Light Scattering and Nanoscale Surface Roughness

Nanostructure Science and Technology

Series Editor: David J. Lockwood, FRSC
National Research Council of Canada
Ottawa, Ontario, Canada

Current volumes in this series:

Alternative Lithography: Unleashing the Potentials of Nanotechnology
Edited by Clivia M. Sotomayor Torres

Controlled Synthesis of Nanoparticles in Microheterogeneous Systems
Vincenzo Turco Liveri

Interfacial Nanochemistry: Molecular Science and Engineering at Liquid-Liquid Interfaces
Edited by Hitoshi Watarai

Introduction to Nanoscale Science and Technology, Vol. 6
Di Ventra, Massimiliano, Evoy Stephane, and James R. Helfin Jr.

Nanoparticles: Building Blocks for Nanotechnology
Edited by Vincent Rotello

Nanoscale Assembly—Chemical Techniques
Edited by Wilhelm T.S. Huck

Nanostructured Catalysts
Edited by Susannah L. Scott, Cathleen M. Crudden, and Christopher W. Jones

Nanotechnology in Catalysis, Volumes 1 and 2
Edited by Bing Zhou, Sophie Hermans, and Gabor A. Somorjai

Ordered Porous Nanostructures and Applications
Edited by Ralf B. Wehrspohn

Polyoxometalate Chemistry for Nano-Composite Design
Edited by Toshihiro Yamase and Michael T. Pope

Self-Assembled Nanostructures
Jin Z. Zhang, Zhong-lin Wang, Jun Liu, Shaowei Chen, and Gang-yu Liu

Self-Organized Nanoscale Materials
Edited by Motonari Adachi and David J. Lockwood

Semiconductor Nanocrystals: From Basic Principles to Applications
Edited by Alexander L. Efros, David J. Lockwood, and Leonid Tsybeskov

Surface Effects in Magnetic Nanoparticles
Dino Fiorani

Light Scattering and Nanoscale Surface Roughness
Edited by Alexei A. Maradudin

Light Scattering and Nanoscale Surface Roughness

Alexei A. Maradudin
Editor

With 197 Figures

Springer

Alexei A. Maradudin
Department of Physics and Astronomy
University of California, Irvine
4129 Frederick Reines Hall
Irvine, CA 92697
USA
aamaradu@uci.edu

Library of Congress Control Number: 2006926878

ISBN-10: 0-387-25580-X e-ISBN-10: 0-387-35659-2
ISBN-13: 978-0387-25580-4 e-ISBN-13: 978-0387-35659-4

Printed on acid-free paper.

9 8 7 6 5 4 3 2 1

springer.com

Preface

All real surfaces, both those occurring naturally, and those fabricated artificially and with great care, are rough to some degree. It is therefore of interest, and often of importance, to know the extent to which this roughness affects physical processes occurring at a surface. A particularly interesting class of physical processes occurring at a rough surface is the scattering of electromagnetic waves from it, or their transmission through it. In this case the degree of the surface roughness is referred to the wavelength of the waves incident on it.

The study of the scattering of electromagnetic waves from rough surfaces has been actively carried out for more than a century now, since Rayleigh's investigations of the scattering of a monochromatic plane wave incident normally on a sinusoidal interface between two different media.[1] The first theoretical treatment of the scattering of an electromagnetic wave from a randomly rough surface was due to Mandel'shtam[2] in the context of the scattering of light from a liquid surface. In these pioneering studies the angular dependence of the intensity of the scattered field was calculated by perturbation theory as an expansion in powers of the surface profile function though the first nonzero term, a single-scattering approximation. For the next 70 years single-scattering approximations, either the small-amplitude perturbation theory introduced by Rayleigh and extended to the scattering of electromagnetic waves from a two-dimensional, randomly rough, perfectly or finitely conducting surface by Rice,[3] or the Kirchhoff approximation,[4] in which scattering from a rough surface is treated as reflection from the plane tangent to the surface at each point, dominated theoretical investigations of rough surface scattering.

The past 20 years have seen many advances in this field. They include improvements in analytic and computational approaches to rough surface scattering. These have simplified the incorporation of multiple scattering into theories of rough surface scattering, which has led to further improvements in analytic and computational methods, and to the prediction and observation of interesting new optical phenomena not captured by single-scattering approximations. There is now an increasing interest in the study of moments of the scattered field higher than the second. The techniques of rough surface scattering theory have been applied to the theory of near-field optical microscopy. Finally, techniques have been developed for the fabrication of one- and two-dimensional randomly rough surfaces with specified statistical properties, and for the characterization of surface roughness.

The development of this field has been driven in part by applications of the scattering of electromagnetic waves from random surfaces encountered in nature, for example, the scattering of electromagnetic waves from the sun and from planets, the propagation of radio waves over the Earth's terrain and over the ocean, and the remote sensing of such features of the Earth's terrain as snow, ice, and vegetation canopy. It has also been driven by applications in which the scale of the surface roughness is comparable to the wavelength of the electromagnetic waves incident on it, as in the transmission characteristics of waveguides with randomly rough walls, the calibration of laser radar standards, the detection of surface defects, the design of microstructured surfaces for directional illumination and thermal control, and *in situ* monitoring of manufacturing processes for the control of such dynamic processes as polishing, etching, film growth, strain relaxation, phase transitions, and interdiffusion.

However, a major driving force for the development of both theory and experiment in the field of rough surface scattering during the past 20 years has been the recognition that the introduction of multiple scattering into the theory of the scattering of electromagnetic waves from randomly rough surfaces yields a variety of effects that have no counterparts in the results obtained on the basis of single-scattering theories. These include enhanced backscattering, the presence of a well-defined peak in the retroreflection direction in the angular dependence of the intensity of the light scattered from a randomly rough surface; enhanced transmission, which is the presence of a well-defined peak in the antispecular direction in the angular dependence of the intensity of the light transmitted through a randomly rough surface; satellite peaks, which are sharp peaks on both sides of the enhanced backscattering and transmission peaks that arise when the scattering system, e.g. a film with a randomly rough surface, supports two or more surface or guided waves; peaks in the angular intensity correlation function of light scattered into the far field from a randomly rough surface; and interesting coherence properties of light scattered or emitted into the near field of a random surface. All of these effects have now been observed experimentally. They are examples of a broader class of multiple-scattering phenomena that go under the name of *weak localization*, and are caused by the coherent interference of multiply-scattered waves, both quantum, and classical.

The initial theoretical studies[5] and subsequent experimental studies[6] of these multiple-scattering effects were carried out for randomly rough surfaces characterized by rms heights of the order of 5–10 nm, and transverse correlation lengths of the order of 100 nm, i.e. surfaces with nanoscale roughness. Subsequent experimental[7] and theoretical[8] work was devoted to the study of surfaces that were significantly rougher than these, e.g. surfaces with microscale roughness. Nanoscale has a somewhat elastic definition. In this volume we have adopted a rather liberal interpretation of this term, extending it in some cases to what purists might consider the microscale regime or beyond. This is because some of the methods developed for treating scattering from surfaces with this larger scale of roughness, especially computational methods, can also be used in the study of scattering from surfaces with nanoscale roughness, and some of the results obtained in studies of surfaces with the larger scale roughness also apply to surfaces with nanoscale roughness.

Theoretical and experimental studies of rough surface scattering can be divided, roughly speaking, into studies of the direct scattering problem and studies of the inverse scattering problem. In the direct problem one is given the surface profile function in the case of a deterministically rough surface, or its statistical properties, in the case of a randomly rough surface, and the task is to solve Maxwell's equations and the associated boundary conditions to obtain the scattered field in response to a prescribed incident field. In the inverse problem one is given experimental data for the angular or spatial dependence of the intensity of the scattered field or, in some cases, of the amplitude and phase of the scattered field, and its dependence on wavelength and polarization, and the task is to invert these data to obtain the surface profile function responsible for them, or some statistical property of the surface profile function such as the power spectrum of the surface roughness or its rms height. Both types of scattering problems are treated in this volume, with the direct problem receiving the majority of the attention, which is simply a reflection of the greater amount of work that has been devoted to this type of problem.

The first several chapters are devoted to the direct scattering problem. In the first chapter, J. M. Bennett introduces definitions of surface roughness, and provides a historical account of the development of various experimental methods for characterizing it, with descriptions of these methods, and describes the forces that stimulated these developments.

Central to the solution of the direct problem is the ability to solve the equations of scattering theory: Maxwell's equations and the associated boundary conditions. There are two general approaches to the solution of this calculational problem: the use of approximate theories of rough surface scattering, usually single-scattering theories, and numerically exact computational approaches that take multiple-scattering into account.

One of the two most frequently used approximate theories of rough surface scattering is the Kirchhoff approximation, a single-scattering approximation. A derivation of this approximation, and of the closely related tangent plane approximation, is presented by A. G. Voronovich in Chapter 2, together with a discussion of some generalizations of it.

In Chapter 3, C. J. R. Sheppard obtains simplified expressions for the scattering of a scalar plane wave from a two-dimensional random surface in the Kirchhoff approximation by introducing the concept of three-dimensional spatial frequencies. The results are used to obtain useful expressions for the bidirectional reflectance distribution function (BRDF) and for the total integrated scattering (TIS).

The other most commonly used approximate approach to the theory of rough surface scattering is small-amplitude perturbation theory. In this theory, the scattering amplitude and the intensity of the scattered field are expanded in powers of the surface profile function, often only up to the lowest order term. Underlying this approximation is the Rayleigh hypothesis, which is the assumption that expressions for the field in the medium of incidence and in the scattering medium outside the selvedge region, which satisfy the boundary conditions at infinity, can be used to satisfy the boundary conditions at the interface between these two regions. This hypothesis is discussed by A. G. Voronovich in Chapter 4, where he argues that, in fact, it can be used even when the rigorous conditions for its validity are not satisfied.

Small-amplitude perturbation theory is not limited to the approximation where only the leading nonzero contribution in powers of the surface profile function is retained. K. A. O'Donnell has used small-amplitude perturbation theory to study the scattering of light from one-dimensional randomly rough surfaces. He has been able to extend such calculations to obtain results that are valid through the eighth order in the surface profile function. Through such high-order calculations he has found new features in the scattering spectrum that are not seen in lower-order calculations. This work is described in Chapter 5. In Chapter 6, G. Berginc describes the application of small-amplitude perturbation theory to the scattering of light from and its transmission through a two-dimensional randomly rough interface between two semi-infinite media, and a film bounded by two random surfaces.

The development of powerful computers with great speed and large memories have enabled calculations of scattering from rough surfaces to be carried out that are largely free from the restrictions that govern the validity of approximate theories such as the Kirchhoff and small-amplitude approximations. This development does not eliminate the need for calculations based on these approximations in regimes where they are applicable, due to their relative simplicity, but affords a means to validate the results obtained by these approximations, and to incorporate multiple-scattering effects into the theory of rough surface scattering in a manner free from approximations, which can lead to improved approximate analytic theories.

In Chapter 7, J. T. Johnson surveys numerically exact approaches that have been developed for the solution of the problem of the scattering of electromagnetic waves from one- and two-dimensional randomly rough surfaces, with recommendations for when such approaches should be used. In Chapter 8, J. A. De Santo describes three kinds of integral equation methods that can be used in solving both scalar and electromagnetic scattering problems when the scattering surface is two-dimensional.

The discussion of the direct scattering problem up to this point is purely theoretical. However, in Chapter 9, K. A. O'Donnell describes experimental studies of angular distributions of light scattered from weakly rough, one-dimensional, random metal surfaces. Fabrication of such surfaces is described, together with the measurement techniques used in the study of the scattering from them. Experimental results for the mean differential reflection coefficient, for angular intensity correlation functions, and second harmonic generation of light scattered from these surfaces are presented. In Chapter 10, T. A. Germer discusses the measurement and interpretation of surface roughness by angle-resolved optical scattering from a single interface and from the two interfaces of a dielectric film. In the latter case the polarization of the scattered light is used to obtain information about the roughness of the two interfaces.

Many of the theoretical studies of the scattering from one- and two-dimensional randomly rough surfaces are based on the assumption that the scattering surface is defined by a surface profile function that is a single-valued function of its argument that is differentiable an arbitrary number of times, and constitutes a stationary, zero-mean, isotropic, Gaussian random process. However, not all randomly rough surfaces of interest are of this type. Surfaces with fractal dimensions are ubiquitous in nature, and are characterized by a divergent root-mean-square slope. In Chapter 11, J. A. Sánchez-Gil *et al.* describe theoretical studies of the light scattered from randomly rough one-dimensional self-affine metal surfaces with a nanoscale lower cutoff. A different type of randomly rough surface is represented by an ensemble of particles with simple geometries seeded onto planar surfaces. Such surfaces are interesting for basic science reasons, because they allow calculations to be carried out in a controlled way for different sizes, shapes, densities, or optical properties of the particles. They are also of interest in applications such as the degradation of mirrors by particle contamination, optical particle sizing, and the fabrication of biosensors. F. Moreno, *et al.* present an overview, in Chapter 12, of experimental and theoretical work on the scattering of light by particles on surfaces, proceeding from the case of a single particle on a surface to the case of an ensemble of particles on a surface. The scattering of light from a randomly rough surface that bounds an inhomogeneous dielectric medium is one of the major unsolved problems of rough surface scattering theory. In Chapter 13, K. K. Tsi *et al.* investigate the multiple scattering of waves by large volume concentrations of random distributions of nanoparticles, and describe ways in which the scattering problem can be solved when the particles are on or are buried in a substrate that has a randomly rough surface.

The mean intensity of a scattered field is a two-point amplitude correlation function of the scattered field in the limit as the two points at which the field and its complex conjugate are determined merge. However, two-point correlation functions of the scattered field for noncoincident points, sometimes called mutual coherence functions, occur in a variety of contexts in rough surface scattering. These include studies of angular and frequency intensity correlation functions, which reveal symmetry properties of speckle patterns and the statistics of the scattered field, and studies of the coherence properties of the scattered field, i.e. the properties of light that are most closely related to interference and diffraction. In Chapter 14, T. A. Leskova and A. A. Maradudin describe how taking into account multiple scattering introduces new features into angular and frequency intensity correlations not present in the results obtained in the lowest order of small-amplitude perturbation theory, some of which have now been observed experimentally. In Chapter 15, J. J. Greffet and R. Carminati study the coherence of the field scattered from a rough surface or of a thermal field generated by an interface, in both the near field and the far field, and discuss the role of surface plasmon polaritons in the structure of the speckle pattern in the near field. They point out that in modeling the detection of an optical near field the role of the tip used as the detector must be taken into account, and present examples to illustrate this.

The direct scattering problem is not the only one of interest in rough surface scattering. The inverse problem, although it is less intensely studied, is another important and challenging one. In Chapter 16, E. R. Méndez and D. Macias describe algorithms for inverting far-field scattering data obtained from measurements on one-dimensional rough surfaces to obtain the surface profile function. As input one of the algorithms uses the angular dependence of the intensity of the scattered light, while another uses the angular dependence of the amplitude and phase of the scattered field. A different type of inverse problem is discussed by A. A. Maradudin in Chapter 17, namely how to design a one- or two-dimensional randomly rough surface that produces a specified angular or wavelength dependence of the radiation scattered from it.

The chapters constituting this book present an up-to-date survey of many aspects of rough surface scattering that are relevant to the scattering of electromagnetic waves from surfaces with nanoscale roughness. Yet, as one reads through them it is clear that there remain several areas of the subject that are in need of further development. These include, for example, improved algorithms for calculating the scattering of electromagnetic waves from, and their transmission through, two-dimensional randomly rough penetrable surfaces; the extension of these algorithms to the case that the random surface bounds an inhomogeneous substrate; the solution of the inverse problem from far-field data for the amplitude and phase or the intensity of light scattered from two-dimensional rough surfaces; the experimental observation of higher-order angular intensity correlation functions; experimental and theoretical studies of the coherence and other correlation functions of light in the near field; and the determination of methods for designing and fabricating surfaces that produce scattered or transmitted light with specified properties that are more complex than those considered to date. If some reader of this volume is motivated to tackle one or another of these problems, all of us involved in producing this volume will feel that the effort was worth it.

Finally, I wish to thank the authors for the thought and care they have put into the preparation of their contributions.

Alexei A. Maradudin
Irvine, California
March, 2006

References

1. L. Rayleigh, *The Theory of Sound*, vol. II, 2nd edn (Macmillan, London, 1896), pp. 89, 297–311.
2. L.I. Mandel'shtam, *Ann. Phys.* **41**, 609 (1913).
3. S.O. Rice, *Commun. Pure Appl. Math.* **4**, 351 (1951).
4. M.A. Isakovich, *Zh. Eksp. Teor. Fiz.* **23**, 305 (1952).

5. A.R. McGurn, A. A. Maradudin, and V. Celli, *Phys. Rev. B* **31**, 4866 (1985).
6. C.S. West and K. A. O'Donnell, *J. Opt. Soc. Am. A* **12**, 390 (1995).
7. E. R. Méndez and K. A. O'Donnell, *Opt. Commun.* **61**, 91 (1987).
8. M. Nieto-Vesperinas and J.M. Soto-Crespo, *Opt. Lett.* **12**, 979 (1987); A.A. Maradudin, E.R. Méndez, and T. Michel, *Opt. Lett.* **14**, 151 (1989).

Contents

Endnote

This volume presents up-to-date surveys of many aspects of rough surface scattering that are relevant to the scattering of electromagnetic waves from surfaces with nanoscale roughness. Both the direct and inverse scattering problems are considered, perturbative and computational approaches to their solution are described, and experimental methods and results are discussed.

Contributors

Jean M. Bennett jbennett@ridgenet.net

Gérard Berginc gerard.berginc@fr.thalesgroup.com

Rémi Carminati remi.carminati@ecp.fr

Chi-Hou Chan eechic@cityu.edu.hk

John A. DeSanto bajdesanto@mac.com

Kung-Hau Ding Kung-Hau.Ding@hanscom.af.mil

José V. Garcia-Ramos imtg160@iem.cfmac.csic.es

Thomas A. Germer germer@nist.gov

Vincenzo Giannini vingenzino@iem.cfmac.csis.es

F. González gonzaleff@unican.es

Jean-Jacques Greffet Jean_Jacques.Greffet@em2c.ecp.fr

Joel T. Johnson johnson@ece.osu.edu

Tamara A. Leskova leskova@duey.ps.uci.edu

Demetrio Macías demetrio.macias@utt.fr

Alexei A. Maradudin aamaradu@uci.edu

Eugenio R. Méndez emendez@cicese.mx

F. Moreno fernando.moreno@unican.es

Kevin O'Donnell odonnell@cicese.mx

J. M. Saiz saizvj@unican.es

José A. Sánchez-Gil j.sanchez@iem.cfmac.csic.es

Colin J.R. Sheppard colin@nus.edu.sg

Leung Tsang tsangl@u.washington.edu

Ka-Ki Tse kktse@ee.cityu.edu.hk

Alexander G. Voronovich Alexander.Voronovich@noaa.gov

1
Characterization of Surface Roughness*

JEAN M. BENNETT
Michelson Laboratory, Naval Air Warfare Center, China Lake, CA 93555

1.1. Introduction

1.1.1. Definition of Nanoscale Roughness

This book is concerned with the interactions between light and matter on the nanoscale level (dimensions of the order of atoms and molecules). These interactions are being studied, measured, and modeled. The particular emphasis is on the interactions of light with surface topography, i.e., surface roughness, not chemical interactions.

It is appropriate to define what range of surface roughness constitutes "nanoscale roughness." There are two parts to surface roughness – the heights of features above and below the mean surface level, and the lateral separations of these height features (Fig. 1.1). The former are generally given in terms of a root-mean-square (rms) value or an average value and are called the "surface roughness," while the latter are distances measured along the surface in the mean surface plane and are called "surface spatial wavelengths." The words "surface spatial" are very important to distinguish this quantity from the wavelength of the incident light beam illuminating a surface in a light scattering measurement or making an interferometric (optical) measurement of surface roughness.

In the nanoscale roughness region, the range of surface heights that can be measured by appropriate instruments varies from subnanometers ($\sim$0.01 nm) (fractions of the spacing between atoms ($\sim$0.4–0.6 nm)) to more than 1 μm (>1,000 nm). Depending on the type of surface, heights larger than the nanoscale region can reach values of millimeters. Many types of surfaces have heights that are a fractal quantity and can be treated by the theory of fractals.[2–4]

Surface spatial wavelengths, on the other hand, are much longer and can vary from $\sim$0.1 μm (100 nm) to 1 mm (1,000 μm or 1,000,000 nm), depending on the surface roughness. The surfaces that have the largest roughness and smallest surface spatial wavelengths will have the steepest slopes. Some special synthetic

* Most of the material in this chapter is taken from the book: Jean M. Bennett and Lars Mattsson, *Introduction to Surface Roughness and Scattering*, 2nd edn. (Optical Society of America, Washington, D.C., 1999)

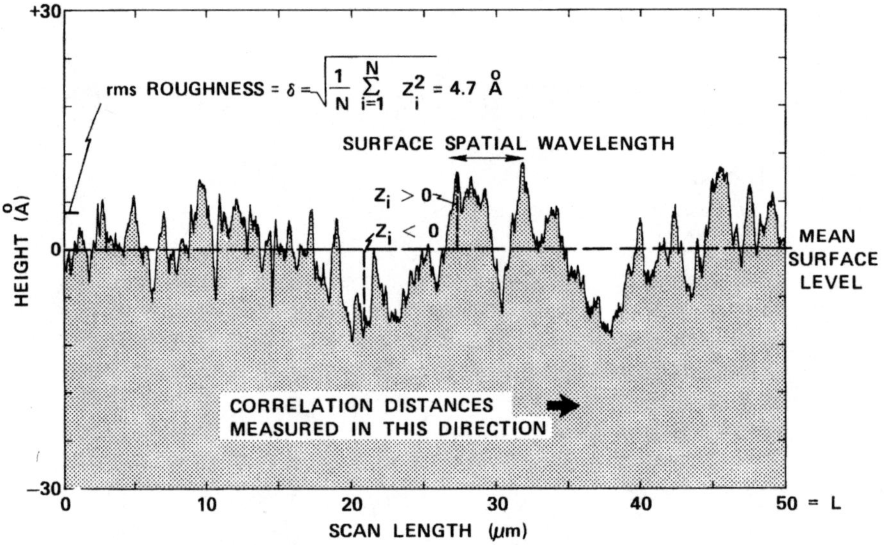

FIGURE 1.1. Schematic representation of a one-dimensional profile of a rough surface showing surface heights and surface spatial wavelengths. The vertical axis has been greatly expanded compared to the horizontal axis (Fig. 23 in Ref. 1).

surfaces that have been made to test various scattering theories have much larger slopes than surfaces normally encountered in the real world. Thus, the ratio between surface spatial wavelengths and surface heights can vary from ~10,000 for the smoothest surfaces to ~10 for rough surfaces with steep slopes.

1.1.2. Early Beginnings: Visual Surface Inspection versus Quantitative Measurements

Prior to the 1940s and World War II, roughness on the surfaces of lenses and other optics was "eyeballed" by looking for a gray cast to the surface, caused by light scattering from residual fine grinding marks. Normally, by the time a lens had the correct shape and focal length, the surface was completely polished out and there was no gray. Now, more sensitive instruments show that the smooth surface contained a network of tiny scratches left by the abrasive particles in the polishing compound (often rouge or iron oxide).

In the same time frame, mass produced, interchangeable parts were starting to be made by the machining industry. Shape and also surface finish became important. Large coordinate measuring machines were being developed that used a ball touch-probe mounted on a movable arm to contact the part and measure its shape. Surface finish was determined by visually comparing the finish on a machined part with one of a set of standards – small blocks of metal machined by different processes (grinding, milling, lapping, buffing, polishing, etc.) that had different roughnesses (groove depths). Sometimes a 10× magnification loupe was used. Often the machinist ran his fingernail over the standard block and machined part to see if they felt similar. This "fingernail test" was in general use prior to and during World War II.

Although visual surface inspection was and is still important, quantitative measurements were essential. Lord Kelvin already recognized this fact in the nineteenth century when he commented: "When you can measure something and express it in numbers, you know something about it." Kelvin's statement is frequently paraphrased as "if you can measure something, you can make it better." This is the primary reason that increasingly sophisticated quantitative surface metrology techniques have been developed over the years – to make optical components "better." In most cases, "better" has meant lower haze and scattering and clearer, sharper images in an instrument. A very important point is that all the improved surface characterization techniques were developed in response to commercial needs. Thus, *industry needs have been the impetus for new fabrication techniques which, in turn, require improved surface metrology techniques*. For example, mass-produced optical or machined parts required sensitive measurements to determine their shapes and performance. As the specifications became tighter, the measuring instruments had to be more accurate. The shape of flat and spherical optical surfaces had been measured since the nineteenth century by a variety of well-known, exceedingly sensitive, interferometric techniques, but the measurement of surface finish was in its infancy.

In the early twentieth century, crude portable "profilometers" were developed to give a semi-quantitative measurement of the surface roughness on machined surfaces. These were small, pocket-sized instruments that had an arm with a diamond-tipped probe sticking out of a box. The probe touched the surface and was drawn across it for a distance ∼ a few millimeters to a few centimeters. As the probe followed the contours of a surface, an analogue signal was recorded. It could be printed out as a wiggly line on a graph paper, to be compared to another wiggly line on the "reference surface," or an internal analogue circuit could average the signal variations to give a "roughness average" (Ra) value. Mechanical profilers have considerably improved in the intervening years, but even today a mechanical profiler must be calibrated in a way that is traceable to NIST (the National Institute of Standards and Technology). NIST certification is essential for calibrating all profilers that are used in industries producing parts with specified surface roughness. NIST will not certify any instruments that measure surface scattering and calculate a corresponding roughness using an appropriate theory because scatter-measuring instruments cannot be directly compared to profile-measuring instruments!

1.1.3. Beginnings of Quantitative Metrology in the 1940s

Prior to World War II, the best optics came primarily from Europe. However, at the start of the War, US optics industries such as Eastman Kodak, Bausch and Lomb, Corning Glass Company, American Optical Company, and others had to gear up to fill US military needs. There were major problems with the quality of military binoculars, cameras, telescopes, periscopes, and other optical instruments. The images in these instruments were hazy and it was often difficult to see the cross hairs or reticules used for aligning the instruments. Mary Banning[5] at the Institute of Optics of the University of Rochester expanded on the earlier work of John Strong[6] and mass-produced single and multilayer antireflection coatings for lenses which greatly reduced the reflections between lens elements in an optical system and thus the ghost reflections in the image plane. She also produced multilayer coatings for other applications.

About the same time, McLeod and Sherwood[7] at Eastman Kodak Company proposed the first semi-quantitative method for characterizing finish of polished optical surfaces. A series of polished optical flats were to be diamond-scribed with scratches of increasing "badness" that would be visually compared *with no magnification* to the sizes of scratches on a finished optical component. The scratch sizes were to go from a barely perceptible #10 to a very large #80. It was hoped that the different "badness" of the scratches would correlate with the measured scratch widths, but the appearance critically depended on the method used to make the scratches. Along with the scratches were a corresponding series of digs, or small pits, with numbers from #10 to #80. A #10 dig was to have a diameter of 100 microns, or 0.1 mm; thus the dig number was to be one-tenth of the dig diameter measured in microns. The sets of standard scratches and standard digs were to be sold in boxes to optical companies and used for checking the surface imperfections on polished optics. The Army adopted the scratches and digs exactly as McLeod and Sherwood had proposed and designated them as the standard MIL-O-13830A, the so-called Scratch and Dig Standard. This standard has had a long and colorful history (see Chap. 9 of Ref. 1) and has created many arguments, much confusion, expense, and frustration among the users. It is no longer a military standard. However, a commercial US standards-writing body, the Opto-Electronic and Optics Standards Council (OEOSC), has adopted the old standard, slightly clarified some parts, and has turned it into a commercial standard. Nothing better has come along to assess the "visual appearance" of an optical surface.

1.1.4. Metrology Advances in the 1950s and 1960s

The US military, particularly ARPA (Advanced Research Projects Agency) and DOE (Department of Energy) provided an impetus for metrology development in the late 1950s and 1960s. The Navy was trying to find a rapid measurement method that could be used to detect wakes of underwater submarines by measuring the scattering of radar waves from ocean surfaces. The hope was that a scattering map of the ocean surface would show submarine wakes. This problem was transferred to optics laboratories – to relate scattering from a surface to its roughness. The simplest scattering measurement was to collect a large solid angle of surface scattering, so-called total integrated scattering (TIS) (see Chap. 4 of Ref. 1). The theory for scattering of radar waves from rough surfaces was already well known but had to be modified for optical measurements because the reflection coefficient for optical surfaces is smaller than the 100% reflection coefficient of radar waves being scattered from water surfaces. The theory was appropriately modified, instruments were built, and many TIS measurements were made on optical and machined surfaces over the last 40 years. TIS has become an ASTM standard measurement protocol and is used in the US and elsewhere (see Chap. 9 of Ref. 1). Even though TIS is now a routine optical surface characterization technique, the original problem – finding a method for detecting submarine wakes – has never been solved!

Other military efforts in this same period involved the production and characterization of large optics to use as beam directors to send intense laser beams to outer space with minimum scattering losses (beam divergence). Large single-point diamond-turning machines were built in several government laboratories and much effort was spent on optimizing methods for producing low scatter mirrors. The optics group at Michelson Laboratory, China Lake, characterized many prototype mirrors produced on these diamond-turning machines by measuring surface profiles and TIS.

A comparable effort was also started to increase the resistance of large multilayer-coated beam director mirrors and other optics subjected to high power laser radiation. Improved polishing and coating techniques were developed along with sophisticated laser damage test facilities to measure failure of mirrors and coatings under intense laser radiation. An annual Laser Damage Symposium was started to enable researchers around the world to share their results. Now, thirty-seven years later, the Boulder Laser Damage Symposium is still going, with some of the original people still on the organizing committee.

1.1.5. Further Metrology Advances in the 1970s and 1980s

Industries were driving the metrology effort during this period. The machining industry needed closer control of surface shape and finish. Improved multiaxis CNC (computer numerically controlled) machines were being built that enabled parts to be completely fabricated automatically based on instructions input on punched paper tape into a rudimentary computer. Modern CNC machines use considerably more sophisticated computer systems. Also, larger, more accurate, coordinate measuring machines checked surface shape and more sophisticated mechanical profilers came on the market. The British company, Rank Taylor Hobson, produced a landmark mechanical profiler, the Talystep step height measuring instrument for quality control measurements (see Fig. 1.2). It was primarily intended to measure small steps on machined surfaces and roughnesses in the microinch range (from fractions of a micron to several microns) as required by the specifications on machined surfaces. The ~40–100 mg loading on the sharp diamond probe always left a track on the surface, as did the probes of all the other profilometers (loadings were often several grams).

Since tracks on high quality optical surfaces were not acceptable, in the late 1960s and early 1970s several groups started quantifying the scratch depths and widths made in thin film coatings by commercial mechanical profilers.[8,9] But it was not until 1981 that comprehensive studies were published showing that a

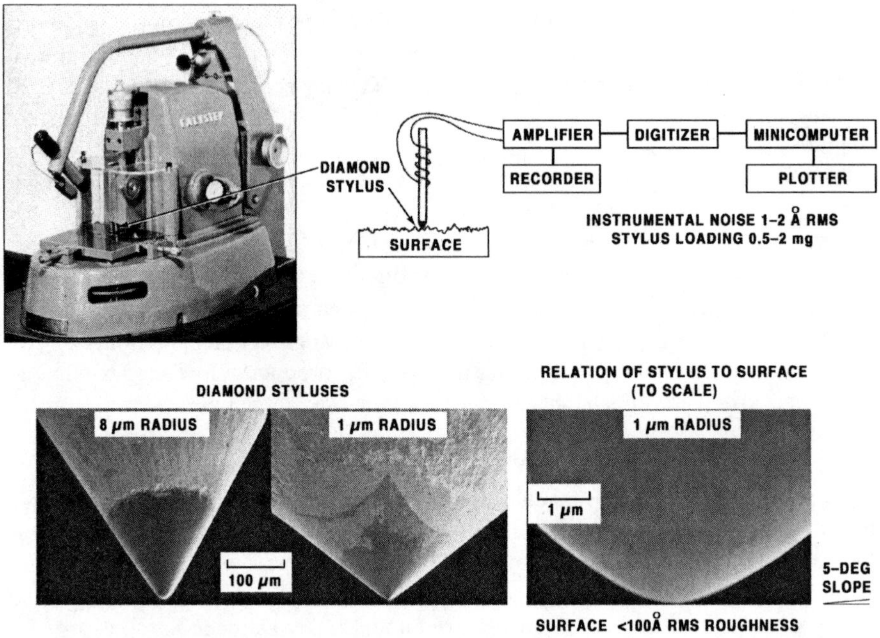

FIGURE 1.2. Talystep mechanical profiler and the associated electronic circuits used to nondestructively profile optical surfaces. A scanning electron micrograph of the standard stylus is shown at the lower left; a micrograph of a specially made, sharp stylus is at the lower center and enlarged at the lower right (From Ref. 10).

mechanical profiler could be used to nondestructively profile soft optical coatings by greatly reducing the stylus loading;[10] also the performance of different mechanical profilers was later compared.[11]

In this same time frame, the microelectronics industry was developing circuits on silicon wafers, data were being stored on floppy disks and magnetic tapes, and the optics industry and the diamond-turning groups were demanding nondestructive methods for quantitatively measuring surface finish. This fueled a two-pronged attack – development of optical, noncontact profilers and formulating scattering theories that would give more statistical information about surface finish. Three main types of optical profilers were developed that later became commercial products and are briefly described in Sect. 1.2.4. The first of these, chronologically, was Sommargren's common path optical heterodyne interferometer.[12] The second was the phase measuring interferometer developed by Wyant and Koliopoulos.[13] About the same time, Eastman and Zavislan[14,15] proposed the third type of interferometer, based on the principle of differential interference contrast.

The demand for more statistical information to be obtained from scattering measurements spurred the development of both scalar and vector scattering theories. The basis for the scalar scattering theories was the work originally published in 1919 by Chinmayanandam,[16] which was considerably expanded and popularized in the scalar theory "Scattering Bible" by Beckmann and Spizzichino in 1963.[17] Church was the recognized expert for both scalar and easy-to-understand vector scattering theories for surfaces having small roughness, beginning in the mid 1970s[18–20] and continuing through the early 1990s.[21] He was primarily concerned with scattering from bare metal surfaces, particularly diamond-turned surfaces. Carniglia[22] proposed a scalar scattering theory for multilayer dielectric coatings, but it had limited use. First-order vector scattering theories were much better for describing polarization effects in angle-resolved scattering and scattering from multilayer-dielectric coatings. Many groups developed vector scattering theories during the 1970s and 1980s. Elson, one of the prolific theorists, first worked on surface plasmon theory[23] along with several others including Kröger and Kretschmann.[24] Later, Elson published a landmark paper on scattering from multilayer-dielectric coatings[25] followed by others.[26,27] Bousquet and coworkers[28] also presented a much-referenced theory about scattering from multilayer films. Starting in the mid 1970s, Maradudin and his colleagues[29,30] published important scattering theory papers (see also the chapters in this book) but they are not generally used for analysis of experimental data. Scattering theories have also been developed in various laboratories and are used for interpreting the angle-resolved-scattering measurements made there. These include two laboratories in Marseilles, France: the laboratory of the Rasignis[31,32] and the one of Amra and Pelletier,[33] plus others elsewhere that were established later. Although many theories deal with scattering from smooth and rougher optical surfaces, some are specifically designed to explain scattering from synthetic surfaces. These cannot be applied to normal optical surfaces.

Another very important event dating back to the early 1980s was the invention and subsequent development of the scanning tunneling microscope (STM) by Binnig and Rohrer[34,35] following the original proof-of-concept instrument demonstrated by Young in 1971.[36–38] Originally developed to look at single atoms or molecules on electrically conducting surfaces, it was later used to look at the structure of gold-coated surfaces by Dragoset *et al.*[39] However, the STM was not practical for measuring most optical surfaces because nearly all metals have a thin, nonconducting surface oxide layer, and dielectrics do not conduct at all unless they are overcoated with gold or platinum. About four years after Binnig and Rohrer's STM was announced, Binnig, Quate, and Gerber[40] proposed an instrument to measure the topography of nonconducting materials with the aid of an STM. After that initial proposal, many other clever people contributed to making a whole family of scanning probe microscopes that can measure, among other things, surface topography, friction, temperature, magnetic domains, and variations of optical properties. The atomic force microscope (AFM) is the most useful of these instruments for looking at fine topographic structure on optical surfaces.

1.1.6. Recent Developments from the 1990s to the Present

The tremendous increase in computing speed and memory of today's computers has made possible the greatly improved surface metrology instruments of today. Second, larger and more sensitive CCD arrays and similar area detectors are now available. Although it took a few years to develop the 1980s demonstration proof-of-principle laboratory instruments into commercial products, with the advent of modern computers they have become rapid, automated, user-friendly instruments. Currently, the phase measuring interferometer, the white light interferometer (the original Michelson interferometer concept applied to surface topography measurements in the 1990s), and the Nomarski-type scanning profiler form the workhorses of today's noncontact surface topography measuring instruments.

Instruments to measure light scattering from optical surfaces are not generally available. However, one-of-a-kind instruments that measure angle-resolved scattering (ARS) and total integrated scattering (TIS) are constantly being built in various research laboratories. An exception is a commercial instrument using ARS at oblique incidence to measure roughness of machined surfaces.[41] A new model with increased sensitivity is now able to read surface roughness in the > 2.5nm rms range. A new instrument to measure scattering and/or emittance (radiation emitted from the sample) has recently been introduced.[42] See Sect. 1.2.7 for more information. Specialized instruments have been developed for inspection of defects and contamination on silicon wafers using light scattering technology (but generally without a sophisticated theory) and are in use in most silicon wafer fabrication laboratories.

Commercial scanning probe microscopes that are capable of measuring many surface properties have been available since the mid 1990s and are greatly improved from earlier models. Now they are automated, user-friendly, i.e., requiring a minimum of skill to operate, and exceedingly rapid. The AFM is still the instrument most used to inspect small areas ($<100\,\mu$m $\times$ $100\,\mu$m) of optical surfaces, particularly surface films. Specialized forms of the AFM have been built to operate remotely in clean rooms of silicon wafer fabrication laboratories to give detailed images of defects found with a light scattering instrument.

1.2. Current Surface Metrology Techniques and Instruments

1.2.1. Questions to Answer Prior to Taking Measurements

Before any measurements are undertaken, various questions need to be answered: (1) What is the purpose of the characterization? (2) What is the sample size and shape? (3) How many samples are there and how much money is available for the measurements? (4) What is the condition of the sample surface?

(1) The person with the sample may not know what the best measurement technique would be. He just wants to know, "What is the roughness of the surface?" Depending on what "the surface roughness" will be used for, there are numerous possible methods for making a measurement; they will not, in general, agree because the instruments measure roughness in different surface spatial wavelength

regions, as will be explained later in this section. For example, if the sample is a prototype of a component to be used in a high performance, low scatter optical system, scattering (not topographic roughness) should be measured. If the sample has a very low scatter, surface contamination is extremely important since a few average size dust particles (diameters $\sim 1\ \mu$m) will produce more scattering than the rest of the surface! If a new surface finishing method is being developed (new polishing method, single-point diamond turning, magnetorheological polishing, etc.), an optical or mechanical profiler measurement would be appropriate. If a new coating technique is being developed, a small area of the coating should be measured with an atomic force microscope to show the "lumpiness" of the coating. If the sample is a tiny replacement plastic eye lens or a soft or hard contact lens, probably the surface should be inspected in a high-quality optical microscope using the differential interference contrast technique.

(2) Depending on the sample size, shape, and quality, some measurement techniques may not be possible. For example, if the sample is a several-meter-size component of a segmented telescope mirror, it will not fit under the measuring heads of normal laboratory instruments. If the project is sufficiently important and there is money available, often special instruments can be built to make the measurement. If the sample is not compatible with a vacuum environment, it cannot be observed or measured in a scanning electron microscope or a transmission electron microscope.

(3) If there are many samples of the same type, all of which need to be measured, careful, individual measurements are not possible. Depending on the importance of the project and the funding available, a special automated instrument can be built that will do all the measurements automatically and only inform the user when samples are out of the surface specification range. If the particular kind of measurement that is desired is very expensive and requires a skilled operator to make the measurement, it is sometimes possible to substitute another kind of measurement that is cheaper and yields similar information.

(4) If the person who has the sample is unaware of the surface condition and wants rapid results, often the metrologist does not take the time to inspect the surface condition, but immediately proceeds to take the desired measurements. If the results are not carefully inspected to see that they look reasonable, it is possible that the entire "roughness value" is caused by contamination on the surface or mishandling marks, so that the measurements are meaningless! In some cases, contamination on the sample surface cannot be removed without damaging the surface. In such a case, the only solution is to provide a clean sample.

1.2.2. Relations Between Surface Metrology Techniques

Figure 1.3 shows examples of certain types of structure that may be present on optical surfaces. The first line shows the structure of a deposited film, whose heights and lateral dimensions (surface spatial wavelengths) are small compared to the illuminating wavelength. (In all the three illustrations, the vertical scale ($\sim$ a

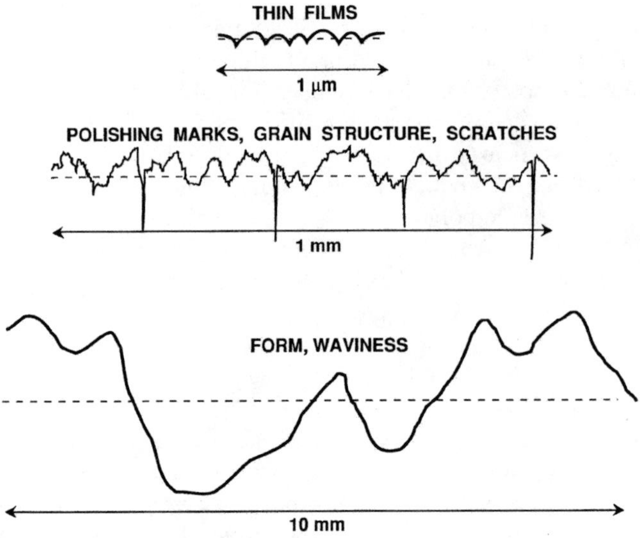

FIGURE 1.3. Schematic representation of different kinds of structure that can be present on optical surfaces. The vertical scale of surface heights has been greatly expanded relative to the horizontal scale of surface spatial wavelengths.

few nanometers or less) is small compared to the horizontal scale.) The instrument used to measure the structure of a film of this sort must be able to completely resolve the structure or else the measurement will be meaningless.

The second line in Fig. 1.3, "Polishing marks, grain structure, scratches" is the most common type of roughness on polished optical surfaces. If the horizontal and vertical scales were equal, the scratches would appear as shallow V's instead of sharp vertical spikes. This type of structure covers the entire surface and produces so-called microirregularity scattering.

There are other kinds of possible surface structure such as a sine wave or pseudo-sine wave in one or more directions, unidirectional grinding marks (most often seen on metal samples), single-point diamond turning marks in the form of concentric circles or arcs of circles (turned off center, i.e., "fly cut"), grids made with square wave or v-shaped grooves, and other types of special surface structure made to test various types of theories.

The third line in Fig. 1.3 is the so-called optical figure, or departure of the actual surface from its ideal shape. This is often called "form" or "waviness" and is the largest component of the surface roughness, with the thin film "lumpiness" being the smallest component. Optical figure is generally measured in some sort of interferometer with results given in fractions of a wavelength. An optical figure of $\lambda/500$ peak–valley, measured at a wavelength in the visible spectral region, is an excellent figure on an optic. However, converted to nm, it is 1.3 nm, while microirregularity roughnesses are sometimes smaller than one-tenth this much. Optical figure is not generally considered part of surface roughness, unless one is

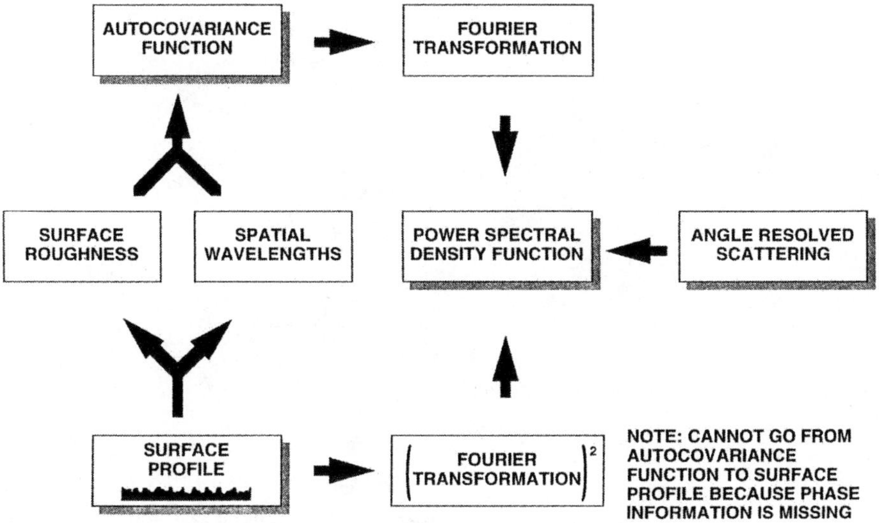

FIGURE 1.4. Relation between different parameters used to characterize optical surfaces. The two primarily measured quantities are surface profiles and angle-resolved scattering (Fig. 19 in Ref. 1).

concerned with mid-spatial frequency (or mid-spatial wavelength) roughness on very large mirrors that can have spatial wavelengths of cm or longer.

The conclusion to draw from this example of the three scales of roughness is that the instrument chosen to take the measurement should have a lateral resolution and profile length (spot size for scattering measurements) appropriate for the kind of roughness that is being measured. All three scales of roughness are normally superposed on a surface. For example, a coated surface would have the tiny roughness bumps of the coating superposed on the longer spatial wavelength roughness composed of polishing marks. Then, the entire surface would be slightly wavy (optical figure), illustrating deviations of the actual surface from the desired shape.

Figure 1.4 shows the relations between different kinds of measurements that can be made on optical surfaces. These blocks assume that the surface heights are small compared to the light wavelength, so that first-order scalar or vector scattering theory is valid. Scalar scattering theory is less useful because polarization information is missing. Normally, the starting measurements are profile (area topography) measurements or scattering measurements. Both types of measurements can be used to determine (band-width limited) parts of the power spectral density function, the so-called master curve for a surface, by the way of intermediate functions and appropriate Fourier transforms. (The lower block marked (Fourier transformation)^2 is shorthand for the square of the magnitude of the Fourier transform.) However, it is not possible to go from a power spectral density function to a surface profile because the phase information about the relation of the various kinds of surface structure to each other is missing.

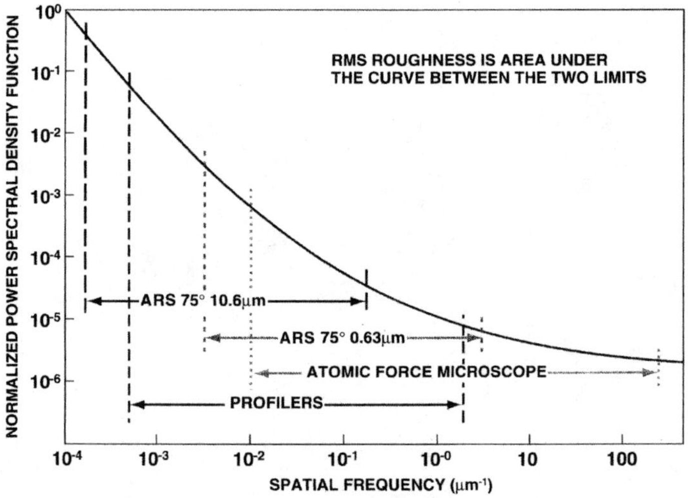

FIGURE 1.5. Idealized power spectral density function (PSD) plotted versus surface spatial frequencies (reciprocal of surface spatial wavelengths). Arrows between vertical dashed lines indicate surface spatial frequency ranges measured by different instruments or techniques (Fig. 35 in Ref. 1).

Figure 1.5 shows a one-dimensional synthetic power spectral density (PSD) function that has the general shape of many that have been measured for isotropic polished optical surfaces. The horizontal (x) axis is surface spatial frequency (reciprocal of the surface spatial wavelength); in this representation the optical figure (long spatial wavelengths or low spatial frequencies) is on the left-hand end of the x-axis. The fine structure from thin films on the surface (short spatial wavelengths or high spatial frequencies) is on the right-hand end of the x-axis. The magnitude of the various roughness components is given on the y-axis. Note that the magnitude of the optical figure is many orders of magnitude larger than the magnitude of the "lumpiness" of a thin film coating. The magnitude of the microirregularity roughness lies between these two limits. Also on this graph are shown various vertical bands (ranges of surface spatial frequencies) that are typically measured by different kinds of surface measuring instruments or techniques. Notice that the profilers have a slightly larger spatial frequency range than do the ARS measurements, but the atomic force microscope has the largest spatial frequency range of all. However, these spatial frequency ranges may be slightly misleading because a single measurement taken with the AFM has a maximum size of $100\,\mu\text{m} \times 100\,\mu\text{m}$ (spatial frequencies $0.01\,\mu\text{m}^{-1} \times 0.01\,\mu\text{m}^{-1}$) but with the typical 1024×1024 data points in the image only resolves features down to a size of $\sim 0.1\,\mu\text{m} \times 0.1\,\mu\text{m}$ (spatial frequencies $10\,\mu\text{m}^{-1} \times 10\,\mu\text{m}^{-1}$). To obtain the extra decade of higher spatial frequencies, smaller image sizes must be measured (the number of pixels in the image does not change because that is the number of pixels on the detector). But smaller image sizes sample even less of the surface area, making it even more imperative to coat the surface uniformly. All surfaces measured with an AFM, optical, or mechanical profilers should be carefully inspected prior to the measurements to make sure that the area being measured represents the overall structure on the surface. The basic problem of generating a PSD from measurements made on the same surface using different instruments that measure in overlapping spatial wavelength bands is that the segments will not necessarily overlap.[43]

The preceding discussion applies to a PSD calculated from line profiles. There are also two-dimensional PSDs that can be calculated from area topographic maps or ARS data. If a surface is isotropic, it is often possible to fit a measured one-dimensional PSD to a relation proposed by Church and Takacs[44] that contains constants A, B, and C. The fitted constants are then used in another relation, also proposed by Church and Takacs, to calculate the two-dimensional form of PSD,[45] as summarized in Sect. 4.D of Ref. 1. The result is an area PSD map of the surface. For an isotropic surface, the PSD map has circular symmetry with a spike at the center (contribution from the optical figure). A radial line taken in any direction can look similar to the curve in Fig. 1.5 except that it is displaced vertically on the y-axis. However, if, for example, a surface has unidirectional structure in two orthogonal directions, the PSD would have structure along narrow bands in two orthogonal directions, such as the one shown in Fig. 34 of Ref. 1.

Figure 1.6 presents more information about specific types of surface measuring instruments and techniques including the AFM, optical and stylus profilers, and ARS and TIS scattering measurements. It is probably misleading to show a range of spatial frequencies for a TIS measurement because this entire band is combined into one TIS measurement. This is why there is no corresponding band on the PSD graph shown in Fig. 1.5. ARS measured at normal incidence would have a narrower spatial frequency band than the one for a large angle of incidence (for a given illumination wavelength). It is also important to realize that the rms roughness values determined by ARS (or TIS) depend on some form of first-order scattering theory that assumes that the surface is covered by many tiny grooves (such as multiple diffraction gratings) that are oriented at different angles to each other. Each grating has some groove spacing and some phase associated with the surface structure. Therefore, the gratings oriented at different angles to each other have different amplitudes and phases. In a profile or areal topography measurement, slices of the structure on the surface are measured and then, through the Fourier transform process, are converted into power spectral density components. An example of how this process works is illustrated in Chap. 4 of Ref. 1.

The range of surface heights (roughnesses) that can be measured (shown in parentheses above the solid bars) depends on the specific instrument or technique. To obtain the wide range of surface roughnesses shown for the optical profiler, the phase measuring interferometer measures small roughnesses and the white light interferometer measures larger roughnesses. For the ARS and TIS techniques, the minimum surface height depends on the sensitivity of the instrument and the amount of scattered light present in the system. The maximum surface height is limited by the theory (the assumption that the surface roughness must be much

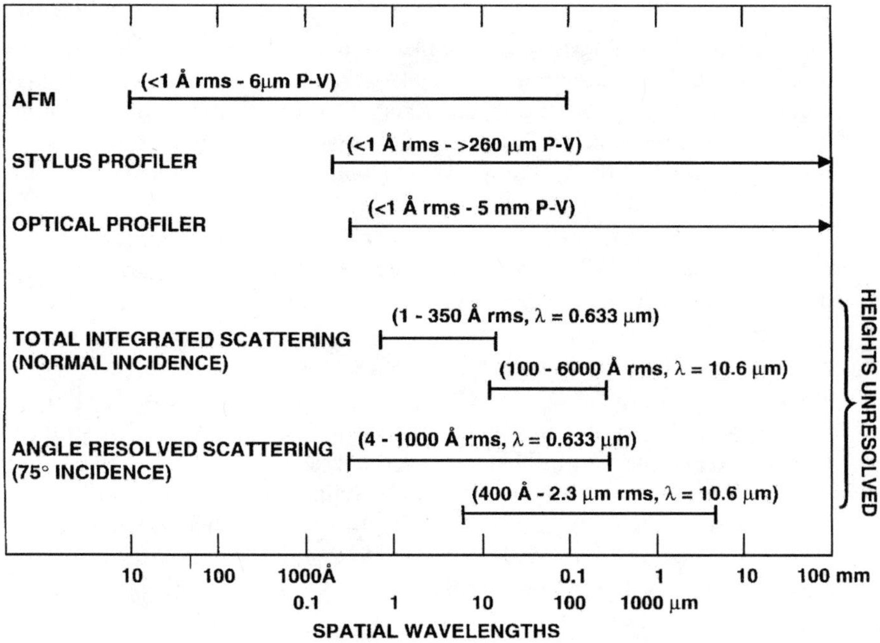

FIGURE 1.6. Ranges of roughness (numbers in parentheses) and surface spatial wavelengths (horizontal bars) accessed by different measuring techniques. The horizontal axis is a log scale of surface spatial wavelengths (Fig. 22 in Ref. 1).

smaller than the illuminating wavelength), so this value depends on the illuminating wavelength and how close to the limit one wishes to push the theory.

1.2.3. Surface Inspection and Imaging

For all kinds of surface characterization, the surface of the sample must be visually inspected to see what is on it. There may be contamination or unexpected structure present that would interfere with the measurement. A decision needs to be made whether to clean the sample to remove contamination or to obtain another sample if the current one has additional unexpected structure.

Visual inspection can be as simple as observing the sample by eye, either under ordinary room light, in bright sunlight, or in a darkened room and using a directed white light beam or laser source. Since what is being observed is light scattered from surface imperfections, the most effective angle is very close to the direction of specular reflection where the scattered light is a maximum. Visual inspection will also show whether a supposedly flat sample has appreciable curvature or waviness that would interfere with the measurement. A convenient way to do this if the sample is reasonably smooth and gives a nice specular reflection is to observe the reflection from an object having parallel bars or a regular grid, such as the pattern formed by a window frame or a pattern in a ceiling. To easily detect waviness, slightly rotate the sample and see if the shape of the lines or bars changes. A piece of a silicon wafer makes a good specimen to demonstrate the effect. The visual observation should always be done in a reasonably dust-free environment (depending on the type of surface being characterized). Remember to always hold the sample tightly in gloved fingers and do not breathe on it or "speak to it," to keep from putting spit marks on the surface. If it is necessary to transport the sample outside of its protective container for any distance, a good way to do it is to hold the sample so the surface is facing down; in this way, dust cannot settle on it.

If visual inspection is not adequate to see the desired surface detail, progressively higher magnifications can be used, starting with a magnifying glass or a 10× loupe. If available, a light microscope used at a low magnification, preferably with differential interference contrast or dark field viewing, is ideal. By observing the sample surface under the microscope and manually translating it, the entire surface of a small sample can be easily scanned. A higher magnification can be used to obtain more detail on a particular defect. Scanning electron microscopes are not helpful because it is very difficult to show detail on smooth surfaces; also, the sample must be coated with a conducting metal film in many SEMs. Furthermore, the surface will become more contaminated for longer inspection times.

An image of the sample surface is often desired. A film or digital camera may be used for low magnification photographs, preferably with oblique illumination to bring out surface detail. For higher magnification (50×–250×), a differential interference contrast (Nomarski) microscope is ideal,[46] (see also Chap. 2B in Ref. 1). Figure 1.7 shows a schematic diagram of a Nomarski microscope with an overall view in Fig. 1.7(a) and a detail of the two sheared images on the sample surface in Fig. 1.7(b). White light from an illuminator (often followed by a green filter) passes through a polarizer and then through a Wollaston prism, where it is split into two beams polarized orthogonally to each other. The microscope objective lens focuses this light into two overlapping spots on the surface whose centers are separated by a small distance, typically $\sim 1\ \mu$m, that depends on the magnification of the lens. Any small defects or slope variations on the surface will introduce a relative phase difference between the two beams. The reflected beams again pass through the lens and the Wollaston prism, interfering in the image plane. Each color or shade in the image is associated with a specific relative phase change between the two beams. By using a retarder/polarizer combination, the background color can be canceled, leaving the part of the image that is caused by surface defects, i.e., any features that have differences in height or optical constants from the surrounding surface. Surface detail is best seen on surfaces having moderate to high reflectance. However, with sufficient care, structure on a bare glass surface (4% reflectance) can also be seen.

Figure 1.8 shows Nomarski micrographs of four typical types of structure on optical surfaces. The image on the left (molybdenum polished) represents a normal polished surface that has different sizes of polishing scratches going in random directions, and also some isolated point defects. The second image (fused quartz bowl feed polished) is of a supersmooth optical surface made by a special bowl

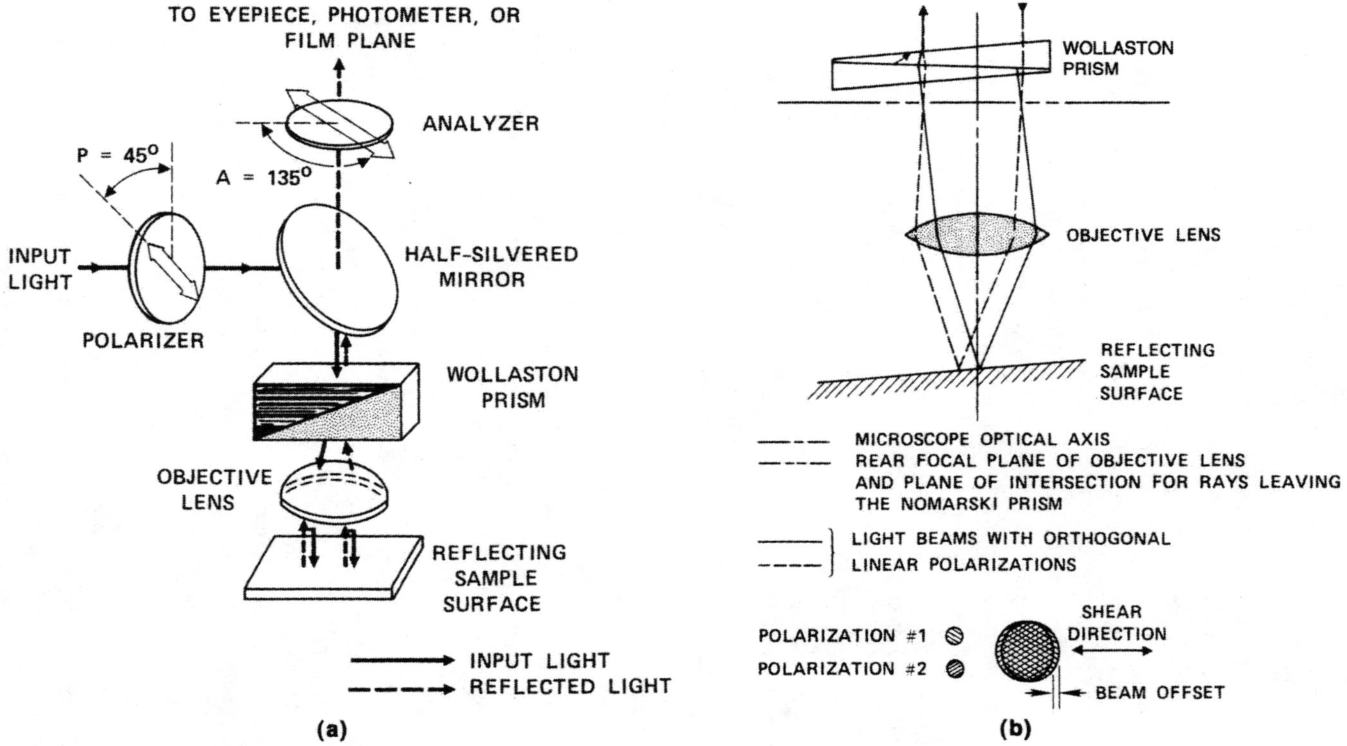

FIGURE 1.7. Schematic diagram of a Nomarski microscope showing (a) overall view and (b) detail of the two sheared images on the sample surface (Fig. 3 in Ref. 1, modified from Ref. 47).

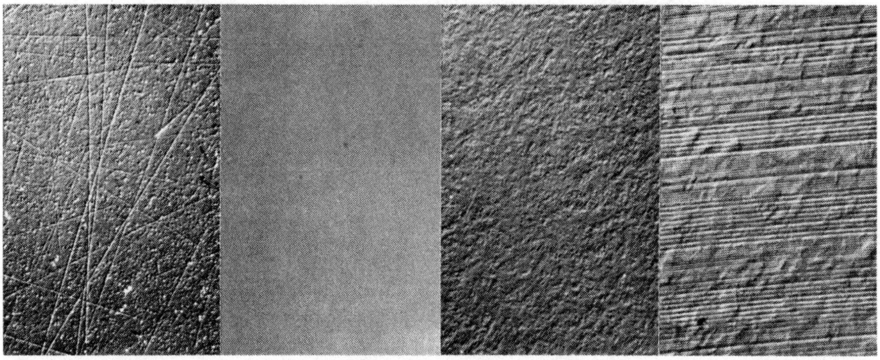

FIGURE 1.8. Nomarski micrographs of various types of surface structure that can occur on optical surfaces.

feed or recirculating feed polishing process[48] that leaves no visible surface defects. The tiny surface scratches are buried beneath a smooth, redeposited layer of fused silica. The third image (molybdenum polished) is of well-polished polycrystalline molybdenum that shows the grain structure on the surface of the material.[49] The fourth image on the right (copper diamond turned) shows a single point diamond-turned copper surface. Structure on this surface is a combination of the adjacent groves of the cutting tool, larger spaced, deeper machine vibration grooves, and grain structure of the polycrystalline copper sample. Many other kinds of surface texture can easily be seen on other surfaces.

Analogue and digital enhancement of Nomarski micrographs bring out even more detail on smooth surfaces, such as that of a silicon wafer shown in Fig. 1.9. In this way, the original featureless Nomarski micrograph can be transformed into the image in Fig. 1.9(f) that shows a structure whose heights are close to the atomic spacing in the silicon lattice. Note, however, that since the lateral resolution is not very large (each image covers $\sim$1 mm $\times$ 1 mm on the surface), individual atoms are not being imaged. Although it is relatively easy to photograph structure on surfaces, it is quite difficult to measure heights of the structure on the microscope images (see Chap. 2B in Ref. 1). One optical profiler can measure these heights using the Nomarksi principle of measuring surface slopes (see Sect. 1.2.4) although without obtaining an image of the surface.

There is a problem using a Nomarski microscope to observe or photograph a surface that is made up of a structure having variations in the optical constants. These produce different phase changes on reflection and thus appear to have different heights on the surface, even if the surface is perfectly flat. Randomly oriented crystallites in a material such as polycrystalline beryllium can also have different

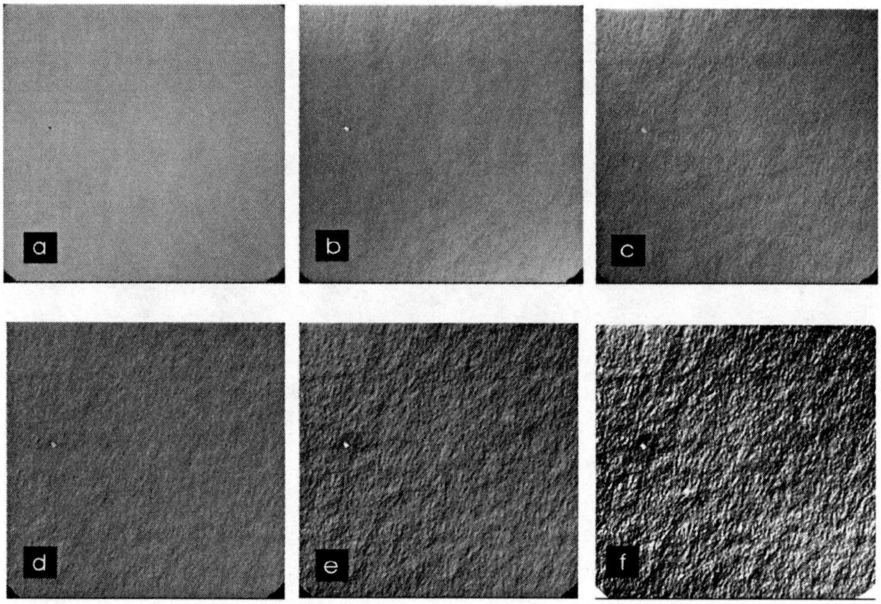

FIGURE 1.9. Nomarski micrographs of the same place on a single crystal silicon wafer surface with the data processed in different ways: (a) bright-field reflection image; (b) Nomarski image, no enhancement; (c) analogue contrast enhancement of (b); (d) background intensity leveling of (c); (e) digital contrast enhancement of (d); and (f) nonlinear digital contrast enhancement of (e) (Fig. 4 in Ref. 1).

phase changes on reflection on a polished surface and produce the same effect as a stepped surface.

Other more specialized types of microscopes can be used to form images of certain kinds of surfaces. The confocal microscope is excellent for showing the structure of fibrous materials such as paper (see Chap. 2E in Ref. 1); a TEM will show structure of multilayer films on cross sections of appropriately prepared samples, and an AFM can be used in various modes to show surface structure at a high lateral resolution (Sect. 1.2.6).

1.2.4. Optical Profilers

To obtain information about the actual topography of a surface, either in a line (profile) or as an area, either an optical profiler or a mechanical profiler may be used. The optical profiler is noncontact, so does not touch the surface, while a mechanical profiler has a diamond probe that contacts the surface and often leaves a line where it has moved across the surface. The optical profilers have either a linear detector array for measuring a profile or an area array for obtaining an area map, so they take data in parallel. Mechanical profilers take data sequentially, so they are much slower. Most surface topography is now measured with optical profilers.

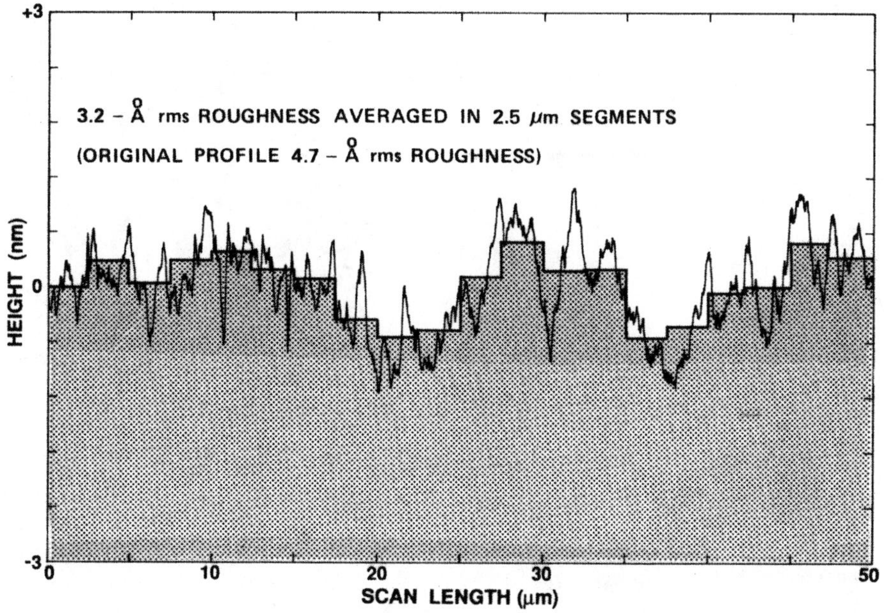

FIGURE 1.10. Schematic representation of the averaging effect of a one-dimensional surface profile made by a noncontact optical profiler (square bars) compared with a profile made with a stylus probe. The vertical axis has been greatly expanded compared to the horizontal axis (Fig. 12 in Ref. 1).

In spite of all the advantages optical profilers have over mechanical profilers, one disadvantage is that optical profilers have a lower lateral resolution for the same linear profile length, as shown in Fig. 1.10. The detector array used in optical profilers has a fixed number of pixels, normally 1024 × 1024, for images taken with all microscope objectives. The surface area corresponding to one pixel in the array depends on the wavelength of the illumination, microscope magnification, diffraction limit of the microscope objective, spacing between pixels, and other factors (see Chap. 3 in Ref. 1). A mechanical profiler has more flexibility and the user can choose the sampling interval, so that more data points can be measured per scan line than the fixed number of pixels in the detector array. Also, the sharp probe of a mechanical profiler has a better lateral resolution than the diffraction limited microscope objectives used for the optical profilers. The net result, as shown in Fig. 1.10, is that the measured rms roughness for one profile made by an optical profiler is smaller than the value measured by a sharp-pointed ($\sim 1\ \mu$m radius) diamond stylus on a mechanical profiler for the same place on the surface. A better lateral resolution can be obtained with an optical profiler by using a higher microscope magnification, but the total profile length will be shorter.

We will now describe the three main types of optical profilers that are currently in use. But first the historically important instrument that still has the lowest noise level of any of the optical profilers will be described. This instrument was

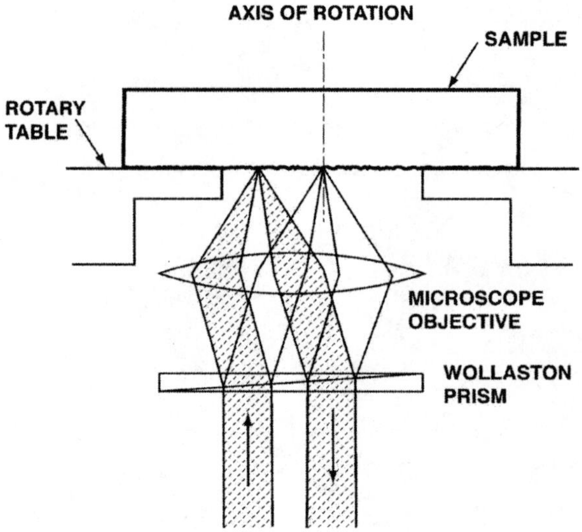

FIGURE 1.11. Schematic diagram of the optical measuring head of the Sommargren interferometer (modified from Ref. 12).

developed by Gary Sommargren while he was at Lawrence Livermore National Laboratory and was described in a 1981 publication.[12] Later, a limited number of instruments were manufactured by the Zygo Corporation.[50] A schematic diagram of this instrument is shown in Fig. 1.11. The common path interferometer uses a single mode He–Ne (632.8 nm) laser beam with its center frequency split by an axial magnetic field on the laser tube into two Zeeman components with a frequency difference of 2 MHz. The two-component, collinear beam is then split into two spatially separated, orthogonally polarized beams by a Wollaston prism which are then focused by a microscope objective onto the sample surface. The reference beam lies on the mechanical axis of a precision rotary table on which the sample is mounted face down. The test beam is offset by $\sim$160 μm. During a measurement, the table slowly rotates through a full circle. The test beam describes a small circle on the sample whose circumference is 1 mm. During the rotation, the two beams are recombined spatially but are slightly displaced from the incoming beam. The small phase differences between the test and reference beams are measured by a heterodyne technique, and are then converted into height differences. The circle is then unwrapped and is plotted as a linear profile. With low noise electronics, a profile repeatability of 0.006 nm can be consistently obtained. (Other optical profilers have a noise/repeatability level of $\sim$0.1 nm.)

A majority of commercial optical profilers use the principle of phase measuring interferometry (PMI) for accurately measuring small surface roughnesses (less than $\lambda/2$). Often the PMI principle is combined with a white light interferometer into one instrument. The original PMI instrument was developed by James C. Wyant and his PhD student Chris Koliopoulos at the Optical Sciences Center,

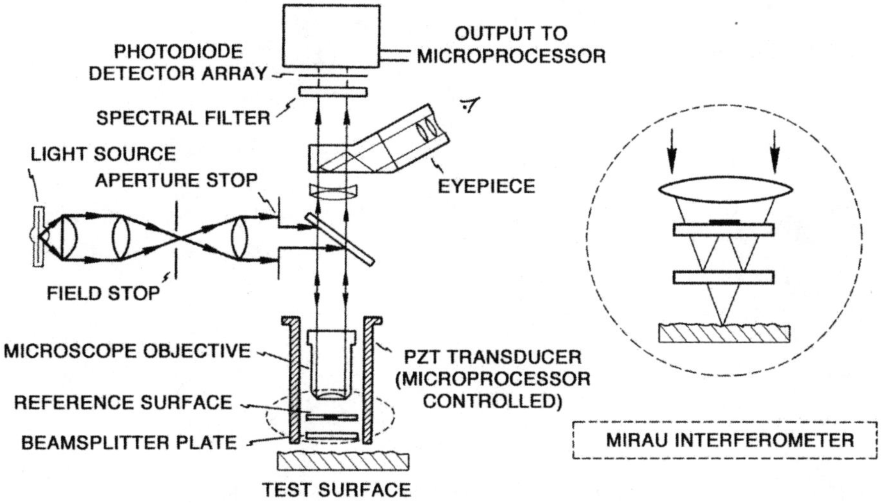

FIGURE 1.12. Schematic diagram of the phase measuring interferometer (modified from Veeco-WYKO literature, Ref. 51).

University of Arizona and described in a paper in 1985.[13] Information about the current PMI family of instruments can be obtained from Veeco Instruments, Inc.[51] A schematic diagram of this instrument is shown in Fig. 1.12 (Chap. 3, Ref. 1). The so-called 2D version that produces a line profile will be described here, although the more common type is now the 3D version that measures area topography on a sample. A beam of white light passes through a narrow-band red filter, is reflected by a beam splitter, and is focused by a microscope objective onto the sample surface (test surface). A beam splitter that is located between the objective lens and the test surface reflects part of the incident beam to a small, aluminized reference surface. After reflection from the test and reference surfaces, the beams recombine at the beam splitter, again pass through the objective lens and form interference fringes on the surface of a photodiode detector array. The fringes can also be seen in the eyepiece. The reference surface is slightly tilted relative to the test surface to form a small number (3 or 4) straight-line fringes. During the profile measurement, the objective lens, beam splitter, and reference surface are oscillated parallel to the optic axis with a piezoelectric transducer that sweeps the interference fringe pattern across the detector array. The array is read four times per oscillation. The number of oscillations is user selected and is normally between 4 and 64. In the analysis, the phase of the interference fringe pattern on each detector pixel is determined by an algorithm known as "averaging of buckets." The phase differences on the pixels are converted into height differences, and the profile is displayed on the monitor screen. The detector has a constant number of pixels, so the profile length and lateral resolution will vary depending on the magnification of the objective lens. The lowest magnification 1.5× lens produces an 8.87-mm-long profile and the highest magnification 200× lens in a Linnik interferometer

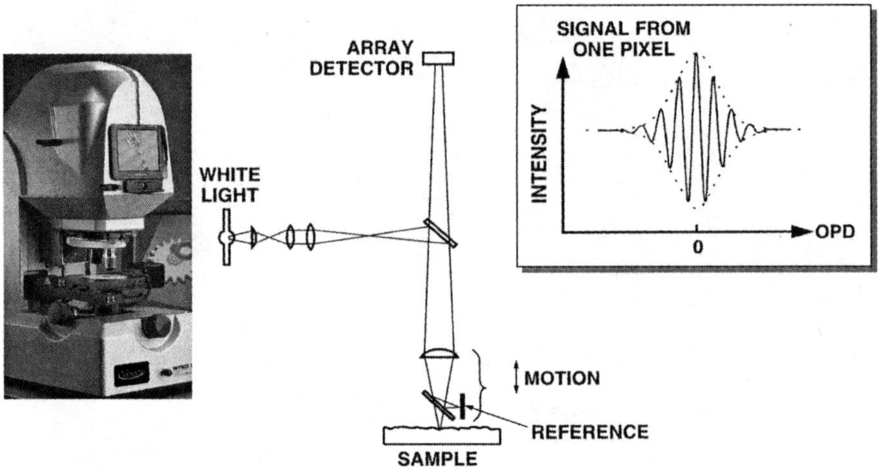

FIGURE 1.13. Schematic diagram of a white light interferometer showing a photograph of the instrument, schematic diagram of the optical components, and a plot of the electrical signal from one pixel as the optical path of the beam splitter–reference surface changes relative to the optical path of the beam splitter-sample – see text (modified from Veeco-WYKO literature, Ref. 51).

configuration yields a 66-μm-long profile. The roughness of the reference surface in the PMI instruments is ~0.6–0.8 nm rms and limits how smooth surfaces can be that are measured with this type of instrument unless one of several correction methods is used to eliminate the reference surface roughness.[52] If the roughness of the reference surface is eliminated, the instrument has a noise level and gives approximately repeatable roughness values ~0.1 nm rms.

The white light interferometer (WLI) follows the Michelson interferometer design but is used for a different application. The first commercial instrument became available around 1993. The WLI, shown schematically in Fig. 1.13, operates on a different principle from the PMI (Chap. 3 in Ref. 1 and references therein.) The white light source contains a continuum of all wavelengths in the visible spectral region and the reference surface is perpendicular to the optic axis of the instrument, in the configuration of a Michelson interferometer. During a measurement, the beam splitter and reference surface move in one direction away from the surface. Consider one tiny spot on the sample surface that is imaged on one pixel of the detector. At the start of the scan, the distance from the sample to the beam splitter is less than the distance from the beam splitter to the reference surface. Because of the white light illumination, there are no interference fringes because the different wavelengths have different equal path length distances and hence do not produce coherent interference fringes unless the path lengths of both arms of the interferometer are equal. Thus, there is a constant signal, as shown in the graph at right that plots the detector signal for one pixel as a function of optical path difference (between the beam splitter and sample surface and beam splitter and reference surface). As the equal path position is approached, the constant signal gradually changes to an oscillatory one because all the wavelengths are producing their interference maxima and minima coherently. When the two paths are exactly equal, the amplitude of the oscillation is a maximum. As the equal path length condition is passed, the oscillations damp out, and the signal again becomes constant. The envelope containing the signal oscillation normally has a Gaussian shape. The maximum of the envelope is analytically determined. Because the height of the beam splitter is correlated with the electrical signal, the equal path condition for the one pixel has been determined and can be saved. To obtain the maximum accuracy, various algorithms are used to determine the height above a preset reference level corresponding to the peak of the oscillatory signal. The same process occurs simultaneously for all the other pixels in the detector array, although the height corresponding to the oscillation maximum height is different because of the topography undulations on the sample surface. The result of the WLI measurement is a topographic map of a small surface area. Since the number of pixels in the detector array is constant, the size of the measured area, lateral resolution, and sampling distance will depend on the magnification and resolution of the microscope objective and other instrumental factors. The WLI does not have the small roughness limitation of the PMI instruments but the original instrument did not have as good height sensitivity because simple algorithms were used to locate the maxima of the oscillation envelope. Now much more sophisticated algorithms are being used so the height sensitivity has considerably improved. The maximum surface heights that can be measured with a WLI depend on the maximum translation of the beam splitter–reference surface unit and the mechanical rigidity of the instrument, and is normally several mm.

The third type of interferometer uses the principle of the differential interference contrast microscope that had been described in 1955 by Georges Nomarski.[46] In the 1984–1985 time frame, Jay Eastman and Jim Zavislan[14,15] proposed a profiling instrument that measures surface slopes (i.e., height differences between adjacent data points) and then integrates them to obtain a surface profile. As described in Chap. 3B of Ref. 1, a linearly polarized laser beam passes through a Wollaston prism where it is split into two beams that are slightly spatially separated and orthogonally polarized. A microscope objective lens focuses the two beams onto the sample surface as in a Nomarski microscope. The beams reflect from the surface, pass again through the Wollaston prism, are spatially combined, and continue to a polarizing beam splitter where interference fringes are formed. The polarizing beam splitter separates the two beams into their two polarized identities and directs them to two detectors. The difference between the detector signals divided by their sum is proportional to the height difference or surface slope between the average levels of the areas illuminated by the two beams. In taking a profile, the microscope objective lens–Wollaston prism combination moves along the surface to a maximum user-set length (currently 100 mm). The surface slopes are then integrated to produce a profile. If a topographic area image is desired, a translation stage steps the sample surface one unit in the y-direction, another scan is taken, and the process is continued. This is not a true topographic map because the surface height is assumed to be zero at the start of each scan.

1.2.5. Mechanical Profilers

Commercial mechanical profilers are primarily used in the machining industry and only a few models are designed specifically for profiling optical surfaces. These have stylus loadings that can be set to a few mg force so the stylus will not damage soft optical surfaces and coatings. Although most mechanical profilers are partly automated and require minimum operator skill, the sample must still be leveled so the profile remains within the height sensitivity range of the detector, a linear variable differential transformer (LVDT). Also, mechanical profilers take data sequentially along a line, so they are considerably slower than optical profilers. If the stylus speed is too fast, the stylus skips over parts of the surface and misses finer surface detail. If the stylus speed is too slow and the sampling distance is short, nonlinear drift can occur during data acquisition that cannot be corrected, yielding a curvature artifact on the surface. Additionally, the large number of data points may overfill the allocated memory and will be too large to process rapidly.

Once a surface has been profiled either with an optical or a mechanical profiler and height differences above or below a mean surface level have been determined, statistical properties of the surface can be calculated from the digitized data including root-mean-square (rms) roughness, peak-to-valley (PV) roughness, height and slope distribution functions, autocovariance function, power spectral density function, and many others (see Chap. 4 in Ref. 1).

1.2.6. Atomic Force Microscopes (AFM)

The AFM was originally developed to study single atoms and molecules, but has been adopted by the optics community to use for studying fine detail on small surface areas ($<100\,\mu$m $\times$ $100\,\mu$m to $\sim 1\,\mu$m $\times$ $1\,\mu$m or smaller). The operating principles of a commercial AFM are illustrated in Chap. 2D in Ref. 1, in Ref. 53, and in company literature. These instruments are able to operate in the contact mode with a tiny diamond probe touching the surface and moving over it in a raster mode and also in the noncontact or "tapping" mode in which the diamond probe oscillates above the surface and only touches it during a small fraction of the oscillation. Other scanning probe technologies have been incorporated into commercial AFMs. The most useful ones for surface characterization are the lateral force (friction) mode to show grain structure on a very smooth, polycrystalline surface, and the phase contrast mode that emphasizes, for example, grain boundaries in a deposited metal film (see Chap. 2D in Ref. 1). The standard AFM probes have radii ~10–50 nm and there are also special sharper probes with radii in the 4–10 nm range that can resolve even smaller lateral structure. A superpolished Zerodur (low expansion glass) surface was scanned with a 10 nm radius probe.[54] The harder β-SiO_2 crystallites with an average measured diameter of 11 nm, could be easily seen because they slightly protruded from the fused silica matrix, giving an overall roughness of 0.16 nm rms. The same surface had been previously profiled with an ~40 nm radius probe but the β-SiO_2 structure was not seen and the measured roughness was 0.065 nm rms. The structure on this Zerodur surface may be the finest surface detail profiled on an optical surface.

1.2.7. Total Integrated Scattering and Angle-Resolved Scattering

There are several commercial ARS instruments including one-of-a-kind instruments that have been built for special purposes. One instrument[41] is intended for measuring machined surfaces but is now being used to measure rougher optical surfaces. It has a near infrared light-emitting diode as a light source at a large angle of incidence and measures light scattered at different angles close to the specularly reflected beam in the plane of incidence. The rms roughness is calculated from first-order scattering theory. The original instrument measured roughness in the range from ~25 nm to ~0.25 μm, although rougher surfaces could be measured with suitable empirical calibration. Recently it has been redesigned to measure smoother surfaces with roughnesses as small as ~2.5 nm rms.

Another scatter-measuring instrument has recently been introduced.[42] ARS or light emitted from a surface can be measured simultaneously at closely spaced angles in the reflection hemisphere (or sphere) at a wavelength of 635 nm. There are six measurement ports in a geodesic structure on the hemisphere, with a pentagon-shaped lens as the first element of a multi-element lens on each port. Six 640 $\times$ 480 pixel video cameras (14 bit digitization increment giving a dynamic range of 1:16,000) take images at the rate of 15 frames/s (60 ms acquisition time). All cameras are synchronized so the images can be stitched together forming a complete hemispherical array of data points, having an angular resolution of 0.15°. A variety of software analysis programs are available.

All scattering measurements for small roughness surfaces rely on a theory to relate the scattered light from the surface to a roughness value. The assumptions of these theories are that the surface heights are small and the lateral correlation lengths are large compared to the illuminating wavelength (i.e., small surface slopes). The scattering is assumed to come from "correlated roughness" in which all the surface microirregularities act together, compared to scattering from isolated particles or point scatterers (see Ref. 1, Chap. 4 including the references). There are many scattering theories in the literature; some of these are used to interpret scattering measurements made in the laboratories of the groups that have developed the theories.

1.2.8. Surface Contamination and Cleaning

As mentioned in Sections 1.2.1 and 1.2.3, surfaces should be visually inspected before they are measured. It is much better to keep a surface clean than to have to clean it later. Thus, surfaces should be stored face down in dust-tight boxes, preferably in clean areas. If any cleaning is necessary, the surfaces should be inspected before, during (if possible), and after they are cleaned.

Cleaning optical surfaces is more of an art than a science. Many techniques are available depending on the type of surface and the type of contamination to be removed (see Chap. 8 in Ref. 1 and the references therein). If a surface is simply dusty, it can often be cleaned by directing a jet of clean air or dry nitrogen onto the surface or by using a hand air blower. Unprotected soft metal surfaces (silver, gold, aluminum, platinum, copper, etc.) will be damaged if they are touched by anything solid during the cleaning process. It is possible to carefully squirt a metal surface, held vertically, with a solvent such as water, alcohol, acetone, tricholorethylene, benzene, or another organic solvent. Alcohol (ethanol or methanol) is the best general purpose solvent. It drains off a clean surface in a thin sheet and drops of liquid can be blotted off the sample edge. Acetone is such an excellent solvent that it will dissolve many kinds of contamination that other solvents will not remove. It is also very fast drying. If drops of acetone are allowed to dry on a surface, white spots will remain that cannot be removed by any solvent, including acetone. A jet of clean air can be used to chase a solvent drop off the surface. However, an isolated drop indicates that there is still contamination remaining on the surface. Isolated particles on a hard surface can sometimes be removed by a cotton swab moistened in alcohol. Care is needed not to leave a larger footprint of dried residue on the surface.

Uncoated glass substrates can be washed in detergent, with or without ultrasonic agitation, rinsed with copious amounts of ultrapure water, spun, dripped, or air blown dry. Any paper, cloth, etc. that touches the wet surface will probably leave a smear that will affect the quality of deposited coatings. Other more specialized cleaning techniques are discussed in Chap. 8 of Ref. 1. When handling optics, unpowdered, polyethylene clean-room gloves should be worn. These gloves dissolve in acetone, so gloves of another material should be worn when using acetone. In summary, *never* handle optics with bare hands because dirt and oils on the fingers can affect the adhesion of coatings. *Never* talk to an optical surface unless a face mask is worn because spit marks can easily strike the surface (they can be removed with water but not with organic solvents). Also, no cleaning is better than bad cleaning.

1.3. Current and Future Surface Metrology Requirements

1.3.1. General Comments

There are many current unsolved surface metrology needs as well as desired future capabilities. These are for optical applications and also for other fields such as microelectronics, communications, medical optics, and military and space applications. They apply to integrated circuit technology, miniaturized components at the micro- and nanolevel such as MEMS (microelectronic and mechanical systems), components for optical fibers, ultraviolet and soft x-ray mirrors, and steep aspherics from normal size to meter size or larger. There are also current requirements to measure surface finish on difficult-to-reach surfaces such as gear teeth, undercut steps, and cavities. Some measurements simply require an accept–reject decision for mass production applications. In other cases, initial inspection of starting components (for example, large, expensive silicon wafers) is mandatory to keep from producing defective circuits.

It may be possible to customize or adapt existing metrology instruments, but new concepts are also urgently needed. Polarization information would be useful in scatter-mapping of components, either to detect defects or for remote sensing applications. New theories are necessary, for example, to fill the gap between first-order scalar and vector theories for small roughness and theories that cover large roughness and steep slopes. There is no microirregularity scattering theory for the region between the small and large roughness theories where the roughness is comparable to the illuminating wavelength. Better theories are needed for scattering from isolated particles (and clusters of particles) to be able to obtain information about sizes, shapes, particle densities, and composition from scattering measurements.

Advances in surface metrology are made by three routes: (a) Graduate students and their professors work on research problems at universities. This research often has a relaxed time frame, and sometimes no direct applications unless it is funded by a grant from a Department of Defense organization. (b) Research on pressing problems to meet industrial and military requirements is done in government laboratories and by industries with government contracts. Industrial laboratories also do short term applied research to develop new products that can be sold in a year or so. The short term research is necessary so the stockholders can see how their money is being spent and see the finished product. (3) Show stoppers are major problems holding up a large military program. Massive amounts of money are poured into such projects because the necessary research and development was not done previously in a timely manner. The bottom line is that military and industrial needs drive most of the development of new surface metrology that is required to measure new state-of-the art products.

New and improved surface metrology instruments are made possible by advances in unrelated fields. Integrated circuits with narrower line widths and miniature components in electronic circuits have made possible greatly increased computing capabilities in tinier packages. Parallel computing capabilities, i.e., linking many small computers together in parallel to perform repetitive operations, has made possible additional increases in computing speed. In the future, optical interconnects and all optical computers will greatly increase computing speed because the information will travel at the speed of light instead of at the speed of electrons. New metrology instruments are benefiting and will benefit in the future from all these advances in computers so they can be automated, faster, more accurate, and more user friendly.

New types of optical components have also become available. New types of detectors include high-density, high-performance array detectors that are spin-offs from the digital camera and video camera industries. Light emitting diodes (LEDs) and liquid crystal displays are cheap and readily available. The traditional cathode-ray tube monitors and low intensity, direction-sensitive, monochrome liquid crystal displays on laptop computers have been replaced with brighter, higher quality flat panel displays. Since the power requirements for these new units are much reduced, battery-operated devices are now possible.

Conventional optical light sources have also greatly improved. The bright, energy efficient quartz-halogen lamps are used for optical applications, for example, as light sources in metrology instruments, for microscope illuminators, and also for other applications such as in searchlights and lighthouses, replacing much larger, lower brightness, tungsten-filament bulbs. Automobiles now have quartz-halogen headlight bulbs and strips of red LEDs as brake light warning devices. Red, yellow, and green LEDs are used in long-lifetime traffic lights and "white" LEDs (combinations of red, yellow, and blue) are found in bright, long-life flashlight bulbs.

Miniature polarization components are being developed as part of nanosized optical systems. Although the extinction ratios of these components are not as good as for conventional, high-quality polarizers, they are more than adequate for the specific applications. There are also many piezo-electric and similar devices, making possible repeatable micromechanical motion.

All of the above components are available to be incorporated into the next generation of surface characterization instruments. The challenge is to interest creative, hardworking, clever people with innovative ideas to develop new types of instruments.

1.3.2. Metrology of Microcomponents

Optical MEMS (microelectronic machines) are interferometers and controllers that primarily use optical components such as tiny interferometers, mirrors, lenses, and polarizers along with mechanical devices to move them. Surfaces of interferometer mirrors need to be smooth, flat, and have low scatter coatings. Currently there are no instruments to test them. Optical fibers used as light pipes for communications, medical optics, and other applications, often have special requirements such as having low dispersion of travelling modes and preserving polarization properties. The surface between the core and the cladding must be as smooth as possible to prevent excessive scattering, leakage into the cladding, and loss in the fiber. Signals should be coupled into and out of fibers with maximized coupling efficiency. This means that the surfaces of the couplers should be smooth and low scatter. Integrated optical components on silicon wafers have tiny features that need to be measured during and after fabrication including undercut steps, film thicknesses, etch depths and line widths. In the machining industry, there are tiny mating surfaces such as gear teeth that must have controlled surface roughness to maximize lubrication and minimize excessive friction. These and other applications require the ability to measure feature dimensions, surface roughness, and scattering of very tiny components.

1.3.3. Metrology in the UV and Soft X-Ray Regions

The timetable for achieving integrated circuits with progressively narrower line widths (now approaching submicron widths), and the corresponding sizes of components is requiring lithographic projection optics to be made of materials that transmit extreme ultraviolet (EUV) and soft x-rays. (Sometimes mirrors must be used instead.) Calcium fluoride, the primary candidate material at these short wavelengths, cannot be polished as well as high quality fused silica. Since ARS is predicted to be inversely proportional to the fourth power of the wavelength, a surface having a given roughness will scatter four times as much at half the wavelength. Thus, techniques used for polishing EUV-transmitting materials must be greatly improved and ARS systems must be able to measure scattering in the EUV. Scattering theories must be further developed. Materials for high reflectance multilayer coatings must be found that can be used to produce low absorption, high reflectance, and low scatter multilayer coatings at EUV wavelengths. Because of the extremely short wavelengths, these multilayer films will be only a few atomic layers thick and must have extremely uniform thicknesses. Properties of these coatings must also be measured.

1.3.4. Metrology of Steeply-Curved Spherical or Aspheric Surfaces

Increasing numbers of optical systems now require steeply curved spherical or aspheric surfaces to increase light gathering power or to reduce the number of optical elements required. There are some new metrology instruments that can measure the optical figure of small areas of these surfaces, and then stitch them together by using appropriate stitching algorithms.[55] The instrument that measures each small area is currently a standard interferometer[50] that measures surface shape but not surface roughness. The current generation of metrology instruments cannot handle very small or very large steeply curved surfaces. In the future, this type of instrument will be improved to meet customer requirements.

1.3.5. Polarization of Scattered Light for Target Discrimination

Current military projects are looking at new methods to distinguish targets from backgrounds by remote sensing from satellites. Passive infrared radiation has been used for this purpose for many years in night vision instruments (for example to detect a camouflaged tank with its motor running in a forest). One of the methods being considered is to detect the polarization of the scattered light that is produced when a signal beamed from an active airborne system strikes a target. To achieve real-time data processing, such a system would have to be extremely sensitive and have high-throughput polarizers and analyzers with moderate extinction ratios. While this application does not require a theory to interpret the results, other applications using polarized scattered light do need a theory.

1.3.6. Automated, Rapid-Response Systems with Accept–Reject Capabilities

There are current requirements, and there will be many more in the future, for instruments and entire systems to operate remotely, without human intervention, to measure certain surface parameters and decide whether to accept or reject a particular item, or to instantly stop production before a catastrophic failure occurs. These systems require rapid data acquisition of topography, dimensions, feature heights, surface or bulk scattering, or other properties. Such systems are starting to look feasible, but clever sample handling, instrument mounting, and extremely fast number crunching of large data files are required. The instrument designer and programmer must take into account all possible scenarios in the automated system. At first the systems will be custom built, but as the demand grows they may later be made into product lines. There will be other pressing surface measurement problems in the future that will be needed to characterize instruments that are currently only ideas in clever people's minds. It has been said that if someone can think up a device that will perform a certain function, someone else will figure out how to build and test it. This is the way of the future for surface characterization instruments.

1.4. Summary

This chapter contains a brief history of the origins of surface characterization instruments along with the circumstances that expedited their development. We have described the current collection of useful, commercially available instruments and have mentioned some current and future requirements for more sophisticated instruments and techniques. Although this is primarily a metrology-oriented presentation, surface scattering theories are essential for interpreting measured scattering data. Surface characterization could be considerably improved if theories were better able to interpret the scattering measurements. Many other surface characterization topics that have been omitted from this brief presentation can be found in Ref. 1. Let us hope that, in another decade, all the current surface metrology problems will have been solved and that the next generation of surface characterization instruments will be much more accurate, user-friendly, faster, and operate remotely, if desired.

References

1. J.M. Bennett and L. Mattsson, *Introduction to Surface Roughness and Scattering*, 2nd edn (Optical Society of America, Washington, D.C., 1999).
2. F. Varnier, A. Llebaria, and G. Rasigni, "Improvement in the method of searching for the fractal nature of rough thin-film top surface," *J. Vac. Sci. Technol. A* **8**, 1554–1559 (1990).
3. F. Varnier, A. Llebaria, and G. Rasigni, "How is the fractal dimension of a thin film top surface connected with the roughness parameters and anisotropy of this surface?" *J. Vac. Sci. Technol. A* **9**, 563–569 (1991).
4. E. L. Church, "Fractal surface finish," *Appl. Opt.* **27**, 1518–1526 (1988).
5. M. Banning, "Practical methods of making and using multilayer filters," *J. Opt. Soc. Am.* **37**, 792–797 (1947).
6. J. Strong, "On a method of decreasing the reflection from nonmetallic substances," *J. Opt. Soc. Am.* **26**, 73–74 (1936).

7. J.H. McLeod and W.T. Sherwood, "A proposed method of specifying appearance defects of optical parts," *J. Opt. Soc. Am.* **35**, 136–138 (1945).
8. W.B. Estill and J.C. Moody, "Deformation caused by stylus tracking on thin gold film," *ISA Trans.* **5**, 373–378 (1966).
9. H.L. Eschbach and F. Verheyen, "Possibilities and limitations of the stylus method for thin film thickness measurements," *Thin Solid Films* **21**, 237–243 (1974).
10. J.M. Bennett and J.H. Dancy, "Stylus profiling instrument for measuring statistical properties of smooth optical surfaces," *Appl. Opt.* **20**, 1785–1802 (1981).
11. J.M. Bennett, "Comparison of instruments for measuring step heights and surface profiles," *Appl. Opt.* **24**, 3766–3772 (1985).
12. G.E. Sommargren, "Optical heterodyne profilometry," *Appl. Opt.* **20**, 610–618 (1981).
13. B. Bushan, J.C. Wyant, and C.I. Koliopoulos, "Measurement of surface topography of magnetic tapes by Mirau interferometry," *Appl. Opt.* **24**, 1489–1497 (1985).
14. J.M. Eastman and J.M. Zavislan, "A new optical surface microprofiling instrument," in *Precision Surface Metrology*, J.C. Wyant, ed., *Proc. Soc. Photo-Opt. Instrum. Eng.* **429**, 56–64 (1984).
15. J.M. Zavislan and J.M. Eastman, "Microprofiling of precision surfaces," in *Measurements and Effects of Surface Defects and Quality of Polish*, L.R. Baker and H.E. Bennett, eds., *Proc. Soc. Photo-Opt. Instrum. Eng.* **525**, 169–173 (1985).
16. T.K. Chinmayanandam, "Specular reflection from rough surfaces," *Phys. Rev.* **13**, 96–101 (1919).
17. P. Beckmann and A. Spizzichino, *The Scattering of Electromagnetic Waves from Rough Surfaces* (Pergamon Press, London, 1963).
18. E.L. Church and J.M. Zavada, "Residual surface roughness of diamond-turned optics," *Appl. Opt.* **14**, 1788–1795 (1975).
19. E.L. Church, H.A. Jenkinson, and J.M. Zavada, "Measurement of the finish of diamond-turned metal surfaces by differential light scattering," *Opt. Eng.* **16**, 360–374 (1977).
20. E.L. Church, H.A. Jenkinson, and J.M. Zavada, "Relationship between surface scattering and microtopographic features," *Opt. Eng.* **18**, 125–136 (1979).
21. E.L. Church and P.Z. Takacs, "BASIC program for power spectrum estimation," *Brookhaven National Laboratory Informal Report* 490.35 (May, 1993).
22. C.K. Carniglia, "Scalar scattering theory for multilayer optical coatings," *Opt. Eng.* **18**, 104–115 (1979).
23. J.M. Elson and R.H. Ritchie, "Photon interactions at a rough metal surface," *Phys. Rev. B* **4** 4129–4138 (1971).
24. E. Kröger and E. Kretschmann, "Scattering of light by slightly rough surfaces or thin films including plasma resonance emission," *Z. Phys.* **237**, 1–15 (1970).
25. J.M. Elson, "Infrared light scattering from surfaces covered with multiple dielectric overlayers," *Appl. Opt.* **16**, 2872–2881(1977).
26. J.M. Elson and J.M. Bennett, "Vector scattering theory," *Opt. Eng.* **18**, 116–124 (1979).
27. J.M. Elson, "Theory of light scattering from a rough surface with an inhomogeneous dielectric permittivity," *Phys. Rev.* **30**, 5460–5480 (1984).
28. P. Bousquet, F. Flory, and P. Roche, "Scattering from multilayer films: theory and experiment," *J. Opt. Soc. Am.* **71**, 1115–1123 (1981).
29. A.A. Maradudin and D.L. Mills, "Scattering and absorption of electromagnetic radiation by a semi-infinite medium in the presence of surface roughness," *Phys. Rev. B* **11**, 1392–1415 (1975).

30. V. Celli, A. Marvin, and F. Toigo, "Light scattering from rough surfaces," *Phys. Rev. B* **4**, 1779–1786 (1975).
31. M. Rasigni, G. Rasigni, J.-P. Palmari, and A. Llebaria, "Study of surface roughness using a microdensitometer analysis of electron micrographs of surface replicas: I. Surface profiles," *J. Opt. Soc. Am.* **71**, 1124–1133 (1981).
32. M. Rasigni, G. Rasigni, J.-P. Palmari, and A. Llebaria, "Study of surface roughness using microdensitometer analysis of electron micrographs of surface replicas. II: Autocovariance functions," *J. Opt. Soc. Am.* **71**, 1230–1237 (1981).
33. C. Amra, P. Roche, and E. Pelletier, "Interface roughness cross-correlation laws deduced from scattering diagram measurements on optical multilayers: effect of the material grain size," *J. Opt. Soc. Am. B* **4**, 1087–1093 (1987).
34. G. Binnig and H. Rohrer, "Scanning tunneling microscopy," *Helv. Phys. Acta* **55**, 726–735 (1982).
35. G. Binnig, H. Rohrer, Ch. Gerber, and E. Weibel, "Surface studies by scanning tunneling microscopy," *Phys. Rev. Lett.* **49**, 57–61 (1982).
36. R. Young, J. Ward, and F. Scire, "Observation of metal–vacuum–metal tunneling, field emission, and the transition region," *Phys. Rev. Lett.* **27**, 922–924 (1971).
37. R.D. Young, "Surface microtopography," *Phys. Today* **24**, 11, 42–49 (Nov. 1971).
38. R. Young, J. Ward, and F. Scire, "The topografiner: an instrument for measuring surface microtopography," *Rev. Sci. Instrum.* **43**, 999–1011 (1972).
39. R.A. Dragoset, R.D. Young, H.P. Layer, S.R. Mielczarek, E.C. Teague, and R.J. Celotta, "Scanning tunneling microscope applied to optical surfaces," *Opt. Lett.* **11**, 560–562 (1986).
40. G. Binnig, C.F. Quate, and Ch. Gerber, "Atomic force microscope," *Phys. Rev. Lett.* **56**, 930–933 (1986).
41. J. Glenn Valliant, Michael P. Foley, and Jean M. Bennett, "Instrument for on-line monitoring of surface roughness of machined surfaces," *Opt. Eng.* **39**, 3247–3254 (2000); www.opticaldimensions.com.
42. Engineering Synthesis Design, Inc., 310 S. Williams Blvd., Suite 210, Tucson, AZ 85711-4483; www.engsynthesis.com.
43. A. Duparré, J. Ferre-Borrull, S. Gliech, G. Notni, J. Steinert, and J.M. Bennett, "Surface characterization techniques for determining the root-mean-square roughness and power spectral densities of optical components," *Appl. Opt.* **41**, 154–171 (2002).
44. E.L. Church and P.Z. Takacs, "The optimal estimation of finish parameters," in *Optical Scatter: Applications, Measurements, and Theory*, J.C. Stover, ed., *Proc. Soc. Photo-Opt. Instrum. Eng.* **1530**, 71–78 (1991).
45. J.M. Elson, J.M. Bennett, and J.C. Stover, "Wavelength and angular dependence of light scattering from beryllium: comparison of theory and experiment," *Appl. Opt.* **32**, 3362–3376 (1993).
46. G. Nomarski, "Microinterféromètre differentiel à ondes polarisées," *J. Phys. Rad.* **16**, 9S–13S (1955); G. Nomarski and A. R. Weill, "Application à la métallographie des méthods interférentielles à deux ondes polarisées," *Rev. Metall. (Paris)* **52**, 121–134 (1955).
47. D.L. Lessor, J.S. Hartman, and R.L. Gordon, "Quantitative surface topography determination by Nomarski reflection microscopy. 1. Theory," *J. Opt. Soc. Am.* **69**, 357–366 (1979).
48. R.W. Dietz and J.M. Bennett, "Bowl feed technique for producing supersmooth optical surfaces," *Appl. Opt.* **5**, 881–882 (1966).

49. J.M. Bennett, S.M. Wong, and G. Krauss, "Relation between the optical and metallurgical properties of polished molybdenum mirrors," *Appl. Opt.* **19**, 3562–3584 (1980).
50. Zygo Corporation, Laurel Brook Road, P. O. Box 448, Middlefield, CT 06455-0448; www.zygo.com; Sommargren interferometer and 4 inch diameter interferometer for measuring optical figure.
51. Veeco Instruments, Inc., WYKO Metrology Group, 2650 E. Elvira Rd., Tucson, AZ 85706-7123; www.veeco.com.
52. K. Creath and J.C. Wyant, "Absolute measurement of surface roughness," *Appl. Opt.* **29**, 3823–3827 (1990).
53. L. Mattsson, "Scanning probe microscopy," in *Surface Characterization: A User's Guide*, D. Brune, R. Hellborg, H.J. Whitlow, and O. Hunderi, eds. (Wiley VCH, Weinheim, Germany, 1997), Chap. 3, pp. 77–81.
54. J.M. Bennett, "Surface roughness measurement," in *Optical Measurement Techniques and Applications*, P.K. Rastogi, ed. (Artech House, Inc., Norwood, MA, 1997), Chap. 12.
55. QED Technologies, Inc., 1040 University Avenue, Rochester, NY 14607; www.qedmrf.com.

2
The Kirchhoff and Related Approximations

A. G. VORONOVICH
NOAA/Earth Systems Research Laboratory, 325 Broadway, Boulder, CO 80305

2.1. Introduction

The Kirchhoff approximation (KA) is one of the two classical approaches in the theory of wave scattering. It represents the short-wavelength limit, which is similar to the semi-classical approximation. It is impossible to encompass in one standard-size paper all the aspects of the KA. Such questions as the criterion for the validity of the KA, shadowing and multiple reflections, statistical aspects, practical applications and comparisons with experiments are not even mentioned here. Instead this paper provides a review of some foundations of this approximation as well as some other approaches that appear naturally as its generalizations.

For the reader's convenience the paper tries to be self-contained; it avoids referencing textbooks for derivations of a few auxiliary relations such as the Helmholtz or Stratton–Chu formulas. They appear naturally in the context of the problem under consideration. Thus, this text can be used by a newcomer to the field. However, the sections related to the treatment of the problem of the scattering of electromagnetic waves seem to contain some novel results, which the author hopes might be of some interest to experts as well.

2.2. The Helmholtz Formula

The basis for developing the Kirchhoff approximation is the Helmholtz formula. It allows calculating the field at any point within a volume based on the values of the field on the boundary of the volume. Both the field itself and its normal derivative at the boundary are assumed to be known in this case. The volume may represent a compact domain or can be a half-space. The latter situation is depicted in Fig. 2.1. Usually the surface $\Sigma_{\mathbf{R}}$ is chosen to coincide with a physical boundary; however, in principle it may be arbitrary. Let the scalar field u satisfy the wave equation

$$(\nabla^2 + K^2)u = Q, \tag{1}$$

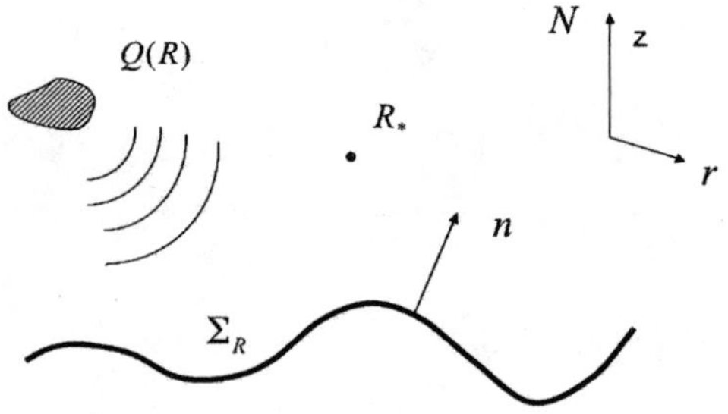

FIGURE 2.1. Geometry of the scattering problem.

where $K = \omega/c$ is a wavenumber and Q represents sources of radiation. Generally, K may have a positive imaginary part which is usually assumed to be small. In this case a solution has to remain bounded at large distances. This is called the limiting absorption principle. The introduction of small absorption allows avoiding singularities and/or making appropriate integrals convergent (as in Eq. (3)). In the final formulae one can usually set the absorption to zero so that K becomes real.

Let us introduce a free space Green function for the wave equation

$$(\nabla^2 + K^2)G = \delta(\mathbf{R} - \mathbf{R}_*). \tag{2}$$

On Fourier transforming this equation one finds

$$G(\mathbf{R} - \mathbf{R}_*) = \int \frac{e^{i\boldsymbol{\kappa}(\mathbf{R}-\mathbf{R}_*)}}{K^2 - \kappa^2} \frac{d\boldsymbol{\kappa}}{(2\pi)^3} = -\frac{e^{iK|\mathbf{R}-\mathbf{R}_*|}}{4\pi\, |\mathbf{R} - \mathbf{R}_*|}, \tag{3}$$

where the integral is easily calculated in spherical coordinates. Another representation for the Green function follows after integration over κ_z using the residue theorem

$$G(\mathbf{R} - \mathbf{R}_*) = -\frac{i}{8\pi^2} \int e^{i\mathbf{k}(\mathbf{r}-\mathbf{r}_*)+iq_k|z-z_*|} \frac{d\mathbf{k}}{q_k}. \tag{4}$$

Here $\mathbf{r}$, $\mathbf{r}_*$ are horizontal- and z, z_* vertical-projections of the 3D vectors $\mathbf{R}$, $\mathbf{R}_*$, respectively, and $\mathbf{k}$ is the horizontal projection of the 3D wave vector $\boldsymbol{\kappa}$. The function q_k represents the vertical component of the wave vector corresponding to a horizontal projection $\mathbf{k}$:

$$q_k = \sqrt{K^2 - k^2}, \qquad \mathrm{Im}q_k \geq 0. \tag{5}$$

The representation of the Green function in terms of a 2D Fourier transform, Eq. (4), is called the Weyl formula. Note that the order of arguments in the integral in Eq. (4) can be changed, and the difference $\mathbf{r} - \mathbf{r}_*$ can be replaced by $\mathbf{r}_* - \mathbf{r}$, without affecting the result.

Let us multiply Eq. (2) by u, Eq. (1) by G and subtract the results:

$$\nabla\,(u\nabla G - G\nabla u) = u(\mathbf{R})\delta(\mathbf{R} - \mathbf{R}_*) - Q(\mathbf{R})G(\mathbf{R} - \mathbf{R}_*). \tag{6}$$

Now we can integrate over the half space above the surface Σ using the Gauss theorem

$$\int \nabla \mathbf{A} d\mathbf{R} = \int \mathbf{A} \cdot \mathbf{n} d\Sigma_{\mathbf{R}}, \tag{7}$$

where the surface integral proceeds over the boundary of the integration volume and $\mathbf{n}$ is an external normal to the boundary. In our case part of the boundary includes a hemi-sphere of a sufficiently large radius encompassing all the sources and the observation point in Fig. 2.1. Now we have to use the limiting absorption principle again. Since K includes an arbitrarily small positive imaginary part, the Green function G decays exponentially at infinity, the solution u remains bounded in the worst case, and despite the fact that the surface area grows like a radius squared the integral over the hemi-sphere tends to zero when its radius goes to infinity. Thus, Eqs. (6),(7) give

$$u(\mathbf{R}_*) = u^{(\text{in})}(\mathbf{R}_*) - \int u \frac{\partial G(\mathbf{R} - \mathbf{R}_*)}{\partial \mathbf{n}_{\mathbf{R}}} d\Sigma_{\mathbf{R}} + \int \frac{\partial u}{\partial \mathbf{n}_{\mathbf{R}}} G(\mathbf{R} - \mathbf{R}_*) \, d\Sigma_{\mathbf{R}}. \tag{8}$$

Here we used usual notation for the normal derivative, namely $\partial/\partial \mathbf{n} = \mathbf{n}\nabla$, and took into account that the normal $\mathbf{n}$ in Fig. 2.1 is in fact the inner normal, which leads to the appropriate changes of sign. The first term in Eq. (8) represents an incident field which would be generated by the sources Q in the homogeneous space,

$$u^{(\text{in})}(\mathbf{R}_*) = \int Q(\mathbf{R}) G(\mathbf{R} - \mathbf{R}_*) \, d\mathbf{R}. \tag{9}$$

In what follows we will simply assume that the incident field represents a plane wave. The second and the third terms in Eq. (8) represent the scattered field generated by distributions of dipole and monopole sources with surface densities equal to $-u$ and $\partial u/\partial \mathbf{n}$, respectively.

Equation (8) is called the Helmholtz formula. It allows calculating the field at any observation point $\mathbf{R}_*$ if the field u and its normal derivative $\partial u/\partial \mathbf{n}$ on the surface Σ are both known. Of course, those two values are not independent. A boundary condition at infinity for our second-order wave equation has already been used: this is the choice of the causal Green function Eq. (3) selected with the help of the limiting absorption principle. Thus, only one value, u or $\partial u/\partial \mathbf{n}$ (or some linear combination of them) at the surface Σ can be assigned; the other value has to be determined in terms of the first one. This requires the solution of a certain integral equation (or a set of equations), and is equivalent to the solution of the scattering problem under investigation.

Let us assume now that the observation point $\mathbf{R}_*$ is in the lower half space. In this case the first term on the right-hand side of Eq. (6) does not contribute to the integral over the upper half space, and the left-hand side of Eq. (8) vanishes. Equation (8) with $\mathbf{R}_*$ belonging to the lower half space and $u(\mathbf{R}_*)$ on the left-hand side replaced by zero is called the extinction theorem, since it claims that the

FIGURE 2.2. To the calculation of the limiting values of the field.

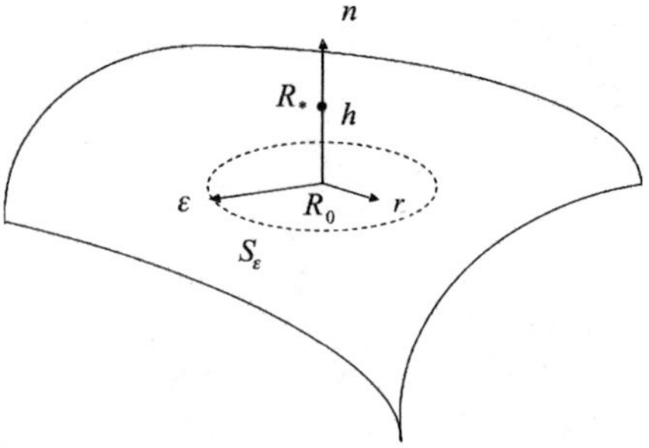

distribution of the dipoles and the monopoles on the surface cancels the incident field beyond the surface.

2.3. The Limit of the Observation Point Tending to the Surface

Another pillar in developing the KA is taking the limit in the Helmholtz formula, Eq. (8), when the observation point tends to the surface: $\mathbf{R}_* \to \Sigma$ (see Fig. 2.2.) For application to electromagnetic waves we will need a slightly more general result when the normal derivative of the Green function is replaced by its gradient:

$$\nabla_{\mathbf{R}} G(\mathbf{R}-\mathbf{R}_*) = \frac{\mathrm{i}K\,|\mathbf{R}-\mathbf{R}_*| - 1}{-4\pi\,|\mathbf{R}-\mathbf{R}_*|^3}(\mathbf{R}-\mathbf{R}_*)\,\mathrm{e}^{\mathrm{i}K|\mathbf{R}-\mathbf{R}_*|}. \tag{10}$$

Let us assume that the observation point $\mathbf{R}_*$ tends to the surface along the normal from outside so that $h \to +0$. One has

$$\int u\nabla_{\mathbf{R}} G(\mathbf{R}-\mathbf{R}_*)\,\mathrm{d}\Sigma_{\mathbf{R}} = \left(\int_{\Sigma/S_\varepsilon} + \int_{S_\varepsilon}\right) u\nabla_{\mathbf{R}} G(\mathbf{R}-\mathbf{R}_*)\,\mathrm{d}\Sigma_{\mathbf{R}}, \tag{11}$$

where S_ε represents a circle on the surface with a small radius ε. In this case Eq. (10) gives

$$\nabla_{\mathbf{R}} G(\mathbf{R}-\mathbf{R}_*) \approx \frac{\mathbf{r} - h\mathbf{n}}{4\pi(r^2+h^2)^{3/2}}. \tag{12}$$

When calculating the integral over S_ε one can replace u by the constant value u_0 corresponding to the center of the circle $\mathbf{R}_0$, and the piece of the generally uneven surface within the circle by a plane, and can integrate using polar coordinates:

$$\int u\nabla_{\mathbf{R}} G(\mathbf{R}-\mathbf{R}_*)\,\mathrm{d}\Sigma_{\mathbf{R}}$$

$$\approx u_0 \int_0^\varepsilon \int_0^{2\pi} \frac{r\cos\varphi\,\mathbf{e}_x + r\sin\varphi\,\mathbf{e}_y}{4\pi(r^2+h^2)^{3/2}} r\,\mathrm{d}r\,\mathrm{d}\varphi - u_0\mathbf{n}\int_0^\varepsilon \frac{h}{4\pi(r^2+h^2)^{3/2}} 2\pi r\,\mathrm{d}r,$$

where $\mathbf{e}_{x,y}$ are unit vectors along the x- and y-axes. The first integral above vanishes due to the angular integration, and we find

$$\int u\nabla_{\mathbf{R}} G(\mathbf{R}-\mathbf{R}_*)\,\mathrm{d}\Sigma_{\mathbf{R}} \approx -u_0\mathbf{n}\int_0^{\varepsilon} \frac{h}{4\pi(r^2+h^2)^{3/2}} 2\pi r\,\mathrm{d}r$$
$$= -\frac{u_0}{2}\mathbf{n}\int_0^{\varepsilon/h} \frac{x\,\mathrm{d}x}{(1+x^2)^{3/2}}. \tag{13}$$

Now we can set the observation point onto the surface taking the limit $h \to +0$ with the value of ε kept fixed. The upper limit of the integral in Eq. (13) tends then to infinity, and the integral itself tends to unity. Note, that if the observation point tended to the surface from below, so that $h \to -0$, then the signs in Eqs. (12),(13) would have changed. Now we can take the limit $\varepsilon \to 0$ which reduces to replacement of the integral over Σ/S_ε in Eq. (11) by a principal value integral. Thus we obtained the relation

$$\lim_{\mathbf{R}_*\to\mathbf{R}_0\in\Sigma} \int u\nabla_{\mathbf{R}} G(\mathbf{R}-\mathbf{R}_*)\,\mathrm{d}\Sigma_{\mathbf{R}} = \mp\frac{u(\mathbf{R}_0)}{2}\mathbf{n} + \text{v.p.}\int u\nabla_{\mathbf{R}} G(\mathbf{R}-\mathbf{R}_0)\,\mathrm{d}\Sigma_{\mathbf{R}}. \tag{14}$$

The upper (minus) sign here corresponds to the case of the observation point tending to the surface from the side to which normal is pointing, and the lower (plus) sign to the case when the observation point tends to the surface from the opposite side.

One can easily see that if the gradient of the Green function in Eq. (11) is replaced by the Green function itself then the integral over S_ε will tend to zero when $\varepsilon \to 0$. Thus, no extra terms appear in this case and one can simply replace $\mathbf{R}_*$ by $\mathbf{R}_0$,

$$\lim_{\mathbf{R}_*\to\mathbf{R}_0\in\Sigma} \int uG(\mathbf{R}-\mathbf{R}_*)\,\mathrm{d}\Sigma_{\mathbf{R}} = \int uG(\mathbf{R}-\mathbf{R}_0)\,\mathrm{d}\Sigma_{\mathbf{R}}. \tag{15}$$

The integrand here is regular since the $|\mathbf{R}-\mathbf{R}_0|^{-1}$ singularity of the Green function is compensated by the surface area element. The limit in Eq. (15) does not depend on the side of the surface from which the point $\mathbf{R}_*$ tends to $\mathbf{R}_0$.

2.4. Kirchhoff Approximation for the Neumann Problem

The simplest problem to analyze is the case of the Neumann boundary condition,

$$\frac{\partial u}{\partial \mathbf{n}} = 0, \qquad \mathbf{R} \in \Sigma. \tag{16}$$

The Helmholtz formula Eq. (8) in this case gives

$$u(\mathbf{R}_*) = u^{(\text{in})}(\mathbf{R}_*) - \int u(\mathbf{R})\frac{\partial G(\mathbf{R}-\mathbf{R}_*)}{\partial \mathbf{n}_{\mathbf{R}}}\,\mathrm{d}\Sigma_{\mathbf{R}}. \tag{17}$$

Now we can take the limit where the observation point $\mathbf{R}_*$ tends to $\mathbf{R}_0$ using Eq. (14) with the upper sign. Thus one finds

$$\frac{1}{2}u(\mathbf{R}_0) = u^{(\text{in})}(\mathbf{R}_0) - \int u(\mathbf{R})\frac{\partial G(\mathbf{R}-\mathbf{R}_0)}{\partial \mathbf{n}_\mathbf{R}}\,\mathrm{d}\Sigma_\mathbf{R}, \qquad \mathbf{R}, \mathbf{R}_0 \in \Sigma. \tag{18}$$

This is an integral equation with respect to the unknown value of the surface field $u(\mathbf{R})$.

If the surface Σ is a plane, then $\partial G(\mathbf{R}-\mathbf{R}_0)/\partial \mathbf{n}_{\mathbf{R}_0} = 0$, and

$$u(\mathbf{R}_0) = 2u^{(\text{in})}(\mathbf{R}_0) \tag{19}$$

provides an exact solution to Eq. (18). In the high-frequency limit one can hope that the solution of the integral equation Eq. (18) at point $\mathbf{R}_0$ is influenced mainly by the points in the vicinity of $\mathbf{R}_0$ (one has to assume, of course, that multiple reflections are absent). If the curvature of the surface at this point is not too large, one can approximate the surface in the vicinity of $\mathbf{R}_0$ by a plane and use Eq. (19) as an approximate solution. The other way to look at this approximation is to solve Eq. (18) by iteration,

$$u(\mathbf{R}_0) = 2u^{(\text{in})}(\mathbf{R}_0) - 4\int u^{(\text{in})}(\mathbf{R})\frac{\partial G(\mathbf{R}-\mathbf{R}_0)}{\partial \mathbf{n}_\mathbf{R}}\,\mathrm{d}\Sigma_\mathbf{R} + \cdots. \tag{20}$$

On neglecting in Eq. (20) all higher-order terms we obtain the KA, Eq. (19).

Now we can explicitly calculate the total field u at any point $\mathbf{R}_*$ by substituting Eq. (19) into Eq. (17)

$$u(\mathbf{R}_*) = u^{(\text{in})}(\mathbf{R}_*) - 2\int u_{\text{in}}(\mathbf{R})\frac{\partial G(\mathbf{R}-\mathbf{R}_*)}{\partial \mathbf{n}_\mathbf{R}}\,\mathrm{d}\Sigma_\mathbf{R}. \tag{21}$$

2.5. Scattering Amplitude

It is convenient to fix a certain basis of incident and scattered waves and describe scattering in terms of a scattering amplitude (SA) with respect to this basis. We will be interested mostly in wave scattering from infinite, uneven, plane on average surfaces located in the area adjacent to the level $z = 0$. In this case a natural choice is the following basis of plane waves:

$$u^{(\text{in})} = \frac{1}{\sqrt{q_0}}\,\mathrm{e}^{\mathrm{i}\mathbf{k}_0\mathbf{r}-\mathrm{i}q_0 z} \tag{22}$$

$$u^{(\text{sc})} = \frac{1}{\sqrt{q_k}}\,\mathrm{e}^{\mathrm{i}\mathbf{k}\mathbf{r}+\mathrm{i}q_k z} \tag{23}$$

so that

$$u(\mathbf{r}, z) = \frac{1}{\sqrt{q_0}}\,\mathrm{e}^{\mathrm{i}\mathbf{k}_0\mathbf{r}-\mathrm{i}q_0 z} + \int S(\mathbf{k}, \mathbf{k}_0)\frac{1}{\sqrt{q_k}}\,\mathrm{e}^{\mathrm{i}\mathbf{k}\mathbf{r}+\mathrm{i}q_k z}\,\mathrm{d}\mathbf{k}. \tag{24}$$

Here we introduced the notation $q_0 = q_{k_0}$, and omitted the asterisk which previously indicated the coordinates of the observation point. The factors $q_0^{-1/2}$, $q_k^{-1/2}$ are introduced here to make the resulting formulae for SA more symmetric. Such a normalization corresponds to waves with a unit energy flux in the vertical direction. The SA $S(\mathbf{k}, \mathbf{k}_0)$ is a complex quantity that represents the amplitude of the process of the scattering of a downward propagating plane wave with the horizontal projection of the wave vector equal to $\mathbf{k}_0$ into an upward propagating plane wave with the horizontal projection of the wave vector equal to $\mathbf{k}$. Since an arbitrary incident field can be expanded in terms of a superposition of plane waves, calculation of SA is essentially equivalent to the solution of the scattering problem. The first statistical moment of the SA gives the so-called average reflection coefficient, and the second moment is directly related to the scattering cross-section, which can be measured experimentally [6].

Let the surface Σ be described by the equation

$$z = h(\mathbf{r}). \tag{25}$$

It is assumed in Eq. (24) that the observation point is in the upper half-space beyond the excursions of the surface: $z > \max h(\mathbf{r})$. Substituting into Eq. (17) the Weyl representation of the Green function Eq. (4) and comparing the result with Eq. (24) one finds

$$S(\mathbf{k}, \mathbf{k}_0) = \frac{1}{8\pi^2} \frac{1}{\sqrt{q_k}} \int u(\mathbf{R})\, (q_k - \mathbf{k}\nabla h)\, \mathrm{e}^{-\mathrm{i}\mathbf{kr} - \mathrm{i}q_k h(\mathbf{r})}\, \mathrm{d}\mathbf{r}, \tag{26}$$

where the surface density $u(\mathbf{R})$ is a solution of Eq. (18) corresponding to the incident plane wave Eq. (22). We have used here that the unit normal $\mathbf{n}_\mathbf{r}$ is given by the formula

$$\mathbf{n}_\mathbf{r} = \frac{\mathbf{N} - \nabla h}{\sqrt{1 + (\nabla h)^2}}, \tag{27}$$

where $\mathbf{N} = (0, 0, 1)$ is a unit vector directed in the positive direction of the z-axis and ∇h is a 2D vector in the horizontal plane, whence

$$\frac{\partial}{\partial \mathbf{n}_\mathbf{r}} \mathrm{d}\Sigma_\mathbf{r} = \left(\frac{\partial}{\partial z} - \nabla h \cdot \nabla \right) \mathrm{d}\mathbf{r}. \tag{28}$$

On substituting for $u(\mathbf{R})$ in Eq. (26) the KA Eq. (19), one finds

$$S(\mathbf{k}, \mathbf{k}_0) = \frac{1}{\sqrt{q_k q_0}} \int (q_k - \mathbf{k}\nabla h)\, \mathrm{e}^{-\mathrm{i}(\mathbf{k}-\mathbf{k}_0)\mathbf{r} - \mathrm{i}(q_k + q_0)h(\mathbf{r})} \frac{\mathrm{d}\mathbf{r}}{(2\pi)^2}. \tag{29}$$

The term proportional to ∇h here can be integrated by parts by the use of the relation

$$\nabla h e^{-i(q_k+q_0)h(\mathbf{r})} = \frac{\mathrm{i}}{q_k + q_0} \nabla \mathrm{e}^{-\mathrm{i}(q_k+q_0)h(\mathbf{r})}. \tag{30}$$

This results in the following replacement in Eq. (29):

$$\nabla h \to -\frac{\mathbf{k}-\mathbf{k}_0}{q_k+q_0}. \tag{31}$$

Finally, one finds

$$S(\mathbf{k},\mathbf{k}_0) = \frac{1}{\sqrt{q_k q_0}} \frac{K^2 + q_k q_0 - \mathbf{k}\mathbf{k}_0}{q_k+q_0} \int \mathrm{e}^{-\mathrm{i}(\mathbf{k}-\mathbf{k}_0)\mathbf{r}-\mathrm{i}(q_k+q_0)h(\mathbf{r})} \frac{\mathbf{dr}}{(2\pi)^2}. \tag{32}$$

Equation (32) is the expression for the SA for the Neumann problem in the KA. Since the KA has a coordinate-invariant nature, the SA in Eq. (32) possesses the following two transformational properties. Namely, if the surface profile is shifted horizontally by a constant vector **d**,

$$h(\mathbf{r}) \to h\left(\mathbf{r}-\mathbf{d}\right), \tag{33}$$

then the SA is transformed according to the relation

$$S_{h(\mathbf{r}-\mathbf{d})}(\mathbf{k},\mathbf{k}_0) = S_{h(\mathbf{r})}(\mathbf{k},\mathbf{k}_0)\cdot \mathrm{e}^{-\mathrm{i}(\mathbf{k}-\mathbf{k}_0)\mathbf{d}}. \tag{34}$$

Similarly, if the surface profile is shifted in the vertical direction by a constant value H,

$$h(\mathbf{r}) \to h\left(\mathbf{r}\right) + H, \tag{35}$$

then

$$S_{h(\mathbf{r})+H}(\mathbf{k},\mathbf{k}_0) = S_{h(\mathbf{r})}(\mathbf{k},\mathbf{k}_0)\cdot \mathrm{e}^{-\mathrm{i}(q_k+q_0)H}. \tag{36}$$

The transformational properties, Eqs. (34),(36), have a purely geometric origin and are exact. They can be easily verified by appropriate shifts of the coordinate system.

2.6. The Tangent Plane Approximation

This approximation is based on the same physical assumptions as the KA and often leads to the identical results. However, its derivation does not require solution of the integral equation, and it can be obtained directly from the Helmholtz formula based on the following considerations. If the plane wave given by Eq. (22) impinges upon the plane $z = 0$, one finds for the resulting total field

$$u = (1 + V(q_0))\, u^{(\mathrm{in})} \tag{37}$$

$$\frac{\partial u}{\partial z} = -\mathrm{i}q_0\,(1 - V(q_0))\, u^{(\mathrm{in})}, \tag{38}$$

where $V(q_0)$ is the reflection coefficient. Now we can try to approximate the uneven surface Σ at each point by a plane tangent to the surface at this point and assume that the total field and its normal derivative at the point are locally given by

Eqs. (37),(38). We substitute those expressions into the Helmholtz formula Eq. (8) and find for the scattered field

$$u_{\mathrm{sc}}(\mathbf{R}_*) = -\int (1 + V(q_t))\, u^{(\mathrm{in})} \frac{\partial G(\mathbf{R} - \mathbf{R}_*)}{\partial \mathbf{n}_{\mathbf{R}}}\, \mathrm{d}\Sigma_{\mathbf{R}} - \mathrm{i}\int q_t\, (1 - V(q_t))\, u^{(\mathrm{in})} G(\mathbf{R} - \mathbf{R}_*)\, \mathrm{d}\Sigma_{\mathbf{R}}. \tag{39}$$

Here $q_t = q_t(\mathbf{R})$ represents the modulus of the projection of the incident wave vector $(\mathbf{k}_0, q_0)$ onto a local normal to the surface. Substituting for u_{in} the expression given by Eq. (22), for the Green function the Weyl representation Eq. (4), and proceeding exactly as in the preceding Section, we find the following expression for the SA:

$$S(\mathbf{k}, \mathbf{k}_0) = -\frac{1}{2\sqrt{q_k q_0}} \int \big[(-q_k + \mathbf{k}\nabla h)(1 + V(q_t)) + q_t\, (1 - V(q_t))\, (1 + (\nabla h)^2)^{1/2} \big]\, \mathrm{e}^{-\mathrm{i}(\mathbf{k}-\mathbf{k}_0)\mathbf{r} - \mathrm{i}(q_k+q_0)h(\mathbf{r})} \frac{\mathrm{d}\mathbf{r}}{(2\pi)^2}, \tag{40}$$

where

$$q_t = \frac{q_0 + \mathbf{k}_0 \nabla h}{\sqrt{1 + (\nabla h)^2}}. \tag{41}$$

As we saw, the KA requires smoothness of the surface profile on the scale of a wavelength. For such surfaces one can try to evaluate the integral in Eq. (40) by the stationary phase method. The stationary points are determined from the requirement that the gradient of the exponent be equal to zero, whence

$$\nabla h = -\frac{\mathbf{k} - \mathbf{k}_0}{q_k + q_0}. \tag{42}$$

We do not need to assume here that there is only one stationary point: Eq. (42) may have multiple solutions. However, all the stationary points correspond to the same values of ∇h and q_t. We will approximate the preexponential factor in Eq. (40) by its value calculated at the stationary points. A simple calculation gives

$$S(\mathbf{k}, \mathbf{k}_0) = V(q_t) \frac{K^2 + q_k q_0 - \mathbf{k}\mathbf{k}_0}{(q_k + q_0)\sqrt{q_k q_0}} \int \mathrm{e}^{-\mathrm{i}(\mathbf{k}-\mathbf{k}_0)\mathbf{r} - \mathrm{i}(q_k+q_0)h(\mathbf{r})} \frac{\mathrm{d}\mathbf{r}}{(2\pi)^2}, \tag{43}$$

where the expression for q_t follows after substitution of Eq. (42) into Eq. (41)

$$q_t = \sqrt{\frac{K^2 + q_k q_0 - \mathbf{k}\mathbf{k}_0}{2}}. \tag{44}$$

Equation (43) provides a general expression for the SA of the scalar waves in the tangent plane approximation. Strictly speaking, the expression (Eq. (40)) is more general. However, the difference between them, probably, exceeds the accuracy of the Kirchhoff-tangent plane approximation itself.

2.7. Scattering of Electromagnetic Waves from the Interface Between Dielectric Half-Spaces

We will assume now that the upper half-space in Fig. 2.1 corresponds to a medium with a (complex) dielectric permittivity ε_1 and the lower half-space to a medium with a dielectric permittivity ε_2. The magnetic permeability of both media is assumed to be equal to unity. In this paper we will be interested in calculating the scattered field in the upper half-space only.

The electromagnetic field in both half-spaces is governed by the Maxwell equations

$$\mathrm{i}\frac{\omega}{c}\mathbf{H} = \nabla \times \mathbf{E} \tag{45}$$

$$-\mathrm{i}\frac{\omega}{c}\varepsilon\mathbf{E} = \nabla \times \mathbf{H}, \tag{46}$$

with the boundary conditions

$$\mathbf{n} \times \mathbf{E}^{(1)} = \mathbf{n} \times \mathbf{E}^{(2)} \tag{47}$$

$$\mathbf{n} \times \mathbf{H}^{(1)} = \mathbf{n} \times \mathbf{H}^{(2)} \tag{48}$$

that express continuity of the tangential components of the electric and magnetic fields across the boundary (superscripts 1 and 2 here and below refer to the upper and the lower half-space, respectively). It follows from Eqs. (45) and (46) that in this case the normal components of the fields at the boundary satisfy the equations

$$\varepsilon_1 \mathbf{n} \cdot \mathbf{E}^{(1)} = \varepsilon_2 \mathbf{n} \cdot \mathbf{E}^{(2)}, \qquad \mathbf{n} \cdot \mathbf{H}^{(1)} = \mathbf{n} \cdot \mathbf{H}^{(2)}. \tag{49}$$

The Maxwell equations (45)–(46) have solutions in the form of plane waves with two polarizations. Similar to Eqs. (22),(23) we will select the following basis for the incident and scattered waves in the upper medium

$$\mathbf{E}^{(\mathrm{in})} = \mathbf{e}_\alpha^{(\mathrm{in})}(\mathbf{k}_0)\left(q_0^{(1)}\right)^{-1/2} \mathrm{e}^{\mathrm{i}\mathbf{k}_0\mathbf{r} - \mathrm{i}q_0^{(1)}z}, \qquad \mathbf{E}^{(\mathrm{sc})} = \mathbf{e}_\alpha^{(\mathrm{sc})}(\mathbf{k})\left(q_k^{(1)}\right)^{-1/2} \mathrm{e}^{\mathrm{i}\mathbf{k}\mathbf{r} + \mathrm{i}q_k^{(1)}z} \tag{50}$$

$$\mathbf{H}^{(\mathrm{in})} = \mathbf{h}_\alpha^{(\mathrm{in})}(\mathbf{k}_0)\left(q_0^{(1)}\right)^{-1/2} \mathrm{e}^{\mathrm{i}\mathbf{k}_0\mathbf{r} - \mathrm{i}q_0^{(1)}z}, \qquad \mathbf{H}^{(\mathrm{sc})} = \mathbf{h}_\alpha^{(\mathrm{sc})}(\mathbf{k})\left(q_k^{(1)}\right)^{-1/2} \mathrm{e}^{\mathrm{i}\mathbf{k}\mathbf{r} + \mathrm{i}q_k^{(1)}z}, \tag{51}$$

where the index $\alpha = 1, 2$ refers to the vertical and horizontal polarization, respectively. The unit polarization vectors $\mathbf{e}_\alpha$, $\mathbf{h}_\alpha$ are given by the following expressions

$$\begin{aligned} \mathbf{e}_1^{(\mathrm{in})}(\mathbf{k}_0) &= \tfrac{1}{K^{(1)}}\left(k_0\mathbf{N} + \tfrac{q_0^{(1)}}{k_0}\mathbf{k}_0\right) \qquad & \mathbf{e}_1^{(\mathrm{sc})}(\mathbf{k}) &= \tfrac{1}{K^{(1)}}\left(k\mathbf{N} - \tfrac{q_k^{(1)}}{k}\mathbf{k}\right) \\ \mathbf{h}_1^{(\mathrm{in})}(\mathbf{k}_0) &= -\tfrac{\mathbf{N}\times\mathbf{k}_0}{k_0} & \mathbf{h}_1^{(\mathrm{sc})}(\mathbf{k}) &= -\tfrac{\mathbf{N}\times\mathbf{k}}{k} \end{aligned} \tag{52}$$

$$\begin{aligned} \mathbf{e}_2^{(\mathrm{in})}(\mathbf{k}_0) &= \tfrac{\mathbf{N}\times\mathbf{k}_0}{k_0} & \mathbf{e}_2^{(\mathrm{sc})}(\mathbf{k}) &= \tfrac{\mathbf{N}\times\mathbf{k}}{k} \\ \mathbf{h}_2^{(\mathrm{in})}(\mathbf{k}_0) &= \tfrac{1}{K^{(1)}}\left(k_0\mathbf{N} + \tfrac{q_0^{(1)}}{k_0}\mathbf{k}_0\right) \qquad & \mathbf{h}_2^{(\mathrm{sc})}(\mathbf{k}) &= \tfrac{1}{K^{(1)}}\left(k\mathbf{N} - \tfrac{q_k^{(1)}}{k}\mathbf{k}\right). \end{aligned} \tag{53}$$

Here $\mathbf{N} = (0, 0, 1)$ is a unit vector in the positive direction of the z-axis, $K^{(1)} = \sqrt{\varepsilon_1}\omega/c$, and

$$q_k^{(1)} = \sqrt{\left[K^{(1)}\right]^2 - k^2}, \qquad \text{Im } q_k^{(1)} \geq 0. \tag{54}$$

One can easily check that the Maxwell equations are satisfied for each of these four sets of plane waves.

The scattering amplitude is introduced by the following representation of the fields in the upper half-space:

$$\mathbf{E} = \mathbf{e}_\beta^{(\text{in})}(\mathbf{k}_0)\frac{e^{i\mathbf{k}_0\mathbf{r} - iq_0^{(1)}z}}{\sqrt{q_0^{(1)}}} + \int S_{\alpha\beta}(\mathbf{k}, \mathbf{k}_0)\mathbf{e}_\alpha^{(\text{sc})}(\mathbf{k})\frac{e^{i\mathbf{k}\mathbf{r} + iq_k^{(1)}z}}{\sqrt{q_k^{(1)}}}\, d\mathbf{k} \tag{55}$$

$$\frac{\mathbf{H}}{\sqrt{\varepsilon_1}} = \mathbf{h}_\beta^{(\text{in})}(\mathbf{k}_0)\frac{e^{i\mathbf{k}_0\mathbf{r} - iq_0^{(1)}z}}{\sqrt{q_0^{(1)}}} + \int S_{\alpha\beta}(\mathbf{k}, \mathbf{k}_0)\mathbf{h}_\alpha^{(\text{sc})}(\mathbf{k})\frac{e^{i\mathbf{k}\mathbf{r} + iq_k^{(1)}z}}{\sqrt{q_k^{(1)}}}\, d\mathbf{k} \tag{56}$$

(here and in what follows summation over a repeated index is assumed). The SA is now a 2×2 matrix $S_{\alpha\beta}(\mathbf{k}, \mathbf{k}_0)$, which describes the amplitude of the process of scattering of the downward propagating incident plane wave of polarization $\beta = 1, 2$ and horizontal projection of the wave vector equal to $\mathbf{k}_0$ into an upward propagating plane wave of polarization $\alpha = 1, 2$ and horizontal projection of the wave vector equal to $\mathbf{k}$. Clearly, Eqs. (55),(56) represent a solution of the Maxwell equations for an arbitrary matrix S. The expression for S follows from the boundary conditions. In the case of the plane boundary $z = 0$ the SA reduces to specular reflection,

$$S_{11}(\mathbf{k}, \mathbf{k}_0) = V_1\delta(\mathbf{k} - \mathbf{k}_0), \qquad S_{22}(\mathbf{k}, \mathbf{k}_0) = V_2\delta(\mathbf{k} - \mathbf{k}_0), \qquad S_{12} = S_{21} = 0, \tag{57}$$

where V_1 and V_2 are the Fresnel reflection coefficients for vertical and horizontal polarizations, correspondingly,

$$V_1 = \frac{\varepsilon_2 q_k^{(1)} - \varepsilon_1 q_k^{(2)}}{\varepsilon_2 q_k^{(1)} + \varepsilon_1 q_k^{(2)}}, \qquad V_2 = \frac{q_k^{(1)} - q_k^{(2)}}{q_k^{(1)} + q_k^{(2)}}, \tag{58}$$

while

$$q_k^{(2)} = \sqrt{\left[K^{(2)}\right]^2 - k^2}, \qquad \text{Im } q_k^{(2)} \geq 0,\ K^{(2)} = \sqrt{\varepsilon_2}\frac{\omega}{c} \tag{59}$$

is the vertical projection of the wave vector in the second (lower) half-space. Apparently, the reflection coefficients can be considered as functions of $q^{(1)}$ only.

To obtain the expression for the SA in the tangent plane approximation we can use the general expression Eq. (43) for each polarization. However, this will be correct only if the waves correspond to the waves of the vertical and horizontal polarization in the basis associated with the local tangent plane. For this reason the original incident polarization vectors in Eqs. (52),(53) should be first projected onto the polarization vectors associated with the local tangent plane, then formula

Eq. (43) is applied to each polarization in the local basis, and then the resulting field is projected again on the polarization vectors of the scattered field in Eqs. (52), (53). This results in the expression Eq. (43), where the scalar reflection coefficient V is replaced by the 2×2 matrix $R(\mathbf{k}, \mathbf{k}_0)$

$$R(\mathbf{k}, \mathbf{k}_0) = T_1(\mathbf{n}, \mathbf{k}) \begin{pmatrix} V_1(q_t) & 0 \\ 0 & V_2(q_t) \end{pmatrix} T_2(\mathbf{n}, \mathbf{k}_0), \tag{60}$$

with

$$T_1(\mathbf{n}, \mathbf{k}) = \frac{1}{k\sqrt{\left[K^{(1)}\right]^2 - q_t^2}} \begin{pmatrix} k^2\mathbf{N}\cdot\mathbf{n} - q_k^{(1)}\mathbf{k}\cdot\mathbf{n} & K^{(1)}\mathbf{k}\cdot\mathbf{N}\times\mathbf{n} \\ -K^{(1)}\mathbf{k}\cdot\mathbf{N}\times\mathbf{n} & k^2\mathbf{N}\cdot\mathbf{n} - q_k^{(1)}\mathbf{k}\cdot\mathbf{n} \end{pmatrix} \tag{61}$$

$$T_2(\mathbf{n}, \mathbf{k}_0) = \frac{1}{k_0\sqrt{\left[K^{(1)}\right]^2 - q_t^2}} \begin{pmatrix} k_0^2\mathbf{N}\cdot\mathbf{n} + q_0^{(1)}\mathbf{k}_0\cdot\mathbf{n} & -K^{(1)}\mathbf{k}_0\cdot\mathbf{N}\times\mathbf{n} \\ K^{(1)}\mathbf{k}_0\cdot\mathbf{N}\times\mathbf{n} & k_0^2\mathbf{N}\cdot\mathbf{n} + q_0^{(1)}\mathbf{k}_0\cdot\mathbf{n} \end{pmatrix}. \tag{62}$$

Here $\mathbf{n}$ is given by Eq. (27), ∇h by Eq. (42) and q_t by Eq. (44) with $K = K^{(1)}$. One can check that $T_{1,2}$ are orthogonal matrices. As a result we obtain the following explicit expression for the SA of the electromagnetic waves in the tangent plane approximation:

$$\begin{aligned} S(\mathbf{k}, \mathbf{k}_0) = {} & \frac{1}{\left(q_k^{(1)} + q_0^{(1)}\right)\sqrt{q_k^{(1)}q_0^{(1)}}\,kk_0} \frac{1}{\left[K^{(1)}\right]^2 - q_k^{(1)}q_0^{(1)} + \mathbf{k}\mathbf{k}_0} \\ & \times \begin{pmatrix} k^2 q_0^{(1)} + q_k^{(1)}\mathbf{k}\mathbf{k}_0 & K^{(1)}\mathbf{N}\cdot\mathbf{k}\times\mathbf{k}_0 \\ -K^{(1)}\mathbf{N}\cdot\mathbf{k}\times\mathbf{k}_0 & k^2 q_0 + q_k^{(1)}\mathbf{k}\mathbf{k}_0 \end{pmatrix} \begin{pmatrix} V_1(q_t) & 0 \\ 0 & V_2(q_t) \end{pmatrix} \\ & \times \begin{pmatrix} k_0^2 q_k^{(1)} + q_0^{(1)}\mathbf{k}\mathbf{k}_0 & -K^{(1)}\mathbf{N}\cdot\mathbf{k}\times\mathbf{k}_0 \\ K^{(1)}\mathbf{N}\cdot\mathbf{k}\times\mathbf{k}_0 & k_0^2 q_k^{(1)} + q_0^{(1)}\mathbf{k}\mathbf{k}_0 \end{pmatrix} \\ & \times \int \mathrm{e}^{-\mathrm{i}(\mathbf{k}-\mathbf{k}_0)\mathbf{r} - \mathrm{i}\left(q_k^{(1)} + q_0^{(1)}\right)h(\mathbf{r})} \frac{d\mathbf{r}}{(2\pi)^2}, \end{aligned} \tag{63}$$

where

$$q_t = \sqrt{\frac{\left[K^{(1)}\right]^2 + q_k^{(1)}q_0^{(1)} - \mathbf{k}\mathbf{k}_0}{2}}. \tag{64}$$

Note, that in the backscattering case $\mathbf{k} = -\mathbf{k}_0$ one has to remove an uncertainty in the preintegral factor of Eq. (63). The reason is that for normal incidence the 3D wave vector and the normal to the plane are parallel and the polarization plane is undetermined.

It is easy to see that expression Eq. (63) satisfies the reciprocity relation [1]:

$$\begin{pmatrix} 1 & 0 \\ 0 & -1 \end{pmatrix} S(\mathbf{k}, \mathbf{k}_0) = S^T(-\mathbf{k}_0, -\mathbf{k}) \begin{pmatrix} 1 & 0 \\ 0 & -1 \end{pmatrix}. \tag{65}$$

Now, as in the case of scalar waves, we will try to build the KA for electromagnetic waves based on the integral equations for the surface fields. To formulate those equations we will use the Stratton–Chu formula.

2.8. The Stratton–Chu Formula

This formula is the Helmholtz formula applied to the case of electromagnetic waves. It allows expressing the electric and magnetic fields in the medium in terms of their surface values. Let us write down the Helmholtz formula Eq. (8) for the α-th component of the electric field:

$$E_\alpha^{(\text{in})} - \int E_\alpha n_k \frac{\partial G(\mathbf{R}-\mathbf{R}_*)}{\partial x_k}\,\mathrm{d}\Sigma_\mathbf{R} + \int n_k \frac{\partial E_\alpha}{\partial x_k} G(\mathbf{R}-\mathbf{R}_*)\,\mathrm{d}\Sigma_\mathbf{R}$$

$$= \begin{cases} E_\alpha(\mathbf{R}_*), & \mathbf{R}_* \in \text{upper half-space} \\ 0 & \mathbf{R}_* \in \text{lower half-space.} \end{cases} \tag{66}$$

One has an obvious identity

$$G n_k \frac{\partial E_\alpha}{\partial x_k} = G n_k \left(\frac{\partial E_\alpha}{\partial x_k} - \frac{\partial E_k}{\partial x_\alpha} \right) + n_k \frac{\partial}{\partial x_\alpha}(E_k G) - n_k E_k \frac{\partial G}{\partial x_\alpha}. \tag{67}$$

Calculating the double vector product it is easy to check that

$$n_k \left(\frac{\partial E_\alpha}{\partial x_k} - \frac{\partial E_k}{\partial x_\alpha} \right) = -\left\{ \mathbf{n} \times (\nabla \times \mathbf{E}) \right\}_\alpha, \tag{68}$$

where the index α on the right-hand side means the α-th component of the corresponding vector expression. To transform the second term in Eq. (67) we will need the following identity:

$$\int \frac{\partial \varphi}{\partial x_\alpha}\,\mathrm{d}\mathbf{R} = \int n_\alpha \varphi \,\mathrm{d}\Sigma_\mathbf{R}, \tag{69}$$

where φ is an arbitrary smooth scalar function. Equation (69) immediately follows after the Gauss theorem Eq. (7) is applied to the vector field $\chi_k = \delta_{k\alpha}\varphi$,

$$\int \frac{\partial}{\partial x_k}(\delta_{k\alpha}\varphi)\,\mathrm{d}\mathbf{R} = \int n_k (\delta_{k\alpha}\varphi)\,\mathrm{d}\Sigma_\mathbf{R}, \tag{70}$$

which coincides with Eq. (69). Applying the Gauss theorem one more time and using Eq. (69) one finds

$$\int n_k \frac{\partial}{\partial x_\alpha}(E_k G)\,\mathrm{d}\Sigma_\mathbf{R} = \int \frac{\partial}{\partial x_k}\frac{\partial}{\partial x_\alpha}(E_k G)\,\mathrm{d}\mathbf{R} = \int \frac{\partial}{\partial x_\alpha}\frac{\partial}{\partial x_k}(E_k G)\,\mathrm{d}\mathbf{R}$$

$$= \int n_\alpha \frac{\partial}{\partial x_k}(E_k G)\,\mathrm{d}\Sigma_\mathbf{R} = \int n_\alpha E_k \frac{\partial G}{\partial x_k}\,\mathrm{d}\Sigma_\mathbf{R}. \tag{71}$$

At the last step we took into account that according to Eq. (45) $\partial E_k/\partial x_k = 0$. Finally, calculating the double vector product we find

$$(\mathbf{n} \times \mathbf{E}) \times \nabla G = \mathbf{E}(\mathbf{n}\nabla G) - \mathbf{n}(\mathbf{E}\nabla G). \tag{72}$$

Substituting Eqs. (67),(68),(71),(72) into Eq. (66), replacing $\nabla \times \mathbf{E}$ in Eq. (68) by the use of the Maxwell equation Eq. (45), and returning to the vector notation, we obtain the Stratton–Chu formula:

$$\mathbf{E}^{(\mathrm{in})} - \frac{\mathrm{i}\omega}{c}\int (\mathbf{n}\times\mathbf{H})\, G\mathrm{d}\Sigma_{\mathbf{R}} - \int (\mathbf{nE})\,\nabla_{\mathbf{R}} G\mathrm{d}\Sigma_{\mathbf{R}} - \int (\mathbf{n}\times\mathbf{E})\times\nabla_{\mathbf{R}} G\mathrm{d}\Sigma_{\mathbf{R}}$$

$$= \begin{cases} \mathbf{E}(\mathbf{R}_*), & \mathbf{R}_* \in \text{upper half-space} \\ 0 & \mathbf{R}_* \in \text{lower half-space,} \end{cases} \tag{73}$$

where $G = G(\mathbf{R} - \mathbf{R}_*)$. Applying the same result to the magnetic field $\mathbf{H}$, and taking into account that in replacing $\nabla \times \mathbf{H}$ we have to use Eq. (46), one finds

$$\mathbf{H}^{(\mathrm{in})} + \varepsilon\frac{\mathrm{i}\omega}{c}\int (\mathbf{n}\times\mathbf{E})\, G\mathrm{d}\Sigma_{\mathbf{R}} - \int (\mathbf{nH})\,\nabla_{\mathbf{R}} G\mathrm{d}\Sigma_{\mathbf{R}} - \int (\mathbf{n}\times\mathbf{H})\times\nabla_{\mathbf{R}} G\mathrm{d}\Sigma_{\mathbf{R}}$$

$$= \begin{cases} \mathbf{H}(\mathbf{R}_*), & \mathbf{R}_* \in \text{upper half-space} \\ 0 & \mathbf{R}_* \in \text{lower half-space.} \end{cases} \tag{74}$$

2.9. The Integral Equations for the Electromagnetic Case

To obtain the equations for the surface values of the fields we have to set the observation point $\mathbf{R}_*$ in Eqs. (73),(74) on the surface using the limiting formulae given by Eqs. (14),(15). Taking into account also that

$$(\mathbf{nE})\,\mathbf{n} + (\mathbf{n}\times\mathbf{E})\times\mathbf{n} = \mathbf{E}, \tag{75}$$

we find the equations

$$\begin{aligned} \frac{1}{2}\mathbf{E}^{(1)}(\mathbf{R}_0) = \mathbf{E}^{(\mathrm{in})}(\mathbf{R}_0) &- \frac{\mathrm{i}\omega}{c}\int (\mathbf{n}\times\mathbf{H})\, G^{(1)}(\mathbf{R}-\mathbf{R}_0)\,\mathrm{d}\Sigma \\ &- \text{v.p.}\int \left(\mathbf{nE}^{(1)}\right)\nabla_{\mathbf{R}} G^{(1)}(\mathbf{R}-\mathbf{R}_0)\,\mathrm{d}\Sigma \\ &- \text{v.p.}\int \left(\mathbf{n}\times\mathbf{E}^{(1)}\right)\times\nabla_{\mathbf{R}} G^{(1)}(\mathbf{R}-\mathbf{R}_0)\,\mathrm{d}\Sigma \end{aligned} \tag{76}$$

$$\begin{aligned} \frac{1}{2}\mathbf{H}(\mathbf{R}_0) = \mathbf{H}^{(\mathrm{in})}(\mathbf{R}_0) &+ \frac{\mathrm{i}\omega}{c}\varepsilon_1\int \left(\mathbf{n}\times\mathbf{E}^{(1)}\right) G^{(1)}(\mathbf{R}-\mathbf{R}_0)\,\mathrm{d}\Sigma \\ &- \text{v.p.}\int (\mathbf{nH})\,\nabla_{\mathbf{R}} G^{(1)}(\mathbf{R}-\mathbf{R}_0)\,\mathrm{d}\Sigma - \text{v.p.}\int (\mathbf{n}\times\mathbf{H}) \\ &\times\nabla_{\mathbf{R}} G^{(1)}(\mathbf{R}-\mathbf{R}_0)\,\mathrm{d}\Sigma \end{aligned} \tag{77}$$

$$\begin{aligned}\frac{1}{2}\mathbf{E}^{(2)}(\mathbf{R}_0) = {} & \frac{\mathrm{i}\omega}{c}\int (\mathbf{n}\times\mathbf{H})\, G^{(2)}(\mathbf{R}-\mathbf{R}_0)\,\mathrm{d}\Sigma \\ & + \text{v.p.}\int \left(\mathbf{n}\mathbf{E}^{(2)}\right)\nabla_{\mathbf{R}} G^{(2)}(\mathbf{R}-\mathbf{R}_0)\,\mathrm{d}\Sigma \\ & + \text{v.p.}\int \left(\mathbf{n}\times\mathbf{E}^{(2)}\right)\times\nabla_{\mathbf{R}} G^{(2)}(\mathbf{R}-\mathbf{R}_0)\,\mathrm{d}\Sigma \end{aligned} \tag{78}$$

$$\begin{aligned}\frac{1}{2}\mathbf{H}(\mathbf{R}_0) = {} & \mathbf{H}^{(\mathrm{in})}(\mathbf{R}_0) - \frac{\mathrm{i}\omega}{c}\varepsilon_2\int \left(\mathbf{n}\times\mathbf{E}^{(2)}\right) G^{(2)}(\mathbf{R}-\mathbf{R}_0)\,\mathrm{d}\Sigma \\ & + \text{v.p.}\int (\mathbf{n}\mathbf{H})\,\nabla_{\mathbf{R}} G^{(2)}(\mathbf{R}-\mathbf{R}_0)\,\mathrm{d}\Sigma \\ & + \text{v.p.}\int (\mathbf{n}\times\mathbf{H})\times\nabla_{\mathbf{R}} G^{(2)}(\mathbf{R}-\mathbf{R}_0)\,\mathrm{d}\Sigma. \end{aligned} \tag{79}$$

The boundary condition, Eq. (47), is

$$\mathbf{n}\times\mathbf{E}^{(1)} = \mathbf{n}\times\mathbf{E}^{(2)}. \tag{80}$$

We took into account in Eqs. (76)–(79) that the magnetic field is continuous at the interface. For this reason a superscript to **H** is not assigned. Although the tangential components of the electric field are also continuous at the interface, the normal components are not, and the superscript to $\mathbf{E}^{(1,2)}$ indicates the field on the upper and lower side of the interface, respectively. When writing Eqs. (78), (79) for the lower medium we used the same Eqs. (73), (74) and took into account that there are no sources in the lower half-space so that $\mathbf{E}^{(\mathrm{in})} = \mathbf{H}^{(\mathrm{in})} = 0$, and the sign of the normal has to be changed. Also, when applying Eq. (14) one has to choose for the lower medium the plus sign.

Apparently, not all of Eqs. (76)–(80) are independent, and this set of equations represents generally an overdetermined system. One might expect that those equations have a unique solution even when the second medium represents a compact body with inner resonances, since surface values of the field represent physical measurables within the first medium.

Let us consider Eqs. (76)–(79) for the case of the plane $z = 0$. One can see that the integral terms do not vanish even in this case. This means that the surface electromagnetic field depends not only on the values of the incident field at the same point, but on the fields in the vicinity of the given point. One possibility would be to evaluate the resulting integrals by the stationary point method. One can see that in the lowest order no derivatives of the interface profile higher than the first enter the result. Essentially this means that the profile is approximated at each point by a plane. It is obvious that this approach is equivalent to the tangent plane approximation.

Another possibility is to make use of the fact that the relation between the surface field and the incident field for the case of the plane $z = 0$ becomes local in the Fourier domain. For surfaces with sufficiently small slopes one can try to approximate the argument of the Green functions in (76)–(79) as follows:

$$|\mathbf{R}-\mathbf{R}_*| = \sqrt{|\mathbf{r}-\mathbf{r}_*|^2 + (h(\mathbf{r}) - h(\mathbf{r}_*))^2} \approx |\mathbf{r}-\mathbf{r}_*| \,. \tag{81}$$

If one also neglects the terms explicitly proportional to ∇h, then all the integrals become convolutions, and the Fourier transform will lead to an explicit solution. This approximation constitutes the Meecham–Lysanov approach, which was suggested for the Dirichlet problem in [2], [3]. We will try to build an approximate solution for the electromagnetic problem based on this idea, keeping track of the higher-order corrections.

Let us Fourier transform Eqs. (76)–(80). First, using the representation given by Eq. (4) one finds

$$\nabla_{\mathbf{R}} G(\mathbf{R}-\mathbf{R}_0) = -\frac{1}{8\pi^2} \times \int (\mathbf{k} + q_k \mathrm{sign}(z_0 - z)\mathbf{N}) \left(1 + \sum_{m=1}^{\infty} \frac{(\mathrm{i} q_k\, |z_0 - z|)^m}{m!}\right) \mathrm{e}^{\mathrm{i}\mathbf{k}(\mathbf{r}_0 - \mathbf{r})} \frac{\mathrm{d}\mathbf{k}}{q_k}. \tag{82}$$

Note that the even powers of $|z - z_*|$ here are always associated with the even powers of q_k which, after integration over $\mathbf{k}$, reduce to the appropriate derivatives of $\delta\,(\mathbf{r}_0 - \mathbf{r})$. Those terms do not contribute to the integrals in Eqs. (76)–(79), which are principal values. The odd powers of $|z - z_*|$ in Eq. (82), being multiplied by the function $\mathrm{sign}(z_0 - z)$, produce the odd powers of $(z_0 - z)$. Let us introduce for any function $\varphi(\mathbf{r})$ of the horizontal vector $\mathbf{r}$ its corresponding Fourier transform $\varphi_{\mathbf{k}}$

$$\varphi_{\mathbf{k}} = \int \varphi(\mathbf{r})\,\mathrm{e}^{-\mathrm{i}\mathbf{k}\mathbf{r}} \frac{\mathrm{d}\mathbf{r}}{(2\pi)^2}, \qquad \varphi(\mathbf{r}) = \int \varphi_{\mathbf{k}}\,\mathrm{e}^{\mathrm{i}\mathbf{k}\mathbf{r}} \mathrm{d}\mathbf{k}. \tag{83}$$

The function φ could be a vector or a scalar; in particular, it can be the elevation $h(\mathbf{r})$. For example, if

$$\varphi(\mathbf{r}_0) = \mathrm{v.p.} \int A(\mathbf{r}) \nabla_{\mathbf{R}} G\,(h(\mathbf{r}) - h\,(\mathbf{r}_0))\,\mathrm{d}\mathbf{r}, \tag{84}$$

and we use for G an approximation that includes only the first term of the series in Eq. (82), one finds

$$\begin{aligned}
\varphi_{\mathbf{k}} &= \int \frac{\mathrm{d}\mathbf{r}_0}{(2\pi)^2}\,\mathrm{e}^{-\mathrm{i}\mathbf{k}\mathbf{r}_0} \int \mathrm{d}\mathbf{k}_1 A_{\mathbf{k}_1} \mathrm{e}^{\mathrm{i}\mathbf{k}_1\mathbf{r}} \\
&\times \left(-\frac{1}{8\pi^2}\right) \int \frac{\mathrm{d}\xi}{q_\xi}\,\mathrm{e}^{\mathrm{i}\xi(\mathbf{r}_0 - r)} \left(\xi + \mathrm{i} q_\xi^2\,(h(\mathbf{r}_0) - h(\mathbf{r}))\,\mathbf{N}\right) \mathrm{d}\mathbf{r} + O(h^2) \\
&= -\frac{\mathbf{k}}{2q_k} A_{\mathbf{k}} - \frac{\mathrm{i}}{2}\mathbf{N} \int A_\xi (q_\xi - q_k) h_{\mathbf{k}-\xi} \mathrm{d}\xi + O(h^2).
\end{aligned} \tag{85}$$

Similarly, if

$$\varphi(\mathbf{r}_0) = \int A(\mathbf{r}) G\left(h(\mathbf{r}) - h\left(\mathbf{r}_0\right)\right) \mathrm{d}\mathbf{r}, \tag{86}$$

then

$$\begin{aligned} \varphi_{\mathbf{k}} &= \int \frac{\mathrm{d}\mathbf{r}_0}{(2\pi)^2} \mathrm{e}^{-\mathrm{i}\mathbf{k}\mathbf{r}_0} \int \mathrm{d}\mathbf{k}_1 A_{\mathbf{k}_1} \mathrm{e}^{\mathrm{i}\mathbf{k}_1\mathbf{r}} \left(-\frac{\mathrm{i}}{8\pi^2}\right) \int \frac{\mathrm{d}\xi}{q_\xi} \mathrm{e}^{\mathrm{i}\xi(\mathbf{r}_0 - r)} \mathrm{d}\mathbf{r} + O(h^2) \\ &= -\frac{\mathrm{i}}{2q_k} A_{\mathbf{k}} + O(h^2). \end{aligned} \tag{87}$$

Using these relations and also taking into account that

$$\mathbf{n}\mathrm{d}\Sigma = (\mathbf{N} - \nabla h)\,\mathrm{d}\mathbf{r}, \tag{88}$$

we transform Eqs. (76)–(79) into the following system of equations:

$$\frac{1}{2}\mathbf{E}_{\mathbf{k}}^{(1)} = \mathbf{E}_{\mathbf{k}}^{(\mathrm{in})} - \frac{\omega}{c}\frac{1}{2q_k^{(1)}}\mathbf{N} \times \mathbf{H}_{\mathbf{k}} + \frac{\mathbf{k}}{2q_k^{(1)}}\mathbf{N}\mathbf{E}_{\mathbf{k}}^{(1)} + \frac{1}{2q_k^{(1)}}\left(\mathbf{N} \times \mathbf{E}_{\mathbf{k}}^{(1)}\right) \times \mathbf{k} + \mathbf{T}_{\mathbf{k}}^{(1)} \tag{89}$$

$$\frac{1}{2}\mathbf{H}_{\mathbf{k}}^{(1)} = \mathbf{H}_{\mathbf{k}}^{(\mathrm{in})} + \frac{\omega}{c}\varepsilon_1\frac{1}{2q_k^{(1)}}\mathbf{N} \times \mathbf{E}_{\mathbf{k}}^{(1)} + \frac{\mathbf{k}}{2q_k^{(1)}}\mathbf{N}\mathbf{H}_{\mathbf{k}} + \frac{1}{2q_k^{(1)}}\left(\mathbf{N} \times \mathbf{H}_{\mathbf{k}}\right) \times \mathbf{k} + \mathbf{U}_{\mathbf{k}}^{(1)} \tag{90}$$

$$\frac{1}{2}\mathbf{E}_{\mathbf{k}}^{(2)} = \frac{\omega}{c}\frac{1}{2q_k^{(2)}}\mathbf{N} \times \mathbf{H}_{\mathbf{k}} - \frac{\mathbf{k}}{2q_k^{(2)}}\mathbf{N}\mathbf{E}_{\mathbf{k}}^{(2)} - \frac{1}{2q_k^{(2)}}\left(\mathbf{N} \times \mathbf{E}_{\mathbf{k}}^{(2)}\right) \times \mathbf{k} + \mathbf{T}_{\mathbf{k}}^{(2)} \tag{91}$$

$$\frac{1}{2}\mathbf{H}_{\mathbf{k}}^{(2)} = -\frac{\omega}{c}\varepsilon_2\frac{1}{2q_k^{(2)}}\mathbf{N} \times \mathbf{E}_{\mathbf{k}}^{(2)} - \frac{\mathbf{k}}{2q_k^{(2)}}\mathbf{N}\mathbf{H}_{\mathbf{k}} - \frac{1}{2q_k^{(2)}}\left(\mathbf{N} \times \mathbf{H}_{\mathbf{k}}\right) \times \mathbf{k} + \mathbf{U}_{\mathbf{k}}^{(2)}. \tag{92}$$

Here the terms $\mathbf{T}_{\mathbf{k}}^{(1,2)}$, $\mathbf{U}_{\mathbf{k}}^{(1,2)}$ include all the terms of $O(h^n)$, with $n \geq 1$. These terms can be represented as

$$\begin{aligned} \mathbf{T}_{\mathbf{k}}^{(1,2)} &= \mathrm{i}\int \mathbf{t}^{(1,2)}(\mathbf{k}, \xi) h_{\mathbf{k}-\xi}\mathrm{d}\xi + O(h^2), \\ \mathbf{U}_{\mathbf{k}}^{(1,2)} &= \mathrm{i}\int \mathbf{u}^{(1,2)}(\mathbf{k}, \xi) h_{\mathbf{k}-\xi}\mathrm{d}\xi + O(h^2), \end{aligned} \tag{93}$$

where the $O(h^2)$ terms are due to the corresponding terms from Eqs. (85) and (87), and

$$\begin{aligned} \mathbf{t}^{(1)}(\mathbf{k}, \xi) = {} & \frac{\omega}{c}\frac{1}{2q_k^{(1)}}(\mathbf{k} - \xi) \times \mathbf{H}_\xi - (\mathbf{k} - \xi)\,\mathbf{E}_\xi^{(1)}\frac{\mathbf{k}}{2q_k^{(1)}} \\ & - \left((\mathbf{k} - \xi) \times \mathbf{E}_\xi^{(1)}\right) \times \frac{\mathbf{k}}{2q_k^{(1)}} + \frac{1}{2}\left(q_\xi^{(1)} - q_k^{(1)}\right)\mathbf{E}_\xi^{(1)} \end{aligned} \tag{94}$$

$$\mathbf{t}^{(2)}(\mathbf{k},\xi) = -\frac{\omega}{c}\frac{1}{2q_k^{(2)}}(\mathbf{k}-\xi)\times\mathbf{H}_\xi + (\mathbf{k}-\xi)\,\mathbf{E}_\xi^{(2)}\frac{\mathbf{k}}{2q_k^{(2)}}$$
$$+\left((\mathbf{k}-\xi)\times\mathbf{E}_\xi^{(2)}\right)\times\frac{\mathbf{k}}{2q_k^{(2)}} - \frac{1}{2}\left(q_\xi^{(2)} - q_k^{(2)}\right)\mathbf{E}_\xi^{(2)} \tag{95}$$

$$\mathbf{u}^{(1)}(\mathbf{k},\xi) = -\frac{\omega}{c}\varepsilon_1\frac{1}{2q_k^{(1)}}(\mathbf{k}-\xi)\times\mathbf{E}_\xi^{(1)} - (\mathbf{k}-\xi)\,\mathbf{H}_\xi\frac{\mathbf{k}}{2q_k^{(1)}}$$
$$-((\mathbf{k}-\xi)\times\mathbf{H}_\xi)\times\frac{\mathbf{k}}{2q_k^{(1)}} + \frac{1}{2}\left(q_\xi^{(1)} - q_k^{(1)}\right)\mathbf{H}_\xi^{(1)} \tag{96}$$

$$\mathbf{u}^{(2)}(\mathbf{k},\xi) = \frac{\omega}{c}\varepsilon_2\frac{1}{2q_k^{(2)}}(\mathbf{k}-\xi)\times\mathbf{E}_\xi^{(2)} + (\mathbf{k}-\xi)\,\mathbf{H}_\xi\frac{\mathbf{k}}{2q_k^{(2)}}$$
$$+((\mathbf{k}-\xi)\times\mathbf{H}_\xi)\times\frac{\mathbf{k}}{2q_k^{(2)}} - \frac{1}{2}\left(q_\xi^{(2)} - q_k^{(2)}\right)\mathbf{H}_\xi^{(1)}. \tag{97}$$

The boundary condition Eq. (80) takes the form

$$\mathbf{N}\times\mathbf{E}_\mathbf{k}^{(1)} - \mathrm{i}\int(\mathbf{k}-\xi)\times\mathbf{E}_\xi^{(1)}h_{\mathbf{k}-\xi}\,\mathrm{d}\xi$$
$$= \mathbf{N}\times\mathbf{E}_\mathbf{k}^{(2)} - \mathrm{i}\int(\mathbf{k}-\xi)\times\mathbf{E}_\xi^{(2)}h_{\mathbf{k}-\xi}\,\mathrm{d}\xi. \tag{98}$$

Equations (89)–(92) can be solved by iteration. In the lowest order one neglects $\mathbf{T}_\mathbf{k}^{(1,2)}$, $\mathbf{U}_\mathbf{k}^{(1,2)}$ and the $O(h)$ terms in Eq. (98). Projecting Eqs. (89)–(92) on $\mathbf{N}$ one calculates first $\mathbf{NE}_\mathbf{k}^{(1,2)}$ and $\mathbf{NH}_k$. Then one multiplies Eqs. (89)–(92) by $\times\mathbf{N}$ and calculates the surface fields:

$$\mathbf{H}_\mathbf{k} = 2\frac{c}{\omega}\frac{q_k^{(1)}}{q_k^{(2)}+q_k^{(1)}}\left(-q_k^{(2)}\mathbf{N}\times\mathbf{E}_\mathbf{k}^{(\mathrm{in})}\right.$$
$$\left.-\frac{\varepsilon_1 q_k^{(1)}+\varepsilon_2 q_k^{(2)}}{\varepsilon_1 q_k^{(2)}+\varepsilon_2 q_k^{(1)}}\left(\mathbf{NE}_\mathbf{k}^{(\mathrm{in})}\right)\mathbf{N}\times\mathbf{k} + \frac{\omega}{c}\left(\mathbf{NH}_\mathbf{k}^{(\mathrm{in})}\right)\mathbf{N}\right) \tag{99}$$

$$\mathbf{E}_\mathbf{k}^{(1)} = 2\frac{c}{\omega}\frac{q_k^{(1)}}{\varepsilon_1 q_k^{(2)}+\varepsilon_2 q_k^{(1)}}\left(q_k^{(2)}\mathbf{N}\times\mathbf{H}_\mathbf{k}^{(\mathrm{in})}\right.$$
$$\left.+\left(\mathbf{NH}_\mathbf{k}^{(\mathrm{in})}\right)\mathbf{N}\times\mathbf{k} + \frac{\omega}{c}\varepsilon_2\left(\mathbf{NE}_\mathbf{k}^{(\mathrm{in})}\right)\mathbf{N}\right). \tag{100}$$

To obtain the corresponding expression for $\mathbf{E}_\mathbf{k}^{(2)}$ one has to replace ε_2 by ε_1 in the last term in Eq. (100).

Let us now calculate the scattering matrix in a manner similar to the calculation of the SA in Sects. 3, 4. We assume that the incident field is given by the plane waves,

Eqs. (50), (51), with polarization index equal to β, and express the Green function in Eqs. (73), (74) by using the Weyl representation Eq. (4). Then we project the electric (or magnetic) spectral component of the field corresponding to the horizontal wave vector $\mathbf{k}$ on the polarization vector $\mathbf{e}_{\alpha}^{(sc)}(\mathbf{k})$ (or $\mathbf{h}_{\alpha}^{(sc)}(\mathbf{k})$). Note, that the third terms in Eqs. (73),(74) do not contribute since they produce longitudinal components of the field, which are, in fact, compensated by appropriate contributions from the second and the fourth terms. Let us neglect first the ∇h term in the expression for the normal $\mathbf{n}$: $\mathbf{n}d\Sigma = (\mathbf{N} - \nabla h) \approx \mathbf{N}d\mathbf{r}$. Then we find

$$S_{\alpha\beta}(\mathbf{k}, \mathbf{k}_0) = \frac{1}{2\sqrt{q_k^{(1)}}} \int \left[-\frac{\omega}{c}\mathbf{N} \times \mathbf{H}_{\xi} + \left(\mathbf{N} \times \mathbf{E}_{\xi}^{(1)}\right) \times \left(\mathbf{k} + q_k^{(1)}\mathbf{N}\right) \right] \mathbf{e}_{\alpha}^{(sc)}(\mathbf{k}) \times \mathrm{e}^{-\mathrm{i}(\mathbf{k}-\xi)\mathbf{r}_1 - \mathrm{i}q_k^{(1)}h(\mathbf{r}_1)} \frac{d\mathbf{r}_1}{(2\pi)^2}\, d\xi. \tag{101}$$

In the lowest-order approximation, according to Eqs. (99), (100), $\mathbf{E}_{\xi}$ and $\mathbf{H}_{\xi}$ are proportional to $\mathbf{E}_{\xi}^{(in)}$, $\mathbf{H}_{\xi}^{(in)}$. Due to Eqs. (50),(51),(83) one has

$$\mathbf{E}_{\xi}^{(in)} = \mathbf{e}_{\beta}^{(in)}(\mathbf{k}_0) \left(q_0^{(1)}\right)^{-1/2} \int \mathrm{e}^{-\mathrm{i}(\xi-\mathbf{k}_0)\mathbf{r}_2 - \mathrm{i}q_0^{(1)}h(\mathbf{r}_2)} \frac{d\mathbf{r}_2}{(2\pi)^2}, \tag{102}$$

and a similar expression for $\mathbf{H}_{\xi}^{(in)}$ with the polarization vector $\mathbf{e}_{\beta}^{(in)}(\mathbf{k}_0)$ replaced by $\mathbf{h}_{\beta}^{(in)}(\mathbf{k}_0)$. Substituting this equation into Eqs. (99), (100) and then into Eq. (101) we find

$$S_{\alpha\beta}(\mathbf{k}, \mathbf{k}_0) = \int \mathrm{e}^{-\mathrm{i}(\mathbf{k}-\xi)\mathbf{r}_1 - \mathrm{i}q_k^{(1)}h(\mathbf{r}_1)} \mathrm{e}^{-\mathrm{i}(\xi-\mathbf{k}_0)\mathbf{r}_2 - \mathrm{i}q_0^{(1)}h(\mathbf{r}_2)} \Phi_{\alpha\beta}(\mathbf{k}, \mathbf{k}_0; \xi) \frac{d\mathbf{r}_1 d\mathbf{r}_2}{(2\pi)^4}\, d\xi, \tag{103}$$

where the explicit expression for matrix Φ can be easily calculated; however, we will not need it in what follows.

To calculate the scattering matrix in the next approximation we have to take into account the $\mathbf{T}$, $\mathbf{U}$ terms in Eqs. (89)–(92), replacing $\mathbf{E}_{\xi}$, $\mathbf{H}_{\xi}$ in Eqs. (94)–(97) by the lowest-order approximation. Also, we have to take into account the $O(\nabla h)$ term in the expression for the normal $\mathbf{n}$ when calculating the scattering matrix from Eq. (73) (or Eq. (74)). The calculations are formidable. However, what we will need is only the general structure of this term. Tracking corresponding terms it is easy to see that the correction term (marked here by a prime) has the following structure:

$$S'_{\alpha\beta}(\mathbf{k}, \mathbf{k}_0) = \int \mathrm{e}^{-\mathrm{i}(\mathbf{k}-\xi)\mathbf{r}_1 - \mathrm{i}q_k^{(1)}h(\mathbf{r}_1)} \mathrm{e}^{-\mathrm{i}(\mathbf{k}_1-\mathbf{k}_0)\mathbf{r}_2 - \mathrm{i}q_0^{(1)}h(\mathbf{r}_2)} \times \Phi'_{\alpha\beta}(\mathbf{k}, \mathbf{k}_0; \xi; \mathbf{k}_1)\, h_{\xi-\mathbf{k}_1} \frac{d\mathbf{r}_1 d\mathbf{r}_2}{(2\pi)^4} d\xi\, d\mathbf{k}_1. \tag{104}$$

Note, that according to Eqs. (93)–(97),

$$\Phi'_{\alpha\beta}(\mathbf{k}, \mathbf{k}_0; \xi; \xi) = 0. \tag{105}$$

Making the change of integration variables $\xi \to \xi + \xi_1/2$, $\mathbf{k}_1 \to \xi - \xi_1/2$, we find

$$S'_{\alpha\beta}(\mathbf{k}, \mathbf{k}_0) = \int e^{-i(\mathbf{k}-\xi)\mathbf{r}_1 - iq_k^{(1)} h(\mathbf{r}_1)} e^{-i(\xi-\mathbf{k}_0)\mathbf{r}_2 - iq_0^{(1)} h(\mathbf{r}_2)}$$

$$\times \Phi'_{\alpha\beta}(\mathbf{k}, \mathbf{k}_0; \xi + \xi_1/2; \xi - \xi_1/2)\, h_{\xi_1} e^{i\xi_1(\mathbf{r}_1+\mathbf{r}_2)/2} \frac{d\mathbf{r}_1 d\mathbf{r}_2}{(2\pi)^4} d\xi d\xi_1. \quad (106)$$

2.10. Nonlocal Small-Slope Approximation

Examination of the higher-order terms shows that continuation of the calculations will result in the following expression for the scattering matrix:

$$S(\mathbf{k}, \mathbf{k}_0) = \int e^{-i(\mathbf{k}-\xi)\mathbf{r}_1 - iq_k h(\mathbf{r}_1)} e^{-i(\xi-\mathbf{k}_0)\mathbf{r}_2 - iq_0 h(\mathbf{r}_2)}$$

$$\times F\left(\mathbf{k}, \mathbf{k}_0; \frac{\mathbf{r}_1 + \mathbf{r}_2}{2}; \xi\right) \frac{d\mathbf{r}_1 d\mathbf{r}_2}{(2\pi)^4} d\xi, \quad (107)$$

where

$$F\left(\mathbf{k}, \mathbf{k}_0; \frac{\mathbf{r}_1 + \mathbf{r}_2}{2}; \xi\right) = \Phi(\mathbf{k}, \mathbf{k}_0; \xi) + \int \Phi_1(\mathbf{k}, \mathbf{k}_0; \xi; \xi_1)\, h_{\xi_1} e^{i\xi_1 \frac{\mathbf{r}_1+\mathbf{r}_2}{2}} d\xi_1$$

$$+ \int \Phi_2(\mathbf{k}, \mathbf{k}_0; \xi; \xi_1, \xi_2)\, h_{\xi_1} h_{\xi_2} e^{i(\xi_1+\xi_2)\frac{\mathbf{r}_1+\mathbf{r}_2}{2}} d\xi_1 d\xi_2 + \cdots \quad (108)$$

is a functional of elevations. We have omitted in Eq. (107) polarization indices, considering the functions F, Φ, Φ_n to be appropriate matrices. We have also omitted the superscripts (1, 2) to the vertical wave numbers $q^{(1,2)}$, assuming they always refer to the upper half-space. Note that the kernels Φ_n can be assumed to be symmetric functions of $\xi_1, \xi_2, \ldots \xi_n$.

Looking at the derivation of the expansion Eq. (108) one can conclude that this expansion is uniquely defined. However, this is not the case, and certain parts of the terms in Eq. (108) can be transferred into each other. The physical reason for this is discussed in [4], and we will also reproduce appropriate transformations used in this work below.

The structure of the higher-order terms in Eq. (108) follows from the transformation property Eq. (34). The horizontal shift $h(\mathbf{r}) \to h(\mathbf{r} - \mathbf{d})$ leads to the substitution $h_\xi \to h_\xi e^{-i\mathbf{d}\xi}$. The change of integration variables $\mathbf{r}_{1,2} - \mathbf{d} \to \mathbf{r}_{1,2}$ in Eq. (107) ensures that Eq. (34) is satisfied.

According to Eqs. (105), (106) one has

$$\Phi_1(\mathbf{k}, \mathbf{k}_0; \xi; 0) = 0. \quad (109)$$

Equation (109) is related to the transformation property Eq. (36). The vertical shift $h(\mathbf{r}) \to h(\mathbf{r}) + H$ leads to the replacement $h_\xi \to h_\xi + H\delta(\xi)$. Since the first two exponents in Eq. (107) ensure fulfillment of Eq. (36), the functional F should be invariant with respect to this transformation. Since H is arbitrary, this requires that

$\Phi_n|_{\xi_k=0} = 0$. Taking into account the symmetry of the kernels Φ_n with respect to ξ_k one can represent Φ_n as follows

$$\Phi_n = \sum_{\alpha_k=x,y} \xi_1^{(\alpha_1)} \dots \xi_n^{(\alpha_n)} \widetilde{\Phi}_n^{(\alpha_1 \dots .\alpha_n)}, \tag{110}$$

where $\widetilde{\Phi}_n$ are nonsingular functions.

Consider, for instance, the contribution from the second term in Eq. (108) using also the representation Eq. (110):

$$S_1(\mathbf{k}, \mathbf{k}_0) = \int \frac{\mathrm{d}\mathbf{r}_1 \mathrm{d}\mathbf{r}_2}{(2\pi)^4} \mathrm{d}\xi \, \mathrm{e}^{-\mathrm{i}(\mathbf{k}-\xi)\mathbf{r}_1 - \mathrm{i}q_k h(\mathbf{r}_1)} \, \mathrm{e}^{-\mathrm{i}(\xi - \mathbf{k}_0)\mathbf{r}_2 - \mathrm{i}q_0 h(\mathbf{r}_2)}$$
$$\times \sum_{\alpha_1=x,y} \widetilde{\Phi}_1^{(\alpha_1)} (\mathbf{k}, \mathbf{k}_0; \xi; \xi_1) \, \xi_1^{(\alpha_1)} h_{\xi_1} \mathrm{e}^{\mathrm{i}\xi_1 \frac{\mathbf{r}_1+\mathbf{r}_2}{2}} \mathrm{d}\xi_1. \tag{111}$$

Let us replace the integration variable ξ by $\xi + \xi_1/2$ and introduce the function

$$\varphi_1^{(\alpha_1)} (\mathbf{k}, \mathbf{k}_0; \xi; \xi_1) = \widetilde{\Phi}_1^{(\alpha_1)} \left(\mathbf{k}, \mathbf{k}_0; \xi + \frac{\xi_1}{2}; \xi_1 \right). \tag{112}$$

This function in turn will be represented as a sum of two terms:

$$\varphi_1^{(\alpha_1)} (\mathbf{k}, \mathbf{k}_0; \xi; \xi_1) = \varphi_1^{(\alpha_1)} (\mathbf{k}, \mathbf{k}_0; \xi; \mathbf{k} - \xi)$$
$$+ \sum_{\beta_1=x,y} \left(k^{(\beta_1)} - \xi^{(\beta_1)} - \xi_1^{(\beta_1)} \right) \Delta\varphi_1^{(\alpha_1, \beta_1)}, \tag{113}$$

where $\Delta\varphi$ is a regular function. Since the first term in Eq. (113) does not depend on ξ_1, its integration over ξ_1 results in the appearance of ∇h:

$$S_1^{(1)}(\mathbf{k}, \mathbf{k}_0) = \int \frac{\mathrm{d}\mathbf{r}_1 \mathrm{d}\mathbf{r}_2}{(2\pi)^4} \mathrm{d}\xi \, \mathrm{e}^{-\mathrm{i}(\mathbf{k}-\xi)\mathbf{r}_1 - \mathrm{i}q_k h(\mathbf{r}_1)} \, \mathrm{e}^{-\mathrm{i}(\xi - \mathbf{k}_0)\mathbf{r}_2 - \mathrm{i}q_0 h(\mathbf{r}_2)}$$
$$\times (-\mathrm{i}) \sum_{\alpha_1=x,y} \varphi_1^{(\alpha_1)} (\mathbf{k}, \mathbf{k}_0; \xi; \mathbf{k} - \xi) \, \nabla^{(\alpha_1)} h(\mathbf{r}_1). \tag{114}$$

Now using a transformation of the type of Eq. (30),

$$(-\mathrm{i}) \nabla h(\mathbf{r}_1) \, \mathrm{e}^{-\mathrm{i}q_k h(\mathbf{r}_1)} = \frac{1}{q_k} \nabla_{\mathbf{r}_1} \mathrm{e}^{-\mathrm{i}q_k h(\mathbf{r}_1)}, \tag{115}$$

in Eq. (114) we can integrate over $\mathbf{r}_1$ by parts:

$$S_1^{(1)}(\mathbf{k}, \mathbf{k}_0) = \int \frac{\mathrm{d}\mathbf{r}_1 \mathrm{d}\mathbf{r}_2}{(2\pi)^4} \mathrm{d}\xi \mathrm{e}^{-\mathrm{i}(\mathbf{k}-\xi)\mathbf{r}_1 - \mathrm{i}q_k h(\mathbf{r}_1)} \, \mathrm{e}^{-\mathrm{i}(\xi - \mathbf{k}_0)\mathbf{r}_2 - \mathrm{i}q_0 h(\mathbf{r}_2)}$$
$$\times \frac{\mathrm{i}}{q_k} \sum_{\alpha_1=x,y}^{(\alpha_1)} \left(k^{(\alpha_1)} - \xi^{(\alpha_1)} \right) \varphi_1^{(\alpha_1)} (\mathbf{k}, \mathbf{k}_0; \xi; \mathbf{k} - \xi). \tag{116}$$

According to Eqs. (110),(112),

$$\sum_{\alpha_1=x,y}^{(\alpha_1)} \left(k^{(\alpha_1)} - \xi^{(\alpha_1)}\right) \varphi_1^{(\alpha_1)}(\mathbf{k}, \mathbf{k}_0; \xi; \mathbf{k} - \xi) = \Phi_1 \left(\mathbf{k}, \mathbf{k}_0; \frac{\mathbf{k} + \xi}{2}; \mathbf{k} - \xi\right). \tag{117}$$

As a result we have found that the contribution from the first term in Eq. (113) has the structure of the first term in Eq. (108) and can be included in it.

The contribution from the second term in Eq. (113) is

$$S_1^{(2)}(\mathbf{k}, \mathbf{k}_0) = \int \frac{d\mathbf{r}_1 d\mathbf{r}_2}{(2\pi)^4} d\xi \, e^{-i(\mathbf{k}-\xi)\mathbf{r}_1 - iq_k h(\mathbf{r}_1)} e^{-i(\xi-\mathbf{k}_0)\mathbf{r}_2 - iq_0 h(\mathbf{r}_2)} \times \sum_{\alpha_1,\beta_1=x,y} \left(k^{(\beta_1)} - \xi^{(\beta_1)} - \xi_1^{(\beta_1)}\right) \Delta\varphi_1^{(\alpha_1,\beta_1)} \xi_1^{(\alpha_1)} h_{\xi_1} e^{i\xi_1 \mathbf{r}_1} d\xi_1. \tag{118}$$

By using the identity

$$(\mathbf{k} - \xi - \xi_1) \, e^{-i(\mathbf{k}-\xi-\xi_1)\mathbf{r}_1} = i\nabla_{\mathbf{r}_1} e^{-i(\mathbf{k}-\xi-\xi_1)\mathbf{r}_1}, \tag{119}$$

we can integrate by parts over $\mathbf{r}_1$ again, which results in the replacement of the factor $(k^{(\beta_1)} - \xi^{(\beta_1)} - \xi_1^{(\beta_1)})$ in Eq. (118) by

$$-q_k \nabla h(\mathbf{r}_1) = -iq_k \int \xi_2 h_{\xi_2} e^{i\xi_2 \mathbf{r}_1} d\xi_2. \tag{120}$$

Now, the change of integration variable $\xi \to \xi - (\xi_1 + \xi_2)/2$ leads to an expression that has the structure of the third term in Eq. (108) with

$$\widetilde{\Phi}_2^{(\alpha_1,\beta_1)}(\mathbf{k}, \mathbf{k}_0; \xi; \xi_1, \xi_2) = -iq_k \Delta\varphi^{(\alpha_1,\beta_1)}\left(\mathbf{k}, \mathbf{k}_0; \xi - \frac{\xi_1 + \xi_2}{2}; \xi_1\right). \tag{121}$$

Thus, the second term in Eq. (108) can be eliminated by incorporating it into the first and the second terms (before its inclusion into the second term the right-hand side of Eq. (121) has to be symmetrized with respect to ξ_1, ξ_2.)

It is shown in [4] that by using similar transformations one can also eliminate the third term in Eq. (108) by transferring appropriate parts of it into the zeroth- and the third-order terms. Neglecting the third- and higher-order terms in Eq. (108) leads to the main ansatz of the nonlocal small-slope approximation:

$$S(\mathbf{k}, \mathbf{k}_0) = \int \Phi(\mathbf{k}, \mathbf{k}_0; \xi) \, e^{-i(\mathbf{k}-\xi)\mathbf{r}_1 - iq_k h(\mathbf{r}_1)} e^{-i(\xi-\mathbf{k}_0)\mathbf{r}_2 - iq_0 h(\mathbf{r}_2)} \frac{d\mathbf{r}_1 d\mathbf{r}_2}{(2\pi)^4} d\xi, \tag{122}$$

where the kernel function Φ is now considered to be independent of the elevations h. This function can be determined by the requirement that in the limit of small elevations Eq. (122) reproduces the two lowest orders of perturbation theory [4]. Perturbation theory provides an expression for the scattering matrix in the following form:

$$S(\mathbf{k}, \mathbf{k}_0) = B(\mathbf{k}, \mathbf{k})\delta(\mathbf{k} - \mathbf{k}_0) - 2i\,(q_k q_0)^{1/2} B(\mathbf{k}, \mathbf{k}_0) h_{\mathbf{k}-\mathbf{k}_0} + + (q_k q_0)^{1/2} \int B_2(\mathbf{k}, \mathbf{k}_0; \xi) h_{\mathbf{k}-\xi} h_{\xi-\mathbf{k}_0} d\xi + \cdots . \tag{123}$$

The expressions for the kernel functions B, B_2 are known for many scattering problems [6]. For the electromagnetic problem they can be found,e.g., in [5]. Expanding Eq. (122) into a power series with respect to h, comparing the result with Eq. (123), and using some relations between B, B_2, one finds the following explicit expression for the kernel Φ:

$$\Phi(\mathbf{k}, \mathbf{k}_0; \xi) = \frac{2(q_k q_0)^{1/2}}{q_k + q_0} \left(B(\mathbf{k}, \mathbf{k}_0) + \frac{B_2(\mathbf{k}, \mathbf{k}_0; \mathbf{k}_0) - B_2(\mathbf{k}, \mathbf{k}_0; \xi)}{2q_0} \right. \\ \left. + \frac{B_2(\mathbf{k}, \mathbf{k}_0; \mathbf{k}) - B_2(\mathbf{k}, \mathbf{k}_0; \xi)}{2q_k} \right). \tag{124}$$

Equations (122),(124) constitute the nonlocal small-slope approximation.

2.11. Relation to other Approaches

The drawback of the nonlocal small-slope approximation is the rather high order of integration. One can try to simplify the representation given by Eq. (122) using the following considerations. Let us represent the contribution from the second term in Eq. (124) as follows:

$$\int (B_2(\mathbf{k}, \mathbf{k}_0; \mathbf{k}_0) - B_2(\mathbf{k}, \mathbf{k}_0; \xi))\, e^{-i(\xi - \mathbf{k}_0)\mathbf{r}_2 - iq_0 h(\mathbf{r}_2)}\, d\mathbf{r}_2 \\ = \int e^{-iq_0 h(\mathbf{r}_2)} (B_2(\mathbf{k}, \mathbf{k}_0; \mathbf{k}_0) - B_2(\mathbf{k}, \mathbf{k}_0; \mathbf{k}_0 + i\nabla_{\mathbf{r}_2})) e^{-i(\xi - \mathbf{k}_0)\mathbf{r}_2}\, d\mathbf{r}_2 \\ = \int e^{-i(\xi - \mathbf{k}_0)\mathbf{r}_2} (B_2(\mathbf{k}, \mathbf{k}_0; \mathbf{k}_0) - B_2(\mathbf{k}, \mathbf{k}_0; \mathbf{k}_0 - i\nabla_{\mathbf{r}_2})) e^{-iq_0 h(\mathbf{r}_2)}\, d\mathbf{r}_2. \tag{125}$$

As a result of the action of the operator in Eq. (125) on $\exp(iq_0 h(\mathbf{r}_2))$ a term linear in h will appear as well as terms proportional to the products of derivatives of h of different orders. Let us make an approximation that neglects all the products of the derivatives:

$$(B_2(\mathbf{k}, \mathbf{k}_0; \mathbf{k}_0) - B_2(\mathbf{k}, \mathbf{k}_0; \mathbf{k}_0 - i\nabla_{\mathbf{r}_2})) e^{iq_0 h(\mathbf{r}_2)} \\ \approx -iq_0 e^{iq_0 h(\mathbf{r}_2)} (B_2(\mathbf{k}, \mathbf{k}_0; \mathbf{k}_0) - B_2(\mathbf{k}, \mathbf{k}_0; \mathbf{k}_0 - i\nabla_{\mathbf{r}_2})) h(\mathbf{r}_2). \tag{126}$$

Similarly

$$\int (B_2(\mathbf{k}, \mathbf{k}_0; \mathbf{k}) - B_2(\mathbf{k}, \mathbf{k}_0; \xi))\, e^{-i(\mathbf{k} - \xi)\mathbf{r}_1 - iq_k h(\mathbf{r}_1)} d\mathbf{r}_1 \\ \approx -iq_k \int e^{-i(\mathbf{k} - \xi)\mathbf{r}_1 - iq_k h(\mathbf{r}_1)} (B_2(\mathbf{k}, \mathbf{k}_0; \mathbf{k}) - B_2(\mathbf{k}, \mathbf{k}_0; \mathbf{k} + i\nabla_{\mathbf{r}_1})) h(\mathbf{r}_1) d\mathbf{r}_1. \tag{127}$$

Now integration over ξ produces $\delta(\mathbf{r}_1 - \mathbf{r}_2)$, and we find in this approximation

$$S(\mathbf{k}, \mathbf{k}_0) = \frac{2(q_k q_0)^{1/2}}{q_k + q_0} \int \frac{d\mathbf{r}}{(2\pi)^2} e^{-i(\mathbf{k} - \mathbf{k}_0)\mathbf{r} - i(q_k + q_0)h(\mathbf{r})} \{B(\mathbf{k}, \mathbf{k}_0) \\ + \frac{i}{4} \int [B_2(\mathbf{k}, \mathbf{k}_0; \mathbf{k} - \xi) + B_2(\mathbf{k}, \mathbf{k}_0; \mathbf{k}_0 + \xi) \\ + 2(q_k + q_0) B(\mathbf{k}, \mathbf{k}_0)]\, h_\xi e^{i\xi\mathbf{r}} d\xi \}. \tag{128}$$

Here the following relation between B, B_2 following from the representation Eq. (123) and the transformation property Eq. (36) was used [4]

$$B_2(\mathbf{k}, \mathbf{k}_0; \mathbf{k}) + B_2(\mathbf{k}, \mathbf{k}_0; \mathbf{k}_0) + 2\,(q_k + q_0)\, B(\mathbf{k}, \mathbf{k}_0) = 0. \tag{129}$$

Equation (128) is called the small-slope approximation [6]. By expanding the exponent in powers of h one verifies that the result matches the perturbation expansion Eq. (123) up to the second-order terms.

The expansion of the expression in the square parentheses in Eq. (128) in powers of ξ starts from the terms of $O(\xi)$ due to Eq. (129). Let us extract from this expression the linear terms representing

$$B_2(\mathbf{k}, \mathbf{k}_0; \mathbf{k} - \xi) + B_2(\mathbf{k}, \mathbf{k}_0; \mathbf{k}_0 + \xi) + 2\,(q_k + q_0)\, B(\mathbf{k}, \mathbf{k}_0)$$
$$= -\left(\left.\frac{\mathrm{d}B_2(\mathbf{k}, \mathbf{k}_0; \eta)}{\mathrm{d}\eta}\right|_{\eta=\mathbf{k}} - \left.\frac{\mathrm{d}B_2(\mathbf{k}, \mathbf{k}_0; \eta)}{\mathrm{d}\eta}\right|_{\eta=\mathbf{k}_0} \right) \xi + T\,(\mathbf{k}, \mathbf{k}_0; \xi)\,. \tag{130}$$

The linear term after integration over ξ produces a term that is proportional to ∇h. This term can be integrated by parts as in Eqs. (29),(30). As a result one finds

$$S(\mathbf{k}, \mathbf{k}_0) = \frac{2\,(q_k q_0)^{1/2}}{q_k + q_0} \int \frac{\mathrm{d}\mathbf{r}}{(2\pi)^2}\, \mathrm{e}^{-\mathrm{i}(\mathbf{k}-\mathbf{k}_0)\mathbf{r} - \mathrm{i}(q_k+q_0)h(\mathbf{r})}$$
$$\times \left[\widetilde{R}(\mathbf{k}, \mathbf{k}_0) + \frac{\mathrm{i}}{4} \int T\,(\mathbf{k}, \mathbf{k}_0; \xi)\, h_\xi \mathrm{e}^{\mathrm{i}\xi\mathbf{r}} \mathrm{d}\xi \right], \tag{131}$$

where

$$\widetilde{R}(\mathbf{k}, \mathbf{k}_0) = B(\mathbf{k}, \mathbf{k}_0)$$
$$+ \frac{1}{4} \left(\left.\frac{\mathrm{d}B_2(\mathbf{k}, \mathbf{k}_0; \eta)}{\mathrm{d}\eta}\right|_{\eta=\mathbf{k}} - \left.\frac{\mathrm{d}B_2(\mathbf{k}, \mathbf{k}_0; \eta)}{\mathrm{d}\eta}\right|_{\eta=\mathbf{k}_0} \right) \frac{\mathbf{k} - \mathbf{k}_0}{q_k + q_0}. \tag{132}$$

If one neglects the second term in Eq. (131), the result will correspond to the general expression for the SA in the KA-tangent plane approximation, Eq. (43), with the preintegral factor replaced by $\widetilde{R}(\mathbf{k}, \mathbf{k}_0)$. Based on the results presented in [5] one can check that the matrix $\widetilde{R}$ coincides with the matrix R in Eq. (60) up to terms of the first order in $(\mathbf{k} - \mathbf{k}_0)$. It is important that application to the case of small roughness, where the exponential can be expanded in a power series, leads generally to the wrong answer even for the terms of $O(h)$, since the factor $\mathrm{i}\,(q_k + q_0)\,\widetilde{R}\,(\mathbf{k}, \mathbf{k}_0)$ generally does not coincide with $2\mathrm{i}\,(q_k q_0)^{1/2}\, B\,(\mathbf{k}, \mathbf{k}_0)$ from Eq. (123). Those factors are close only in the near-specular direction. Thus, the KA does not correctly describe scattering at large angles, i.e. the Bragg scattering (which is consistent with the assumption of the smoothness of the roughness). This drawback induced development of different "unifying" theories, which are able to handle both the Kirchhoff- and the Bragg scattering within a single theoretical scheme.

The expansion of the kernel $T\,(\mathbf{k}, \mathbf{k}_0; \xi)$ in powers of ξ starts from the terms of order $O(\xi^2)$. This means that the contribution of this term in Eq. (131) depends on the curvature and higher-order derivatives of the surface profile. The term associated with T in Eq. (131) can be called a curvature correction. In [7,8] another choice of matrices $\widetilde{R}$, T was suggested, with the matrix $\widetilde{R}$ essentially coinciding with R from Eq. (60). The matrix T is modified to match the perturbation expansion Eq. (123) up to the first-order terms $O(h)$. This approach is called the local curvature approximation. Some numerical tests and comparisons of it with other approximations can be found in [9].

Another possible approach is called the tilt-invariant approximation. It was suggested in [10] for the Dirichlet problem. This approach can be illustrated as follows. Consider the Neumann problem and represent the solution of Eq. (18) as an iterative series:

$$u(\mathbf{R}) = 2u^{(\mathrm{in})}(\mathbf{R}) + 2\sum_{n=1}^{\infty}(-2)^n \int \frac{\partial G(\mathbf{R}-\mathbf{R}_n)}{\partial \mathbf{n}_{\mathbf{R}_n}} \cdots \frac{\partial G(\mathbf{R}_2-\mathbf{R}_1)}{\partial \mathbf{n}_{\mathbf{R}_1}} u^{(\mathrm{in})}(\mathbf{R}_1) d\Sigma_{\mathbf{R}_{1,\ldots,n}}. \tag{133}$$

Assuming that the incident field corresponds to a point source $u^{(\mathrm{in})}(\mathbf{R}) = G(\mathbf{R}-\mathbf{R}_0)$, and calculating the field at the point $\mathbf{R}$ using Eq. (21), we obtain the following representation for the Green function for the Neumann problem G_{Neum}:

$$G_{\mathrm{Neum}}(\mathbf{R}) = G(\mathbf{R}-\mathbf{R}_0) + \sum_{n=1}^{\infty}(-2)^n \int \frac{\partial G(\mathbf{R}-\mathbf{R}_n)}{\partial \mathbf{n}_{\mathbf{R}_n}} \cdots \frac{\partial G(\mathbf{R}_2-\mathbf{R}_1)}{\partial \mathbf{n}_{\mathbf{R}_1}} G(\mathbf{R}_1-\mathbf{R}_0)\, d\Sigma_{\mathbf{R}_{1,\ldots,n}}. \tag{134}$$

The Green function here is represented in a coordinate-invariant form. The expression for the SA follows after substitution for the first and the last Green functions in Eq. (134) the Weyl expansion Eq. (4). To make the result tractable the rest of the Green functions are represented as power series in the elevations h. At this point the coordinate-invariant property is lost. Since the normal derivatives of the Green functions on the surface vanish in the case of a plane surface, all the expansions depend in fact on the curvature and higher derivatives of the surface profile and start from the terms of $O(h)$. Thus, the n-th term in the sum in Eq. (134) is of $O(h^{n-1})$. The first terms of the resulting expansion were calculated for the Dirichlet problem in [10].

2.12. Conclusion

We presented here derivations of the Kirchhoff approximation for the problem of wave scattering from rough surfaces, using as examples the scalar Neumann and the vector electromagnetic cases. It was demonstrated how extensions of the KA lead to other approaches, such as the small-slope, tilt-invariant, and the local curvature approximations. Other theories not mentioned in this paper were also tried in the literature; they were reviewed, in particular, in a recent paper[11]. It is clear, however, that not all the possibilities are exhausted yet. The specific geometry of the problem, with scattering being confined to a 2D plane-like manifold, opens additional possibilities which are not present in the general case of 3D scattering. If this paper encourages the reader to try his/her own ideas in this area, the author will consider his task fulfilled.

References

1. A. Voronovich, "Small-slope approximation for electromagnetic wave scattering at a rough interface of two dielectric half-spaces," *Waves Random Media* **4**, 337–367 (1994).
2. W.C. Meecham, "Fourier transform method for the treatment of the problem of the reflection of radiation from irregular surfaces," *J. Acoust. Soc. Am.* **28**, 370–377 (1956).
3. Yu.P. Lysanov, "About one approximate solution of the problem of acoustic wave scattering by a rough interface," *Sov. Phys. – Acoust.* **2**, 190–198 (1956).
4. A.G. Voronovich, "Non-local small-slope approximation," *Waves Random Media* **6**, 151–167 (1996).
5. A.G. Voronovich, "The effect of the modulation of Bragg scattering in small-slope approximation," *Waves Random Media* **12**, 341–349 (2002).
6. A.G. Voronovich, *Wave Scattering from Rough Surfaces*, 2nd edn (Springer-Verlag, New York, 1999).
7. T. Elfouhaily, S. Guignard, R. Awadallah, and D. Thompson, "Local and non-local curvature approximation: a new asymptotic theory for wave scattering," *Waves Random Media* **13**, 321–337 (2003).
8. T. Elfouhaily, S. Guignard, and D. Thompson, "Formal tilt invariance of the local curvature approximation," *Waves Random Media* **13**, L7–L11 (2003).
9. C.-A. Guerin, G. Soriano, and T. Elfouhaily, "Weighted curvature approximation: numerical tests for 2D dielectric surfaces," *Waves Random Media* **14**, 349–363 (2004).
10. M.I. Charnotskii and V.I. Tatarskii, "Tilt-invariant theory of rough-surface scattering: I," *Waves Random Media* **5**, 361–380 (1995).
11. T. Elfouhaily and C.-A. Guerin, "A critical survey of approximate scattering wave theories from random rough surfaces," *Waves Random Media* **14**, R1–R40 (2004).

3
Scattering and the Spatial Frequency Representation

COLIN JR SHEPPARD
Division of Bioengineering, 9 Engineering Drive 1, National University of Singapore, Singapore, 117576, and Department of Diagnostic Radiology, National University of Singapore, 5 Lower Kent Ridge Road, Singapore 119074.

3.1. Introduction

This chapter describes surface scattering in terms of direction cosines, or, equivalently, in terms of three-dimensional (3D) spatial frequencies.[1] This approach results in simplified expressions and provides a physical interpretation. It is also directly applicable to investigation of 3D imaging (including holography, tomography, microscopy, interferometry, surface profiling and shape measurement). It is notable that different areas of optics tend to have their own adherents with their own literature, and few connections between the areas are exploited. Imaging is usually based on diffraction theory, which is distinguished from scattering theory mainly on the basis that diffraction takes place from objects large compared with the wavelength and scattering from structures of the order of the wavelength in size. This distinction breaks down for microscopic imaging where the resolution limit can be sub-wavelength. Both diffraction and scattering also have a geometrical optics limit for large structures, smooth surfaces, and so on.

3.2. Plane Waves

The amplitude of a scalar plane wave at the point $\mathbf{r} = x\mathbf{i} + y\mathbf{j} + z\mathbf{k}$ can be written as (the time dependence $\exp(-i\omega t)$ is suppressed)

$$U(\mathbf{r}) = \exp[in_0k(px + qy + sz)] = \exp(in_0k\mathbf{p} \cdot \mathbf{r}), \tag{1}$$

where $k = 2\pi/\lambda$, n_0 is the refractive index of the medium, and the vector $\mathbf{p}$ is

$$\mathbf{p} = p\mathbf{i} + q\mathbf{j} + s\mathbf{k}, \tag{2}$$

with p, q, s direction cosines so that

$$s = \pm\sqrt{1 - (p^2 + q^2)}, \tag{3}$$

the plus and minus signs representing forward and backward propagating waves, respectively. The vector **p** can be regarded as a normalized three-dimensional spatial frequency vector. Then

$$p^2 + q^2 + s^2 = 1 \tag{4}$$

represents the surface of a sphere, called the Ewald sphere.

3.3. Scattering

If a scalar plane wave is scattered by some scattering structure, the scattered angular spectrum of plane waves can in general be written in spherical coordinates

$$\tilde{U}_2(\theta_2, \phi_2) = S(\theta_1, \phi_1; \theta_2, \phi_2)\, \tilde{U}_1(\theta_1, \phi_1), \tag{5}$$

where $S(\theta_1, \phi_1; \theta_2, \phi_2)$ is the scattering function. The incident and scattered waves are each specified by two degrees of freedom, so that the scattering function has in general four degrees of freedom. The scattering structure can be, for example, a surface, a particle, e.g. a sphere, a medium with varying refractive index or a turbid medium. Scattering can occur in either the forward or the backward direction. The scattering can be expressed in terms of the direction cosines as

$$\tilde{U}_2(p_2, q_2) = [S_f(p_1, q_1; p_2, q_2) + S_b(p_1, q_1; p_2, q_2)]\tilde{U}_1(p_1, q_1), \tag{6}$$

where S_f, S_b are the scattering functions for forward and backward scattering, respectively.

Scattering can be calculated rigorously by a number of alternative techniques such as the modal method, the coupled-wave theory, the waveguide model and by integral equations. These different approaches are equivalent, but one or another technique may be numerically superior according to the geometry or material of the scattering structure. However, there are several limitations of rigorous scattering theory. These methods tend to break down for large angles of incidence or scattering, or for very deep structures. Investigation of total integrated scatter (TIS) and also image modeling are computationally intensive, because the scattered waves must be integrated over angles of incidence or scattering, or both. It is difficult to identify trends in the scattering behavior. In general the inverse problem of reconstructing the scattering object from scattering data seems intractable.

So we are led to consider an approximate theory. For forward scattering we can use the Born approximation or the Rytov approximation. For backward scattering we can use the Kirchhoff approximation or the Rayleigh approximation. In the Rayleigh approximation, the height change is assumed small, but polarization effects are easily incorporated. For the present work, we may be interested in large height changes, so this approximation is not considered further. In the Kirchhoff approximation for surface scattering, the surface is considered to be made up of planar patches, the curvature of which is negligible. The slope of the surface is assumed to be not too large, so that multiple scattering and shadowing effects can be neglected. But the height change can be large, as long as the conditions mentioned are satisfied. The Kirchhoff approximation does not predict enhanced backscattering, but for TIS or imaging modeling the contribution of enhanced backscattering is small after the integration process.

The standard derivation for surface scattering in the Kirchhoff approximation is that of Beckmann and Spizzichino.[2] They found that the reflection coefficient ρ for surface scattering from a surface of infinite conductivity is

$$\rho = \frac{F_3}{A} \iint_A \exp(i\mathbf{v} \cdot \mathbf{r})\, dx\, dy, \tag{7}$$

where A is the area of projection of the surface,

$$\mathbf{r} = x\mathbf{i} + y\mathbf{j} + \zeta(x, y)\mathbf{k} \tag{8}$$

represents the surface profile $\zeta(x, y)$, and $\mathbf{v}$ and the "F-factor" F_3 are expressed in coordinates θ_1, θ_2, ϕ as

$$\mathbf{v} = n_0 k[(\sin\theta_1 - \sin\theta_2 \cos\phi)\mathbf{i} - \sin\theta_2 \sin\phi \mathbf{j} - (\cos\theta_1 + \cos\theta_2)\mathbf{k}] \tag{9}$$

$$F_3(\theta_1; \theta_2, \phi) = \frac{1 + \cos\theta_1 \cos\theta_2 - \sin\theta_1 \sin\theta_2 \cos\phi}{\cos\theta_1(\cos\theta_1 + \cos\theta_2)}. \tag{10}$$

Here θ_1, θ_2 are the angles of incidence and scattering, and ϕ is the meridional angle (the angle of the plane of scattering from the plane of incidence). The geometric significance of $\mathbf{v}$ and F_3 is not too clear from these expressions.

For scattering by a surface, we are only concerned with backward scattering, S_b, so we will omit the suffix in the following. We introduce the direction cosines $p_1, q_1, s_1; p_2, q_2, s_2$ for the incident and scattered waves.[3] We also introduce the scattering vector $kn_0\mathbf{p} = kn_0(p\mathbf{i} + q\mathbf{j} + s\mathbf{k})$, where the normalized spatial frequencies are given by

$$\begin{aligned} p &= p_2 - p_1, \\ q &= q_2 - q_1, \\ s &= s_2 - s_1. \end{aligned} \tag{11}$$

Note that s_1 is negative, and for scattering in the backward direction, s_2 and therefore s, are positive. Then, in some cases, the scattering function can be expressed in terms of p, q, s only, as $S = T(\mathbf{p}) = T(p, q, s)$. The cases when this occurs include those when the Born or Kirchhoff approximation is valid, for the symmetrical case of a sphere, or for a thin grating. In general the scattering function has four degrees of freedom, but, if the Kirchhoff approximation is valid, these reduce to three degrees of freedom. Thus $T(\mathbf{p})$ is constant for all angles of incidence, providing p, q, s are kept constant. The vectors $\mathbf{p}_1, \mathbf{p}_2$ represent Ewald spheres, with centers separated by the vector $\mathbf{p}$, which intersect in a circle, the scattering function in the Kirchhoff approximation being constant for incident and scattered waves represented by points on this circle. The scattering function is given in terms of the reflection coefficient by

$$T = \rho A \cos\theta_1, \tag{12}$$

so, substituting in for the angles of incidence and scattering

$$T(\mathbf{p}) = \frac{\mathbf{p}^2}{2s} \iint_A \zeta(x, y) \exp(-in_0 k \mathbf{p} \cdot \mathbf{r}) \mathrm{d}x \, \mathrm{d}y. \tag{13}$$

We immediately appreciate that this expression is much simpler than that of Beckmann and Spizzichino.[2] It also has the advantage that it is invariant under a rotation about the axis. However, we can continue by expressing the double integral as a triple integral over a δ-function:[4–7]

$$T(\mathbf{p}) = \frac{\mathbf{p}^2}{2s} \iiint_{-\infty}^{\infty} \delta[z - \zeta(x, y)] \exp(-in_0 k \mathbf{p} \cdot \mathbf{r}) \mathrm{d}^3\mathbf{r}. \tag{14}$$

This can be recognized as the 3D Fourier transform of the surface profile:

$$T(\mathbf{p}) = \frac{\mathbf{p}^2}{2s} F\{\delta[z - \zeta(x, y)]\}, \tag{15}$$

where $F\{\cdot\}$ denotes the Fourier transformation operation, that is the scattering function is given by the 3D Fourier transform of the profile, multiplied by a geometric factor that is independent of the form of the surface. Then taking the factor $1/s$ inside the Fourier transform, we have

$$T(\mathbf{p}) = -\frac{\mathrm{i} n_0 k \mathbf{p}^2}{4\pi} F\{b(\mathbf{r})\}, \tag{16}$$

where $b(\mathbf{r})$ represents the bulk of the surface. Finally, by the differentiation theorem of Fourier transforms, we have

$$T(\mathbf{p}) = \frac{\mathrm{i}\pi}{n_0 k} F\{\nabla^2 b(\mathbf{r})\}. \tag{17}$$

Thus we see that the scattering function is given simply by the Fourier transform of a scattering potential[8]

$$V(\mathbf{r}) = \frac{\mathrm{i}\pi}{n_0 k} \nabla^2 b(\mathbf{r}). \tag{18}$$

Expressing the scattering function in terms of the normalized spatial frequencies has resulted in a much simpler form for the scattering function. It also provides a simple explanation of the so-called memory effect. The scattering function for a surface that is not too rough is almost independent of s, as the Fourier transform of the small height is stretched out. The relationships between $\mathbf{p}$ and the coordinates θ_1, θ_2, ϕ are

$$\begin{aligned} p &= \sin\theta_2 \cos\phi - \sin\theta_1, \\ q &= \sin\theta_2 \sin\phi, \\ s &= \cos\theta_2 + \cos\theta_1. \end{aligned} \tag{19}$$

For a particular plane of incidence there are only three degrees of freedom so that there is a unique relationship between θ_1, θ_2, ϕ, and $\mathbf{p}$. We can simply obtain an expression for θ_1 and ϕ in terms of $\mathbf{p}$. However, explicit expressions for θ_1, θ_2, ϕ in terms of $\mathbf{p}$ are quite complicated. Introducing the radial spatial frequency ℓ

$$\ell^2 = p^2 + q^2, \tag{20}$$

we then have

$$\ell^2 = \sin^2\theta_1 + \sin^2\theta_2 - 2\sin\theta_1 \sin\theta_2 \cos\phi. \tag{21}$$

For an isotropic surface, the statistics of $T(\mathbf{p})$ reduce to a function of two variables, ℓ and s, and θ_1, θ_2, ϕ are related for a constant value of ℓ.

3.4. Significance of the Three-Dimensional Spatial Frequencies

The vector $\mathbf{p}$ can be regarded as a vector formed from the three direction cosines, or as a normalized spatial frequency vector. The geometrical significance is illustrated in Fig. 3.1, which shows a section through the 3D spatial frequency space.[9–12] For normal incidence, the spatial frequencies lie on the surface of a sphere of radius unity that passes through the origin, $\ell^2 + s(s-2) = 0$. For specular reflection, $p = q = 0$, and so the spatial frequencies lie on the s-axis. Thus the factor $\mathbf{p}^2/2s$ in (15) reduces to $s/2$. This is consistent with the well-known property first described by Darwin[13] in the connection of X-ray diffraction, that a plane of atoms (which can be represented by a δ-function) does not reflect radiation independently of

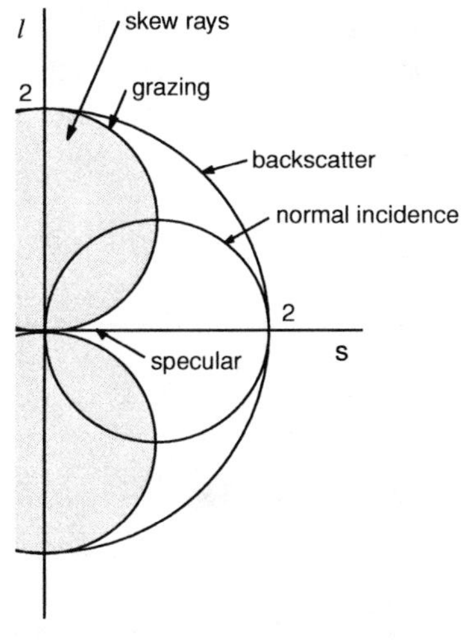

FIGURE 3.1. The geometrical significance of the normalized spatial frequency vector $\mathbf{p}$.

angle but as $1/\cos\theta$. Thus a perfectly conducting plane does not behave as a perfect reflector, but the surface of a perfectly conducting solid does.

All scattered waves, for any angle of incidence, have a value of **p** that lies within a sphere of radius 2, centered at the origin. Backscatter is represented by spatial frequencies on the surface of the sphere $\ell^2 + s^2 = 4$. The region outside of this sphere cannot represent propagating waves. The figure shows a shaded region of toroidal shape. This represents skew rays, which are scattered out of the plane of incidence. The boundary of this volume, $\ell(2 - \ell) = s^2$, represents illumination at grazing incidence.

It should be noted that for the assumed case of a surface with isotropic statistics, scattering in the Kirchhoff approximation reduces to a function of only two variables ℓ, s instead of the three trigonometric variables. Thus use of the 3D spatial frequencies shows very simply how the scattering data change with angle of incidence. Further, measurements out of the plane of incidence give in principle no further information if the Kirchhoff approximation is satisfied. This is not true if the ranges of illumination and scattering angles are restricted to a finite range, say to $\pm\alpha$. Then for measurements in the plane of incidence only, for information to be obtained for a large continuous range of transverse spatial frequencies, s must be in the range between $1 + \cos\alpha = 2\cos^2(\alpha/2)$ and 2, giving $|\ell|$ in the range between 0 and $4\sin^2(\alpha/2)(1 + \cos^2(\alpha/2))$. On the other hand, if measurements out of the plane of incidence are included, the range of s is increased to cover from $2\cos\alpha$ to 2, with $|\ell|$ in the range from 0 to $2\sin\alpha$.

3.5. Polarization Effects

So far we have been concerned with the case of scalar waves. For light waves no depolarization occurs for scattering by 1D rough surfaces for either TE or TM polarized waves, with plane of incidence normal to the direction of constant surface height.

In general, the depolarization can be calculated in terms of the polarization angles β_1,β_2, being the angles of the electric field vector of the incident and scattered radiation, respectively, from the intersection of the wavefront with the reflecting plane (Ref. [2], p.169). It is straightforward to show that

$$\sin\beta_{1,2} = \frac{pq_{1,2} - qp_{1,2}}{\ell_{1,2}\,|\mathbf{p}|\,\sqrt{1 - \mathbf{p}^2/4}}. \tag{22}$$

It is seen that if the conditions,

$$\begin{aligned} p &\ll p_1, \\ q &\ll q_1, \end{aligned} \tag{23}$$

are satisfied, then $\beta_1 \approx \beta_2$. In this case, for given values of p_1 and q_1, s is related to them by

$$p_1^2 + q_1^2 = 4 - s^2. \tag{24}$$

The polarization factor P_2 of the scattered wave can then be written in terms of the polarization factor P_1 of the incident wave (Ref. [2], p.171)

$$P_2 = \frac{\tan 2\beta_1 + P_1}{P_1 \tan 2\beta_1 - 1}. \tag{25}$$

The depolarization thus does not depend only on p, q, and s, but also on the direction cosines of the incident wave.

3.6. Random Surfaces

3.6.1. Statistics of Surface Scattering

We take our surface as that generated by a continuous stationary random process with a height ζ which is distributed normally with mean value

$$\langle \zeta \rangle = 0 \tag{26}$$

and distribution

$$w(\zeta) = \frac{1}{\sigma\sqrt{2\pi}} \exp\left(-\frac{\zeta^2}{2\sigma^2}\right), \tag{27}$$

where σ is the standard deviation (the root-mean-square value of ζ). The characteristic function associated with the height variation is

$$\chi(s) = \exp\left[-\frac{1}{2}(n_0 k s \sigma)^2\right] = \exp(-h^2/2), \tag{28}$$

where

$$h = n_0 k s \sigma. \tag{29}$$

The autocorrelation coefficient is taken as $C(x, y)$. The 2D normal distribution of the two random variables ζ_1 and ζ_2 at points 1 and 2 is, with C_{12} the autocorrelation coefficient between the points,

$$w(\zeta_1, \zeta_2) = \frac{1}{2\pi\sigma^2\sqrt{1 - C_{12}^2}} \exp\left[-\frac{\zeta_1^2 - 2C\zeta_1\zeta_2 + \zeta_2^2}{2\sigma^2(1 - C_{12}^2)}\right], \tag{30}$$

so that the joint characteristic function of the distribution is

$$\chi_2(s, -s) = \exp\left[-h^2\left(1 - C_{12}^2\right)\right]. \tag{31}$$

The mean value of the scattering function for a large surface is[2,14]

$$\langle T(\mathbf{p})\rangle = \frac{\mathbf{p}^2}{2s} \iint_{-\infty}^{\infty} \langle \exp(-in_0 k s \zeta)\rangle \exp[-in_0 k(px + qy)]\, \mathrm{d}x\, \mathrm{d}y$$

$$= \frac{s\lambda^2}{2n_0^2}\chi(s)\delta(p)\delta(q)$$
$$= \frac{s\lambda^2}{2n_0^2}\exp(-h^2/2)\delta(p)\delta(q). \tag{32}$$

Thus the mean of the scattering function, which is in general a complex quantity, vanishes except for $p = q = 0$, corresponding to specular reflection. For p or $q \neq 0$, the mean is zero, and the scattering function, which is in general complex, corresponds to diffuse scattering.

The modulus square of the mean value is for a surface of large area A

$$|\langle T(\mathbf{p})\rangle|^2 = A\left(\frac{\mathbf{p}^2}{2s}\right)^2 \iint_{-\infty}^{\infty} \chi(s)\chi^*(s)\exp\left[-in_0k(px+qy)\right]\mathrm{d}x\,\mathrm{d}y$$
$$= \frac{A\lambda^2}{n_0^2}\left(\frac{\mathbf{p}^2}{2s}\right)^2 \exp(-h^2)\delta(p)\delta(q). \tag{33}$$

The mean-square value of the scattering function is

$$\langle TT^*\rangle = A\left(\frac{\mathbf{p}^2}{2s}\right)^2 \iint_{-\infty}^{\infty} \langle\exp\left[-in_0ks(\zeta_1-\zeta_2)\right]\rangle \times \exp\left[-in_0k(px+qy)\right]\mathrm{d}x\,\mathrm{d}y$$
$$= A\left(\frac{\mathbf{p}^2}{2s}\right)^2 \iint_{-\infty}^{\infty} \chi_2(s,-s) \times \exp\left[-in_0k(px+qy)\right]\mathrm{d}x\,\mathrm{d}y. \tag{34}$$

The variance of the scattering function is then[2,14]

$$D\{T\} = A\left(\frac{\mathbf{p}^2}{2s}\right)^2 \iint_{-\infty}^{\infty} \left[\chi_2(s,-s)-\chi(s)\chi^*(s)\right] \times \exp\left[-in_0k\,(px+qy)\right]\mathrm{d}x\,\mathrm{d}y, \tag{35}$$

where χ and χ_2 are the characteristic functions describing the surface.

The usual measure of scattering from a rough surface is the BRDF (bidirectional reflectance distribution function), defined as the scattered power per unit solid angle divided by the input power.[15] Then, the total BRDF is[7]

$$\mathrm{BRDF} = \frac{n_0^2}{A\lambda^2}\langle TT^*\rangle. \tag{36}$$

Neglecting the coherent component to the BRDF, we have for the incoherent part

$$\mathrm{BRDF}_{\mathrm{inc}} = \frac{n_0^2}{A\lambda^2}D\{T\}, \tag{37}$$

so that

$$\mathrm{BRDF}_{\mathrm{inc}} = \frac{n_0^2}{\lambda^2}\left(\frac{\mathbf{p}^2}{2s}\right)^2 \iint_{-\infty}^{\infty} [\chi_2(s,-s) - \chi(s)\chi^*(s)] \times \exp[-in_0k(px+qy)]\,\mathrm{d}x\,\mathrm{d}y. \tag{38}$$

Thus the BRDF is dimensionless and independent of the area of the surface.

For a surface described by a normal distribution function, using Eq. (35) we have

$$\mathrm{BRDF}_{\mathrm{inc}} = \frac{n_0^2}{\lambda^2}\left(\frac{\mathbf{p}^2}{2s}\right)^2 \exp(-h^2) \times \iint_{-\infty}^{\infty} \{\exp[h^2C(x,y)] - 1\} \times \exp[-in_0k(px+qy)]\,\mathrm{d}x\,\mathrm{d}y. \tag{39}$$

For an isotropic surface roughness, let us assume that the autocorrelation coefficient can be written as $C(Lt)$ where $t = \rho/L$, and L is some characteristic distance. Then we have

$$\begin{aligned}\mathrm{BRDF}_{\mathrm{inc}} &= \frac{c_0^2}{2\pi}\left(\frac{\mathbf{p}^2}{2s}\right)^2 \exp(-h^2) \times \int_0^{\infty} \{\exp[h^2C(Lt)] - 1\} J_0(ct)t\,\mathrm{d}t \\ &= \frac{(n_0kL)^2}{4\pi}\left(\frac{\mathbf{p}^2}{2s}\right)^2 \exp(-h^2) T_3(c,h),\end{aligned} \tag{40}$$

where $c_0 = n_0kL$ and

$$T_3(c,h) = 2\int_0^{\infty} \{\exp[h^2C(Lt)] - 1\} J_0(ct)t\,\mathrm{d}t, \tag{41}$$

the suffix 3 refers to the 3D case.

3.6.1.1. Smooth Surface

For a smooth surface for which $h \ll 1$, expanding the exponential function of the autocorrelation coefficient as a power series and retaining the first two terms[2]

$$\begin{aligned}\mathrm{BRDF}_{\mathrm{inc}} &= \frac{n_0^2}{\lambda^2}\left(\frac{\mathbf{p}^2}{2s}\right)^2 h^2 \exp(-h^2) \times \iint_{-\infty}^{\infty} C(x,y) \exp[-in_0k(px+qy)]\mathrm{d}x\,\mathrm{d}y \\ &= \frac{n_0^4\pi^2}{\lambda^4}\mathbf{p}^4 \exp(-h^2) G(p,q),\end{aligned} \tag{42}$$

where $G(p, q)$ is the power spectral density of the surface profile

$$G(p, q) = \sigma^2 \iint_{-\infty}^{\infty} C(x, y) \exp[-in_0 k(px + qy)] \mathrm{d}x\, \mathrm{d}y. \tag{43}$$

If $h \ll 1$ then the exponential term in Eq. 42 is also approximately unity giving

$$\mathrm{BRDF_{inc}} = \frac{n_0^4 \pi^2}{\lambda^4} \mathbf{p}^4 G(p, q), \tag{44}$$

so that investigation of the image spectrum in the appropriate region can give the power spectral density.

It should be noted that this last equation is true for any distribution function of surface heights. We have for expansion of the characteristic function in terms of the moments

$$\chi(s) = \sum_j \frac{(in_0 ks)^j}{j!} \overline{\zeta^j}, \tag{45}$$

where $\overline{\zeta^j}$ is the jth moment of the surface height variation. If $n_0 ks\zeta$ is small, then expanding as a power series

$$\chi(s) = 1 + in_0 ks\overline{\zeta} - \frac{(n_0 ks)^2 \overline{\zeta^2}}{2} + \cdots \tag{46}$$

so that

$$\chi(s)\chi^*(s) = 1 - h^2 + \cdots. \tag{47}$$

Similarly for the joint characteristic function

$$\begin{aligned} \chi_2(s, -s) &= \sum_j \sum_l \frac{(in_0 ks)^j (-in_0 ks)^l}{j!l!} \left\langle \zeta_1^j \zeta_2^l \right\rangle \\ &= 1 - h^2[1 - C(x, y) + \cdots] \end{aligned} \tag{48}$$

giving

$$\chi_2(s, -s) - \chi(s)\chi^*(s) = h^2 C(x, y) + \cdots. \tag{49}$$

This is true for any distribution function, the mean height having canceled out so that the distribution function does not need to be symmetric.

3.6.1.2. Rough Surface

If on the other hand we have a rough surface $h \gg 1$, the only contributions are from the region where $x = y = 0$, and, assuming an isotropic surface, the autocorrelation coefficient can be expanded as a power series, assuming it is continuous in slope, to give

$$C(\rho) = 1 - \frac{\rho^2}{L^2} + \cdots, \tag{50}$$

where L is identified in terms of the power series expansion. Then

$$\mathrm{BRDF}_{\mathrm{inc}} = \frac{1}{4\pi}\left(\frac{\mathbf{p}^2}{2s}\right)^2\left(\frac{c}{\ell h}\right)^2 \exp\left[-\left(\frac{c}{2h}\right)^2\right], \tag{51}$$

where

$$c = c_0\ell = n_0 k\ell L. \tag{52}$$

Thus for a fixed h, the BRDF depends on the ratio h/c, or equivalently σ/L. However, inverting (43)

$$C(x, y) = \left(\frac{n_0}{\sigma\lambda}\right)^2 \iint G(p, q)\exp\left[in_0 k(px + qy)\right]\mathrm{d}p\,\mathrm{d}q. \tag{53}$$

For an isotropic surface, we can transform to polar coordinates, and by expanding the resulting Bessel function kernel in a power series

$$C(\rho) = 1 - \frac{n_0^4 k^4 \sigma_G^2}{8\pi\sigma^2}\rho^2, \tag{54}$$

where σ_G^2 is the second moment of the power spectral density, so that

$$\frac{\sigma}{L} = \frac{n_0^2 k^2 \sigma_G}{2\sqrt{2\pi}}, \tag{55}$$

and is seen to be directly related to σ_G.

3.6.2. Gaussian Autocorrelation Coefficient

The autocorrelation coefficient is now assumed to be normal:

$$C(\rho) = \exp\left(-\frac{\rho^2}{L^2}\right), \tag{56}$$

where L is the correlation length. Then the BRDF is[2,7]

$$\begin{aligned}\mathrm{BRDF}_{\mathrm{inc}} &= \frac{c_0^2}{2\pi}\left(\frac{\mathbf{p}^2}{2s}\right)^2 \exp(-h^2) \\ &\quad\times \int_0^\infty \{\exp[h^2\exp(-t^2)] - 1\}J_0(ct)t\,\mathrm{d}t \\ &= \frac{c_0^2}{4\pi}\left(\frac{\mathbf{p}^2}{2s}\right)^2 \exp(-h^2)T_3(c, h),\end{aligned} \tag{57}$$

where

$$T_3(c, h) = 2\int_0^\infty \{\exp[h^2\exp(-t^2)] - 1\}J_0(ct)t\,\mathrm{d}t, \tag{58}$$

and the suffix 3 refers to the 3D case. Again we can investigate the case of a smooth surface ($h \ll 1$), giving

$$\mathrm{BRDF}_{\mathrm{inc}} = \frac{c_0^2}{4\pi}\left(\frac{\mathbf{p}^2}{2s}\right)^2 h^2 \exp\left(-\frac{c^2}{4}\right). \tag{59}$$

Also for reasonably small values of h, expanding the outer exponential[2]

$$T_3\left(c, h\right) = \sum_{j=1}^{\infty} \frac{h^{2j}}{j!j} \exp\left(-\frac{c^2}{4j}\right), \tag{60}$$

so that

$$\mathrm{BRDF}_{\mathrm{inc}} = \frac{c_0^2}{4\pi}\left(\frac{\mathbf{p}^2}{2s}\right)^2 \exp(-h^2) \sum_{j=1}^{\infty} \frac{h^{2j}}{j!j} \exp\left(-\frac{c^2}{4j}\right). \tag{61}$$

In the specular direction, for $\ell = 0$, the incoherent part of the BRDF is[16]

$$\mathrm{BRDF}_{\mathrm{inc}} = \frac{c_0^2 s^2}{16\pi} \exp(-h^2)[\mathrm{Ei}(h^2) - \gamma - \ln(h^2)], \tag{62}$$

where Ei is an exponential integral and γ is Euler's constant. For large h,

$$\exp(-h^2)[\mathrm{Ei}(h^2) - \gamma - \ln(h^2)] \to 1/h^2. \tag{63}$$

The function T_3 can be written as

$$T_3\left(c, h\right) = P_3\left(c, h\right) + Q_3\left(c, h\right), \tag{64}$$

where[16]

$$\begin{aligned} P_3\left(c, h\right) &= \sum_{j=1}^{\infty} \frac{(h^2)^{2j}}{(2j)!2j} \exp\left(-\frac{c^2}{4(2j)}\right), \\ Q_3\left(c, h\right) &= \sum_{j=0}^{\infty} \frac{(h^2)^{2j+1}}{(2j+1)!\,(2j+1)} \exp\left(-\frac{c^2}{4(2j+1)}\right). \end{aligned} \tag{65}$$

Thus P, Q are even and odd functions in h^2, respectively. Then

$$\begin{aligned} P_3\left(0, h\right) &= \mathrm{Chi}(h^2) - \gamma - \ln(h^2), \\ Q_3\left(0, h\right) &= \mathrm{Shi}(h^2), \end{aligned} \tag{66}$$

where Chi, Shi are hyperbolic cosine and sine integrals, respectively.

The root-mean-square value of the scattering function is, from (34) and (31)

$$T_{\mathrm{RMS}} = L\sqrt{\pi A}\left(\frac{\mathbf{p}^2}{2s}\right)\exp(-h^2/2)\{T_3(c^2, h^2)\}^{1/2}, \qquad \ell \neq 0,$$
$$= \frac{As}{2}\exp(-h^2/2), \qquad \ell = 0. \tag{67}$$

The condition which must be satisfied for the second of these expressions to be valid is that the contribution from the incoherent component is negligible compared with the coherent component, that is when

$$\frac{A}{\pi L^2} \gg T_3(0, h^2)$$
$$\gg \frac{\exp(h^2)}{h^2}, \tag{68}$$

or, if $A/(\pi L^2) \ll \exp(h^2)/h^2$, then it is valid even for $\ell = 0$.

The total BRDF, including the coherent part is, for an infinite surface,

$$\mathrm{BRDF} = \frac{1}{8\pi}\exp(-h^2)$$
$$\times\left[2\left(\frac{\mathbf{p}^2}{2s}\right)^2 c_0^2 T_3(c, h) + s^2\delta(\ell)/\ell\right], \tag{69}$$

where $\delta(\ell)$ is a radial delta function in cylindrical coordinates. The first component is the diffuse component, and the second component is the coherent specular component. The relative strength of the coherent specular and scattered components is for large surfaces independent of the area of the surface.

3.6.3. Measurement of Surface Roughness

Surface roughness can be studied either by direct investigation of the scattering statistics, or by measuring the surface profile and investigating the statistics of the profile. The scattering statistics can be measured by angular-resolved scattering (ARS).[17,15] In this case, we measure the BRDF as a function of incidence and scattering angles. The data can be presented as a function of the normalized spatial frequency vector $\mathbf{p}$. Only values of $\mathbf{p}^2 \leq 4$ represent real angles of incidence and scattering. For incident and scattering angles limited within an angle α relative to the surface normal, we also have $2\cos\alpha \leq s \leq 2$.

For an isotropic surface described by a continuous stationary random process with a normal distribution of surface heights and a Gaussian autocorrelation coefficient, the BRDF (57) is given by a product of a function of $\mathbf{p}$ and a function of c, h. For different values of σ, L, the scale of the function of c, h varies relative to the value of $\mathbf{p}$. It should be noted that there are thus scaling parameters in both longitudinal (height) and transverse directions. Thus we distinguish between smooth and rough surfaces, and between coarse and fine roughness.

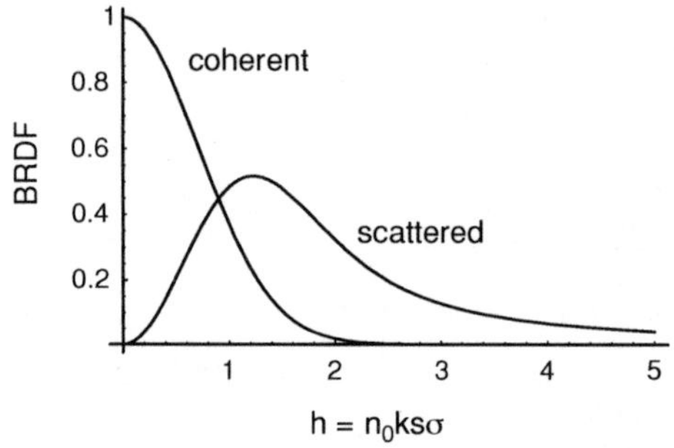

FIGURE 3.2. The surface dependent parts of the scattered and coherent components of the BRDF, for $\ell = 0$, corresponding to the specular direction, given by $\exp(-h^2)[\mathrm{Ei}(h^2) - \gamma - \ln(h^2)]$ and $\exp(-h^2)$, respectively. The surface exhibits a normal distribution of surface heights and a Gaussian autocorrelation coefficient. Constant numerical factors and terms dependent on **p** are suppressed.

Figure 3.2 shows the variation in the function of h for the coherent and scattered components of the BRDF in the specular direction, given by $\exp(-h^2)$ and $\exp(-h^2)[\mathrm{Ei}(h^2) - \gamma - \ln(h^2)]$, respectively, from (32) and (62). We see that the coherent component decays monotonically with h. The scattered component exhibits a maximum value at $h \approx 1$, and then decays as $1/h$. Thus when this is multiplied by the function of $\mathbf{p}$ (i.e. by $s/2$), it will tend to a constant value. Figure 3.3 shows the variation of the scattered component of the BRDF given by the function of c, h, given by $\exp(-h^2)T_3(c, h)$ from (57). It exhibits a maximum on the $\ell = 0$ axis. The importance of this function is that it is valid, within the Kirchhoff approximation, for surfaces of any magnitude of roughness or correlation length. Alteration of the values of σ and L just changes the scale of this function relative to the observed spatial frequencies $\mathbf{p}$.

We now investigate the behavior for some special cases.

3.6.3.1. Smooth Surface, $n_0 k\sigma = h_0 \ll 0.5$

In this case the condition for a smooth surface ($h \ll 1$) applies for all $s \leq 2$. Then from (69) and (42), we have

$$\mathrm{BRDF} = \frac{1}{16\pi}\left\{\mathbf{p}^2(c_0 h_0)^2 \exp\left(-\frac{c^2}{4}\right) + 2s^2\delta(\ell)/\ell\right\}, \tag{70}$$

where $h_0 = n_0 k\sigma$. It is seen that s appears only in a term that is a property of the optical system only, and not of the particular surface. Integrating over axial and transverse spatial frequencies we can obtain expressions for the power in the specular and diffuse components. This is analogous to the total integrated scatter (TIS) method of roughness measurement.

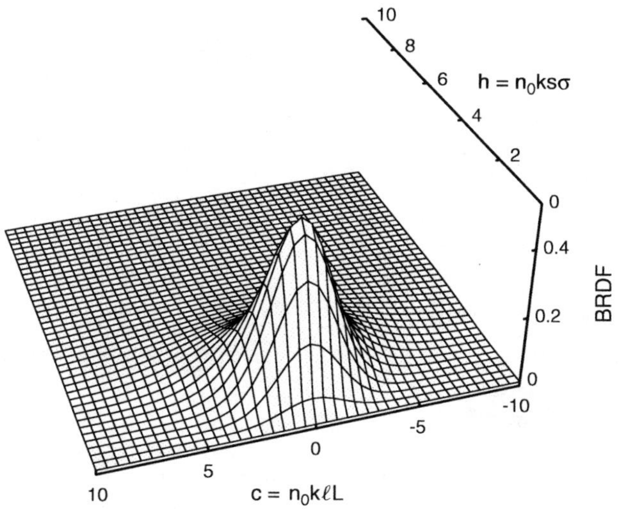

FIGURE 3.3. The surface dependent part of the scattered component of the BRDF, given by $\exp\left(-h^2\right) T_3\left(c, h\right)$. The surface exhibits a normal distribution of surface heights and a Gaussian autocorrelation coefficient. A constant numerical factor and term dependent on $\mathbf{p}$ are suppressed.

3.6.3.2. Rough Surface, $h_0 \gg 0.5 \sec \alpha$

Now the surface is so rough that the monotonically decaying region of the BRDF completely fills the observed range of spatial frequencies. We now have from (69) and (51)

$$\text{BRDF} = \frac{1}{8\pi}\left\{2\left(\frac{\mathbf{p}^2}{2s}\right)^2\left(\frac{L}{s\sigma}\right)^2 \exp\left[-\frac{1}{4}\left(\frac{\ell L}{s\sigma}\right)^2\right] + s^2 \exp(-h^2)\frac{\delta\left(\ell\right)}{\ell}\right\}. \tag{71}$$

For large h the specular component becomes weak. Observation of the variation of the BRDF allows σ/L to be determined.

3.6.3.3. Fine Surface, $c_0 \ll 1/\sin\alpha$

In the case of a small correlation length, we can assume $c \ll 1$ over the whole observable region of spatial frequencies, so that from (40) and (62), we have

$$\text{BRDF} = \frac{1}{8\pi}\exp\left(-h^2\right) \times \left\{2\left(\frac{\mathbf{p}^2}{2s}\right)^2 c_0^2[\text{Ei}(h^2) - \gamma - \ln(h^2)] + s^2\delta(\ell)/\ell\right\}, \tag{72}$$

and it is seen that for the scattered component ℓ appears only in the term that is a property of the optical system only, so we can integrate over ℓ. For a fine and smooth surface

$$\text{BRDF} = \frac{s^2}{16\pi}\{\mathbf{p}^4(c_0 h_0)^2 + 2\delta(\ell)/\ell\}. \tag{73}$$

For a fine and rough surface, the condition that must be satisfied is now $L/\sigma \ll 0.5\cot(\alpha/2)$, and then

$$\text{BRDF} = \frac{1}{8\pi}\left\{2\left(\frac{\mathbf{p}^2}{2s}\right)^2\left(\frac{L}{s\sigma}\right)^2 + s^2\exp(-h^2)\delta(\ell)/\ell\right\}. \tag{74}$$

3.6.3.4. Coarse Surface, $c_0 \gg 1/\sin\alpha$

If the correlation length of the surface is large, then variation in ℓ^2 can be neglected over the region where the surface spectrum is appreciable:

$$\text{BRDF} = \frac{s^2}{16\pi}\exp(-h^2)[c_0^2 T_3(c, h) + 2\delta(\ell)/\ell]. \tag{75}$$

Then for a coarse and smooth surface

$$\text{BRDF} = \frac{s^2}{16\pi}\left\{c_0^2 h^2 \exp\left(-\frac{c^2}{4}\right) + 2\delta(\ell)/\ell\right\}. \tag{76}$$

For a coarse and rough surface, for $L/\sigma \gg \cot(\alpha/2)/2$, we have

$$\text{BRDF} = \frac{s^2}{16\pi}\left\{\left(\frac{L}{s\sigma}\right)^2 \exp\left[-\frac{1}{4}\left(\frac{\ell L}{s\sigma}\right)^2\right] + 2\exp(-h^2)\frac{\delta(\ell)}{\ell}\right\}. \tag{77}$$

3.6.4. Imaging of Surface Roughness

Measurement of the BRDF loses the relative phase of the scattered components, and hence the surface profile cannot be recovered without using phase retrieval techniques. On the other hand, the surface profile can be recovered by various techniques such as confocal profilometry[18] and interferometry. These imaging techniques can be characterized by a coherent transfer function (CTF), which also has a support $\mathbf{p}^2 \leq 4$, $2\cos\alpha \leq s \leq 2$ for an objective lens of semi-angular aperture α. The image amplitude is then given by the 3D Fourier transform of the spatial frequencies (scattered components) transmitted through the system. Thus the relative phase of these components is retained in the imaging process. In interferometry, the phase of the image is measured. In confocal microscopy the image intensity only is measured, but nevertheless the profile can be extracted from these 3D intensity image data.[19]

3.6.5. Inversion of Scattering Data

For a smooth surface the power spectral density, and hence the autocorrelation coefficient, can be determined directly from the scattering data. Even for rougher surfaces, for a normally distributed surface height the simple form of BRDF can be recognized as a Fourier transform, so that the scattering data, measured by altering both incidence and scattering directions so that s is kept constant, can be inverse transformed to determine the autocorrelation coefficient. For a fixed value of s we can write, by inverting the BRDF,

$$\exp\{-h^2[1 - C(x, y)]\} - \exp[-h^2]$$
$$= \iint_{-\infty}^{\infty} \mathrm{BRDF}_{\mathrm{inc}} \left(\frac{2s}{\mathbf{m}^2}\right)^2 \exp[-in_0k(px + qy)]\mathrm{d}p\,\mathrm{d}q. \qquad (78)$$

Denoting the integral in this equation by $I(x, y)$, the condition $C(0, 0) = 1$ allows us to determine the RMS surface height

$$h^2 = -\ln[1 - I(0, 0)]. \qquad (79)$$

It should be noted that this requires measurement of the absolute value of the BRDF. Then the autocorrelation coefficient is given by

$$C(x, y) = 1 - \frac{\ln[1 - I(0, 0) + I(x, y)]}{\ln[1 - I(0, 0)]}. \qquad (80)$$

Whilst this equation is exact within the Kirchhoff approximation, the approach is only applicable for intermediate values of roughness, as for very rough surfaces it breaks down because the scattering becomes insensitive to the form of the autocorrelation coefficient except for values of x, y very close to zero. In practice, the value of h can be reduced by increasing the value of s, which requires an increased range of measurement angles or a longer wavelength.

For very rough surfaces it is still possible to obtain information concerning the behavior of the autocorrelation coefficient for small distances. Fractal behavior, associated with a cusp in the autocorrelation coefficient, is indicated by a cusp in $I(x, y)$. Hence the existence and properties of an inner scale of fractal behavior can in some cases be determined. For determination of statistics rather than the actual profile, the autocorrelation coefficient and BRDF are both real and symmetric functions, so that knowledge of phase information is unnecessary.

3.6.6. Statistics of the Scattered Field

Let us now consider the statistics obeyed by the scattered field. Beckmann and Spizzichino[2] show that the statistics satisfy a distribution $p(u)$, where u is a normalized amplitude, which depends on two parameters, B representing the relative strength of the constant to random components, and K, the asymmetry factor, which is the square root of the ratio of the imaginary to real components of the random part. For normally distributed surface heights we find that, using the approximation of Beckmann and Spizzichino,[2]

$$K = [\coth(h^2/2)]^{1/2}, \qquad \ell = 0,$$
$$= 1, \qquad \ell \neq 0. \qquad (81)$$

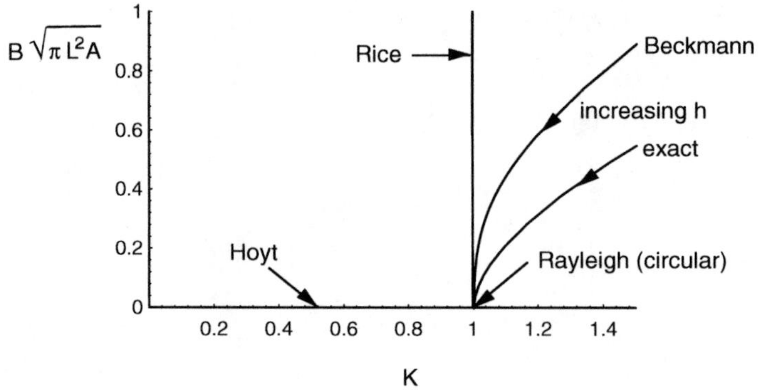

FIGURE 3.4. The relationship between B and K, with normalized roughness h as parameter, for a surface exhibiting a normal distribution of surface heights and a Gaussian autocorrelation coefficient, predicted by the exact Kirchhoff theory and the Beckmann approximation. The behavior of the Rayleigh, Rice and Hoyt distributions are also shown.

and for the 3D case

$$B = (\pi L^2 A)^{-1}[\mathrm{Ei}(h^2) - \gamma - \ln(h^2)]^{-1}, \qquad \ell = 0,$$
$$= 0, \qquad \ell \neq 0. \tag{82}$$

Beckmann calculated the value of K by neglecting the interdependence of the real and imaginary components of the variance of the scattering function. Goodman[16] has calculated the variance of the real and imaginary parts of the scattered field directly, so that for $\ell = 0$

$$D\{T_r\} = \frac{\pi L^2 A s^2}{4} \exp(-h^2)[\mathrm{Chi}(h^2) - \gamma - \ln(h^2)]$$
$$D\{T_i\} = \frac{\pi L^2 A s^2}{4} \exp(-h^2)\mathrm{Shi}(h^2). \tag{83}$$

We obtain

$$K = \left[\frac{\mathrm{Shi}(h^2)}{\mathrm{Chi}(h^2) - \gamma - \ln(h^2)}\right]^{1/2}, \qquad \ell = 0,$$
$$= 1, \qquad \ell \neq 0. \tag{84}$$

Thus for $\ell \neq 0$, when there is no specular component, the statistics are circular (Rayleigh). For $\ell = 0$ there are both specular and random components. The values of B and K given by the exact theory and using the Beckmann approximation are illustrated in Fig. 3.4. For large values of h the statistics become circular, but for values of h larger than about 1.5 they satisfy approximately a Rice distribution

$$p(u) = 2u \exp[-(B^2 + u^2)] I_0(2Bu), \tag{85}$$

where I_0 is a modified Bessel function of order zero.

3.6.7. Limitations of the Kirchhoff Approximation

The usual theory for image plane speckle[16] assumes that there are at least several correlation areas within the point spread function of the imaging system, so that the central limit theorem can be invoked to give an image field which is a Gaussian random variable. We take the number of correlation areas within the point spread function to be

$$\begin{aligned} N &= \frac{4}{n_0 k L \sin\alpha} \\ &= \frac{4}{n_0 k \ell_0 L} = \frac{4}{c_0}, \end{aligned} \tag{86}$$

where ℓ_0, c_0 refer to values for the transverse cut-off frequency for coherent imaging. The RMS value of the scattering function varies as $\exp(-c^2/8)$, i.e. at the cut-off for confocal imaging it has a value $\exp(-2/N^2)$, so that if N is large the scattering function varies slowly with transverse spatial frequency over the pass-band of the optical system. It should be noted that the treatment presented earlier makes no assumption as to the value of L, although the total area of the surface is assumed to contain many correlation areas.

The application of the Kirchhoff approximation relies on the assumption that the radius of curvature of the surface is large compared with the wavelength. The curvature is always smaller than the second derivative of the surface profile. From the theory of stochastic processes, the distribution of the second derivative of a normal process is itself normal. The distribution function of the second derivative is

$$w(\zeta'') = \frac{l^2}{2\sigma\sqrt{2\pi}} \exp\left[-\frac{(\zeta'')^2 L^4}{8\sigma^2}\right] \tag{87}$$

so that the standard deviation of the second derivative is

$$\sigma'' = \frac{2\sigma}{L^2}. \tag{88}$$

The second derivative is unlikely to have a value greater than two standard deviations so that the Kirchhoff approximation will be valid for almost all points on the surface if

$$\frac{(n_0 k L)^2}{n_0 k \sigma} \gg 8\pi. \tag{89}$$

This can be written as

$$\frac{c^2}{h} \gg \frac{8\pi \ell^2}{s}, \tag{90}$$

and as the maximum value of ℓ is $2\sin\alpha$ and the minimum value of s is $2\cos\alpha$, the approximation is valid over the whole of the nonzero region of the transfer function if

$$\frac{c^2}{h} \gg 16\pi \sin\alpha \tan\alpha, \tag{91}$$

which is more restrictive the larger is the angular aperture.

3.7. Fractal Surfaces with an Outer Scale

In early work on the scattering of light by rough surfaces, a Gaussian autocorrelation coefficient was assumed. However, autocorrelation coefficients of exponential form have been experimentally observed. Elson and Bennett[20] pointed out that an exponential correlation is physically unacceptable as it results in an infinite mean square slope. This is an indication of fractal behavior, and of course in practice there has to be a limitation in the range of scales over which the fractal nature exists.

We therefore explore scattering by surfaces with exponential correlation. These are, in fact, special cases of a more general surface with correlation given by a modified Bessel function of the second kind.

Thus an exponential autocorrelation coefficient, for separation ρ, is of the form

$$C(\rho) = \exp(-\rho/L). \tag{92}$$

Still assuming rotational symmetry, this can be generalized to

$$C(\rho) = \frac{2}{\Gamma(m-1)} \left(\frac{\rho}{2L}\right)^{m-1} K_{m-1}\left(\frac{\rho}{L}\right), \qquad \rho \geq 0, \tag{93}$$

where K is a modified Bessel function of the second kind and m is a constant ≥ 1. This expression has been normalized to give a value of unity for the autocorrelation coefficient for zero separation. The power spectral density (PSD) is then

$$G(\ell) = \frac{4\pi\sigma^2 L^2(m-1)}{(1+c^2)^m}, \tag{94}$$

where σ is the rms surface height and $c = n_0 k\ell L$ can be regarded as a normalized radial spatial frequency. For high spatial frequencies we see that the PSD is proportional to c^{-2m}. It thus behaves in this range as a fractal surface, and the fractal dimension is

$$D = 4 - m. \tag{95}$$

For $1 < m < 2$, it is a fractal with an outer scale L. For $m = 3/2$, corresponding to $D = 5/2$, this reduces to the exponential correlation, the so-called Brownian fractal, and for $m = 2$, to a marginal fractal,

$$C(\rho) = \frac{\rho}{L} K_1\left(\frac{\rho}{L}\right). \tag{96}$$

For $m = 4/3$, corresponding to $D = 8/3$, i.e. the case of Kolmogorov turbulence,[21] the correlation function is

$$C(\rho) = \frac{2\pi 3^{1/6}}{\gamma} \mathrm{Ai}\left[\left(\frac{3\rho}{2L}\right)^{2/3}\right], \tag{97}$$

where Ai is an Airy function. For $m > 2$, the surface becomes a subfractal. For large m, the power spectral density tends to a Gaussian

$$G(\ell) = 4\pi\sigma^2 L^2(m-1)\exp[-(m-1)c^2]. \tag{98}$$

Scattering by a fractal structure with an outer scale has been investigated by Uscinski *et al.*,[22] and also by Jakeman,[23] although they did not consider rough surface scattering specifically, or 2D surfaces.

We can normalize L so that the power spectral density has the same value for small spatial frequencies for all values of m. We thus introduce

$$T = 2L\sqrt{m-1}. \tag{99}$$

Plots of the PSD and the correlation function for different values of m are shown in Fig. 3.5.

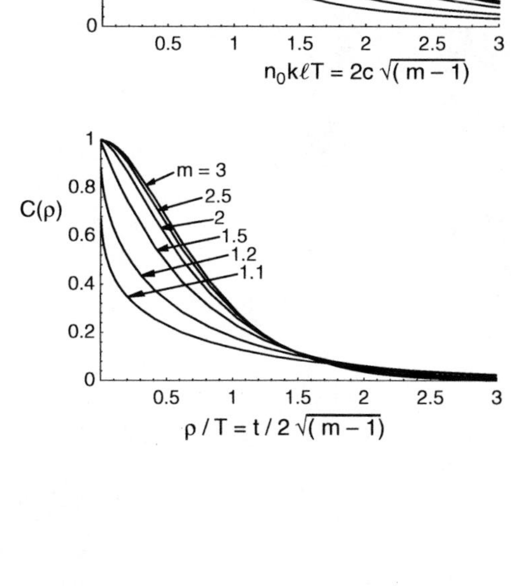

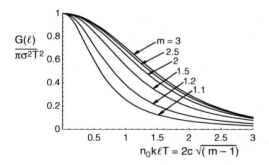

FIGURE 3.5. The power spectral density and the autocorrelation coefficient for fractal surfaces with an outer scale.

3.7.1. Scattering by a Fractal Surface with an Outer Scale

For scattering by a surface with Gaussian distribution of surface heights and a fractal power spectrum, we can calculate the BRDF as previously. For the particular value of $m = 3/2$ we then have[24,25]

$$T_3(c, h) = 2\int_0^\infty \{\exp[h^2 \exp(-t)] - 1\} J_0(ct)t\,dt$$

$$= \sum_{r=1}^{\infty} \frac{h^{2r}}{r!r^2(1 + c^2/r^2)^{3/2}}. \tag{100}$$

For $\ell = 0$, corresponding to the specular direction

$$T_3(c, h) = \sum_{r=1}^{\infty} \frac{h^{2r}}{r!r^2}. \tag{101}$$

Fig. 3.6 shows the form of the function $\exp(-h^2)T_3(c, h)$. The behavior is broadly similar to that described earlier for Gaussian correlation functions. Fig. 3.7 shows the behavior for $\ell = 0$, corresponding to the specular direction, for different values of m. The curves tend to that for a Gaussian correlation function for large values of m, whereas for smaller values of m they decay more quickly with h.

We now turn to the important case of rough surfaces, corresponding to $h \gg 1$. Then the dominant contribution to the integral of (100) is from small values of t. Thus the correlation function can be expanded as a power series, and the second term of the integral neglected, so taking

$$C(\rho) = 1 - \frac{\rho}{L} + \cdots, \qquad \rho \geq 0, \tag{102}$$

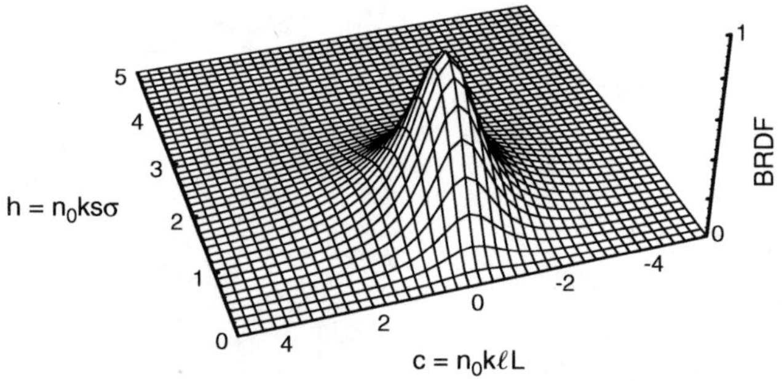

FIGURE 3.6. The surface-dependent part of the scattered component of the BRDF for a surface exhibiting a normal distribution of surface heights and an exponential autocorrelation coefficient. A constant numerical factor and term dependent on **p** are suppressed.

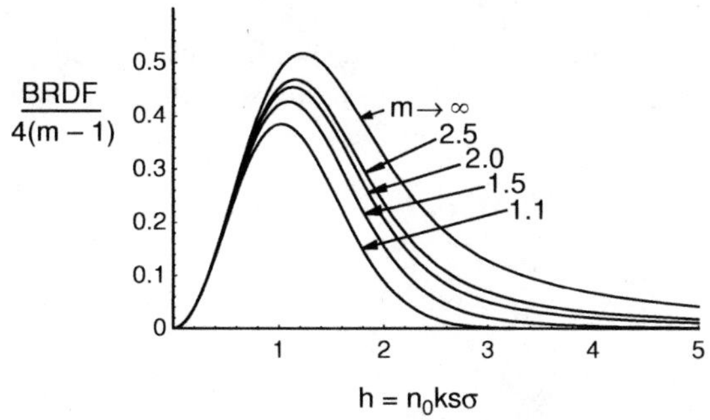

FIGURE 3.7. The surface-dependent parts of the scattered component of the BRDF, for $\ell = 0$, corresponding to the specular direction, for a surface exhibiting a normal distribution of surface heights and a K-distribution autocorrelation coefficient, for different values of the parameter m. Constant numerical factors and terms dependent on $\mathbf{p}$ are suppressed.

the BRDF is

$$\begin{aligned}\text{BRDF} &= \frac{(n_0 kL)^2}{4\pi}\left(\frac{\mathbf{p}^2}{2s}\right)^2 \int_0^\infty \exp(-h^2 t) J_0(ct)t\,\mathrm{d}t \\ &= \frac{(n_0 kL)^2}{2\pi}\left(\frac{\mathbf{p}^2}{2s}\right)^2 \frac{(c/h^2)^2}{[1+(c/h^2)^2]^{3/2}}. \end{aligned} \tag{103}$$

It is seen that for all rough surfaces with correlation functions that have a nonzero first power term in their power series expansion, the same form of BRDF results. The behavior is different from that for surfaces with Gaussian correlation function as the BRDF is a function of c/h^2 rather than c/h.

The surface for $m = 3/2$ exhibits a simple cusp at the origin, but for other values of m in the range $1 < m < 2$ the behavior is more complicated. For small t we can use the approximation, valid for $0 < \nu < 1$ and small x,

$$K_\nu(x) = 2^{\nu-1}\Gamma(\nu)\left(\frac{1}{x^\nu} - \mathrm{d}^2 x^\nu\right), \tag{104}$$

where

$$d = \frac{1}{\Gamma(\nu)2^\nu}\sqrt{\frac{\pi}{\nu\sin\nu\pi}}, \tag{105}$$

so that

$$C(\rho) = 1 - d^2\left(\frac{\rho}{L}\right)^{2(m-1)}, \tag{106}$$

and for $h \gg 1$

$$T_3(c,h) = 2\exp(h^2)\int_0^\infty \exp\left\{-h^2\mathrm{d}^2 t^{2(m-1)}\right\} J_0(ct)t\,\mathrm{d}t. \tag{107}$$

It is noticed that the exponential term is of the same form as a super-Gaussian,[26] but that as the exponent is less than unity it should really be called a sub-Gaussian.

For $c = 0$, corresponding to scattering in the specular direction,

$$T_3(0,h) = \exp(h^2)\frac{1}{(hd)^{2/(m-1)}}\frac{1}{(m-1)}\Gamma\left(\frac{1}{m-1}\right)$$
$$= \left[\frac{a(m)}{h}\right]^{2/(m-1)}. \tag{108}$$

The value of $a\,(m)$ does not change appreciably in magnitude until m approaches either unity, corresponding to an extreme fractal, or two, corresponding to a marginal fractal. The power law of the decay of T_3 with h increases strongly as m decreases.

The behavior for large h can be described by

$$T_3(c,h) = \exp(h^2)\frac{1}{(hd)^{2/(m-1)}}\frac{1}{(m-1)}\Gamma\left(\frac{1}{m-1}\right)f(w), \tag{109}$$

where

$$f(w) = \frac{(m-1)}{8\Gamma\left(\frac{1}{m-1}\right)}\int_0^\infty \exp\{-q^{2(m-1)}\}J_0(2(m-1)^{1/2(m-1)}wq)q\,dq \tag{110}$$

with

$$w = \frac{2c}{(\sqrt{m-1}hd)^{1/(m-1)}}$$

and

$$q = \frac{1}{2}(hd)^{1/(m-1)}t. \tag{111}$$

For rational values for m, the integral $f\,(w)$ can be evaluated in terms of generalized hypergeometric functions. It is found that as $m \to 1$,

$$\mathrm{Lt}_{m\to 1}\{f(w)\} = \exp\left(-\frac{w}{4}\right)I_0\left(\frac{w}{4}\right). \tag{112}$$

For particular values the hypergeometric functions may reduce to simpler forms. As before, for the Brownian fractal $m = 3/2$

$$f\,(w) = \frac{1}{\left[1+\left(\frac{w}{4}\right)^2\right]^{3/2}}. \tag{113}$$

For the marginal fractal $m = 2$, the cusp disappears and the behavior for large h varies similarly to a Gaussian correlation function:

$$f(w) = \exp\left[-\left(\frac{w}{4}\right)^2\right].$$

As m decreases, the peak becomes sharper relative to the decay.

The behavior for large h and small c can be determined by expanding the Bessel function in Eq. (110) in a power series and retaining the first two terms. We obtain

$$
\begin{aligned}
f(w) &= \frac{1}{16} - \frac{(m-1)^{1/(m-1)}}{16} \frac{\Gamma\left(\frac{2}{m-1}\right)}{\Gamma\left(\frac{1}{m-1}\right)} w^2 + \cdots, \\
&= \frac{1}{16} - \frac{1}{4(hd)^{2/(m-1)}} \frac{\Gamma\left(\frac{2}{m-1}\right)}{\Gamma\left(\frac{1}{m-1}\right)} c^2 + \cdots.
\end{aligned} \tag{114}
$$

These equations predict that for fixed and large h the scattering varies as c^{m-1}/h. Thus the scattering spreads out more quickly with h for higher fractal dimension, with proportionally more power being scattered through high angles.

3.8. Total Integrated Scatter (TIS)

Total integrated scatter (TIS) measurements are an established technique for measuring the roughness of surfaces.[17] The surface is illuminated with a plane wave, and the scattered light collected. Then, if the surface is illuminated normally, the TIS, defined as the total scattered power normalized by the total reflected power, has been shown to be related to the rms surface height by[25,27]

$$
\mathrm{TIS} = 1 - \exp[-4(n_0 k \sigma)^2]. \tag{115}
$$

The TIS can be calculated by integrating over the power scattered at different angles. The angular scattering can be described by the bidirectional reflectance distribution function (BRDF)[15]. As before, under certain circumstances, scattering can be calculated by the Kirchhoff approximation,[2,28] which assumes that the radius of curvature of the surface is large compared with the wavelength, and neglects multiple scattering and shadowing effects.

For normal incidence the factor

$$
\left(\frac{\mathbf{p}^2}{2s}\right) = 1, \tag{116}
$$

giving

$$
s^2 = 2 - \ell^2 + 2\sqrt{1-\ell^2}, \tag{117}
$$

and we have the relationships between the differentials in solid angle ω and the normalized spatial frequencies

$$
\mathrm{d}\omega = \frac{2\pi \ell \mathrm{d}\ell}{\sqrt{1-\ell^2}} = -2\pi \mathrm{d}s. \tag{118}
$$

Integrating the coherent specular component over the solid angle around the axis, the normalized coherent power is, for any finite angular semi-aperture α

$$P_c = \int_0^{\sin\alpha} \frac{2\pi \ell d\ell}{\sqrt{1-\ell^2}} \exp[-(n_0 k\sigma)^2]\ell d\ell$$
$$= \exp[-4(n_0 k\sigma)^2], \tag{119}$$

thus agreeing with the results of Bennett and Porteus.[27]

Integrating the scattered contribution over semi-angular apertures from α_0 to α, we have for the normalized scattered power[30]

$$P_s = \int_{1+\cos\alpha}^{1+\cos\alpha_0} \frac{c_0^2}{2} \exp[-(h_0 s)^2] T_3 \left[c_0\sqrt{s\,(2-s)},\, h_0 s\right] \mathrm{d}s, \tag{120}$$

where $c_0 = c/\ell = n_0 kL$ and $h_0 = h/s = n_0 k\sigma$ are independent of s.

For the particular case of a long correlation length, the light is scattered over small angles so we can put, for the case when $\alpha_0 = 0$,

$$s = 2 - \delta\,, \tag{121}$$

with δ small, so that

$$P_s = \int_0^{\infty} \frac{c_0^2}{2} \exp(-4h_0^2) T_3(c_0\sqrt{2\delta}, 2h_0)\mathrm{d}s \tag{122}$$

where, as the exponential decays quickly, the upper limit of the integral has been replaced by infinity, and after performing the integral and then the summation in T_3, we obtain

$$P_s = 1 - \exp(-4h_0^2). \tag{123}$$

The total power, specular and scattered, is thus from (119) and (123) unity, satisfying conservation of energy. Thus (123) also gives the TIS, agreeing with (8.1).We note that if the collection aperture is not high enough to collect all the scattered light, the total scattered plus specular power is not unity. In addition, for shorter correlation lengths, some of the scattered power will go into evanescent waves, which cannot be collected. Thus it is more convenient to define TIS as the ratio of the scattered power to the reflected incident power,[15] although defining it as the ratio of the scattered power to the sum of the coherent and scattered powers[17] has the advantage of being defined in terms of measurable quantities.

We can continue to consider various other special cases. For a smooth surface, such that $h_0 \ll 1$, we can retain only the first team of the expansion of T_3 in (120), giving

$$P_s = 4h_0^2 \int_{1+\cos\alpha}^{1+\cos\alpha_0} \frac{(c_0 s)^2}{8} \exp\left[-\frac{1}{4}c_0^2 s(2-s)\right] \mathrm{d}s. \tag{124}$$

The integral tends to unity for large c_0, so that the scattered power is given by $4h_0^2$, agreeing with (115) for small h_0. Thus the factor

$$f = \frac{P_s}{4h_0^2} = \int_{1+\cos\alpha}^{1+\cos\alpha_0} \frac{(c_0 s)^2}{8} \exp\left[-\frac{1}{4}c_0^2 s(2-s)\right] \mathrm{d}s \tag{125}$$

can be considered as a correction factor, describing the effects of the finite range of collection angles on the measured surface roughness. This factor can be expressed in terms of the error function. For high apertures the correction factor is close to unity if $c_0 = n_0 kL$ is greater than about 4.

For a very rough surface, $h_0 \gg 1$,

$$\begin{aligned} P_s &= \int_{1+\cos\alpha}^{1+\cos\alpha_0} \frac{c_0^2}{2(h_0 s)^2} \exp\left[-\frac{(2-s)c_0^2}{4h_0^2 s}\right] \mathrm{d}s \\ &= \exp\left[-\frac{1}{4}\left(\frac{c_0}{h_0}\right)^2 \tan^2\frac{\alpha_0}{2}\right] - \exp\left[-\frac{1}{4}\left(\frac{c_0}{h_0}\right)^2 \tan^2\frac{\alpha}{2}\right]. \end{aligned} \tag{126}$$

It is seen, as has been pointed out many times previously, that the scattering depends on the ratio c_0/h_0 rather than on the parameters separately.

The correction factor is in general

$$f = \frac{P_s}{4h_0^2} = \int_{1+\cos\alpha}^{1+\cos\alpha_0} \frac{1}{8}\left(\frac{c_0}{h_0}\right)^2 \exp\left[-(h_0 s)^2\right] T_3\left[c_0\sqrt{s\,(2-s)},\, h_0 s\right] \mathrm{d}s, \tag{127}$$

which is shown in Fig. 3.8 for $\alpha_0 = 0$ and $\alpha = \pi/2$. If the correlation length is too small or the RMS surface height too large for the TIS to be proportional to the roughness, the correction factor allows the correct roughness to be predicted.

3.9. Dielectric Medium

So far we have assumed that the surface is a perfect conductor. Consider now scattering by the rough surface of a dielectric medium. First consider reflection from a plane interface between media of refractive indices n_0 and n_2. The angular reflectivity can be derived from Maxwell's equations giving, for parallel and perpendicular polarizations, respectively

$$\begin{aligned} \rho_{\parallel} &= \frac{n_2 c - n_0 c_2}{n_2 c + n_0 c_2}, \\ \rho_{\perp} &= \frac{n_0 c - n_2 c_2}{n_0 c + n_2 c_2}, \end{aligned} \tag{128}$$

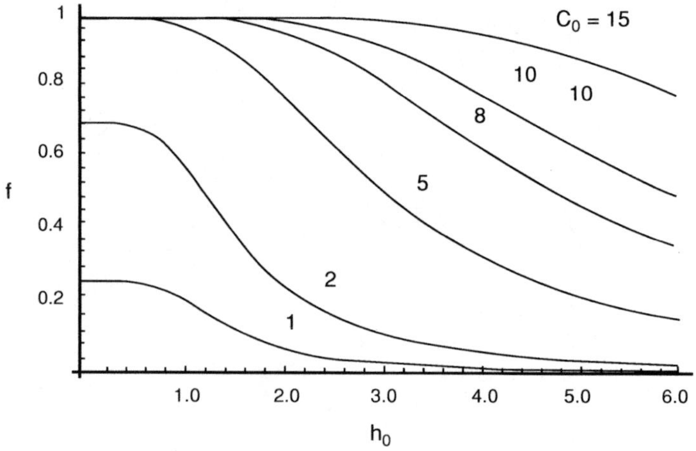

FIGURE 3.8. The correction factor $f = P_s/4h_0^2$ as a function of normalized rms height h_0, with normalized correlation length c_0 as parameter, for total integrated scatter for a surface exhibiting a normal distribution of surface heights and a Gaussian autocorrelation coefficient.

where c, c_2 are the values of $\cos\theta$ in the two media. This can be expressed exactly as[29,30]

$$\rho_{\parallel} = \tanh\left[\frac{1}{2}\ln\left(\frac{n_2 c}{n_0 c_2}\right)\right],$$
$$\rho_{\perp} = \tanh\left[\frac{1}{2}\ln\left(\frac{n_0 c}{n_2 c_2}\right)\right]. \tag{129}$$

Putting $n = n_2/n_0$ and introducing Snell's law

$$c_2 = \frac{1}{n}\sqrt{n^2 - 1 + c^2},$$

we have[30,31]

$$\rho_{\parallel} = \tanh\left[\frac{1}{4}\ln\left(\frac{n^4 c}{n^2 - 1 + c^2}\right)\right],$$
$$\rho_{\perp} = -\tanh\left[\frac{1}{4}\ln\left(\frac{n^2 - 1 + c^2}{c}\right)\right]. \tag{130}$$

Next, assuming the reflection coefficient is small, we linearize by approximating the hyperbolic tangent $\tanh x$ by x. We then use a binomial expansion for values of the refractive index ratio n close to unity, to give

$$\rho_{\parallel} = \left(\frac{2c^2 - 1}{2c^2}\right)\ln n = \left(\frac{s^2 - 2}{s^2}\right)\ln n,$$
$$\rho_{\perp} = -\frac{1}{2c^2}\ln n = -\frac{2}{s^2}\ln n, \tag{131}$$

where $s = 2\cos\theta$ is the axial normalized spatial frequency. It is found that the reflection coefficient for normal incidence predicted by this approximation agrees remarkably well with that calculated from the Fresnel equations, in fact being correct to better than 2% for the large refractive index ratio of 1.5. We also find that the approximation gives a good prediction for the angular variation when the reflection coefficient is small, as long as the angle of incidence is not large. The range of validity is determined by the condition

$$\left|n^2 - 1\right| \ll c^2. \tag{132}$$

We note that in the approximate expressions the effects of refractive index and angle of incidence are separated, and that the reflectivity depends on the ratio of the refractive indices rather than on the individual values.

The Brewster angle according to the approximation occurs at 45°. A change in refractive index of 0.1 gives rise, according to the exact theory, to a shift of less than 3° from this value. These approximate values for the reflection coefficients have been applied to multilayer thin film stacks.[30]

The reflectivity is given by the modulus squared of the reflection coefficient. For unpolarized light, the reflectivity R is thus

$$R = \frac{\rho_\parallel^2 + \rho_\perp^2}{2} = \left(\frac{1}{2}\ln n\right)^2 \left(\frac{2c^4 - 2c^2 + 1}{c^4}\right) = \left(\frac{1}{2}\ln n\right)^2 \left(\frac{s^4 - 4s^2 + 8}{s^4}\right). \tag{133}$$

The agreement is good, even for comparatively large angles, for a refractive index ratio of 1.01. For a refractive index ratio of 1.1 the agreement is still good for small angles of incidence.

We can also consider quantities $\rho_1 = (\rho_\parallel - \rho_\perp)/2$ and $\rho_2 = -(\rho_\parallel + \rho_\perp)/2$. For small angles of incidence, the reflection coefficients have been defined so that $\rho_\perp$ is negative, and $\rho_\parallel$ and $\rho_\perp$ are of opposite sign. Thus ρ_1 can be regarded as the mean reflection coefficient. We obtain

$$\begin{aligned} \rho_1 &= \frac{1}{2}\ln n, \\ \rho_2 &= \frac{1}{2}\ln n \left(\frac{1 - c^2}{c^2}\right) = \frac{1}{2}\ln n \tan^2\theta = \frac{1}{2}\ln n \left(\frac{4 - s^2}{s^2}\right). \end{aligned} \tag{134}$$

The approximation for ρ_1 shows no angular dependence and is a good description of the behavior for small refractive index ratios and moderate angles of incidence. For ρ_2 there is again good prediction of the exact behavior for weak reflectors and angles of incidence which are not too large.

The scattering function for a rough surface on a dielectric medium can be now obtained by substituting (131) into (17):

$$T_\parallel(\mathbf{p}) = \frac{i\pi}{n_0 k}\left(\frac{s^2 - 2}{s^2}\right)\ln n F\{\nabla^2 b(\mathbf{r})\} \tag{135}$$

$$T_\perp(\mathbf{p}) = -\frac{i\pi}{n_0 k}\frac{2}{s^2}\ln n F\{\nabla^2 b(\mathbf{r})\}. \tag{136}$$

For small angles of incidence ($s \approx 2$), these become equal, but of opposite sign. Taking the mean reflection coefficient as $T(\mathbf{p}) = [T_{\|}(\mathbf{p}) - T_{\perp}(\mathbf{p})]/2$, we obtain

$$T(\mathbf{p}) = \frac{\mathrm{i}\pi}{2n_0 k} \ln n F\{\nabla^2 b(\mathbf{r})\}. \tag{137}$$

Effectively, we have assumed that the angle of incidence is constant over the surface profile. The mean reflection coefficient is then equivalent to

$$T(\mathbf{p}) = \frac{\mathrm{i}\pi}{n_0 k} F\left\{\nabla^2\left[\frac{1}{2}\ln n(\mathbf{r})\right]\right\}. \tag{138}$$

We thus recognize the scattering potential as

$$V(\mathbf{r}) = \frac{\mathrm{i}\pi}{n_0 k}\nabla^2\left[\frac{1}{2}\ln n(\mathbf{r})\right]. \tag{139}$$

We have discussed elsewhere[31] how these expressions can be applied for scattering from a general spatial variation of refractive index.

3.10. Conclusions

We have shown how the expressions for scattering by a rough surface in the Kirchhoff approximation can be expressed in a very simple form by introduction of direction cosines, or equivalently, of 3D normalized spatial frequencies **p**. According to the Kirchhoff approximation, the scattering function can be expressed as a function of **p**. Thus different values of incidence and scattering angles that correspond to the same value of **p** have the same value of the scattering function. This can be used as a test of the validity of the Kirchhoff approximation.

A convenient way of illuminating at different angles of incidence is by using a microscope objective with a mask in its back focal plane. Then investigation of the scattered radiation in the back focal plane allows the scattering function to be measured. As the phase is maintained across the back focal plane, in principle the phase of the scattering function can also be measured using an interferometric technique.

The approach described in this chapter brings together the concepts of scattering and imaging. Although we do not have space to consider these connections in detail here, we mention that we can consider the imaging of surface roughness, including the effects of speckle and the properties of surface profiling methods. Further, consideration of surface profiling techniques has suggested algorithms[19,32] for reconstruction of surface profiles from scattering data.

References

1. C. J. R. Sheppard, "Imaging of random surfaces and inverse scattering in the Kirchhoff approximation," *Waves Random Media* **8**, 33-66 (1998).
2. P. Beckmann and A. Spizzichino, *The Scattering of Electromagnetic Waves from Rough Surfaces* (Pergamon, London, 1963).

3. S. Chandrasekhar, "Stochastic problems in physics and astronomy," *Rev. Mod. Phys.* **15**, 1-89 (1943).
4. R. J. Wombell and J. A. DeSanto, "Reconstruction of rough-surface profiles with the Kirchhoff approximation," *J. Opt. Soc. Am. A* **8**, 1892-1897 (1991).
5. C. J. R. Sheppard, T. J. Connolly, and M. Gu, "Scattering by a one-dimensional rough surface and surface reconstruction by confocal imaging," *Phys. Rev. Lett.* **70**, 1409-1412 (1993).
6. C. J. R. Sheppard, T. J. Connolly, and M. Gu, "Imaging and reconstruction for rough surface scattering in the Kirchhoff approximation by confocal microscopy," *J. Mod. Opt.* **40**, 2407-2421 (1993).
7. C. J. R. Sheppard, "Simplified expressions for the bidirectional reflectance distribution function, and inverse scattering of surface roughness," *J. Mod. Opt.* **43**, 373-380 (1996).
8. C. J. R. Sheppard, T. J. Connolly, and M. Gu, "The scattering potential for imaging in the reflection geometry," *Opt. Commun.* **117**, 16-19 (1995).
9. C. J. R. Sheppard, "The spatial frequency cut-off in three-dimensional imaging," *Optik* **72**, 131-133 (1986).
10. C. J. R. Sheppard, "The spatial frequency cut-off in three-dimensional imaging II," *Optik* **74**, 128-129 (1986).
11. C. J. R. Sheppard, M. Gu, and X. Q. Mao, "Three-dimensional coherent transfer function in a reflection-mode confocal scanning microscope," *Opt. Commun.* **81**, 281-284 (1991).
12. C. J. R. Sheppard, M. Gu, Y. Kawata, and S. Kawata, "Three-dimensional transfer functions for high aperture systems," *J. Opt. Soc. Am. A* **11**, 593-598 (1994).
13. C. G. Darwin, "The theory of X-ray reflexion," *Phil. Mag.* **27**, 211-213, 313-333, 675-690 (1914).
14. C. J. R. Sheppard and T. J. Connolly, "Imaging of random surfaces," *J. Mod. Opt.* **42**, 861-881 (1995).
15. J. C. Stover, *Optical Scattering: Measurement and Analysis* (McGraw-Hill, New York, 1990).
16. J. W. Goodman, "Dependence of image speckle contrast on surface roughness," *Opt. Commun.* **14**, 324-327 (1975).
17. J. M. Bennett and L. Mattson, *Introduction to Surface Roughness and Scattering* (Optical Society of America, Washington DC, 1989).
18. I. J. Cox and C. J. R. Sheppard, "Digital image processing of confocal images," *Image Vis. Comput.*, **1**, 52-56 (1983).
19. J. C. Quartel and C. J. R. Sheppard, "Surface reconstruction using an algorithm based on confocal imaging," *J. Mod. Opt.* **43**, 469-486 (1996).
20. J. M. Elson and J. M. Bennett, "Relation between the angular dependence of scattering and the statistical properties of optical surfaces," *J. Opt. Soc. Am.* **69**, 31-47 (1979).
21. V. I. Tatarski, *Wave Propagation in Turbulent Media* (McGraw-Hill, New York, 1961).
22. B. J. Uscinski, H. G. Booker, and M. Marians, "Intensity fluctuations due to a deep phase screen with a power-law spectrum," *Proc. Roy. Soc. Lond. A* **374**, 503-530 (1981).
23. E. Jakeman, "Fresnel scattering by a corrugated random surface with fractal slope," *J. Opt. Soc. Am. A* **72**, 1034-1041 (1982).
24. C. J. R. Sheppard, "Scattering by fractal surfaces with an outer scale," *Opt. Commun.* **122**, 178-188 (1996).
25. H. Davies, "The reflection of electromagnetic waves from a rough surface," *Proc. I.E.E.* **101**, 209-214 (1954).
26. S. de Silverstri, P. La Porta, V. Magni, O. Svelto, and B. Majocchi, "Unstable laser resonators with super-Gaussian mirrors," *Opt. Lett.* **13**, 201-203 (1988).

27. H. E. Bennett and J. O. Porteus, "Relation between surface roughness and specular reflectance at normal incidence," *J. Opt. Soc. Am.* **61**, 123-129 (1961).
28. C. J. R. Sheppard and M. Roy, "Total integrated scatter in the Kirchhoff approximation," *Optik* **103**, 42-44 (1996).
29. C. H. Greenewald, W. Brandt, and D. D. Friel, "Irridescent colors of hummingbird feathers," *J. Opt. Soc. Am.* **50**, 1005-1013 (1960).
30. C. J. R. Sheppard, "Approximate calculation of the reflection coefficient from a stratified medium," *Pure and Applied Optics* **4**, 669-669 (1995).
31. C. J. R. Sheppard and F. Aguilar "Fresnel coefficients for weak reflection, and the scattering potential for three-dimensional imaging," *Opt. Commun.* **162**, 182-186 (1999).
32. J. C. Quartel and C. J. R. Sheppard, "A surface reconstruction algorithm based on confocal interferometric profiling," *J. Mod. Opt.* **43**, 591-605 (1996).

4
Rayleigh Hypothesis

A. G. VORONOVICH
NOAA/Earth Systems Research Laboratory, 325 Broadway, Boulder, CO 80305

4.1. Introduction

Consider a wave incident onto a rough surface Σ from above (see Fig. 4.1). We will assume that the wave is a scalar and that the Dirichlet boundary condition

$$u = 0, \qquad \mathbf{R} \in \Sigma \tag{1}$$

for the total field at the surface holds, although the results presented here should not depend on the nature of the field or on the boundary condition. From the Helmholtz formula [1], Eq. (8), the following representation for the field

$$u^{(\text{in})}(\mathbf{R}_*) + \int \mu\,(\mathbf{R})\,G\,(\mathbf{R} - \mathbf{R}_*)\,\mathrm{d}\Sigma_{\mathbf{R}} = \begin{cases} u(\mathbf{R}_*), & \mathbf{R}_* \in \text{upper half-space} \\ 0 & \mathbf{R}_* \in \text{lower half-space} \end{cases} \tag{2}$$

is obtained, where

$$\mu\,(\mathbf{R}) = \frac{\partial u}{\partial \mathbf{n}_{\mathbf{R}}}, \qquad \mathbf{R} \in \Sigma$$

is the normal derivative of the total field and G is the free-space Green function

$$G(\mathbf{R}) = -\frac{\mathrm{e}^{\mathrm{i}KR}}{4\pi R} = -\frac{\mathrm{i}}{8\pi^2} \int \mathrm{e}^{\mathrm{i}\mathbf{kr} + \mathrm{i}q_k|z|} \frac{\mathrm{d}\mathbf{k}}{q_k}. \tag{3}$$

Here $\mathbf{R} = (\mathbf{r}, z)$, where $\mathbf{r}$ and z are the horizontal and vertical components of the radius vector, while $\mathbf{k}$ is the horizontal, and q_k

$$q_k = \sqrt{K^2 - k^2}, \qquad \operatorname{Im} q_k \geq 0 \tag{4}$$

is the vertical component of the wave vector (see Eqs. (3),(4) from [1]). The second term in Eq. (2) represents the scattered field generated by the monopole sources with surface density μ located on the surface:

$$u^{(\text{sc})}(\mathbf{R}_*) = \int \mu\,(\mathbf{R})\,G\,(\mathbf{R} - \mathbf{R}_*)\,\mathrm{d}\Sigma_{\mathbf{R}}. \tag{5}$$

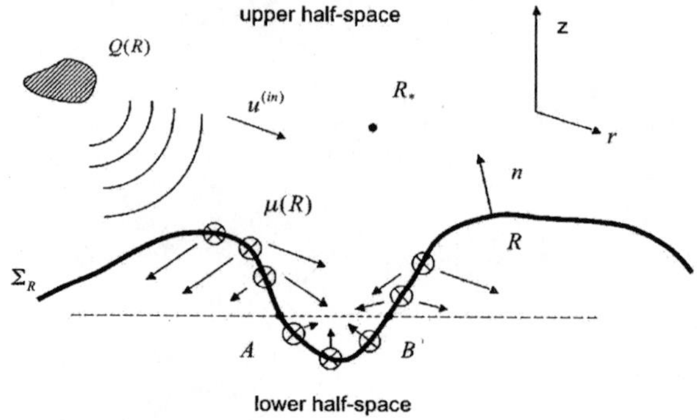

FIGURE 4.1. Illustration of the geometry of the problem. Secondary monopole sources located on the surface radiate in all directions.

On substituting into Eq. (2) the Weyl representation Eq. (3), and assuming that $z_* > \max h(\mathbf{r})$, where

$$z = h(\mathbf{r}) \tag{6}$$

is the equation of the boundary, one finds

$$u^{(\mathrm{sc})}(\mathbf{R}_*) = \int S(\mathbf{k})\, \mathrm{e}^{\mathrm{i}\mathbf{k}\mathbf{r}_* + \mathrm{i}q_k z_*}\, \mathbf{dk}, \tag{7}$$

where

$$S(\mathbf{k}) = -\frac{\mathrm{i}}{8\pi^2 q_k} \int \mu(\mathbf{R})\, \mathrm{e}^{-\mathrm{i}\mathbf{k}\mathbf{r} - \mathrm{i}q_k h(\mathbf{r})} \left(1 + (\nabla h)^2\right)^{1/2} \mathbf{dr}. \tag{8}$$

The scattering amplitude (SA) S introduced here is slightly different from the one used in [1]. However the definition Eq. (8) simplifies the formulas used in this chapter by removing normalization factors that are insignificant in the present context. It is important that the observation point $\mathbf{R}_*$ was assumed to be located above the excursions of the roughness. If this point were located within the roughness (e.g., $\mathbf{R}_* \in AB$, see Fig. 4.1), then we would find

$$u^{(\mathrm{sc})}(\mathbf{R}_*) = \int S_{\mathrm{up}}(\mathbf{k})\, \mathrm{e}^{\mathrm{i}\mathbf{k}\mathbf{r}_* + \mathrm{i}q_k z_*} \mathbf{dk} + \int S_{\mathrm{dn}}(\mathbf{k})\, \mathrm{e}^{\mathrm{i}\mathbf{k}\mathbf{r}_* - \mathrm{i}q_k z_*}\, \mathbf{dk}, \tag{9}$$

where

$$S_{\mathrm{up}}(\mathbf{k}) = -\frac{\mathrm{i}}{8\pi^2 q_k} \int_{h(\mathbf{r}) < z_*} \mu(\mathbf{R})\, \mathrm{e}^{-\mathrm{i}\mathbf{k}\mathbf{r} - \mathrm{i}q_k h(\mathbf{r})} \left(1 + (\nabla h)^2\right)^{1/2} \mathbf{dr} \tag{10}$$

and

$$S_{\mathrm{dn}}(\mathbf{k}) = -\frac{\mathrm{i}}{8\pi^2 q_k} \int_{h(\mathbf{r}) > z_*} \mu(\mathbf{R})\, \mathrm{e}^{-\mathrm{i}\mathbf{k}\mathbf{r} + \mathrm{i}q_k h(\mathbf{r})} \left(1 + (\nabla h)^2\right)^{1/2} \mathbf{dr}. \tag{11}$$

Seemingly, in this case the field consists of both upward and downward propagating waves, corresponding to the first and the second terms in Eq. (9), respectively.

Nevertheless, when calculating scattering from small surface roughness, Rayleigh[2] used the representation given by Eq. (7), which consists of waves propagating only upwards, to calculate the scattered field within the roughness as well, in particular at the points belonging to the boundary. Such a possibility is called the Rayleigh hypothesis (RH). Since then the RH has been considered in a large number of papers. The RH has a few different aspects. First, could the RH be justified, at least in certain cases or, in view of the representation Eq. (9), since it ignores the downgoing component of the scattered field, is it fundamentally wrong? Second, if the RH is at least sometimes correct, could one use the representation given by Eq. (7) to calculate the expansion of the SA in a power series of elevations h? By this we mean the calculation of the coefficient functions of the expansion (i.e., functions B, $B_2, \ldots$ from [1], Eq. (123)), and not the issue of the convergence of the resulting series since, in practical applications, we always use it as an asymptotic expansion. Third, could one use the RH for a numerical evaluation of the SA? These issues will be addressed below. The reference list to this paper is not intended to be complete; an interested reader will easily track the chain of references starting from the most recent ones.

4.2. Is the Representation Given by Eq. (7) Fundamentally Wrong?

At first sight it is: the accurate calculation given by Eq. (9) clearly shows that within the roughness there are both upward and downward propagating waves, and Eq. (7) contains only upward propagating waves [3]. This conclusion was supported by a finding of Petit and Cadilhac[4], who demonstrated that for sinusoidal elevations

$$h = a \cos px \tag{12}$$

of sufficient steepness, $ap > 0.448$, the integral in Eq. (7) diverges. More precisely, since they considered periodic 2D undulations, the integral reduces to the corresponding infinite sum

$$u^{(\mathrm{sc})}(x_*, z_*) = \sum_{n=-\infty}^{n=\infty} S_n \, \mathrm{e}^{\mathrm{i}k_n x_* + \mathrm{i}q_n z_*}, \tag{13}$$

where

$$k_n = k_0 + pn, \qquad q_n = \sqrt{K^2 - k_n^2},$$

and the divergence of the sum Eq. (13) was proven. Thus, it looked like the RH was proven to be wrong, both from physical and mathematical perspectives. On the other hand, three years later Millar proved that the sum in Eq. (13) converges for *all* points of the upper half space (including the boundary) for sufficiently gentle undulations [5], and two years later [6] he established that convergence takes place for slopes smaller than the critical slope found by Petit and Cadilhac:

$$ap < 0.448. \tag{14}$$

If the series in Eq. (13) converges for all points above the roughness, the corresponding expression certainly provides an exact solution of the wave equation and the representation Eq. (13) appears to be *exact* in this case. This seems strange, since the downward propagating waves in Eq. (9) in principle exist for gentle undulations as well, and should not crop up only after the slope exceeds the critical value. How can one resolve this quandary?

In fact, the reasoning regarding up- and downward propagating waves implicitly assumes that the up- and downward propagating waves are physical entities, and the representation of the field in the form of Eq. (9) is unique, so that the second term representing downgoing waves cannot be reexpanded in terms of the upward propagating waves. However, this is not necessarily correct. Let us assume now that the observation point $(\mathbf{r}_*, z_*)$ is located at the same level z_*, but is beneath the surface (e.g., it lies on the ray $(-\infty, A)$ or (B, ∞) in Fig. 4.1.) According to the extinction theorem, Eq. (2), on the lower line one has

$$u^{(\text{in})}(\mathbf{r}_*, z_*) + \int S_{\text{dn}}(\mathbf{k})\, \mathrm{e}^{\mathrm{i}\mathbf{kr}_* - \mathrm{i}q_k z_*}\, \mathrm{d}\mathbf{k} = -\int S_{\text{up}}(\mathbf{k})\, \mathrm{e}^{\mathrm{i}\mathbf{kr}_* + \mathrm{i}q_k z_*}\, \mathrm{d}\mathbf{k}, \tag{15}$$

and a certain linear combination of *downward* propagating waves on the left-hand side is represented on the right-hand side in terms of a linear combination of *upward* propagating waves. The up- and downward propagating waves are defined uniquely if one can measure the field in the vicinity of the *whole* plane $z_* = \text{const}$. However, in the case shown in Fig. 4.1, part of the plane, namely the segment AB, is excluded. In other words, if the field (and its gradient) is measured only on a part of the infinite plane $z_* = \text{const}$, there are situations when one cannot figure out whether it is coming from above or from below. This seems to be counterintuitive: if we see something, we know that it is in front of us, and not behind. Still, Eq. (15) shows that confusing situations may exist. Thus, the representation Eqs. (7) and (9) do not contradict each other, since the downward propagating waves in Eq. (9) can be reexpanded in terms of upward propagating waves.

Now let us also look at this issue from another perspective. Note, that a solution of the Helmholtz equation is an analytic function of the coordinates. This immediately follows from the Helmholtz formula Eq. (2): the Green function in the integrand can be expanded into an absolutely convergent series in powers of the coordinates $(x_* - x_0)^{n_1}(y_* - y_0)^{n_2}(z_* - z_0)^{n_3}$ at any observation point $\mathbf{R}_0$ belonging to the upper half-space provided $|\mathbf{R}_* - \mathbf{R}_0| < r_0 = \min_{\mathbf{R}\in\Sigma} |\mathbf{R} - \mathbf{R}_0|$, i.e. if the sphere with the center at $\mathbf{R}_0$ does not intersect the boundary Σ. This series converges uniformly with respect to $\mathbf{R} \in \Sigma$. For this reason the series according to Eq. (2) can be integrated term-by-term over $\mathrm{d}\Sigma$ with the weight $\mu(\mathbf{R})$ thus providing within the sphere $|\mathbf{R}_* - \mathbf{R}_0| < r_0$ the representation for $u(\mathbf{R}_*)$ in the form of a convergent power series.

The simplest estimate based on Eq. (8) readily gives

$$|S_n| < C\, |n|^{-1/2}\, \mathrm{e}^{\mathrm{ph}_{\max}|n|}, \tag{16}$$

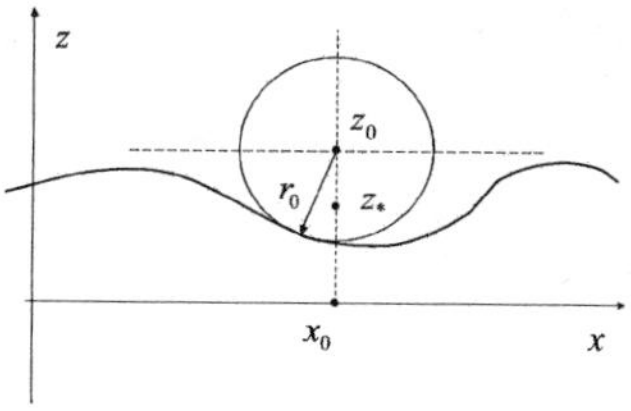

FIGURE 4.2. The circle of convergence of the expansion of the solution into powers of coordinates.

where

$$h_{\max} = \max h(\mathbf{r}), \tag{17}$$

whence

$$\left| S_n \, \mathrm{e}^{\mathrm{i}k_n x_* + \mathrm{i}q_n z_*} \right| < C \, |n|^{-1/2} \, \mathrm{e}^{-p(z_* - h_{\max})|n|} \tag{18}$$

so that the series in Eq. (13) converges absolutely for $z_* > h_{\max}$. Let us assume for simplicity that the points $\mathbf{R}_*$ and $\mathbf{R}_0$ lie on the same vertical $x_* = x_0$ and the point z_0 is located slightly above the excursions of the undulations: $z_0 > h_{\max}$. Let us expand the exponentials in Eq. (13) in powers of $(z_* - z_0)$ and let us assume first that the point z_* is also above the undulations: $z_0 > z_* > h_{\max}$. Using the absolute convergence of the resulting series as well as the functional series in Eq. (13), we can exchange the orders of summation and find

$$\sum_{n=-\infty}^{n=\infty} S_n \, \mathrm{e}^{\mathrm{i}k_n x_* + \mathrm{i}q_n z_*} = \sum_{k=0}^{\infty} \left[\frac{1}{k!} \sum_{n=-\infty}^{n=\infty} S_n \, \mathrm{e}^{\mathrm{i}k_n x_0 + \mathrm{i}q_n z_0} \, (\mathrm{i}q_n)^k \right] (z_* - z_0)^k . \tag{19}$$

The internal series in the square parentheses here converges due to the estimate given by Eq. (18). Although we have assumed that $z_0 > h_{\max}$, in fact due to the analyticity of the field the series over k on the right-hand side of Eq. (19) converges for $|z_0 - z_*| < r_0$, where r_0 is the radius of the circle which is centered at (x_0, z_0) and touches the boundary (see Fig. 4.2.)

Although both the series over n on the left-hand side and the series over k on the right-hand side represent the same value of the scattered field at the point (x_*, z_*), the areas of convergence of both series are different. In contrast to the right-hand side, the series over plane waves on the left-hand side converges for $z_* > h_{\max}$; however, it may diverge for $z_* < h_{\max}$. A formal interchange of the orders of summation over k and n on the right-hand side of Eq. (19) will result in the expression on the left-hand side accompanied by a change of the area of convergence. Thus, the representation of the scattered field Eq. (13) should, strictly speaking, include a recipe for how the corresponding expression is supposed to be calculated. The RH tacitly assumes that Eq. (13) is calculated as a functional series, i.e. in the most natural way: by sequentially summing the exponentials. However, this is not the only possibility. If one is allowed to rearrange the terms one way or another, the convergence area may change. This emphasizes that divergence of the series in Eq. (13) does not correspond to an inadequacy of the field representation; it is simply a matter of the proper way of calculating the functional series.

4.3. Convergence of the Rayleigh Series

The estimate Eq. (16) is too rough; it leaves open the question of convergence of the Rayleigh series Eq. (13). However, this estimate can be easily refined and the asymptotic behavior of the spectral amplitudes S_n can be determined without solving the scattering problem. Assuming that the surface profile is an analytic function one can estimate the integral in Eq. (8) by the saddle point method. For large k the exponent in the integrand of Eq. (8) becomes

$$-\mathrm{i}kx - \mathrm{i}q_k h(x) \to -\mathrm{i}kx + |k|\, h(x). \tag{20}$$

Consider the case $k \to +\infty$. The equation for the stationary point x_s is

$$h'(x_s) = \mathrm{i}. \tag{21}$$

Using the standard formula of the saddle point method and omitting the pre-exponential factor, which is unimportant for determining the convergence criterion, one finds the following asymptotic form for the spectral amplitudes:

$$S_k \sim \mathrm{e}^{k(-\mathrm{i}x_s + h(x_s))}, \qquad k \to +\infty. \tag{22}$$

Hence

$$S_k\, \mathrm{e}^{\mathrm{i}kx_* + \mathrm{i}q_k z_*} \sim \mathrm{e}^{\mathrm{i}kx_*}\, \mathrm{e}^{k(-\mathrm{i}x_s + h(x_s) - z_*)}, \qquad k \to +\infty. \tag{23}$$

Thus, the integral in Eq. (7) converges for all boundary points provided

$$\mathrm{Re}[-\mathrm{i}x_s + h(x_s)] < \min h(x), \tag{24}$$

and diverges otherwise. Consideration of the case $k \to -\infty$ results in the same condition. Applying this criterion to the sinusoidal profile Eq. (12) gives

$$-ap \sin px_s = \mathrm{i}, \tag{25}$$

whence

$$px_s = -\mathrm{i} \ln \left(\frac{1}{ap} + \sqrt{\frac{1}{(ap)^2} + 1} \right) \tag{26}$$

and

$$\mathrm{Re}[-\mathrm{i}x_s + a \cos px_s] = -\frac{1}{p} \ln \left(\frac{1}{ap} + \sqrt{\frac{1}{(ap)^2} + 1} \right) + \frac{\sqrt{1 + (ap)^2}}{p} < -a. \tag{27}$$

Simple analysis shows that inequality (27) holds when

$$ap < \frac{\xi^2 - 1}{2\xi} = 0.448, \tag{28}$$

where $\xi = 1.544$ is the root of the equation $e^{\xi} = (\xi + 1)/(\xi - 1)$. Many other examples of the application of the criterion (24) can be found in [7].

Let us consider now elevations of the type

$$ah(x), \tag{29}$$

where $h(x)$ is an analytic function and a is some real amplitude parameter. For $a \to 0$ the solution of Eq. (21) in the lower half-space of the complex plane will tend to infinity: $\mathrm{Im}\, x_s \to -\infty$ and $\mathrm{Re}(-\mathrm{i}x_s) \to -\infty$. Let us assume that the function $h(x)$ is such that $h(x)/h'(x)$ remains bounded when $\mathrm{Im}\, x_s \to -\infty$ (this condition will hold, for instance, if $h(x)$ is a sum of an arbitrary finite number of sinusoids). Then

$$\mathrm{Re}[-\mathrm{i}x_s + ah(x_s)] = \mathrm{Re}\left[-\mathrm{i}x_s + \mathrm{i}\frac{h(x_s)}{h'(x_s)}\right] \to -\infty, \tag{30}$$

and for sufficiently small a criterion (24) will be satisfied [8].

When estimating the integral in Eq. (8) by the saddle point method we assumed that the function μ can be analytically continued into the complex domain. Such a possibility was proven in [5, 6]. Note also, that criterion (24) does not depend on the wavenumber in the medium K; it applies equally to the Laplace equation.

4.4. Rayleigh Hypothesis and the Perturbative Expansion of the SA

The possibility of applying representation (7) for boundary points greatly simplifies calculation of the expansion of the SA in a power series with respect to elevations h. The problem simply reduces to substitution of the representation into the boundary condition. For example, in the case of the Dirichlet problem (Eq. (1)) and an incident plane wave one finds

$$e^{\mathrm{i}\mathbf{k}_0\mathbf{r} - \mathrm{i}q_0 h(\mathbf{r})} + \int S(\mathbf{k})\, e^{\mathrm{i}\mathbf{k}\mathbf{r} + \mathrm{i}q_k h(\mathbf{r})}\, d\mathbf{k} = 0. \tag{31}$$

We can represent this equation as follows:

$$S(\mathbf{k}) + \int S(\mathbf{k}')\, e^{\mathrm{i}(\mathbf{k}'-\mathbf{k})\mathbf{r}} \left(e^{\mathrm{i}q_{k'} h(\mathbf{r})} - 1\right) \frac{d\mathbf{r}}{(2\pi)^2}\, d\mathbf{k}'$$
$$= -\int e^{\mathrm{i}(\mathbf{k}_0 - \mathbf{k})\mathbf{r} - \mathrm{i}q_0 h(\mathbf{r})} \frac{d\mathbf{r}}{(2\pi)^2}. \tag{32}$$

One can solve this equation by iteration, expanding all exponentials in power series; the calculation is very simple. A more cautious way of calculating the power series expansion of the SA, not relying on representation (7), would be application of a perturbative analysis to the corresponding boundary integral equation. The calculations in this case are much more extensive; this is especially true in the case of electromagnetic waves [1].

The question arises: will the coefficient functions of the power expansion of the SA (i.e., the functions B, B_2, ... from [1], Eq. (123)) calculated by those two different approaches coincide? The answer is affirmative [10]. One way to make sure this is true is to notice that when calculating the field at the boundary based on the scattering amplitude one rather uses the analytical dependence of the field on coordinates, i.e. the right-hand side of Eq. (19). Moreover, only a finite number of k-terms is required to calculate the expansion of the SA for any finite power of h, and it does not matter whether the right- or left-hand side of Eq. (19) is used. Another way to prove this is as follows. Let us consider elevations of the type Eq. (29). Then use of representation (7) is justified, and one can legitimately substitute it into the boundary conditions obtaining equations of the type of (31). As demonstrated above, this allows an unambiguous calculation of the power series expansion of SA. The assumption of the elevations being analytical does not impose any restrictions here.

One can make sure that Eq. (31) holds *per se* and provides correct expressions for coefficient functions B, $B_2, \ldots$ by a direct calculation also. Introducing

$$\widetilde{\mu}(\mathbf{r}) = -\frac{\mathrm{i}}{8\pi^2}\left(1+(\nabla h)^2\right)^{1/2}\mu(\mathbf{r}), \tag{33}$$

according to Eq. (8), we find

$$S(\mathbf{k}) = \frac{1}{q_k}\int \widetilde{\mu}(\mathbf{r})\,\mathrm{e}^{-\mathrm{i}\mathbf{kr}-\mathrm{i}q_k h(\mathbf{r})}\,\mathrm{d}\mathbf{r}. \tag{34}$$

On the other hand, substituting into Eq. (2) the spectral representation of the Green function Eq. (3), and setting $\mathbf{R}_* = \mathbf{R} \in \Sigma$, we obtain

$$\int \mathrm{d}\mathbf{r}'\widetilde{\mu}(\mathbf{r}')\int \mathrm{d}\mathbf{k}\frac{1}{q_k}\,\mathrm{e}^{\mathrm{i}\mathbf{k}(\mathbf{r}-\mathbf{r}')+\mathrm{i}q_k|h(\mathbf{r})-h(\mathbf{r}')|} = -\mathrm{e}^{\mathrm{i}\mathbf{k}_0\mathbf{r}-\mathrm{i}q_0 h(\mathbf{r})}. \tag{35}$$

Let us represent this equation as follows:

$$\begin{aligned}&\int \mathrm{d}\mathbf{r}'\widetilde{\mu}(\mathbf{r}')\int \mathrm{d}\mathbf{k}\frac{\cos\left[q_k(h(\mathbf{r})-h(\mathbf{r}'))\right]}{q_k}\,\mathrm{e}^{\mathrm{i}\mathbf{k}(\mathbf{r}-\mathbf{r}')}\\ &\quad+\mathrm{i}\int \mathrm{d}\mathbf{r}'\widetilde{\mu}(\mathbf{r}')\int \mathrm{d}\mathbf{k}\frac{\sin\left[q_k\left|h(\mathbf{r})-h(\mathbf{r}')\right|\right]}{q_k}\,\mathrm{e}^{\mathrm{i}\mathbf{k}(\mathbf{r}-\mathbf{r}')} = -\mathrm{e}^{\mathrm{i}\mathbf{k}_0\mathbf{r}-\mathrm{i}q_0 h(\mathbf{r})},\end{aligned} \tag{36}$$

The modulus sign in the first term associated with the cosine function can be dropped since cosine is an even function. Let us consider the second term associated with the sine function on the left-hand side of this equation:

$$
\begin{aligned}
&\int d\mathbf{r}'\widetilde{\mu}(\mathbf{r}') \int d\mathbf{k} \frac{\sin\left[q_k \left|h(\mathbf{r}) - h(\mathbf{r}')\right|\right]}{q_k} e^{i\mathbf{k}(\mathbf{r}-\mathbf{r}')} \\
&= \int d\mathbf{r}'\widetilde{\mu}(\mathbf{r}') \int d\mathbf{k} \sum_{n=0}^{\infty} \frac{(-1)^n}{(2n+1)!} q_k^{2n} \left|h(\mathbf{r}) - h(\mathbf{r}')\right|^{2n+1} e^{i\mathbf{k}(\mathbf{r}-\mathbf{r}')} \\
&= \sum_{n=0}^{\infty} \frac{(-1)^n}{(2n+1)!} \int d\mathbf{r}'\widetilde{\mu}(\mathbf{r}')\left|h(\mathbf{r}) - h(\mathbf{r}')\right|^{2n+1} \left(K^2 + \nabla_{\mathbf{r}'}^2\right)^n \int d\mathbf{k}\, e^{i\mathbf{k}(\mathbf{r}-\mathbf{r}')}.
\end{aligned}
\tag{37}
$$

Integrating by parts with respect to $\mathbf{r}'$ and then integrating with respect to $\mathbf{k}$ and $\mathbf{r}'$ we obtain

$$
(37) = (2\pi)^2 \sum_{n=0}^{\infty} \frac{(-1)^n}{(2n+1)!} \left[\left(K^2 + \nabla_{\mathbf{r}'}^2\right)^n \widetilde{\mu}(\mathbf{r}') \left|h(\mathbf{r}) - h(\mathbf{r}')\right|^{2n+1}\right]_{\mathbf{r}'=\mathbf{r}}. \tag{38}
$$

For smooth profiles one has

$$
\widetilde{\mu}(\mathbf{r}') \left|h(\mathbf{r}) - h(\mathbf{r}')\right|^{2n+1} \sim \left(\mathbf{r} - \mathbf{r}'\right)^{2n} \left|\mathbf{r} - \mathbf{r}'\right|, \qquad \mathbf{r}' \to \mathbf{r}. \tag{39}
$$

On the other hand, the differential operator in Eq. (38) includes no more than $2n$ spatial derivatives. For this reason all expressions in the square parentheses in Eq. (38) vanish. Apparently, the presence of the modulus sign in Eqs. (37)–(39) does not matter, and we find

$$
\begin{aligned}
&\int d\mathbf{r}'\widetilde{\mu}(\mathbf{r}') \int d\mathbf{k} \frac{\sin\left[q_k \left|h(\mathbf{r}) - h(\mathbf{r}')\right|\right]}{q_k} e^{i\mathbf{k}(\mathbf{r}-\mathbf{r}')} \\
&= \int d\mathbf{r}'\widetilde{\mu}(\mathbf{r}') \int d\mathbf{k} \frac{\sin\left[q_k \left(h(\mathbf{r}) - h(\mathbf{r}')\right)\right]}{q_k} e^{i\mathbf{k}(\mathbf{r}-\mathbf{r}')} = 0
\end{aligned}
\tag{40}
$$

for any smooth functions $\widetilde{\mu}$, h. Thus, the modulus sign in the second term on the left-hand side of Eq. (36) can be omitted, and as a result it can be omitted in Eq. (35), too. In this case Eq. (35) transforms into Eq. (31). One can compare the transformations above with those made in Eq. (82) of [1].

The calculation above assumed that both integrals associated with the sine and cosine functions in Eq. (36) exist. This will be the case if criterion (24) holds. Alternatively, we can suppose that both $\widetilde{\mu}$ and h are represented as asymptotic expansions with respect to h and assume that the transformations above involve only powers of elevations not exceeding a certain finite value. Then the calculation above proves that Eq. (31) holds for any finite power of h.

4.5. Application to Numerical Analysis

Direct substitution of representation (7) into boundary conditions could also be used for the numerical solution of the corresponding scattering problem. The question arises: is this approach limited to the situations where the RH holds? The answer to this question is negative, and representation (7) can be used beyond the domain of validity of the RH.

The set of functions

$$\varphi_{\mathbf{k}}(\mathbf{r}) = \mathrm{e}^{\mathrm{i}\mathbf{kr}+\mathrm{i}q_k h(\mathbf{r})}, \tag{41}$$

which are called metaharmonic functions, or a topological basis, is complete in L_2 [11]. To prove this it is sufficient to demonstrate that if the functions $\varphi_{\mathbf{k}}(\mathbf{r})$ for all $\mathbf{k}$ are orthogonal to a function $f(\mathbf{r})$ from L_2, then $f = 0$. Let

$$\int \overline{f(\mathbf{r})}\mathrm{e}^{\mathrm{i}\mathbf{kr}+\mathrm{i}q_k h(\mathbf{r})}\,\mathrm{d}\mathbf{r} = 0, \tag{42}$$

where the bar stands for complex conjugation. We will multiply this equation by the functions

$$-\frac{\mathrm{i}}{8\pi^2}\frac{1}{q_k}\mathrm{e}^{-\mathrm{i}\mathbf{kr}_*-\mathrm{i}q_k z_*}, \tag{43}$$

where $(\mathbf{r}_*, z_*) = \mathbf{R}_*$ are parameters, and we will assume that

$$z_* < \min h(\mathbf{r}). \tag{44}$$

Due to Eq. (44) the integrand in Eq. (42) exponentially decays for $k \to \infty$, and we can integrate over $\mathbf{k}$, interchanging orders of integration over $\mathbf{k}$ and $\mathbf{r}$. By replacing $\mathbf{k}$ by $-\mathbf{k}$ and using Eq. (3), we find

$$u(\mathbf{R}_*) = \int \overline{f(\mathbf{r})}G(\mathbf{R} - \mathbf{R}_*)\,\mathrm{d}\mathbf{r} = 0. \tag{45}$$

Apparently, $u(\mathbf{R}_*)$ as a function of $\mathbf{R}_*$ represents a solution of the wave equation. Since $u(\mathbf{R}_*)$ is an analytic function and vanishes for $z_* < \min h(\mathbf{r})$, it is identically equal to zero in the entire lower half space $z \le h(\mathbf{r})$, including the boundary. The field generated by monopole sources (with density $\overline{f(\mathbf{r})}$ in the present instance) is continuous through the boundary (see, e.g., [1]); hence, $u(\mathbf{R}_*)$ equals zero on the upper side of the surface $z = h(\mathbf{r})$ also. A solution of the wave equation consisting of outgoing waves only vanishing on the boundary does not radiate energy and vanishes identically in the entire upper half-space:

$$\int_{z>h(\mathbf{r})} |\nabla u|^2\,\mathrm{d}\mathbf{R}_* = -\int u\frac{\partial \overline{u}}{\partial \mathbf{n_R}}\mathrm{d}\Sigma_{\mathbf{R}} = 0. \tag{46}$$

The integral over an infinite hemisphere closing the volume vanishes because due to the limiting absorption principle outgoing waves decay exponentially at infinity. Thus, $u(\mathbf{R}_*)$ vanishes identically, and the same is true for the monopole density $\overline{f(\mathbf{r})}$.

If the boundary value μ in Eq. (5) is matched in the L_2 sense, it follows immediately from this equation that the scattered field at any internal point $\mathbf{R}_*$ is calculated exactly. Using the completeness of the set of outgoing waves, Eq. (41), on the boundary in L_2 one can try to calculate the SA for a periodic surface with period L from the equations [12]

$$\int_0^L \left| \int \sum_{n=-\infty}^{\infty} S_n \, \mathrm{e}^{\mathrm{i}k_n x + \mathrm{i}q_n h(x)} - \mathrm{e}^{\mathrm{i}k_0 x - \mathrm{i}q_0 h(x)} \right|^2 \mathrm{d}x = 0. \tag{47}$$

Strictly speaking, a unique solution of these equations should exist. However, numerical realization of the appropriate algorithm in the cases when a significant number of inhomogeneous harmonics are included in the sum in Eq. (47) requires accurate evaluation of integrals with exponentially large integrands, and will generally lead to linear sets with large condition numbers. For this reason consideration of steep undulations and cases with significant shadowing usually cannot be treated by this method.

Using the collocation method instead of estimating the boundary condition mismatch in L_2, Eq. (47) leads to the linear set

$$\sum_{n=-N}^{N} S_n \, \mathrm{e}^{\mathrm{i}k_n x_j + \mathrm{i}q_n h(x_j)} = \mathrm{e}^{\mathrm{i}k_0 x_j - \mathrm{i}q_0 h(x_j)}, \qquad j = 1, 2, \ldots, 2N+1. \tag{48}$$

The existence of converged solutions should generally require validity of the RH. If, however, instead of plane waves, Eq. (41), one uses other basis functions, which are certain linear combinations of the former, convergence may be recovered and the collocation method will give correct results [11,13]. Note also that the collocation method should not necessarily fail even when the RH is not valid [14]. If the scattered field corresponds mostly of homogeneous modes, the solution of Eqs. (48) will give accurate enough results. However, if one increases N including sufficiently strong inhomogeneous modes, the solution provided by Eqs. (48) will fail sooner or later. Performance of the method depends also on the positions of the collocation points [13].

Extensive comparisons between the numerical performance of different basis functions in the context of different methods of solution of the scattering problem were made in [15,16].

4.6. Conclusion

Representation of the scattered field in the form Eq. (13), which looks like it consists of outgoing waves only, in fact places no such restriction on the field. The Rayleigh hypothesis consists not only of ansatz (13) but also (and this is even more important) of the statement that series (13) converges if it is calculated by sequentially adding the exponentials. Such convergence is guaranteed above the roughness; however, it may or may not take place if the observation point is within the roughness. However, if one is allowed to identically transform and rearrange the terms in Eq. (13), this expression will give a correct result at any point. This is quite similar to analytic continuation: different representations of the same function may have different areas of validity. For instance, if one is bound to calculate the series

$$u(x) = 1 - \mathrm{e}^{\mathrm{i}x} + \mathrm{e}^{2\mathrm{i}x} - \mathrm{e}^{3\mathrm{i}x} + \cdots \tag{49}$$

by summing up the exponentials, the result exists in the upper half space $\mathrm{Im}\, x > 0$ only. If one represents $u(x)$ as a power series of x, the series will converge within the circle $|x| < \pi$. Finally, in the form

$$u(x) = \frac{1}{1 + \mathrm{e}^{\mathrm{i}x}}, \tag{50}$$

this function can be evaluated everywhere. When calculating the coefficients in the expansion of the SA in powers of elevations one never sums up infinite series, and the issue of convergence is irrelevant. The Rayleigh hypothesis can always be used for such calculations.

References

1. A.G. Voronovich, "The Kirchhoff and related approximations," in *Light Scattering and Nanoscale Roughness*, A.A. Maradudin (ed.) (Springer, New York, 2006), pp. 34–59.
2. Lord Rayleigh, "On the dynamical theory of gratings," *Proc. R. Soc. Lond., Ser. A*, **79**, 399–416 (1907).
3. B.A. Lippmann, "Note on theory of gratings," *J. Opt. Soc. Am.* **43**, 408 (1953).
4. R. Petit and M. Cadilhac, "Sur la diffraction d'une onde plane par un reseau infiniment conducteur," *C.R. Acad. Sci. Paris, Ser. A–B* **262**, 468–471 (1966).
5. R.F. Millar, "On the Rayleigh assumption in scattering by a periodic surface," *Proc. Camb. Phil. Soc.* **65**, 773–791 (1969).
6. R.F. Millar, "On the Rayleigh assumption in scattering by a periodic surface: II," *Proc. Camb. Phil. Soc.* **69**, 217–225 (1971).
7. J.A. DeSanto, "Scattering from a perfectly reflecting arbitrary periodic surface: an exact theory," *Radio Sci.* **16**, 1315–1326 (1981).
8. P.M. van den Berg and J.T. Fokkema, "The Rayleigh hypothesis in the theory of reflection by a grating," *J. Opt. Soc. Am.* **69**, 27–31 (1979).
9. A.G. Voronovich, *Wave Scattering from Rough Surfaces*, 2nd edn (Springer Series on Wave Phenomena, Springer Verlag, New York 1999).
10. M.L. Burrows, "Equivalence of the Rayleigh solution and the extended-bounday-condition solution for scattering problems," *Electron. Lett.* **5**, 277–278 (1969).
11. R.F. Millar, "The Rayleigh hypothesis and a related least-squares solution to scattering problems for periodic surfaces and other scatterers," *Radio Sci.* **8**, 785–796 (1973).
12. H. Ikuno and K. Yasuura, "Improved point-matching method with application to scattering from a periodic surface," *IEEE Trans. Ant. Prop.* **AP-21**, 657–662 (1973).
13. A.I. Kleev and A.B. Malenkov, "The convergence of point-matching techniques," *IEEE Trans. Ant. Prop.* **AP-37**, 50–54 (1989).
14. S. Christiansen and R. Kleinman, "On a misconception involving point collocation and the Rayleigh hypothesis," *IEEE Trans. Ant. Prop.* **AP-44**, 1309–1316 (1996).

15. J. DeSanto, G. Erdmann, W. Hereman, and M. Misra, "Theoretical and computational aspects of scattering from rough surfaces: one dimensional perfectly reflecting surfaces," *Waves Random Media* **8**, 385–414 (1998).
16. J. DeSanto, G. Erdmann, W. Hereman, and M. Misra, "Theoretical and computational aspects of scattering from periodic surfaces: one dimensional transmission interface," *Waves Random Media* **11**, 425–453 (2001).

5
Small-Amplitude Perturbation Theory for One-Dimensionally Rough Surfaces

K. A. O'DONNELL
División de Física Aplicada, Centro de Investigación Científica y de Educación Superior de Ensenada, Apartado Postal 2732, Ensenada, Baja California, 22800 México

5.1. Introduction

There has been considerable interest in the scattering of light from rough surfaces having surface height fluctuations small compared to the illumination wavelength. This case of weak roughness is important in practical applications such as scattering from the residual roughness of polished optical surfaces.[1] There has also been interest in scattering from weakly rough metal surfaces when the excitation of surface plasmon polaritons is significant.[2,3] Under appropriate conditions, a vertical roughness of only a few nanometers can produce remarkably strong scattering effects arising from polariton excitation. This line of research has been driven more by fundamental interest, and studies have often addressed the unusual features such as backscattering enhancement that appear in the diffuse scattering distributions.

A theoretical approach that is suitable for scattering from weak roughness is perturbation theory, in which the perturbation parameter is the ratio of the surface height fluctuation to the illumination wavelength. The leading-order terms of this theory have been known for many years and have been widely employed. It is only more recently that theoretical approaches have developed sufficient sophistication to produce perturbation terms beyond the leading order. One such theory is infinite-order perturbation theory, which includes a number of approximations, as was developed in the original works on polariton-related backscattering enhancement.[2,3] Other perturbation methods have also appeared that can determine, in principle at least, the exact perturbation term of any particular order.[4,5] It is stressed that the exact terms are quite different from their counterparts in the approximate infinite-order theories, and that their sheer complexity can make it difficult to evaluate them in high perturbation orders.

In the work presented here, the objective is simply stated: to determine the mean diffuse intensity scattered by a rough surface to as high an order as possible, while keeping exact perturbation terms in all calculated orders. This objective is appealing because no scattering mechanisms will be omitted and no artifacts will be introduced, apart from whatever effects arise from the inevitable truncation of the series. The theoretical approach is described in Sect. 5.2 and is an extension of the fourth-order formulation of Ref. 4 to sixth and eighth order in the surface roughness parameter.[6] It may be anticipated that the perturbation terms will be complicated in the higher orders, and that they will require considerable computational effort to evaluate. To simplify the analysis, the random roughness will be assumed to be a one-dimensional Gaussian process. Still, this assumption is not unrealistic, and comparisons with a number of related experimental results[7,8] will later be presented.

The theoretical approach described here may be applied to a surface having an arbitrary dielectric constant and roughness power spectrum. However, our main interest is in the consequences of surface wave excitation, so that calculations are presented for metal surfaces with roughness power spectra that produce significant coupling of the incident wave to surface plasmon polaritons. This interest provides justification for taking the perturbation theory to a high order because surface plasmon polaritons may be scattered many times by the surface roughness, until the excitation finally emerges from the surface as a contribution to the diffusely scattered light. We ask: what features will these multiple-scattering processes produce in the diffuse scattering distributions? Will there be any unusual or unanticipated effects revealed by the new perturbation terms?

These questions are addressed here as follows. Many of the perturbation terms contain contributions to the backscattering enhancement peak and, in calculations in Sect. 5.3 for a Gaussian roughness spectrum, it is shown that they predict a broadening of the peak with increasing surface roughness. In Sect. 5.4, it is discussed that the theory does indeed predict an entirely unexpected effect: a peak in the diffuse scatter that is centered at the specular angle. This remarkable peak arises from eighth-order perturbation terms and is produced by multiple scattering processes resembling those producing backscattering enhancement. Finally, in Sect. 5.5, the theory is employed to elucidate the origin of backscattering peaks that have appeared in experiments.

5.2. Theory

As shown in Fig. 5.1, a plane wave of frequency ω is incident at angle θ_i on the rough surface of a medium having dielectric constant $\varepsilon = \varepsilon_1 + \mathrm{i}\,\varepsilon_2$. The surface profile is given by the one-dimensional random process $\zeta(x)$ and, with the incident wave vector in the plane of Fig. 5.1, all scattered wave vectors lie in the plane of incidence. The objective is to determine the mean diffuse intensity $I(\theta_s|\theta_i)$, which is a function of the scattering angle θ_s of Fig. 5.1. The case of p polarization is considered throughout the analysis because this is the case in which surface plasmon polaritons may be excited; the surface waves have wavenumber $\pm k_{\mathrm{sp}} = \pm(\omega/c)\sqrt{\varepsilon_1/(\varepsilon_1+1)}$ for polaritons traveling to the right (+) or left (−) along the surface.

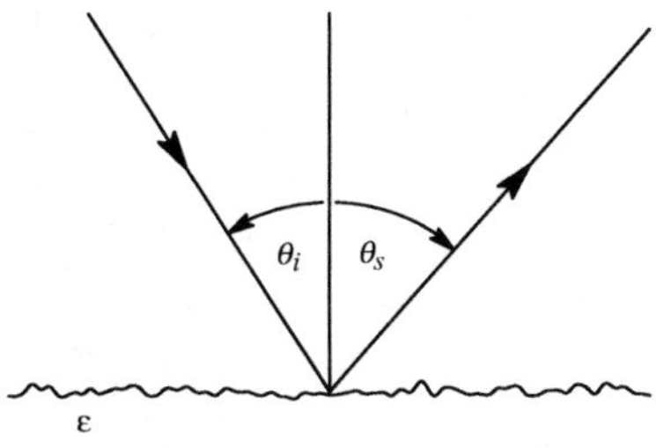

FIGURE 5.1. Scattering by a rough surface. The incident and scattering angles θ_i and θ_s are positive as shown.

The profile $\zeta(x)$ is taken to be a stationary Gaussian random process of zero mean, with the Fourier transform

$$\hat{\zeta}(k) = \int_{-\infty}^{\infty} \mathrm{d}x\, \zeta(x) \exp(-\mathrm{i}kx)\,. \tag{1}$$

It is then readily shown that

$$\langle \hat{\zeta}^*(k)\, \hat{\zeta}(k') \rangle = 2\pi\, \sigma^2\, \delta(k - k')\, g(k)\,, \tag{2}$$

where $\langle \cdot \rangle$ denotes an ensemble average, σ is the standard deviation of the roughness, and $g(k)$ is the power spectrum of $\zeta(x)$.

The amplitude scattered by the surface is simply related to the transition matrix $T(q|k)$ of scattering theory, with $q = (\omega/c)\sin\theta_s$ and $k = (\omega/c)\sin\theta_i$. In direct perturbation theory, $T(q|k)$ is expanded into a power series

$$T(q|k) = \sum_{n=1}^{\infty} \frac{(-\mathrm{i})^n}{n!}\, T^{(n)}(q|k) \tag{3}$$

where $T^{(n)}(q|k)$ is of nth order in $\zeta(x)$. In particular, $T^{(n)}(q|k)$ may be expressed in the form

$$T^{(1)}(q|k) = \mathcal{A}^{(1)}(q|k)\, \hat{\zeta}(q - k)\,, \tag{4}$$

$$T^{(2)}(q|k) = \frac{1}{2\pi} \int_{-\infty}^{\infty} \mathrm{d}p\, \mathcal{A}^{(2)}(q|p|k)\, \hat{\zeta}(q - p)\, \hat{\zeta}(p - k)\,, \tag{5}$$

$$T^{(3)}(q|k) = \frac{1}{(2\pi)^2} \int^{-\infty}\!\!\int_{\infty} \mathrm{d}p\, \mathrm{d}r\, \mathcal{A}^{(3)}(q|p|r|k)\, \hat{\zeta}(q - p)\, \hat{\zeta}(p - r)\, \hat{\zeta}(r - k)\,, \tag{6}$$

and so on. The quantities $\mathcal{A}^{(n)}$ may be determined from rigorous scattering theory which, in the present case, is the reduced Rayleigh equations for p polarization.[4] This task is a substantial one and is not reproduced here, but the terms $\mathcal{A}^{(n)}$ follow in a straightforward way from the recursion relations of Ref. 4. It is notable that the algebraic forms of the $\mathcal{A}^{(n)}$ are neither unique nor reciprocal, and it is necessary to transform them to exhibit their manifestly reciprocal forms.[4] However, when calculating a physically meaningful quantity such as a scattered amplitude or intensity, the numerical result is unique and, largely for convenience, we employ the nonreciprocal forms obtained from Eqs. (3.19) and (3.22) of Ref. 4. The interpretation of Eqs. (4)–(6) is that $T^{(1)}(q|k)$ represents the transfer of the incident state k directly to the scattered state q, $T^{(2)}(q|k)$ represents the process $k \rightarrow p \rightarrow q$ via an intermediate state p, $T^{(3)}(q|k)$ includes processes $k \rightarrow r \rightarrow p \rightarrow q$ via states r and p, and other terms introduce higher scattering processes. Integration over all intermediate states then serves to obtain the total contribution from k to q.

In order to determine $I(\theta_s|\theta_i)$, one must first square and then ensemble average Eq. (3). Including all necessary factors, the mean diffuse intensity $I(\theta_s|\theta_i)$ is given by

$$I(\theta_s|\theta_i) = \frac{2}{\pi}\left(\frac{\omega}{c}\right)^3 \cos^2(\theta_s)\,\cos(\theta_i)\;|G_o(q)|^2\;\langle|\Delta T(q|k)|^2\rangle\;|G_o(k)|^2\,, \tag{7}$$

where, assuming that $\zeta(x)$ is Gaussian, we have that

$$\begin{aligned}\langle|\Delta T(q|k)|^2\rangle = {} & \left\langle|T^{(1)}(q|k)|^2\right\rangle \\ & + \frac{1}{2!^2}\left\langle|\Delta T^{(2)}(q|k)|^2\right\rangle - \frac{2}{1!\,3!}\mathrm{Re}\left\langle T^{(1)*}(q|k)\,T^{(3)}(q|k)\right\rangle \\ & + \frac{1}{3!^2}\left\langle|T^{(3)}(q|k)|^2\right\rangle - \frac{2}{2!\,4!}\mathrm{Re}\left\langle \Delta T^{(2)*}(q|k)\,\Delta T^{(4)}(q|k)\right\rangle \\ & + \frac{2}{1!\,5!}\mathrm{Re}\left\langle T^{(1)*}(q|k)\,T^{(5)}(q|k)\right\rangle \\ & + \frac{1}{4!^2}\left\langle|\Delta T^{(4)}(q|k)|^2\right\rangle - \frac{2}{3!\,5!}\mathrm{Re}\left\langle T^{(3)*}(q|k)\,T^{(5)}(q|k)\right\rangle \\ & + \frac{2}{2!\,6!}\mathrm{Re}\left\langle \Delta T^{(2)*}(q|k)\,\Delta T^{(6)}(q|k)\right\rangle \\ & - \frac{2}{1!\,7!}\mathrm{Re}\left\langle T^{(1)*}(q|k)\,T^{(7)}(q|k)\right\rangle\,, \end{aligned} \tag{8}$$

and the flat surface Green's function $G_o(k)$ is given by

$$G_o(k) = \frac{\mathrm{i}\varepsilon}{\varepsilon\,\alpha_o(k) + \alpha(k)}\,, \tag{9}$$

with $\alpha_o(k) = \sqrt{(\omega/c)^2 - k^2}$ and $\alpha(k) = \sqrt{\varepsilon(\omega/c)^2 - k^2}$. In Eq. (8) we have retained all nonzero terms of the form $\langle T^{(n)*}(q|k)\,T^{(m)}(q|k)\rangle$ for $(n+m) \le 8$. The assumption of Gaussian statistics guarantees that only terms with even $(n+m)$ survive in Eq. (8) because all odd-ordered moments of the Gaussian process $\zeta(x)$ vanish.[9] The specular reflection is associated with $\langle T(q|k)\rangle$ and, in Eq. (7), we have taken $\Delta T(q|k) \equiv T(q|k) - \langle T(q|k)\rangle$ to isolate pure diffuse scatter. The Gaussian surface assumption also similarly implies that $\langle T^{(n)}(q|k)\rangle$ is nonzero only for even n and, consequently, it is essential to take $\Delta T^{(n)}(q|k) = T^{(n)}(q|k) - \langle T^{(n)}(q|k)\rangle$ in Eq. (8) for even n to remove the specular reflection.

It is now necessary to substitute expressions of the form of Eqs. (4)–(6) into Eq. (8), substitute for the Fourier transforms as in Eq. (1) and, after averaging, apply the Gaussian moment theorem to the moments of the profile function $\zeta(x)$. It is then found that the contribution of $\langle T^{(n)*}(q|k)\,T^{(m)}(q|k)\rangle$ to Eq. (7) is proportional to $(\sigma/\lambda)^{n+m}$ where λ is the illumination wavelength, so that Eq. (8)

exhibits all contributions of second order (the 1–1 term), fourth order (2–2 and 1–3 terms), sixth order (3–3, 2–4, and 1–5 terms), and eighth order (4–4, 3–5, 2–6, and 1–7 terms) in σ/λ. The physical interpretation of the n–m term of Eq. (8), based on the discussion of Eqs. (4)–(6), is that it represents the interference term between the scattering contributions of n-fold scattering sequences and m-fold scattering sequences.

There are integrations over Dirac delta functions as well as variable changes that must be performed to put the terms into a simplified form; these manipulations are lengthy and were carried out using computer-based symbolic manipulation. The resulting full set of terms is too long to be presented but may be found in Appendix A of Ref. 6, and here the discussion is confined to general comments. The number of terms in each order in Eq. (8) (taken throughout our discussions to be the number of terms produced by application of the Gaussian moment theorem to the moments of $\zeta(x)$) are 1, 5, 42, and 396 in, respectively, the second, fourth, sixth, and eighth orders in σ (and there would be 4575 and 60,885 terms in, respectively, the tenth and twelfth orders). The terms contain one-dimensional integrals in the fourth order in σ and, at most, two- and three-dimensional integrals in, respectively, the sixth and eighth orders. The integrals cannot be done analytically and numerical quadrature is necessary for exact term evaluation.

In general, the n–m term integrand contains expressions involving $\mathcal{A}^{(n)*}$ and $\mathcal{A}^{(m)}$, accompanied by factors of the roughness spectrum $g(\cdot)$. To perform the integrals numerically, it is first necessary to evaluate the integrands efficiently. However, directly coding the recursion relations for $\mathcal{A}^{(n)}$ is inefficient because many redundant calculations are performed. Thus the recursion relations were rewritten so as to remove this redundancy. Complications also arise because the integrands often present narrow peaks. In the numerical integration, when the integration variables bring an intermediate argument of any $\mathcal{A}^{(\cdot)}$ to within a few resonance widths $\Delta_\varepsilon = k_{sp}\,\varepsilon_2/[2\varepsilon_1(\varepsilon_1+1)]$ of $\pm k_{sp}$, the integrand usually becomes large. Thus an adaptive integrator was developed, based on a modified Genz–Malik algorithm,[10] that continued sampling the integrands in such regions until the total residual error calculated was at most 10^{-3} of the integral value.

5.3. Gaussian Roughness Spectrum

We now consider a case in which the roughness power spectrum has the Gaussian form

$$g(k) = \sqrt{\pi}\, a \exp(-a^2k^2/4)\,, \tag{10}$$

which has been employed in a number of previous theoretical works,[2–4,6,11] even though no related experimental works have appeared. As in the theoretical works, we assume that $\sigma = 5$ nm, $\varepsilon = -7.5 + 0.24\,\mathrm{i}$, $\lambda = 457.9$ nm, and $a = 100$ nm is the surface correlation length. This spectrum is sufficiently broad to produce coupling of the incident wave to plasmon polaritons ($g(k_{sp})/g(0) \simeq 0.58$), as well as coupling of counterpropagating polaritons ($g(2k_{sp})/g(0) \simeq 0.11$).

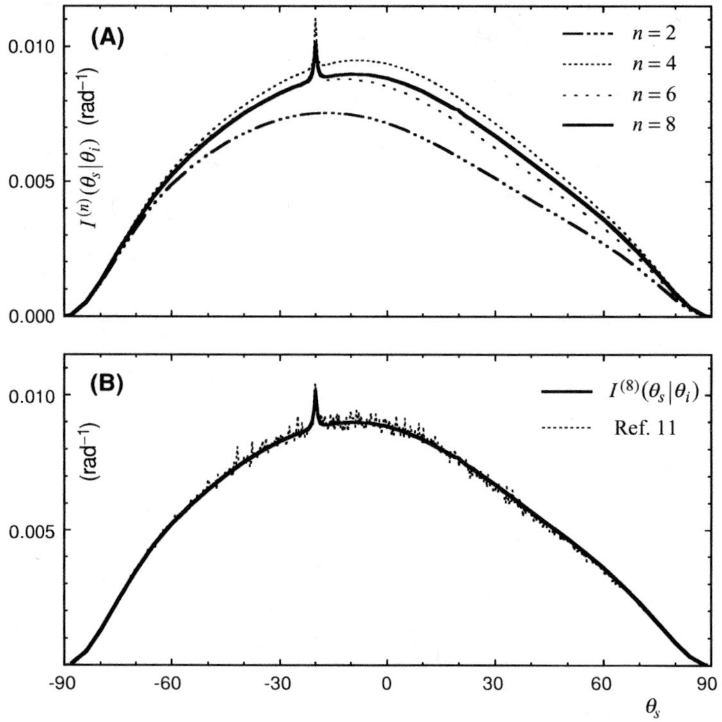

FIGURE 5.2. Scattering from a surface having a Gaussian roughness spectrum with $\sigma = 5$ nm, $\varepsilon = -7.5 + 0.24\,i$, $\lambda = 457.9$ nm, and $a = 100$ nm. Upper plot: $I^{(n)}(\theta_s|\theta_i)$ for n as indicated. Lower plot: comparison of $I^{(8)}(\theta_s|\theta_i)$ with results from Ref. 11.

We denote the mean diffuse intensity obtained by truncating Eq. (8) at order n by $I^{(n)}(\theta_s|\theta_i)$. With $\theta_i = 20°$, Fig. 5.2A shows $I^{(n)}(\theta_s|\theta_i)$ for the cases $n = 2$, 4, 6, and 8. A broad, featureless distribution is seen in $I^{(2)}(\theta_s|\theta_i)$, while $I^{(4)}(\theta_s|\theta_i)$ is upwardly corrected and shows a distinct peak at the backscattering angle ($\theta_s = -\theta_i$). Both of these results are consistent with similar calculations reported earlier in Ref. 4. In the high-order contributions it is seen that, with respect to $I^{(4)}(\theta_s|\theta_i)$, $I^{(6)}(\theta_s|\theta_i)$ is downwardly corrected, and that $I^{(8)}(\theta_s|\theta_i)$ then receives a modest upward correction. Significant contributions arise from the 3–3 and 2–4 terms of $I^{(6)}(\theta_s|\theta_i)$ and from the 4–4, 3–5, and 2–6 terms of $I^{(8)}(\theta_s|\theta_i)$.

Figure 5.2B shows comparisons of $I^{(8)}(\theta_s|\theta_i)$ with the results of Michel,[11] who employed a Monte Carlo technique. In particular, in Ref. 11 the intensity scattered by a realization of $\zeta(x)$ was calculated from exact numerical solution of the reduced Rayleigh equations, and the result was averaged over an ensemble of $N = 2000$ realizations of $\zeta(x)$. Due to the finite value of N, there is noise in the results analogous to the speckle noise of experimental data. Nevertheless, the Monte Carlo results appear to fluctuate randomly about the curve of $I^{(8)}(\theta_s|\theta_i)$, so that the agreement is excellent.

The detailed shape of the backscattering peak is a subtle issue for which the higher-order corrections are significant. The peak width is the diffraction width of the plasmon–polariton travel length along the surface and is affected by two mechanisms.[2] First, the polariton is attenuated at the rate $\Delta_\varepsilon = k_{\text{sp}}\,\varepsilon_2/[2\varepsilon_1(\varepsilon_1 + 1)]$ due to Joule losses in the metal. In addition, as the polariton travels it may be scattered by the roughness, thereby increasing its effective attenuation. It is found that only the first mechanism contributes to $I^{(4)}(\theta_s|\theta_i)$ and, in the pole approximation, it produces a backscattering peak of Lorentzian form in the quantity $(q - k)$ with angular width (taken as the full width at half maximum throughout this chapter) $\Delta\theta_\varepsilon = 4\Delta_\varepsilon/\cos\theta_i$.[4] In Ref. 2, the second mechanism was included by writing the total decay rate as $\Delta_{\text{tot}} = \Delta_\varepsilon + \Delta_{\text{sp}}$, where Δ_{sp} was of order σ^2 and was an estimate of the rate of roughness-induced damping; this theory then predicts a Lorentzian peak having width $4\Delta_{\text{tot}}/\cos\theta_i$. However, it is pointed out here that Δ_{sp} is only an estimate and that the shape need not be Lorentzian. Further, Monte Carlo approaches such as Ref. 11 have speckle noise and it can be difficult to discern the structure of the peak. Thus the peak shape is an important issue that is well-addressed by the high-order perturbation theory developed here. While previous works have estimated the polariton attenuation explicitly, here the broadening of the backscattering peak arises naturally and without approximation as higher-order terms are introduced.

The details of the peak of Fig. 5.2A are shown in Fig. 5.3A, where the nearly Lorentzian form appears in $I^{(4)}(\theta_s|\theta_i)$. It is also seen in that a strong sixth-order broadening correction has produced an unphysical double peak in $I^{(6)}(\theta_s|\theta_i)$. In the next order, however, $I^{(8)}(\theta_s|\theta_i)$ presents a more reasonable single peak once again. The peak shape is not Lorentzian but instead has a sharp central region and a broad base, which arise in a subtle way from the sum of the intensity contributions of multiple scattering processes (the 2–2, 3–3, and 4–4 terms of Eq. (8)), as well as from the interference between processes of different order (the 2–4, 3–5, and 2–6 terms); only the 1–1, 1–3, 1–5, and 1–7 terms do not contribute to the peak.

We now consider the broadening of the peak with increasing σ. In these calculations, the perturbation terms employed earlier are simply scaled by the appropriate power of σ and recombined to determine the intensity. The background is then subtracted by curve-fitting the broad envelope and the width of the isolated backscattering peak is determined. The results are shown in Fig. 5.3B. The width $\Delta\theta_\varepsilon = 0.64^\circ$ is essentially identical to the σ-independent peak width of $I^{(4)}(\theta_s|\theta_i)$. The peak width of $I^{(6)}(\theta_s|\theta_i)$ rises with increasing σ, and the width from $I^{(8)}(\theta_s|\theta_i)$ shares the same initial increase, but then rises less rapidly. The arrows indicate the value of σ for which the peaks become unphysical (i.e., contain multiple maxima, as noted for $I^{(6)}(\theta_s|\theta_i)$ in Fig. 5.3A) so that, for larger values in the range of σ plotted, higher-order perturbation terms are required and the curves are probably not accurate. If the pattern of behavior continues, the tenth order would increase the peak width and we speculate that an exact calculation would lie somewhat above the curve for $I^{(8)}(\theta_s|\theta_i)$. For comparison, the prediction of the approximate theory of Ref. 2 is also shown in Fig. 5.3B and, relative to

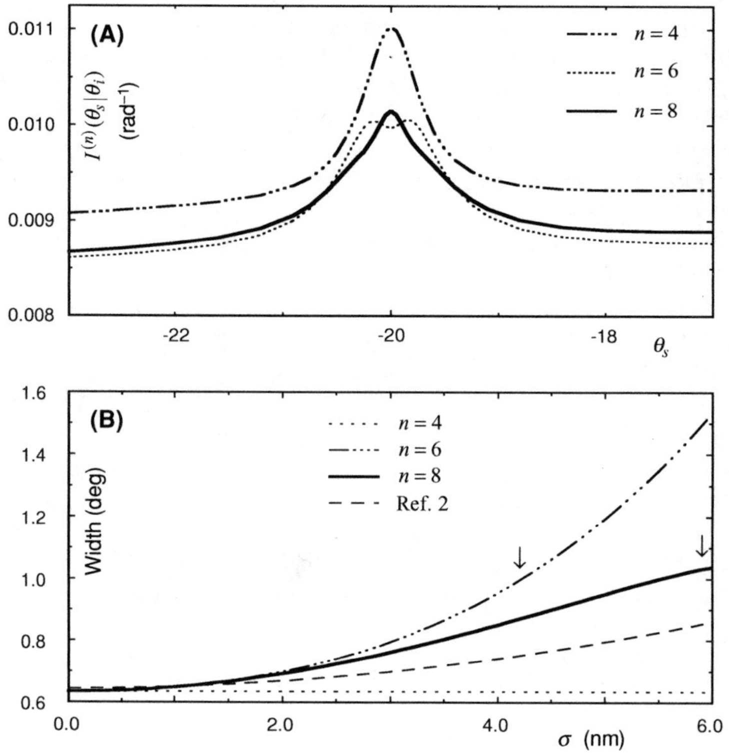

FIGURE 5.3. Upper plot: backscattering peaks in $I^{(n)}(\theta_s|\theta_i)$ from Fig. 5.2 for n as indicated. Lower plot: peak width from $I^{(n)}(\theta_s|\theta_i)$ as a function of surface roughness σ. Arrows indicate points where peak shape becomes unphysical for larger σ.

the baseline provided by $\Delta\theta_\varepsilon$, it predicts only one-half of the broadening seen in $I^{(8)}(\theta_s|\theta_i)$.

Thus the approach utilized here provides a direct means of studying the shape of the backscattering peak. In passing, it is notable that Michel[11] exhibits one remarkable case of this Gaussian spectrum for $\sigma = 16$ nm with peak width approximately eight times $\Delta\theta_\varepsilon$. Such strong effects are well beyond the limitations of the eighth-order calculations developed here.

5.4. Enhanced Specular Peaks

There is an unexpected and remarkable effect that is predicted by the eighth-order perturbation terms. In particular, the new effect is a peak in $I(\theta_s|\theta_i)$ that is centered on the specular angle.[6,12] It is stressed that this peak has no relation whatsoever to the specular reflection from the surface because, as noted in writing Eq. (8), the specular reflection has been subtracted to isolate pure diffuse scatter in $I(\theta_s|\theta_i)$.

The enhanced specular peak is instead a feature of the diffuse scattering distribution that arises from constructive interference of multiple scattering contributions, in a manner similar to the mechanisms that produce backscattering enhancement. It originates in the 4–4 term of Eq. (8) and thus, in its leading order, relies on a 4-fold scattering sequence.

There do not seem to have been experimental observations of this enhanced specular peak although, in principle, it would be a straightforward matter to observe it by blocking the specular reflection itself and scanning a detector near specular to resolve the peak (the specular reflection has angular width inversely proportional to the illuminated surface area, and may thus be made arbitrarily narrow). In their respective leading orders, the specular peak has twice the angular width of the backscattering peak.

In this section we present a discussion of this specular peak. It is studied with the perturbation theory and its physical origins are described. In cases presented here, the peak becomes distinct when the perturbation theory is on the verge of producing unphysical results, presumably due to the truncation of the series of Eq. (8) at eighth order. Thus, to further support the existence of the peak, we also present calculations for stronger roughness based on nonperturbative Monte Carlo techniques.

5.4.1. Gaussian Spectra

The specular peak is indeed present in $I^{(8)}(\theta_s|\theta_i)$ for the case of the Gaussian spectrum such as the one discussed in Sect. 5.3, although it is of such low contrast in Fig. 5.2A for $\sigma = 5$ nm that it is not clearly seen. By simply increasing σ to 7.5 nm, while keeping all other surface parameters fixed, the specular peak rises distinctly to 5% of the height of the surrounding distribution. This case is shown in Fig. 5.4A where a nearly Lorentzian peak is seen at the specular angle. It has a width of 1.25°, which is essentially $2\,\Delta\theta_\varepsilon$, and is due to the 4–4 term of Eq. (8). It can also be seen that the backscattering peak appears with unphysical secondary minima in Fig. 5.4A, indicating that yet higher-order terms are needed to produce an accurate distribution. These terms may also modify the height and shape of the specular peak, but they could not possibly remove it for all σ. Thus, despite the unphysical nature of $I^{(8)}(\theta_s|\theta_i)$ in Fig. 5.4A, the specular peak is a real effect. However, it has not been possible to find values of σ and a that produce both a distinct specular peak and a physically reasonable backscattering peak.

These two criteria may by satisfied by instead employing a nonperturbative Monte Carlo approach[12] similar to the work of Michel[11] discussed earlier. The roughness σ has been increased to 10 nm and a has been reduced to 40 nm to produce the modest but distinct specular peak in the distribution of Fig. 5.4B. The backscattering peak of Fig. 5.4B is considerably broadened and has a width of 4.8°, well above the leading-order perturbation width of 0.61°. The width of the low specular peak is difficult to estimate but is approximately 6°, so it too is considerably wider than the leading-order width of 1.2°. While the values of σ and

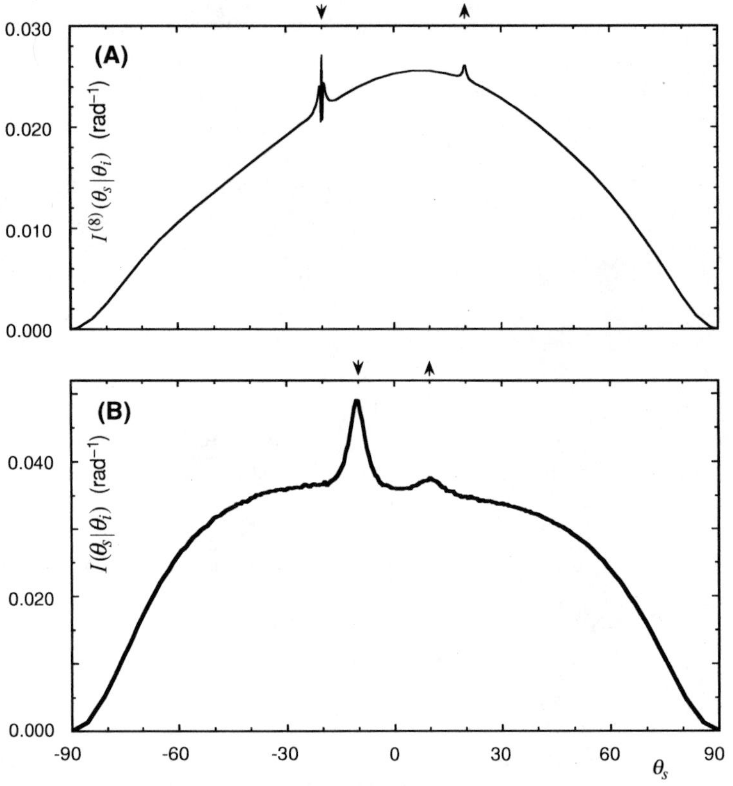

FIGURE 5.4. Results for enhanced specular peaks in cases with Gaussian roughness spectra. Upper plot: $I^{(8)}(\theta_s|\theta_i)$ for $\theta_i = 20°$, $\sigma = 7.5$ nm, and $a = 100$ nm. Lower plot: from Monte Carlo calculations, $I(\theta_s|\theta_i)$ for $\theta_i = 10°$, $\sigma = 10$ nm, and $a = 40$ nm.

a were chosen through trial and error to produce a distinct specular peak, specular peaks are still apparent for considerable range of these parameters.

5.4.2. Physical Origins

The 4–4 term of Eq. (8) contains a total of 96 distinct perturbation terms, of which only four are associated with the specular peak. This may be shown by discarding terms and recomputing results until only the critical terms remain, as discussed in Ref. 6. The key contributions to the 4–4 term may be written as

$$\frac{\sigma^8}{(2\pi)^3} \int_{-\infty}^{\infty} \mathrm{d}p \, [\mathcal{F}(q, p, k) + \mathcal{G}(q, p, k)] \, g(q - p) g(p - k) , \tag{11}$$

with

$$\mathcal{F}(q, p, k) = [F^*(q, p, k) + F^*(q, q + k - p, k)]F(q, p, k) \tag{12}$$

and

$$\mathcal{G}(q, p, k) = [G^*(q, p, k) + G^*(q, q + k - p, k)]G(q, p, k), \tag{13}$$

where

$$F(q, p, k) = \int_{-\infty}^{\infty} \mathrm{d}r \, \mathcal{A}^{(4)}(q|q + r|p + r|k + r|k) \, g(r) \tag{14}$$

and

$$G(q, p, k) = \int_{-\infty}^{\infty} \mathrm{d}r \, \mathcal{A}^{(4)}(q|p|p + r|p|k) \, g(r). \tag{15}$$

The contribution to the 4–4 terms is unusual in that integrals over the *amplitude* $\mathcal{A}^{(4)}$ are first performed (Eqs. (14)–(15)), and products of these integrals are then integrated again in Eq. (11) to determine the 4–4 term contribution.

It remains to ask what scattering processes are described by these terms. This requires more effort and the arguments are subtle; thus we do not describe the analysis here and the reader is referred to Ref. 6 for details. In particular, numerical studies show that the significant contributions to the integrals occur when the arguments of the integrands read either as $\mathcal{A}^{(4)}(q| - k_{\mathrm{sp}}| + k_{\mathrm{sp}}| - k_{\mathrm{sp}}|k)$ or $\mathcal{A}^{(4)}(q|+k_{\mathrm{sp}}|-k_{\mathrm{sp}}|+k_{\mathrm{sp}}|k)$ which implies that, respectively, the scattering processes $k \to -k_{\mathrm{sp}} \to +k_{\mathrm{sp}} \to -k_{\mathrm{sp}} \to q$ or $k \to +k_{\mathrm{sp}} \to -k_{\mathrm{sp}} \to +k_{\mathrm{sp}} \to q$ are the contributing sequences. Further, it is shown that the first process or the second process *alone* is sufficient to produce a specular peak. However, if both are present, the *interference* between the processes produces an additional contribution to the height of the specular peak.

Inasmuch as the theory employed here is formulated in terms of wavenumber, the significance of the scattering mechanisms just described is clear and unambiguous. However, it is still an open issue as to what is occurring in the surface itself, and here we put forth the following model that uses the appropriate wavenumber sequences. In particular, one must search for scattering paths that interfere constructively at the specular angle. Thus consider the scattering paths that involve four points on the surface as shown in Fig. 5.5. Paths A and B lie in the plane of the mean surface, and are each linked by $k \to -k_{\mathrm{sp}} \to +k_{\mathrm{sp}} \to -k_{\mathrm{sp}} \to q$. In the usual far-field approximation, the phase delay along path A is

$$\phi_{\mathrm{A}} = k\, x_1 - q\, x_4 + \beta_{1-4}, \tag{16}$$

where x_n is the coordinate of point n, and β_{1-4} is the total phase delay along the path $1 \to 2 \to 3 \to 4$. Path B is chosen with the same points 2 and 3 but new points $1'$ and $4'$, producing the phase delay

$$\phi_{\mathrm{B}} = k\, x_{1'} - q\, x_{4'} + \beta_{1'-4'}. \tag{17}$$

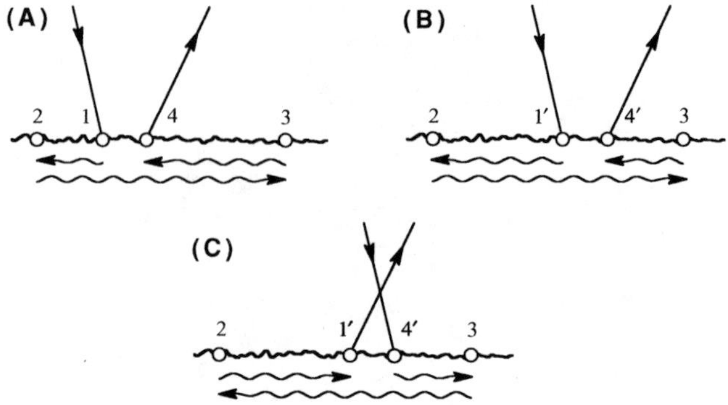

FIGURE 5.5. The three types of scattering processes that contribute to the enhanced specular peak.

Now if path B is chosen with $(x_{4'} - x_{1'}) = (x_4 - x_1)$, so that $\beta_{1'-4'} = \beta_{1-4}$, the phase difference $\Delta\phi_{\mathrm{BA}} \equiv \phi_{\mathrm{B}} - \phi_{\mathrm{A}}$ is then

$$\Delta\phi_{\mathrm{BA}} = (q - k)(x_1 - x_{1'})\,. \tag{18}$$

It is clear that $\Delta\phi_{\mathrm{BA}}$ vanishes for $q = k$ and the contributions interfere constructively at specular. There are many possible paths of type B; for all possible points $1'$ and $4'$ between points 2 and 3, constructive interference occurs as long as the condition $(x_4 - x_1) = (x_{4'} - x_{1'})$ is met.

By symmetry, one may also link the four points of Fig. 5.5 with the second sequence $k \to +k_{\mathrm{sp}} \to -k_{\mathrm{sp}} \to +k_{\mathrm{sp}} \to q$ and draw identical conclusions. Thus if the surface supports this second sequence, as is the case for the Gaussian roughness spectrum, there is constructive interference and another contribution to the specular peak is produced. However, it was pointed out in Ref. 6 that the perturbation theory also indicates that there is *constructive interference* between the contributions of the two scattering sequences. Thus consider path C of Fig. 5.5, which is similar to path B, but is linked with the second sequence. It is readily shown that the phase difference $\Delta\phi_{\mathrm{CA}} \equiv \phi_{\mathrm{C}} - \phi_{\mathrm{A}}$ is

$$\Delta\phi_{\mathrm{CA}} = (q - k)(x_1 - x_{1'}) + (q + k)\Delta x\,, \tag{19}$$

where $\Delta x \equiv (x_4 - x_1) = (x_{4'} - x_{1'})$. It is only when the path closes on itself ($\Delta x = 0$) that the phase difference takes the form of Eq. (18). Thus, for a recurrent scattering path, the interference between processes A and C is fully consistent with a specular peak contribution. The specular peak is thereby seen to arise from three sources: paths linked by $k \to -k_{\mathrm{sp}} \to +k_{\mathrm{sp}} \to -k_{\mathrm{sp}} \to q$, others linked by $k \to +k_{\mathrm{sp}} \to -k_{\mathrm{sp}} \to +k_{\mathrm{sp}} \to q$, as well as the interference between contributions arising from the two wavenumber sequences.

Thus it is plausible that these are the physical scattering paths on the surface that produce the enhanced specular peak. It is to be stressed that the specular effect is quite remarkable for the same reasons that backscattering enhancement has attracted so much attention. In particular, the constructive interference between the contributions of these multiple scattering paths prevails to the extent that, even after averaging over the randomness of a rough surface, the specular peak rises distinctly in the diffuse scatter.

5.4.3. Rectangular Spectra

With the intent of improving the contrast of the enhanced specular peak, we now consider another form of the surface power spectrum $g(k)$. The approach taken is to construct a spectrum that is nonzero only in regions essential to the couplings associated with the peak. It is hoped that this approach will maintain these couplings, while it will discourage other unessential scattering processes, thereby leading to an improved peak contrast.

5.4.3.1. The Spectrum

We now consider the wavenumbers of the surface roughness required to support the essential sequence $k \to -k_{\text{sp}} \to +k_{\text{sp}} \to -k_{\text{sp}} \to q$. The first coupling of $k = (\omega/c)\,\sin\theta_i$ to $-k_{\text{sp}}$ may be represented by the grating equation

$$-k_{\text{sp}} = k + k_{r1}\,, \tag{20}$$

where k_{r1} represents a wavenumber that is present in the surface roughness spectrum. Similarly, the coupling of $-k_{\text{sp}}$ to $+k_{\text{sp}}$ requires a roughness wavenumber k_{r2} satisfying the coupling equation

$$+k_{\text{sp}} = -k_{\text{sp}} + k_{r2}\,, \tag{21}$$

while the coupling of $+k_{\text{sp}}$ to $-k_{\text{sp}}$ requires a roughness wavenumber k_{r3} satisfying the equation

$$-k_{\text{sp}} = +k_{\text{sp}} + k_{r3}\,. \tag{22}$$

Finally, the outward coupling of $-k_{\text{sp}}$ to the escaping wave $q = (\omega/c)\,\sin\theta_s$ is expressed by

$$q = -k_{\text{sp}} + k_{r4}\,, \tag{23}$$

where k_{r4} is the essential roughness wavenumber.

The proposed spectrum is shown in Fig. 5.6, where $g(k)$ consists of two rectangles of height g_1 and half-width Δ_1 centered on $\pm k_{\text{sp}}$, and two rectangles of height g_2 and half-width Δ_2 centered on $\pm 2\,k_{\text{sp}}$. The rectangles at $\pm k_{\text{sp}}$ provide roughness wavenumbers for the inward and outward coupling of Eqs. (20) and (23); it is readily shown that the width Δ_1 maintains coupling as long as $|\theta_i|$ and $|\theta_s|$ are less than $\theta_{\max} = \arcsin[(c/\omega)\Delta_1]$. The rectangles at $\pm 2\,k_{\text{sp}}$ provide roughness wavenumbers to satisfy Eqs. (21) and (22). The ratio g_2/g_1 may be considered a parameter that controls the relative amounts of counterpropagating polariton coupling and inward/outward coupling.

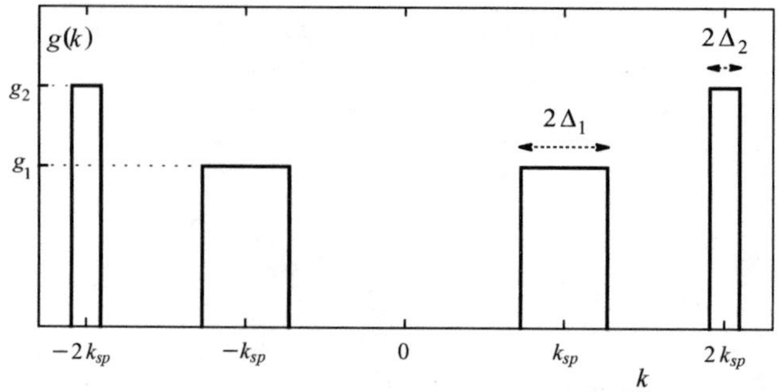

FIGURE 5.6. Surface power spectrum $g(k)$ proposed to produce the enhanced specular peak.

The preceding arguments have considered only the sequence $k \to -k_{sp} \to +k_{sp} \to -k_{sp} \to q$. However, if $g(k)$ of Fig. 5.6 supports this sequence, it is readily shown that the sequence $k \to +k_{sp} \to -k_{sp} \to +k_{sp} \to q$ is also supported. This may be proven by inverting the signs of all plasmon polariton wavenumbers in Eqs. (20)–(23), and then noting that all necessary roughness wavenumbers are still present in $g(k)$. It is concluded that the spectrum of Fig. 5.6 produces the essential couplings, and is then a prime candidate for producing a distinct specular peak.

5.4.3.2. Results

We now consider a case with $\sigma = 2$ nm and, once again, $\lambda = 457.9$ nm and $\varepsilon = -7.5 + 0.24i$. It follows that $k_{sp} = 1.074\,(\omega/c)$, and the spectral parameters chosen are $\Delta_1 = (\omega/c)\,\sin(17°)$ (which maintains coupling to $\theta_{max} = 17°$), $\Delta_2 = 0.10\,(\omega/c)$, and $g_2/g_1 = 1.5$. The diffuse intensity $I^{(8)}(\theta_s|\theta_i)$ is shown in Fig. 5.7A for $\theta_i = 9°$ where, due to the outward polariton coupling, a distinct distribution appears for $|\theta_s| \leq \theta_{max}$. Not only does a peak fall at backscattering but also, more notably, a clear specular enhancement peak appears at $\theta_s = \theta_i$. This specular peak appears with good contrast, rising to a height of one-fifth of that of the surrounding distribution. In this case both peaks have a physically reasonable appearance, unlike what was seen in Fig. 5.4A. The scatter at large $|\theta_s|$ is mostly single scatter as in $q = k + k_r$, where k_r are roughness wavenumbers from the two rectangles centered at $\pm k_{sp}$ in $g(k)$ of Fig. 5.6. For the most part, this scatter arises from the 1–1 term of Eq. (8), although higher terms make small corrections. It is clear that the single scatter is well-isolated from the plasmon polariton coupling region with $|\theta_s| \leq \theta_{max}$, and that this isolation significantly improves the contrast of the specular peak.

However, if σ is increased, the backscattering peak soon becomes unphysical, presumably due to the higher-order terms missing in Eq. (8). In order to show a result for stronger roughness where the eighth-order theory is inappropriate, we now present a result obtained from the Monte Carlo approach described in

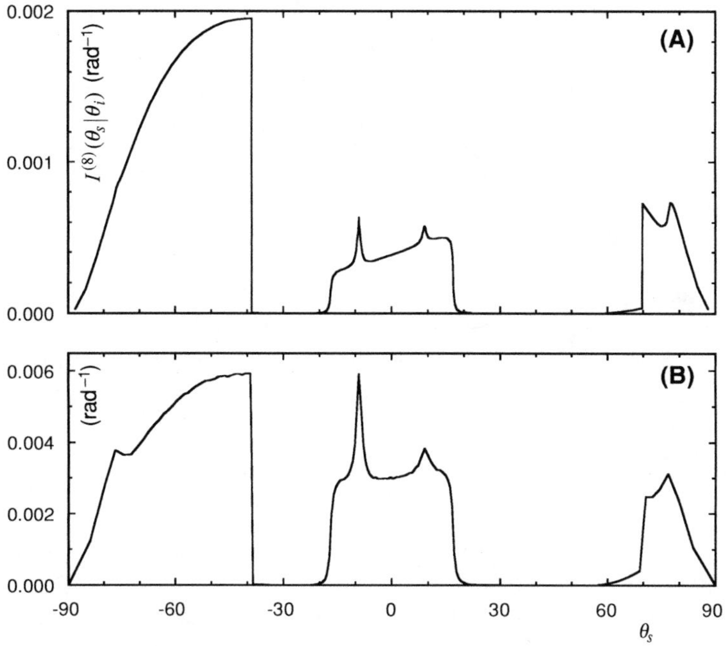

FIGURE 5.7. Mean diffuse intensities exhibiting specular peaks at $\theta_s = \theta_i$ for the spectrum $G(k)$ of Fig. 5.6, $\theta_i = 9^\circ$, and other parameters as in main text. Upper plot: $I^{(8)}(\theta_s|\theta_i)$ from perturbation theory for $\sigma = 2$ nm. Lower plot: $I(\theta_s|\theta_i)$ from Monte Carlo calculations with $\sigma = 3.5$ nm.

Sect. 5.3. The result shown in Fig. 5.7B has $\sigma = 3.5$ nm but all other parameters are as in Fig. 5.7A. The general levels of $I(\theta_s|\theta_i)$ have increased and show clear backscattering and specular enhancement peaks, with the specular enhancement being quite distinct. Peak widths are also considerably wider than in Fig. 5.7A, and the 1.1° backscattering peak width of Fig. 5.7A has increased to 1.7° in Fig. 5.7B. In addition, the specular peak width of 1.2° in Fig. 5.7A has increased to 2.7° and has thus been roughness broadened to 2.2 times its leading-order width. This broadening indicates that terms of tenth and higher orders produce large broadening contributions to the specular peak, even for a modest roughness of only $\sigma = 3.5$ nm.

In summary, the spectrum of Fig. 5.6 does indeed produce enhanced specular peaks having improved contrast. More generally, the high-order perturbation theory has served well in predicting this new effect and in suggesting this form of $g(k)$, while Monte Carlo calculations are more appropriate for cases with stronger roughness where the specular peak appears with higher contrast. While we are unaware of experimental observations of the enhanced specular peak, there are experimental techniques available that have produced surfaces with rectangular power spectra.[7] An extension of these techniques could be used to produce a spectrum of the form of Fig. 5.6, which would be an important step in attempts to produce the first experimental observation of this remarkable effect.

5.5. Comparisons with Experiments

It is perhaps counterintuitive that the experimental observations of polariton-related backscattering enhancement that have appeared have not employed the "natural" Gaussian spectrum of Eq. (10) but, because of the relative ease of surface fabrication, they have instead used rectangular spectra.[7,8] In this section, calculations for such spectra are presented for comparison. Two cases are considered of backscattering effects that rely on, respectively, the 2–2 and the 4–4 perturbation terms. In passing, it is notable that an effect that relies on the 3–3 term is also possible,[6,13] but this is not discussed here.

5.5.1. The 2–2 Effect

We now consider the case when $g(k)$ is composed of only two rectangles centered at $\pm k_{\rm sp}$ (as in Fig. 5.6, but with the rectangles at $\pm 2k_{\rm sp}$ removed). In couplings as in Eq. (20), the rectangle at positive (or negative) k guarantees that $+k_{\rm sp}$ (or $-k_{\rm sp}$) is excited, as long as $|\theta_i| \leq \theta_{\rm max}$. Similarly, as in Eq. (23), $g(k)$ guarantees that both $+k_{\rm sp}$ and $-k_{\rm sp}$ are coupled to diffuse scatter emerging for $|\theta_s| \leq \theta_{\rm max}$. For this spectrum, a backscattering peak should appear that arises from interference between the wavenumber sequences $k \to +k_{\rm sp} \to q$ and $k \to -k_{\rm sp} \to q$; these processes represent double scattering and the backscattering peak should thus, in the lowest order, appear in the 2–2 term of Eq. (8).[14]

The particular parameters employed here are chosen to be consistent with the first experimental observations of polariton-related backscattering enhancement.[7] Specifically, we set the limits of the rectangles of $g(k)$ at $\pm 0.83\,\omega/c$ and $\pm 1.29\,\omega/c$, and we assume that $\lambda = 612$ nm and $\varepsilon = -9.0 + 1.29\,\mathrm{i}$ (gold), so that $k_{\rm sp} \simeq 1.06\,\omega/c$ is centered in $g(k)$. The results are shown in Fig. 5.8 for $\theta_i = 10°$ and $\sigma = 8.3$ nm. Here $I^{(4)}(\theta_s|\theta_i)$ exhibits a clear backscattering peak appearing in a distribution for $|\theta_s| \leq 13.5°$ that arises from the 2–2 term. The physical mechanism is double scattering involving a pair of points 1 and 2 on the surface; as shown in Fig. 5.8, for the scattering sequence $k \to +k_{\rm sp} \to q$ linking them from 1 to 2, there is a time-reversed sequence $k \to -k_{\rm sp} \to q$ linking them from 2 to 1. It is straightforward to show that the two associated contributions have zero phase difference at backscattering and, upon averaging over all such pairs of points on a random surface, the scattering contributions interfere constructively to produce the peak.[14] The limited bandwidth of $g(k)$ constrains the outward coupling $\pm k_{\rm sp} \to q$ to $|\theta_s| \leq 13.5°$, and similarly the polariton launching $k \to \pm k_{\rm sp}$ only occurs for $|\theta_i| \leq 13.5°$. The scattering for $\theta_s \leq -41°$ is single scatter $k \to q$ from the 1–1 term; the 1–3 term makes a small downward correction to this distribution. The result for $I^{(2)}(\theta_s|\theta_i)$ is not shown,

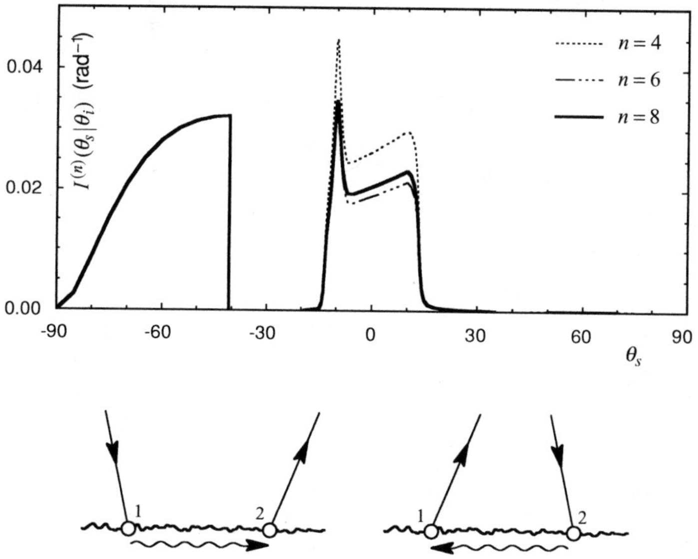

FIGURE 5.8. Upper plot: for a rectangular spectrum spanning $0.83\,\omega/c$ to $1.29\,\omega/c$, $I^{(n)}(\theta_s|\theta_i)$ for $\theta_i = 10^\circ$, $\sigma = 8.3$ nm, $\lambda = 612$ nm, and $\varepsilon = -9.0 + 1.29$i. Lower plot: the two lowest-order scattering processes that contribute to the backscattering peak.

but contains only the distribution for $\theta_s \le -41^\circ$ and remains zero for all other θ_s in Fig. 5.8.

It is seen in Fig. 5.8 that $I^{(6)}(\theta_s|\theta_i)$ has a central distribution of reduced height as compared to $I^{(4)}(\theta_s|\theta_i)$, which is due to a negative 2–4 term, while $I^{(8)}(\theta_s|\theta_i)$ receives positive contributions there that arise from the 4–4 and 2–6 terms. For $\theta_s \le -41^\circ$, a number of high-order terms contribute (1–3, 3–3, 1–5, 3–5, and 1–7) but the overall contribution is small. Upon comparing $I^{(8)}(\theta_s|\theta_i)$ to $I^{(4)}(\theta_s|\theta_i)$, there is a slight increase in the peak width, and the steep walls at $|\theta_s| = 13.5^\circ$ receive reductions in slope, although these subtle effects are not easily seen in Fig. 5.8. It is thus clear that the effects remain much as in $I^{(4)}(\theta_s|\theta_i)$, and that terms of higher order modify the height of the distribution but otherwise produce only subtle changes in its shape.

The qualitative comparison with the experimental result is extremely good (see Fig. 7 of Ref. 7). The experimental curve is somewhat higher, although such discrepancies can easily arise from differences in ε or from errors in the experimentally determined value of σ. The fact that $I^{(8)}(\theta_s|\theta_i)$ receives small corrections, as compared to $I^{(6)}(\theta_s|\theta_i)$, suggests that the numerical convergence of the series is rapid enough so that $I^{(8)}(\theta_s|\theta_i)$ should be close to the exact result. Indeed, we do not compare with the other results of Ref. 7 taken with $\sigma = 10.9$ nm; here $I^{(8)}(\theta_s|\theta_i)$ receives a large correction and consequently it is probably not close to the exact result.

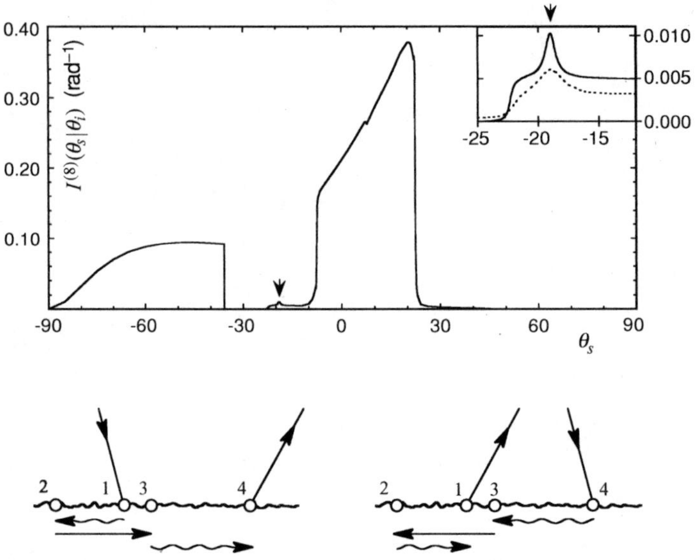

FIGURE 5.9. Upper plot: for a rectangular spectrum spanning $0.91\,\omega/c$ to $1.42\,\omega/c$, $I^{(8)}(\theta_s|\theta_i)$ for $\theta_i = 19°$, $\sigma = 14.8$ nm, $\lambda = 674$ nm, and $\varepsilon = -12.6 + 1.16$i. The inset shows the backscattering region, where $I^{(8)}(\theta_s|\theta_i)$ (solid curve) is compared to experimental results (dashed curve). Lower plots: the two fourth-order scattering processes that contribute to the backscattering peak. The solid line below the surface denotes the nonresonant wave.

5.5.2. *The 4–4 Effect*

We now consider a backscattering effect that arises from the 4–4 term of the eighth-order theory. An experimental observation of this effect has indeed appeared,[8] although the origin of the effect was not recognized. For consistency with these experiments, we assume that the limits of $g(k)$ lie at $\pm 0.91\,\omega/c$ and $\pm 1.42\,\omega/c$, and that $\theta_i = 19°$, $\lambda = 674$ nm, $\sigma = 14.8$ nm, and $\varepsilon = -12.6 + 1.16$i (gold). Thus $k_{\text{sp}} \simeq 1.04\,\omega/c$ is well off-center in the rectangular region, which implies that the direct excitations of $+k_{\text{sp}}$ and $-k_{\text{sp}}$ are no longer simultaneous. In particular, from coupling equations such as Eq. (20), it is verified that $+k_{\text{sp}}$ (or $-k_{\text{sp}}$) is directly excited for θ_i between $-22°$ and $7°$ (or $-7°$ and $22°$).

The intensity $I^{(8)}(\theta_s|\theta_i)$ is shown in Fig. 5.9. In a manner similar to Fig. 5.4A, there is a single-scatter contribution for $\theta_s \leq -36°$ from the 1–1 term. The 2–2 term also makes a strong contribution, but it appears far from backscattering at $-7° \leq \theta_s \leq 22°$. These angular limits are the outward coupling limits of $-k_{\text{sp}}$ as may be readily verified from $q = -k_{\text{sp}} + k_r^{(+)}$ where the superscript indicates that k_r is taken from the positive-k part of $g(k)$, so it is clear that $-k_{\text{sp}}$ is strongly excited. The 1–1 and 2–2 contributions receive corrections from higher-order terms in the same manner discussed in Sect. 5.5.1.

The inset in Fig. 5.9 shows the backscattering region in more detail. There is a backscattering peak that arises from the 4–4 term (more specifically, from the triple integral of Eq. (A7) of Ref. 6). The experimental results of Ref. 8 are also shown for comparison. The experimental peak is somewhat lower and wider then the calculation; these differences can, in a high-order term, easily arise from small errors in σ or ε. Generally, the qualitative agreement is good and, in both theory and experiment, the effect is well-isolated and unambiguous.

The scattering process involves two polaritons and a nonresonant wave as also shown in Fig. 5.9. The incident wave is first roughness-coupled to $-k_{\mathrm{sp}}$ as in $-k_{\mathrm{sp}} = k + k_{r1}^{(-)}$. The polariton is scattered into a nonresonant wave k_{nr}, $k_{\mathrm{nr}} = -k_{\mathrm{sp}} + k_{r2}^{(+)}$, which is then coupled to $+k_{\mathrm{sp}}$ as $+k_{\mathrm{sp}} = k_{\mathrm{nr}} + k_{r3}^{(+)}$; thus the counterpropagating polaritons are linked via a two-step process. Finally $+k_{\mathrm{sp}}$ is outwardly coupled to a propagating wave as $q = +k_{\mathrm{sp}} + k_{r4}^{(-)}$. For the process involving points 1 through 4 in Fig. 5.9, there is another process as shown in which these points are linked in the reverse order; the interference of the contributions from these two processes produces the peak at backscattering.

This interpretation may be verified by using a pole approximation[4] in the 4–4 integrand for the two resonant polariton excitations and evaluating the remaining one-dimensional integral numerically, which produces accurate results. It follows that the width of the peak should be $\Delta\theta_{\varepsilon} \simeq 1.00°$, which is close to the $1.02°$ width of the numerical results of Fig. 5.9. However, these calculations are lengthy and are not presented here.

5.6. Conclusions

A perturbation theory has been presented for the mean diffuse intensity scattered by a surface with one-dimensional Gaussian roughness. All perturbation terms through eighth order in the surface roughness parameter σ have been evaluated without approximation. This approach has the advantage that no physical processes will be omitted and no artifacts will be introduced, apart from whatever effects arise from the unavoidable truncation of the perturbation series.

The theory has been applied to rough metals when σ is a few nanometers and there is strong surface plasmon–polariton excitation. Under these conditions there is often a backscattering enhancement peak, to which many of the perturbation terms contribute. In calculations for a Gaussian roughness power spectrum, it has been shown that the terms of sixth and eighth order predict a broadening of the backscattering peak with increasing surface roughness. For rectangular roughness spectra such as those employed in experiments, it has been discussed that the theory predicts distinct backscattering peaks that first appear in the fourth or even in the eighth perturbation order, depending on the parameters of the spectrum.

The eighth-order terms of the perturbation theory also predict an entirely new effect: a second peak in the mean diffuse scatter that falls at the specular angle. This specular peak, which bears no relation to the surface's specular reflection, appears with twice the angular width of the backscattering peak in their respective leading orders. It has been discussed that the specular peak arises from constructive interference of multiple-scattering contributions, in a manner that resembles the physical processes producing backscattering enhancement. Through supporting calculations based on Monte Carlo techniques, suggestions are made for surface parameters that may lead to the first experimental observations of this new effect.

Acknowledgments. This research was supported by CICESE Internal Project 7329 and CONACYT grant 41947F.

References

1. J.M. Bennett and L. Mattsson, *Introduction to Surface Roughness and Scattering* (Optical Society of America, Washington, D.C., 1989) and references therein.
2. A.R. McGurn, A.A. Maradudin, and V. Celli, "Localization effects in the scattering of light from a randomly rough grating," *Phys. Rev. B* **31**, 4866–4871 (1985).
3. V. Celli, A.A. Maradudin, A.M. Marvin, A.R. McGurn, "Some aspects of light scattering from a randomly rough metal surface," *J. Opt. Soc. Am. A* **2**, 2225–2239 (1985).
4. A.A. Maradudin and E.R. Méndez, "Enhanced backscattering of light from weakly rough, random metal surfaces," *Appl. Opt.* **32**, 3335–3343 (1993).
5. A.R. McGurn and A.A. Maradudin, "Perturbation theory results for the diffuse scattering of light from two-dimensional randomly rough metal surfaces," *Waves Random Media* **6**, 251–267 (1996).
6. K.A. O'Donnell, "High-order perturbation theory for light scattering from a rough metal surface," *J. Opt. Soc. Am. A* **18**, 1507–1518 (2001).
7. C.S. West and K.A. O'Donnell, "Observations of backscattering enhancement from polaritons on a rough metal surface," *J. Opt. Soc. Am. A* **12**, 390–397 (1995).
8. C.S. West and K.A. O'Donnell, "Scattering by plasmon polaritons on a metal surface with a detuned roughness spectrum," *Opt. Lett.* **21**, 1–3 (1996).
9. J.W. Goodman, *Statistical Optics* (Wiley, New York, 1985).
10. A.C. Genz and A.A. Malik, "An adaptive algorithm for numerical integration over an *N*-dimensional rectangular region," *J. Comput. Appl. Math.* **6**, 295–302 (1980).
11. T.R. Michel, "Resonant light scattering from weakly rough random surfaces and imperfect gratings," *J. Opt. Soc. Am. A* **11**, 1874–1885 (1994).
12. K.A. O'Donnell and E. R. Méndez, "Enhanced specular peaks in diffuse light scattering from weakly rough metal surfaces," *J. Opt. Soc. Am. A* **20**, 2338–2346 (2003).
13. K.A. O'Donnell, C.S. West, and E.R. Méndez, "Backscattering enhancement from polariton-polariton coupling on a rough metal surface," *Phys. Rev. B* **57**, 13209–13219 (1998).
14. A.A. Maradudin, A.R. McGurn, and E.R. Méndez, "Surface plasmon polariton mechanism for enhanced backscattering of light from one-dimensional randomly rough metal surfaces," *J. Opt. Soc. Am. A* **12**, 2500–2506 (1995).

6
Small-Amplitude Perturbation Theory for Two-Dimensional Surfaces

GÉRARD BERGINC
Thales Optronique, BP 55, 78233 Guyancourt, France

6.1. Introduction

Electromagnetic wave scattering from randomly rough surfaces and films has emerged as a distinct discipline of considerable research in many areas including radio-physics, geophysical remote sensing, ocean acoustics, surface optics, and plasmonics.[1–14] Wave behavior on rough surfaces is an old subject that has undergone a tremendous transformation in the past ten years. In plasmonics, we study optical processus and their applications for building ultra-small waveguides and highly efficient sensors. Basic building blocks in plasmonics research are metal nanoscale structures that confine light. A main tool for studying such structures is wave scattering simulation for nanoscale objects. The purposes of this chapter are to delineate a coherent outline of the small-amplitude perturbation theory for two-dimensional surfaces and to present both the physical and mathematical framework and the relevant theoretical technique. The small-amplitude perturbation theory was one of the earlier theories used, it was originally developed by Rice.[15] This theory still remains of interest[14,16–19] because perturbative terms of order higher than one can produce enhanced backscattering or improve the accuracy of predictions in an emission model. The Rice method can be used to determine all orders in the perturbative development, but very few works use terms of higher-order for the scattering of electromagnetic waves from two-dimensional rough surfaces due to the calculation complexity. The standard method developed by Rice is difficult to apply when we consider second or third-order of scattered fields as a function of the surface height. The second order has been given in a compact form by Voronovich[20] in his work on the small-slope approximation, and only recently the third-order has been presented.[16] However, a different way[18] exists to obtain small-amplitude perturbation which dates back from the work of Brown *et al.*[21] Using both the Rayleigh hypothesis and the extinction theorem, we can obtain an integral equation, called the reduced Rayleigh equation, which only involves the incident and the scattered field alone. In this method the field transmitted through the surface is eliminated and the scattered field can be expressed as a function of only the incident field. This reduced equation has been extensively used by Maradudin *et al.* to study localization effects by a conducting surface,[22–24] coherent effects in reflection factor,[25] scattering by one-dimensional[26] and two-dimensional conducting surfaces in terms of Green function's development.[27] The phenomenon of backscattering enhancement which has been predicted manifests itself as a well-defined peak in the retro-reflection direction when looking at the angular dependence of the intensity related to the diffuse component of the scattered field. In the case of small-roughness metallic surfaces this peak was explained by the infinite perturbation theory,[22–24] and further developments[26] have shown that the major contribution for the enhanced backscattering peak comes from the second-order term in the field perturbation. Observations[28] of surface backscattering enhancement phenomena from metallic randomly rough surfaces have been reported experimentally and stimulated critical discussions.

Similar studies have been done in the case of thin films or slabs bounded by rough surfaces,[14,29] but only in the two-dimensional case. Early works mainly focused on two-dimensional randomly rough slabs due to the calculation complexity.* Fruitful results were reported in numerical and experimental studies. But to address practical application-oriented situations, scattering theory of three-dimensional rough slabs must be studied. In order to calculate the three-dimensional case it becomes necessary to derive an extension of the reduced Rayleigh equation for this system. In this chapter, we show the existence of a set of four equations, which we also call reduced Rayleigh equation. We derive the corresponding equations for a slab where the boundaries are rough surfaces. Next, a perturbative development up to third-order is obtained in a compact matrix form for this three-dimensional system. This third-order term is mandatory if we need the expression of the scattering cross-section up to fourth-order approximation. As in the one-dimensional case, the results of the incoherent scattering cross-section show a well defined peak in the retroreflection direction.

We determine the perturbative development up to the third-order term in the surface height, for a rough surface alone, and for slabs with rough surfaces. We introduce the Mueller matrix and the definition of the statistical parameters for the rough surface. We then obtain the bistatic matrix in terms of a perturbative development.

Using these results, we show by means of numerical simulations, what are the different mechanisms responsible of the enhanced backscattering. For a metallic surface, the phenomenon is produced by the interference of waves which excite a surface plasmon polariton along a certain path and then follow the same path but in the reverse direction. For a dielectric bounded by a metallic plate with one rough surface, the enhanced backscattering[14] is produced by a similar mechanism where the surface plasmon polariton is replaced by a slab guided wave. In this case the incident wave excites first a guided mode due to surface roughness, and then the roughness transforms the surface wave into a bulk wave. Furthermore,

---

* We mention the work[30] where they consider a three-dimensional film. Their method has in principle the same domain of validity as the small-amplitude perturbation theory, the difference comes from their use of a sophisticated development called Wiener–Hermite expansion to write down the scattered fields.

if the slab supports several guided modes, recent investigations[14,18] have shown the presence of additional peaks, called satellite peaks, in the angular distribution of the incoherent intensity. When the slab is bounded by two rough boundaries, the enhancement of backscattering can be produced by two kinds of interaction: the classic one, where the wave is scattered two times by the same boundary, and another one where the wave is transmitted by the first rough surface, then scattered by the second rough surface and finally by the first one. Before closing this introduction, we want just to add a remark: in this chapter we also try to demonstrate that in metals, such as silver or gold, plasmon–polariton resonances are responsible for unique and remarkable optical phenomena.

6.2. Derivation of the Reduced Rayleigh Equations

The reduced Rayleigh equation was obtained for a two-dimensional surface by Brown *et al.*[21] using the extinction theorem and the Rayleigh hypothesis. This equation allows us to calculate only the scattered field from the rough surface. In order to compute the transmitted field from the rough surface, we use another reduced Rayleigh equation derived by Greffet.[31] However, because these two equations were established in the case where there is no up-going field inside the medium, we cannot use them to obtain the field scattered by a slab with a rough surface on its upper side. To generalize these equations to a slab, we consider all the fields shown in Fig. 6.1. In this case, we derive four reduced Rayleigh equations, which involve the following fields $\boldsymbol{E}_0^-$, $\boldsymbol{E}_0^+$, $\boldsymbol{E}_1^-$, $\boldsymbol{E}_1^+$. We consider that the electromagnetic waves propagate with a frequency ω, and in the following the factor $\exp(-\mathrm{i}\omega t)$ will be omitted. We work with both polarizations and we use a Cartesian coordinate system $\boldsymbol{r} = (\boldsymbol{x}, z) = (x, y, z)$, where the z-axis is directed upwards, and we consider a boundary of the form $z = h(\boldsymbol{x})$. Moreover, we suppose that there exists a length L for which $h(x, y) = 0$, if $|x| > L/2$ or $|y| > L/2$; L may be arbitrary large but finite.

6.2.1. Propagation Equations and Boundary Conditions

The electric field $\boldsymbol{E}$ satisfies the Helmholtz equations in the two media:

$$\left(\nabla^2 + \epsilon_0 K_0^2\right) \boldsymbol{E}^0(\boldsymbol{r}) = 0 \qquad \text{for } z > h(\boldsymbol{x}), \tag{1}$$

$$\left(\nabla^2 + \epsilon_1 K_0^2\right) \boldsymbol{E}^1(\boldsymbol{r}) = 0 \qquad \text{for } z < h(\boldsymbol{x}), \tag{2}$$

where $K_0 = \omega/c$. Since the system is homogeneous in the $\boldsymbol{x} = (x, y)$ directions, we can represent the electric field by its Fourier transform. Using the Helmholtz equation, we deduce the following expression for the electric field[9,13] in the medium 0:

$$\boldsymbol{E}^0(\boldsymbol{r}) = \int \frac{\mathrm{d}^2\boldsymbol{p}}{(2\pi)^2} \boldsymbol{E}^{0-}(\boldsymbol{p}) \exp\left(\mathrm{i}\boldsymbol{k}_p^{0-} \cdot \boldsymbol{r}\right) + \int \frac{\mathrm{d}^2\boldsymbol{p}}{(2\pi)^2} \boldsymbol{E}^{0+}(\boldsymbol{p}) \exp\left(\mathrm{i}\boldsymbol{k}_p^{0+} \cdot \boldsymbol{r}\right), \tag{3}$$

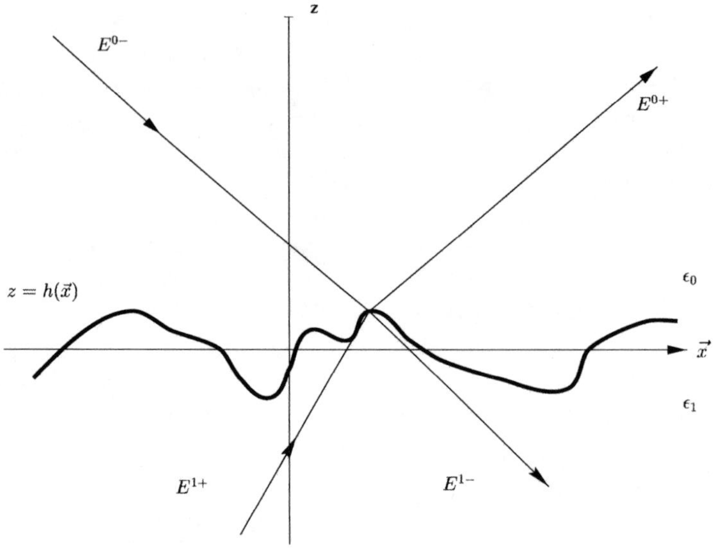

FIGURE 6.1. A randomly rough surface with a system of incident waves coming from both sides of the interface (medium 0 and 1).

where (see Fig. 6.2), we have the expressions

$$\alpha_0(\boldsymbol{p}) \equiv \left(\epsilon_0 K_0^2 - \boldsymbol{p}^2\right)^{\frac{1}{2}}, \tag{4}$$

$$\boldsymbol{k}_{\boldsymbol{p}}^{0\pm} \equiv \boldsymbol{p} \pm \alpha_0(\boldsymbol{p})\hat{\boldsymbol{e}}_z. \tag{5}$$

With this definition, we made implicitly the assumption that the Rayleigh hypothesis is correct. This representation is only valid when $z > \max[h(\boldsymbol{x})]$ and in that case $\boldsymbol{E}^{0-}(\boldsymbol{p})$ represents the incident wave amplitude. In order to be correct we need to add an explicit dependence on the z-coordinate. However, explicit calculations in the case of infinite conducting surfaces,[32] and for a dielectric medium[5] without this hypothesis, have shown that the perturbative developments are identical. The validity of this hypothesis is by no doubt a matter of convergence domain as discussed by Voronovich.[9,20]

In the medium 1, we have a similar expression:

$$\boldsymbol{E}^1(\boldsymbol{r}) = \int \frac{\mathrm{d}^2\boldsymbol{p}}{(2\pi)^2}\boldsymbol{E}^{1-}(\boldsymbol{p})\exp\left(\mathrm{i}\boldsymbol{k}_{\boldsymbol{p}}^{1-}\cdot\boldsymbol{r}\right) + \int \frac{\mathrm{d}^2\boldsymbol{p}}{(2\pi)^2}\boldsymbol{E}^{1+}(\boldsymbol{p})\exp\left(\mathrm{i}\boldsymbol{k}_{\boldsymbol{p}}^{1+}\cdot\boldsymbol{r}\right), \tag{6}$$

where we have

$$\alpha_1(\boldsymbol{p}) \equiv \left(\epsilon_1 K_0^2 - \boldsymbol{p}^2\right)^{\frac{1}{2}}, \tag{7}$$

$$\boldsymbol{k}_{\boldsymbol{p}}^{1\pm} \equiv \boldsymbol{p} \pm \alpha_1(\boldsymbol{p})\hat{\boldsymbol{e}}_z. \tag{8}$$

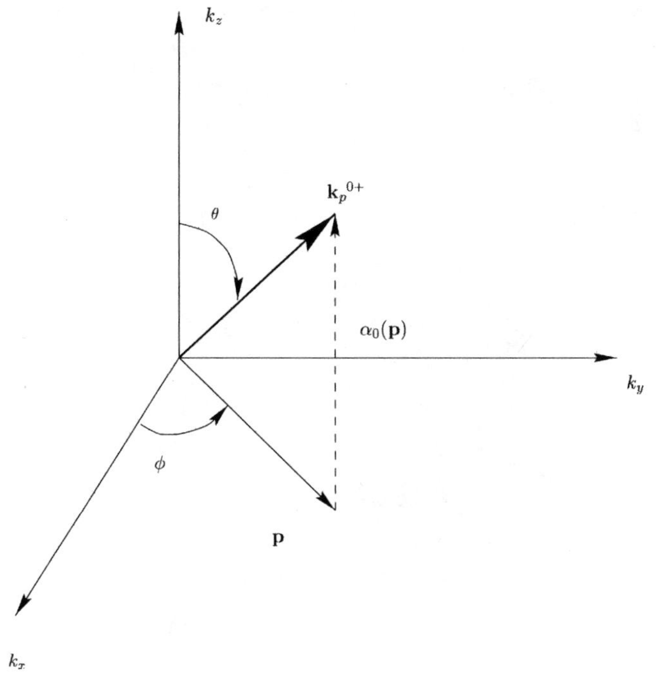

FIGURE 6.2. Decomposition of the wave vector $\boldsymbol{k}_p^{0+}$.

The vectors $\boldsymbol{E}(\boldsymbol{p})$ are decomposed on a two-dimensional basis due to the fact that $\nabla\cdot\boldsymbol{E}(\boldsymbol{r}) = 0$. This leads to the conditions

$$\boldsymbol{k}_p^{0\pm}\cdot\boldsymbol{E}^{0\pm}(\boldsymbol{p}) = 0, \qquad \boldsymbol{k}_p^{1\pm}\cdot\boldsymbol{E}^{1\pm}(\boldsymbol{p}) = 0. \tag{9}$$

We define the horizontal polarization vectors H for TE and V for TM in medium 0 by

$$\hat{\boldsymbol{e}}_H(\boldsymbol{p}) \equiv \frac{\hat{\boldsymbol{e}}_z \times \boldsymbol{k}_p^{0\pm}}{||\hat{\boldsymbol{e}}_z \times \boldsymbol{k}_p^{0\pm}||} = \hat{\boldsymbol{e}}_z \times \hat{\boldsymbol{p}}, \tag{10}$$

$$\hat{\boldsymbol{e}}_V^{0\pm}(\boldsymbol{p}) \equiv \frac{\hat{\boldsymbol{e}}_H(\boldsymbol{p}) \times \boldsymbol{k}_p^{0\pm}}{||\hat{\boldsymbol{e}}_H(\boldsymbol{p}_0) \times \boldsymbol{k}_p^{0\pm}||} = \pm\frac{\alpha_0(\boldsymbol{p})}{\sqrt{\epsilon_0}K_0}\hat{\boldsymbol{p}} - \frac{||\boldsymbol{p}||}{\sqrt{\epsilon_0}K_0}\hat{\boldsymbol{e}}_z, \tag{11}$$

with similar expressions for the medium 1:

$$\hat{\boldsymbol{e}}_H(\boldsymbol{p}) \equiv \frac{\hat{\boldsymbol{e}}_z \times \boldsymbol{k}_p^{1\pm}}{||\hat{\boldsymbol{e}}_z \times \boldsymbol{k}_p^{1\pm}||} = \hat{\boldsymbol{e}}_z \times \hat{\boldsymbol{p}}, \tag{12}$$

$$\hat{\boldsymbol{e}}_V^{1\pm}(\boldsymbol{p}) \equiv \frac{\hat{\boldsymbol{e}}_H(\boldsymbol{p}) \times \boldsymbol{k}_p^{1\pm}}{||\hat{\boldsymbol{e}}_H(\boldsymbol{p}_0) \times \boldsymbol{k}_p^{1\pm}||} = \pm\frac{\alpha_1(\boldsymbol{p})}{\sqrt{\epsilon_1}K_0}\hat{\boldsymbol{p}} - \frac{||\boldsymbol{p}||}{\sqrt{\epsilon_1}K_0}\hat{\boldsymbol{e}}_z. \tag{13}$$

We decompose the waves in medium 0 on the basis $[p]^{0-} \equiv (\hat{\boldsymbol{e}}_V^{0-}(\boldsymbol{p}), \hat{\boldsymbol{e}}_H(\boldsymbol{p}))$, and $[p]^{0+} \equiv (\hat{\boldsymbol{e}}_V^{0+}(\boldsymbol{p}), \hat{\boldsymbol{e}}_H(\boldsymbol{p}))$. In this scheme, the expressions are:

$$\boldsymbol{E}^{0-}(\boldsymbol{p}) = \begin{pmatrix} E_V^{0-}(\boldsymbol{p}) \\ E_H^{0-}(\boldsymbol{p}) \end{pmatrix}_{[p]^{0-}}, \qquad \boldsymbol{E}^{0+}(\boldsymbol{p}) = \begin{pmatrix} E_V^{0+}(\boldsymbol{p}) \\ E_H^{0+}(\boldsymbol{p}) \end{pmatrix}_{[p]^{0+}}, \tag{14}$$

and for medium 1 on the basis $[p]^{1-} \equiv (\hat{\boldsymbol{e}}_V^{1-}(\boldsymbol{p}), \hat{\boldsymbol{e}}_H(\boldsymbol{p}))$ and $[p]^{1+} \equiv (\hat{\boldsymbol{e}}_V^{1+}(\boldsymbol{p}), \hat{\boldsymbol{e}}_H(\boldsymbol{p}))$, the expression are

$$\boldsymbol{E}^{1-}(\boldsymbol{p}) = \begin{pmatrix} E_V^{1-}(\boldsymbol{p}) \\ E_H^{1-}(\boldsymbol{p}) \end{pmatrix}_{[p]^{1-}}, \qquad \boldsymbol{E}^{1+}(\boldsymbol{p}) = \begin{pmatrix} E_V^{1+}(\boldsymbol{p}) \\ E_H^{1+}(\boldsymbol{p}) \end{pmatrix}_{[p]^{1+}}. \tag{15}$$

The electric $\boldsymbol{E}(\boldsymbol{x}, z)$ and magnetic fields $\boldsymbol{B}(\boldsymbol{x}, z) = \frac{c}{i\omega} \nabla \times \boldsymbol{E}(\boldsymbol{x}, z)$ obey the following boundary conditions:

$$\boldsymbol{n}(\boldsymbol{x}) \times \left[\boldsymbol{E}^0(\boldsymbol{x}, h(\boldsymbol{x})) - \boldsymbol{E}^1(\boldsymbol{x}, h(\boldsymbol{x}))\right] = 0, \tag{16}$$

$$\boldsymbol{n}(\boldsymbol{x}) \cdot \left[\epsilon_0 \boldsymbol{E}^0(\boldsymbol{x}, h(\boldsymbol{x})) - \epsilon_1 \boldsymbol{E}^1(\boldsymbol{x}, h(\boldsymbol{x}))\right] = 0, \tag{17}$$

$$\boldsymbol{n}(\boldsymbol{x}) \times \left[\boldsymbol{B}^0(\boldsymbol{x}, h(\boldsymbol{x})) - \boldsymbol{B}^1(\boldsymbol{x}, h(\boldsymbol{x}))\right] = 0, \tag{18}$$

$$\boldsymbol{n}(\boldsymbol{x}) \equiv \hat{\boldsymbol{e}}_z - \nabla h(\boldsymbol{x}) .$$

Substituting the field Fourier transforms, Eq. (3) and Eq. (6), into the boundary conditions Eqs. (16)–(18), we obtain the following expressions:

$$\sum_{a=\pm} \int \frac{d^2 \boldsymbol{p}}{(2\pi)^2} \boldsymbol{n}(\boldsymbol{x}) \times \boldsymbol{E}^{0a}(\boldsymbol{p}) \exp\left(i\boldsymbol{k}_{\boldsymbol{p}}^{0a} \cdot \boldsymbol{r}_{\boldsymbol{x}}\right)$$
$$= \sum_{a=\pm} \int \frac{d^2 \boldsymbol{p}}{(2\pi)^2} \boldsymbol{n}(\boldsymbol{x}) \times \boldsymbol{E}^{1a}(\boldsymbol{p}) \exp\left(i\boldsymbol{k}_{\boldsymbol{p}}^{1a} \cdot \boldsymbol{r}_{\boldsymbol{x}}\right), \tag{19}$$

$$\frac{\epsilon_0}{\epsilon_1} \sum_{a=\pm} \int \frac{d^2 \boldsymbol{p}}{(2\pi)^2} \boldsymbol{n}(\boldsymbol{x}) \cdot \boldsymbol{E}^{0a}(\boldsymbol{p}) \exp\left(i\boldsymbol{k}_{\boldsymbol{p}}^{0a} \cdot \boldsymbol{r}_{\boldsymbol{x}}\right)$$
$$= \sum_{a=\pm} \int \frac{d^2 \boldsymbol{p}}{(2\pi)^2} \boldsymbol{n}(\boldsymbol{x}) \cdot \boldsymbol{E}^{1a}(\boldsymbol{p}) \exp\left(i\boldsymbol{k}_{\boldsymbol{p}}^{1a} \cdot \boldsymbol{r}_{\boldsymbol{x}}\right), \tag{20}$$

$$\sum_{a=\pm} \int \frac{d^2 \boldsymbol{p}}{(2\pi)^2} \boldsymbol{n}(\boldsymbol{x}) \times \left[\boldsymbol{k}_{\boldsymbol{p}}^{0a} \times \boldsymbol{E}^{0a}(\boldsymbol{p})\right] \exp\left(i\boldsymbol{k}_{\boldsymbol{p}}^{0a} \cdot \boldsymbol{r}_{\boldsymbol{x}}\right)$$
$$= \sum_{a=\pm} \int \frac{d^2 \boldsymbol{p}}{(2\pi)^2} \boldsymbol{n}(\boldsymbol{x}) \times \left[\boldsymbol{k}_{\boldsymbol{p}}^{1a} \times \boldsymbol{E}^{1a}(\boldsymbol{p})\right] \exp\left(i\boldsymbol{k}_{\boldsymbol{p}}^{1a} \cdot \boldsymbol{r}_{\boldsymbol{x}}\right), \tag{21}$$

$$\boldsymbol{r}_{\boldsymbol{x}} = \boldsymbol{x} + h(\boldsymbol{x})\hat{\boldsymbol{e}}_z, \qquad \boldsymbol{k}_{\boldsymbol{p}}^{0a} \equiv \boldsymbol{p} + a\alpha_0(\boldsymbol{p})\hat{\boldsymbol{e}}_z,$$
$$\boldsymbol{k}_{\boldsymbol{p}}^{1a} \equiv \boldsymbol{p} + a\alpha_1(\boldsymbol{p})\hat{\boldsymbol{e}}_z, \tag{22}$$

where the summation includes the two possible signs, $a = \pm$, linked to the propagation directions. We will also use the condition $\nabla \cdot \boldsymbol{E}^0(\boldsymbol{x}, z) = \nabla \cdot \boldsymbol{E}^1(\boldsymbol{x}, z)$ which yields the relation

$$\sum_{a=\pm}\int \frac{d^2p}{(2\pi)^2}\boldsymbol{k}_p^{0a}\cdot \boldsymbol{E}^{0a}(\boldsymbol{p})\exp\left(i\boldsymbol{k}_p^{0a}\cdot\boldsymbol{r}_x\right)=\sum_{a=\pm}\int \frac{d^2p}{(2\pi)^2}\boldsymbol{k}_p^{1a}\cdot \boldsymbol{E}^{1a}(\boldsymbol{p})\exp\left(i\boldsymbol{k}_p^{1a}\cdot\boldsymbol{r}_x\right). \tag{23}$$

6.2.2. Field Elimination

Equations (19)–(21) and (23) are linear in the fields $\boldsymbol{E}^{0-}$, $\boldsymbol{E}^{0+}$, $\boldsymbol{E}^{1-}$, $\boldsymbol{E}^{1+}$. To eliminate $\boldsymbol{E}^{1-}$ or $\boldsymbol{E}^{1+}$ in these equations, we shall take the following linear combination of their left and right members:

$$\int d^2x\,\left[\boldsymbol{k}_u^{1b}\times(\text{Eq. (19)})+(\text{Eq. (21)})-\boldsymbol{k}_u^{1b}(\text{Eq. (20)})\right.$$
$$\left.-\boldsymbol{n}(\boldsymbol{x})(\text{Eq. (23)})\right]\exp\left(-i\boldsymbol{k}_u^{1b}\cdot\boldsymbol{r}_x\right), \tag{24}$$

with $\boldsymbol{k}_u^{1b}\equiv\boldsymbol{u}+b\alpha_1(\boldsymbol{u})\hat{\boldsymbol{e}}_z$, and where we must fix $b=\pm$ according to the choice of the field we want to eliminate. Using the vectorial identity, $\boldsymbol{a}\times(\boldsymbol{b}\times\boldsymbol{c})=\boldsymbol{b}(\boldsymbol{a}\cdot\boldsymbol{c})-\boldsymbol{c}(\boldsymbol{a}\cdot\boldsymbol{b})$, we get for the right member of Eq. (24):

$$\sum_{a=\pm}\iint d^2x\frac{d^2p}{(2\pi)^2}\left[-\left(\boldsymbol{k}_u^{1b}+\boldsymbol{k}_p^{1a}\right)\cdot\boldsymbol{n}(\boldsymbol{x})\,\boldsymbol{E}^{1a}(\boldsymbol{p})+\left(\boldsymbol{k}_u^{1b}-\boldsymbol{k}_p^{1a}\right)\cdot\boldsymbol{E}^{1a}(\boldsymbol{p})\,\boldsymbol{n}(\boldsymbol{x})\right.$$
$$\left.-\boldsymbol{n}(\boldsymbol{x})\cdot\boldsymbol{E}^{1a}(\boldsymbol{p})\left(\boldsymbol{k}_u^{1b}-\boldsymbol{k}_p^{1a}\right)\right]\exp\left(-i\left(\boldsymbol{k}_u^{1b}-\boldsymbol{k}_p^{1a}\right)\cdot\boldsymbol{r}_x\right). \tag{25}$$

We discuss now the different cases depending on the relative signs of a and b:

(1) If $a=-b$, we can use an integration by parts (given in Appendix 6.7) to evaluate $\boldsymbol{n}(\boldsymbol{x})\equiv\hat{\boldsymbol{e}}_z-\nabla h(\boldsymbol{x})$. This leads to

$$\boldsymbol{n}(\boldsymbol{x})=\hat{\boldsymbol{e}}_z-\nabla h(\boldsymbol{x})\longleftrightarrow\boldsymbol{n}(\boldsymbol{x})=\hat{\boldsymbol{e}}_z+\frac{(\boldsymbol{u}-\boldsymbol{p})}{(b\alpha_1(\boldsymbol{u})-a\alpha_1(\boldsymbol{p}))}. \tag{26}$$

We can notice that the denominator $(b\alpha_1(\boldsymbol{u})-a\alpha_1(\boldsymbol{p}))$ does not present any singularity because $a=-b$. For the first term in integral (25) we get

$$-\left(\boldsymbol{k}_u^{1b}+\boldsymbol{k}_p^{1a}\right)\cdot\boldsymbol{n}(\boldsymbol{x})\,\boldsymbol{E}^{1a}(\boldsymbol{p})=\frac{-b\boldsymbol{E}^{1a}(\boldsymbol{p})}{(\alpha_1(\boldsymbol{u})+\alpha_1(\boldsymbol{p}))}\left[\boldsymbol{u}^2-\boldsymbol{p}^2+\alpha_1(\boldsymbol{u})^2-\alpha_1(\boldsymbol{p})^2\right],$$
$$=0. \tag{27}$$

Using Eq. (7), the last equality can be easily verified. For the sum of the second and third terms of Eq. (25), we obtain

$$\left(\boldsymbol{k}_u^{1b}-\boldsymbol{k}_p^{1a}\right)\cdot\boldsymbol{E}^{1a}(\boldsymbol{p})\,\boldsymbol{n}(\boldsymbol{x})-\boldsymbol{n}(\boldsymbol{x})\cdot\boldsymbol{E}^{1a}(\boldsymbol{p})\left(\boldsymbol{k}_u^{1b}-\boldsymbol{k}_p^{1a}\right)=0, \tag{28}$$

due to

$$\boldsymbol{n}(\boldsymbol{x})=\frac{\boldsymbol{k}_u^{1b}-\boldsymbol{k}_p^{1a}}{b\alpha_1(\boldsymbol{u})-a\alpha_1(\boldsymbol{p})}. \tag{29}$$

(2) If $a=b$, we can use the integration by parts only if $\alpha_1(\boldsymbol{u})\neq\alpha_1(\boldsymbol{p})$. And we have to consider three cases:
   (a) $\boldsymbol{u}\neq\boldsymbol{p}$ and $\boldsymbol{u}\neq-\boldsymbol{p}$, similarly as in the first case by using an integration by parts, we show that Eq. (25) is zero.

(b) $\boldsymbol{u} = \boldsymbol{p}$, then $\boldsymbol{k}_{\boldsymbol{u}}^{1b} = \boldsymbol{k}_{\boldsymbol{p}}^{1a}$:

$$\begin{aligned}
&-\int \mathrm{d}^2\boldsymbol{x}\left(\boldsymbol{k}_{\boldsymbol{u}}^{1b}+\boldsymbol{k}_{\boldsymbol{p}}^{1a}\right)\cdot\boldsymbol{n}(\boldsymbol{x})\,\boldsymbol{E}^{1a}(\boldsymbol{p})\exp\left(-i\left(\boldsymbol{k}_{\boldsymbol{u}}^{1b}-\boldsymbol{k}_{\boldsymbol{p}}^{1a}\right)\cdot\boldsymbol{r}_{\boldsymbol{x}}\right)\\
&\quad=-\int \mathrm{d}^2\boldsymbol{x}2\boldsymbol{k}_{\boldsymbol{u}}^{1b}\cdot\boldsymbol{n}(\boldsymbol{x})\,\boldsymbol{E}^{1a}(\boldsymbol{p})\\
&\quad=-2b\alpha_1(\boldsymbol{u})\int \mathrm{d}^2\boldsymbol{x}\boldsymbol{E}^{1a}(\boldsymbol{p})\\
&\quad=-2b\alpha_1(\boldsymbol{u})\,L^2\boldsymbol{E}^{1a}(\boldsymbol{p}),
\end{aligned}\tag{30}$$

because $\int \mathrm{d}^2\boldsymbol{x}\nabla h(\boldsymbol{x}) = 0$, and

$$\left(\boldsymbol{k}_{\boldsymbol{u}}^{1b}-\boldsymbol{k}_{\boldsymbol{p}}^{1a}\right)\cdot\boldsymbol{E}^{1a}(\boldsymbol{p})\boldsymbol{n}(\boldsymbol{x})-\boldsymbol{n}(\boldsymbol{x})\cdot\boldsymbol{E}^{1a}(\boldsymbol{p})\left(\boldsymbol{k}_{\boldsymbol{u}}^{1b}-\boldsymbol{k}_{\boldsymbol{p}}^{1a}\right)=0. \tag{31}$$

(c) $\boldsymbol{u} = -\boldsymbol{p} \neq 0$, then $\boldsymbol{k}_{\boldsymbol{u}}^{1b} - \boldsymbol{k}_{\boldsymbol{p}}^{1a} = 2\boldsymbol{u}$,

$$\begin{aligned}
&\int \mathrm{d}^2\boldsymbol{x}\,\boldsymbol{n}(\boldsymbol{x})\exp\left(-\mathrm{i}\left(\boldsymbol{k}_{\boldsymbol{u}}^{1b}-\boldsymbol{k}_{\boldsymbol{p}}^{1a}\right)\cdot\boldsymbol{r}_{\boldsymbol{x}}\right)=\int \mathrm{d}\boldsymbol{x}\,(\hat{\boldsymbol{e}}_z-\nabla h(\boldsymbol{x}))\exp(-2\mathrm{i}\boldsymbol{u}\cdot\boldsymbol{x})\\
&=\hat{\boldsymbol{e}}_z\int \mathrm{d}^2\boldsymbol{x}\exp(-2\mathrm{i}\boldsymbol{u}\cdot\boldsymbol{x})\boldsymbol{x}-\hat{\boldsymbol{e}}_x\int \mathrm{d}y\left[\exp(-2\mathrm{i}\boldsymbol{u}\cdot\boldsymbol{x})h(x,y)\right]_{x=-L/2}^{x=L/2}\\
&\quad-\hat{\boldsymbol{e}}_y\int \mathrm{d}x\left[\exp(-2i\boldsymbol{u}\cdot\boldsymbol{x})h(x,y)\right]_{y=-L/2}^{y=L/2}\\
&=\hat{\boldsymbol{e}}_z\delta(\boldsymbol{u})\qquad\text{when}\quad L\to+\infty\\
&=0\qquad\text{since}\qquad \boldsymbol{u}\neq 0.
\end{aligned}\tag{32}$$

This result shows us that expression (25) is also zero in that case.

We can now write the above results in the following form:

$$\begin{aligned}
&-\int \mathrm{d}^2\boldsymbol{x}\left(\boldsymbol{k}_{\boldsymbol{u}}^{1b}+\boldsymbol{k}_{\boldsymbol{p}}^{1a}\right)\cdot\boldsymbol{n}(\boldsymbol{x})\,\boldsymbol{E}^{1a}(\boldsymbol{p})\exp\left(-\mathrm{i}\left(\boldsymbol{k}_{\boldsymbol{u}}^{1b}-\boldsymbol{k}_{\boldsymbol{p}}^{1a}\right)\cdot\boldsymbol{r}_{\boldsymbol{x}}\right)\\
&=-2b\alpha_1(\boldsymbol{u})\,\delta_{a,b}\delta_{\boldsymbol{u},\boldsymbol{p}}L^2\boldsymbol{E}^{1a}(\boldsymbol{p})\\
&=-2b\alpha_1(\boldsymbol{u})\,\delta_{a,b}\,(2\pi)^2\delta(\boldsymbol{u}-\boldsymbol{p})\boldsymbol{E}^{1b}(\boldsymbol{u})\\
&\quad\text{when}\quad L\to+\infty,
\end{aligned}\tag{33}$$

where $\delta_{\boldsymbol{u},\boldsymbol{p}}$ is the kronecker symbol, and $\delta(\boldsymbol{u}-\boldsymbol{p}) = L^2/(2\pi)^2\,\delta_{\boldsymbol{u},\boldsymbol{p},}$ is the Dirac function.

After an integration on $\boldsymbol{p}$ and a summation on a, expression (25) leads to

$$-2b\alpha_1(\boldsymbol{u})\boldsymbol{E}^{1b}(\boldsymbol{u}). \tag{34}$$

We see that we can eliminate the field $\boldsymbol{E}^{1-}(\boldsymbol{u})$ or $\boldsymbol{E}^{1+}(\boldsymbol{u})$ depending on the choice made for $b = \pm$.

Now, we consider the left member of Eq. (24), we can write

$$\sum_{a=\pm}\iint \mathrm{d}^2\boldsymbol{x}\frac{\mathrm{d}^2\boldsymbol{p}}{(2\pi)^2}\ \Big[-\left(\boldsymbol{k}_{\boldsymbol{u}}^{1b}+\boldsymbol{k}_{\boldsymbol{p}}^{0a}\right)\cdot\boldsymbol{n}(\boldsymbol{x})\,\boldsymbol{E}^{0a}(\boldsymbol{p})+\left(\boldsymbol{k}_{\boldsymbol{u}}^{1b}-\boldsymbol{k}_{\boldsymbol{p}}^{0a}\right)\cdot\boldsymbol{E}^{0a}(\boldsymbol{p})\,\boldsymbol{n}(\boldsymbol{x})$$
$$-\boldsymbol{n}(\boldsymbol{x})\cdot\boldsymbol{E}^{0a}(\boldsymbol{p})\left(\frac{\epsilon_0}{\epsilon_1}\boldsymbol{k}_{\boldsymbol{u}}^{1b}-\boldsymbol{k}_{\boldsymbol{p}}^{0a}\right)\Big]\exp\left(-i\left(\boldsymbol{k}_{\boldsymbol{u}}^{1b}-\boldsymbol{k}_{\boldsymbol{p}}^{0a}\right)\cdot\boldsymbol{r}_{\boldsymbol{x}}\right). \tag{35}$$

Using an integration by parts, we replace $\boldsymbol{n}(\boldsymbol{x})$ by (26), and we get

$$\boldsymbol{n}(\boldsymbol{x})\longleftrightarrow\hat{\boldsymbol{e}}_z+\frac{(\boldsymbol{u}-\boldsymbol{p})}{(b\alpha_1(\boldsymbol{u})-a\alpha_0(\boldsymbol{p}))}=\frac{\boldsymbol{k}_{\boldsymbol{u}}^{1b}-\boldsymbol{k}_{\boldsymbol{p}}^{0a}}{(b\alpha_1(\boldsymbol{u})-a\alpha_0(\boldsymbol{p}))}. \tag{36}$$

In this case, we do not need to discuss the relative sign between a and b because we have $b\alpha_1(\boldsymbol{u})-a\alpha_0(\boldsymbol{p})\neq 0$, because $\epsilon_0\neq\epsilon_1$. We obtain

$$\begin{aligned}
-\left(\boldsymbol{k}_{\boldsymbol{u}}^{1b}+\boldsymbol{k}_{\boldsymbol{p}}^{1a}\right)\cdot\boldsymbol{n}(\boldsymbol{x})\,\boldsymbol{E}^{0a}(\boldsymbol{p}) &= -\frac{\boldsymbol{u}^2-\boldsymbol{p}^2+\alpha_1(\boldsymbol{u})^2-\alpha_0(\boldsymbol{p})^2}{b\alpha_1(\boldsymbol{u})-a\alpha_0(\boldsymbol{p})}\,\boldsymbol{E}^{0a}(\boldsymbol{p}),\\
&= -\frac{(\epsilon_1-\epsilon_0)K_0^2}{b\alpha_1(\boldsymbol{u})-a\alpha_0(\boldsymbol{p})}\,\boldsymbol{E}^{0a}(\boldsymbol{p}),
\end{aligned} \tag{37}$$

where we use definitions (4) and (7).

The remaining terms of Eq. (35) lead to

$$\begin{aligned}
&\left(\boldsymbol{k}_{\boldsymbol{u}}^{1b}-\boldsymbol{k}_{\boldsymbol{p}}^{0a}\right)\cdot\boldsymbol{E}^{0a}(\boldsymbol{p})\,\boldsymbol{n}(\boldsymbol{x})-\boldsymbol{n}(\boldsymbol{x})\cdot\boldsymbol{E}^{0a}(\boldsymbol{p})\left(\frac{\epsilon_0}{\epsilon_1}\boldsymbol{k}_{\boldsymbol{u}}^{1b}-\boldsymbol{k}_{\boldsymbol{p}}^{0a}\right)\\
&=\left(\boldsymbol{k}_{\boldsymbol{u}}^{1b}-\boldsymbol{k}_{\boldsymbol{p}}^{0a}\right)\cdot\boldsymbol{E}^{0a}(\boldsymbol{p})\,\boldsymbol{n}(\boldsymbol{x})-\boldsymbol{n}(\boldsymbol{x})\cdot\boldsymbol{E}^{0a}(\boldsymbol{p})\left(\boldsymbol{k}_{\boldsymbol{u}}^{1b}-\boldsymbol{k}_{\boldsymbol{p}}^{0a}\right)\\
&\quad+\boldsymbol{n}(\boldsymbol{x})\cdot\boldsymbol{E}^{0a}(\boldsymbol{p})\left(\boldsymbol{k}_{\boldsymbol{u}}^{1b}-\frac{\epsilon_0}{\epsilon_1}\boldsymbol{k}_{\boldsymbol{p}}^{1b}\right)\\
&=\frac{\boldsymbol{k}_{\boldsymbol{u}}^{1b}-\boldsymbol{k}_{\boldsymbol{p}}^{0a}}{b\alpha_1(\boldsymbol{u})-a\alpha_0(\boldsymbol{p})}\cdot\boldsymbol{E}^{0a}(\boldsymbol{p})\frac{(\epsilon_1-\epsilon_0)}{\epsilon_1}\boldsymbol{k}_{\boldsymbol{u}}^{1b}.
\end{aligned} \tag{38}$$

With the following expression:

$$I(\alpha|\boldsymbol{p})\equiv\int\mathrm{d}^2\boldsymbol{x}\ \exp(-\mathrm{i}\boldsymbol{p}\cdot\boldsymbol{x}-\mathrm{i}\alpha\,h(\boldsymbol{x})), \tag{39}$$

and taking into account expressions (34), (37)–(38), the resulting linear combination (24) leads to

$$\begin{aligned}
&\sum_{a=\pm}\int\frac{d^2p}{(2\pi)^2}\ \frac{I(b\alpha_1(\boldsymbol{u})-a\alpha_0(\boldsymbol{p})|\boldsymbol{u}-\boldsymbol{p})}{b\alpha_1(\boldsymbol{u})-a\alpha_0(\boldsymbol{p})}\\
&\qquad\times\left[K_0^2\,\boldsymbol{E}^{0a}(\boldsymbol{p})-\frac{\boldsymbol{k}_{\boldsymbol{u}}^{1b}}{\epsilon_1}\left(\boldsymbol{k}_{\boldsymbol{u}}^{1b}-\boldsymbol{k}_{\boldsymbol{p}}^{0a}\right)\cdot\boldsymbol{E}^{0a}(\boldsymbol{p})\right]\\
&\qquad=\frac{2\,b\,\alpha_1(\boldsymbol{u})}{(\epsilon_1-\epsilon_0)}\,\boldsymbol{E}^{1b}(\boldsymbol{u}).
\end{aligned} \tag{40}$$

This expression represents in fact four equations, depending on the choice for a and $b=\pm$. The last step is to project Eq. (40) on the natural basis of $\boldsymbol{E}^{1b}(\boldsymbol{u})$, which is given by $[\boldsymbol{u}]^{1b} \equiv (\hat{\boldsymbol{e}}_V^{1b}(\boldsymbol{u}), \hat{\boldsymbol{e}}_H(\boldsymbol{u}))$. This basis has the property to be orthogonal to $\boldsymbol{k}_{\boldsymbol{u}}^{1b}$ and to eliminate the second term of Eg. (40). Let us notice, that in order to decompose $\boldsymbol{E}^{0a}(\boldsymbol{p})$ on $[\boldsymbol{p}]^{0a}$, we have to define a matrix $\overline{\boldsymbol{M}}^{1b,0a}(\boldsymbol{u}|\boldsymbol{p})$ transforming a vector expressed on the basis $[\boldsymbol{p}]^{0a}$ into a vector on the basis $[\boldsymbol{u}]^{1b}$, multiplied by a numerical factor $(\epsilon_0\,\epsilon_1)^{\frac{1}{2}}\,K_0^2$. This matrix is given by the expression

$$\overline{\boldsymbol{M}}^{1b,0a}(\boldsymbol{u}|\boldsymbol{p}) \equiv (\epsilon_0\,\epsilon_1)^{\frac{1}{2}}\,K_0^2 \begin{pmatrix} \hat{\boldsymbol{e}}_V^{1b}(\boldsymbol{u})\cdot\hat{\boldsymbol{e}}_V^{0a}(\boldsymbol{p}) & \hat{\boldsymbol{e}}_V^{1b}(\boldsymbol{u})\cdot\hat{\boldsymbol{e}}_H(\boldsymbol{p}) \\ \hat{\boldsymbol{e}}_H(\boldsymbol{u})\cdot\hat{\boldsymbol{e}}_V^{0a}(\boldsymbol{p}) & \hat{\boldsymbol{e}}_H(\boldsymbol{u})\cdot\hat{\boldsymbol{e}}_H(\boldsymbol{p}) \end{pmatrix}. \tag{41}$$

6.2.3. *The Reduced Rayleigh Equations*

Now, we will go into more detailed discussion of the reduced Rayleigh equations. Equations (10)–(13) imply that the matrix $\overline{\boldsymbol{M}}$ takes the form

$$\overline{\boldsymbol{M}}^{1b,0a}(\boldsymbol{u}|\boldsymbol{p}) = \begin{pmatrix} ||\boldsymbol{u}||\,||\boldsymbol{p}|| + ab\alpha_1(\boldsymbol{u})\alpha_0(\boldsymbol{p})\,\hat{\boldsymbol{u}}\cdot\hat{\boldsymbol{p}} & -b\,\epsilon_0^{\frac{1}{2}}\,K_0\,\alpha_1(\boldsymbol{u})\,(\hat{\boldsymbol{u}}\times\hat{\boldsymbol{p}})_z \\ a\,\epsilon_1^{\frac{1}{2}}\,K_0\,\alpha_0(\boldsymbol{p})\,(\hat{\boldsymbol{u}}\times\hat{\boldsymbol{p}})_z & (\epsilon_0\,\epsilon_1)^{\frac{1}{2}}\,K_0^2\,\hat{\boldsymbol{u}}\cdot\hat{\boldsymbol{p}} \end{pmatrix}. \tag{42}$$

The reduced Rayleigh equation resulting from Eq. (40) can be expanded as follows:

$$\sum_{a=\pm}\int \frac{d^2\boldsymbol{p}}{(2\pi)^2}\frac{I(b\alpha_1(\boldsymbol{u}) - a\alpha_0(\boldsymbol{p})|\boldsymbol{u}-\boldsymbol{p})}{b\alpha_1(\boldsymbol{u}) - a\alpha_0(\boldsymbol{p})}\overline{\boldsymbol{M}}^{1b,0a}(\boldsymbol{u}|\boldsymbol{p})\,\boldsymbol{E}^{0a}(\boldsymbol{p})$$
$$= \frac{2\,b\,(\epsilon_0\,\epsilon_1)^{\frac{1}{2}}\,\alpha_1(\boldsymbol{u})}{(\epsilon_1 - \epsilon_0)}\,\boldsymbol{E}^{1b}(\boldsymbol{u}), \tag{43}$$

where we suppose that $\boldsymbol{E}^{0a}(\boldsymbol{p})$, $\boldsymbol{E}^{1b}(\boldsymbol{u})$ are respectively decomposed on the basis $[\boldsymbol{p}]^{0a}$ and $[\boldsymbol{u}]^{1b}$. We can derive a similar equation where $\boldsymbol{E}^{0b}$ is now eliminated, by exchanging ϵ_0 and ϵ_1 in (42) and (43), owing to the symmetry of the equation, we get

$$\sum_{a=\pm}\int \frac{d^2\boldsymbol{p}}{(2\pi)^2}\frac{I(b\alpha_0(\boldsymbol{u}) - a\alpha_1(\boldsymbol{p})|\boldsymbol{u}-\boldsymbol{p})}{b\alpha_0(\boldsymbol{u}) - a\alpha_1(\boldsymbol{p})}\overline{\boldsymbol{M}}^{0b,1a}(\boldsymbol{u}|\boldsymbol{p})\,\boldsymbol{E}^{1a}(\boldsymbol{p})$$
$$= -\frac{2\,b\,(\epsilon_0\,\epsilon_1)^{\frac{1}{2}}\,\alpha_1(\boldsymbol{u})}{(\epsilon_1 - \epsilon_0)}\,\boldsymbol{E}^{0b}(\boldsymbol{u}), \tag{44}$$

$$\overline{\boldsymbol{M}}^{0b,1a}(\boldsymbol{u}|\boldsymbol{p}) = \begin{pmatrix} ||\boldsymbol{u}||\,||\boldsymbol{p}|| + ab\alpha_0(\boldsymbol{u})\alpha_1(\boldsymbol{p})\,\hat{\boldsymbol{u}}\cdot\hat{\boldsymbol{p}} & -b\,\epsilon_1^{\frac{1}{2}}\,K_0\,\alpha_0(\boldsymbol{u})\,(\hat{\boldsymbol{u}}\times\hat{\boldsymbol{p}})_z \\ a\,\epsilon_0^{\frac{1}{2}}\,K_0\,\alpha_1(\boldsymbol{p})\,(\hat{\boldsymbol{u}}\times\hat{\boldsymbol{p}})_z & (\epsilon_0\,\epsilon_1)^{\frac{1}{2}}\,K_0^2\,\hat{\boldsymbol{u}}\cdot\hat{\boldsymbol{p}} \end{pmatrix}. \tag{45}$$

We will show how these equations greatly simplify the perturbative calculations of plane wave scattering by a rough surface. Another helpful compact notation is given by the following new matrices $\overline{\boldsymbol{M}}_h$:

$$\overline{\boldsymbol{M}}_h^{1b,0a}(\boldsymbol{u}|\boldsymbol{p}) \equiv \frac{I(b\alpha_1(\boldsymbol{u}) - a\alpha_0(\boldsymbol{p})|\boldsymbol{u}-\boldsymbol{p})}{b\alpha_1(\boldsymbol{u}) - a\alpha_0(\boldsymbol{p})}\overline{\boldsymbol{M}}^{1b,0a}(\boldsymbol{u}|\boldsymbol{p}), \tag{46}$$

$$\overline{\boldsymbol{M}}_h^{0b,1a}(\boldsymbol{u}|\boldsymbol{p}) \equiv \frac{I(b\alpha_0(\boldsymbol{u}) - a\alpha_1(\boldsymbol{p})|\boldsymbol{u}-\boldsymbol{p})}{b\alpha_0(\boldsymbol{u}) - a\alpha_1(\boldsymbol{p})}\overline{\boldsymbol{M}}^{0b,1a}(\boldsymbol{u}|\boldsymbol{p}). \tag{47}$$

6.3. The Diffusion Matrix

We consider the scattering of an incident plane wave by a rough surface. From the previous formalism, we define an incident plane wave defined by a wave vector $\boldsymbol{k}_{\boldsymbol{p}_0}^{0-}$:

$$\boldsymbol{E}^{0-}(\boldsymbol{p}) = (2\pi)^2\,\delta(\boldsymbol{p} - \boldsymbol{p}_0)\boldsymbol{E}^i(\boldsymbol{p}_0). \tag{48}$$

We define the diffusion operator $\overline{\boldsymbol{R}}$:

$$\boldsymbol{E}^{0+}(\boldsymbol{p}) \equiv \overline{\boldsymbol{R}}(\boldsymbol{p}\,|\,\boldsymbol{p}_0)\cdot\boldsymbol{E}^i(\boldsymbol{p}_0). \tag{49}$$

We represent the diffusion operator in a matrix form, using the vectorial basis described above. This leads to the expression

$$\overline{\boldsymbol{R}}(\boldsymbol{p}\,|\,\boldsymbol{p}_0) = \begin{pmatrix} R_{\mathrm{VV}}(\boldsymbol{p}\,|\,\boldsymbol{p}_0) & R_{\mathrm{VH}}(\boldsymbol{p}\,|\,\boldsymbol{p}_0) \\ R_{\mathrm{HV}}(\boldsymbol{p}\,|\,\boldsymbol{p}_0) & R_{\mathrm{HH}}(\boldsymbol{p}\,|\,\boldsymbol{p}_0) \end{pmatrix}_{[p_0^-]\to[p^+]}.$$

We can now write the field in medium 0 using decomposition (3):

$$\boldsymbol{E}^0(\boldsymbol{r}) = \boldsymbol{E}^i(\boldsymbol{p}_0)\exp\left(\mathrm{i}\boldsymbol{k}_{\boldsymbol{p}_0}^{0-}\cdot\boldsymbol{r}\right) + \int \frac{\mathrm{d}^2\boldsymbol{p}}{(2\pi)^2}\overline{\boldsymbol{R}}(\boldsymbol{p}\,|\,\boldsymbol{p}_0)\cdot\boldsymbol{E}^i(\boldsymbol{p}_0)\exp\left(\mathrm{i}\boldsymbol{k}_{\boldsymbol{p}}^{0+}\cdot\boldsymbol{r}\right). \tag{50}$$

6.4. A Perturbative Development

Our aim now is to obtain a perturbative development, thus we have to determine a perturbative expansion of the given boundary problem. A direct approach which uses an exact integral equation named the extended boundary condition (EBC) (see Ref.[5,32]) requires tedious calculations. Another issue is to use the Rayleigh hypothesis in the boundary conditions. This is the method generally used to obtain the small-amplitude development.[15,16,20] It is very important to notice that a great deal of simplifications can be achieved if we are only interested by the field outside the slab. Brown *et al.*[21] show that an exact integral equation can be obtained (under the Rayleigh hypothesis), which only involves the scattering matrix $\overline{\boldsymbol{R}}(\boldsymbol{p}\,|\,\boldsymbol{p}_0)$. The proof is based on the extinction theorem which decouples the fields inside and outside the media. In this section, this integral equation will be obtained from the previous development, including a generalization to the case of bounded media.

6.4.1. *Case of One Rough Surface*

Working in the framework of a small perturbation regime, we look for an asymptotic perturbative development of $\overline{\boldsymbol{R}}$ in powers of the height h:

$$\overline{\boldsymbol{R}}(\boldsymbol{p}\,|\,\boldsymbol{p}_0) = \overline{\boldsymbol{R}}^{(0)}(\boldsymbol{p}\,|\,\boldsymbol{p}_0) + \overline{\boldsymbol{R}}^{(1)}(\boldsymbol{p}\,|\,\boldsymbol{p}_0) + \overline{\boldsymbol{R}}^{(2)}(\boldsymbol{p}\,|\,\boldsymbol{p}_0) + \overline{\boldsymbol{R}}^{(3)}(\boldsymbol{p}\,|\,\boldsymbol{p}_0) + \cdots. \tag{51}$$

We can easily prove that this development takes the following form (see Appendix B):

$$\begin{aligned}\overline{\boldsymbol{R}}(\boldsymbol{p}\,|\,\boldsymbol{p}_0) &= (2\pi)^2\delta(\boldsymbol{p}-\boldsymbol{p}_0)\,\overline{\boldsymbol{X}}^{(0)}(\boldsymbol{p}_0) + \alpha_0(\boldsymbol{p}_0)\,\overline{\boldsymbol{X}}^{(1)}(\boldsymbol{p}\,|\,\boldsymbol{p}_0)\,h(\boldsymbol{p}-\boldsymbol{p}_0)\\ &+\alpha_0(\boldsymbol{p}_0)\int\frac{\mathrm{d}^2\boldsymbol{p}_1}{(2\pi)^2}\,\overline{\boldsymbol{X}}^{(2)}(\boldsymbol{p}\,|\,\boldsymbol{p}_1\,|\,\boldsymbol{p}_0)\,h(\boldsymbol{p}-\boldsymbol{p}_1)h(\boldsymbol{p}_1-\boldsymbol{p}_0)\\ &+\alpha_0(\boldsymbol{p}_0)\iint\frac{\mathrm{d}^2\boldsymbol{p}_1}{(2\pi)^2}\frac{\mathrm{d}\boldsymbol{p}_2}{(2\pi)^2}\,\overline{\boldsymbol{X}}^{(3)}(\boldsymbol{p}\,|\,\boldsymbol{p}_1\,|\,\boldsymbol{p}_2\,|\,\boldsymbol{p}_0)\,h(\boldsymbol{p}-\boldsymbol{p}_1)\\ &\times h(\boldsymbol{p}_1-\boldsymbol{p}_2)h(\boldsymbol{p}_2-\boldsymbol{p}_0),\end{aligned} \tag{52}$$

where $h(\boldsymbol{p})$ is the Fourier transform† of $h(\boldsymbol{x})$:

$$h(\boldsymbol{p}) \equiv \int \mathrm{d}\boldsymbol{x}\,\exp(-\mathrm{i}\boldsymbol{p}\cdot\boldsymbol{x})\,h(\boldsymbol{x}). \tag{53}$$

We will now prove how powerful are the reduced Rayleigh equations for the three following configurations. The $\overline{\boldsymbol{R}}$ matrix satisfies an exact integral equation under the Rayleigh hypothesis, and we get compact formulas suitable for numerical simulations.

6.4.1.1. A Rough Surface Separating Two Different Media

We consider a rough surface delimiting two media which are semi infinite, (see Fig. 6.3). We suppose that there is no upward field propagating in the medium 1, so we get $\boldsymbol{E}^{1+}=0$. With the choice, $b=+$, in Eq. (43), we obtain the following integral equation for the scattering matrix $\overline{\boldsymbol{R}}_{s\,\epsilon_0,\epsilon_1}(\boldsymbol{p}\,|\,\boldsymbol{p}_0)$ for a single surface‡ (the subscript s means a single surface located at $z=0$)

$$\int\frac{\mathrm{d}^2\boldsymbol{p}}{(2\pi)^2}\overline{\boldsymbol{M}}_h^{1+,0+}(\boldsymbol{u}\,|\,\boldsymbol{p})\,\overline{\boldsymbol{R}}_{s\,\epsilon_0,\epsilon_1}(\boldsymbol{p}\,|\,\boldsymbol{p}_0)+\overline{\boldsymbol{M}}_h^{1+,0-}(\boldsymbol{u}\,|\,\boldsymbol{p}_0)=0. \tag{54}$$

This equation has been already obtained making use of the extinction theorem.[21] We can notice that, since the right member of the Eq. (54) is null, we can simplify the second line of the matrices, $\overline{\boldsymbol{M}}^{1+,0+}$,$\overline{\boldsymbol{M}}^{1+,0-}$, by a factor $(\epsilon_1)^{\frac{1}{2}}$, then they coincide with the $\overline{\boldsymbol{M}}$,$\overline{\boldsymbol{N}}$ matrices derived by Celli *et al.*[21]

To construct a perturbative development, we expand in Taylor series the term $\exp(\mathrm{i}\alpha\, h(\boldsymbol{x}))$ inside $I(\alpha\,|\,\boldsymbol{p})$ (Eq. (39)), which yields

$$I(\alpha\,|\,\boldsymbol{p}) = (2\pi)^2\,\delta(\boldsymbol{p}) - \mathrm{i}\alpha\,h^{(1)}(\boldsymbol{p}) - \frac{\alpha^2}{2}\,h^{(2)}(\boldsymbol{p}) + \frac{\mathrm{i}\alpha^3}{3!}\,h^{(3)}(\boldsymbol{p}) + \cdots, \tag{55}$$

$$h^{(n)}(\boldsymbol{p}) \equiv \int \mathrm{d}^2\boldsymbol{x}\,\exp(-\mathrm{i}\boldsymbol{p}\cdot\boldsymbol{x})\,h^n(\boldsymbol{x}), \tag{56}$$

† We use the same symbol for a function and its Fourier transform, their arguments denoting the difference.

‡ We denote explicitly the permittivity dependance of $\overline{\boldsymbol{R}}_s$ by the subscript ϵ_0, ϵ_1 because in the following section we will use the $\overline{\boldsymbol{R}}_s$ matrix with different permittivity values.

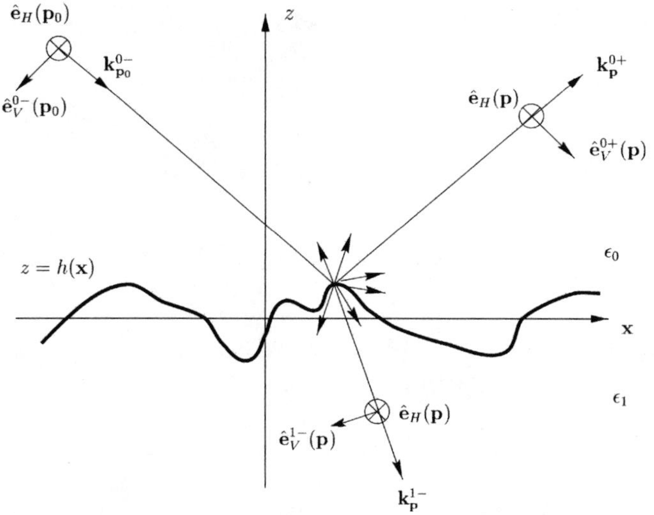

FIGURE 6.3. A two-dimensional random rough surface delimiting two dielectric media 0 and 1.

and we collect the terms of the same order in $h(\boldsymbol{x})$. Let us define the matrix

$$\overline{\boldsymbol{D}}^{\pm}_{10}(\boldsymbol{p}_0) \equiv \begin{pmatrix} \epsilon_1\,\alpha_0(\boldsymbol{p}_0) \pm \epsilon_0\,\alpha_1(\boldsymbol{p}_0) & 0 \\ 0 & \alpha_0(\boldsymbol{p}_0) \pm \alpha_1(\boldsymbol{p}_0) \end{pmatrix}, \tag{57}$$

the classical specular reflection coefficients for (TM) and (TE) waves are given by the diagonal elements of the matrix

$$\overline{\boldsymbol{V}}^{10}(\boldsymbol{p}_0) \equiv \overline{\boldsymbol{D}}^{-}_{10}(\boldsymbol{p}_0)\left[\overline{\boldsymbol{D}}^{+}_{10}(\boldsymbol{p}_0)\right]^{-1}. \tag{58}$$

Substituting Eq. (55) into Eq. (54), we construct for $\overline{\boldsymbol{R}}_{s\,\epsilon_0,\epsilon_1}$ a perturbative development of the well known form (52), and the coefficients of the perturbative development are given by

$$\begin{aligned} \overline{\boldsymbol{X}}^{(0)}(\boldsymbol{p}_0) &= -\frac{\alpha_1(\boldsymbol{p}_0) - \alpha_0(\boldsymbol{p}_0)}{\alpha_1(\boldsymbol{p}_0) + \alpha_0(\boldsymbol{p}_0)}\left[\overline{\boldsymbol{M}}^{1+,0+}(\boldsymbol{p}_0|\boldsymbol{p}_0)\right]^{-1} \cdot \overline{\boldsymbol{M}}^{1+,0-}(\boldsymbol{p}_0|\boldsymbol{p}_0) \\ &= \overline{\boldsymbol{V}}^{10}(\boldsymbol{p}_0), \end{aligned} \tag{59}$$

and

$$\overline{\boldsymbol{X}}^{(1)}_{s\,\epsilon_0,\epsilon_1}(\boldsymbol{u}|\boldsymbol{p}_0) = 2\mathrm{i}\,\overline{\boldsymbol{Q}}^{+}(\boldsymbol{u}|\boldsymbol{p}_0), \tag{60}$$

$$\begin{aligned} \overline{\boldsymbol{X}}^{(2)}_{s\,\epsilon_0,\epsilon_1}(\boldsymbol{u}|\boldsymbol{p}_1|\boldsymbol{p}_0) &= \alpha_1(\boldsymbol{u})\,\overline{\boldsymbol{Q}}^{+}(\boldsymbol{u}|\boldsymbol{p}_0) + \alpha_0(\boldsymbol{p}_0)\,\overline{\boldsymbol{Q}}^{-}(\boldsymbol{u}|\boldsymbol{p}_0) \\ &\quad -2\,\overline{\boldsymbol{P}}(\boldsymbol{u}|\boldsymbol{p}_1) \cdot \overline{\boldsymbol{Q}}^{+}(\boldsymbol{p}_1|\boldsymbol{p}_0), \end{aligned} \tag{61}$$

$$\overline{\boldsymbol{X}}^{(3)}_{s\,\epsilon_0,\epsilon_1}(\boldsymbol{u}|\boldsymbol{p}_1|\boldsymbol{p}_2|\boldsymbol{p}_0) = -\frac{\mathrm{i}}{3}\Big[\left(\alpha_1^2(\boldsymbol{u})+\alpha_0^2(\boldsymbol{p}_0)\right)\overline{\boldsymbol{Q}}^{+}(\boldsymbol{u}|\boldsymbol{p}_0) + 2\,\alpha_1(\boldsymbol{u})\,\alpha_0(\boldsymbol{p}_0)\,\overline{\boldsymbol{Q}}^{-}(\boldsymbol{u}|\boldsymbol{p}_0)\Big] + \mathrm{i}\,\overline{\boldsymbol{P}}(\boldsymbol{u}|\boldsymbol{p}_1)\,\overline{\boldsymbol{X}}^{(2)}(\boldsymbol{p}_1|\boldsymbol{p}_2|\boldsymbol{p}_0) + \mathrm{i}\,(\alpha_1(\boldsymbol{u}) - \alpha_0(\boldsymbol{p}_2))\,\overline{\boldsymbol{P}}(\boldsymbol{u}|\boldsymbol{p}_2)\cdot\overline{\boldsymbol{Q}}^{+}(\boldsymbol{p}_2|\boldsymbol{p}_0), \tag{62}$$

where we have the expression

$$\overline{\boldsymbol{Q}}^{\pm}(\boldsymbol{u}|\boldsymbol{p}_0) \equiv \frac{\alpha_1(\boldsymbol{u})-\alpha_0(\boldsymbol{u})}{2\,\alpha_0(\boldsymbol{p}_0)}\Big[\overline{\boldsymbol{M}}^{1+,0+}(\boldsymbol{u}|\boldsymbol{u})]^{-1}\cdot[\overline{\boldsymbol{M}}^{1+,0-}(\boldsymbol{u}|\boldsymbol{p}_0) \pm \overline{\boldsymbol{M}}^{1+,0+}(\boldsymbol{u}|\boldsymbol{p}_0)\cdot\overline{\boldsymbol{X}}^{(0)}(\boldsymbol{p}_0)\Big], \tag{63}$$

and, after some simple algebra, we obtain

$$\overline{\boldsymbol{Q}}^{+}(\boldsymbol{u}|\boldsymbol{p}_0) = (\epsilon_1-\epsilon_0)\left[\overline{\boldsymbol{D}}^{+}_{10}(\boldsymbol{u})\right]^{-1} \times \begin{pmatrix} \epsilon_1\,||\boldsymbol{u}||\,||\boldsymbol{p}_0|| - \epsilon_0\,\alpha_1(\boldsymbol{u})\,\alpha_1(\boldsymbol{p}_0)\,\hat{\boldsymbol{u}}\cdot\hat{\boldsymbol{p}}_0 & -\epsilon_0^{\frac{1}{2}}\,K_0\,\alpha_1(\boldsymbol{u})\,(\hat{\boldsymbol{u}}\times\hat{\boldsymbol{p}}_0)_z \\ -\epsilon_0^{\frac{1}{2}}\,K_0\,\alpha_1(\boldsymbol{p}_0)\,(\hat{\boldsymbol{u}}\times\hat{\boldsymbol{p}}_0)_z & \epsilon_0\,K_0^2\,\hat{\boldsymbol{u}}\cdot\hat{\boldsymbol{p}}_0 \end{pmatrix} \times \left[\overline{\boldsymbol{D}}^{+}_{10}(\boldsymbol{p}_0)\right]^{-1}, \tag{64}$$

$$\overline{\boldsymbol{Q}}^{-}(\boldsymbol{u}|\boldsymbol{p}_0) = \frac{(\epsilon_1-\epsilon_0)}{\alpha_0(\boldsymbol{p}_0)}\left[\overline{\boldsymbol{D}}^{+}_{10}(\boldsymbol{u})\right]^{-1} \times \begin{pmatrix} \epsilon_0\,\alpha_1(\boldsymbol{p}_0)\,||\boldsymbol{u}||\,||\boldsymbol{p}_0|| - \epsilon_1\,\alpha_1(\boldsymbol{u})\,\alpha_0^2(\boldsymbol{p}_0)\,\hat{\boldsymbol{u}}\cdot\hat{\boldsymbol{p}}_0 & -\epsilon_0^{\frac{1}{2}}\,K_0\,\alpha_1(\boldsymbol{u})\,\alpha_1(\boldsymbol{p}_0)\,(\hat{\boldsymbol{u}}\times\hat{\boldsymbol{p}}_0)_z \\ -\epsilon_0^{\frac{1}{2}}\,K_0\,\epsilon_1\,\alpha_0^2(\boldsymbol{p}_0)\,(\hat{\boldsymbol{u}}\times\hat{\boldsymbol{p}}_0)_z & \epsilon_0\,K_0^2\,\alpha_1(\boldsymbol{p}_0)\,\hat{\boldsymbol{u}}\cdot\hat{\boldsymbol{p}}_0 \end{pmatrix} \times \left[\overline{\boldsymbol{D}}^{+}_{10}(\boldsymbol{p}_0)\right]^{-1}, \tag{65}$$

$$\overline{\boldsymbol{P}}(\boldsymbol{u}|\boldsymbol{p}_1) \equiv (\alpha_1(\boldsymbol{u})-\alpha_0(\boldsymbol{u}))\left[\overline{\boldsymbol{M}}^{1+,0+}(\boldsymbol{u}|\boldsymbol{u})\right]^{-1}\overline{\boldsymbol{M}}^{1+,0+}(\boldsymbol{u}|\boldsymbol{p}_1) \tag{66}$$

$$= (\epsilon_1-\epsilon_0)\left[\overline{\boldsymbol{D}}^{+}_{10}(\boldsymbol{u})\right]^{-1} \times \begin{pmatrix} ||\boldsymbol{u}||\,||\boldsymbol{p}|| + \alpha_1(\boldsymbol{u})\,\alpha_0(\boldsymbol{p})\,\hat{\boldsymbol{u}}\cdot\hat{\boldsymbol{p}}_1 & -\epsilon_0^{\frac{1}{2}}\,K_0\,\alpha_1(\boldsymbol{u})\,(\hat{\boldsymbol{u}}\times\hat{\boldsymbol{p}}_1)_z \\ \epsilon_0^{-\frac{1}{2}}\,K_0\,\alpha_0(\boldsymbol{p})\,(\hat{\boldsymbol{u}}\times\hat{\boldsymbol{p}}_1)_z & K_0^2\,\hat{\boldsymbol{u}}\cdot\hat{\boldsymbol{p}}_1 \end{pmatrix}. \tag{67}$$

It can be easily shown that $\overline{\boldsymbol{X}}^{(1)}$ is the well known first-order term in perturbation theory which was obtained by Rice.[15] We can prove after some lengthy calculations that Eqs. (61),(62), are identical to those found by Johnson.[16] Thus expressions Eqs. (61),(62), are a compact manner to write the second- and third-order terms of the perturbative expansion; moreover, they are well adapted for numerical computations. However, we notice that only the first term $\overline{\boldsymbol{X}}^{(1)}$ is reciprocal. Since

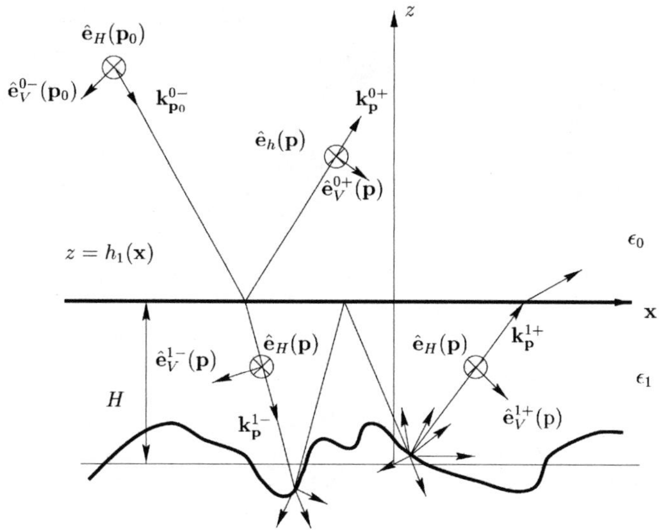

FIGURE 6.4. A slab formed with a bottom two-dimensional randomly rough surface and an upper planar surface.

the second- and third-order perturbative terms are included in an integral, the coefficient $\overline{X}^{(2)}$, $\overline{X}^{(3)}$, are not unique; however they can be put into a reciprocal form (see Appendix B).

It is worth to notice that we can follow an analogous procedure to calculate the transmitted field. By taking $b = -$ in Eq. (44), we get

$$\int \frac{d^2 p}{(2\pi)^2} \overline{M}_h^{0-,1-}(u|p)\, E^{1-}(p) = -\frac{-2\,(\epsilon_0\,\epsilon_1)^{\frac{1}{2}}\,\alpha_1(u)}{(\epsilon_1 - \epsilon_0)}\, E^{0-}(u). \tag{68}$$

This equation was already obtained with the extinction theorem.[31]

6.4.1.2. A Slab with a Rough Surface on the Bottom Side

We consider a slab delimited on the upper side by a planar surface and on the bottom side by a rough surface, (see Fig. 6.4). Since there is no incident upward field in medium 2, the scattering matrix we have obtained in the previous section is sufficient to determine the scattering matrix of the present configuration. In order to get a proof, let us give some definitions as explained in Fig. 6.5. The scattering matrix for an incident plane wave coming from the medium 0, and scattered in the medium 1, is given by the following expression:

$$\overline{V}^{0}(p\,|\,p_0) = (2\pi)^2\,\delta(p - p_0)\,\overline{V}^{10}(p_0)\,, \tag{69}$$

where $\overline{V}^{10}$ is defined by Eq. (58). The transmitted wave in medium 1 is given by

$$\overline{T}^{0}(p\,|\,p_0) = (2\pi)^2\,\delta(p - p_0)\,\frac{\alpha_0(p_0)}{\alpha_1(p_0)}\,\overline{T}^{10}(p_0), \tag{70}$$

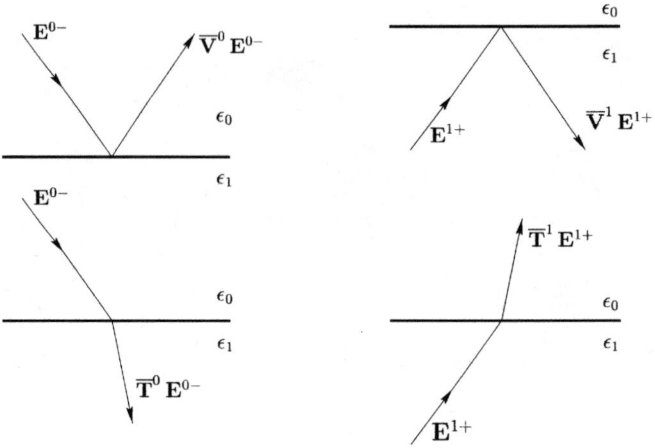

FIGURE 6.5. Definitions of scattering matrices for a planar surface.

$$\overline{\boldsymbol{T}}^{10}(\boldsymbol{p}_0) \equiv 2\alpha_1(\boldsymbol{p}_0)\begin{pmatrix}(\epsilon_0\,\epsilon_1)^{\frac{1}{2}} & 0\\ 0 & 1\end{pmatrix}.\left[\overline{\boldsymbol{D}}_{10}^{+}(\boldsymbol{p}_0)\right]^{-1}. \tag{71}$$

When the incident wave is coming from medium 1, we have similarly the following equations:

$$\overline{\boldsymbol{V}}^{1}(\boldsymbol{p}\,|\,\boldsymbol{p}_0) = -(2\pi)^2\,\delta(\boldsymbol{p}-\boldsymbol{p}_0)\,\overline{\boldsymbol{V}}^{10}(\boldsymbol{p}_0), \tag{72}$$

$$\overline{\boldsymbol{T}}^{1}(\boldsymbol{p}\,|\,\boldsymbol{p}_0) = (2\pi)^2\,\delta(\boldsymbol{p}-\boldsymbol{p}_0)\,\overline{\boldsymbol{T}}^{10}(\boldsymbol{p}_0). \tag{73}$$

The scattering matrix $\overline{\boldsymbol{R}}^{H}_{s\,\epsilon_1,\epsilon_2}$ for the rough surface h which is located at $z=-H$[§], and separating respectively two media of permittivity ϵ_1 and ϵ_2 is given by

$$\overline{\boldsymbol{R}}^{H}_{s\,\epsilon_1,\epsilon_2}(\boldsymbol{p}\,|\,\boldsymbol{p}_0) = \exp(\mathrm{i}(\alpha_1(\boldsymbol{p})+\alpha_1(\boldsymbol{p}_0))\,H)\,\overline{\boldsymbol{R}}_{s\,\epsilon_1,\epsilon_2}(\boldsymbol{p}\,|\,\boldsymbol{p}_0). \tag{74}$$

It is worth noticing that the phase term comes from the translation $z=-H$, (see Eq. (217)), and $\overline{\boldsymbol{R}}_{s\,\epsilon_1,\epsilon_2}$ denotes the scattering matrix $\overline{\boldsymbol{R}}_{s\,\epsilon_0,\epsilon_1}$ of the previous section, where we have replaced ϵ_0 by ϵ_1, and ϵ_1 by ϵ_2. Furthermore, we define the product of two operator $\overline{\boldsymbol{A}}$ and $\overline{\boldsymbol{B}}$ by

$$(\overline{\boldsymbol{A}}\cdot\overline{\boldsymbol{B}})(\boldsymbol{p}\,|\,\boldsymbol{p}_0) \equiv \int \frac{\mathrm{d}^2\boldsymbol{p}_1}{(2\pi)^2}\,\overline{\boldsymbol{A}}(\boldsymbol{p}\,|\,\boldsymbol{p}_1)\cdot\overline{\boldsymbol{B}}(\boldsymbol{p}_1\,|\,\boldsymbol{p}_0), \tag{75}$$

then we easily prove for the configuration shown in Fig. 6.4 that (we use for the fields the notation of Fig. 6.1),

$$\boldsymbol{E}^{1+} = \overline{\boldsymbol{R}}^{H}_{s\,\epsilon_1,\epsilon_2}\cdot\overline{\boldsymbol{T}}^{0}\cdot\boldsymbol{E}^{0-} + \overline{\boldsymbol{R}}^{H}_{s\,\epsilon_1,\epsilon_2}\cdot\overline{\boldsymbol{V}}^{1}\cdot\boldsymbol{E}^{1+}, \tag{76}$$

[§] Throughout this chapter, the symbol H in upper index is related to the height of a surface, in lower index to the polarization of the electromagnetic field.

$$\boldsymbol{E}^{0+} = \overline{\boldsymbol{V}}^{0} \cdot \boldsymbol{E}^{0-} + \overline{\boldsymbol{T}}^{1} \cdot \boldsymbol{E}^{1+}, \tag{77}$$

where $\boldsymbol{E}^{0-}(\boldsymbol{p}) = (2\pi)^2\,\delta(\boldsymbol{p}-\boldsymbol{p}_0)\boldsymbol{E}^{i}(\boldsymbol{p}_0)$. These equations have been recently used by Fuks[17] to calculate in the first order the field scattered by a layered medium. In fact, as we shall see below, these equations allow us to obtain all orders of the field perturbation. Expression (76) is analogous to the Dyson equation usually used in random media.[5] So, we are led to define the scattering operator $\overline{\boldsymbol{U}}$

$$\boldsymbol{E}^{1+} = \overline{\boldsymbol{R}}^{H}_{s\,\epsilon_1,\epsilon_2} \cdot \overline{\boldsymbol{U}} \cdot \overline{\boldsymbol{T}}^{0} \cdot \boldsymbol{E}^{0-}, \tag{78}$$

which satisfies the equation

$$\overline{\boldsymbol{U}} = \overline{\mathbf{1}} + \overline{\boldsymbol{V}}^{1} \cdot \overline{\boldsymbol{R}}^{H}_{s\,\epsilon_1,\epsilon_2} \cdot \overline{\boldsymbol{U}}\,. \tag{79}$$

$\overline{\boldsymbol{R}}_d(\boldsymbol{p}\,|\,\boldsymbol{p}_0)$ is the expression of the global scattering matrix for the upper planar surface and the bottom rough surface and we get

$$\boldsymbol{E}^{0+}(\boldsymbol{p}) = \overline{\boldsymbol{R}}_d(\boldsymbol{p}\,|\,\boldsymbol{p}_0) \cdot \boldsymbol{E}^{i}(\boldsymbol{p}_0). \tag{80}$$

With Eqs. (77),(78), the scattering matrix can be written as follows:

$$\overline{\boldsymbol{R}}_d = \overline{\boldsymbol{V}}^{0} + \overline{\boldsymbol{T}}^{1} \cdot \overline{\boldsymbol{R}}^{H}_{s\,\epsilon_1,\epsilon_2} \cdot \overline{\boldsymbol{U}} \cdot \overline{\boldsymbol{T}}^{0}. \tag{81}$$

We improve development (79), by summing all the specular reflections inside the slab. This is done by defining the operator $\overline{\boldsymbol{U}}^{(0)}$ which satisfies the equation

$$\overline{\boldsymbol{U}}^{(0)} = \overline{\mathbf{1}} + \overline{\boldsymbol{V}}^{1} \cdot \overline{\boldsymbol{R}}^{H\,(0)}_{s\,\epsilon_1,\epsilon_2} \cdot \overline{\boldsymbol{U}}^{(0)}, \tag{82}$$

where $\overline{\boldsymbol{R}}^{H\,(0)}_{s\,\epsilon_1,\epsilon_2}$ is the zeroth-order term of the perturbative development given by

$$\overline{\boldsymbol{R}}^{H\,(0)}_{s\,\epsilon_1,\epsilon_2}(\boldsymbol{p}\,|\,\boldsymbol{p}_0) = (2\pi)^2\,\delta(\boldsymbol{p}-\boldsymbol{p}_0)\,\overline{\boldsymbol{V}}^{H\,21}(\boldsymbol{p}_0), \tag{83}$$

where $\overline{\boldsymbol{V}}^{H\,21}$ is the scattering matrix for a planar surface located at the height $z = -H$

$$\overline{\boldsymbol{V}}^{H\,21}(\boldsymbol{u}) \equiv \exp(2\,i\,\alpha_1(\boldsymbol{u})\,H)\,\overline{\boldsymbol{D}}^{-}_{21}(\boldsymbol{p}_0)\left[\overline{\boldsymbol{D}}^{+}_{21}(\boldsymbol{p}_0)\right]^{-1}, \tag{84}$$

$$\overline{\boldsymbol{D}}^{\pm}_{21}(\boldsymbol{p}_0) \equiv \begin{pmatrix} \epsilon_2\,\alpha_1(\boldsymbol{p}_0) \pm \epsilon_1\,\alpha_2(\boldsymbol{p}_0) & 0 \\ 0 & \alpha_1(\boldsymbol{p}_0) \pm \alpha_2(\boldsymbol{p}_0) \end{pmatrix}. \tag{85}$$

The term $\exp(2\,i\,\alpha_1(\boldsymbol{u})\,H)$ comes from the phase shift induced by the translation of the planar surface from the height $z = 0$ to $z = -H$ (see Appendix B).The diagrammatic representation of Eq. (82) is shown in Fig. 6.6. It is in fact a geometric series which can be summed, and we get the expressions:

$$\overline{\boldsymbol{U}}^{(0)}(\boldsymbol{p}\,|\,\boldsymbol{p}_0) = (2\pi)^2\,\delta(\boldsymbol{p}-\boldsymbol{p}_0)\,\overline{\boldsymbol{U}}^{(0)}(\boldsymbol{p}_0), \tag{86}$$

$$\overline{\boldsymbol{U}}^{(0)}(\boldsymbol{p}_0) \equiv \left[\overline{\mathbf{1}} + \overline{\boldsymbol{V}}^{10}(\boldsymbol{p}_0) \cdot \overline{\boldsymbol{V}}^{H\,21}(\boldsymbol{p}_0)\right]^{-1}. \tag{87}$$

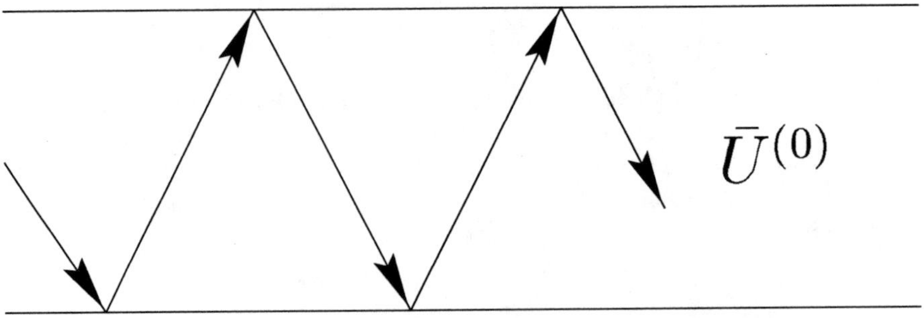

FIGURE 6.6. Diagrammatic representation of the operator $\overline{\boldsymbol{U}}^{(0)}$.

From the previous results, Eq. (79) can be written as follows :

$$\overline{\boldsymbol{U}} = \overline{\boldsymbol{U}}^{(0)} + \overline{\boldsymbol{U}}^{(0)} \cdot \overline{\boldsymbol{V}}^{1} \cdot \Delta\overline{\boldsymbol{R}}^{H}_{S\,\epsilon_1,\epsilon_2} \cdot \overline{\boldsymbol{U}}\,, \tag{88}$$

where

$$\Delta\overline{\boldsymbol{R}}^{H}_{S\,\epsilon_1,\epsilon_2} \equiv \overline{\boldsymbol{R}}^{H}_{S\,\epsilon_1,\epsilon_2} - \overline{\boldsymbol{R}}^{H\,(0)}_{S\,\epsilon_1,\epsilon_2}. \tag{89}$$

In order to obtain the perturbative development of $\overline{\boldsymbol{R}}_d$, we substitute the expansion

$$\overline{\boldsymbol{R}}^{H}_{S\,\epsilon_1,\epsilon_2} = \overline{\boldsymbol{R}}^{H\,(0)}_{S\,\epsilon_1,\epsilon_2} + \overline{\boldsymbol{R}}^{H\,(1)}_{S\,\epsilon_1,\epsilon_2} + \overline{\boldsymbol{R}}^{H\,(2)}_{S\,\epsilon_1,\epsilon_2} + \overline{\boldsymbol{R}}^{H\,(3)}_{S\,\epsilon_1,\epsilon_2}, \tag{90}$$

into Eqs. (81) and (88), and we get the following terms:

$$\overline{\boldsymbol{R}}^{(0)}_{d} = \overline{\boldsymbol{V}}^{0} + \overline{\boldsymbol{T}}^{1} \cdot \overline{\boldsymbol{R}}^{H\,(0)}_{S\,\epsilon_1,\epsilon_2} \cdot \overline{\boldsymbol{U}}^{(0)} \cdot \overline{\boldsymbol{T}}^{0}, \tag{91}$$

$$\overline{\boldsymbol{R}}^{(1)}_{d} = \overline{\boldsymbol{T}}^{1} \cdot \overline{\boldsymbol{U}}^{(0)} \cdot \overline{\boldsymbol{R}}^{H\,(1)}_{S\,\epsilon_1,\epsilon_2} \cdot \overline{\boldsymbol{U}}^{(0)} \cdot \overline{\boldsymbol{T}}^{0}, \tag{92}$$

$$\overline{\boldsymbol{R}}^{(2)}_{d} = \overline{\boldsymbol{T}}^{1} \cdot \overline{\boldsymbol{U}}^{(0)} \cdot \left[\overline{\boldsymbol{R}}^{H\,(2)}_{S\,\epsilon_1,\epsilon_2} + \overline{\boldsymbol{R}}^{H\,(1)}_{S\,\epsilon_1,\epsilon_2} \cdot \overline{\boldsymbol{U}}^{(0)} \cdot \overline{\boldsymbol{V}}^{1} \cdot \overline{\boldsymbol{R}}^{H\,(1)}_{S\,\epsilon_1,\epsilon_2}\right] \cdot \overline{\boldsymbol{U}}^{(0)} \cdot \overline{\boldsymbol{T}}^{0}, \tag{93}$$

$$\begin{aligned}\overline{\boldsymbol{R}}^{(3)}_{d} = \overline{\boldsymbol{T}}^{1} \cdot \overline{\boldsymbol{U}}^{(0)} \cdot \Big[&\overline{\boldsymbol{R}}^{H\,(3)}_{S\,\epsilon_1,\epsilon_2} + \overline{\boldsymbol{R}}^{H\,(2)}_{S\,\epsilon_1,\epsilon_2} \cdot \overline{\boldsymbol{U}}^{(0)} \cdot \overline{\boldsymbol{V}}^{1} \cdot \overline{\boldsymbol{R}}^{H\,(1)}_{S\,\epsilon_1,\epsilon_2} \\ &+ \overline{\boldsymbol{R}}^{H\,(1)}_{S\,\epsilon_1,\epsilon_2} \cdot \overline{\boldsymbol{U}}^{(0)} \cdot \overline{\boldsymbol{V}}^{1} \cdot \overline{\boldsymbol{R}}^{H\,(2)}_{S\,\epsilon_1,\epsilon_2} \\ &+ \overline{\boldsymbol{R}}^{H\,(1)}_{S\,\epsilon_1,\epsilon_2} \cdot \overline{\boldsymbol{U}}^{(0)} \cdot \overline{\boldsymbol{V}}^{1} \cdot \overline{\boldsymbol{R}}^{H\,(1)}_{S\,\epsilon_1,\epsilon_2} \cdot \overline{\boldsymbol{U}}^{(0)} \cdot \overline{\boldsymbol{V}}^{1} \cdot \overline{\boldsymbol{R}}^{H\,(1)}_{S\,\epsilon_1,\epsilon_2}\Big] \cdot \overline{\boldsymbol{U}}^{(0)} \cdot \overline{\boldsymbol{T}}^{0}. \end{aligned} \tag{94}$$

Using development (52) for $\overline{\boldsymbol{R}}^{H}_{S\,\epsilon_1,\epsilon_2}$, and definitions (69)–(70), (72)–(73), we obtain after some calculations a development of form (52) for $\overline{\boldsymbol{R}}_d$ with the following coefficients:

$$\overline{\boldsymbol{X}}^{(0)}_{b}(\boldsymbol{p}_0) = \left(\overline{\boldsymbol{V}}^{10}(\boldsymbol{p}_0) + \overline{\boldsymbol{V}}^{H\,21}(\boldsymbol{p}_0)\right)\left[\overline{\boldsymbol{1}} + \overline{\boldsymbol{V}}^{10}(\boldsymbol{p}_0) \cdot \overline{\boldsymbol{V}}^{H\,21}(\boldsymbol{p}_0)\right]^{-1}. \tag{95}$$

This matrix is diagonal, and its coefficients are identical to the reflection coefficients for a planar slab.[5] The other coefficients are

$$\overline{\boldsymbol{X}}^{(1)}_{d}(\boldsymbol{p}\,|\,\boldsymbol{p}_0) = \overline{\boldsymbol{T}}^{10}(\boldsymbol{p}) \cdot \overline{\boldsymbol{U}}^{(0)}(\boldsymbol{p}) \cdot \overline{\boldsymbol{X}}^{H\,(1)}_{S\,\epsilon_1,\epsilon_2}(\boldsymbol{p}\,|\,\boldsymbol{p}_0) \cdot \overline{\boldsymbol{U}}^{(0)}(\boldsymbol{p}_0) \cdot \overline{\boldsymbol{T}}^{10}(\boldsymbol{p}_0), \tag{96}$$

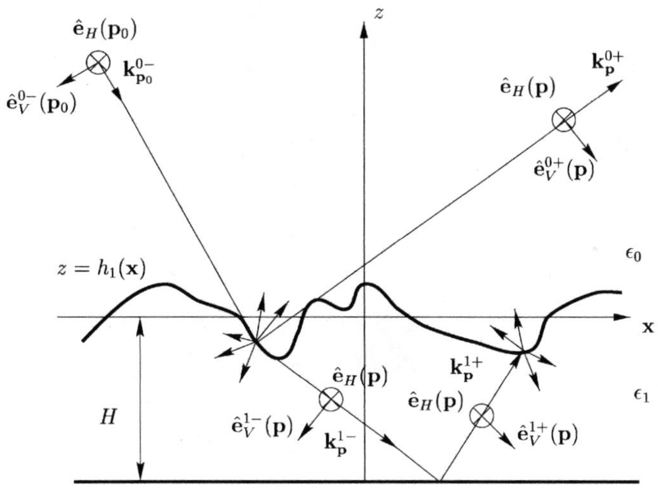

FIGURE 6.7. A slab formed with an upper two-dimensional randomly rough surface and a bottom planar surface.

$$\begin{aligned}\overline{\boldsymbol{X}}_d^{(2)}(\boldsymbol{p}|\boldsymbol{p}_1|\boldsymbol{p}_0) = \overline{\boldsymbol{T}}^{10}(\boldsymbol{p})\cdot\overline{\boldsymbol{U}}^{(0)}(\boldsymbol{p})\cdot\Big[&\overline{\boldsymbol{X}}^{H\,(2)}_{s\,\epsilon_1,\epsilon_2}(\boldsymbol{p}|\boldsymbol{p}_1|\boldsymbol{p}_0)\\ &-\alpha_1(\boldsymbol{p}_1)\overline{\boldsymbol{X}}^{H\,(1)}_{s\,\epsilon_1,\epsilon_2}(\boldsymbol{p}|\boldsymbol{p}_1)\cdot\overline{\boldsymbol{U}}^{(0)}(\boldsymbol{p}_1)\cdot\overline{\boldsymbol{V}}^{10}(\boldsymbol{p}_1)\cdot\overline{\boldsymbol{X}}^{H\,(1)}_{s\,\epsilon_1,\epsilon_2}(\boldsymbol{p}_1|\boldsymbol{p}_0)\Big]\\ &\times\overline{\boldsymbol{U}}^{(0)}(\boldsymbol{p}_0)\cdot\overline{\boldsymbol{T}}^{10}(\boldsymbol{p}_0),\end{aligned} \tag{97}$$

$$\begin{aligned}\overline{\boldsymbol{X}}_d^{(3)}(\boldsymbol{p}|\boldsymbol{p}_1|\boldsymbol{p}_2|\boldsymbol{p}_0) = \overline{\boldsymbol{T}}^{10}(\boldsymbol{p})\cdot\overline{\boldsymbol{U}}^{(0)}(\boldsymbol{p})\cdot\Big[&\overline{\boldsymbol{X}}^{H\,(3)}_{s\,\epsilon_1,\epsilon_2}(\boldsymbol{p}|\boldsymbol{p}_1|\boldsymbol{p}_2|\boldsymbol{p}_0)\\ -\alpha_1(\boldsymbol{p}_2)&\overline{\boldsymbol{X}}^{H\,(2)}_{s\,\epsilon_1,\epsilon_2}(\boldsymbol{p}|\boldsymbol{p}_1|\boldsymbol{p}_2)\cdot\overline{\boldsymbol{U}}^{(0)}(\boldsymbol{p}_2)\cdot\overline{\boldsymbol{V}}^{10}(\boldsymbol{p}_2)\cdot\overline{\boldsymbol{X}}^{H\,(1)}_{s\,\epsilon_1,\epsilon_2}(\boldsymbol{p}_2|\boldsymbol{p}_0)\\ -\alpha_1(\boldsymbol{p}_1)&\overline{\boldsymbol{X}}^{H\,(1)}_{s\,\epsilon_1,\epsilon_2}(\boldsymbol{p}|\boldsymbol{p}_1)\cdot\overline{\boldsymbol{U}}^{(0)}(\boldsymbol{p}_1)\cdot\overline{\boldsymbol{V}}^{10}(\boldsymbol{p}_1)\cdot\overline{\boldsymbol{X}}^{H\,(2)}_{s\,\epsilon_1,\epsilon_2}(\boldsymbol{p}_1|\boldsymbol{p}_2|\boldsymbol{p}_0)\\ +\alpha_1(\boldsymbol{p}_1)\alpha_1(\boldsymbol{p}_2)&\overline{\boldsymbol{X}}^{H\,(1)}_{s\,\epsilon_1,\epsilon_2}(\boldsymbol{p}|\boldsymbol{p}_1)\cdot\overline{\boldsymbol{U}}^{(0)}(\boldsymbol{p}_1)\cdot\overline{\boldsymbol{V}}^{10}(\boldsymbol{p}_1)\cdot\overline{\boldsymbol{X}}^{H\,(1)}_{s\,\epsilon_1,\epsilon_2}(\boldsymbol{p}_1|\boldsymbol{p}_2)\\ \times\overline{\boldsymbol{U}}^{(0)}(\boldsymbol{p}_2)\cdot\overline{\boldsymbol{V}}^{10}(\boldsymbol{p}_2)\cdot&\overline{\boldsymbol{X}}^{H\,(1)}_{s\,\epsilon_1,\epsilon_2}(\boldsymbol{p}_2|\boldsymbol{p}_0)\Big]\cdot\overline{\boldsymbol{U}}^{(0)}(\boldsymbol{p}_0)\cdot\overline{\boldsymbol{T}}^{10}(\boldsymbol{p}_0).\end{aligned} \tag{98}$$

In these expressions, we have

$$\overline{\boldsymbol{X}}^{H\,(n)}_{s\,\epsilon_1,\epsilon_2}(\boldsymbol{p}|\boldsymbol{p}_0) \equiv \exp(\mathrm{i}(\alpha_1(\boldsymbol{p})+\alpha_1(\boldsymbol{p}_0))\,H)\,\overline{\boldsymbol{X}}^{(n)}_{s\,\epsilon_1,\epsilon_2}(\boldsymbol{p}|\boldsymbol{p}_0)\,, \tag{99}$$

and the subscripts ϵ_1, ϵ_2 in $\overline{\boldsymbol{X}}_{s\,\epsilon_1,\epsilon_2}$ denote that we substitute ϵ_0 by ϵ_1, and ϵ_1 by ϵ_2 in (60)–(62).

6.4.1.3. A Slab with a Rough Surface on the Upper Side

We consider a slab delimited on the upper side by a two-dimensional rough surface, and on the bottom side by a planar surface. This system is described in Fig. 6.7. To derive the reduced Rayleigh equation in this case, we combine the two following equations:

$$\int \frac{d^2p}{(2\pi)^2} \overline{M}_h^{1+,0+}(u|p)\,\overline{R}_u(p|p_0)\,E^{0-}(p_0) + \overline{M}_h^{1+,0-}(u|p_0)\,E^{0-}(p_0) = \frac{2\,(\epsilon_0\,\epsilon_1)^{\frac{1}{2}}\,\alpha_1(u)}{(\epsilon_1-\epsilon_0)}\,E^{1+}(u), \tag{100}$$

$$\int \frac{d^2p}{(2\pi)^2} \overline{M}_h^{1-,0+}(u|p)\,\overline{R}_u(p|p_0)\,E^{0-}(p_0) + \overline{M}_h^{1-,0-}(u|p_0)\,E^{0-}(p_0) = -\frac{2\,(\epsilon_0\,\epsilon_1)^{\frac{1}{2}}\,\alpha_1(u)}{(\epsilon_1-\epsilon_0)}\,E^{1-}(u), \tag{101}$$

with

$$E^{1+}(u) = \overline{V}^{H\,21}(u)\,E^{1-}(u), \tag{102}$$

where $\overline{V}^{H\,21}$ is given by (85), and $\overline{R}_u$ is defined as the global scattering matrix for the upper rough surface and the bottom planar surface.

The reduced Rayleigh equation for the scattering matrix $\overline{R}_u$ is given by the following expression:

$$\int \frac{d^2p}{(2\pi)^2}\left[\overline{M}_h^{1+,0+}(u|p) + \overline{V}^{H\,21}(u)\cdot\overline{M}_h^{1-,0+}(u|p)\right]\overline{R}_u(p|p_0) = -\left[\overline{M}_h^{1+,0-}(u|p_0) + \overline{V}^{H\,21}(u)\cdot\overline{M}_h^{1-,0-}(u|p_0)\right]. \tag{103}$$

By using the expansion of $I(\alpha|p)$ in power series, we can then write the perturbative development

$$\begin{aligned}\overline{X}^{(0)}_{u\,\epsilon_0,\epsilon_1}(p_0) &= -\left[\frac{\overline{M}^{1+,0+}(p_0|p_0)}{\alpha_1(p_0)-\alpha_0(p_0)} - \overline{V}^{H\,21}(p_0)\cdot\frac{\overline{M}^{1-,0+}(p_0|p_0)}{\alpha_1(p_0)+\alpha_0(p_0)}\right]^{-1}\\ &\quad\times\left[\frac{\overline{M}^{1+,0-}(p_0|p_0)}{\alpha_1(p_0)+\alpha_0(p_0)} + \overline{V}^{H\,21}(p_0)\cdot\frac{\overline{M}^{1-,0-}(p_0|p_0)}{-\alpha_1(p_0)+\alpha_0(p_0)}\right]\\ &= \left(\overline{V}^{10}(p_0) + \overline{V}^{H\,21}(p_0)\right)\left[\overline{1} + \overline{V}^{10}(p_0)\cdot\overline{V}^{H\,21}(p_0)\right]^{-1},\end{aligned} \tag{104}$$

$$\overline{X}^{(1)}_u(u|p_0) \equiv 2\,i\,\overline{Q}^{++}(u|p_0) \tag{105}$$

$$\overline{X}^{(2)}_u(u|p_1|p_0) = \alpha_1(u)\,\overline{Q}^{-+}(u|p_0) + \alpha_0(p_0)\,\overline{Q}^{+-}(u|p_0) - 2\,\overline{P}^{+}(u|p_1)\cdot\overline{Q}^{++}(p_1|p_0) \tag{106}$$

$$\begin{aligned}\overline{X}^{(3)}_u(u|p_1|p_2|p_0) &= -\frac{i}{3}\Big[\left(\alpha_1^2(u)+\alpha_0^2(p_0)\right)\overline{Q}^{++}(u|p_0)\\ &\quad + 2\,\alpha_1(u)\,\alpha_0(p_0)\,\overline{Q}^{--}(u|p_0)\Big]\\ &\quad + i\,\overline{P}^{+}(u|p_1)\,\overline{X}^{(2)}(p_1|p_2|p_0)\\ &\quad + i\left[\alpha_1(u)\,\overline{P}^{-}(u|p_2) - \alpha_0(p_2)\,\overline{P}^{+}(u|p_2)\right]\cdot\overline{Q}^{++}(p_2|p_0)\end{aligned} \tag{107}$$

with the following expressions:

$$\overline{Q}^{ba}(u|p_0) \equiv \frac{\mathrm{i}}{2\,\alpha_0(p_0)}\left[\frac{\overline{M}^{1+,0+}(u|u)}{\alpha_1(u)-\alpha_0(u)} - \overline{V}^{H\,21}(u)\cdot\frac{\overline{M}^{1-,0+}(u|u)}{\alpha_1(u)+\alpha_0(u)}\right]^{-1}$$
$$\times\left[\overline{M}^{1+,0+}(u|p_0)\cdot\overline{X}^{(0)}_{s\,\epsilon_0,\epsilon_1}(p_0) + a\,\overline{M}^{1+,0-}(u|p_0)\right.$$
$$\left. + b\,\overline{V}^{H\,21}(u)\cdot\left(\overline{M}^{1-,0+}(u|p_0)\cdot\overline{X}^{(0)}_{s\,\epsilon_0,\epsilon_1}(p_0) + a\,\overline{M}^{1-,0-}(u|p_0)\right)\right], \tag{108}$$

$$\overline{P}^{\pm}(u|p_1) \equiv \left[\frac{\overline{M}^{1+,0+}(u|u)}{\alpha_1(u)-\alpha_0(u)} - \overline{V}^{H\,21}(u)\cdot\frac{\overline{M}^{1-,0+}(u|u)}{\alpha_1(u)+\alpha_0(u)}\right]^{-1}$$
$$\times\left[\overline{M}^{1+,0+}(u|p_1) \pm \overline{V}^{H\,21}(u)\cdot\overline{M}^{1-,0+}(u|p_1)\right], \tag{109}$$

where, $a=\pm$, $b=\pm$ denote the sign indices. After some calculations, we obtain

$$\overline{Q}^{++}(u|p_0) = (\epsilon_1-\epsilon_0)\left[\overline{D}^{+}_{10}(u)\right]^{-1}$$
$$\times\begin{pmatrix} \epsilon_1\,\overline{A}^{++} - \epsilon_0\,\alpha_1(u)\alpha_1(p_0)\,\overline{B}^{--} & -\epsilon_0^{\frac{1}{2}}\,\alpha_1(u)\,\overline{J}^{-+} \\ -\epsilon_0^{\frac{1}{2}}\,\alpha_1(p_0)\,\overline{G}^{+-} & \epsilon_0\,K_0^2\,\overline{C}^{++} \end{pmatrix}\cdot\left[\overline{D}^{+}_{10}(p_0)\right]^{-1}, \tag{110}$$

$$\overline{Q}^{-+}(u|p_0) = (\epsilon_1-\epsilon_0)\left[\overline{D}^{+}_{10}(u)\right]^{-1}$$
$$\times\begin{pmatrix} \epsilon_1\,\overline{A}^{-+} - \epsilon_0\,\alpha_1(u)\alpha_1(p_0)\,F_V^{+}(u)\,\overline{B}^{+-} & -\epsilon_0^{\frac{1}{2}}\,\alpha_1(u)\,\overline{J}^{++} \\ -\epsilon_0^{\frac{1}{2}}\,\alpha_1(p_0)\,\overline{G}^{--} & \epsilon_0\,K_0^2\,\overline{C}^{-+} \end{pmatrix}\cdot\left[\overline{D}^{+}_{10}(p_0)\right]^{-1}, \tag{111}$$

$$\overline{Q}^{+-}(u|p_0) = \frac{(\epsilon_1-\epsilon_0)}{\alpha_0(p_0)}\,[\overline{D}^{+}_{10}(u)]^{-1}$$
$$\times\begin{pmatrix} \epsilon_0\,\alpha_1(p_0)\,\overline{A}^{+-} - \epsilon_1\,\alpha_1(u)\,\alpha_0^2(p_0)\,\overline{B}^{-+} & -\epsilon_0^{\frac{1}{2}}\,\alpha_1(u)\,\alpha_1(p_0)\,\overline{J}^{--} \\ -\epsilon_0^{\frac{1}{2}}\,\epsilon_1\,\alpha_0^2(p_0)\,\overline{G}^{++} & \epsilon_0\,K_0^2\,\alpha_1(p_0)\,\overline{C}^{+-} \end{pmatrix}$$
$$\cdot\left[\overline{D}^{+}_{10}(p_0)\right]^{-1}, \tag{112}$$

$$\overline{Q}^{--}(u|p_0) = \frac{(\epsilon_1-\epsilon_0)}{\alpha_0(p_0)}\,[\overline{D}^{+}_{10}(u)]^{-1}$$
$$\times\begin{pmatrix} \epsilon_0\,\alpha_1(p_0)\,\overline{A}^{--} - \epsilon_1\,\alpha_1(u)\,\alpha_0^2(p_0)\,\overline{B}^{++} & -\epsilon_0^{\frac{1}{2}}\,\alpha_1(u)\,\alpha_1(p_0)\,\overline{J}^{+-} \\ -\epsilon_0^{\frac{1}{2}}\,\epsilon_1\,\alpha_0^2(p_0)\,\overline{G}^{-+} & \epsilon_0\,K_0^2\,\alpha_1(p_0)\,\overline{C}^{--} \end{pmatrix}$$
$$\cdot\left[\overline{D}^{+}_{10}(p_0)\right]^{-1}, \tag{113}$$

where

$$\overline{A}^{a,b} = ||u||||p_0||\, F_V^a(u)F_V^b(p_0), \tag{114}$$

$$\overline{B}^{a,b} = F_V^a(u)\, F_V^b(p_0)\hat{u}\cdot\hat{p}_0, \tag{115}$$

$$\overline{C}^{a.b} = F_H^a(u)F_H^b(p_0)\hat{u}\cdot\hat{p}_0, \tag{116}$$

$$\overline{J}^{a,b} = F_V^a(u)F_H^b(p_0)\,(\hat{u}\times\hat{p}_0)_z, \tag{117}$$

$$\overline{G}^{a,b} = F_H^a(u)F_V^b(p_0)\,(\hat{u}\times\hat{p}_0)_z, \tag{118}$$

$$\begin{pmatrix} F_V^{\pm}(p_0) & 0 \\ 0 & F_H^{\pm}(p_0)\end{pmatrix} = \left(\overline{1} \pm \overline{V}^{H\,21}(p_0)\right)\left(\overline{1} + \overline{V}^{10}(p_0)\cdot\overline{V}^{H\,21}(p_0)\right)^{-1}. \tag{119}$$

The explicit expression of the matrix $\overline{P}^{\pm}$ is the following one:

$$\overline{P}^{+}(u|p_0) = (\epsilon_1-\epsilon_0)\left[\overline{D}_{10}^{+}(u)\right]^{-1}\cdot \begin{pmatrix} ||u||||p||F_V^{+}(u) + \alpha_1(u)\alpha_0(p)\,F_V^{-}(u)\hat{u}\cdot\hat{p}_1 & -\epsilon_0^{\frac{1}{2}}\,K_0\,\alpha_1(u)\,F_V^{-}(u)\,(\hat{u}\times\hat{p}_1)_z \\ \epsilon_0^{-\frac{1}{2}}\,K_0\,\alpha_0(p)\,F_H^{+}(u)\,(\hat{u}\times\hat{p}_1)_z & K_0^2\,F_H^{+}(u)\hat{u}\cdot\hat{p}_1 \end{pmatrix} \tag{120}$$

$$\overline{P}^{-}(u|p_0) = (\epsilon_1-\epsilon_0)\left[\overline{D}_{10}^{+}(u)\right]^{-1}\cdot \begin{pmatrix} ||u||||p||F_V^{-}(u) + \alpha_1(u)\alpha_0(p)\,F_V^{+}(u)\hat{u}\cdot\hat{p}_1 & -\epsilon_0^{\frac{1}{2}}\,K_0\,\alpha_1(u)\,F_V^{+}(u)\,(\hat{u}\times\hat{p}_1)_z \\ \epsilon_0^{-\frac{1}{2}}\,K_0\,\alpha_0(p)\,F_H^{-}(u)\,(\hat{u}\times\hat{p}_1)_z & K_0^2\,F_H^{-}(u)\hat{u}\cdot\hat{p}_1 \end{pmatrix}. \tag{121}$$

The first-order term was derived by Fuks *et al.*[17] They have noticed that for this order, the matrix differs from the one obtained for a surface separating two semi-infinite media by only the factors $F^{\pm}$. Likewise, for higher-orders, we see that Eqs. (110),(111) differ from Eq. (64) by only $F^{\pm}$, similarly for Eqs. (112),(113) with respect to Eq. (65), and Eqs. (120),(121) with respect to Eq. (67). So when the thickness H becomes infinite, and the absorption $\mathrm{Im}(\epsilon_1) \neq 0$, or if $\epsilon_1 = \epsilon_2$, we have $\overline{V}^{H\,21} = 0$, thus $F^{\pm} = 1$, and in that case we recover matrix (64)–(67) for a rough surface between two semi-infinite media.

6.4.2. Case of Two Rough Surfaces

The slab is bounded by two weakly rough surfaces. The structure is shown in Fig. 6.8, where the two rough surfaces separate three media.

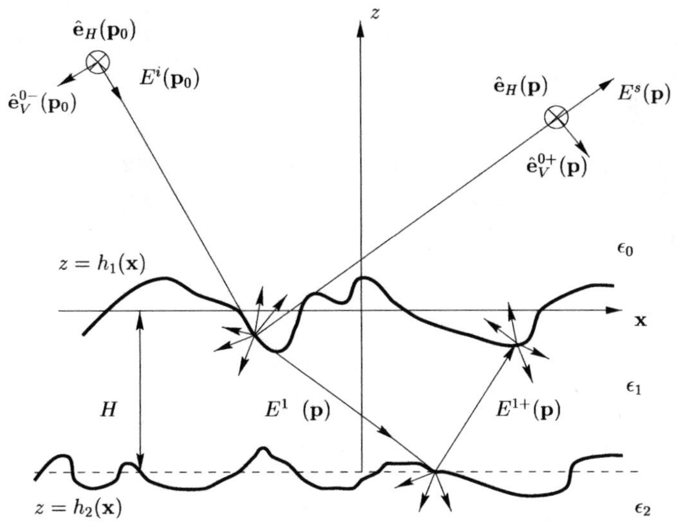

FIGURE 6.8. A rough surface with an incident wave coming from the medium 0 and scattered by a slab with two randomly rough surfaces.

These three media are characterized by isotropic, homogeneous dielectric constants ϵ_0, ϵ_1 and ϵ_2, respectively. The two boundaries of the rough surfaces are located at the heights $z = h_1(\boldsymbol{x})$, $z = -H + h_2(\boldsymbol{x})$, $\boldsymbol{x} = (x, y)$. When the slab has two rough boundaries, the enhancement of backscattering can be produced by two kinds of interaction. The classic one, where the wave is scattered twice by the same boundary, and another[19](see Fig. 6.9) where the wave is transmitted at a

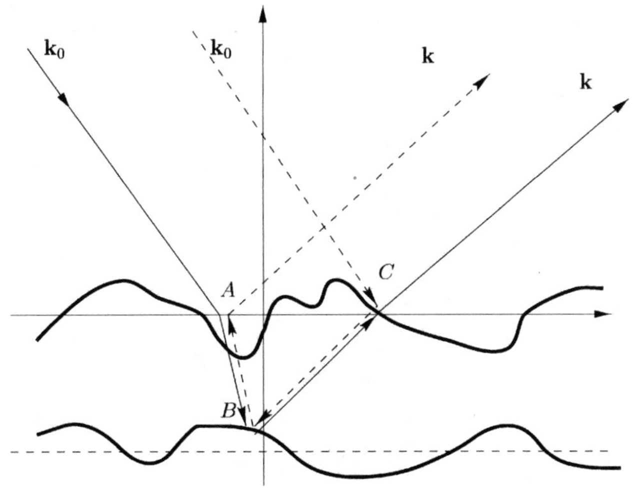

FIGURE 6.9. Mechanisms responsible of enhanced backscattering.

point A of the surface without being diffused, then the wave is scattered by the second rough surface at a point B and by the first one at C. If the wave follows the same path in the reverse direction and excites the same guided mode, then the phase difference between the two paths is $\Delta\phi = \boldsymbol{r}_{BC} \cdot (\boldsymbol{k} + \boldsymbol{k}_i)$, where $\boldsymbol{r}_{BC}$ is the distance between B and C. This phase difference between these two waves is zero and independent of the random position of the points B and C in the antispecular direction ($\boldsymbol{k} = -\boldsymbol{k}_i$) which produces the backscattering peak.

In this section, we study the perturbative development of the scattered fields as a function of the surface elevations h_1 and h_2. In a perturbative expansion of the scattering matrix, the terms which contain an expression such as $h_1^n h_2^m$ will be denoted $\overline{\boldsymbol{R}}^{(nm)}$. The perturbative development of $\overline{\boldsymbol{R}}$ can be written as follows:

$$\overline{\boldsymbol{R}} = \overline{\boldsymbol{R}}^{(00)} + \overline{\boldsymbol{R}}^{(10)} + \overline{\boldsymbol{R}}^{(01)} + \overline{\boldsymbol{R}}^{(11)} + \overline{\boldsymbol{R}}^{(20)} + \overline{\boldsymbol{R}}^{(02)} + \overline{\boldsymbol{R}}^{(21)} + \overline{\boldsymbol{R}}^{(12)} + \overline{\boldsymbol{R}}^{(22)} + \overline{\boldsymbol{R}}^{(30)} + \overline{\boldsymbol{R}}^{(03)} + \cdots \tag{122}$$

To obtain the reduced Rayleigh equations, we follow the same calculation procedure we have given in the case of one rough surface, whose equations are

$$\int \frac{\mathrm{d}^2 \boldsymbol{p}}{(2\pi)^2} \overline{\boldsymbol{M}}_h^{1+,0+}(\boldsymbol{u}|\boldsymbol{p}) \cdot \overline{\boldsymbol{R}}(\boldsymbol{p}|\boldsymbol{p}_0) \cdot \boldsymbol{E}^i(\boldsymbol{p}_0) + \overline{\boldsymbol{M}}_h^{1+,0-}(\boldsymbol{u}|\boldsymbol{p}_0) \cdot \boldsymbol{E}^i(\boldsymbol{p}_0) = \frac{2\,(\epsilon_0\,\epsilon_1)^{\frac{1}{2}}\,\alpha_1(\boldsymbol{u})}{(\epsilon_1 - \epsilon_0)} \boldsymbol{E}^{1+}(\boldsymbol{u}), \tag{123}$$

$$\int \frac{\mathrm{d}^2 \boldsymbol{p}}{(2\pi)^2} \overline{\boldsymbol{M}}_h^{1-,0+}(\boldsymbol{u}|\boldsymbol{p}) \cdot \overline{\boldsymbol{R}}(\boldsymbol{p}|\boldsymbol{p}_0) \cdot \boldsymbol{E}^i(\boldsymbol{p}_0) + \overline{\boldsymbol{M}}_h^{1-,0-}(\boldsymbol{u}|\boldsymbol{p}_0) \cdot \boldsymbol{E}^i(\boldsymbol{p}_0) = -\frac{2\,(\epsilon_0\,\epsilon_1)^{\frac{1}{2}}\,\alpha_1(\boldsymbol{u})}{(\epsilon_1 - \epsilon_0)} \boldsymbol{E}^{1-}(\boldsymbol{u}), \tag{124}$$

where $\overline{\boldsymbol{M}}_h^{1b,0a}$ are given by the Eqs. (45)–(46) . In order to obtain a single equation for $\overline{\boldsymbol{R}}(\boldsymbol{p}\,|\,\boldsymbol{p}_0)$, we find a relation between $\boldsymbol{E}^{1-}$ and $\boldsymbol{E}^{1+}$. We already know the expression of the scattering matrix for a single rough surface separating two homogenous media of permittivity ϵ_1 and ϵ_2, translated along the z-axis to the height $z = -H$. This matrix is denoted by $\overline{\boldsymbol{R}}_{s\,\epsilon_1,\epsilon_2}(\boldsymbol{p}\,|\,\boldsymbol{p}_0)$ (see Eq. (74)), and we have the following expressions:

$$\overline{\boldsymbol{R}}^H_{s\,\epsilon_1,\epsilon_2}(\boldsymbol{p}\,|\,\boldsymbol{p}_0) = \exp(\mathrm{i}(\alpha_1(\boldsymbol{p}) + \alpha_1(\boldsymbol{p}_0))\,H)\,\overline{\boldsymbol{R}}_{s\,\epsilon_1,\epsilon_2}(\boldsymbol{p}\,|\,\boldsymbol{p}_0). \tag{125}$$

The phase term in Eq. (125) is given by the translation $z = -H$ of the rough surface $h_2(\boldsymbol{x})$. We obtain the following relation:

$$\boldsymbol{E}^{1+}(\boldsymbol{u}) = \int \frac{\mathrm{d}^2 \boldsymbol{u}_1}{(2\pi)^2} \overline{\boldsymbol{R}}^H_{s\,\epsilon_1,\epsilon_2}(\boldsymbol{u}\,|\,\boldsymbol{u}_1) \cdot \boldsymbol{E}^{1-}(\boldsymbol{u}_1). \tag{126}$$

Combining Eq. (126) with Eqs. (123)–(124), we obtain the integral equation for the operator $\overline{\boldsymbol{R}}(\boldsymbol{p}|\boldsymbol{p}_0)$

$$\int \frac{d^2\boldsymbol{p}}{(2\pi)^2}\left[\overline{\boldsymbol{M}}_h^{1+,0+}(\boldsymbol{u}|\boldsymbol{p}) + \int \frac{d^2\boldsymbol{u}_1}{(2\pi)^2}\frac{\alpha_1(\boldsymbol{u})}{\alpha_1(\boldsymbol{u}_1)}\overline{\boldsymbol{R}}^H_{s\,\epsilon_1,\epsilon_2}(\boldsymbol{u}|\boldsymbol{u}_1)\cdot\overline{\boldsymbol{M}}_h^{1-,0+}(\boldsymbol{u}_1|\boldsymbol{p})\right]$$
$$\times\overline{\boldsymbol{R}}(\boldsymbol{p}|\boldsymbol{p}_0) = -\left[\overline{\boldsymbol{M}}_h^{1+,0-}(\boldsymbol{u}|\boldsymbol{p}_0) + \int \frac{d^2\boldsymbol{u}_1}{(2\pi)^2}\frac{\alpha_1(\boldsymbol{u})}{\alpha_1(\boldsymbol{u}_1)}\overline{\boldsymbol{R}}^H_{s\,\epsilon_1,\epsilon_2}(\boldsymbol{u}|\boldsymbol{u}_1)\cdot\overline{\boldsymbol{M}}_h^{1-,0-}(\boldsymbol{u}|\boldsymbol{p}_0)\right]. \tag{127}$$

Expanding $I(\alpha|\boldsymbol{p})$ in powers of h_1 (Eqs. (55),(56)) and using the perturbative development of $\overline{\boldsymbol{R}}^H_{s\,\epsilon_1,\epsilon_2}$ in powers of h_2 (Eq. (52)), we get the expression

$$\begin{aligned}\overline{\boldsymbol{R}}^H_{s\,\epsilon_1,\epsilon_2}(\boldsymbol{p}|\boldsymbol{p}_0) &= (2\pi)^2\delta(\boldsymbol{p}-\boldsymbol{p}_0)\,\overline{\boldsymbol{X}}^{H(0)}_{s\,\epsilon_1,\epsilon_2}(\boldsymbol{p}_0)\\ &\quad+\alpha_0(\boldsymbol{p}_0)\,\overline{\boldsymbol{X}}^{H(1)}_{s\,\epsilon_1,\epsilon_2}(\boldsymbol{p}|\boldsymbol{p}_0)\,h_2(\boldsymbol{p}-\boldsymbol{p}_0)\\ &\quad+\alpha_0(\boldsymbol{p}_0)\int\frac{d^2\boldsymbol{p}_1}{(2\pi)^2}\,\overline{\boldsymbol{X}}^{H(2)}_{s\,\epsilon_1,\epsilon_2}(\boldsymbol{p}|\boldsymbol{p}_1|\boldsymbol{p}_0)\,h_2(\boldsymbol{p}-\boldsymbol{p}_1)h_2(\boldsymbol{p}_1-\boldsymbol{p}_0)\\ &\quad+\alpha_0(\boldsymbol{p}_0)\iint\frac{d^2\boldsymbol{p}_1}{(2\pi)^2}\frac{d\boldsymbol{p}_2}{(2\pi)^2}\,\overline{\boldsymbol{X}}^{H(3)}_{s\,\epsilon_1,\epsilon_2}(\boldsymbol{p}|\boldsymbol{p}_1|\boldsymbol{p}_2|\boldsymbol{p}_0)\\ &\quad\times h_2(\boldsymbol{p}-\boldsymbol{p}_1)h_2(\boldsymbol{p}_1-\boldsymbol{p}_2)h_2(\boldsymbol{p}_2-\boldsymbol{p}_0),\end{aligned} \tag{128}$$

where $\overline{\boldsymbol{X}}^{H(0)}$ is given by Eq. (99). We define the following notation:

$$\overline{\boldsymbol{R}}^{(n0)}(\boldsymbol{p}|\boldsymbol{p}_0) = \overline{\boldsymbol{R}}^{(n)}_u(\boldsymbol{p}|\boldsymbol{p}_0), \tag{129}$$
$$\overline{\boldsymbol{R}}^{(0n)}(\boldsymbol{p}|\boldsymbol{p}_0) = \overline{\boldsymbol{R}}^{(n)}_d(\boldsymbol{p}|\boldsymbol{p}_0), \tag{130}$$
$$\overline{\boldsymbol{X}}^{(n0)}(\boldsymbol{p}|\boldsymbol{p}_0) = \overline{\boldsymbol{X}}^{(n)}_u(\boldsymbol{p}|\boldsymbol{p}_0), \tag{131}$$
$$\overline{\boldsymbol{X}}^{(0n)}(\boldsymbol{p}|\boldsymbol{p}_0) = \overline{\boldsymbol{X}}^{(n)}_d(\boldsymbol{p}|\boldsymbol{p}_0), \tag{132}$$

where $\overline{\boldsymbol{R}}^{(n)}$ is given by Eqs. (91)–(94), and $\overline{\boldsymbol{X}}^{(n)}$ by Eqs. (96)–(98) and Eqs. (105)–(107).

We finally obtain the perturbative development (122), and we have

$$\overline{\boldsymbol{R}}^{(10)}(\boldsymbol{p}|\boldsymbol{p}_0) = \alpha_0(\boldsymbol{p}_0)\,\overline{\boldsymbol{X}}^{(10)}(\boldsymbol{p}|\boldsymbol{p}_0)\,h_1(\boldsymbol{p}-\boldsymbol{p}_0), \tag{133}$$
$$\overline{\boldsymbol{R}}^{(01)}(\boldsymbol{p}|\boldsymbol{p}_0) = \alpha_0(\boldsymbol{p}_0)\,\overline{\boldsymbol{X}}^{(01)}(\boldsymbol{p}|\boldsymbol{p}_0)\,h_2(\boldsymbol{p}-\boldsymbol{p}_0), \tag{134}$$

$$\begin{aligned}\overline{\boldsymbol{R}}^{(11)}(\boldsymbol{p}|\boldsymbol{p}_0) &= \alpha_0(\boldsymbol{p}_0)\int\frac{d^2\boldsymbol{p}_1}{(2\pi)^2}\left[\overline{\boldsymbol{X}}^{(11)12}(\boldsymbol{p}|\boldsymbol{p}_1|\boldsymbol{p}_0)\,h_1(\boldsymbol{p}-\boldsymbol{p}_1)\,h_2(\boldsymbol{p}-\boldsymbol{p}_0)\right.\\ &\quad\left.+\overline{\boldsymbol{X}}^{(11)21}(\boldsymbol{p}|\boldsymbol{p}_1|\boldsymbol{p}_0)\,h_2(\boldsymbol{p}-\boldsymbol{p}_1)\,h_1(\boldsymbol{p}-\boldsymbol{p}_0)\right],\end{aligned} \tag{135}$$

$$\overline{\boldsymbol{R}}^{(20)}(\boldsymbol{p}\,|\,\boldsymbol{p}_0) = \alpha_0(\boldsymbol{p}_0) \int \frac{\mathrm{d}^2 \boldsymbol{p}_1}{(2\pi)^2}\, \overline{\boldsymbol{X}}^{(20)}(\boldsymbol{p}\,|\,\boldsymbol{p}_1|\,\boldsymbol{p}_0)\, h_1(\boldsymbol{p} - \boldsymbol{p}_1)\, h_1(\boldsymbol{p}_1 - \boldsymbol{p}_0)\,, \tag{136}$$

$$\overline{\boldsymbol{R}}^{(02)}(\boldsymbol{p}\,|\,\boldsymbol{p}_0) = \alpha_0(\boldsymbol{p}_0) \int \frac{\mathrm{d}^2 \boldsymbol{p}_1}{(2\pi)^2}\, \overline{\boldsymbol{X}}^{(02)}(\boldsymbol{p}\,|\,\boldsymbol{p}_1|\,\boldsymbol{p}_0)\, h_2(\boldsymbol{p} - \boldsymbol{p}_1)\, h_2(\boldsymbol{p}_1 - \boldsymbol{p}_0)\,, \tag{137}$$

$$\begin{aligned}\overline{\boldsymbol{R}}^{(21)}(\boldsymbol{p}\,|\,\boldsymbol{p}_0) = \alpha_0(\boldsymbol{p}_0) \iint \frac{\mathrm{d}^2 \boldsymbol{p}_1}{(2\pi)^2}\frac{\mathrm{d}\boldsymbol{p}_2}{(2\pi)^2}\\ \times \Big[\overline{\boldsymbol{X}}^{(21)112}(\boldsymbol{p}\,|\,\boldsymbol{p}_1|\,\boldsymbol{p}_2|\,\boldsymbol{p}_0)\, h_1(\boldsymbol{p} - \boldsymbol{p}_1)\, h_1(\boldsymbol{p}_1 - \boldsymbol{p}_2)\, h_2(\boldsymbol{p}_2 - \boldsymbol{p}_0)\\ + \overline{\boldsymbol{X}}^{(21)121}(\boldsymbol{p}\,|\,\boldsymbol{p}_1|\,\boldsymbol{p}_2|\,\boldsymbol{p}_0)\, h_1(\boldsymbol{p} - \boldsymbol{p}_1)\, h_2(\boldsymbol{p}_1 - \boldsymbol{p}_2)\, h_1(\boldsymbol{p}_2 - \boldsymbol{p}_0)\\ + \overline{\boldsymbol{X}}^{(21)211}(\boldsymbol{p}\,|\,\boldsymbol{p}_1|\,\boldsymbol{p}_2|\,\boldsymbol{p}_0)\, h_2(\boldsymbol{p} - \boldsymbol{p}_1)\, h_1(\boldsymbol{p}_1 - \boldsymbol{p}_2)\, h_1(\boldsymbol{p}_2 - \boldsymbol{p}_0)\Big]\,,\end{aligned} \tag{138}$$

$$\begin{aligned}\overline{\boldsymbol{R}}^{(12)}(\boldsymbol{p}\,|\,\boldsymbol{p}_0) = \alpha_0(\boldsymbol{p}_0) \iint \frac{\mathrm{d}^2 \boldsymbol{p}_1}{(2\pi)^2}\frac{\mathrm{d}\boldsymbol{p}_2}{(2\pi)^2}\\ \times \Big[\overline{\boldsymbol{X}}^{(12)221}(\boldsymbol{p}\,|\,\boldsymbol{p}_1|\,\boldsymbol{p}_2|\,\boldsymbol{p}_0)\, h_2(\boldsymbol{p} - \boldsymbol{p}_1)\, h_2(\boldsymbol{p}_1 - \boldsymbol{p}_2)\, h_1(\boldsymbol{p}_2 - \boldsymbol{p}_0)\\ + \overline{\boldsymbol{X}}^{(12)212}(\boldsymbol{p}\,|\,\boldsymbol{p}_1|\,\boldsymbol{p}_2|\,\boldsymbol{p}_0)\, h_2(\boldsymbol{p} - \boldsymbol{p}_1)\, h_1(\boldsymbol{p}_1 - \boldsymbol{p}_2)\, h_2(\boldsymbol{p}_2 - \boldsymbol{p}_0)\\ + \overline{\boldsymbol{X}}^{(12)122}(\boldsymbol{p}\,|\,\boldsymbol{p}_1|\,\boldsymbol{p}_2|\,\boldsymbol{p}_0)\, h_1(\boldsymbol{p} - \boldsymbol{p}_1)\, h_2(\boldsymbol{p}_1 - \boldsymbol{p}_2)\, h_2(\boldsymbol{p}_2 - \boldsymbol{p}_0)\Big]\,,\end{aligned} \tag{139}$$

$$\begin{aligned}\overline{\boldsymbol{R}}^{(30)}(\boldsymbol{p}\,|\,\boldsymbol{p}_0) = \alpha_0(\boldsymbol{p}_0) \iint \frac{\mathrm{d}^2 \boldsymbol{p}_1}{(2\pi)^2}\frac{\mathrm{d}\boldsymbol{p}_2}{(2\pi)^2}\\ \times \overline{\boldsymbol{X}}^{(30)}(\boldsymbol{p}\,|\,\boldsymbol{p}_1|\,\boldsymbol{p}_2|\,\boldsymbol{p}_0)\, h_1(\boldsymbol{p} - \boldsymbol{p}_1)\, h_1(\boldsymbol{p}_1 - \boldsymbol{p}_2)\, h_1(\boldsymbol{p}_2 - \boldsymbol{p}_0)\end{aligned} \tag{140}$$

$$\begin{aligned}\overline{\boldsymbol{R}}^{(03)}(\boldsymbol{p}\,|\,\boldsymbol{p}_0) = \alpha_0(\boldsymbol{p}_0) \iint \frac{\mathrm{d}^2 \boldsymbol{p}_1}{(2\pi)^2}\frac{\mathrm{d}\boldsymbol{p}_2}{(2\pi)^2}\\ \times \overline{\boldsymbol{X}}^{(03)}(\boldsymbol{p}\,|\,\boldsymbol{p}_1|\,\boldsymbol{p}_2|\,\boldsymbol{p}_0)\, h_2(\boldsymbol{p} - \boldsymbol{p}_1)\, h_2(\boldsymbol{p}_1 - \boldsymbol{p}_2)\, h_2(\boldsymbol{p}_2 - \boldsymbol{p}_0)\,.\end{aligned} \tag{141}$$

In these expressions, superscripts in some terms indicate the order of appearance of the functions h_1 and h_2. For example, in $\overline{\boldsymbol{X}}^{(21)121}$ the superscript 121 indicates that this coefficient is associated with the product $h_1(\boldsymbol{p} - \boldsymbol{p}_1)\, h_2(\boldsymbol{p}_1 - \boldsymbol{p}_2)\, h_1(\boldsymbol{p}_2 - \boldsymbol{p}_0)$. We have the following expressions:

$$\begin{aligned}\overline{\boldsymbol{X}}^{(11)21}(\boldsymbol{p}\,|\,\boldsymbol{p}_1|\,\boldsymbol{p}_0) = \overline{\boldsymbol{T}}^{10}(\boldsymbol{p}) \cdot \overline{\boldsymbol{U}}^{(0)}(\boldsymbol{p}) \cdot \overline{\boldsymbol{X}}^{H(1)}_{s\,\epsilon_1,\epsilon_2}(\boldsymbol{p}\,|\,\boldsymbol{p}_1)\\ \times \Big[-\overline{\epsilon} \cdot \overline{\boldsymbol{D}}^{-}_{10}(\boldsymbol{p}_1) \cdot \overline{\boldsymbol{X}}^{(10)}(\boldsymbol{p}_1|\,\boldsymbol{p}_0) + \mathrm{i}\overline{\boldsymbol{S}}^{+}(\boldsymbol{p}_1|\,\boldsymbol{p}_0)\Big]\end{aligned} \tag{142}$$

$$\overline{\boldsymbol{X}}^{(11)12}(\boldsymbol{p}|\boldsymbol{p}_1|\boldsymbol{p}_0) = \mathrm{i}\,\overline{\boldsymbol{P}}^{+}(\boldsymbol{p}|\boldsymbol{p}_1)\cdot\overline{\boldsymbol{X}}^{(01)}(\boldsymbol{p}_1|\boldsymbol{p}_0), \tag{143}$$

$$\overline{\boldsymbol{X}}^{(21)112}(\boldsymbol{p}|\boldsymbol{p}_1|\boldsymbol{p}_2|\boldsymbol{p}_0) = \mathrm{i}\,\overline{\boldsymbol{P}}^{+}(\boldsymbol{p}|\boldsymbol{p}_1)\cdot\overline{\boldsymbol{X}}^{(11)12}(\boldsymbol{p}_1|\boldsymbol{p}_2|\boldsymbol{p}_0) + \frac{1}{2}\Big[\alpha_1(\boldsymbol{p})\overline{\boldsymbol{P}}^{-}(\boldsymbol{p}|\boldsymbol{p}_2) - \alpha_0(\boldsymbol{p}_2)\overline{\boldsymbol{P}}^{+}(\boldsymbol{p}|\boldsymbol{p}_2)\Big]\cdot\overline{\boldsymbol{X}}^{(01)}(\boldsymbol{p}_2|\boldsymbol{p}_0) \tag{144}$$

$$\overline{\boldsymbol{X}}^{(21)121}(\boldsymbol{p}|\boldsymbol{p}_1|\boldsymbol{p}_2|\boldsymbol{p}_0) = \mathrm{i}\,\overline{\boldsymbol{P}}^{+}(\boldsymbol{p}|\boldsymbol{p}_1)\cdot\overline{\boldsymbol{X}}^{(11)21}(\boldsymbol{p}_1|\boldsymbol{p}_2|\boldsymbol{p}_0), \tag{145}$$

$$\begin{aligned}\overline{\boldsymbol{X}}^{(21)211}(\boldsymbol{p}|\boldsymbol{p}_1|\boldsymbol{p}_2|\boldsymbol{p}_0) = {}& \overline{\boldsymbol{T}}^{10}(\boldsymbol{p})\cdot\overline{\boldsymbol{U}}^{(0)}(\boldsymbol{p})\cdot\overline{\boldsymbol{X}}^{H(1)}_{s\,\epsilon_1,\epsilon_2}(\boldsymbol{p}|\boldsymbol{p}_1)\\ &\times\Big[-\overline{\epsilon}\cdot\overline{\boldsymbol{D}}^{-}_{10}(\boldsymbol{p}_1)\cdot\overline{\boldsymbol{X}}^{(20)}(\boldsymbol{p}_1|\boldsymbol{p}_2|\boldsymbol{p}_0)\\ &+\frac{\mathrm{i}\,(\epsilon_1-\epsilon_0)}{2\,(\epsilon_0\,\epsilon_1)^{1/2}}\overline{\boldsymbol{M}}^{1-,0+}(\boldsymbol{p}_1|\boldsymbol{p}_2)\cdot\overline{\boldsymbol{X}}^{(10)}(\boldsymbol{p}_2|\boldsymbol{p}_0)\\ &-\frac{1}{2}\Big(\alpha_1(\boldsymbol{p}_1)\overline{\boldsymbol{S}}^{+}(\boldsymbol{p}_1|\boldsymbol{p}_0)+\alpha_0(\boldsymbol{p}_0)\overline{\boldsymbol{S}}^{-}(\boldsymbol{p}_1|\boldsymbol{p}_0)\Big)\Big]\end{aligned} \tag{146}$$

$$\overline{\boldsymbol{X}}^{(12)122}(\boldsymbol{p}|\boldsymbol{p}_1|\boldsymbol{p}_2|\boldsymbol{p}_0) = \mathrm{i}\,\overline{\boldsymbol{P}}^{+}(\boldsymbol{p}|\boldsymbol{p}_1)\cdot\overline{\boldsymbol{X}}^{(02)}(\boldsymbol{p}_1|\boldsymbol{p}_2|\boldsymbol{p}_0), \tag{147}$$

$$\begin{aligned}\overline{\boldsymbol{X}}^{(12)212}(\boldsymbol{p}|\boldsymbol{p}_1|\boldsymbol{p}_2|\boldsymbol{p}_0) = {}& \overline{\boldsymbol{T}}^{10}(\boldsymbol{p})\cdot\overline{\boldsymbol{U}}^{(0)}(\boldsymbol{p})\cdot\overline{\boldsymbol{X}}^{H(1)}_{s\,\epsilon_1,\epsilon_2}(\boldsymbol{p}|\boldsymbol{p}_1)\\ &\times\Big[-\overline{\epsilon}\cdot\overline{\boldsymbol{D}}^{-}_{10}(\boldsymbol{p}_1)\times\overline{\boldsymbol{X}}^{(11)12}(\boldsymbol{p}_1|\boldsymbol{p}_2|\boldsymbol{p}_0)\\ &+\frac{\mathrm{i}\,(\epsilon_1-\epsilon_0)}{2\,(\epsilon_0\,\epsilon_1)^{1/2}}\overline{\boldsymbol{M}}^{1-,0+}(\boldsymbol{p}_1|\boldsymbol{p}_2)\cdot\overline{\boldsymbol{X}}^{(01)}(\boldsymbol{p}_2|\boldsymbol{p}_0)\Big],\end{aligned} \tag{148}$$

$$\begin{aligned}\overline{\boldsymbol{X}}^{(12)221}(\boldsymbol{p}|\boldsymbol{p}_1|\boldsymbol{p}_2|\boldsymbol{p}_0) = {}& \overline{\boldsymbol{T}}^{10}(\boldsymbol{p})\cdot\overline{\boldsymbol{U}}^{(0)}(\boldsymbol{p})\Big[-\overline{\boldsymbol{X}}^{H(1)}_{s\,\epsilon_1,\epsilon_2}(\boldsymbol{p}|\boldsymbol{p}_1)\cdot\overline{\epsilon}\cdot\overline{\boldsymbol{D}}^{-}_{10}(\boldsymbol{p}_1)\\ &\times\overline{\boldsymbol{X}}^{(11)21}(\boldsymbol{p}_1|\boldsymbol{p}_2|\boldsymbol{p}_0)+\overline{\boldsymbol{X}}^{H(2)}_{s\,\epsilon_1,\epsilon_2}(\boldsymbol{p}|\boldsymbol{p}_1|\boldsymbol{p}_2)\\ &\times\Big(-\overline{\epsilon}\cdot\overline{\boldsymbol{D}}^{-}_{10}(\boldsymbol{p}_2)\cdot\overline{\boldsymbol{X}}^{(10)}(\boldsymbol{p}_2|\boldsymbol{p}_0)+\mathrm{i}\,\overline{\boldsymbol{S}}^{+}(\boldsymbol{p}_2|\boldsymbol{p}_0)\Big)\Big].\end{aligned} \tag{149}$$

The matrices $\overline{\epsilon}$ and $\overline{\boldsymbol{S}}^{\pm}(\boldsymbol{p}_1|\boldsymbol{p}_0)$ are given in Appendix C, $\overline{\boldsymbol{P}}^{\pm}$ by Eqs. (120),(121), and $\overline{\boldsymbol{U}}^{(0)}$ by Eq. (87).

6.5. The Mueller Matrix Cross-Section and the Surface Statistics

When we consider an observation point in the far-field limit, the saddle-point method yields an asymptotic expression for the scattered field $\boldsymbol{E}^s \equiv \boldsymbol{E}^{0+}$ given by Eq. (50) :

$$\boldsymbol{E}^s(\boldsymbol{x}, z) = \frac{\exp(\mathrm{i}K_0||\boldsymbol{r}||)}{||\boldsymbol{r}||}\overline{\boldsymbol{f}}(\boldsymbol{p}|\boldsymbol{p}_0)\cdot\boldsymbol{E}^i(\boldsymbol{p}_0), \tag{150}$$

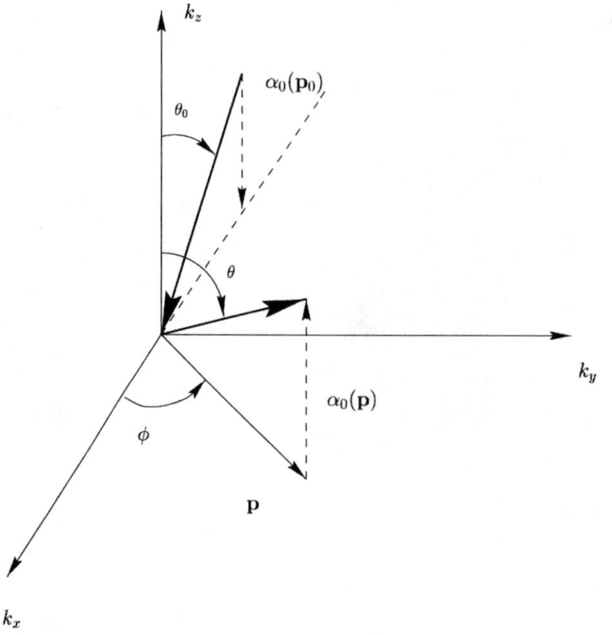

FIGURE 6.10. Definition of the scattering vector.

with

$$\overline{f}(\boldsymbol{p}\,|\,\boldsymbol{p}_0) \equiv \frac{K_0 \cos\theta}{2\pi \mathrm{i}}\,\overline{\boldsymbol{R}}(\boldsymbol{p}\,|\,\boldsymbol{p}_0), \tag{151}$$

$$\boldsymbol{p} = K_0 \frac{\boldsymbol{x}}{||\boldsymbol{r}||}, \tag{152}$$

where θ is the angle between $\hat{\boldsymbol{e}}_z$ and the scattering direction (see Fig. 6.10). In order to describe the incident and the scattered waves, we define the modified Stokes parameters:

$$\boldsymbol{I}^s(\boldsymbol{p}) \equiv \begin{pmatrix} |E_V^s(\boldsymbol{p})|^2 \\ |E_H^s(\boldsymbol{p})|^2 \\ 2Re(E_V^s(\boldsymbol{p})E_H^{s*}(\boldsymbol{p})) \\ 2Im(E_V^s(\boldsymbol{p})E_H^{s*}(\boldsymbol{p})) \end{pmatrix}, \qquad \boldsymbol{I}^i(\boldsymbol{p}_0) \equiv \begin{pmatrix} |E_V^i(\boldsymbol{p}_0)|^2 \\ |E_H^i(\boldsymbol{p}_0)|^2 \\ 2Re(E_V^i(\boldsymbol{p}_0)E_H^{i*}(\boldsymbol{p}_0)) \\ 2Im(E_V^i(\boldsymbol{p}_0)E_H^{i*}(\boldsymbol{p}_0)) \end{pmatrix}. \tag{153}$$

The analogue of the scattering matrix for these parameters is the Mueller matrix, defined[2] by

$$\boldsymbol{I}^s(\boldsymbol{p}) \equiv \frac{1}{||\boldsymbol{r}||^2}\overline{\boldsymbol{M}}(\boldsymbol{p}\,|\,\boldsymbol{p}_0) \cdot \boldsymbol{I}^i(\boldsymbol{p}_0). \tag{154}$$

This matrix can be expressed as a function[2] of $\overline{f}(\boldsymbol{p}\,|\,\boldsymbol{p}_0)$. To keep a matrix formulation in the following calculations, we define a new product between two-dimensional matrices with the definition:

$$\overline{f} \odot \overline{g} \equiv \begin{pmatrix} f_{VV} & f_{VH} \\ f_{HV} & f_{HH} \end{pmatrix} \odot \begin{pmatrix} g_{VV} & g_{VH} \\ g_{HV} & g_{HH} \end{pmatrix}$$

$$= \begin{pmatrix} f_{VV}g^*_{VV} & f_{VH}g^*_{VH} & \mathrm{Re}(f_{VV}g^*_{VH}) & -\mathrm{Im}(f_{VV}g^*_{VH}) \\ f_{HV}g^*_{HV} & f_{HH}g^*_{HH} & \mathrm{Re}(f_{HV}g^*_{HH}) & -\mathrm{Im}(f_{HV}g^*_{VH}) \\ 2\mathrm{Re}(f_{VV}g^*_{HV}) & 2\mathrm{Re}(f_{VH}g^*_{HH}) & \mathrm{Re}(f_{VV}g^*_{VV} + f_{HV}g^*_{VH}) & -\mathrm{Im}(f_{VV}g^*_{HH} - f_{VH}g^*_{HV}) \\ 2\mathrm{Im}(f_{VV}g^*_{HV}) & 2\mathrm{Im}(f_{VH}g^*_{HH}) & \mathrm{Im}(f_{VV}g^*_{VV} + f_{HV}g^*_{VH}) & \mathrm{Re}(f_{VV}g^*_{HH} - f_{VH}g^*_{HV}) \end{pmatrix}. \tag{155}$$

This product allows us to express the matrix $\overline{\boldsymbol{M}}$ as:

$$\overline{\boldsymbol{M}}(\boldsymbol{p}\,|\,\boldsymbol{p}_0) = \overline{\boldsymbol{f}}(\boldsymbol{p}\,|\,\boldsymbol{p}_0) \odot \overline{\boldsymbol{f}}(\boldsymbol{p}\,|\,\boldsymbol{p}_0), \tag{156}$$

$$= \frac{K_0^2 \cos^2\theta}{(2\pi)^2}\, \overline{\boldsymbol{R}}(\boldsymbol{p}\,|\,\boldsymbol{p}_0) \odot \overline{\boldsymbol{R}}(\boldsymbol{p}\,|\,\boldsymbol{p}_0). \tag{157}$$

Following Ishimaru *et al.*,[33] we define the Mueller matrix cross-section per unit area $\overline{\boldsymbol{\sigma}} = (\sigma_{ij})$:

$$\overline{\boldsymbol{\sigma}} \equiv \frac{4\pi}{A}\,\overline{\boldsymbol{M}}, \tag{158}$$

and the bistatic Mueller matrix[5] $\overline{\gamma} = (\gamma_{ij})$:

$$\overline{\gamma} \equiv \frac{1}{A\cos\theta_0}\,\overline{\boldsymbol{M}}, \tag{159}$$

where A is the unit area and θ_0 is the incidence angle. These matrices are the generalization of the classical coefficients. In fact, if we assume, for example, that the incident wave is vertically polarized we have

$$\frac{1}{A\,\cos\theta_0}\left|E_V^s(\boldsymbol{p})\right|^2 = \frac{1}{||r||^2}\gamma_{11}(\boldsymbol{p}\,|\,\boldsymbol{p}_0)\left|E_V^i(\boldsymbol{p})\right|^2, \tag{160}$$

$$\frac{1}{A\,\cos\theta_0}\left|E_H^s(\boldsymbol{p})\right|^2 = \frac{1}{||r||^2}\gamma_{21}(\boldsymbol{p}\,|\,\boldsymbol{p}_0)\left|E_V^i(\boldsymbol{p})\right|^2. \tag{161}$$

Thus γ_{11} and γ_{21} are respectively the classical bistatic coefficients γ_{VV} and γ_{HV}. We remark that we have computed the full Mueller matrix, giving us all the information on the diffuse process, in particular, for circular polarization. Thus, we can also define the cross-section and the bistatic coefficients for an incident circular polarization. As an example, taking the incident wave right circularly polarized, we have

$$\boldsymbol{I}^i(\boldsymbol{p}_0) = \frac{1}{2}(1\ 1\ 0\ {-2})^t. \tag{162}$$

If we put a right-hand side polarizer at the receiver, we get

$$\frac{1}{4}\begin{pmatrix} 1 & 1 & 0 & 1 \\ 1 & 1 & 0 & 1 \\ 0 & 0 & 0 & 0 \\ 2 & 2 & 0 & 2 \end{pmatrix}. \tag{163}$$

The right to right bistatic coefficient γ_{rr} is

$$\gamma_{rr} = \tfrac{1}{4}\left(\gamma_{11} + \gamma_{12} + 2\,\gamma_{14} + \gamma_{21} + \gamma_{22} + 2\,\gamma_{24} + \gamma_{41} + \gamma_{42} + 2\,\gamma_{44}\right), \tag{164}$$

where γ_{ij} are coefficients of the matrix $\overline{\gamma} = (\gamma_{ij})$.

In a similar way, we can obtain the right to left bistatic coefficient:

$$\gamma_{lr} = \tfrac{1}{4}\left(\gamma_{11} + \gamma_{12} + 2\,\gamma_{14} + \gamma_{21} + \gamma_{22} + 2\,\gamma_{24} - \gamma_{41} - \gamma_{42} - 2\,\gamma_{44}\right). \tag{165}$$

Up to now, we have made no hypothesis on the nature of the rough surface. Let us define the statistical characteristics of the function $h(\boldsymbol{x})$. We suppose that it is a stationary, isotropic Gaussian random process defined by its moments. We suppose that, in the case of two surfaces, $h_1(\boldsymbol{x})$ and $h_2(\boldsymbol{x})$, they are given by

$$\langle h_i(\boldsymbol{x})\rangle = 0, \tag{166}$$

$$\langle h_i(\boldsymbol{x})\, h_i(\boldsymbol{x}')\rangle = W_i(\boldsymbol{x} - \boldsymbol{x}'), \tag{167}$$

$$\langle h_1(\boldsymbol{x})\, h_2(\boldsymbol{x}')\rangle = 0, \tag{168}$$

where $i = 1, 2$, and the brackets denote an average over the ensemble of realizations of the functions $h_1(\boldsymbol{x})$ and $h_2(\boldsymbol{x})$. We note that the surface height correlation function $W_i(\boldsymbol{x})$ is not necessarily a Gaussian function. In the following sections, we use a Gaussian form for the surface-height correlation function $W_i(\boldsymbol{x})$:

$$W_i(\boldsymbol{x}) = \sigma_i^2 \, \exp(-\boldsymbol{x}^2/l_i^2), \tag{169}$$

where σ is the root mean square height of the surface, and l is the transverse correlation length. In momentum space, eqs. (166)–(168) lead to

$$\langle h_i(\boldsymbol{p})\rangle = 0, \tag{170}$$

$$\langle h_i(\boldsymbol{p})\, h_i(\boldsymbol{p}')\rangle = (2\pi)^2\, \delta(\boldsymbol{p} + \boldsymbol{p}')\, W_i(\boldsymbol{p}), \tag{171}$$

$$\langle h_1(\boldsymbol{p})\, h_2(\boldsymbol{p}')\rangle = 0 \tag{172}$$

$$\left\langle h_1^{2p+1}(\boldsymbol{x})\right\rangle = 0, \quad \text{with } p \text{ being a positive integer} \tag{173}$$

$$\left\langle h_2^{2p+1}(\boldsymbol{x})\right\rangle = 0\,, \tag{174}$$

with

$$W_i(\boldsymbol{p}) \equiv \int \mathrm{d}^2\boldsymbol{x}\; W(\boldsymbol{x}) \, \exp(-\mathrm{i}\boldsymbol{p}\cdot\boldsymbol{x}) \tag{175}$$

$$= \pi\, \sigma_i^2\, l_i^2 \, \exp(-\boldsymbol{p}^2\, l_i^2/4). \tag{176}$$

We can now define the bistatic coherent matrix:

$$\overline{\gamma}^{\rm coh} \equiv \frac{1}{A\,\cos\theta_0}\,\langle \overline{\boldsymbol{f}}(\boldsymbol{p}\,|\,\boldsymbol{p}_0)\rangle \odot \langle \overline{\boldsymbol{f}}(\boldsymbol{p}\,|\,\boldsymbol{p}_0)\rangle$$
$$= \frac{K_0^2\cos^2\theta}{A\,(2\pi)^2\,\cos\theta_0}\,\langle \overline{\boldsymbol{R}}(\boldsymbol{p}\,|\,\boldsymbol{p}_0)\rangle \odot \langle \overline{\boldsymbol{R}}(\boldsymbol{p}\,|\,\boldsymbol{p}_0)\rangle, \tag{177}$$

and the incoherent bistatic matrix

$$\overline{\gamma}^{\rm incoh}(\boldsymbol{p}\,|\,\boldsymbol{p}_0) \equiv \frac{1}{A\,\cos\theta_0}$$
$$\times[\langle \overline{\boldsymbol{f}}(\boldsymbol{p}\,|\,\boldsymbol{p}_0) \odot \overline{\boldsymbol{f}}(\boldsymbol{p}\,|\,\boldsymbol{p}_0)\rangle - \langle \overline{\boldsymbol{f}}(\boldsymbol{p}\,|\,\boldsymbol{p}_0)\rangle \odot \langle \overline{\boldsymbol{f}}(\boldsymbol{p}\,|\,\boldsymbol{p}_0)\rangle],$$
$$= \frac{K_0^2\cos^2\theta}{A\,(2\pi)^2\,\cos\theta_0}$$
$$\times\left[\langle \overline{\boldsymbol{R}}(\boldsymbol{p}\,|\,\boldsymbol{p}_0) \odot \overline{\boldsymbol{R}}(\boldsymbol{p}\,|\,\boldsymbol{p}_0)\rangle - \langle \overline{\boldsymbol{R}}(\boldsymbol{p}\,|\,\boldsymbol{p}_0)\rangle \odot \langle \overline{\boldsymbol{R}}(\boldsymbol{p}\,|\,\boldsymbol{p}_0)\rangle\right]. \tag{178}$$

6.5.1. *Case of One Randomly Rough Surface*

From Eq. (52), and the property of a Gaussian random process, we obtain

$$\overline{\gamma}^{\rm coh}(\boldsymbol{p}\,|\,\boldsymbol{p}_0) = \frac{K_0^2\cos^2\theta}{\cos\theta_0}\,\delta(\boldsymbol{p}-\boldsymbol{p}_0)\,\overline{\boldsymbol{R}}^{\rm coh}(\boldsymbol{p}_0) \odot \overline{\boldsymbol{R}}^{\rm coh}(\boldsymbol{p}_0) \tag{179}$$

$$\overline{\boldsymbol{R}}^{\rm coh}(\boldsymbol{p}_0) \equiv \overline{\boldsymbol{X}}^{(0)}(\boldsymbol{p}_0) + K_0\,\cos\theta_0 \int \frac{{\rm d}^2\boldsymbol{p}_1}{(2\pi)^2}\,W(\boldsymbol{p}_1-\boldsymbol{p}_0)$$
$$\overline{\boldsymbol{X}}^{(2)}(\boldsymbol{p}_0|\boldsymbol{p}_1|\boldsymbol{p}_0) + \cdots, \tag{180}$$

where $\overline{\boldsymbol{R}}^{\rm coh}(\boldsymbol{p}_0)$ is a diagonal matrix describing the reflection coefficients of the coherent waves. For the incoherent part of the scattered waves, we have

$$\overline{\gamma}^{\rm incoh}(\boldsymbol{p}\,|\,\boldsymbol{p}_0) = \frac{K_0^4\cos^2\theta\,\cos\theta_0}{(2\pi)^2}\left[\overline{\boldsymbol{I}}^{(1-1)}(\boldsymbol{p}\,|\,\boldsymbol{p}_0) + \overline{\boldsymbol{I}}^{(2-2)}(\boldsymbol{p}\,|\,\boldsymbol{p}_0) + \overline{\boldsymbol{I}}^{(3-1)}(\boldsymbol{p}\,|\,\boldsymbol{p}_0)\right], \tag{181}$$

where

$$\overline{\boldsymbol{I}}^{(1-1)}(\boldsymbol{p}\,|\,\boldsymbol{p}_0) \equiv W(\boldsymbol{p}-\boldsymbol{p}_0)\,\overline{\boldsymbol{X}}^{(1)}(\boldsymbol{p}\,|\,\boldsymbol{p}_0) \odot \overline{\boldsymbol{X}}^{(1)}(\boldsymbol{p}\,|\,\boldsymbol{p}_0) \tag{182}$$

$$\overline{\boldsymbol{I}}^{(2-2)}(\boldsymbol{p}\,|\,\boldsymbol{p}_0) \equiv \int \frac{{\rm d}^2\boldsymbol{p}_1}{(2\pi)^2}\,W(\boldsymbol{p}-\boldsymbol{p}_1)W(\boldsymbol{p}_1-\boldsymbol{p}_0)\,\overline{\boldsymbol{X}}^{(2)}(\boldsymbol{p}\,|\,\boldsymbol{p}_1|\,\boldsymbol{p}_0)$$
$$\odot\left[\overline{\boldsymbol{X}}^{(2)}(\boldsymbol{p}\,|\,\boldsymbol{p}_1|\,\boldsymbol{p}_0) + \overline{\boldsymbol{X}}^{(2)}(\boldsymbol{p}\,|\,\boldsymbol{p}+\boldsymbol{p}_0-\boldsymbol{p}_1|\,\boldsymbol{p}_0)\right] \tag{183}$$

$$\overline{\boldsymbol{I}}^{(3-1)}(\boldsymbol{p}\,|\,\boldsymbol{p}_0) \equiv W(\boldsymbol{p}-\boldsymbol{p}_0)\left[\overline{\boldsymbol{X}}^{(1)}(\boldsymbol{p}\,|\,\boldsymbol{p}_0) \odot \overline{\boldsymbol{X}}^{(3)}(\boldsymbol{p}\,|\,\boldsymbol{p}_0)\right.$$
$$\left.+\,\overline{\boldsymbol{X}}^{(3)}(\boldsymbol{p}\,|\,\boldsymbol{p}_0) \odot \overline{\boldsymbol{X}}^{(1)}(\boldsymbol{p}\,|\,\boldsymbol{p}_0)\right], \tag{184}$$

with

$$\overline{\boldsymbol{X}}^{(3)}(\boldsymbol{p}|\boldsymbol{p}_0) \equiv \int \frac{\mathrm{d}^2\boldsymbol{p}_1}{(2\pi)^2}\Big[W(\boldsymbol{p}_1-\boldsymbol{p}_0)\overline{\boldsymbol{X}}^{(3)}(\boldsymbol{p}|\boldsymbol{p}_0|\boldsymbol{p}_1|\boldsymbol{p}_0)$$
$$+W(\boldsymbol{p}-\boldsymbol{p}_1)\Big(\overline{\boldsymbol{X}}^{(3)}(\boldsymbol{p}|\boldsymbol{p}_1|\boldsymbol{p}_0-\boldsymbol{p}+\boldsymbol{p}_1|\boldsymbol{p}_0)$$
$$+\overline{\boldsymbol{X}}^{(3)}(\boldsymbol{p}|\boldsymbol{p}_1|\boldsymbol{p}|\boldsymbol{p}_0)\Big)\Big]. \tag{185}$$

6.5.2. *Case of Two Randomly Rough Surfaces*

In this section, we give the procedure to get the incoherent bistatic scattering matrix for a three-dimensional structure which combines two randomly rough surfaces. The incoherent bistatic matrix is given by the contribution of three terms:

$$\overline{\gamma}^{\text{incoh}}(\boldsymbol{p}|\boldsymbol{p}_0) = \overline{\gamma}_u^{\text{incoh}}(\boldsymbol{p}|\boldsymbol{p}_0) + \overline{\gamma}_d^{\text{incoh}}(\boldsymbol{p}|\boldsymbol{p}_0) + \overline{\gamma}_{up}^{\text{incoh}}(\boldsymbol{p}|\boldsymbol{p}_0), \tag{186}$$

where

$$\overline{\gamma}_u^{\text{incoh}}(\boldsymbol{p}|\boldsymbol{p}_0) = \frac{K_0^2\cos^2\theta}{A\,(2\pi)^2\,\cos\theta_0}\Big[\Big\langle\overline{\boldsymbol{R}}^{(10)}\odot\overline{\boldsymbol{R}}^{(10)}\Big\rangle$$
$$+\Big\langle\overline{\boldsymbol{R}}^{(20)}\odot\overline{\boldsymbol{R}}^{(20)}\Big\rangle+\Big\langle\overline{\boldsymbol{R}}^{(30)}\odot\overline{\boldsymbol{R}}^{(10)}\Big\rangle\Big] \tag{187}$$

corresponds to the incoherent bistatic matrix for the slab where only the upper surface has a roughness ($h_2(\boldsymbol{x}) = 0$), its expansion is developed up to order four in the rms-height elevation σ_1. Similarly, we define the matrix

$$\overline{\gamma}_d^{\text{incoh}}(\boldsymbol{p}|\boldsymbol{p}_0) = \frac{K_0^2\cos^2\theta}{A\,(2\pi)^2\,\cos\theta_0}$$
$$\times\Big[\Big\langle\overline{\boldsymbol{R}}^{(01)}\odot\overline{\boldsymbol{R}}^{(01)}\Big\rangle+\Big\langle\overline{\boldsymbol{R}}^{(02)}\odot\overline{\boldsymbol{R}}^{(02)}\Big\rangle+\Big\langle\overline{\boldsymbol{R}}^{(03)}\odot\overline{\boldsymbol{R}}^{(01)}\Big\rangle\Big], \tag{188}$$

where only the bottom surface is rough ($h_1(\boldsymbol{x}) = 0$), and the perturbative development is calculated up to order four in terms of σ_2. The last contribution $\overline{\gamma}_{ud}^{\text{incoh}}$ contains terms which describe the scattering process between the two rough surfaces, in this case only the leading terms are retained:

$$\overline{\gamma}_{ud}^{\text{incoh}}(\boldsymbol{p}|\boldsymbol{p}_0) = \frac{K_0^2\cos^2\theta}{A\,(2\pi)^2\,\cos\theta_0}\Big[\Big\langle\overline{\boldsymbol{R}}^{(10)}\odot\overline{\boldsymbol{R}}^{(12)}\Big\rangle+\Big\langle\overline{\boldsymbol{R}}^{(12)}\odot\overline{\boldsymbol{R}}^{(10)}\Big\rangle$$
$$+\Big\langle\overline{\boldsymbol{R}}^{(01)}\odot\overline{\boldsymbol{R}}^{(21)}\Big\rangle+\Big\langle\overline{\boldsymbol{R}}^{(21)}\odot\overline{\boldsymbol{R}}^{(01)}\Big\rangle$$
$$+\Big\langle\overline{\boldsymbol{R}}^{(11)}\odot\overline{\boldsymbol{R}}^{(11)}\Big\rangle+\cdots\Big]. \tag{189}$$

All terms up to the order $\sigma_1^2\sigma_2^2$ are included. If the values of σ_1 and σ_2 are of the same order of magnitude, these terms will be comparable to the order four in expressions (187), (188). Thus we can suppose that in expansion (189) the following terms which are of order $\sigma_1^4\sigma_2^2$, $\sigma_1^2\sigma_2^4$, and $\sigma_1^4\sigma_2^4$ will be negligible compared to those retained. However, due to the complexity of the perturbative development, these terms of order six have not been calculated.

When we combine Eq. (171), and $\delta(0) = A/(2\pi)^2$, with the previous development, we obtain the following expressions for eqs. (187)–(189):

$$\overline{\gamma}_u^{\text{incoh}}(\boldsymbol{p}\,|\,\boldsymbol{p}_0) = \frac{K_0^4 \cos^2\theta \, \cos\theta_0}{(2\pi)^2} \times \left[\overline{\boldsymbol{I}}^{(10-10)}(\boldsymbol{p}\,|\,\boldsymbol{p}_0) + \overline{\boldsymbol{I}}^{(20-20)}(\boldsymbol{p}\,|\,\boldsymbol{p}_0) + \overline{\boldsymbol{I}}^{(30-10)}(\boldsymbol{p}\,|\,\boldsymbol{p}_0)\right], \tag{190}$$

$$\overline{\gamma}_d^{\text{incoh}}(\boldsymbol{p}\,|\,\boldsymbol{p}_0) = \frac{K_0^4 \cos^2\theta \, \cos\theta_0}{(2\pi)^2} \times \left[\overline{\boldsymbol{I}}^{(01-01)}(\boldsymbol{p}\,|\,\boldsymbol{p}_0) + \overline{\boldsymbol{I}}^{(02-02)}(\boldsymbol{p}\,|\,\boldsymbol{p}_0) + \overline{\boldsymbol{I}}^{(03-01)}(\boldsymbol{p}\,|\,\boldsymbol{p}_0)\right], \tag{191}$$

$$\overline{\gamma}_{ud}^{\text{incoh}}(\boldsymbol{p}\,|\,\boldsymbol{p}_0) = \frac{K_0^4 \cos^2\theta \, \cos\theta_0}{(2\pi)^2} \times \left[\overline{\boldsymbol{I}}^{(12-10)}(\boldsymbol{p}\,|\,\boldsymbol{p}_0) + \overline{\boldsymbol{I}}^{(11-11)}(\boldsymbol{p}\,|\,\boldsymbol{p}_0) + \overline{\boldsymbol{I}}^{(21-01)}(\boldsymbol{p}\,|\,\boldsymbol{p}_0)\right], \tag{192}$$

where

$$\overline{\boldsymbol{I}}^{(10-10)}(\boldsymbol{p}\,|\,\boldsymbol{p}_0) = W_1(\boldsymbol{p} - \boldsymbol{p}_0)\,\overline{\boldsymbol{X}}^{(10)}(\boldsymbol{p}\,|\,\boldsymbol{p}_0) \odot \overline{\boldsymbol{X}}^{(10)}(\boldsymbol{p}\,|\,\boldsymbol{p}_0) \tag{193}$$

$$\overline{\boldsymbol{I}}^{(20-20)}(\boldsymbol{p}\,|\,\boldsymbol{p}_0) = \int \frac{\mathrm{d}^2\boldsymbol{p}_1}{(2\pi)^2}\, W_1(\boldsymbol{p} - \boldsymbol{p}_1) W_1(\boldsymbol{p}_1 - \boldsymbol{p}_0)\,\overline{\boldsymbol{X}}^{(20)}(\boldsymbol{p}\,|\,\boldsymbol{p}_1\,|\,\boldsymbol{p}_0) \odot \left[\overline{\boldsymbol{X}}^{(20)}(\boldsymbol{p}\,|\,\boldsymbol{p}_1\,|\,\boldsymbol{p}_0) + \overline{\boldsymbol{X}}^{(20)}(\boldsymbol{p}\,|\,\boldsymbol{p} + \boldsymbol{p}_0 - \boldsymbol{p}_1\,|\,\boldsymbol{p}_0)\right] \tag{194}$$

$$\overline{\boldsymbol{I}}^{(30-10)}(\boldsymbol{p}\,|\,\boldsymbol{p}_0) = W_1(\boldsymbol{p} - \boldsymbol{p}_0)\left[\overline{\boldsymbol{X}}^{(10)}(\boldsymbol{p}\,|\,\boldsymbol{p}_0) \odot \overline{\boldsymbol{X}}^{(30)}(\boldsymbol{p}\,|\,\boldsymbol{p}_0) + \overline{\boldsymbol{X}}^{(30)}(\boldsymbol{p}\,|\,\boldsymbol{p}_0) \odot \overline{\boldsymbol{X}}^{(10)}(\boldsymbol{p}\,|\,\boldsymbol{p}_0)\right], \tag{195}$$

$$\overline{\boldsymbol{I}}^{(01-01)}(\boldsymbol{p}\,|\,\boldsymbol{p}_0) = W_2(\boldsymbol{p} - \boldsymbol{p}_0)\,\overline{\boldsymbol{X}}^{(01)}(\boldsymbol{p}\,|\,\boldsymbol{p}_0) \odot \overline{\boldsymbol{X}}^{(01)}(\boldsymbol{p}\,|\,\boldsymbol{p}_0) \tag{196}$$

$$\overline{\boldsymbol{I}}^{(02-02)}(\boldsymbol{p}\,|\,\boldsymbol{p}_0) = \int \frac{\mathrm{d}^2\boldsymbol{p}_1}{(2\pi)^2}\, W_2(\boldsymbol{p} - \boldsymbol{p}_1) W_2(\boldsymbol{p}_1 - \boldsymbol{p}_0)\,\overline{\boldsymbol{X}}^{(02)}(\boldsymbol{p}\,|\,\boldsymbol{p}_1\,|\,\boldsymbol{p}_0) \odot \left[\overline{\boldsymbol{X}}^{(02)}(\boldsymbol{p}\,|\,\boldsymbol{p}_1\,|\,\boldsymbol{p}_0) + \overline{\boldsymbol{X}}^{(02)}(\boldsymbol{p}\,|\,\boldsymbol{p} + \boldsymbol{p}_0 - \boldsymbol{p}_1\,|\,\boldsymbol{p}_0)\right] \tag{197}$$

$$\overline{\boldsymbol{I}}^{(03-01)}(\boldsymbol{p}\,|\,\boldsymbol{p}_0) = W_2(\boldsymbol{p} - \boldsymbol{p}_0)\left[\overline{\boldsymbol{X}}^{(01)}(\boldsymbol{p}\,|\,\boldsymbol{p}_0) \odot \overline{\boldsymbol{X}}^{(03)}(\boldsymbol{p}\,|\,\boldsymbol{p}_0) + \overline{\boldsymbol{X}}^{(03)}(\boldsymbol{p}\,|\,\boldsymbol{p}_0) \odot \overline{\boldsymbol{X}}^{(01)}(\boldsymbol{p}\,|\,\boldsymbol{p}_0)\right], \tag{198}$$

$$\overline{\boldsymbol{I}}^{(12-10)}(\boldsymbol{p}\,|\,\boldsymbol{p}_0) = W_1(\boldsymbol{p} - \boldsymbol{p}_0)\left[\overline{\boldsymbol{X}}^{(12)}(\boldsymbol{p}\,|\,\boldsymbol{p}_0) \odot \overline{\boldsymbol{X}}^{(10)}(\boldsymbol{p}\,|\,\boldsymbol{p}_0) + \overline{\boldsymbol{X}}^{(10)}(\boldsymbol{p}\,|\,\boldsymbol{p}_0) \odot \overline{\boldsymbol{X}}^{(12)}(\boldsymbol{p}\,|\,\boldsymbol{p}_0)\right], \tag{199}$$

$$\overline{\boldsymbol{I}}^{(21-01)}(\boldsymbol{p}\,|\,\boldsymbol{p}_0) = W_2(\boldsymbol{p} - \boldsymbol{p}_0)\left[\overline{\boldsymbol{X}}^{(21)}(\boldsymbol{p}\,|\,\boldsymbol{p}_0) \odot \overline{\boldsymbol{X}}^{(01)}(\boldsymbol{p}\,|\,\boldsymbol{p}_0) + \overline{\boldsymbol{X}}^{(01)}(\boldsymbol{p}\,|\,\boldsymbol{p}_0) \odot \overline{\boldsymbol{X}}^{(21)}(\boldsymbol{p}\,|\,\boldsymbol{p}_0)\right], \tag{200}$$

$$
\begin{aligned}
\overline{\boldsymbol{I}}^{(11-11)}(\boldsymbol{p}|\boldsymbol{p}_0) = \int \frac{\mathrm{d}^2\boldsymbol{p}_1}{(2\pi)^2} \Big[& W_1(\boldsymbol{p}-\boldsymbol{p}_1)\, W_2(\boldsymbol{p}_1-\boldsymbol{p}_0)\, \overline{\boldsymbol{X}}^{(11)12}(\boldsymbol{p}|\boldsymbol{p}_1|\boldsymbol{p}_0) \\
& \odot \Big(\overline{\boldsymbol{X}}^{(11)12}(\boldsymbol{p}|\boldsymbol{p}_1|\boldsymbol{p}_0) + \overline{\boldsymbol{X}}^{(11)21}(\boldsymbol{p}|\boldsymbol{p}+\boldsymbol{p}_0-\boldsymbol{p}_1|\boldsymbol{p}_0) \Big) \\
& + W_2(\boldsymbol{p}-\boldsymbol{p}_1)\, W_1(\boldsymbol{p}_1-\boldsymbol{p}_0)\, \overline{\boldsymbol{X}}^{(11)21}(\boldsymbol{p}|\boldsymbol{p}_1|\boldsymbol{p}_0) \\
& \odot \Big(\overline{\boldsymbol{X}}^{(11)21}(\boldsymbol{p}|\boldsymbol{p}_1|\boldsymbol{p}_0) + \overline{\boldsymbol{X}}^{(11)12}(\boldsymbol{p}|\boldsymbol{p}+\boldsymbol{p}_0-\boldsymbol{p}_1|\boldsymbol{p}_0) \Big) \Big]
\end{aligned} \tag{201}
$$

with

$$
\begin{aligned}
\overline{\boldsymbol{X}}^{(30)}(\boldsymbol{p}|\boldsymbol{p}_0) = \int \frac{\mathrm{d}^2\boldsymbol{p}_1}{(2\pi)^2} \Big[& W_1(\boldsymbol{p}_1-\boldsymbol{p}_0)\, \overline{\boldsymbol{X}}^{(30)}(\boldsymbol{p}|\boldsymbol{p}_0|\boldsymbol{p}_1|\boldsymbol{p}_0) \\
& + W_1(\boldsymbol{p}-\boldsymbol{p}_1) \Big(\overline{\boldsymbol{X}}^{(30)}(\boldsymbol{p}|\boldsymbol{p}_1|\boldsymbol{p}_0-\boldsymbol{p}+\boldsymbol{p}_1|\boldsymbol{p}_0) \\
& + \overline{\boldsymbol{X}}^{(30)}(\boldsymbol{p}|\boldsymbol{p}_1|\boldsymbol{p}|\boldsymbol{p}_0) \Big) \Big],
\end{aligned} \tag{202}
$$

$$
\begin{aligned}
\overline{\boldsymbol{X}}^{(03)}(\boldsymbol{p}|\boldsymbol{p}_0) = \int \frac{\mathrm{d}^2\boldsymbol{p}_1}{(2\pi)^2} \Big[& W_2(\boldsymbol{p}_1-\boldsymbol{p}_0)\, \overline{\boldsymbol{X}}^{(03)}(\boldsymbol{p}|\boldsymbol{p}_0|\boldsymbol{p}_1|\boldsymbol{p}_0) \\
& + W_2(\boldsymbol{p}-\boldsymbol{p}_1) \Big(\overline{\boldsymbol{X}}^{(03)}(\boldsymbol{p}|\boldsymbol{p}_1|\boldsymbol{p}_0-\boldsymbol{p}+\boldsymbol{p}_1|\boldsymbol{p}_0) \\
& + \overline{\boldsymbol{X}}^{(03)}(\boldsymbol{p}|\boldsymbol{p}_1|\boldsymbol{p}|\boldsymbol{p}_0) \Big) \Big],
\end{aligned} \tag{203}
$$

$$
\begin{aligned}
\overline{\boldsymbol{X}}^{(12)}(\boldsymbol{p}|\boldsymbol{p}_0) = \int \frac{\mathrm{d}^2\boldsymbol{p}_1}{(2\pi)^2} & [W_2(\boldsymbol{p}-\boldsymbol{p}_1) \\
& \Big(\overline{\boldsymbol{X}}^{(12)221}(\boldsymbol{p}|\boldsymbol{p}_1|\boldsymbol{p}|\boldsymbol{p}_0) + \overline{\boldsymbol{X}}^{(12)212}(\boldsymbol{p}|\boldsymbol{p}_1|\boldsymbol{p}_0-\boldsymbol{p}+\boldsymbol{p}_1|\boldsymbol{p}_0) \Big) \\
& + W_2(\boldsymbol{p}_1-\boldsymbol{p}_0)\, \overline{\boldsymbol{X}}^{(12)122}(\boldsymbol{p}|\boldsymbol{p}_0|\boldsymbol{p}_1|\boldsymbol{p}_0) \Big],
\end{aligned} \tag{204}
$$

$$
\begin{aligned}
\overline{\boldsymbol{X}}^{(21)}(\boldsymbol{p}|\boldsymbol{p}_0) = \int \frac{\mathrm{d}^2\boldsymbol{p}_1}{(2\pi)^2} & [W_1(\boldsymbol{p}-\boldsymbol{p}_1) \\
& \Big(\overline{\boldsymbol{X}}^{(21)112}(\boldsymbol{p}|\boldsymbol{p}_1|\boldsymbol{p}|\boldsymbol{p}_0) + \overline{\boldsymbol{X}}^{(21)121}(\boldsymbol{p}|\boldsymbol{p}_1|\boldsymbol{p}_0-\boldsymbol{p}+\boldsymbol{p}_1|\boldsymbol{p}_0) \Big) \\
& + W_1(\boldsymbol{p}_1-\boldsymbol{p}_0)\, \overline{\boldsymbol{X}}^{(21)211}(\boldsymbol{p}|\boldsymbol{p}_0|\boldsymbol{p}_1|\boldsymbol{p}_0) \Big].
\end{aligned} \tag{205}
$$

6.6. Numerical Examples and Analysis of the Phenomena

In the preceding sections we have described a small-amplitude perturbation method to compute the scattering matrices for a weakly rough surface between two media, and for slabs which include one or two weakly rough surfaces. The surface variations are assumed to be much smaller than the incident wavelength and the slope of the randomly rough surface are relatively small. In this paragraph, we evaluate numerically the incoherent bistatic coefficients given by Eqs. (181)–(184) and (190)–(201) for different values of the parameters which characterize the configurations presented in the preceding sections, and we analyze the different wave phenomena. In the following sections, medium 0 will be the vacuum ($\epsilon_0 = 1$).

6.6.1. A Randomly Rough Surface Separating Two Different Semi-Infinite Media

The rough surface is regarded as a random Gaussian process. The distribution of surface heights is supposed to be described by a Gaussian function. And we assume that the surface correlation function is also Gaussian, the correlation length is defined by the distance over which the correlation function falls by $1/e$.

We consider polarized light of wavelength $\lambda = 457.9$ nm which is normally incident ($\theta_0 = 0^o$, $\phi_0 = 0^o$) on a two-dimensional weakly rough silver surface (see Fig. 6.3) characterized by the roughness parameters, respectively, the root mean square height of the surface and the correlation length $\sigma = 20$ nm, $l = 200$ nm, $\epsilon = -7.5 + i0.24$. The perturbative development is given by Eqs. (60),(67). In Fig. 6.11, we present the results for an incident wave linearly polarized, the scattered field being observed in the plane of incidence ($\phi = 0^o$). The single scattering contribution associated with the term $\overline{\boldsymbol{I}}^{(1-1)}$ is plotted as a dotted line, the double-scattering contribution $\overline{\boldsymbol{I}}^{(2-2)}$ as a dashed line, the scattering term $\overline{\boldsymbol{I}}^{(3-1)}$ as a dash-dotted line, and the sum of all these terms $\overline{\gamma}^{\text{incoh}}$ by the solid curve.

We observed an enhancement of the backscattering which corresponds to the physical process in which the incident light excites a surface electromagnetic wave. In fact, the surface polariton propagates along the rough surface, then it is scattered into a bulk wave due to the roughness, at the same time, a reverse partner exists with a path traveling in the opposite direction. These two paths can interfere constructively near the backscattering direction to produce a peak.[22–24] It is worth noticing that for a two-dimensional randomly rough surface, the surface plasmon polariton exists for a TM incident polarization. A TE polarized incident wave can excite a TM surface wave, and this surface wave is scattered into bulk wave with both polarization modes as can be seen in Fig. 6.11. This phenomenon can be easily explained. In the one-dimensional case, we know that the enhanced backscattering for a metallic weakly rough surface appears only for a TM incident wave due to the fact that plasmon polaritons can only be propagated with TM modes. In the two-dimensional case, we work with both polarization modes and the depolarization phenomenon implies that an incident TE wave can excite a TM plasmon mode which can be transformed into a TE or TM bulk electromagnetic wave. Thus, the enhanced backscattering is present independently of the polarization mode of the incident and scattered waves.[18] Let us notice that the backscattering enhancement is only produced by the term $\overline{\boldsymbol{I}}^{(2-2)}$, because only in that case, we have a constructive interference of the wave propagating along a path and its time reversal.

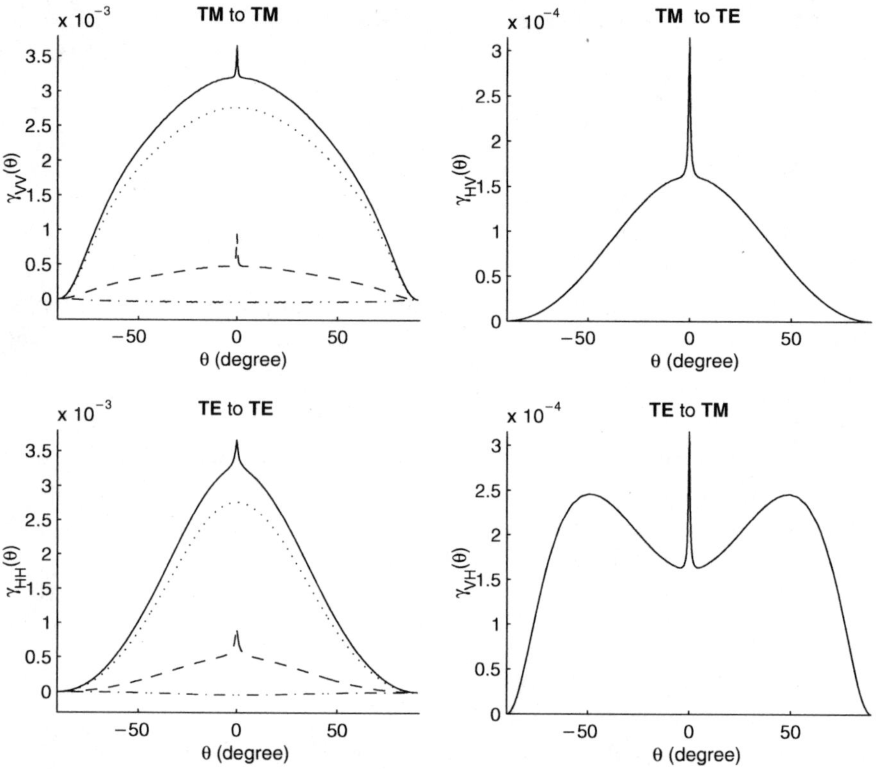

FIGURE 6.11. The bistatic incoherent scattering coefficients for a horizontal TE and a vertical TM polarized incident optical wave on a two-dimensional Gaussian randomly rough silver surface. The scattered field is observed in the incident plane. The total incoherent scattering coefficient $\overline{\gamma}^{\text{incoh}}$ (solid curve), the first order is given by $\overline{\boldsymbol{I}}^{(1-1)}$ (dotted line), the second order by $\overline{\boldsymbol{I}}^{(2-2)}$ (dashed line), and the third order by $\overline{\boldsymbol{I}}^{(3-1)}$ (dash-dotted line).

In expression (181), the peak is produced by the term $\overline{\boldsymbol{I}}^{(2-2)}$. We see that the term $\overline{\boldsymbol{P}}(\boldsymbol{u}|\boldsymbol{p}_1) \cdot \overline{\boldsymbol{Q}}^{+}(\boldsymbol{p}_1|\boldsymbol{p}_0)$ in $\overline{\boldsymbol{X}}^{(2)}_{s\,\epsilon_0,\epsilon_1}(\boldsymbol{p}|\boldsymbol{p}_1|\boldsymbol{p}_0)$ contains a factor of the form

$$\left[D^{+}_{10\,V}(\boldsymbol{p}_1)\right]^{-1} = \frac{1}{\epsilon_1\alpha_0(\boldsymbol{p}_1) + \epsilon_0\alpha_1(\boldsymbol{p}_1)}, \tag{206}$$

which is close to zero except when $\boldsymbol{p}_1$ is near the polariton resonance mode $\boldsymbol{p}_r$, which is given by $D^{+}_{10\,V}(\boldsymbol{p}_r) = 0$. When we observe the field scattered far away from the backscattering direction ($\boldsymbol{p} + \boldsymbol{p}_0 \neq 0$), the terms $\overline{\boldsymbol{X}}^{(2)}_{s\,\epsilon_0,\epsilon_1}(\boldsymbol{p}|\boldsymbol{p}_1|\boldsymbol{p}_0)$ and $\overline{\boldsymbol{X}}^{(2)}_{s\,\epsilon_0,\epsilon_1}(\boldsymbol{p}|\boldsymbol{p} + \boldsymbol{p}_0 - \boldsymbol{p}_1|\boldsymbol{p}_0)$ containing $D^{+}_{10\,V}$ are nonzero when $\boldsymbol{p}_1 \approx \boldsymbol{p}_r$ and $\boldsymbol{p} + \boldsymbol{p}_0 - \boldsymbol{p}_1 \approx \boldsymbol{p}_r$. Since these domains are disjoint, the product $\odot$ of these two terms is approximatively zero. When we are near the backscattering direction ($\boldsymbol{p}+\boldsymbol{p}_0 \approx 0$), the terms inside brackets are almost equal and produce the backscattering enhancement. The enhancement factor is not equal to 2 because the matrices

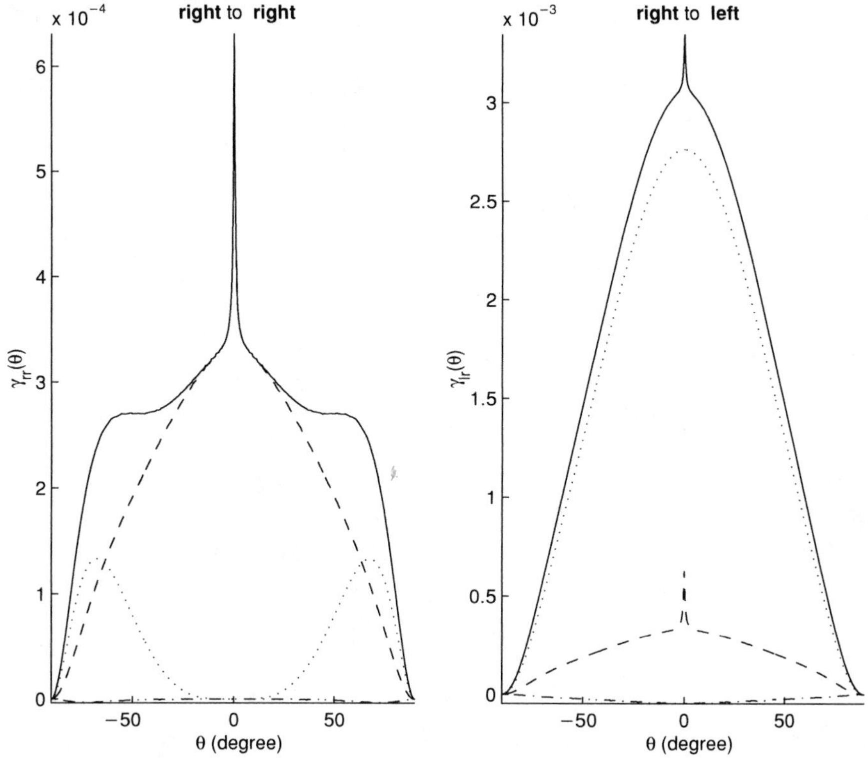

FIGURE 6.12. The bistatic incoherent scattering coefficient for the same configuration as Fig. 6.11, but with a right incident circularly polarized optical wave, and a right to right (or left to left), right to left (or left to right) observed polarizations.

$\overline{Q}^{+}(\boldsymbol{p}|\boldsymbol{p}_0)$ and $\overline{Q}^{-}(\boldsymbol{p}|\boldsymbol{p}_0)$ in $\overline{X}^{(2)}_{s\,\epsilon_0,\epsilon_1}$ do not contain $[D^{+}_{10\,V}(\boldsymbol{p}_1)]^{-1}$, they produce a significant contribution whatever the scattering angle is.

If the incident wave is now circularly polarized, we see in Fig. 6.12 that enhanced backscattering also takes place. We have not displayed the left to left, and left to right polarizations because the media are not optically active. As a consequence, the results are the same whether the incident wave is right or left polarized.

6.6.2. *A Film with a Randomly Rough Surface on the Upper Side*

We consider a dielectric slab (see Fig. 6.7) with a mean thickness $H = 500$ nm, a dielectric constant $\epsilon_1 = 2.6896 + 0.0075\text{i}$, deposited on a planar perfectly conducting substrate ($\epsilon_2 = -\infty$) and illuminated by a linearly polarized light of wavelength $\lambda = 632.8$ nm, which is incident normally ($\phi_0 = 0^{\circ}$, $\theta_0 = 0^{\circ}$). The upper two-dimensional Gaussian randomly rough surface is characterized by the

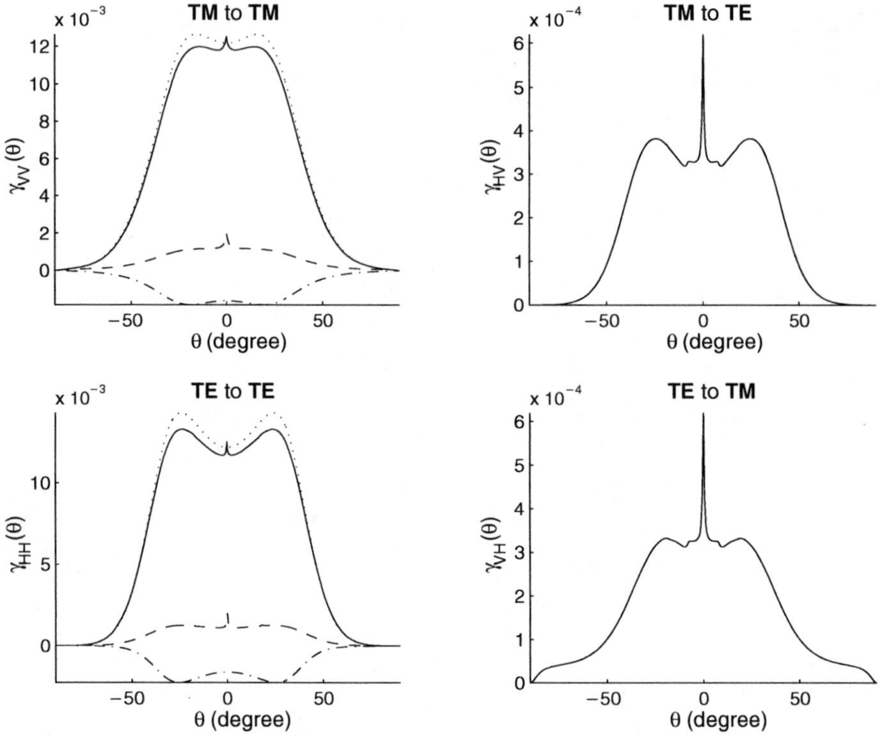

FIGURE 6.13. The bistatic incoherent scattering coefficients for a horizontal TE and a vertical TM polarized incident optical wave on a slab with an upper two-dimensional Gaussian randomly rough surface deposited on an infinite conducting plane. The scattered field is observed in the incident plane. The total incoherent scattering $\overline{\gamma}^{\text{incoh}}$ (solid curve), the first order is given by $\overline{\boldsymbol{I}}^{(1-1)}$ (dotted line), the second order by $\overline{\boldsymbol{I}}^{(2-2)}$ (dashed line), and the third-order by $\overline{\boldsymbol{I}}^{(3-1)}$ (dash-dotted line).

statistical parameters $\sigma = 15$ nm and $l = 100$ nm. For this three-dimensional structure, there exists an up-going wave in the material, which is not present in the case of the two-dimensional structure. The scattering diagrams are shown in Fig. 6.13 with the same curve labeling as before. The field is observed in the incident plane. The perturbative development is given by Eqs. (105),(107) and Eqs. (110),(121). For the first case, we choose an infinite conducting plane ($\epsilon_2 = -\infty$). Thus the coefficients $F^{\pm}$ (Eq. (119)) take the following expression:

$$F_V^{\pm}(\boldsymbol{p}_0) = \frac{1 \pm \exp^{2\mathrm{i}\,\alpha_0(\boldsymbol{p}_0)\,H}}{(\epsilon_1\,\alpha_0(\boldsymbol{p}_0) + \epsilon_0\,\alpha_1(\boldsymbol{p}_0)) + (\epsilon_1\,\alpha_0(\boldsymbol{p}_0) - \epsilon_0\,\alpha_1(\boldsymbol{p}_0))e^{2\mathrm{i}\,\alpha_0(\boldsymbol{p}_0)\,H}}, \tag{207}$$

$$F_H^{\pm}(\boldsymbol{p}_0) = \frac{1 \mp \exp^{2\mathrm{i}\,\alpha_0(\boldsymbol{p}_0)\,H}}{(\alpha_0(\boldsymbol{p}_0) + \alpha_1(\boldsymbol{p}_0)) + (\alpha_0(\boldsymbol{p}_0) - \alpha_1(\boldsymbol{p}_0))e^{2\mathrm{i}\,\alpha_0(\boldsymbol{p}_0)\,H}}. \tag{208}$$

We choose the thickness of the slab so that the slab supports only two guided wave modes, $p^1_{\mathrm{TE}} = 1.5466\, K_0$, and $p^2_{\mathrm{TE}} = 1.2423\, K_0$ for the TE polarization. These modes are resonance modes, they verify $[F^{\pm}_H]^{-1}(p^{1,2}_{\mathrm{TE}}) = 0$.

For the TM case, we have three modes given by the roots of the following equation $[F^{\pm}_V]^{-1}(p_{\mathrm{TM}}) = 0$, these modes are $p^1_{\mathrm{TM}} = 1.6126\, K_0$, $p^2_{\mathrm{TM}} = 1.3823\, K_0$ and $p^3_{\mathrm{TM}} = 1.0030\, K_0$. As described in Ref,[14, 29, 30] these guided modes can produce a classical enhanced backscattering with satellite peaks which are symmetrically positioned. The satellite peak angles are given by the equation

$$\sin\theta^{nm}_{\pm} = -\sin\theta_0 \pm \frac{1}{K_0}\,[p^n - p^m], \tag{209}$$

where p^n, p^m describe one of the guided modes. When $n = m$, we recover the classical enhanced backscattering. For the TE polarization, we have only two guided waves and the satellite peaks can only exist at the angles $\theta^{12}_{\pm}(TE) = \pm 17.7^{\mathrm{o}}$. For the TM polarization we have three angles: $\theta^{12}_{\pm}(TM) = \pm 13.3^{\mathrm{o}}$, $\theta^{13}_{\pm}(TM) = \pm 37.6^{\mathrm{o}}$, and $\theta^{23}_{\pm}(TM) = \pm 22.3^{\mathrm{o}}$. The satellite peaks are produced by the term $\overline{\boldsymbol{I}}^{(2-2)}$. In the case of TM polarization we do not obtain any significant contribution to satellite peaks. However, for the TE to TE

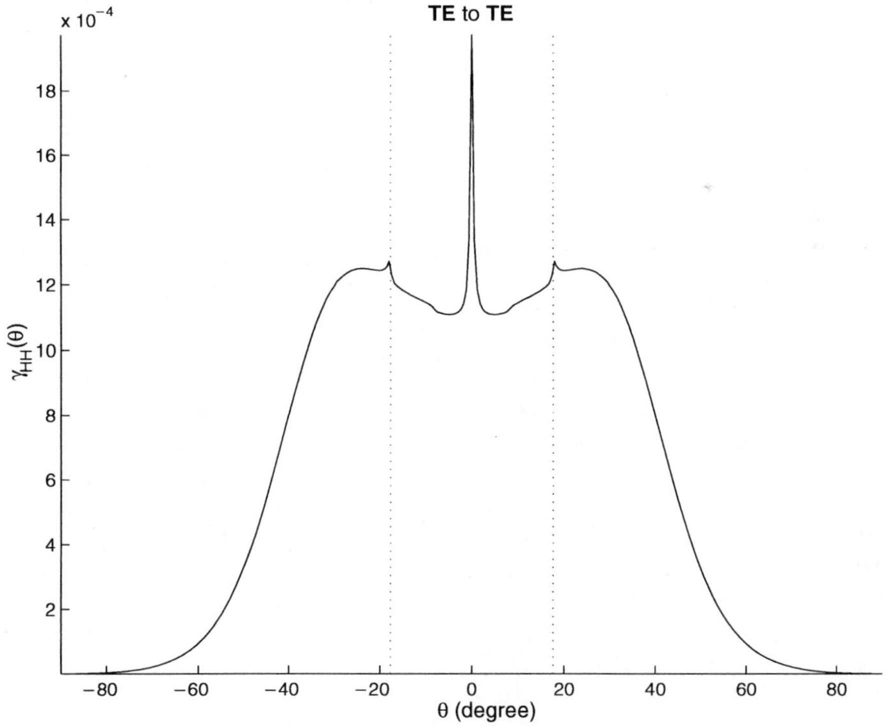

FIGURE 6.14. Details of second-order TE to TE contribution to the scattering shown in Fig. 6.13. The dotted-lines indicate the angle position of the two satellite peaks.

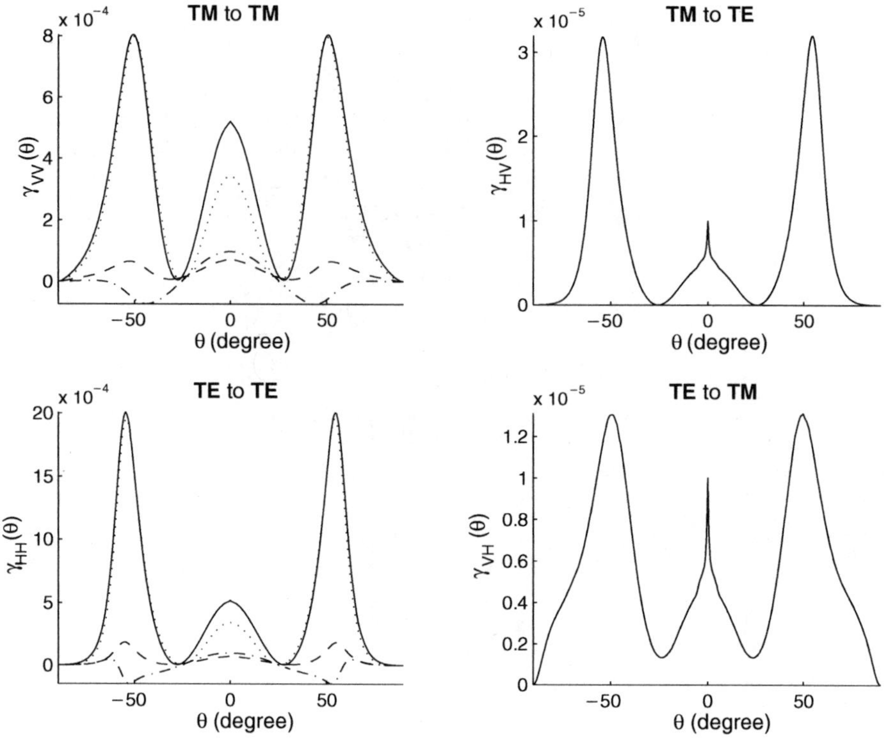

FIGURE 6.15. Selényi fringes. The slab thickness is $H = 1000$ nm for the configuration shown in Fig. 6.13.

scattering shown in Fig. 6.14, we prove the existence of the satellite peaks at the angle $\theta_{\pm}^{12}(TE) = \pm 17.7^{\circ}$ positioned along the dotted line in the figure. The peaks are attenuated because the polarization modes TM and TE are propagating at the same time, so the incident wave is split between the two polarization modes. In the case of TM polarization mode, we find more propagating modes than in TE polarization mode, as a consequence the corresponding peaks for TM polarization exist but they are too attenuated to be visible in the scattering figure.

Now, by increasing the slab thickness (see Fig. 6.15), the satellite peaks disappear for all the polarization modes, but we see a new phenomenon called the Selényi fringes.[30,34,35] For a slightly random rough surface, the slab produces fringes similar to those obtained with a Fabry–Perot interferometer illuminated by an extended source. The roughness modulates amplitude fringes but their localization remains the same as for the interferometer. Their position is mainly given by classical optics. Next, instead of doubling the slab thickness, we have replaced the infinitely conducting plane by a silver plane ($\epsilon_2 = -18.3 + 0.55i$). We see in Fig. 6.16 that the enhancement of backscattering is also decreased, and that there is no more satellite peak corresponding to TE to TE scattering. For a rough dielectric

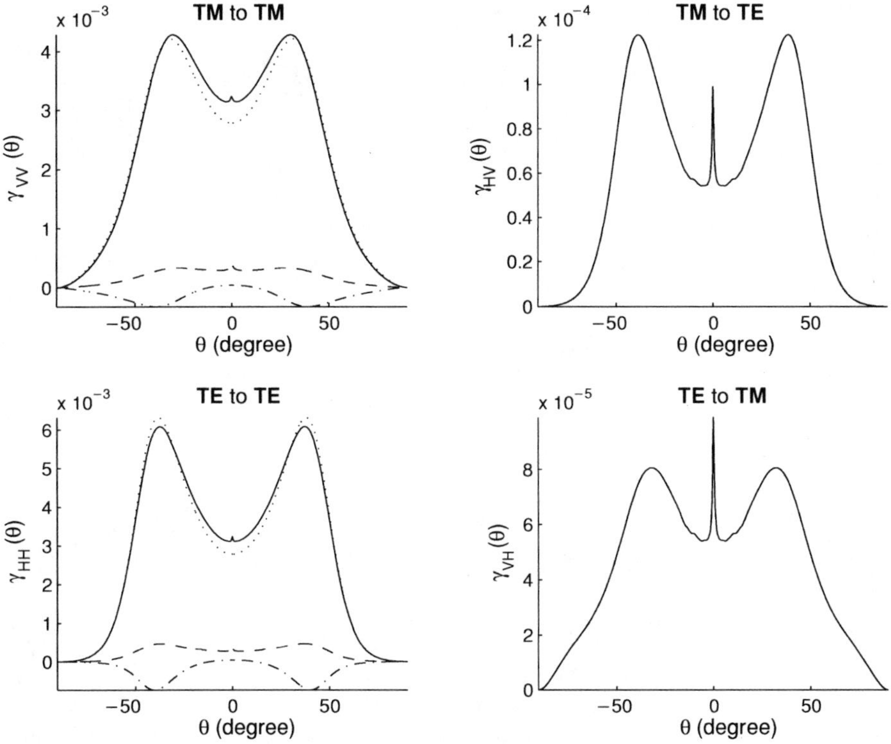

FIGURE 6.16. Same parameters as in Fig. 6.13, but with a silver plane.

slab bounded by a conducting plane the enhanced backscattering is present for both TE and TM incident waves, even for a two-dimensional structure because guided waves exist for these two polarization modes. The qualitative effect of the two-dimensional randomly rough surface is particularly sensitive when we study a three-dimensional thin film. For instance, in the case of two-dimensional slabs, satellite peaks[14,29] appear on each side of the enhanced backscattering peaks. However, for three-dimensional structures, the coupling between TE and TM modes drastically attenuates these peaks.

We showed that a three-dimensional system with a two-dimensional randomly rough surface deposited on a silver surface can display satellite peaks in scattering of electromagnetic waves when the randomness is introduced into the slab owing to a two-dimensional weakly rough surface. The position of these peaks are predicted accurately, the computer simulations of the small-amplitude perturbation theory up to fourth-order in the surface profile function corroborate these peaks as multiple-scattering phenomena for a three-dimensional slab. We proved that the slab thickness and the conductivity of the plane are key parameters in order to observe satellite peaks.

6.6.3. A Slab with Two Randomly Rough Surfaces

We consider a three-dimensional system, which is constituted of an air-dielectric film whose dielectric constant is $\epsilon_1 = 2.6896 + \mathrm{i}\,0.0075(\epsilon_0 = 1)$, deposited on a silver surface with $\epsilon_2 = -18.3 + \mathrm{i}\,0.55$. The vacuum–dielectric interface is a two-dimensional Gaussian randomly rough surface, whose parameters are $\sigma_1 = 15$ nm and $l_1 = 100$ nm. The dielectric–silver boundary has a Gaussian roughness defined by $\sigma_2 = 5$ nm and $l_2 = 100$ nm. The incident wave has an arbitrary polarization and its wavelength is $\lambda = 632.8$ nm. The conditions of validity of the small-perturbation theory are satisfied[7] with the following set of parameters:

$$2\pi \left|\frac{\epsilon_1}{\epsilon_0}\right|^{1/2} \frac{\sigma_1}{\lambda} \ll 1, \qquad 2\pi \left|\frac{\epsilon_2}{\epsilon_1}\right|^{1/2} \frac{\sigma_2}{\lambda} \ll 1 \tag{210}$$

$$\frac{\sigma_1}{l_1} \ll 1, \qquad \frac{\sigma_2}{l_2} \ll 1\,. \tag{211}$$

The thickness of the film is $H = 500$ nm and supports two-guided wave polaritons for the TE polarizations at $p_{\mathrm{TE}}^1 = 1.5534\,K_0$, and $p_{\mathrm{TE}}^2 = 1.2727\,K_0$, and three guided-modes for the TM polarizations at $p_{\mathrm{TM}}^1 = 1.7752\,K_0$, $p_{\mathrm{TM}}^2 = 1.4577\,K_0$ and $p_{\mathrm{TM}}^3 = 1.034\,K_0$.

In this paragraph, we use the theory developed in the previous sections to compute the incoherent bistatic coefficient $\gamma^{\mathrm{incoh}}(\boldsymbol{p}\,|\,\boldsymbol{p}_0)$ given by Eqs. (194), (197), (201), (202)–(205) where the integrals involved in these expressions are evaluated using Legendre quadrature. The results are shown in Figs. (6.17–6.20), where the incoherent bistatic coefficients are drawn as functions of the scattering angle θ for two different angles of incidence and a linearly polarized incident wave. In Fig. 6.17, the wave is normally incident and the scattered field is observed in the incident plane ($\phi = 0^{\mathrm{o}}$). The single scattering contribution on each surface, associated with the terms $\overline{\boldsymbol{I}}^{(10-10)} + \overline{\boldsymbol{I}}^{(01-01)}$, is plotted as a dotted line, the double-scattering contribution $\overline{\boldsymbol{I}}^{(20-20)} + \overline{\boldsymbol{I}}^{(02-02)} + \overline{\boldsymbol{I}}^{(11-11)}$ as a dashed line, the other terms $\overline{\boldsymbol{I}}^{(30-10)} + \overline{\boldsymbol{I}}^{(03-01)} + \overline{\boldsymbol{I}}^{(12-10)} + \overline{\boldsymbol{I}}^{(21-01)}$ as a dash-dotted line, and the total contribution γ^{incoh} by the solid curve. We observe an enhancement of the backscattering which corresponds to the classical physical process according to which the incident light excites a guided-mode through the roughness of the slab and is then scattered into a bulk wave which is also due to the roughness effect. During the same process, the light can follow this path in the opposite direction where one possible configuration is shown in Fig. 6.9. These two paths can interfere constructively near the backscattering direction to produce a peak. These paths are identical for the two waves under consideration, they have the same degree of interaction with the rough surface. A term such as $\overline{\boldsymbol{I}}^{(30-10)}$ cannot produce the peak because the first wave interacts three times with the upper rough surface while the second wave interacts only once. This analysis shows that the effect giving rise to backscattering enhancement comes only from the terms $\overline{\boldsymbol{I}}^{(20-20)} + \overline{\boldsymbol{I}}^{(02-02)} + \overline{\boldsymbol{I}}^{(11-11)}$, which contain the paths indicated in Fig. 6.9. However, it is worth noticing that these

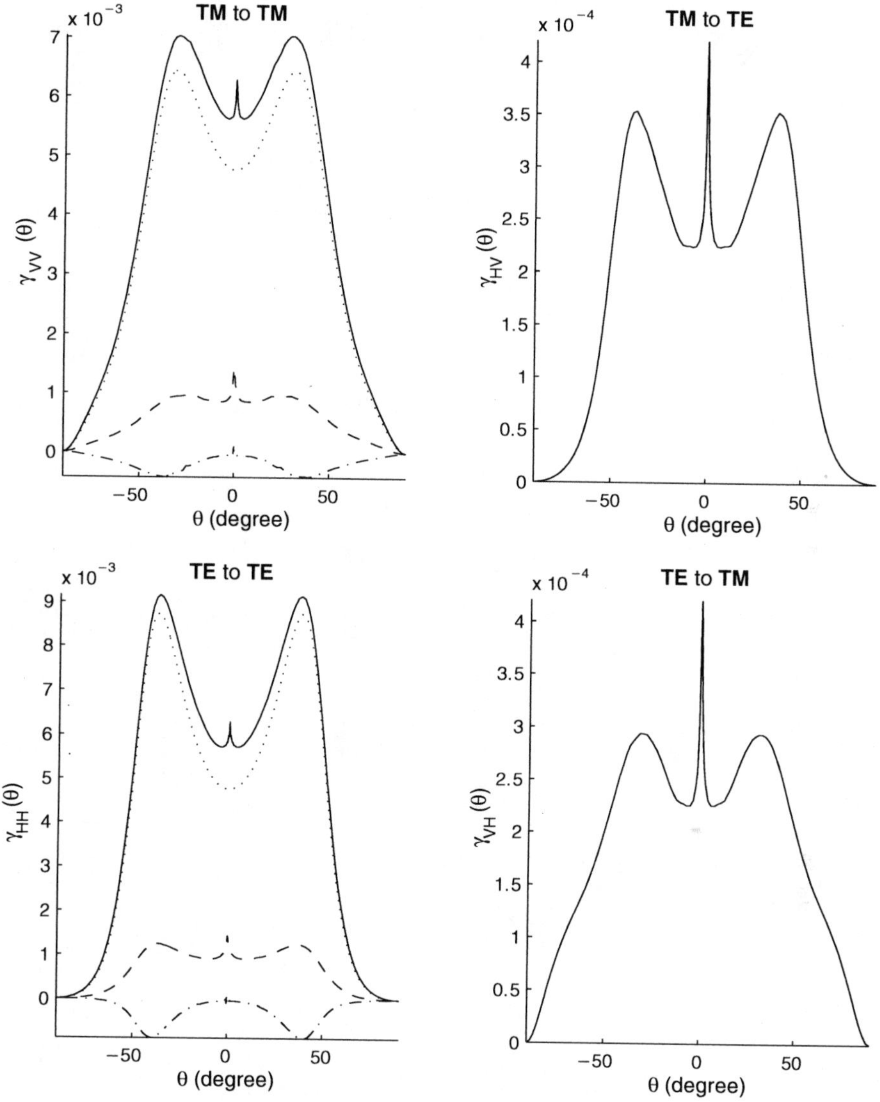

FIGURE 6.17. The bistatic incoherent scattering coefficients for a horizontal TE and a vertical TM polarized normally incident optical wave on a slab bounded by two two-dimensional randomly rough surfaces. The scattered field is observed in the incident plane. The total incoherent scattering $\overline{\gamma}^{\text{incoh}}$ (solid curve), the first order is given by $\overline{\boldsymbol{I}}^{(10-10)} + \overline{\boldsymbol{I}}^{(01-01)}$ (dotted curve), the second order by $\overline{\boldsymbol{I}}^{(20-20)} + \overline{\boldsymbol{I}}^{(02-02)} + \overline{\boldsymbol{I}}^{(11-11)}$ (dashed curve), and the third order by $\overline{\boldsymbol{I}}^{(30-10)} + \overline{\boldsymbol{I}}^{(03-01)} + \overline{\boldsymbol{I}}^{(12-10)} + \overline{\boldsymbol{I}}^{(21-01)}$ (dash-dotted curve).

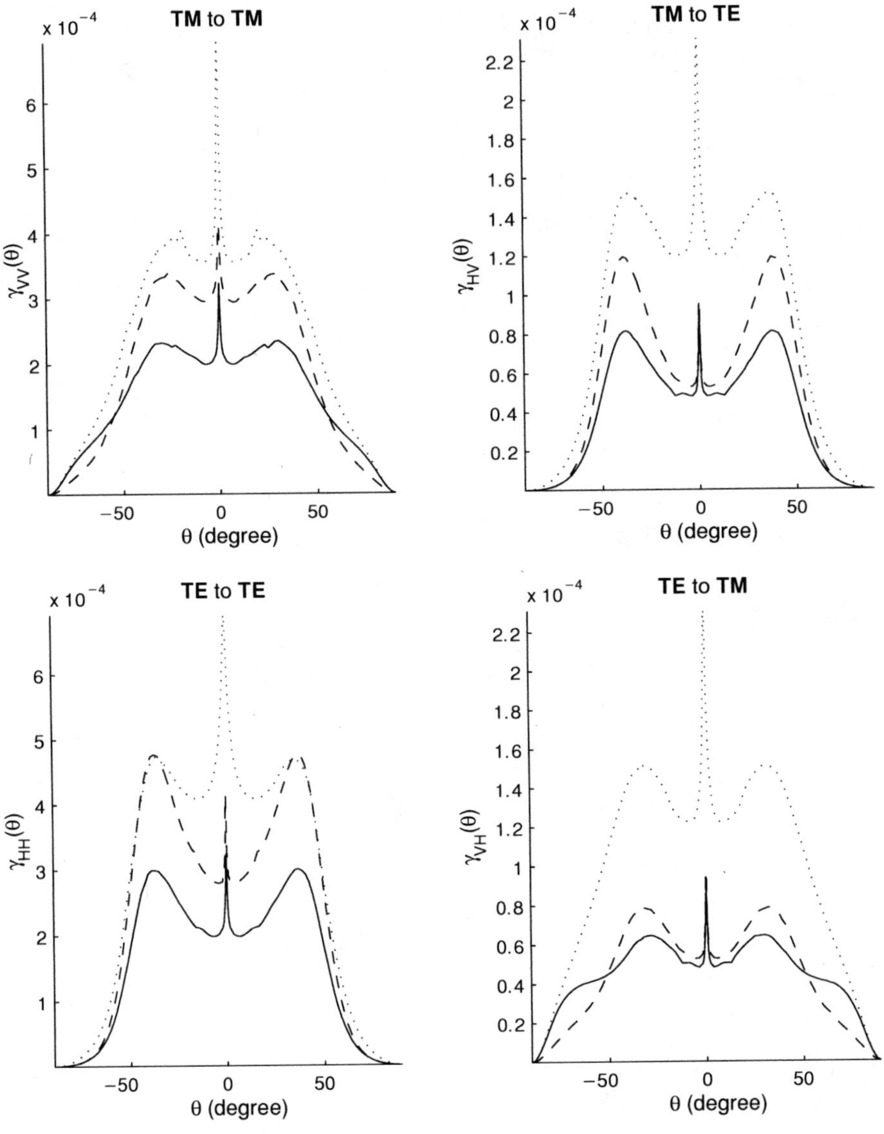

FIGURE 6.18. Details of the second-order contributions of Fig. 6.17. $\overline{\boldsymbol{I}}^{(11-11)}$ is the solid curve, $\overline{\boldsymbol{I}}^{(20-20)}$ is the dashed curve and $\overline{\boldsymbol{I}}^{(02-02)}$ is the dotted curve.

terms also contain paths which do not produce the enhanced backscattering, for instance the term $\overline{\boldsymbol{I}}^{(20-20)}$ contains a scattering process where the incident wave is only scattered once by the upper rough surface although the scattering process is of order two in h_1. This is the reason the terms $\overline{\boldsymbol{I}}^{(20-20)} + \overline{\boldsymbol{I}}^{(02-02)} + \overline{\boldsymbol{I}}^{(11-11)}$ are not null away from the antispecular direction. We see here the interest of the

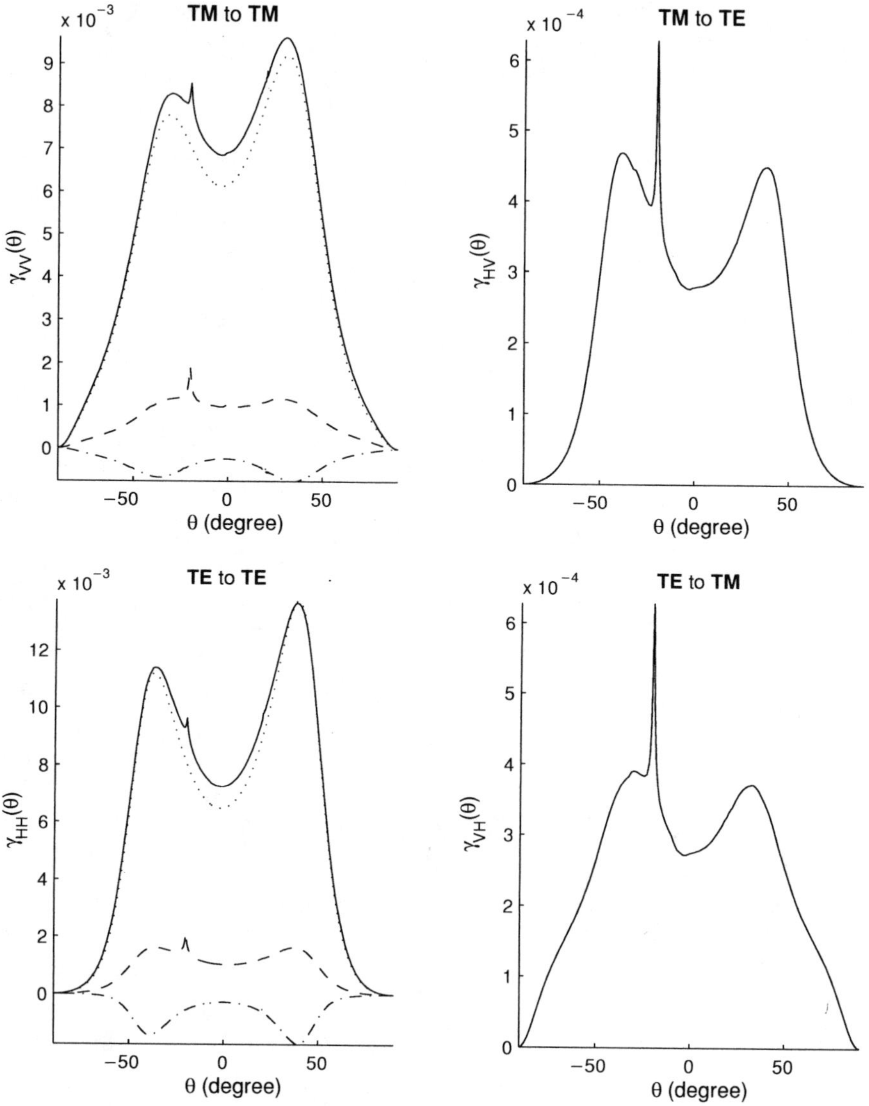

FIGURE 6.19. Same parameters as in Fig. 6.17, the incident angle is $\theta_0 = 20^{\circ}$.

perturbation method, which allows us to analyze the different contributions due to the analytical calculation. Thus we can explain the mechanisms of the backscattering enhancement.

In order to separate the different contributions to the backscattering peak, we have drawn the contributions $\overline{\boldsymbol{I}}^{(20-20)}$, $\overline{\boldsymbol{I}}^{(02-02)}$, $\overline{\boldsymbol{I}}^{(11-11)}$ separately in Fig. 6.18 as a dashed-line, dotted-line and solid curve, respectively. We see that each term

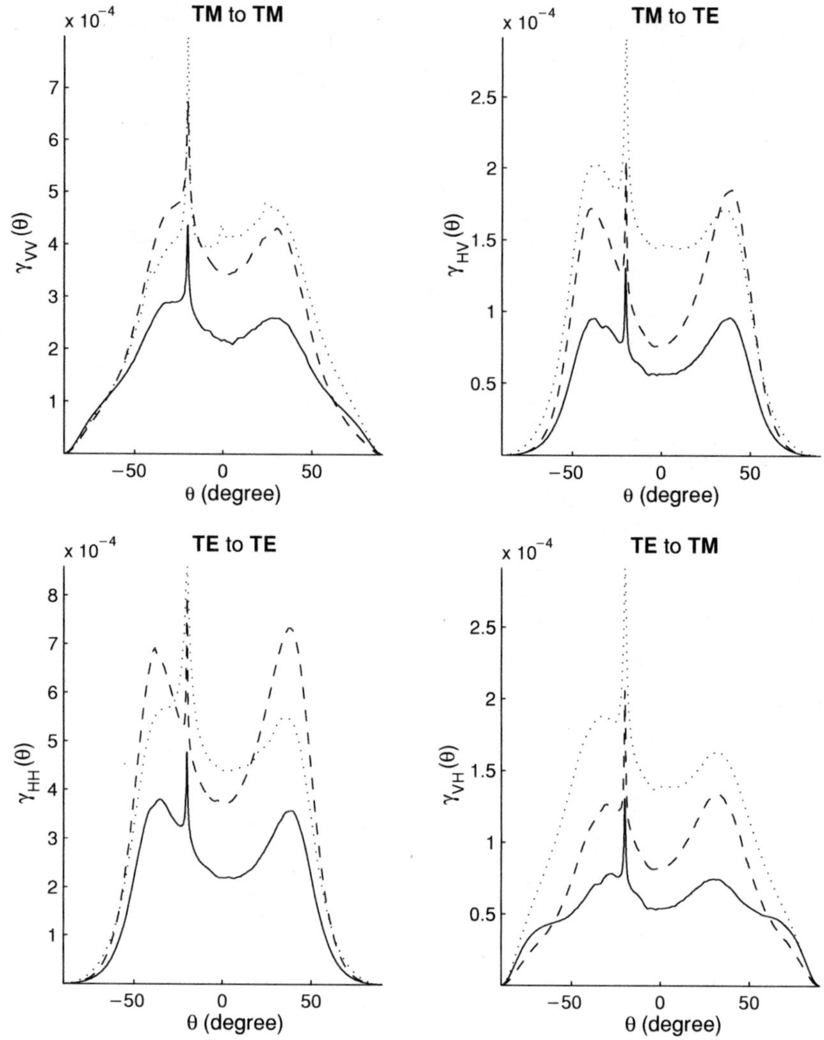

FIGURE 6.20. Second-order contributions to Fig. 6.19.

produces an enhancement near the antispecular direction. The terms $\overline{\boldsymbol{I}}^{(20-20)}$, $\overline{\boldsymbol{I}}^{(02-02)}$ are the classic ones,[14,18] their associated fields do not interact with both rough surfaces; however, they produce a peak due to the scattering on the same rough surface: the upper one for $\overline{\boldsymbol{I}}^{(20-20)}$ and the bottom one for $\overline{\boldsymbol{I}}^{(02-02)}$. Using the computation simulation, we prove that the contribution $\overline{\boldsymbol{I}}^{(11-11)}$[19] takes place and has the same magnitude as the other terms for the chosen input parameters. The mechanism of this interaction is shown in Fig. 6.9. Recent papers[14] have also explored the satellite peaks phenomenon which occurs when the wave follows two reverse paths but with different guided-mode excitations. In Fig. 6.18, this phenomenon appears for the TM to TM polarization due to the term $\overline{\boldsymbol{I}}^{(02-02)}$, and is produced by the roughness of the lower surface. A similar phenomenon occurs for the other terms, but it is too low to give a significant contribution.

We show in Figs. (6.19) and (6.20) the numerical results of the small-amplitude perturbation method for $\theta_0 = 20^{\mathrm{o}}$. We clearly see the peak which is now located at $\theta_s = -20^{\mathrm{o}}$. Due to the incidence angle, we observe a dissymmetry in the small-order contributions.

In the presented structure, we have chosen a smaller roughness for the second randomly rough surface, however we see that the second surface has a scattering effect comparable to the first one.

These computer simulations allow us to prove the existence of the different phenomena occurring in rough films or slabs.

6.7. Discussion

In this chapter, we have presented four generalized reduced Rayleigh equations which are exact integral equations. These equations offer a systematic method to compute the small-amplitude development without lengthy calculations in the cases of two-dimensional or three-dimensional systems, and the scattering matrices are only two-dimensional. All the theoretical calculations have been made up to order three in the height elevation, which allows us to obtain all the fourth-order cross-section terms. The method of reduced Rayleigh equations provides a convenient starting point for the small-amplitude perturbation theory, but also for nonperturbative numerical solutions because the equations we find are exact integral equation. Reduced Rayleigh equations leads to a computer simulation approach which is free from the limitations imposed by the Rayleigh hypothesis.

We have calculated the perturbative development for different structures composed of rough surfaces separating homogeneous semi-infinite media. The numerical results show an enhancement of the backscattering for co- and cross-polarizations for all the structures we have presented. For three-dimensional rough slabs, we have focused on less-known mechanisms that occur in the cases that the randomly rough surfaces enclose bounded structures. These structures possess a discrete number of guided waves at the frequency of the incident electromagnetic field. In the slab cases, for some configurations and definite polarizations, we have determined the physical mechanisms of the creation of satellite peaks which result from interference of different waveguide modes. This general formulation is extended to the configuration including two randomly rough surfaces. We can conclude in this case that up to order four in the perturbative development, the backscattering enhancement peak is produced by a double scattering mechanism and also when the wave is scattered by one of the rough surfaces and then by the other.

We have illustrated both analytically and numerically the occurrence of backscattering enhancement peaks and satellite peaks in three-dimensional structure with one or two weakly rough surfaces. The simulation computation can give some favorable experimental conditions and it will be very interesting to observe these results experimentally. In the very near future, we can hope that highly integrated optical devices will use metallic or metallo-dielectric microstructures which enable propagation of plasmon polariton waves, strong guidance, and manipulation of light.

Acknowledgments. It is my pleasure to thank my coworkers, Claude Bourrely and Antoine Soubret, without whom the results described above, would not have possible.

Appendix

A Expression of $\nabla h(\boldsymbol{x})$

We need to calculate the following integral:

$$\int \mathrm{d}^2\boldsymbol{x} \exp(-\mathrm{i}\left(\boldsymbol{k}_{\boldsymbol{u}}^{1b} - \boldsymbol{k}_{\boldsymbol{p}}^{1a}\right) \cdot \boldsymbol{r}_{\boldsymbol{x}}) \nabla h(\boldsymbol{x}). \tag{212}$$

Since $\nabla h(x, y)$ is zero for $|x| > L/2$ or $|y| > L/2$, we can fix the integration limits. We choose the boundary limits x_l in x such that $|x_l| > L/2$, and $(\boldsymbol{u} - \boldsymbol{p})_x\, x_l = 2\pi m_x$, with $m_x \in \mathrm{Z}$. Similarly, we choose the boundary y_l in y such that $|y_l| > L/2$ and $(\boldsymbol{u} - \boldsymbol{p})_y\, y_l = 2\pi m_y$, with $m_y \in \mathrm{Z}$. Thus, integral (212) is

$$\begin{aligned}
&\int_{-x_l}^{x_l}\int_{-y_l}^{y_l} \mathrm{d}x\, \mathrm{d}y \exp(-\mathrm{i}(\boldsymbol{u} - \boldsymbol{p}) \cdot \boldsymbol{x}) \nabla h(\boldsymbol{x}) \exp(-\mathrm{i}(b\alpha_1(\boldsymbol{u}) - a\alpha_1(\boldsymbol{p}))h(\boldsymbol{x})) \\
&= \hat{\boldsymbol{e}}_x \int_{-y_l}^{y_l} \mathrm{d}y \left[\frac{\exp(-\mathrm{i}(\boldsymbol{u} - \boldsymbol{p}) \cdot \boldsymbol{x} - \mathrm{i}(b\alpha_1(\boldsymbol{u}) - a\alpha_1(\boldsymbol{p}))h(\boldsymbol{x}))}{-\mathrm{i}(b\alpha_1(\boldsymbol{u}) - a\alpha_1(\boldsymbol{p}))}\right]_{x=-x_l}^{x=+x_l} \\
&\quad + \hat{\boldsymbol{e}}_y \int_{-x_l}^{x_l} \mathrm{d}x \left[\frac{\exp(-\mathrm{i}(\boldsymbol{u} - \boldsymbol{p}) \cdot \boldsymbol{x} - \mathrm{i}(b\alpha_1(\boldsymbol{u}) - a\alpha_1(\boldsymbol{p}))h(\boldsymbol{x}))}{-\mathrm{i}(b\alpha_1(\boldsymbol{u}) - a\alpha_1(\boldsymbol{p}))}\right]_{y=-y_l}^{y=+y_l} \\
&\quad - \int_{-x_l}^{x_l} \mathrm{d}x \int_{-y_l}^{y_l} \mathrm{d}y \frac{-\mathrm{i}(\boldsymbol{u} - \boldsymbol{p})}{-\mathrm{i}(b\alpha_1(\boldsymbol{u}) - a\alpha_1(\boldsymbol{p}))} \\
&\quad \times \exp(-\mathrm{i}(\boldsymbol{u} - \boldsymbol{p}) \cdot \boldsymbol{x} - \mathrm{i}(b\alpha_1(\boldsymbol{u}) - a\alpha_1(\boldsymbol{p}))h(\boldsymbol{x})) \qquad (213) \\
&= -\int \mathrm{d}^2\boldsymbol{x} \frac{(\boldsymbol{u} - \boldsymbol{p})}{(b\alpha_1(\boldsymbol{u}) - a\alpha_1(\boldsymbol{p}))} \exp(-\mathrm{i}(\boldsymbol{k}_{\boldsymbol{u}}^{1b} - \boldsymbol{k}_{\boldsymbol{p}}^{1a}) \cdot \boldsymbol{r}_{\boldsymbol{x}}). \qquad (214)
\end{aligned}$$

The term in the square bracket cancels due to the choice made for x_l and y_l. From the previous calculations, we can now substitute $\nabla h(\boldsymbol{x})$ by

$$\nabla h(\boldsymbol{x}) \longleftrightarrow -\frac{(\boldsymbol{u} - \boldsymbol{p})}{(b\alpha_1(\boldsymbol{u}) - a\alpha_1(\boldsymbol{p}))}. \tag{215}$$

B Perturbative Development and Reciprocity Condition

As noticed by Voronovich,[9] the scattering operator $\overline{\boldsymbol{R}}$ has a very simple law of transformation when we shift the boundary in the horizontal direction by a vector $\boldsymbol{d}$:

$$\overline{\boldsymbol{R}}_{\boldsymbol{x}\to h(\boldsymbol{x}-\boldsymbol{d})}(\boldsymbol{p}\,|\,\boldsymbol{p}_0) = \exp[-\mathrm{i}(\boldsymbol{p} - \boldsymbol{p}_0)\cdot\boldsymbol{d}]\,\overline{\boldsymbol{R}}_{\boldsymbol{x}\to h(\boldsymbol{x})}(\boldsymbol{p}\,|\,\boldsymbol{p}_0), \tag{216}$$

when we translate the surface by a vertical shift $H\,\hat{\boldsymbol{e}}_z$:

$$\overline{\boldsymbol{R}}_{h+H}(\boldsymbol{p}\,|\,\boldsymbol{p}_0) = \exp[-\mathrm{i}(\alpha_0(\boldsymbol{p}) + \alpha_0(\boldsymbol{p}_0))\,H]\,\overline{\boldsymbol{R}}_h\,(\boldsymbol{p}\,|\,\boldsymbol{p}_0). \tag{217}$$

Now, using Eq. (216), we can deduce some properties for the perturbative development of the scattering operator. The generalization of the Taylor expansion for a function depending on a real variable to an expansion depending on a function, which is in fact a functional can be expressed in the following form:

$$\overline{\boldsymbol{R}}(\boldsymbol{p}\,|\,\boldsymbol{p}_0) = \overline{\boldsymbol{R}}^{(0)}(\boldsymbol{p}\,|\,\boldsymbol{p}_0) + \overline{\boldsymbol{R}}^{(1)}(\boldsymbol{p}\,|\,\boldsymbol{p}_0) + \overline{\boldsymbol{R}}^{(2)}(\boldsymbol{p}\,|\,\boldsymbol{p}_0) + \overline{\boldsymbol{R}}^{(3)}(\boldsymbol{p}\,|\,\boldsymbol{p}_0) + \cdots, \tag{218}$$

where we have

$$\overline{\boldsymbol{R}}^{(1)}(\boldsymbol{p}\,|\,\boldsymbol{p}_0) = \int \frac{\mathrm{d}^2\boldsymbol{p}_1}{(2\pi)^2}\,\overline{\boldsymbol{R}}^{(1)}(\boldsymbol{p}\,|\,\boldsymbol{p}_1\,|\,\boldsymbol{p}_0)\,h(\boldsymbol{p}_1), \tag{219}$$

$$\overline{\boldsymbol{R}}^{(2)}(\boldsymbol{p}\,|\,\boldsymbol{p}_0) = \iint \frac{\mathrm{d}^2\boldsymbol{p}_1}{(2\pi)^2}\frac{\mathrm{d}\boldsymbol{p}_2}{(2\pi)^2}\,\overline{\boldsymbol{R}}^{(2)}(\boldsymbol{p}\,|\,\boldsymbol{p}_1\,|\,\boldsymbol{p}_2\,|\,\boldsymbol{p}_0)\,h(\boldsymbol{p}_1)\,h(\boldsymbol{p}_2), \tag{220}$$

$$\overline{\boldsymbol{R}}^{(3)}(\boldsymbol{p}\,|\,\boldsymbol{p}_0) = \iiint \frac{\mathrm{d}^2\boldsymbol{p}_1}{(2\pi)^2}\frac{\mathrm{d}^2\boldsymbol{p}_2}{(2\pi)^2}\frac{\mathrm{d}^2\boldsymbol{p}_3}{(2\pi)^2}\,\overline{\boldsymbol{R}}^{(3)}(\boldsymbol{p}\,|\,\boldsymbol{p}_1\,|\,\boldsymbol{p}_2\,|\,\boldsymbol{p}_3\,|\,\boldsymbol{p}_0)\,h(\boldsymbol{p}_1)\,h(\boldsymbol{p}_2)\,h(\boldsymbol{p}_3). \tag{221}$$

$$\vdots$$

Applying this perturbative development on each side of (216), and taking their functional derivative defined by

$$\frac{\delta^{(n)}}{\delta h(\boldsymbol{q}_1)\ldots\delta h(\boldsymbol{q}_n)}, \tag{222}$$

we obtain for all $n \geq 0$ in the limit $h = 0$:

$$\overline{\boldsymbol{R}}^{(n)}(\boldsymbol{p}\,|\boldsymbol{q}_1|\cdots|\boldsymbol{q}_n|\,\boldsymbol{p}_0) = \exp(-\mathrm{i}(\boldsymbol{p} - \boldsymbol{q}_1.\cdots - \boldsymbol{q}_n - \boldsymbol{p}_0)\cdot\boldsymbol{d})\,\overline{\boldsymbol{R}}^{(n)}(\boldsymbol{p}\,|\boldsymbol{q}_1|\cdots|\boldsymbol{q}_n|\,\boldsymbol{p}_0). \tag{223}$$

We get

$$\overline{\boldsymbol{R}}^{(n)}(\boldsymbol{p}\,|\boldsymbol{q}_1|\cdots|\boldsymbol{q}_n|\,\boldsymbol{p}_0) \propto \delta(\boldsymbol{p} - \boldsymbol{q}_1.\ldots - \boldsymbol{q}_n - \boldsymbol{p}_0), \tag{224}$$

and we define the $\overline{\boldsymbol{X}}$ matrices by the relations

$$\overline{\boldsymbol{R}}^{(0)}(\boldsymbol{p}\,|\,\boldsymbol{p}_0) = (2\pi)^2\,\delta(\boldsymbol{p} - \boldsymbol{p}_0)\,\overline{\boldsymbol{X}}^{(0)}(\boldsymbol{p}_0), \tag{225}$$

$$\overline{\boldsymbol{R}}^{(1)}(\boldsymbol{p}\,|\,\boldsymbol{p}_0) = \alpha_0(\boldsymbol{p}_0)\,\overline{\boldsymbol{X}}^{(1)}(\boldsymbol{p}\,|\,\boldsymbol{p}_0)\,h(\boldsymbol{p} - \boldsymbol{p}_0) \tag{226}$$

$$\overline{\boldsymbol{R}}^{(2)}(\boldsymbol{p}|\boldsymbol{p}_0) = \alpha_0(\boldsymbol{p}_0) \int \frac{\mathrm{d}^2\boldsymbol{p}_1}{(2\pi)^2} \overline{\boldsymbol{X}}^{(2)}(\boldsymbol{p}|\boldsymbol{p}_1|\boldsymbol{p}_0)\, h(\boldsymbol{p}-\boldsymbol{p}_1)\, h(\boldsymbol{p}_1-\boldsymbol{p}_0), \quad (227)$$

$$\overline{\boldsymbol{R}}^{(3)}(\boldsymbol{p}|\boldsymbol{p}_0) = \alpha_0(\boldsymbol{p}_0) \iint \frac{\mathrm{d}^2\boldsymbol{p}_1}{(2\pi)^2} \frac{\mathrm{d}\boldsymbol{p}_2}{(2\pi)^2} \overline{\boldsymbol{X}}^{(3)}(\boldsymbol{p}|\boldsymbol{p}_1|\boldsymbol{p}_2|\boldsymbol{p}_0)\, h(\boldsymbol{p}-\boldsymbol{p}_1) \times h(\boldsymbol{p}_1-\boldsymbol{p}_2)\, h(\boldsymbol{p}_2-\boldsymbol{p}_0), \quad (228)$$

$$\vdots$$

where $\alpha_0(\boldsymbol{p}_0)$ is defined for a matter of convenience.

Let us now discuss the reciprocity condition. If we define the antitranspose operation by

$$\begin{pmatrix} a & b \\ c & d \end{pmatrix}^{aT} = \begin{pmatrix} a & -c \\ -b & d \end{pmatrix}, \quad (229)$$

the reciprocity condition for an incident and a scattered waves in the medium 0 reads:[9]

$$\frac{\overline{\boldsymbol{R}}^{aT}(\boldsymbol{p}|\boldsymbol{p}_0)}{\alpha_0(\boldsymbol{p}_0)} = \frac{\overline{\boldsymbol{R}}(-\boldsymbol{p}_0|-\boldsymbol{p})}{\alpha_0(\boldsymbol{p})}. \quad (230)$$

Using the previous functional derivative, we would like to prove that each order of the perturbative development satisfies this condition. It is easy to show that

$$\left[\overline{\boldsymbol{X}}^{(1)}(\boldsymbol{p}|\boldsymbol{p}_0)\right]^{aT} = \overline{\boldsymbol{X}}^{(1)}(-\boldsymbol{p}_0|-\boldsymbol{p}), \quad (231)$$

thus $\overline{\boldsymbol{X}}^{(1)}$ is reciprocal, but the same conclusion cannot be extended to $\overline{\boldsymbol{X}}^{(n)}$ when $n \geq 2$. For example, in the case $n = 2$, using Eq. (230), we can only deduce that

$$\int \frac{\mathrm{d}^2\boldsymbol{p}_1}{(2\pi)^2} \left[\overline{\boldsymbol{X}}^{(2)}(\boldsymbol{p}|\boldsymbol{p}_1|\boldsymbol{p}_0)\right]^{aT} h(\boldsymbol{p}-\boldsymbol{p}_1)\, h(\boldsymbol{p}_1-\boldsymbol{p}_0) = \int \frac{\mathrm{d}^2\boldsymbol{p}_1}{(2\pi)^2} \overline{\boldsymbol{X}}^{(2)}(-\boldsymbol{p}_0|-\boldsymbol{p}_1|-\boldsymbol{p}) h(\boldsymbol{p}-\boldsymbol{p}_1)\, h(\boldsymbol{p}_1-\boldsymbol{p}_0). \quad (232)$$

From this, we cannot deduce a result similar to Eq. (231) for $\overline{\boldsymbol{X}}^{(2)}$. This fact is well illustrated with the following identity, which can be demonstrated with a transformation of the integration variables.

$$\int \frac{\mathrm{d}^2\boldsymbol{p}_1}{(2\pi)^2} (\boldsymbol{p}+\boldsymbol{p}_0-2\,\boldsymbol{p}_1)\, h(\boldsymbol{p}-\boldsymbol{p}_1)\, h(\boldsymbol{p}_1-\boldsymbol{p}_0) = 0. \quad (233)$$

We see that $\boldsymbol{p}_1 \longrightarrow \boldsymbol{p}+\boldsymbol{p}_0-2\,\boldsymbol{p}_1$ is not the null function although the integral is null. From this, we demonstrate that $\overline{\boldsymbol{X}}^{(n)}$ for $n > 1$ are not unique. Moreover in using Eq. (233) we can transform $\overline{\boldsymbol{X}}^{(n)}$ in a reciprocal form. This procedure is illustrated in the one-dimensional case in[26] and the results for the second-order in the electromagnetic case are given in[9] and[27].

C Scattering Matrix Coefficients

We give the expression of the matrices in the case of two randomly rough surfaces

$$\overline{\epsilon} \equiv \frac{1}{2}\begin{pmatrix} (\epsilon_0\,\epsilon_1)^{-1/2} & 0 \\ 0 & 1 \end{pmatrix} \tag{234}$$

$$\overline{S}^{\pm}(\boldsymbol{p}\,|\boldsymbol{p}_0) \equiv \frac{\epsilon_1 - \epsilon_0}{2\,\alpha_0(\boldsymbol{p}_0)\,(\epsilon_0\,\epsilon_1)^{1/2}}\left[\overline{\boldsymbol{M}}^{1-,0+}(\boldsymbol{p}\,|\boldsymbol{p}_0).\cdot\overline{\boldsymbol{X}}^{(00)}(\boldsymbol{p}\,|\boldsymbol{p}_0) \pm \overline{\boldsymbol{M}}^{1-,0-}(\boldsymbol{p}\,|\boldsymbol{p}_0)\right]. \tag{235}$$

After some calculations, we get

$$\begin{aligned}\overline{S}^{+}(\boldsymbol{p}\,|\boldsymbol{p}_0) &= \frac{(\epsilon_1 - \epsilon_0)}{(\epsilon_0\,\epsilon_1)^{1/2}} \\ &\times \begin{pmatrix} \begin{matrix}\epsilon_1\,||\boldsymbol{p}||\,||\boldsymbol{p}_0||\,F_V^{+}(\boldsymbol{p}_0) \\ +\epsilon_0\,\alpha_1(\boldsymbol{p})\,\alpha_1(\boldsymbol{p}_0)\,F_V^{-}(\boldsymbol{p}_0)\,\hat{\boldsymbol{p}}\cdot\hat{\boldsymbol{p}}_0\end{matrix} & \epsilon_0^{1/2}\,K_0\,\alpha_1(\boldsymbol{p})\,F_H^{+}(\boldsymbol{p}_0)\,(\hat{\boldsymbol{p}}\times\hat{\boldsymbol{p}}_0)_z \\ -\epsilon_0\,\epsilon_1^{1/2}\,K_0\,\alpha_1(\boldsymbol{p}_0)\,F_V^{-}(\boldsymbol{p}_0)\,(\hat{\boldsymbol{p}}\times\hat{\boldsymbol{p}}_0)_z & (\epsilon_0\,\epsilon_1)^{1/2}\,K_0^2\,F_H^{+}(\boldsymbol{p}_0)\,\hat{\boldsymbol{p}}\cdot\hat{\boldsymbol{p}}_0 \end{pmatrix} \\ &\times\left[\overline{\boldsymbol{D}}_{10}^{+}(\boldsymbol{p}_0)\right]^{-1},\end{aligned} \tag{236}$$

$$\begin{aligned}\overline{S}^{-}(\boldsymbol{p}\,|\boldsymbol{p}_0) &= \frac{(\epsilon_1 - \epsilon_0)}{\alpha_0(\boldsymbol{p}_0)\,(\epsilon_0\,\epsilon_1)^{1/2}} \\ &\times \begin{pmatrix} \begin{matrix}-\epsilon_0\,\alpha_1(\boldsymbol{p}_0)\,||\boldsymbol{p}||\,||\boldsymbol{p}_0||\,F_V^{-}(\boldsymbol{p}_0) \\ -\epsilon_1\,\alpha_1(\boldsymbol{p})\,\alpha^2 0(\boldsymbol{p}_0)\,F_V^{+}(\boldsymbol{p}_0)\,\hat{\boldsymbol{p}}\cdot\hat{\boldsymbol{p}}_0\end{matrix} & -\epsilon_0^{1/2}\,K_0\,\alpha_1(\boldsymbol{p})\,\alpha_1(\boldsymbol{p}_0)\,F_H^{-}(\boldsymbol{p}_0)\,(\hat{\boldsymbol{p}}\times\hat{\boldsymbol{p}}_0)_z \\ \epsilon_1^{3/2}\,K_0\,\alpha_0^2 0\,F_V^{+}(\boldsymbol{p}_0)\,(\hat{\boldsymbol{p}}\times\hat{\boldsymbol{p}}_0)_z & -(\epsilon_0\,\epsilon_1)^{1/2}\,K_0^2\,\alpha_1(\boldsymbol{p}_0)\,F_H^{-}(\boldsymbol{p}_0)\,\hat{\boldsymbol{p}}\cdot\hat{\boldsymbol{p}}_0 \end{pmatrix} \\ &\times[\overline{\boldsymbol{D}}_{10}^{+}(\boldsymbol{p}_0)]^{-1}.\end{aligned} \tag{237}$$

References

1. F.G. Bass and I.M. Fuks, *Wave Scattering from Statistically Rough Surfaces* (Pergamon, Oxford, U.K., 1979).
2. A. Ishimaru, *Wave Propagation and Scattering in Random Media* (Academic Press, New York, 1978).
3. P. Bousquet, F. Flory, and P. Roche, "Scattering from multilayer thin films: theory and experiment," *J. Opt. Soc. Am. A* **71**, 1115–23 (1981).
4. F.T. Ulaby, R.K. Moore, A.K. Fung, *Microwave Remote Sensing*, vol. 2 (Addison-Wesley, Reading, MA, 1986).
5. L. Tsang, G.T.J. Kong, R. Shin, *Theory of Microwave Remote Sensing* (Wiley-Interscience, New York, 1985).
6. J.A. DeSanto, G.S. Brown, "Analytical techniques for multiple scattering from rough surfaces," in *Progress in Optics XXIII*, E. Wolf, ed. (Elsevier Science Publishers B. V., Amsterdam, 1986).
7. J.A. Ogilvy, *Theory of Wave Scattering From Random Rough Surfaces* (Adam Hilger, Bristol, 1991).
8. S.M. Rytov, Y.A. Kravtsov, and V.I. Tatarskii, *Principle of Statistical Radiophysics* vol. 3–4 (Springer-Verlag, New York, 1989).

9. A.G. Voronovich, *Wave Scattering from Rough Surfaces* (Springer, Berlin, 1994).
10. I. Ohlídal, K. Navrátil, M. Ohlídal, "Scattering of light from multilayer systems with rough boundaries," *Prog. Opt.* **34**, 251–334 (1995).
11. J.M. Elson, "Multilayer-coated optics: guided-wave coupling and scattering of interface random roughness," *J. Opt. Soc. Am. A* **12**, 729–742 (1995).
12. A.K. Fung, *Microwave Scattering and Emission Models and their Applications* (Artech House, Boston, 1994).
13. M. Nieto-Veperinas, *Scattering and Diffraction in Physical Optics* (Wiley, New York, 1991).
14. V. Freilikher, E. Kanzieper, A.A. Maradudin, "Coherent scattering enhancement in systems bounded by rough surfaces," *Phys. Rep.* **288**, 127–204 (1997).
15. S.O. Rice, "Reflection of electromagnetic waves from slightly rough surfaces," *Commun. Pure Appl. Math.* **4**, 351–378 (1951).
16. J.T. Johnson, "Third-order small-perturbation method for scattering from dielectric rough surfaces," *J. Opt. Soc. Am. A* **16**, 2720–2736 (1999).
17. I.M. Fuks, A.G. Voronovich, "Wave diffraction by rough interfaces in an arbitrary plane-layered medium," *Waves Random Media* **10**, 253–272 (2000).
18. A. Soubret, G. Berginc, and C. Bourrely, "Application of reduced Rayleigh equation to electromagnetic wave scattering by two-dimensional randomly rough surfaces," *Phys. Rev.* B **63**, 245411–245431 (2001).
19. A. Soubret, G. Berginc, and C. Bourrely, "Backscattering enhancement of an electromagnetic wave by two-dimensional rough layers," *J. Opt. Soc. Am. A* **18**, 2778–2788 (2001).
20. A.G. Voronovich, "Small-slope approximation for electromagnetic wave scattering at a rough interface of two dielectric half-spaces," *Waves Random Media* **4**, 337–367 (1994).
21. G.C. Brown, V. Celli, M. Haller, and A. Marvin, "Vector theory of light scattering from a rough surface: unitary and reciprocal expansions," *Surf. Sci.* **136**, 381–397 (1984).
22. A.R. McGurn, A.A. Maradudin, V. Celli, "Localization effects in the scattering of light from a randomly rough grating," *Phys. Rev. B* **31**, 4866 (1985).
23. V. Celli, A.A. Maradudin, A.M. Marvin, A.R. McGurn, "Some aspects of light scattering from a randomly rough metal surface," *J. Opt. Soc. Am. A* **2**, 2225–2239 (1985).
24. A.R. McGurn, A.A. Maradudin, "Localization effects in the elastic scattering of light from a randomly rough surface," *J. Opt. Soc. Am. B* **4**, 910–926 (1987).
25. C. Baylard, J.J. Greffet, A.A. Maradudin, "Coherent reflection factor of random rough surface: applications," *J. Opt. Soc. Am. A* **10**, 2637–2647 (1993).
26. A.A. Maradudin, E.R. Méndez, "Enhanced backscattering of light from weakly rough random metal surfaces," *Appl. Opt.* **32**, 3335–3343 (1993).
27. A.R. McGurn, A.A. Maradudin, "Perturbation theory results for the diffuse scattering of light from two-dimensional randomly rough metal surfaces," *Waves Random Media* **6**, 251–267 (1996).
28. C.S. West, K.A. O'Donnell "Observations of backscattering enhancement from polaritons on a rough metal surface," *J. Opt. Soc. Am. A* **12**, 390–397 (1995).
29. J.A. Sánchez-Gil, A.A. Maradudin, J.Q. Lu, V.D. Freilikher, M. Pustilnik, and I. Yurkevich, "Scattering of electromagnetic waves from a bounded medium with a random surface," *Phys. Rev. B* **50**, 15353–15368 (1994).
30. T. Kawanishi, H. Ogura, Z.L. Wang, "Scattering of an electromagnetic wave from a planar waveguide structure with a slightly 2D random surface," *Waves Random Media* **7**, 35–64 (1997).

31. J.J. Greffet, "Scattering of electromagnetic waves by rough dielectric surfaces," *Phys. Rev. B* **37**, 6436–6441 (1988).
32. M. Nieto-Veperinas, "Depolarization of electromagnetic waves scattered from slightly rough random surfaces: a study by means of the extinction theorem," *J. Opt. Soc. Am. A* **72**, 539–547 (1982).
33. A. Ishimaru, C. Le, Y. Kuga, L.A. Sengers, and T.K. Chan, "Polarimetric scattering theory for high slope rough surfaces," *PIER* **14**, 1–36 (1996).
34. J.Q. Lu, J.A. Sánchez-Gil, E.R. Méndez, Z. Gu, and A.A. Maradudin, "Scattering of light from a rough dielectric film on a reflecting substrate: diffuse fringes," *J. Opt. Soc. Am. A* **15**, 185–195 (1998).
35. Y.S. Kaganovskii, V.D. Freilikher, E. Kanzieper, Y. Nafcha, M. Rosenbluh, and I.M. Fuks, "Light scattering from slightly rough dielectric films," *J. Opt. Soc. Am. A* **16**, 331–338 (1999).

7
Computer Simulations of Rough Surface Scattering

JOEL T. JOHNSON

Department of Electrical and Computer Engineering and ElectroScience Laboratory, The Ohio State University, 1320 Kinnear Rd., Columbus, OH 43212

7.1. Introduction

"Numerically exact" methods for rough surface scattering have increased in their relevance to surface scattering studies over the past quarter century. This increase in relevance follows the increase in computational power that has become readily available over the same period, along with developments in approaches for reducing the computational complexity of such methods. However, while computational power has increased by several orders of magnitude, and the range of surface scattering problems that can be studied numerically has followed directly, the overall impact of numerical studies has been decidedly less dramatic, while still of some import, as will be discussed throughout this chapter.

Several excellent review articles have been previously written on the subject of numerical algorithms for rough surface scattering[1–3], including two recent contributions[4,5]; all these reviews are strongly recommended to the reader as excellent surveys of the variety of studies and approaches that have been used. Given these articles, the current chapter is written as a more specific review of the author's research in this area, with particular examples and recommendations provided from the author's experiences. The discussions provided therefore should not be taken as representative of all possible studies and techniques; again the reader is referred to the cited review articles for broader information.

The next section provides a description of a few fundamental issues involved in describing a rough surface scattering problem, while specific integral equation formulation and matrix solution methods utilized in previous studies are described in Sects. 7.3 and 7.4. Sample results are illustrated in Sect. 7.5 to demonstrate a range of problems, and Sect. 7.6 provides summary recommendations from the author's experience regarding use of numerical methods in future studies.

7.2. Fundamental Issues

Any numerical surface scattering simulation begins with a description of the problem to be solved, including properties of the surface and the background media

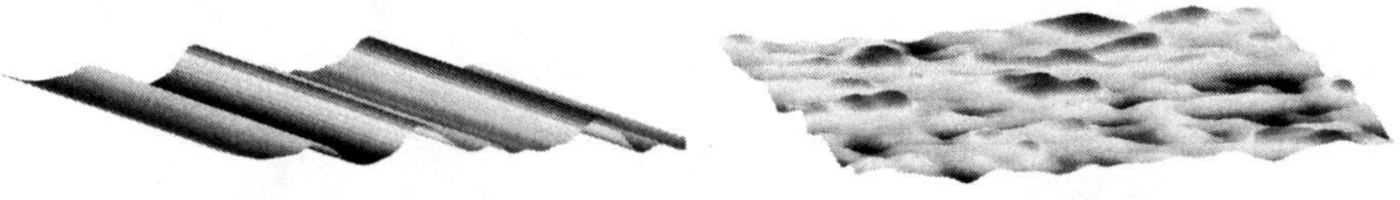

FIGURE 7.1. Geometries of one- and two-dimensional rough surfaces.

involved. Here the basic problem considered involves scattering from a rough interface between two simple and homogeneous media. Typically the coordinate system is defined so that the incident field approaches from above the interface, and the resulting scattered fields are determined above the interface as well as below the interface in the case of a penetrable lower medium. Typically it is the scattered and/or transmitted fields in the far field of the surface that are of interest, although it is quite simple to compute the resulting fields in the near field of the rough surface, and a few studies have taken this route.

7.2.1. One- and Two-Dimensional Surfaces

A major factor in modeling rough surface scattering problems involves the choice of a "one-" or "two"-dimensional surface model. Here a one-dimensional surface refers to a surface with variations along one horizontal coordinate only, with the surface profile constant along the other horizontal coordinate. Two-dimensional surfaces alternatively have variations along both horizontal coordinates, as illustrated in Fig. 7.1.

The major advantage of the one-dimensional geometry is the fact that discretization for a numerical method is required only in the coordinate along which the surface varies. This results in a tremendous computational savings compared to the two-dimensional case. Typically it is found that results using one-dimensional surfaces show nearly identical physical behaviors to those from simulations with two-dimensional surfaces, and in the majority of cases, use of one-dimensional surfaces is found reasonable. The author is not aware of any cases where significantly different physical behaviors were obtained from two-dimensional versus one-dimensional surface predictions, in cases where one-dimensional simulations are relevant.

However there exists a set of physical scattering effects for which one-dimensional models are certainly inadequate. These include studies of cross-polarized scattering as well as studies of scattering outside the plane of incidence (i.e. when the scattered field propagation direction in the far field lies outside the plane formed by the incident field propagation direction and the normal to the mean surface). For scattering within the plane of incidence, one-dimensional models always predict zero cross-polarized fields, while it is well known that such fields, though small, can result from two-dimensional surfaces. Although it is possible to compute bistatically scattered fields outside the plane of incidence for a one-dimensional surface, the resulting fields are representative of bistatic scattering only when the surface of interest is truly one-dimensional.

Due to the reduced computational complexity of numerical methods for one-dimensional surfaces, the majority of rough surface scattering studies have utilized this model. Only in the last decade have significant numerical studies using two-dimensional surfaces been performed. The results to be shown in Sect. 7.5 will include both one- and two-dimensional surface cases.

7.2.2. Description of the Rough Surface

A second fundamental issue involves the description of the rough surface to be simulated. Although it is certainly possible to specify a deterministic rough surface profile function if one is completely known, it is far more common to utilize a stochastic model of the surface profile. In this case, the surface scattering problem itself becomes stochastic, and it is statistics of the scattered fields that are of interest.

Assuming a stochastic surface model, it is noted that stochastic process descriptions of a rough surface profile function can take a wide variety of forms, and exhibit an extremely wide range of behaviors. A general stochastic process model assumes knowledge of the infinite-dimensional probability density function of all surface profile height random variables. Although a few alternate models have been investigated, the majority of numerical studies make use of a stationary Gaussian random process description of the rough surface; in this case, surface statistics are completely specified through knowledge of the two-point covariance function alone. While interest in the use of non-Gaussian process models is increasing, the simplicity of the Gaussian random process model makes it highly preferable. It is also found that many surfaces created through natural or uncontrolled methods are reasonably modeled as Gaussian random processes, at least to an initial approximation. Simple algorithms for generating realizations of a Gaussian random process with a specified covariance function are available and have been widely used.

Note that one of the alternate models that has been explored involves modeling the rough surface through a quasi-fractal description, where the fractal behaviors extend only over a finite range of length scales. However, a subset of these quasi-fractal surfaces can also be described as Gaussian random processes, and at present, the author is not convinced of the advantages of a quasi-fractal model over these similar Gaussian process descriptions.

Once a Gaussian random process model is adopted, the covariance function must be specified, or equivalently, its Fourier transform, the power spectral density function. It is important to recognize that it is often roughness in length scales comparable to the electromagnetic wavelength that strongly influences the scattering process. Therefore it is important to describe surface roughness on these scales accurately. The development of surface statistical models in terms of the power spectral density is then often most useful, because the energy in varying surface length scales is explicitly treated, rather than somewhat masked in a spatial covariance function description.

Although simple two-parameter covariance function descriptions of surface properties have been highly desired, the rms height and correlation length parameters involved are usually woefully inadequate for capturing real surface behaviors. Again it is far better to attempt to describe the power spectral density function over as wide a range of length scales as possible, to the extent that information on the surface to be modeled is available. For purely theoretical studies, the two parameter covariance functions remain widely used due to their simplicity, but again, when modeling a real surface one must ensure that the description used is realistic. Errors in the description of surface statistics can be much more important than errors in any scattering models used, numerical or otherwise.

The boundary condition at the surface interface must also be considered; only electromagnetic scattering is considered in this chapter, resulting in boundary condition possibilities of perfectly conducting, finitely conducting, or penetrable. The finitely conducting case here refers to treatment of the boundary by an "impedance boundary condition", while penetrable includes the possibility of lossy media modeled without such an impedance boundary condition. Perfectly conducting cases with one-dimensional surfaces and in-plane scattering reduce to the Dirichlet and Neumann boundary conditions depending on the polarization considered.

7.2.3. Finite Size Surface Effects

The numerical approaches described in this chapter all consider the surfaces modeled to be of finite horizontal size, as shown in Fig. 7.2. Because typically scattering

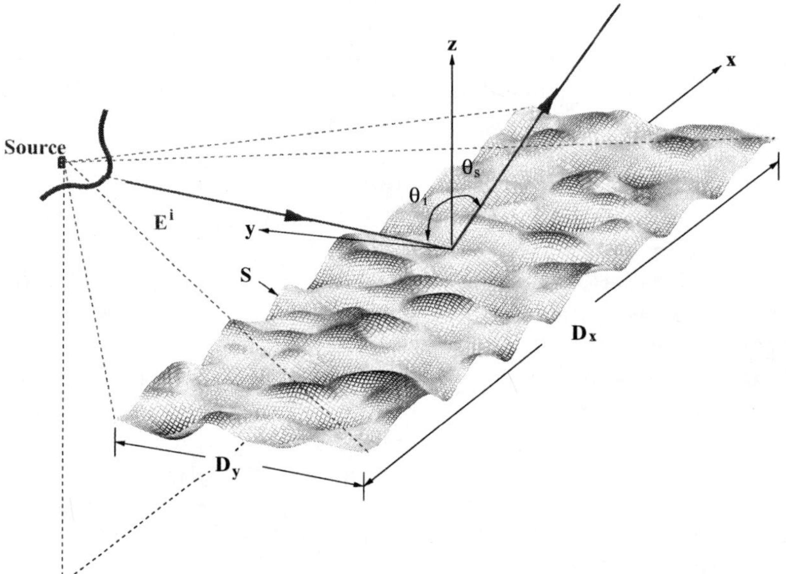

FIGURE 7.2. Two-dimensional canonical surface scattering problem including finite surface size effects.

from a surface of infinite horizontal extent is of interest, numerically computed results must be carefully examined to determine if finite surface size is influencing the results. The obvious means for making such an assessment involves generating numerical results for progressively larger horizontal surface sizes while all other parameters remain the same, and examining scattered field predictions for convergence[6].

Finite surface size results in two clear problems: a reduction in the angular resolution of scattered fields, as well as the potential influence of scattering and diffraction from surface edges. The former case results simply by considering scattered fields obtained from integrating surface fields over finite surface size as having originated from an integration of surface fields over an infinite surface multiplied within the integration by a rectangular "window" function. Because the integration to obtain far-field results is essentially a Fourier transform between space domain surface fields and angular domain scattered fields, inclusion of the window function results in a convolution of the scattered field angular pattern with the Fourier transform of the spatial domain window function. Smaller surface sizes (or smaller windows) therefore result in a greater degree of angular averaging, reducing the ultimate scattered field angular resolution obtained.

Because the average incoherent scattered power obtained from a stochastic surface is typically a relatively smooth function of angle, such reduction in angular resolution often has only a minor influence on the obtained results. However, in some cases, a reduction in angular resolution is extremely problematic. For example, when surface backscattering near low grazing angles under plane wave illumination is considered (i.e. with the incident field approaching the surface near horizontally), the rapid decrease in backscattering with angle that can occur as grazing is approached becomes easily smoothed out as angular resolution decreases. The need for high angular resolution in this case requires use of large surface sizes if true plane wave scattering is to be simulated. Reference [7] provides a discussion of the relationship between angular resolution and surface size.

The second problem with finite surface size involves diffraction effects from surface edges. The use of a "tapered wave" instead of a plane wave is most commonly followed to address this issue. Tapered waves reduce the incident field intensity on surface edges, and thereby attempt to reduce the overall effect of surface edge contributions. A number of tapered wave formulations are available[8–11], and generally it is found that this method is successful at reducing edge contributions to manageable levels. An alternate approach treats surfaces edges with a material of gradually changing resistivity (so called "R-cards"); although this method has been utilized less frequently, it has also been shown to yield reasonable reduction in surface edge contributions[12,13]. One advantage of the R-card technique is that a larger percentage of surface profile function can potentially remain illuminated, whereas the tapered wave method reduces the surface profile illumination near surface edges. Again using the concept of a windowed Fourier transform, the tapered wave approach thus further reduces the angular resolution of the scattered field due to the effective additional reduction in surface size.

For two-dimensional surfaces, tapered wave descriptions must be chosen carefully to minimize corruption of field polarization properties, particularly in cross-polarized quantities. If it is considered that a tapered wave can be regarded as a superposition of incident plane waves at varying incidence angles, the backscattered field becomes an integration of the surface bistatic scattering coefficient over a range of incidence angles. This is particularly important for cross-polarization, because cross-polarized scattering near the backscatter direction is typically minimum within the plane of incidence and rises rapidly out of the plane of incidence. Thus, any angular averaging over bistatic scattering coefficients outside the plane of incidence can cause significant increases in the obtained cross-polarized backscattering. Tapered wave parameters must be chosen appropriately to minimize this effect; typically this sets a limit on the cross range surface size that must be simulated if cross-polarized scattering is of interest. Often, two-dimensional surface numerical simulations that produce reasonable copolarized scattering coefficients remain inaccurate for cross-polarized quantities.

7.2.4. Other Physical Parameters

In formulating a rough surface scattering problem, it is also important that the sensor to be utilized is modeled accurately. Properties such as the polarizations of interest (here referred to as "horizontal" or "H" pol for that polarization perpendicular to both the mean surface normal and the wave propagation direction, and "vertical" or "V" for the remaining polarization), the relevant incidence and scattering angles, and the frequencies of interest must be known.

If examination of scattered fields as a function of frequency is of interest, it must be decided whether any surface profile discretization is to be varying or fixed versus frequency. The former can allow more efficient computations, since typically the number of unknown quantities involved in a numerical simulation decreases as the frequency decreases, but the latter can often result in a more clear and simple description of the surface profiles involved. Such a decision is likely to be made on the basis of the range of frequencies involved; for a moderate range (around an octave approximately), using a fixed discretization may remain desirable.

A brief discussion of time domain algorithms for simulating rough surface scattering, though they are not the focus of this chapter, is appropriate at this point. With the methods to be described in Sect. 7.3, only time harmonic problems are considered, so that results versus frequency are computed only through repeated independent runs of the numerical model at each frequency of interest. Time domain methods seemingly can provide responses at multiple frequencies from a single simulation, and thereby may appear more appealing in this case. However, it should be considered that all numerical algorithms, whether in the frequency or time domain, will require sampling the surface on the scale of the shortest electromagnetic wavelength for which results are desired. In addition, rough surface problems will also typically require that the surface profiles considered be at least as long as several of the longest wavelengths to be considered. Therefore, because the size of the problem involved for the time domain algorithm also scales with the range of frequencies to be investigated, the advantages of time domain algorithms for typical rough surface scattering problems are reduced, even when wide frequency ranges are of interest. The possibility of varying sample rates with frequency in the time harmonic methods further reduces any potential gains from a time domain approach.

The antenna pattern of any sources or receivers involved in a particular experiment are also of importance. While plane wave illumination and reception is usually of interest in theoretical studies, practical problems may involve specified transmit and receive antenna patterns. These antenna patterns then influence surface size needs and the angular resolution obtained as described in the previous section.

7.2.5. Other "Numerical" Parameters

The methods to be described also involve an additional numerical parameter based on the sampling rate utilized in discretizing the surface fields. Typical expectations for the methods discussed are that sampling at approximately 4 to 16 points per electromagnetic wavelength in the incident medium is needed for the "point matching" approaches to be described in Sect. 7.3. The wide range specified allows for varying amounts of accuracy, as well as the fact that differing surface profiles contain varying amounts of energy in small length scales, which in turn influences the discretization rate needed. The sampling rate to be chosen should be appropriate both for the surface profile of interest and for the electromagnetic field. However, unless there are known sharp features in a specific surface profile, it is expected that surface variations can be smoothed to a scale of approximately one quarter electromagnetic wavelength without overly influencing the obtained scattering behavior[14]. Surfaces without roughness on these scales are then treated reasonably with the lower sampling rate, while those with roughness on these scales can require a finer sampling to obtain accurate predictions.

A second issue involved in determining the discretization rate for surface fields is related to the wavelength in the lower medium. In cases where the wavelength in the lower medium is much smaller than that in the incident medium, questions can arise regarding which wavelength to use in determining the sample rate. However, it is clear that in the limit of highly conducting surfaces under an impedance boundary approximation, it is the wavelength in the incident medium that matters. This is also apparent if it is considered that any surface field variations on scales much more rapid that those of the incident medium wavelength will result in evanescent scattered fields in the incident medium. Thus if scattering into the incident medium is of primary interest, such rapid variations in the surface fields are not important. However it remains important to ensure that the required integrations within an integral equation formulation are computed accurately. In this case, integrations of matrix elements may be required, but no increase in the number of unknowns is necessarily needed. References [15–17] provide further discussion of this point. As with surface size, it is generally a good practice to ensure convergence with the discretization rate by examining numerical results as the spatial resolution is increased.

For stochastic problems, it is also commonplace to report averages of scattered fields obtained through repeated computations over an ensemble of surface realizations. These "Monte Carlo" simulations are the primary tool for reporting information on scattered field statistical properties. An issue then involves convergence of results with respect to the number of realizations in the ensemble. Again this convergence is usually examined through comparisons of results from independent ensembles, and also by directly examining the statistics obtained. It is to be expected usually that incoherent scattered fields in time harmonic problems will exhibit real and imaginary parts that are uncorrelated (i.e. random phase), with each following a Gaussian probability density function. The resulting incoherent power is then an exponential random variable, with a standard deviation equal to the mean power. Deviation from these behaviors however is certainly possible in unusual scattering situations[6], so that direct examinations of convergence are usually appropriate.

7.2.6. Use of Approximate Theories

Given the wide range of "artificial" parameters involved in numerical studies of surface scattering (i.e. finite surface length, discretization rate, use of tapered waves, etc.), it can be very difficult to have complete confidence in the results achieved without extensive convergence studies. While several basic expected behaviors have been described in this section to assist in this process, another approach to help build confidence in a particular method involves comparing results with predictions from approximate theories in limits where the approximate theory is known to be accurate.

While some users of numerical algorithms may feel that numerical approaches are inherently superior to the approximate theories, and that therefore approximate theories have no place in verifying their algorithms, in fact the opposite is true. Approximate theories when used appropriately usually will provide far higher accuracy than that obtained from a numerical method, due to the superiority of analytical computations when possible over discretized methods. Therefore it is highly recommended to any user of a numerical method that the method be validated in an appropriate limit by comparing with an approximate theory. If agreement cannot be obtained, it is almost certainly the numerical method that is yielding inaccurate predictions, likely due to the one of the myriad possible sources of error discussed.

An ideal approximate theory for this purpose is the small perturbation method. Once the problem description and choice of numerical parameters has been completed, the surface profile of interest can simply be substituted by a similar surface profile with small surface height and short length scale variations compared to the electromagnetic wavelength. Comparison of the numerically obtained results to the analytical predictions of perturbation theory will yield some idea of the level of accuracy of the numerical simulations.

A second similar approach involves comparison with predictions from the physical optics theory. In this case, the numerical solution is abandoned and replaced with a numerically evaluated physical optics theory. Comparisons are made of numerically computed physical optics scattered powers with those predicted analytically by the physical optics theory. The numerical computation of physical optics predictions is performed using discretization rates, surface sizes, and incident fields identical to those to be used with the numerical method. Because the physical optics approximation is used both in the "numerical" and analytical techniques in this case, the results should be identical if effects of the numerical parameters are negligible. Again these comparisons can provide some idea of the influence of the numerical parameters, and allow improvements in parameter choices to be determined.

Once it is shown that the numerical configuration can yield appropriate results in these tests, the user can have increased confidence for use of the numerical solution, although every attempt to validate the predictions by other means should continue to be pursued as well.

7.3. Integral Equation Formulations

The methods described in this chapter are all integral equation approaches for time harmonic problems. The basic integral equation formulations have been described in the literature numerous times (for example, [18–19] and [20] for the one- and two-dimensional cases, respectively), and therefore are not repeated here.

Unknowns in these integral equations are the tangential electric and/or magnetic fields on the interface. In two-dimensional surface problems, these fields are two component vectors, resulting in four unknown scalar functions to be determined in the penetrable case. This number is reduced to two scalar functions for a two-dimensional perfectly conducting or impedance surfaces or for in-plane scattering with a one-dimensional penetrable surface, and further reduces to a single scalar function for a one-dimensional conducting or impedance surface. In some two-dimensional surface studies, normal components of the surface fields at the interface have also been retained as unknowns[21,22], but this is not required if these quantities are represented in terms of derivatives of tangential fields.

A variety of integral equations are available, and appropriate choices can improve computational efficiency[23,24]. For penetrable surfaces, integral equations must be formulated for fields in the media above and below the surface profile. The integral equations utilized in works by the author are the standard electric field and magnetic field integral equations (labeled the EFIE and MFIE, respectively), with the MFIE typically used in two-dimensional problems in the region above the interface. Although combinations of these integral equations into a "combined field integral equation" (CFIE) are sometimes suggested as providing improved equation conditioning, preconditioning techniques for accomplishing the same goal will be discussed in the next Section.

Discretization of integral equations into a matrix equation is accomplished through the choice of "basis" expansion functions for the unknown scalar functions, as well as "weighting" functions over which scalar products of the integral equations are performed. A variety of these expansion functions are available, ranging from simple "pulse" expansions which discretize the unknown function in terms of a set of locally constant values, versus "entire domain" functions which span the entire surface. The accuracy of pulse expansions is often questioned, but it is clear that the effects of this somewhat crude discretization can always be reduced by increasing the number of points utilized. The advantage of the pulse expansion is that it greatly simplifies computation of the required inner product integrals in formulating a matrix equation. Such simplicity of the matrix elements is highly advantageous in the efficient solution algorithms to be discussed in the next section.

Entire domain basis methods will always require numerical integrations for the computation of the required matrix elements; if the problem becomes sufficiently large that these integrations cannot be stored in memory, then such integrations will need to be repeated multiple times in a iterative matrix equation solution, resulting in low computational efficiency. Integrations can sometimes be avoided through the use of approximations for some matrix elements, but developing such approximations requires extensive analytical work. While it is often argued that entire domain functions may be able to be formulated such that only a small set of such functions is required to accurately represent surface fields, this is only likely to be true in cases where the range of roughness length scales is not large, for example in cases involving high frequency examination of surfaces with roughness on scales much larger than the electromagnetic wavelength. Because multi-scale roughness is usually the norm rather than the exception when numerical solutions of surface scattering are called for, the author's expectations are that pulse basis methods will typically be at least competitive if not far more efficient than entire domain basis methods. For this reason, pulse expansions are used exclusively in the results to be illustrated in Sect. 7.5.

Weighting functions can still be chosen arbitrarily once an expansion function is selected, although the accuracy of the formulation is typically expected to be improved when the weighting and testing functions are chosen to be identical. Given choice of pulse expansion functions, the corresponding weighting functions are then pulses as well (the so-called point matching approach), and the inner product integration involves integration over a pulse (i.e. constant over a small portion of the surface) function. This integration is often simplified in terms of only the value of the function at the center of the pulse multiplied by the pulse width, i.e. a single point approximation of the integral. The accuracy of this approximation degrades as the distance between the weighting and expansion functions decreases, but typically remains acceptable for all but the "self" term (i.e. when the weighting and expansion functions overlap). In cases where increased accuracy is needed, numerical integration over the weighting pulse functions can be performed for a subset of matrix elements; it is important that sufficient memory be available to store the integrated results to avoid the need for repeated integrations. The need for integrating matrix elements in the penetrable medium case discussed in Sect. 7.2.5 is an example of this requirement.

Integrations remain necessary for computing self terms in the formulation, and are typically obtained through an analytical approximation of the Green's function. Appropriate analytical self terms are well known in the one-dimensional case (see [14]). In the two-dimensional case, analytical forms are not easily available, and numerical integrations are required[20]. Recently, an analytical formulation has been developed for including second derivatives in two-dimensional surface self terms[25], analogous to the one-dimensional "curvature" term discussed in [14]. However numerical integrations of any remainders are still required even after the analytical forms are extracted.

7.4. Matrix Solution Methods

Once a matrix equation is formulated, its solution is required in order to determine the unknown coefficients of the surface field expansion. When these are known it is simple to compute whatever field quantities are desired (usually far field scattering). The solution of a linear system of equations is a standard problem in computational physics, and is discussed in numerous references. It is well known that a direct solution of an N by N matrix equation requires an operations count proportional to N^3. Although parallel implementations of direct solution methods are available for poorly conditioned matrix equations [26], iterative matrix equation solutions are generally for more efficient, and preferred in cases with reasonably conditioned equations.

7.4.1. Iterative Solution of Matrix Equations

A variety of iterative matrix inversion algorithms are available[27], including "stationary" and "non-stationary" algorithms. All iterative algorithms are based on a repeated process for generating guesses at the unknown solution vector, along with continued multiplication of these guesses by the matrix in order to determine the degree of error in matching the right-hand side of the matrix equation. Iterative matrix solutions therefore require an operations count proportional to N^2 times the number of iterations; if the number of iterations can be made independent of N, the overall solution is order N^2, and therefore dramatically more efficient than the direct solution.

Stationary iterative algorithms are those in which the same operation is performed on the current solution guess at each iteration to generate a new solution guess. These algorithms are usually easier to understand and physically interpret. Nonstationary algorithms on the other hand can be thought of as having "iteration-dependent coefficients," and are typically based on attempting to generate a sequence of orthogonal search vectors in the solution space. Nonstationary algorithms usually provide better performance over a range of possible matrices, although stationary algorithms can converge more rapidly in some cases if the matrix condition number is low.

Several stationary algorithms have been utilized for rough surface scattering, particularly in the one-dimensional case[28]. The "banded matrix iterative approach" (BMIA)[29] is a stationary algorithm based on iteration on the contributions of matrix elements outside of a banded region from the matrix diagonal. The "forward–backward" or "method of ordered multiple interactions" approach[30–34] is a stationary algorithm known as "successive over relaxation" in which the iterative process results in a forward- and backward-sweeping process for interactions among points on the surface. A Neumann iterative approach[22,35,36] is also a stationary algorithm; in this case, surface fields are written as a series of multiple contributions from nonself points, and an iterative process used to compute these multiple interactions. Generally all of these stationary algorithms are found to work well for one-dimensional surface scattering, although convergence problems can be found in cases where multiple interactions become important. Although physical interpretations of these algorithms may be easily available, essentially the algorithms used are stationary iterative approaches based on repeated matrix multiplies, and are therefore order N^2 algorithms.

Nonstationary algorithms have also been widely used. Due to the fact that most matrices describing rough surface problems are nonsymmetric, the standard conjugate gradient algorithm is not applicable. However other algorithms such as GMRES, Bi-CGstab, and QMR[27] all are applicable to this case, with GMRES having the best analytical convergence properties. However GMRES also requires a significant amount of storage, while the Bi-CGstab algorithm avoids this requirement and is generally found to have properties similar to GMRES. For these reasons, the Bi-CGstab method is the most commonly used nonstationary iterative solution algorithm in rough surface studies.

In some cases, nested iterative algorithms are utilized, typically with a nonstationary algorithm inside a stationary algorithm. This occurs when the original matrix is approximated by a matrix whose matrix-vector multiply is more easily performed. The approximate matrix equation is then solved by an "inner" nonstationary algorithm. The "outer" stationary algorithm then consists of correcting for the approximation of the matrix. This involves modifying the right-hand side of the inner iterative equation to include the difference between the exact and approximated matrices multiplied by the current solution vector. This approach can be advantageous if convergence of the inner algorithm is slow but convergence of the outer algorithm is rapid. The "banded matrix flat surface iterative approach" (BMFSIA)[18] and "sparse matrix flat surface iterative approach" (SMFSIA)[37] are based on such nested algorithms.

7.4.2. *Preconditioning*

The rate at which iterative solutions converge is typically determined by the conditioning properties of the matrix of interest. Better conditioned matrices are generally those in which the near diagonal elements of the matrix are much larger than far off-diagonal elements, so choices of integral equation formulations to produce this behavior are usually preferable. Convergence can also be improved through the use of preconditioning matrices[27,38]; when used, the matrix equation is essentially left multiplied by the inverse of the preconditioner matrix. If the preconditioner matrix is chosen to approximate the original matrix while remaining easily invertible, the condition properties of the new product matrix are improved. Preconditioners can be utilized with both stationary and nonstationary methods, although in the case of physically based stationary methods, the iterative solution itself essentially is based on identification of a preconditioner matrix[38]. Within the iterative algorithm, a step involving solution of a preconditioner matrix equation is required (i.e. solution of an equation specifying the result of the preconditioner multiplied by an unknown vector.) The right-hand side of this preconditioner equation typically varies as the iterative algorithm progresses.

While standard mathematical techniques exist for formulating preconditioner matrices[27], the author believes that preconditioners for rough surface scattering are best identified on the basis of physical insight. One insight that has been utilized involves the expectation that coupling between points on a surface is strongest for points located close to each other. Under this assumption, a preconditioner is formulated that retains matrix elements corresponding to a specified distance between the "source" and "test" points involved. The resulting preconditioner is either banded or sparse, and methods for computing the inverse of such matrices exist. For sparse matrices (i.e. the two-dimensional surface case), the storage required for the sparse matrix inverse can become large, so an iterative solution of the preconditioner matrix equation may be required.

7.4.3. Physically Based Preconditioning

While these methods have been utilized in the literature, the author recommends instead preconditioners that retain contributions from all pairs of surface points, but under some approximation. This is because even though these contributions may individually be weak, their sum when added over all surface points remains important. This can be demonstrated by considering the case of a flat surface profile.

Because the preconditioner matrix is designed to approximate the original matrix equation, and because the original matrix equation fundamentally represents a physical process for determining induced surface fields on a rough surface given a specified incident field, solution of the preconditioner equation can be considered as an approximate solution of the rough surface scattering problem. However due to the varying right-hand side of the preconditioner equation at each iteration, the incident field to be utilized in the approximate surface scattering problem is distinct from that of the original problem. In cases involving penetrable media, an arbitrary right-hand side vector can represent fields impinging on the surface from both above and below. In any case, a preconditioner can be formulated simply by attempting to approximately solve the rough surface scattering problem with the incident fields corresponding to the specified right-hand side. As numerous methods are available for these approximations, there are numerous potential preconditioning algorithms.

A particular technique that is useful is physical optics. Once the right-hand side vector is expressed in terms of a set of plane wave incident fields, surface fields for each plane wave can be evaluated under the physical optics approximation, and summed to produce a preconditioner solution. Note no explicit formulation of a preconditioner matrix is required in this case, but the linear nature of the solution process ensures that such a matrix exists. Because the right-hand side vector is expressed in the space domain, a quasi-Fourier transform operation is necessary to interpret the incident field in terms of a set of plane waves. One limitation of this step is that it must be assumed that the surface is flat to utilize an FFT algorithm for this process. However even with this approximation, a preconditioner based on this approach has been found effective in improving Bi-CGstab convergence in penetrable two-dimensional surface problems for moderate height surfaces[20]. The resulting preconditioner solution requires only a small set of FFT operations, and therefore remains computationally efficient. Use of other approximate rough surface scattering methods is also possible for this purpose.

An alternative technique for generating and solving preconditioners is based on use of a flat-surface approximation for matrix elements, in which all arguments involving surface heights within the Green's function exponential terms are replaced with zero. In this case, the resulting matrix equation assumes a Toeplitz form, and therefore matrix-vector multiplies can be performed through a set of FFT operations. A generalization of this idea results in the "canonical grid" method to be discussed in Sect. 7.4.4.1, but if the approximation is utilized for all nondiagonal matrix points, an effective preconditioner can result for surfaces of moderate height. This technique has also been proposed as a stand-alone approximate method[39]. Although in the one-dimensional perfectly conducting case it is possible to formulate an analytical solution of this preconditioner matrix equation in terms of FFT operations[40], more typically it is required to utilize a nonstationary algorithm to solve the preconditioner equation. However, the efficiency with which matrix multiplies can be performed within the preconditioner solution can still make this an effective choice if the original matrix equation is poorly conditioned.

7.4.4. Accelerating the Matrix-Vector Multiply Operation

If reduction in the operation count below N^2 operations is desired, then more efficient means must be developed for computing the matrix-vector multiply. Several "acceleration" approaches have been proposed for this purpose in the past decade, including the more general "fast-multipole method" which can be used for arbitrary electromagnetic scattering problems[41–45], as well as the more recent "UV" technique based on singular-value decompositions of the matrix equation[46]. Typically these algorithms claim to achieve an operations count proportional to N or $N \log N$ as N becomes large. This proportionality however is often achieved only in the limit of very large N due to the addition of a fixed number of operations to the computational workload in order to utilize the acceleration method.

Standard acceleration algorithms are based on approximations to the interactions among surface points which are separated horizontally by more than a minimum distance, called the "neighborhood distance" in what follows; this distance is then a parameter of the algorithm. Note the use of subspace domain basis functions is implied in this separation. Coupling between points within the neighborhood distance is usually computed exactly, meaning that a banded (one-dimensional) or sparse (two-dimensional) matrix multiply operation is part of the complete matrix-vector multiply. The efficiency of the algorithm depends on efficiently computing interactions among points separated by distances larger than the neighborhood distance, as well as retaining accuracy in evaluating these interactions when the neighborhood distance is much smaller than the overall surface length. While it is possible to formulate such algorithms for general kinds of objects (as in the standard fast multipole method), acceleration methods for rough surface scattering can further exploit the assumed quasi-planar nature of the scatterer.

Here the focus is placed on two specific acceleration algorithms: the "canonical grid" method[47,48] and the "novel spectral approach"[49,50].

7.4.4.1. Canonical Grid Method

In the canonical grid (CAG) method, interactions among points separated by more than the neighborhood distance are computed using a Taylor series expansion of the Green's function in the difference between source- and observation point heights. When this expansion is used along with a rectangular grid for resolving the surface height, the resulting series terms involve translationally invariant functions (i.e. functions of differences in horizontal coordinates only) multiplied by polynomial expansions in surface height. An FFT algorithm can then be used to simultaneously compute coupling between all surface points for a given term in the Green's function expansion. The approach achieves an order $N \log N$ matrix multiply, but can also require a large number of series terms for large height and/or large slope surfaces. For this reason the canonical grid method is best used with surfaces whose heights and/or slopes are moderate compared to the electromagnetic wavelength and unity, respectively.

Reference [40] derives the canonical grid expansion to arbitrary order for one-dimensional surfaces, and also provides an efficient pseudo-code implementation of the matrix multiplication process. It is found that the canonical grid expansion can be interpreted as a power series expansion of the Fresnel phase term in the Green's function, when the horizontal distance is taken as the leading order range term. The quantity expanded is found to be proportional to the height difference between two surface points relative to the wavelength, multiplied by the slope of the line joining the two points. This expansion term can always be made smaller by increasing the neighborhood distance, and thereby decreasing this slope, but at the expense of increased within-neighborhood computations. Similar statements apply for two-dimensional surfaces, although a simple form for series terms at arbitrary order is not as easily available. Reference [48] provides the first three series terms, while an algorithm for generating higher order terms to arbitrary order has been developed in [25].

A simple method for estimating the number of canonical grid series terms needed can be developed. First, the largest possible value of the Fresnel expansion term is estimated, typically using a multiple of the surface rms height (possibly times five or ten) in the Fresnel term numerator and the neighborhood distance in its denominator. Given this Fresnel phase value, the number of power series terms needed in an exponential phase function evaluated at the Fresnel phase value is found to estimate the number of canonical grid series terms needed. Note in surfaces with large heights but small slopes, use of the surface correlation distance may be more appropriate in the denominator of this estimation.

Typically, when more than two or three canonical grid series terms are required, the number of FFT's needed by the algorithm becomes excessive, and use of the method is not recommended. Thus the canonical grid method is usually ineffective for large height and/or large slope surfaces. However, when it is applicable, the method is likely to be among the most efficient means possible for numerical simulations of this type. A multi-level expansion of the canonical grid series has been proposed to address these limitations[24,25], but remains expensive due to the requirement of a three-dimension FFT computation with two-dimensional surfaces.

Note the expansion utilized in the canonical grid algorithm is also used in analytical approximations of rough surface scattering, including the operator expansion method (OEM)[52] and the small slope approximation (SSA)[53]. These three techniques can be successively classified as purely analytical (SSA), quasi-numerical (OEM), and purely numerical (CAG) algorithms essentially making use of the same Green's function expansion.

7.4.4.2. Spectral Approach

One algorithm that can address the limitations of the canonical grid technique for large height surfaces is based on a "novel spectral approach" (NSA) for computing interactions among widely separated points on the surface. In this case, a spectral form of the Green's function is used for points outside the neighborhood distance; the simplicity of the spectral form allows contributions from multiple source points to a single observation point to be expressed in terms of a single spectral integration, and this integration is easily updated as the observation point is modified. A critical factor in the success of this technique is the computational requirements for computing the spectral integration. These are reduced by deforming the spectral integration onto a complex contour, with the contour derived to approach a steepest descent path for the flat surface case when possible. The resulting discretization and distance along the spectral contour are made manageable by this process, and the number of points needed in the spectral integration is found to remain fixed as the surface size increases. The overall algorithm is then order N. Reference [54] provides a detailed review of the spectral parameters for the one-dimensional surface case. Although the number of points needed for the spectral integration increases with surface height, the increase is relatively slow and the algorithm remains highly efficient in this case.

While the novel spectral approach has been widely used in conjunction with the "forward–backward" iterative matrix solution, it remains simply an algorithm for computing a matrix-vector multiply and can therefore be used in any iterative matrix solution method. For one-dimensional surfaces, the approach is highly efficient, competitive with the canonical grid method for moderate height surfaces, and remains efficient as surface heights increase. Use of this algorithm is recommended for studies of one-dimensional surface scattering, although careful choice of the spectral integration parameters should be made to ensure accurate computations[54].

Unfortunately, the gains of the NSA are not as dramatic in the two-dimensional case. This is because the algorithm still requires a beneficial discretization of the spectral path for coupling between all pairs of points in order to remain efficient. The spectral path can be interpreted in terms of an angle space, with one angle representing vertical deviations between points, and a second angle representing horizontal deviations in the two-dimensional surface case. The expansion is typically developed by assuming a "forward propagating" direction, which is usually parallel to one of the surface edges, and labeled as zero degrees in horizontal angle. As coupling between all possible pairs of points is considered, the horizontal angle then varies over both positive and negative angles that can become large as the "width" of the surface increases. Developing an appropriate spectral discretization over this wide range of angles results in a greatly increased number of spectral integration points required. References [55–56] describe means for choosing the spectral integration parameters in this case. Essentially the algorithm treats the width of the surface as a roughness parameter, and thereby is most efficient for rectangular surface sizes with small cross-range dimensions. This can be acceptable for some situations, although edge scatter effects and the corruption of cross-polarized scattering cross-sections must be considered carefully.

Reference [57] describes a method for modifying the NSA algorithm in the two-dimensional case to improve computational efficiency. In this case, three distinct spectral integrations are used, one being the original "along range" spectral path, and the remaining two being utilized in "cross range" for points within a moderate distance of the observation point. This algorithm has been further extended in the work of [58]. These techniques make the NSA method competitive with other approaches for large height and/or large slope surfaces, although no detailed comparisons of computational loads with other algorithms (for example, the fast multipole method) have been reported to date.

7.4.5. Parallelization

Even with the use of matrix-vector multiply acceleration algorithms, problems involving very large surfaces and/or Monte Carlo averaging will often require the use of supercomputing resources to allow results to be obtained in reasonable times. The wide availability of parallel computing resources makes their use of interest, although approaches for dividing the computational load among multiple processors must be developed.

The applicability of parallel computers in Monte Carlo simulations is obvious: if a simulation for a single surface realization can be designed to run on a single processor, then multiple processors can be utilized in a parallel Monte Carlo simulation. No communication among processors (except perhaps minimal control communication) is required, and parallelization is perfectly efficient. Similar statements apply to simulations involving frequency swept computations.

In some cases, the size of the problem of interest becomes such that it can no longer be run on a single processor. Simple algorithms for parallelization can still be devised, based on division of the within-neighborhood multiply operation among processors[18], as well as on parallel FFT[59] or spectral integration computations. Cases requiring matrix element integrations can also benefit from parallel performance of these computations, particularly if a multiple incidence angle simulation is of interest, within which all processors utilize the same matrix but varying right-hand sides[16]. Although development of the codes necessary for these operations may seem challenging, in general the availability of numerous software libraries for parallel communications makes code completion relatively straightforward.

7.4.6. Storage Issues

A final issue of importance for numerical techniques involves storage requirements. Even with parallel computers, it remains highly advantageous to keep a single surface simulation within the memory limitations of single processor if possible. Storage limitations are another motivation for resorting to iterative solution of a matrix equation: consider that with $N = 10^4$, the RAM storage needs of a complex matrix approach 1.5 gigabytes. Iterative methods do not necessarily require that the matrix be stored, only that an algorithm exist for performing the matrix-vector multiply. In cases where the matrix elements are sufficiently simple (no integration), these elements are easily regenerated when needed without a requirement for storage. In cases where at least some fraction of the matrix elements must be integrated, it is advantageous to store the integration results if sufficient memory is available; memory limitations should therefore be considered when designing the simulation of interest.

Both the CAG and NSA techniques have additional storage requirements. In the CAG method, it is desirable to store all translationally invariant terms (actually their FFT's) to avoid repeated computations. This storage requirement increases with the number of canonical grid series terms, and again becomes impractical as the number of series terms becomes large. The NSA method requires storage of "source" terms on the spectral integration path. This is only a modest requirement and typically is manageable.

7.5. Sample Results

To illustrate a range of problems which can be studied using numerical methods, several examples from the author's previous work are discussed in this section, with particular attention to the numerical parameters utilized in a given problem. Two one-dimensional surface and three two-dimensional surface cases are shown.

Figure 7.3 is a plot of normalized incoherent backscattering cross-sections (defined in [7]) from a one-dimensional surface study of low grazing angle backscattering from impedance power-law surfaces. The surfaces are Gaussian random processes with a power-law spectrum, originally intended to approximate

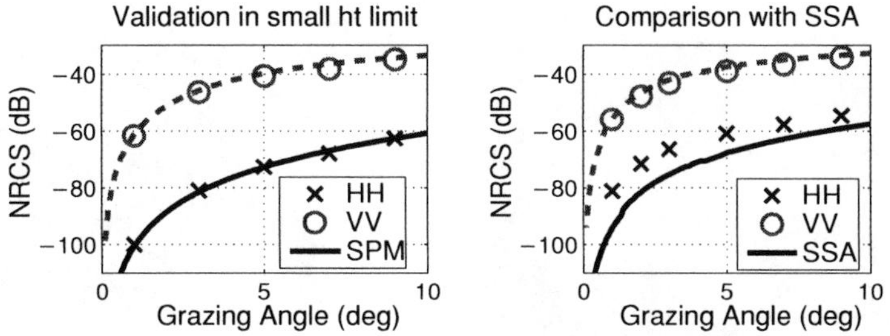

FIGURE 7.3. Normalized backscattering cross-sections obtained from a one-dimensional surface simulation of low grazing angle backscattering with power-law impedance surfaces[7]. Results illustrated are averages over 64 realizations with surface length 8192 wavelengths and 65 536 total unknowns. Left plot illustrates verification of the numerical method through comparison with the small perturbation method for small height surfaces, while right plot illustrates a test of the "small slope approximation" (SSA) analytical theory for larger height surfaces.

the sea surface, but also relevant for other near-fractal surface types. Backscattering results down to 1° grazing incidence were of interest, so a large surface size was clearly called for in order to retain sufficient angular resolution. Studies of this angular resolution suggested that a surface size of 8192 wavelengths was sufficient with the tapered incident wave utilized, and a discretization of eight point per wavelength resulted in 65 536 surface unknowns. Although these parameters were chosen from analytical expectations, further confirmation of their use was provided through comparison with predictions of the small perturbation method (SPM) in the small height limit. The left plot in Fig. 7.1 illustrates this comparison, and shows that the numerical results (averages over 64 realizations, and marked by symbols in the plots) are matching the SPM well in this case, where the surface rms height was set to 0.014 wavelengths through elimination of the low spatial frequency content of the spectrum.

Results for larger height surfaces (with surface rms height 2.27 wavelengths) were then computed given the confidence obtained from the SPM test; these results are compared with the SSA theory in the right plot. The results confirm the well-known limitations of the SSA method at low grazing angles, although SSA performance for vertical polarization was found to remain acceptable. Results were computed using the CAG approach, which required 15 series terms in the larger height case. Later simulations of these same results using the NSA algorithm[49] were far more efficient in the larger height case.

Given the smaller computational requirements in the one-dimensional surface case, it is possible to extend surface scattering simulations beyond the typical Monte Carlo studies of average scattered powers. Studies involving time evolving surfaces, frequency responses, or other physical behaviors can be easily performed. Another example involves the formation of images of rough surface scattering through the use of both frequency and angle swept computations[60].

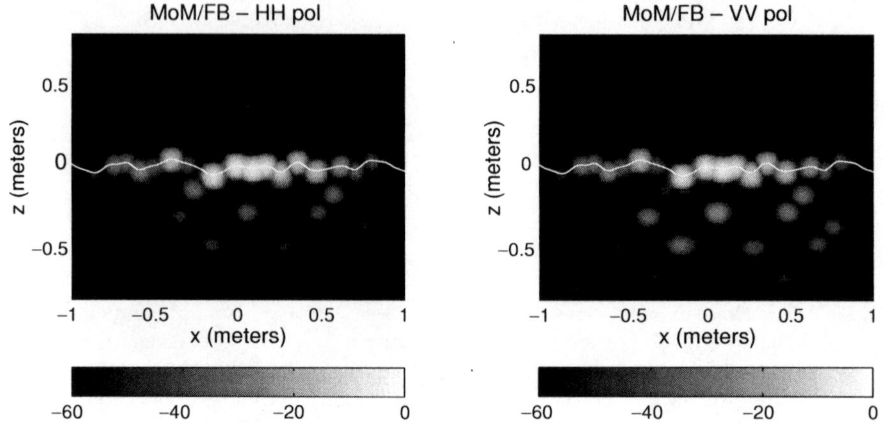

FIGURE 7.4. Radar image of rough surface NRCS values for a single realization of a Gaussian correlated impedance surface[60]. Image amplitude in decibels is plotted. Surface rms height is 0.8 wavelengths, correlation length is three wavelengths, and total surface length is 64 wavelengths, at the center frequency of the image. Image features below the surface profile indicate multiple scattering effects.

Figure 7.4 illustrates such an image, obtained using a single Gaussian random process realization with a Gaussian correlation function, rms height 0.8 center frequency wavelengths, and correlation length three center frequency wavelengths. Images were formed from a frequency sweep over $\pm$ 16% of the center frequency, using 80 frequencies, and from the backscattering response at $\pm 10^\circ$ from normal incidence, with angular step size 0.2°. The required 8000 evaluations of scattering from this surface realization were performed with the NSA method. Because only near normal incidence scattering was of interest, a modest surface size of 64 center frequency wavelengths was possible, along with tapered wave incidence.

Although results are shown from only a single surface realization, the insight provided by such images can be useful in assessing physical scattering effects. In this case, most image features are associated with near specular points on the surface profile (overlaid on the image), although some image features below the surface profile are also observed. A ray tracing analysis confirmed that these features result from multiple scattering interactions among surface points. Comparisons of these images with those predicted by approximate surface scattering theories were performed in [60] in order to assess the approximate theories.

Figure 7.5 presents incoherent normalized radar cross-section (NRCS) values obtained from a two-dimensional surface backscattering study using penetrable surfaces[61]. In this case, the boundary considered is between free space and a medium with relative permittivity $4 + \mathrm{i}$, and again a Gaussian random process surface with an isotropic Gaussian correlation function is used. The results shown use surface statistics of 0.08 wavelengths rms height, 0.48 wavelengths correlation length, ($K\sigma = 0.5$, $KL = 3$, with K the electromagnetic wavenumber) or 0.16

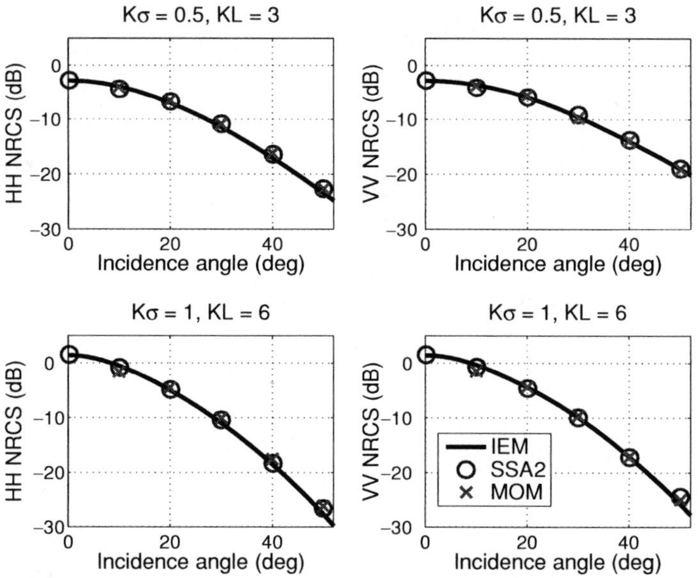

FIGURE 7.5. Normalized backscattering cross-sections from penetrable dielectric surfaces ($\epsilon = 4 + i$) from a two-dimensional surface simulation (numerical results labeled "MOM"). Gaussian correlated surface statistics indicated in figure titles. Results are averaged over 32 realizations using surface sizes of 16 by 16 wavelengths in the Canonical grid method[61], and compared to predictions from the "small slope approximation" (SSA) and "integral equation method" (IEM) theories.

wavelengths rms height, 0.96 wavelengths correlation length ($K\sigma = 1$, $KL = 6$) as indicated in the plot titles. Though these surfaces may seem applicable to classical models, the relatively large slopes obtained make both perturbation theory and physical optics inapplicable.

Numerical results for both HH and VV scattering combinations are shown (labeled "MOM" for method of moments in the legend), and are compared with predictions of the SSA and "integral equation method" (IEM)[62] analytical theories. Due to the moderate range of incidence angles considered here (only to up to 50° incidence) a surface size of 16 by 16 wavelengths was found appropriate; each of four surface field components was sampled into 128 by 128 points, resulting in 65 536 unknowns in the simulation. Results shown are averaged over 32 realizations, found sufficient due to the moderate surface heights considered. The canonical grid method with two terms in the expansion was used, and found to be of acceptable accuracy. Results from all three theories are found to be in good agreement for these results.

One limitation of this study was the poor quality of the numerically computed cross-polarized cross-sections, due to the extremely small expected values of cross-polarization and the small surface size utilized. These problems resulted

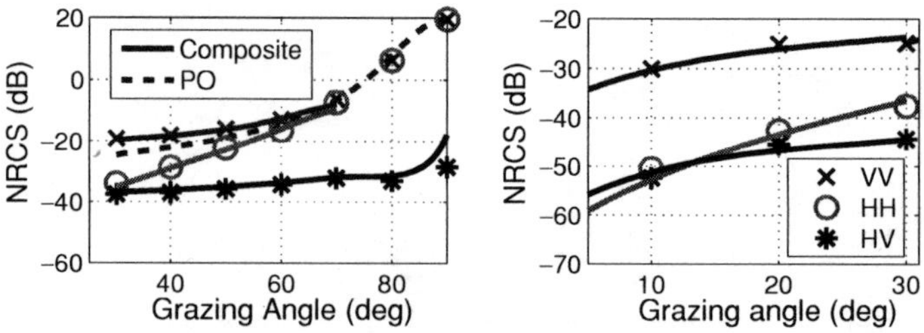

FIGURE 7.6. Normalized backscattering cross sections from two dimensional simulations with perfectly conducting (left) and impedance (right) power-law surfaces using the canonical grid method. Results in the left plot utilized surfaces of 64 by 64 electromagnetic wavelengths (524 K total unknowns), and are averaged over 128 realizations. Results in the right plot required surfaces of 128 by 32 wavelengths and finer sampling (2 M total unknowns) to achieve sufficient accuracy, and are averaged over 32 realizations. Numerical results are compared with predictions of the composite surface and physical optics (PO) theories.

in significant out-of-plane bistatic averaging due to the incident tapered field, and an overestimation of cross-polarized backscattering. This was confirmed through comparison with SPM cross-polarized predictions in the small height limit.

A second limitation involves computations near normal incidence. Given the strong specular response of these moderate height surfaces, a strong coherent response is obtained at near normal angles. This specular response is not the perfect plane wave obtained in analytical theories, but is rather spread over an angular width due to the finite surface size of the numerical simulation, as well as any further angular averaging due to tapered wave incidence. Computation of the incoherent NRCS value requires removing these coherent contributions, but, with only a small number of surface realizations available, removing the coherent term to sufficient accuracy is not possible in these results. Therefore no result is shown for backscattering near normal incidence. Generally it is found that accurate computation of incoherent responses at angles where strong coherent responses exist is difficult in Monte Carlo simulations.

Results from two distinct two-dimensional surface studies of backscattering from power-law surfaces are shown in Fig. 7.6. Both cases utilize the same surface description: Gaussian random processes with an isotropic power-law spectrum, similar to those used in Fig. 7.3. Results in the left plot [63] consider grazing angles from 30 to 90° (90° is normal incidence), and utilize perfectly conducting surfaces, while those in the right plot are for impedance surfaces and are at grazing angles less than[64] 30°. Numerically computed results are indicated by the symbols in the plots, while predictions of analytical theories are represented by lines. Surface sizes used to generate results in the left plot were 64 by 64 wavelengths, and with the two unknown scalar functions sampled into 512 by 512 points, a total of 512 K unknowns results. Averages over 128 realizations were performed using a parallel Monte Carlo simulation, and the canonical grid method with two series terms was found applicable in this case for which the surface rms height was 0.73 wavelengths. Again comparisons with the small perturbation method were utilized to ensure proper choice of numerical parameters for this case, and the surface size used was found acceptable down to 30° grazing incidence. The larger surface sizes utilized in this simulation resulted in improved accuracy in cross-polarized predictions, which are illustrated in the plots.

The smaller grazing angles of the right plot, however, required a modification of surface size to extend the along range dimension. The resulting surface size utilized was 128 by 32 wavelengths, which was found still to retain accuracy in computation of cross-polarized cross sections. In addition, the smaller cross-section values obtained necessitated a finer sampling rate of 16 unknowns per wavelength, so that a total of 2 million surface unknowns resulted. The canonical grid method with two series terms remained applicable. To reduce the total computational time, a parallel algorithm was developed in which a group of four processors was used to perform both the set of CAG method FFT operations as well as the within-neighborhood coupling computation. Multiple sets of four processors were then used in a parallel Monte Carlo simulation with 32 surface realizations to obtain the averaged results shown.

Note that these highly expensive numerical simulations were well matched by the approximate composite surface (or "two-scale" theory) of scattering from a power-law type surface, even for the cross-polarization results. Use of the composite surface model provided these results with around a factor of 10^5 reduction in computations. Although the goal of these simulations was to seek an evaluation of the composite surface theory, the results demonstrate that the approximate theories are often highly accurate, and that the use of numerical methods is an extreme waste of computational and researcher resources when reasonable approximate methods are available. It is also noted that the conclusions obtained regarding the use of approximate theories for copolarized scattering in this case were consistent with those obtained from one-dimensional surface studies[7,63].

A final result illustrated in Fig. 7.7 demonstrates a case to which approximate theories have only recently been extended, involving the "backscattering enhancement" phenomenon of rough surface scattering for large slope surfaces[65–68]. The surfaces considered in these plots are perfectly conducting, and are realizations of a Gaussian random process with an isotropic Gaussian correlation function. An rms height of one wavelength and correlation length of $\sqrt{2}$ wavelengths are specified, so that the surface has very large slopes. In-plane incoherent bistatic scattering patterns are shown as the incidence angle (indicated in the legend) is varied from normal incidence to 70° from normal incidence; the scattering angle (θ_S) of the plots is defined so that specular returns occur at positive angles while backscattering occurs at negative angles. The strong backscattering enhancement peak is observed in the smaller incidence angle cases, but is reduced dramatically as the incidence angle increases. The surface size used in this case was 128 by 16 wavelengths in order to allow simulations down to 70° incidence while avoiding cross-polarized cross-section corruption. The two-dimensional surface NSA method was used to generate these results in a parallel Monte Carlo simulation

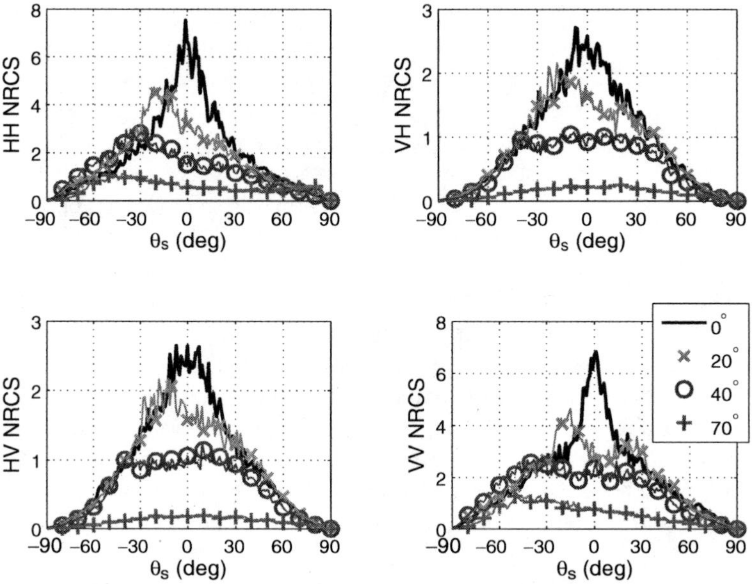

FIGURE 7.7. Normalized bistatic scattering cross-sections from two-dimensional simulations with perfectly conducting surfaces[68]. Results shown are averaged over 150 realizations, and computed using the FB/NSA method for surfaces 128 by 16 wavelengths using 262 K total unknowns. Gaussian correlated surfaces with rms height one wavelength and correlation length $\sqrt{2}$ wavelengths are utilized.

over 150 surface realizations. For this large slope case, the CAG method could not provide efficient computations. Simulations involving impedance surfaces and for other surface statistics are reported in [68]; these results have been utilized in testing new analytical theories of the backscattering enhancement effect[69].

7.6. Conclusions and Recommendations for the Use of Numerical Methods

Given the issues discussed here for numerical simulations of rough surface scattering, it is appropriate to provide a few summary recommendations for readers interested in this area.

First, it is recommended that the problem of interest be formulated thoroughly before determining whether an approximate or numerical approach is warranted for its solution. In this regard, all relevant sensor and surface parameters should be specified, including particularly the statistics of the rough surface itself. In purely theoretical studies surface statistics can usually be specified as desired, but when attempting to model particular surfaces, specification of surface properties is likely to be the major problem of the model regardless of the electromagnetic method used. The influence of any transmit or receive antenna patterns should also be considered if attempting to model a practical measurement. It should also be decided what an acceptable level of accuracy is; in measurements involving surface scattering, an accuracy of approximately a factor of 2 in normalized cross-sections is usually within the level of uncertainty of the measurement.

Once the problem has been formulated, it is recommended to attempt to identify any available approximate theory of surface scattering that may be applicable. The review articles cited, as well as a more recent review article[70], can be of assistance in this process. In addition, numerous studies (similar to those reported in Sect. 7.5) of the accuracy of several approximate theories have been performed, including several recent contributions[71–73]. If a reasonable approximate theory is available, it will likely provide answers within the desired accuracy with far less commitment of computational and personnel resources. Therefore it is important for surface scattering researchers, even those specifically interested in numerical methods, to remain aware of the most recent developments in approximate theories. Recent years have seen great expansion in the range of problems that can be treated by approximate methods[70]. Such methods can also be used for frequency sweep, image formation, or other advanced simulations.

If a numerical solution is called for, then all attempts to produce an accurate solution should be made. The methods described in this chapter and in the literature can be used to construct an efficient numerical algorithm, so long as the limitations of the methods used are known to the developer. The convergence and other tests described in this chapter should be used to ensure that choice of numerical parameters or other issues are not interfering with the results obtained.

Given the current state of affairs in surface scattering studies, it appears that only a subset of current problems of interest should be treated with the numerical approach. The main cases are those that involve either strong shadowing (i.e. low grazing angle problems) or multiple-scattering (i.e. large slope) effects, for which analytical theories continue to have difficulty. Another topic of interest involves studies of scattering from objects in the presence of rough surfaces, for which many one-dimensional and two-dimensional simulations have been performed recently[74–79]. Note that in all these studies, the ultimate goal is improved physical understanding of the scattering process, so that new analytical techniques can be developed, rather than continued use of the numerical methods themselves.

References

1. J.A. DeSanto and R.J. Wombell, "Rough surface scattering," *Waves Random Media* **1**, S41–S56 (1991).
2. D. Maystre, M. Saillard, and J. Ingers, "Scattering by one or two dimensional randomly rough surfaces," *Waves Random Media* **3**, S143–S155 (1991).
3. E.I. Thorsos and D.R. Jackson, "Studies of scattering theory using numerical methods," *Waves Random Media* **3**, S165–S190 (1991).
4. M. Saillard and A. Sentenac, "Rigorous solutions for electromagnetic scattering from rough surfaces," *Waves Random Media* **11**, R103–R137 (2001).

5. K.M. Warnick and W.C. Chew, "Numerical simulation methods for rough surface scattering," *Waves Random Media* **11**, R1–R30 (2001).
6. R.T. Marchand and G.S. Brown, "On the use of finite surfaces in the numerical prediction of rough surface scattering," *IEEE Trans. Antennas Propag.* **47**, 600–604 (1999).
7. J.T. Johnson, "A numerical study of low grazing angle backscatter from ocean-like impedance surfaces with the canonical grid method," *IEEE Trans. Antennas Propag.* **46**, 114–120 (1998).
8. E.I. Thorsos and D. Jackson, "The validity of the Kirchhoff approximation for rough surface scattering using a Gaussian roughness spectrum," *J. Acoust. Soc. Am.* **83**, 78–92 (1988 Jan.).
9. J.V. Toporkov, R. Awadallah, and G.S. Brown, "Issues related to the use of a Gaussian-like incident field for low-grazing-angle scattering," *J. Opt. Soc. Am. A* **16**, 176–187 (1999).
10. H. Braunisch, Y. Zhang, C.O. Ao, S.E. Shih, Y.E. Yang, K.H. Ding, J.A. Kong, and L. Tsang, "Tapered wave with dominant polarization state for all angles of incidence," *IEEE Trans. Antennas Propag.* **48**, 1086–1096 (2000).
11. H.X. Ye and Y.Q. Jin, "Parameterization of the tapered incident wave for numerical simulation of scattering from rough surfaces," *IEEE Trans. Antennas Propag.* **53**, 1234–1237 (2005).
12. J.C. West, "On the control of edge diffraction in numerical rough surface scattering using resistive tapering," *IEEE Trans. Antennas Propag.* **51**, 3180–3183 (2003).
13. Z.Q. Zhao and J.C. West, "Resistive suppression of edge effects in MLFMA scattering from finite conductivity surfaces," *IEEE Trans. Antennas Propag.* **53**, 1848–1852 (2005).
14. J.V. Toporkov, R.T. Marchand, and G.S. Brown, "On the discretization of the integral equation describing scattering by rough conducting surfaces," *IEEE Trans. Antennas Propag.* **46**, 150–161 (1998).
15. Q. Li, C.H. Chan, and L. Tsang, "Monte Carlo simulations of wave scattering from lossy dielectric random rough surfaces using the physics-based two-grid method and the canonical grid method," *IEEE Trans. Antennas Propag.* **47**, 752–769 (1999).
16. J.T. Johnson, R.T. Shin, J.A. Kong, L. Tsang, and K. Pak, "A numerical study of ocean polarimetric thermal emission," *IEEE Trans. Geosci. Remote Sens.* **37**, 8–20 (1999).
17. J.T. Johnson, "Surface currents induced on a dielectric halfspace by a Gaussian beam: a useful validation for MOM codes," *Radio Sci.* **32**, 923–934 (1997).
18. J.T. Johnson, R.T. Shin, J. Eidson, L. Tsang, and J.A. Kong, "A method of moments model for VHF propagation," *IEEE Trans. Antennas Propag.* **45**, 115–125 (1997).
19. C.H. Chan, L. Tsang, and Q. Li, "Monte-Carlo simulation of large-scale one-dimensional random rough-surface scattering at near-grazing incidence: penetrable case," *IEEE Trans. Antennas Propag.* **46**, 142–149 (1998 Jan.).
20. J.T. Johnson and R.J. Burkholder, "Coupled canonical grid/discrete dipole approach for computing scattering from objects above or below a rough interface," *IEEE Trans. Geosci. Remote Sens.* **39**, 1214–1220 (2001).
21. K. Pak, L. Tsang, and J.T. Johnson, "Numerical simulations and backscattering enhancement of electromagnetic waves from two dimensional dielectric random rough surfaces with sparse matrix canonical grid method," *J. Opt. Soc. Am. A.* **14**, 1515–1529 (1997).
22. P. Tran and A.A. Maradudin, "The scattering of electromagnetic waves from a randomly rough 2-D metallic surface," *Opt. Commun.* **110**, 269–273 (1994).

23. J.C. West, "Integral equation formulation for iterative calculation of scattering from lossy rough surfaces," *IEEE Trans. Geosci. Remote Sens.* **38**, 1609–1615 (2000).
24. M.Y. Xia, C.H. Chan, S.Q. Li, B. Zhang, and L. Tsang, "An efficient algorithm for electromagnetic scattering from rough surfaces using a single integral equation and multilevel sparse-matrix canonical grid method," *IEEE Trans. Antennas Propag.* **51**, 1142–1149 (2003).
25. J.T. Johnson, unpublished notes (2003).
26. L.S. Blackford, J. Choi, A. Cleary, E. D'Azevedo, J. Demmel, I. Dhillon, J. Dogarra, S. Hammarling, G. Henry, A. Petitet, K. Stanley, D. Walker, and R.C. Whaley, *Scalapack Users' Guide* (SIAM Publications, Philadelphia, 1997).
27. R. Barrett, M. Berry, T. Chan, J. Demmel, J. Donato, J. Dongarra, V. Eljkhout, R. Pozo, C. Romine, and H. van der Vorst, *Templates for the Solution of Linear Systems: Building Blocks for Iterative Methods*, available at http://netlib2.cs.utk.edu/linalg/html_templates/Templates.html.
28. J.C. West and J.M. Sturm, "On iterative approaches for electromagnetic rough surface scattering problems," *IEEE Trans. Antennas Propag.* **47**, 1281–1288 (1999).
29. L. Tsang, C.H. Chan, H. Sangani, A. Ishimaru, and P. Phu, "A banded matrix iterative approach to Monte Carlo simulations of scattering of waves by large-scale random rough surface scattering," *J. Electr. Waves Appl.* **7**, 185–200 (1993).
30. D. Holliday, L.L. Deraad, and G.J. St-Cyr, "Forward-backward: a new method for computing low grazing angle scattering," *IEEE Trans. Antennas Propag.* **44**, 722–729 (1996).
31. D.A. Kapp and G.S. Brown, "A new numerical method for rough surface scattering calculations," *IEEE Trans. Antennas Propag.* **44**, 711–721 (1996).
32. D. Holliday, L.L. Deraad, and G.J. St-Cyr, "Forward-backward method for scattering from imperfect conductors," *IEEE Trans. Antennas Propag.* **46**, 101–107 (1998).
33. P. Tran, "Calculation of the scattering of electromagnetic waves from a two-dimensional perfectly conducting surface using the method of order multiple interactions," *Waves Random Media*, **7**, 295–302 (1997).
34. R.J. Adams and G.S. Brown, "An iterative solution of one-dimensional rough surface scattering problems based on factorization of the Helmholtz equation," *IEEE Trans. Antennas Propag.* **47**, 765–767 (1999).
35. P. Tran, V. Celli, and A.A. Maradudin, "Electromagnetic scattering from a 2-dimensional randomly rough perfectly conducting surface: iterative methods," *J. Opt. Soc. Am. A.* **11**, 1686–1689 (1994).
36. D.J. Wingham and R.H. Devayya, "A note on the use of the Neumann expansion in calculating the scatter from rough surfaces," *IEEE Trans. Antennas Propag.* **40**, 560–563 (1992).
37. L. Tsang, C.H. Chan, and K. Pak, "Monte Carlo simulations of a two-dimensional random rough surface using the sparse-matrix flat surface iterative approach," *Electron. Lett.* **29**, 1153–1154 (1993).
38. J.C. West, "Preconditioned iterative solution of scattering from rough surfaces," *IEEE Trans. Antennas Propag.* **48**, 1001–1002 (2000).
39. M. Saillard and G. Soriano, "Fast numerical solution for scattering from rough surfaces with small slopes," *IEEE Trans. Antennas Propag.* **52**, 2799–2802 (2004).
40. J.T. Johnson, "On the canonical grid method for two dimensional scattering problems," *IEEE Trans. Antennas Propag.* **46**, 297–302 (1998).

41. N. Engheta, W.D. Murphy, V. Rokhlin, and M.S. Vassiliou, "The fast multipole method for electromagnetic scattering problems," *IEEE Trans. Antennas Propag.* **40**, 634–641 (1992).
42. R.L. Wagner, J.M. Song and W.C. Chew, "Monte Carlo Simulation of Electromagnetic Scattering from Two-Dimensional Random Rough Surfaces," *IEEE Trans. Antennas Propag.* **45**, 235–245 (1997).
43. R.J. Adams and G.S. Brown, "Use of fast multipole method with method of ordered multiple interactions," *Electron. Lett.* **34**, 2219–2220 (1998).
44. V. Jandhyala, E. Michielssen, S. Balasubramaniam and W.C. Chew, "A combined steepest descent-fast multipole algorithm for the fast analysis of three-dimensional scattering by rough surfaces," *IEEE Trans. Geosci. Remote Sens.* **36**, 738–748 (1998).
45. V. Jandhyala, B. Shanker, E. Michielssen and W.C. Chew, "Fast algorithm for the analysis of scattering by dielectric rough surfaces," *J. Opt. Soc. Am. A.* **15**, 1877–1885 (1998b).
46. L. Tsang, D. Chen, P. Xu, Q. Li, and V. Jandhyala, "Wave scattering with the UV multilevel partitioning method: parts 1 and 2," *Radio Sci.* **39**, RS5010–RS5011 (2004).
47. L. Tsang, C.H. Chan, K. Pak, H. Sangani, "Monte Carlo simulations of large scale problems of random rough surface scattering and applications to grazing incidence with the BMIA/canonical grid method," *IEEE Trans. Antennas Propag.* **43**, 851–859 (1995).
48. K. Pak, L. Tsang, C.H. Chan, and J.T. Johnson, "Backscattering enhancement of electromagnetic waves from two-dimensional perfectly conducting random rough surfaces based on Monte Carlo simulations," *J. Opt. Soc. Am. A.* **12**, 2491–2499 (1995).
49. H.T. Chou and J.T. Johnson, "A novel acceleration algorithm for the computation of scattering from rough surfaces with the forward-backward method," *Radio Sci.* **33** (5), 1277–1287 (1998).
50. H.T. Chou and J.T. Johnson, "Formulation of forward-backward method using novel spectral acceleration for the modeling of scattering from impedance rough rough surfaces," *IEEE Trans. Geosci. Remote Sens.* **38** (1), 605–607 (2000).
51. S.Q. Li, C.H. Chan, M.Y. Xia, B. Zhang, and L. Tsang, "Multilevel expansion of the sparse-matrix canonical grid method for two-dimensional random rough surfaces," *IEEE Trans. Antennas Propag.* **49**, 1579–1589 (2001).
52. D.M. Milder, "An improved formalism for wave scattering from rough surfaces," *J. Acoust. Soc. Am.* **89**, 529–541 (1991).
53. A.G. Voronovich, *Wave Scattering from Rough Surfaces* (Springer-Verlag, Berlin, 1994).
54. D. Torrungrueng, J.T. Johnson, and H.T. Chou, "Some issues related to the novel spectral acceleration method for the fast computation of radiation/scattering from one dimensional extremely large scale quasi-planar structures," *Radio Sci.* **37**, 3(1)–3(20) (2002).
55. D. Torrungrueng, H.T. Chou, and J.T. Johnson, "A novel spectral acceleration algorithm for the computation of scattering from two-dimensional rough surfaces with the forward-backward method," *IEEE Trans. Geosci. Remote Sens.* **38** (4), 1656–1668 (2000).
56. D. Torrungrueng and J.T. Johnson, "The forward-backward method with a novel acceleration algorithm (FB/NSA) for the computation of scattering from two-dimensional large-scale impedance random rough surfaces," *Microwave Opt. Techol. Lett.* **29**, 232–236 (2001).

57. D. Torrungrueng and J.T. Johnson, "An improved FB/NSA algorithm for the computation of scattering from two-dimensional large-scale rough surfaces," *J. Electromagn. Waves Appl.* **15**, 1337–1362 (2001).
58. H.C. Ku, R. Awadallah, R.L. McDonald, and N.E. Woods, "Fast and accurate algorithm for scattering from large-scale 2-D dielectric ocean surfaces," *Int'l Antennas and Propagation Symposium*, conference proceedings, **3A**, 454–457 (2005).
59. S.Q. Li, C.H. Chan, L. Tsang, Q. Li, and L. Zhou, "Parallel implementation of the sparse-matrix canonical grid method for the analysis of two-dimensional random rough surfaces (3-D scattering problem) on a Beowulf system," *IEEE Trans. Geosci. Remote Sens.* **38**, 1600–1608 (2000).
60. H. Kim and J.T. Johnson, "Radar images of rough surface scattering: comparison of numerical and analytical models," *IEEE Trans. Antennas Propag.* **50**, 94–100 (2002).
61. H.T. Ewe, J.T. Johnson, and K.S. Chen, "A comparison study of surface scattering models," *Int'l Geoscience and Remote Sensing Symposium (IGARSS 01)*, conference proceedings, **6**, 2692–2694 (2001).
62. T. Wu, K.S. Chen, J. Shi, and A.K. Fung, "A transition model for the reflection coefficient in surface scattering," *IEEE Trans. Geosci. Remote Sens.* **39**, 2040–2050 (2001).
63. J.T. Johnson, R.T. Shin, J.A. Kong, L. Tsang, and K. Pak, "A numerical study of the composite surface model for ocean scattering," *IEEE Trans. Geosci. Remote Sens.* **36** (1), 72–83 (1998).
64. J.T. Johnson and H.T. Chou, "Numerical studies of low grazing angle backscatter from 1-D and 2-D impedance surfaces," *IGARSS'98 Conference Proceedings* **4**, 2295–2297 (1998).
65. A.A. Maradudin, E.R. Mendez, and T. Michel, "Backscattering effects in the elastic scattering of P-polarized light from a large amplitude random metallic grating," *Opt. Lett.* **14**, 151 (1989).
66. A.A. Maradudin, T. Michel, A.R. McGurn, and E.R. Mendez, "Enhanced backscattering of light from a random grating," *Ann. Phys.* **203**, 255–307 (1990).
67. J.T. Johnson, L. Tsang, R.T. Shin, K. Pak, C.H. Chan, A. Ishimaru, and Y. Kuga, "Backscattering enhancement of electromagnetic waves from two dimensional perfectly conducting random rough surfaces: a comparison of Monte Carlo simulations with experimental data," *IEEE Trans. Antennas Propag.* **44**, 748–756 (1996).
68. D. Torrungrueng and J.T. Johnson, "Numerical studies of backscattering enhancement of electromagnetic waves from two-dimensional random rough surfaces using the FB/NSA method," *J. Opt. Soc. Am. A.* **18**, 2518–2526 (2001).
69. C. Bourlier and G. Berginc, "Multiple scattering in the high-frequency limit with second order shadowing function from 2-D anisotropic rough dielectric surfaces: parts I and II," *Waves Random Media* **14**, 229–276 (2004).
70. T. Elfouhaily and C.A. Guerin, "A critical survey of approximate scattering theories from random rough surfaces," *Wave Random Media* **14**, R1–R40 (2004).
71. G. Soriano, C.A. Guerin, and M. Saillard, "Scattering by two-dimensional rough surfaces: comparison between the method of moments, Kirchhoff, and small-slope approximations," *Waves Random Media* **12**, 63–83 (2002).
72. C.A. Guerin, G. Soriano, and T. Elfouhaily, "Weight curvature approximation: numerical tests for 2-D dielectric surfaces," *Waves Random Media* **14**, 349–363 (2004).
73. F.W. Millet and K.L. Warnick, "Validity of rough surface backscattering models," *Waves Random Media* **14**, 327–347 (2004).

74. K. O'Neill, R.F. Lussky, and K.D. Paulsen, "Scattering from a metallic object embedded near the randomly rough surface of a lossy dielectric," *IEEE Trans. Geosci. Remote Sens.* **34**, 367–376 (1996).
75. G. F. Zhang, L. Tsang, and K. Pak, "Angular correlation function and scattering coefficient of electromagnetic waves scattered by a buried object under a two-dimensional rough surface," *J. Opt. Soc. Am. A.* **15**, 2995–3002 (1998).
76. M. El-Shenawee, C. Rappaport, and M. Silevitch, "Monte Carlo simulations of electromagnetic wave scattering from a random rough surface with three dimensional buried object," *J. Opt. Soc. Am. A.* **18**, 3077–3084 (2001).
77. J.T. Johnson, "A numerical study of scattering from an object above a rough surface," *IEEE Trans. Antennas Propag.* **40**, 1361–1367 (2002).
78. J.T. Johnson and R.J. Burkholder, "A study of scattering from an object below a rough interface," *IEEE Trans. Geosci. Remote Sens.* **42**, 59–66 (2004).
79. M. El-Shenawee, "Polarimetric scattering from two-layered two-dimensional random rough surfaces with and without buried objects," *IEEE Trans. Geosci. Remote Sens.* **42**, 67–76 (2004).

8 Overview of Rough Surface Scattering

JOHN A. DESANTO
Department of Physics, Colorado School of Mines, Golden, CO 80401

8.1. Introduction

All real surfaces are rough, the roughness being relative to the interrogating wavelength λ for any physical situation. We confine our discussion here to acoustic and electromagnetic problems, and limit the surface description to vertical height h and one or two horizontal scales L. Practical problems in manufacturing or measuring surfaces can include and often completely depend upon topics such as ripples, polishing marks, and behavior of surface extrema. We thus limit our consideration to acoustic and electromagnetic scattering from surfaces with vertical scales kh ($k = 2\pi/\lambda$)) and horizontal scales kL or tangent scales h/L. This is broad enough to discuss scattering from sea surfaces, terrain, atomic systems, and nanoscale roughness, as well as both deterministic and random problems characterizing the surface behavior.

Various mathematical techniques are used to model the scattering problem. Approximations such as perturbation theory, the Kirchhoff approximation, and Born or distorted-Born methods are popular. Although they have limited validity they are comparatively simple to implement. With the advances in computational ability rigorous exact methods have become dominant. The most popular of these are integral equations, and we confine this overview to these methods as well as some approximations. Even with this restriction, there are several variants as well as diverse solution methods. We present this review paper as a summary of the variety of such methods.

In Sect. 8.2 we treat coordinate-space based integral equation methods for both scalar and electromagnetic scattering problems. To distinguish them from other methods below we characterize them relative to the two arguments in the kernel of the integral term or equivalently the discretized sample spaces resulting in the rows and columns of the matrix system or equivalently the fields and sources of the problem. Here both are in coordinate space and the development is referred to as coordinate coordinate or CC. This is the most common method and we label it to distinguish it from other methods we introduce later. The rough surface separates two homogeneous media. For the scalar problem we have standard boundary integral equations where the boundary unknowns are the field and its normal derivative. For the electromagnetic problem several integral equation versions are available. The first version draws an analogy with the scalar case, and uses the boundary unknowns of the electric field $\boldsymbol{E}$ and its normal derivative $\boldsymbol{N}$ and results in a system of equations which is highly sparse compared to the usual Stratton–Chu formulation which follows from this version with some algebraic manipulations. The second version is the Stratton–Chu equations where the boundary unknowns are the magnetic surface current $\boldsymbol{n} \times \boldsymbol{E}$ and the electric surface current $-\boldsymbol{n} \times \boldsymbol{H}$ where n is the surface normal and $\boldsymbol{H}$ is the magnetic field. This formulation uses the scalar free-space Green's function as does the first. If we take the curl of the Stratton–Chu equations we get the third version, the Franz equations, which also follow directly from Green's theorem using the free-space dyadic Green's function.

In Sect. 8.3 we discuss spectral-space methods for both acoustic and electromagnetic problems. Again we refer to the field-source or row-column interpretation. Here either field or source or both are in spectral space (S) and the formalisms are referred to as SC, CS, and SS. Each formalism has an advantage.

In Sect. 8.4 we treat the topic of inverse surface reconstruction, that is, the reconstruction of the scattering surface from scattering data. Two scalar algorithms for the Dirichlet problem are presented and some other methods are briefly discussed.

In Sect. 8.5 we briefly present some solution methods. Since we only discuss the integral equation formulation in this paper the most important methods are formally exact numerical methods such as collocation or the Galerkin approach, and approximate iterative techniques which often involve factorization of the integral equation kernel.

Sect. 8.6 is a brief discussion of other approaches to this problem.

8.2. Coordinate-Space Methods

8.2.1. Scalar Problems

We first develop integral equations in coordinate-space (CC equations) for scalar problems. The rough surface $h(\boldsymbol{x}_t)$, $(\boldsymbol{x}_t = (x, y))$, separates two homogeneous regions $V_j (j = 1, 2)$ of different wavenumber k_j and density ρ_j. For $j = 1$, $z > h$, and for $j = 2$, $z < h$. For the moment consider the surface to be infinite. In region 1 we have incident $(\psi^{\text{in}}(\boldsymbol{x}))$ and scattered $(\psi_1^{\text{sc}}(\boldsymbol{x}))$ fields (where $\boldsymbol{x} = (x, y, z)$) and in region 2 a transmitted field $(\psi_2(\boldsymbol{x}))$. Fields in the jth region satisfy the scalar Helmholtz equation (in index/summation notation with $i = 1 - 3$)

$$\left(\partial_i \partial_i + k_j^2\right) \phi(\boldsymbol{x}) = 0, \tag{1}$$

where

$$\phi(\boldsymbol{x}) = \begin{cases} \psi_1^{\text{sc}}(\boldsymbol{x}) & z > h,\ j = 1 \\ \psi_2(\boldsymbol{x}) & z < h,\ j = 2. \end{cases} \tag{2}$$

An incident plane wave satisfies the same equation in region 1. The free-space Green's functions G_j satisfy an analogous equation with a delta-function source

$$\left(\partial_i \partial_i + k_j^2\right) G_j(\boldsymbol{x}, \boldsymbol{x}') = -\delta(\boldsymbol{x} - \boldsymbol{x}'), \tag{3}$$

and are explicitly given by

$$G_j(\boldsymbol{x}, \boldsymbol{x}') = \frac{\exp(\mathrm{i}k_j r)}{4\pi r}, \qquad r = |\boldsymbol{x} - \boldsymbol{x}'|. \tag{4}$$

To derive the integral equations we use Green's theorem on ψ_1^{sc} and G_1 in region 1, and ψ_2 and G_2 in region 2, each integration thus in a region bounded by the surface and a semicircle of radius R. We also use single(S)- and double(D)-layer potential integrals on the surface h with $\boldsymbol{x}_h = (x, y, h)$ given by

$$(S_j u)(\boldsymbol{x}) = \iint_h G_j(\boldsymbol{x}, \boldsymbol{x}'_h) u(\boldsymbol{x}'_h)\, \mathrm{d}\boldsymbol{x}'_t, \tag{5}$$

and

$$(D_j v)(\boldsymbol{x}) = \iint_h \partial'_n G_j(\boldsymbol{x}, \boldsymbol{x}'_h) v(\boldsymbol{x}'_h)\, \mathrm{d}\boldsymbol{x}'_t, \tag{6}$$

where u and v are densities and ∂_n is the normal derivative. Since the fields satisfy a radiation condition, the results can be written as

$$\Theta_1(\boldsymbol{x})\psi_1^{\mathrm{sc}}(\boldsymbol{x}) = \left(D_1\psi_1^{\mathrm{sc}}\right)(\boldsymbol{x}) - \left(S_1 N_1^{\mathrm{sc}}\right)(\boldsymbol{x}), \tag{7}$$

and

$$-\Theta_2(\boldsymbol{x})\psi_2(\boldsymbol{x}) = (D_2\psi_2)(\boldsymbol{x}) - (S_2 N_2)(\boldsymbol{x}), \tag{8}$$

where Θ_j are the characteristic functions of the regions and N_j the normal derivatives of the appropriate fields. These are field representations, i.e. given the boundary values, the scattered field in region 1 is given by (7) and the transmitted field in region 2 by (8). The boundary values on the surface h are related using the total field ψ_1 in region 1

$$\psi_1(\boldsymbol{x}) = \psi^{\mathrm{in}}(\boldsymbol{x}) + \psi^{\mathrm{sc}}(\boldsymbol{x}), \tag{9}$$

where ψ^{in} is the incident field, and they are for the total fields

$$\psi_1(\boldsymbol{x}_h) = \rho\psi_2(\boldsymbol{x}_h), \tag{10}$$

where $\rho = \rho_2/\rho_1$, and, for the normal derivatives,

$$N_1(\boldsymbol{x}_h) = N_2(\boldsymbol{x}_h). \tag{11}$$

Define the boundary field

$$F(\boldsymbol{x}_h) = \psi_1(\boldsymbol{x}_h), \tag{12}$$

and normal derivative

$$N(\boldsymbol{x}_h) = N_1(\boldsymbol{x}_h), \tag{13}$$

then (7) and (8) become

$$\Theta_1(\boldsymbol{x})\psi_1(\boldsymbol{x}) = W(\boldsymbol{x}) + (D_1 F)(\boldsymbol{x}) - (S_1 N)(\boldsymbol{x}), \tag{14}$$

where

$$W(\boldsymbol{x}) = \Theta_1(\boldsymbol{x})\psi^{\text{in}}(\boldsymbol{x}) - (D_1 \psi^{\text{in}})(\boldsymbol{x}) + (S_1 N^{\text{in}})(\boldsymbol{x}), \tag{15}$$

and

$$-\Theta_2(\boldsymbol{x})\psi_2(\boldsymbol{x}) = \rho^{-1}(D_2 F)(\boldsymbol{x}) - (S_2 N)(\boldsymbol{x}). \tag{16}$$

In the limit as $\boldsymbol{x}$ approaches the surface from above in (14) and from below in (16), we use the continuity of the single-layer potential and the discontinuity of the double-layer potential to form coupled integral equations on F and N. They are

$$\tfrac{1}{2}F(\boldsymbol{x}_h) = W(\boldsymbol{x}_h) + P(D_1 F)(\boldsymbol{x}_h) - (S_1 N)(\boldsymbol{x}_h), \tag{17}$$

and

$$-\tfrac{1}{2}F(\boldsymbol{x}_h) = P(D_2 F)(\boldsymbol{x}_h) - \rho(S_2 N)(\boldsymbol{x}_h), \tag{18}$$

where the double-layer integrals must be evaluated using principal values (P). These are the basic equations which must be solved for the general interface problem. There are two equations for the two boundary unknowns F and N. Using F and N the fields in each region are evaluated using (14) and (16). It is possible to generate additional equations by taking the (exterior) normal derivative of (14) and (16) and then passing to the surface limit. For bounded obstacles, combinations of these latter equations with (17) and (18) are used to eliminate interior resonance effects and W reduces to the incident field.

Perfectly reflecting Dirichlet (D) ($F = 0$ and $\rho = 0$) and Neumann (N) ($N = 0$ and ρ approach ∞) problems follow from (17) and (18). The Neumann problem yields a second-kind equation for F (here F_N) written symbolically as

$$\left(\tfrac{1}{2}I - PD_1\right) F_N = W, \tag{19}$$

and a first-kind equation for the Dirichlet problem for N (here N_D) given by

$$S_1 N_D = W. \tag{20}$$

A second-kind equation for N_D follows by taking the exterior normal derivative of (14) and then the boundary limit. Again, symbolically the result is

$$\left(\tfrac{1}{2}I + \partial_n S_1\right) N_D = \partial_n W. \tag{21}$$

The field value in region 1 follows from (14). Since the surface is perfectly reflecting there is no field in region 2.

Finally, note that we have been discussing infinite surfaces. For the integrations to proceed as we have presented, no plane waves horizontal to the surface can be present since otherwise the usual radiation condition on the scattered field is not satisfied. In this case the equations must be modified or a more restrictive radiation condition is necessary.[19–21]

8.2.2. Electromagnetic Problems

Each ith component of the scattered (sc) field in region 1, E_i^{sc}, satisfies (1) so that Green's theorem proceeds as in the scalar case component by component. Using the Silver–Mueller radiation condition the results are a straightforward generalization of the scalar field equations from the upper and lower regions (7) and (8) for the scattered electric field and its normal derivative N_i^{sc} and the transmitted electric field $E_i^{(2)}$ and its normal derivative $N_i^{(2)}$. They are

$$\Theta_1(\boldsymbol{x})E_i^{\mathrm{sc}}(\boldsymbol{x}) = \big(D_1 E_i^{\mathrm{sc}}\big)(\boldsymbol{x}) - \big(S_1 N_i^{\mathrm{sc}}\big)(\boldsymbol{x}), \tag{22}$$

and

$$-\Theta_2(\boldsymbol{x})E_i^{(2)}(\boldsymbol{x}) = \big(D_2 E_i^{(2)}\big)(\boldsymbol{x}) - \big(S_2 N_i^{(2)}\big)(\boldsymbol{x}), \tag{23}$$

using the same single and double layer potentials (5) and (6) (and with the same scalar Green's functions) which now however operate on vector densities. The total field in region 1 is the sum of incident (in) and scattered (sc) fields

$$E_i^{(1)}(\boldsymbol{x}) = E_i^{\mathrm{in}}(\boldsymbol{x}) + E_i^{\mathrm{sc}}(\boldsymbol{x}), \tag{24}$$

and using (24) and its normal derivative in (22) we can write an equation on the total field

$$\Theta_1(\boldsymbol{x})E_i^{(1)}(\boldsymbol{x}) = W_i(\boldsymbol{x}) + \big(D_1 E_i^{(1)}\big)(\boldsymbol{x}) - \big(S_1 N_i^{(1)}\big)(\boldsymbol{x}), \tag{25}$$

where

$$W_i(\boldsymbol{x}) = \Theta_1(\boldsymbol{x})E_i^{\mathrm{in}}(\boldsymbol{x}) - \big(D_1 E_i^{\mathrm{in}}\big)(\boldsymbol{x}) + \big(S_1 N_i^{\mathrm{in}}\big)(\boldsymbol{x}), \tag{26}$$

is the analogue of (15). Coupled integral equations result if we let $\boldsymbol{x}$ approach the boundary from above in (25) and from below in (23) to yield

$$\tfrac{1}{2}E_i^{(1)}(\boldsymbol{x}_h) = W_i(\boldsymbol{x}_h) + P\big(D_1 E_i^{(1)}\big)(\boldsymbol{x}_h) - \big(S_1 N_i^{(1)}\big)(\boldsymbol{x}_h), \tag{27}$$

and

$$-\tfrac{1}{2}E_i^{(2)}(\boldsymbol{x}_h) = P\big(D_2 E_i^{(2)}\big)(\boldsymbol{x}_h) - \big(S_2 N_i^{(2)}\big)(\boldsymbol{x}_h). \tag{28}$$

To combine (27) and (28) we need boundary conditions on the electric field[10,28,29,34] and its normal derivative.[18] The usual transmission boundary conditions on the electric field can be written as the continuity of the normal component of the displacement field which is permeability times the electric field as

$$\epsilon \boldsymbol{n}\cdot \boldsymbol{E}^{(2)}(\boldsymbol{x}_h) = \boldsymbol{n}\cdot \boldsymbol{E}^{(1)}(\boldsymbol{x}_h), \tag{29}$$

(where ϵ is the permeability ratio ϵ_2/ϵ_1) and the continuity of the tangential magnetic currents $\boldsymbol{J} = \boldsymbol{n}\times \boldsymbol{E}$

$$\boldsymbol{n}\times \boldsymbol{E}^{(2)}(\boldsymbol{x}_h) = \boldsymbol{n}\times \boldsymbol{E}^{(1)}(\boldsymbol{x}_h). \tag{30}$$

These are four equations for the three independent boundary unknowns which we choose as the electric field in region 1. They can be algebraically solved using (29) and two of the three equations (30) or the four equations solved using a Moore–Penrose pseudo-inverse to express the boundary fields in region 2 in terms of those in region 1 as

$$E_i^{(2)}(\boldsymbol{x}_h) = A_{ij}(\boldsymbol{x}_t)E_j^{(1)}(\boldsymbol{x}_h), \tag{31}$$

with summation over j from 1 to 3 where

$$A_{ij}(\boldsymbol{x}_t) = \delta_{ij} + (\epsilon^{-1} - 1)\hat{n}_i\hat{n}_j, \tag{32}$$

and where δ_{ij} is the Kronecker delta and the normals are unit normals. This is an exact result and will be used in (27) and (28), but first we need boundary conditions on the normal derivatives of the electric fields.

The boundary conditions on the normal derivatives of the electric fields follow from the continuity conditions of the electric surface currents $\boldsymbol{K} = -\boldsymbol{n} \times \boldsymbol{H}$, which are

$$K_i^{(2)}(\boldsymbol{x}_h) = K_i^{(1)}(\boldsymbol{x}_h). \tag{33}$$

Using Maxwell's equations with no primary (free) sources we derive the following vector identity which holds for either region:

$$n_j\partial_j E_i = n_m\partial_i E_m + \mathrm{i}(\omega\tilde{\mu})K_i, \tag{34}$$

where ω is circular frequency, repeated indices are summed from 1 to 3, and $\tilde{\mu}$ will be set equal to μ_1 or μ_2 below. In (23) and (25) we have normal derivative terms such as

$$n_j\partial_j E_i(\boldsymbol{x}_h), \tag{35}$$

which means to first take the normal derivative of E_i and then set the result on the surface. We want to use the opposite procedure, that is to first set the field E_i on the surface and then differentiate it. To accomplish this, introduce the bracket notation

$$\{E_m\} = E_m(\boldsymbol{x}_h), \tag{36}$$

which means to set the field on the surface first, so that the quantity is only a function of x and y. If we now differentiate with respect to x and y we get

$$\partial_x\{E_m\} = \{\partial_x E_m\} + h_x\{\partial_z E_m\}, \tag{37}$$

and

$$\partial_y\{E_m\} = \{\partial_y E_m\} + h_y\{\partial_z E_m\}, \tag{38}$$

where h_x and h_y are surface slopes. The equations follow if we note for example that $\{E_m\}$ is a function of x as both an independent variable and a dependent variable through $h(\boldsymbol{x}_t)$. Equations (37) and (38) are just the chain rule of calculus. Obviously (34) becomes with this bracket notation

$$n_j\{\partial_j E_m\} = n_m\{\partial_i E_m\} + i(\omega\tilde{\mu})\{K_i\}. \tag{39}$$

We can thus write the first term on the rhs of (39)

$$n_m\{\partial_i E_m\} = n_m \partial_{it}\{E_m\} + n_m n_i \{\partial_z E_m\}, \tag{40}$$

where the it subscript refers to only transverse variables x and y or $i = 1$ and 2. Finally, the divergence condition $\partial_m E_m = 0$ also holds on the boundary which yields

$$\{\partial_z E_3\} = -\{\partial_x E_1\} - \{\partial_y E_2\}. \tag{41}$$

Combining all the results we are able to write (39) as

$$n_j\{\partial_j E_i\} = n_m \partial_{it}\{E_m\} - n_i \partial_{pt}\{E_{pt}\} + i(\omega\tilde{\mu})\{K_i\}. \tag{42}$$

The latter equation holds in both regions to give

$$n_j\{\partial_j E_i^{(1)}\} = n_m \partial_{it}\{E_m^{(1)}\} - n_i \partial_{pt}\{E_{pt}^{(1)}\} + \mathrm{i}\omega\mu_1\{K_i^{(1)}\}, \tag{43}$$

and

$$n_j\{\partial_j E_i^{(2)}\} = n_m \partial_{it}\{E_m^{(2)}\} - n_i \partial_{pt}\{E_{pt}^{(2)}\} + \mathrm{i}\omega\mu_2\{K_i^{(2)}\}. \tag{44}$$

The continuity condition on electric currents (33) thus yields a discontinuity condition on the normal derivatives

$$n_j\{\partial_j E_i^{(2)}\} = \mu n_j\{\partial_j E_i^{(1)}\} + V_i(\boldsymbol{x}_t), \tag{45}$$

where

$$\begin{aligned} V_i(\boldsymbol{x}_t) = {} & n_m \partial_{it}\{E_m^{(2)}\} - n_i \partial_{pt}\{E_{pt}^{(2)}\} \\ & -\mu n_m \partial_{it}\{E_m^{(1)}\} + \mu n_i \partial_{pt}\{E_{pt}^{(1)}\}. \end{aligned} \tag{46}$$

The boundary fields in region 2 in (46) can be replaced by boundary fields in region 1 using (31), and all the terms in (46) can be integrated by parts.

Choose the boundary unknowns as

$$E_i(\boldsymbol{x}_h) = E_i^{(1)}(\boldsymbol{x}_h), \tag{47}$$

and

$$N_i(\boldsymbol{x}_h) = n_j\{\partial_j E_i^{(1)}\}. \tag{48}$$

Then, (27) and (28) become

$$\tfrac{1}{2}E_i(\boldsymbol{x}_h) = W_i(\boldsymbol{x}_h) + P(D_1 E_i)(\boldsymbol{x}_h) - (S_1 N_i)(\boldsymbol{x}_h), \tag{49}$$

and

$$-\tfrac{1}{2}A_{ij}(\boldsymbol{x}_t)E_j(\boldsymbol{x}_h) = P(D_2 A_{ij} E_j)(\boldsymbol{x}_h) - \mu(S_2 N_i)(\boldsymbol{x}_h) - (S_2 V_i)(\boldsymbol{x}_h). \tag{50}$$

The term involving V_i in (50) can be integrated by parts to yield terms only on the electric field. Defining the vector of six boundary unknowns

$$\boldsymbol{X} = (E_i, N_i)^{\mathrm{T}}, \tag{51}$$

the resulting matrix system of equations from (49) and (50) can be written in the form

$$\mathrm{M}\boldsymbol{X} = \boldsymbol{W}, \tag{52}$$

where three of the four blocks of the matrix M are in diagonal form. Only the fourth block arising from the electric field coupling in (50) is not diagonal. The result is an exact 50% sparse matrix which has proved successful in computations.[25]

Finally, note that these equations were developed for the electric field and its normal derivative. Corresponding equations can be developed for the magnetic field and its normal derivative.

The development above using boundary unknowns of the vector field and its normal derivative is not the standard version of electromagnetic integral equations which appear in the literature. These are referred to as the Stratton–Chu equations. They follow from the previous series of equations by using vector identities in each region V_j involving the two types of terms which appear in integrals such as (49) and (50). They are

$$Gn_j\partial_j E_i = \mathrm{i}\omega\mu G K_i - n_m E_m \partial_i G + n_m \partial_i (G E_m), \tag{53}$$

and

$$(n_j\partial_j G)E_i = n_i \partial_m (G E_m) - [\boldsymbol{\nabla} G \times \boldsymbol{J}]_i, \tag{54}$$

which are written in mixed notation and hold in each region with the appropriate choice of parameters. The integral of the difference of terms involving the vector derivative of GE_m vanishes. Analogous vector identities on the magnetic fields are found via the replacement $\boldsymbol{E} \Longleftrightarrow \boldsymbol{H}$ and $\mu \Longleftrightarrow -\epsilon$. The resulting electric field equations in each region are written as

$$\begin{aligned}\Theta_1(\boldsymbol{x})E_i^{(1)}(\boldsymbol{x}) = {} & W_i(\boldsymbol{x}) - \mathrm{i}\omega\mu_1\big(S_1 K_i^{(1)}\big)(\boldsymbol{x}) \\ & + \iint_h \partial_i G_1(\boldsymbol{x}, \boldsymbol{x}'_h)\left(n_m E_m^{(1)}\right)(\boldsymbol{x}'_h)\,\mathrm{d}\boldsymbol{x}'_t \\ & - \iint_h \left[\boldsymbol{\nabla} G_1(\boldsymbol{x}, \boldsymbol{x}'_h) \times \boldsymbol{J}^{(1)}(\boldsymbol{x}'_h)\right]_i \mathrm{d}\boldsymbol{x}'_t,\end{aligned} \tag{55}$$

and

$$\begin{aligned}\Theta_2(\boldsymbol{x})E_i^{(2)}(\boldsymbol{x}) = {} & \mathrm{i}\omega\mu_2\left(S_2 K_i^{(2)}\right)(\boldsymbol{x}) \\ & + \iint_h \partial_i G_2(\boldsymbol{x}, \boldsymbol{x}'_h)(n_m E_m^{(2)})(\boldsymbol{x}'_h)\,\mathrm{d}\boldsymbol{x}'_t \\ & - \iint_h \left[\boldsymbol{\nabla} G_2(\boldsymbol{x}, \boldsymbol{x}'_h) \times \boldsymbol{J}^{(2)}(\boldsymbol{x}'_h)\right]_i \mathrm{d}\boldsymbol{x}'_t.\end{aligned} \tag{56}$$

Terms such as $n_m E_m$ are related to surface charge. Here we set them to zero consistent with our choice of $\boldsymbol{\nabla} \cdot \boldsymbol{E} = 0$. (Both could be included with $n_m E_m$ related to the divergence of the current.) The boundary conditions on the electric fields are (31) and the boundary conditions on the currents are (30) and (33) and, taking the surface limits as $\boldsymbol{x}$ approaches the surface from above in (55) and from below in (56), the result is

$$\frac{1}{2}E_i(\boldsymbol{x}_h) = W_i(\boldsymbol{x}_h) - \mathrm{i}\omega\mu_1(S_1 K_i)(\boldsymbol{x}_h) - \oint\!\!\oint_h [\boldsymbol{\nabla} G_1(\boldsymbol{x}_h, \boldsymbol{x}'_h) \times \boldsymbol{J}(\boldsymbol{x}'_h)]_i \,\mathrm{d}\boldsymbol{x}'_t, \tag{57}$$

and

$$\frac{1}{2}A_{ij}E_j(\boldsymbol{x}_h) = \mathrm{i}\omega\mu_2(S_2 K_i)(\boldsymbol{x}_h) - \oint\!\!\oint_h [\boldsymbol{\nabla} G_2(\boldsymbol{x}_h, \boldsymbol{x}'_h) \times \boldsymbol{J}(\boldsymbol{x}'_h)]_i \,\mathrm{d}\boldsymbol{x}'_t. \tag{58}$$

If the currents $\boldsymbol{J}$ are expressed in terms of the electric fields, (57) and (58) are boundary equations on the electric fields and the currents $\boldsymbol{K}$. To find coupled equations on only the currents it is necessary to first write equations analogous to (55) and (56) for the magnetic field $\boldsymbol{H}$ and then to take the cross product of those equations with the exterior normal and then take the surface limits of the result. Computational applications have been developed.[25] Another application to two-dimensional problems is,[45] and fast solution methods for problems of this type can be found in.[9]

As an example, if we assume that the electric surface current is known, then we can generate two different equations on the magnetic surface current. The first is an equation of first kind of the general form

$$\mathrm{i}\omega\mu_1\varepsilon_{pqi}n_q \iint_h G(\boldsymbol{x}_h, \boldsymbol{x}'_h)K_i(\boldsymbol{x}'_h)\,\mathrm{d}\boldsymbol{x}'_t = U_p(\boldsymbol{x}_h), \tag{59}$$

where U_p contains incident field terms and the known surface current values, and ε_{pqi} is the antisymmetric Levi-Civita symbol. This general type of equation is referred to as an electric field integral equation (EFIE). It is also possible to derive an equation of second kind of the general form

$$\frac{1}{2}K_p(\boldsymbol{x}_h) + \varepsilon_{pqi}n_q \oint\!\!\oint_h [\boldsymbol{\nabla} G_1(\boldsymbol{x}_h, \boldsymbol{x}'_h) \times \boldsymbol{K}(\boldsymbol{x}'_h)]_i \,\mathrm{d}\boldsymbol{x}'_t = Q_p(\boldsymbol{x}_h), \tag{60}$$

where the Q_p contains the incident field terms and the known surface current values. This general type of equation is referred to as a magnetic field integral equation (MFIE). Combinations of (59) and (60) are used for bounded body scattering problems in order to eliminate the effects of internal resonances.

There is a third integral equation version of the electromagnetic scattering equations. They can be derived from the Stratton–Chu equations for the magnetic field. Take the curl of these magnetic field equations and, using the Maxwell equations and the fact that the gradient terms in the Stratton–Chu equations vanish under the curl operation, we get equations on the electric field. The equations have the form

$$\Theta_1(\boldsymbol{x})E_i^{(1)}(\boldsymbol{x}) = E_i^{(0)}(\boldsymbol{x}) + \left\{\nabla \times \iint_h G_1(\boldsymbol{x}, \boldsymbol{x}'_h)\boldsymbol{J}^{(1)}(\boldsymbol{x}'_h)\right\}_i \mathrm{d}\boldsymbol{x}'_t$$
$$- (i/\omega\epsilon_1)\left\{\nabla \times \left[\nabla \times \iint_h G_1(\boldsymbol{x}, \boldsymbol{x}'_h)\boldsymbol{K}^{(1)}(\boldsymbol{x}'_h)\right]\right\}_i \mathrm{d}\boldsymbol{x}'_t, \quad (61)$$

and

$$\Theta_2(\boldsymbol{x})E_i^{(2)}(\boldsymbol{x}) = -\left\{\nabla \times \iint_h G_2(\boldsymbol{x}, \boldsymbol{x}'_h)\boldsymbol{J}^{(1)}(\boldsymbol{x}'_h)\right\}_i \mathrm{d}\boldsymbol{x}'_t$$
$$+ (\mathrm{i}/\omega\epsilon_2)\left\{\nabla \times \left[\nabla \times \iint_h G_2(\boldsymbol{x}, \boldsymbol{x}'_h)\boldsymbol{K}^{(1)}(\boldsymbol{x}'_h)\right]\right\}_i \mathrm{d}\boldsymbol{x}'_t, \quad (62)$$

where $E_i^{(0)}$ contains the incident field terms. The general question of conditioning the equations has been discussed in [12]. Similarly, by taking the curl of (55) and (56) we derive equations on the magnetic field. They also follow from (61) and (62) under the replacement $\epsilon \Longleftrightarrow -\mu$ and $\boldsymbol{E} \Longleftrightarrow \boldsymbol{H}$. The surface limits of these equations follows the same development as for the Stratton–Chu equations depending on whether we want to generate equations on say the electric fields and magnetic currents or fully on the currents. The equations also follow directly from Green's theorem on the electric field and the tensor free-space Green's function Γ_{ij} [14] which satisfies the differential equation

$$\partial_l \partial_l \Gamma_{in}(\boldsymbol{x}, \boldsymbol{x}') - \partial_i \partial_m \Gamma_{mn}(\boldsymbol{x}, \boldsymbol{x}') + k^2 \Gamma_{in}(\boldsymbol{x}, \boldsymbol{x}') = -\delta_{ij}\delta(\boldsymbol{x} - \boldsymbol{x}'), \quad (63)$$

and is explicitly given by

$$\Gamma_{ij}(\boldsymbol{x}, \boldsymbol{x}') = \delta_{ij} G(\boldsymbol{x}, \boldsymbol{x}') + k^{-2}\partial_i \partial_i G(\boldsymbol{x}, \boldsymbol{x}'). \quad (64)$$

8.3. Spectral-Space Methods

The development in Sect. 8.3 yielded coordinate-space integral equations where the arguments of the kernels of the equations were both in coordinate-space (CC). Here we discuss in some detail a method which generates integral equations on the same boundary unknowns, but where the "field" variable is in spectral-space, which is the SC version of the equations.[16] These will also be shown to be useful later when we look at the problem of surface reconstruction. Another method, called the CS method, which reverses the behavior of the variables, has also been developed in [38]. It is useful when there are extended surface waves present.

8.3.1. Scalar Problems

It is possible to derive the SC method by using the Weyl representation for the free-space Green's function in (14) and (16), but it is simpler to use Green's theorem on the fields ψ_1 and ψ_2 in regions 1 and 2 along with appropriate up- and down-going wave states for the regions, the latter of which in fact form the kernel of the Weyl representation. The plane wave states in region 1 are

$$\phi_1^{\pm}(\boldsymbol{x}) = \exp(\mathrm{i}k_1 \boldsymbol{M}^{\pm} \cdot \boldsymbol{x}), \tag{65}$$

which satisfies the Helmholtz equation (1) for $j = 1$ where

$$\boldsymbol{M}^{\pm} = (-\boldsymbol{M}_t, \pm M_z), \tag{66}$$

where $\boldsymbol{M}_t = (M_x, M_y)$, $M_z = m_1(\boldsymbol{M}_t)$, and $M_z = \sqrt{1 - M_t^2}$ if $M_t^2 \leq 1$ and $M_z = \mathrm{i}\sqrt{M_t^2 - 1}$ if $M_t^2 \geq 1$. The $\pm$ designation indicates plane waves which propagate in an upward (downward) direction from the surface $z = h(\boldsymbol{x}_t)$. We write Green's theorem on ψ_1 and $\phi_1^{\pm}$ in the domain V_{1L} defined by $z \geq h(\boldsymbol{x}_t)$, $-L/2 \leq (x, y) \leq L/2$, and $z \leq S_1$ where S_1 is a constant and $S_1 \geq \max(h(\boldsymbol{x}_t))$. The characteristic function for this region is

$$\Theta_1(\boldsymbol{x}) = \theta(z - h(\boldsymbol{x}_t))\theta(S_1 - z)\Theta_+(\boldsymbol{x}_t), \tag{67}$$

where

$$\Theta_+(\boldsymbol{x}_t) = \theta(x + L/2)\theta(L/2 - x)\theta(y + L/2)\theta(L/2 - y), \tag{68}$$

where θ is the Heaviside function. The result is

$$\mathcal{U}^{\pm}(L) = S_1^{\pm}(L) - E_1^{\pm}(L), \tag{69}$$

where

$$\mathcal{U}^{\pm}(L) = L^{-2} \iint_{-L/2}^{L/2} [n_l \partial_l \phi_1^{\pm}(\boldsymbol{x}_h)\psi_1(\boldsymbol{x}_h) - \phi_1^{\pm}(\boldsymbol{x}_h) n_l \partial_l \psi_1(\boldsymbol{x}_h)] \,\mathrm{d}\boldsymbol{x}_t, \tag{70}$$

are integrals on the upper side of h,

$$S_1^{\pm}(L) = L^{-2} \iint_{-L/2}^{L/2} [\partial_z \phi_1^{\pm}(\boldsymbol{x}_1)\psi_1(\boldsymbol{x}_1) - \phi_1^{\pm}(\boldsymbol{x}_1)\partial_z \psi_1(\boldsymbol{x}_1)] \,\mathrm{d}\boldsymbol{x}_t, \tag{71}$$

are integrals on S_1 and

$$E_1^{\pm}(L) = L^{-2} \int_{h(\boldsymbol{x}_t)}^{S_1} \mathrm{d}z \iint_{-L/2}^{L/2} \big[\partial_l \phi_1^{\pm}(\boldsymbol{x})\psi_1(\boldsymbol{x}) - \phi_1^{\pm}(\boldsymbol{x})\partial_l \psi_1(\boldsymbol{x})\big] \,\partial_l \Theta_+(\boldsymbol{x}_t) \,\mathrm{d}\boldsymbol{x}_t, \tag{72}$$

are integrals on the side or edge of the domain. In (72) one of the x or y integrations can be carried out using the delta function which arises from $\partial_l \Theta_+$ and, for L large, bounded surface fields, and bounded h we can write the asymptotic estimate $E_1^{\pm}(L) \sim O(L^{-1})$ where O is the order symbol. The integrals in (70) and (71) behave like $O(1)$ for large L, so that asymptotically the side integrals vanish faster than the integrals on h and S_1 and are dropped. (For periodic surfaces the integrals cancel exactly due to the Floquet periodicity.) If we define the limits

$$\mathcal{U}^{\pm} = \lim_{L \to \infty} L^2 \mathcal{U}^{\pm}(L), \tag{73}$$

and

$$S_1^{\pm} = \lim_{L\to\infty} L^2 S_1^{\pm}(L), \tag{74}$$

then (69) becomes just

$$\mathcal{U}^{\pm} = S_1^{\pm}, \tag{75}$$

relating an integral on h to results on S_1 and where the integrals in (75) are now over the domain $(-\infty, \infty)$. They are

$$\mathcal{U}^{\pm} = \iint_{-\infty}^{\infty} \left[n_l \partial_l \phi_1^{\pm}(\boldsymbol{x}_h) F(\boldsymbol{x}_h) - ik_1 \phi_1^{\pm}(\boldsymbol{x}_h) \widetilde{N}(\boldsymbol{x}_h)\right] \mathrm{d}\boldsymbol{x}_t, \tag{76}$$

and

$$S_1^{\pm} = \iint_{-\infty}^{\infty} \left[\partial_z \phi_1^{\pm}(\boldsymbol{x}_1)\psi_1(\boldsymbol{x}_1) - \phi_1^{\pm}(\boldsymbol{x}_1)\partial_z \psi_1(\boldsymbol{x}_1)\right] \mathrm{d}\boldsymbol{x}_t, \tag{77}$$

where $F(\boldsymbol{x}_h)$ is the boundary value defined by (12), and $\widetilde{N}(\boldsymbol{x}_h)$ is the scaled boundary value $(\mathrm{i}k_1)^{-1}N(\boldsymbol{x}_h)$ with $N(\boldsymbol{x}_h)$ given by (13). Note that (75) relates the unknown boundary values on h to values of the field and its normal derivative on S_1 and represents an analytic continuation. Further, the terms are functions of the spectral parameter $\boldsymbol{M}^{\pm}$ from (65). Differentiating $\phi_1^{\pm}$ we get

$$n_l \partial_l \phi_1^{\pm}(\boldsymbol{x}_h) = \mathrm{i}k_1 n_l M_l^{\pm} \phi_1^{\pm}(\boldsymbol{x}_h), \tag{78}$$

and

$$\partial_z \phi_1^{\pm}(\boldsymbol{x}_1) = \pm \mathrm{i}k_1 m_1(\boldsymbol{M}_t)\phi_1^{\pm}(\boldsymbol{x}_1), \tag{79}$$

so that (76) and (77) reduce to

$$\mathcal{U}^{\pm} = \mathrm{i}k_1 \iint_{-\infty}^{\infty} \phi_1^{\pm}(\boldsymbol{x}_h)\left[n_l M_l^{\pm} F(\boldsymbol{x}_h) - \widetilde{N}(\boldsymbol{x}_h)\right] \mathrm{d}\boldsymbol{x}_t, \tag{80}$$

and

$$S_1^{\pm} = \mathrm{i}k_1 \iint_{-\infty}^{\infty} \phi_1^{\pm}(\boldsymbol{x}_1)[\pm m_1(\boldsymbol{M}_t)\psi_1(\boldsymbol{x}_1) - (ik_1)^{-1}\partial_z \psi_1(\boldsymbol{x}_1)]\, \mathrm{d}\boldsymbol{x}_t. \tag{81}$$

Using (9) and a further plane wave decomposition of the incident ψ^{i} and scattered ψ^{sc} fields as

$$\psi^{\mathrm{i}}(\boldsymbol{x}) = \iint I(\beta_t)\exp(\mathrm{i}k_1 \boldsymbol{\beta}\cdot\boldsymbol{x})\,\mathrm{d}\boldsymbol{\beta}_t, \tag{82}$$

where $\boldsymbol{\beta} = (\boldsymbol{\beta}_t, -\beta_z)$ and $\beta_z = \sqrt{1-\beta_t^2}$ if $\beta_t^2 \le 1$, and $\beta_z = \mathrm{i}\sqrt{\beta_t^2 - 1}$ if $\beta_t^2 \ge 1$, which represents downgoing plane waves (or decaying waves), and

$$\psi^{\mathrm{sc}}(\boldsymbol{x}) = \iint R(\boldsymbol{\alpha}_t)\exp(\mathrm{i}k_1 \boldsymbol{\alpha}\cdot\boldsymbol{x})\,\mathrm{d}\boldsymbol{\alpha}_t, \tag{83}$$

where $\boldsymbol{\alpha} = (\boldsymbol{\alpha}_t, \alpha_z)$, and α_z represents either upward propagating plane waves $\alpha_z = \sqrt{1-\alpha_t^2}$ if $\alpha_t^2 \le 1$, or decaying waves in the positive z-direction,

$\alpha_z = \mathrm{i}\sqrt{\alpha_t^2 - 1}$ when $\alpha_t^2 \geq 1$. Note that (83) will only be used in the S_1 integral which is above the highest surface excursion. It is *not* a use of the Rayleigh Hypothesis. Using (9), (82), and (83) to evaluate (81), we can write (75) as

$$\mathcal{U}^{\pm} = 8\pi^2 \mathrm{i} k_1^{-1} m_1(\boldsymbol{M}_t) \begin{cases} I(\boldsymbol{M}_t) \\ -R(\boldsymbol{M}_t), \end{cases} \tag{84}$$

where we use (80) for $\mathcal{U}^{\pm}$. The + equation in (84) uses ϕ_1^+ which is an upgoing plane (or decaying) wave and effectively projects out the downgoing incident field spectral amplitude I. Correspondingly, the − equation in (84) uses ϕ_1^- which is downgoing, and it projects out the upgoing reflected field spectral amplitude R. The incident spectral amplitude I can be used to create incident fields composed of plane waves, cylindrical waves, spherical waves, or beams. For a single plane wave, for example, we have

$$I(\boldsymbol{M}_t) = \delta(\boldsymbol{M}_t - \boldsymbol{M}_t^{\mathrm{i}}), \tag{85}$$

where $M_1^{\mathrm{i}} = \cos\varphi^{\mathrm{i}} \sin\vartheta^{\mathrm{i}}$, $M_2^{\mathrm{i}} = \sin\varphi^{\mathrm{i}} \cos\vartheta^{\mathrm{i}}$, and $M_3^{\mathrm{i}} = m_1(\boldsymbol{M}_t^{\mathrm{i}}) = \cos\vartheta^{\mathrm{i}}$ in terms of incident polar angle ϑ^{i} and azimuthal angle φ^{i}. The usual interpretation of (84) for direct scattering is that given the incident field I, the resulting equation involving I is one equation used to solve for the two boundary unknowns in $\mathcal{U}^{\pm}$. A second equation is generated from the transmitted field discussed below. Given the two boundary unknowns F and $\widetilde{N}$ the term R can be evaluated, and the scattered field can be found from (83). Alternatively if both I and R are known we have in (84) two equations to find the boundary unknowns, or to find the (now) unknown surface h as we discuss later.

An analogous development can be carried out for the transmitted field. We use Green's theorem on the transmitted field ψ_2 and the (transmitted) plane wave states

$$\phi_2^{\pm}(\boldsymbol{x}) = \exp(\mathrm{i}k_2 \boldsymbol{P}^{\pm} \cdot \boldsymbol{x}), \tag{86}$$

where both fields satisfy (1) for $j = 2$. Here, $\boldsymbol{P}^{\pm} = (-\boldsymbol{P}_t, \pm P_z)$ where $P_z = \sqrt{1 - P_t^2}$ for $P_t^2 \leq 1$ and $P_z = \mathrm{i}\sqrt{P_t^2 - 1}$ if $P_t^2 \geq 1$. The domain is V_{2L} where $z \leq h(\boldsymbol{x}_t)$, $-L/2 \leq (x, y) \leq L/2$, and $z \geq S_2$ where S_2 is a constant below the lowest surface excursion, $S_2 \leq \min(h(\boldsymbol{x}_t))$. Again, as $L \to \infty$, the side integrals are of lower asymptotic order, and we get the representation

$$\mathcal{L}^{\pm} = S_2^{\pm}, \tag{87}$$

where

$$\mathcal{L}^{\pm} = \iint_{-\infty}^{\infty} \left[n_l \partial_l \phi_2^{\pm}(\boldsymbol{x}_h) \psi_2(\boldsymbol{x}_h) - \phi_2^{\pm}(\boldsymbol{x}_h) n_l \partial_l \psi_2(\boldsymbol{x}_h) \right] \mathrm{d}\boldsymbol{x}_t \tag{88}$$

is an integral on the lower side of h and

$$S_2^{\pm} = \iint_{-\infty}^{\infty} \left[\partial_z \phi_2^{\pm}(\boldsymbol{x}_2) \psi_2(\boldsymbol{x}_2) - \phi_2^{\pm}(\boldsymbol{x}_2) \partial_z \psi_2(\boldsymbol{x}_2) \right] \mathrm{d}\boldsymbol{x}_t \tag{89}$$

is an integral along S_2. To proceed, differentiate the $\phi_2^{\pm}$ in (88) and (89), use the boundary conditions (10) and (11) and definitions (12) and (13) in (88), and the spectral representation for the transmitted field (for $z \leq S_2$)

$$\psi_2(\boldsymbol{x}) = K^2 \iint T(\gamma_t)\exp(\mathrm{i}k_2\gamma\cdot\boldsymbol{x})\,\mathrm{d}\gamma_t, \tag{90}$$

where, $\gamma = (\gamma_t, -\gamma_z)$ with $\gamma_z = \sqrt{1-\gamma_t^2}$ when $\gamma_t^2 \leq 1$ and $\gamma_z = \mathrm{i}\sqrt{\gamma_t^2 - 1}$ when $\gamma_t^2 \geq 1$, and $K = k_2/k_1$. Further, using Snell's law of conservation of ray parameter

$$\boldsymbol{M}_t = K\boldsymbol{P}_t \tag{91}$$

yields the following pair of equations from (87), (88), and (89)

$$\iint_{-\infty}^{\infty} \phi_2^{\pm}(\boldsymbol{x}_h)\left[\rho^{-1}\left[\pm\sqrt{K^2 - M_t^2} + h_x M_x + h_y M_y\right]F(\boldsymbol{x}_h) - \widetilde{N}(\boldsymbol{x}_h)\right]\mathrm{d}\boldsymbol{x}_t$$
$$= 8\pi^2 k_1^{-2}\sqrt{K^2 - M_t^2}\begin{cases} T(K^{-1}\boldsymbol{M}_t) \\ 0. \end{cases} \tag{92}$$

Again, the + equation in (92) contains ϕ_2^+ which is an upgoing plane wave which projects out the downgoing wave spectral amplitude T. The $-$ equation in (92) contains ϕ_2^- which is downgoing and the zero on the right-hand side illustrates that there are no upgoing (incident) waves from region 2. For direct scattering problems there are thus two equations to solve for the boundary unknowns F and $\widetilde{N}$, the + equation (84) using (80) and the $-$ equation from (92). The kernels of both integral equations have arguments in a mixed spectral-coordinate domain and the equations are referred to as SC equations. The other two equations in (84) and (92) are used to evaluate the reflection R and transmission T spectral amplitudes. For an arbitrary incident field spectral amplitude (not a delta function) there is an overall energy flux conservation relation of the form

$$\iint |I(\boldsymbol{\beta}_t)|^2 \Re m_1(\boldsymbol{\beta}_t)\,\mathrm{d}\boldsymbol{\beta}_t = \iint |R(\boldsymbol{\alpha}_t)|^2 \Re m_1(\boldsymbol{\alpha}_t)\,\mathrm{d}\boldsymbol{\alpha}_t$$
$$+\rho\iint |T(\boldsymbol{M}_t)|^2 \Re m_2(\boldsymbol{M}_t)\,\mathrm{d}\boldsymbol{M}_t, \tag{93}$$

where $\Re$ represents "real part". This is used as a computational check.

For the Dirichlet problem $F(\boldsymbol{x}_h) = 0$ and there is no transmitted field so that (80) and (84) reduce to

$$\iint_{-\infty}^{\infty} k_D^{\pm}(\boldsymbol{M}_t, \boldsymbol{x}_t)\widetilde{N}(\boldsymbol{x}_h)\,\mathrm{d}\boldsymbol{x}_t = D^{\pm}(\boldsymbol{M}_t), \tag{94}$$

where we have explicitly written the Dirichlet kernel (which is just $\phi_1^{\pm}$) as

$$k_D^{\pm}(\boldsymbol{M}_t, \boldsymbol{x}_t) = \exp[-\mathrm{i}k_1\boldsymbol{M}_t\cdot\boldsymbol{x}_t \pm \mathrm{i}k_1 m_1(\boldsymbol{M}_t)h(\boldsymbol{x}_t)], \tag{95}$$

to emphasize its spectral (row, field) and coordinate (column, source) arguments and

$$D^{\pm}(\boldsymbol{M}_t) = 8\pi^2 k_1^{-2} m_1(\boldsymbol{M}_t)\begin{cases} -I(\boldsymbol{M}_t) \\ R^{\mathrm{D}}(\boldsymbol{M}_t), \end{cases} \tag{96}$$

with R^{D} referring to the spectral amplitude for the Dirichlet problem.

For the Neumann problem $\widetilde{N}(\boldsymbol{x}_h) = 0$, again there is no transmitted field and (80) and (84) reduce to

$$\iint_{-\infty}^{\infty} k_N^{\pm}(\boldsymbol{M}_t, \boldsymbol{x}_t) F(\boldsymbol{x}_h)\, \mathrm{d}\boldsymbol{x}_t = \mathrm{N}^{\pm}(\boldsymbol{M}_t), \tag{97}$$

using the Neumann kernel

$$k_N^{\pm}(\boldsymbol{M}_t, \boldsymbol{x}_t) = n_l M_l^{\pm} k_D^{\pm}(\boldsymbol{M}_t, \boldsymbol{x}_t), \tag{98}$$

and

$$N^{\pm}(\boldsymbol{M}_t) = 8\pi^2 k_1^{-2} m_1(\boldsymbol{M}_t) \begin{cases} I(\boldsymbol{M}_t) \\ -R^N(\boldsymbol{M}_t), \end{cases} \tag{99}$$

where R^N is the Neumann reflection spectral amplitude.

8.3.2. Electromagnetic Problems, Infinite Surface

For each component of the electric field the spectral-coordinate development proceeds in an analogous way to the scalar case. In particular, using Green's theorem on $E_i^{(1)}$ and $\phi_1^{\pm}$ (each of which satisfies (1) for $j = 1$) in the domain V_{1L} yields, after we let $L \to \infty$ so that the side integrals drop, a relation analogous to (75) but with vector fields

$$\mathcal{U}_i^{\pm} = S_{1i}^{\pm}, \tag{100}$$

where

$$\mathcal{U}_i^{\pm} = \iint_{-\infty}^{\infty} \left[n_l \partial_l \phi_1^{\pm}(\boldsymbol{x}_h) E_i(\boldsymbol{x}_h) - \mathrm{i}k_1 \phi_1^{\pm}(\boldsymbol{x}_h) \widetilde{N}_i(\boldsymbol{x}_h) \right] \mathrm{d}\boldsymbol{x}_t \tag{101}$$

is an integral over h and

$$S_{1i}^{\pm} = \iint_{-\infty}^{\infty} \left[\partial_z \phi_1^{\pm}(\boldsymbol{x}_1) E_i^{(1)}(\boldsymbol{x}_1) - \phi_1^{\pm}(\boldsymbol{x}_1) \partial_z E_i^{(1)}(\boldsymbol{x}_1) \right] \mathrm{d}\boldsymbol{x}_t, \tag{102}$$

which is an integral along S_1 where E_i and $\widetilde{N}_i$ are the obvious generalizations of F and $\widetilde{N}$. Differentiating the $\phi_1^{\pm}$ terms in (101) and (102) yields a relation similar to (84)

$$\mathcal{U}_i^{\pm} = 8\pi^2 \mathrm{i} k_1^{-1} m_1(\boldsymbol{M}_t) \begin{cases} I_i(\boldsymbol{M}_t) \\ -R_i(\boldsymbol{M}_t), \end{cases} \tag{103}$$

where

$$\mathcal{U}_i^{\pm} = \mathrm{i}k_1 \iint_{-\infty}^{\infty} \phi_1^{\pm}(\boldsymbol{x}_h) \left[n_l M_l^{\pm} E_i(\boldsymbol{x}_h) - \widetilde{N}_i(\boldsymbol{x}_h) \right] \mathrm{d}\boldsymbol{x}_t, \tag{104}$$

and where the vector spectral amplitudes follow from generalizations of (82) and (83)

$$E_i^{\mathrm{in}}(\boldsymbol{x}) = \iint I_i(\boldsymbol{\beta}_t) \exp(\mathrm{i}k_1 \boldsymbol{\beta}_t \cdot \boldsymbol{x})\, \mathrm{d}\boldsymbol{\beta}_t, \tag{105}$$

and

$$E_i^{\mathrm{sc}}(\boldsymbol{x}) = \iint R_i(\boldsymbol{\alpha}_t)\exp(\mathrm{i}k_1\boldsymbol{\alpha}\cdot\boldsymbol{x})\,\mathrm{d}\boldsymbol{\alpha}_t. \tag{106}$$

Green's theorem on $E_i^{(2)}$ and $\phi_1^{\pm}$ in region V_{2L} with $L \to \infty$ yields the generalization of (87)

$$\mathcal{L}_i^{\pm} = S_{2i}^{\pm}, \tag{107}$$

with

$$\mathcal{L}_i^{\pm} = \iint_{-\infty}^{\infty} \left[n_j \partial_j \phi_2^{\pm}(\boldsymbol{x}_h) E_i^{(2)}(\boldsymbol{x}_h) - \phi_2^{\pm}(\boldsymbol{x}_h) n_j \partial_j E_i^{(2)}(\boldsymbol{x}_h)\right] \mathrm{d}\boldsymbol{x}_t, \tag{108}$$

which is an integral over the lower part of the surface h and

$$S_{2i}^{\pm} = \iint_{-\infty}^{\infty} \left[\partial_z \phi_2^{\pm}(\boldsymbol{x}_2) E_i^{(2)}(\boldsymbol{x}_2) - \phi_2^{\pm}(\boldsymbol{x}_2)\partial_z E_i^{(2)}(\boldsymbol{x}_2)\right] \mathrm{d}\boldsymbol{x}_t, \tag{109}$$

an equation on S_2.

Using the boundary conditions (31) and (45) and differentiating the $\phi_2^{\pm}$ in (108) and (109) we first get

$$\begin{aligned} \mathcal{L}_i^{\pm} = \mathrm{i}k_2 \iint_{-\infty}^{\infty} & \phi_2^{\pm}(\boldsymbol{x}_h)\left[n_j P_j^{\pm} A_{im}(\boldsymbol{x}_t) E_m(\boldsymbol{x}_h) - \mu K^{-1} \widetilde{N}_i(\boldsymbol{x}_h)\right. \\ & \left. - (\mathrm{i}k_2)^{-1} V_i(\boldsymbol{x}_t)\right] \mathrm{d}\boldsymbol{x}_t, \end{aligned} \tag{110}$$

and the latter term, using (31), becomes

$$\begin{aligned} V_i(\boldsymbol{x}_t) = {} & n_m \partial_{it}\{A_{mr} E_r\} - n_i \partial_{pt}\{A_{pt} E_{pt}\} \\ & - \mu n_m \partial_{it}\{E_m\} + \mu n_i \partial_{pt}\{E_{pt}\}. \end{aligned} \tag{111}$$

Without going into details, each of the terms in (111) can be integrated by parts and (110) has the interpretation as nondiagonal (coupled) in the E_i fields and diagonal in the $\widetilde{N}_i$. Further, evaluating (109) using the spectral representation of the transmitted field

$$E_i^{(2)}(\boldsymbol{x}) = K^2 \iint T_i(\boldsymbol{\gamma}_t)\exp(\mathrm{i}k_2\boldsymbol{\gamma}\cdot\boldsymbol{x})\,\mathrm{d}\boldsymbol{\gamma}_t, \tag{112}$$

which is a generalization of (90), and using Snell's law (91) we get the result

$$\mathcal{L}_i^{\pm} = 8\pi^2 \mathrm{i}k_1^{-1}\sqrt{K^2 - M_t^2}\begin{cases} T_i(K^{-1}\boldsymbol{M}_t) \\ 0. \end{cases} \tag{113}$$

Equations (113) have an analogous projection interpretation as (92). The result for the direct scattering problem is a total of six equations to solve for the boundary unknowns E_i and $\widetilde{N}_i$ given by the + equation in (103) and the − equation in (113). Again note that, as discussed in Sect. 2.2, a block matrix solution of the resulting system has three of the four blocks diagonal, with the only coupling occurring in the lower region integration in the E_i field in (110). The − equation in (103) is used to evaluate the scattered vector spectral amplitude R_i and the scattered field by (106). The + equation in (113) yields the transmitted spectral density and thus the transmitted field by (112). For a single plane wave of the form

$$E_i^{\text{in}}(\boldsymbol{x}) = E_0 e_i \exp(\mathrm{i}k_1 \boldsymbol{\alpha} \cdot \boldsymbol{x}). \tag{114}$$

If $e_i = \delta_{i2}$ we have TE polarization and $e_i = a\delta_{i1} + b\delta_{i3}$ we have TH polarization, where a and b are related to the angle of incidence.

8.4. Surface Inversion

The problems which have been discussed are direct scattering problems, i.e. given the surface h find the scattered field. To do the latter we need the boundary values of the field and its normal derivative (or the currents). To find them we solve the appropriate integral equations. The inverse problem of surface reconstruction assumes something else is known and its knowledge will be used to infer the surface. What is generally assumed known is some information which is gathered away from the surface, i.e. remotely, for example the scattered field. The incident field is assumed known. We illustrate two algorithms to reconstruct the surface from these data. Both use the scalar Dirichlet problem from (94), (95), and (96)[48,49].

The first algorithm uses perturbation theory in surface height or, more accurately, perturbation theory in the phase of $k_D^{\pm}$ from (95), so that the small quantity is $t = k_1 m_1(\boldsymbol{M}_t) h(\boldsymbol{x}_t)$, and we assume $t \ll 1$. The difference of terms in (95) to first order in t is

$$k_D^{+}(\boldsymbol{M}_t, \boldsymbol{x}_t) - k_D^{-}(\boldsymbol{M}_t, \boldsymbol{x}_t) = 2\mathrm{i}k_1 m_1(\boldsymbol{M}_t) h(\boldsymbol{x}_t) \exp(-\mathrm{i}k_1 \boldsymbol{M}_t \cdot \boldsymbol{x}_t). \tag{115}$$

Using this result, the difference of the two equations (94) becomes with (96)

$$\mathcal{F}[h(\boldsymbol{x}_t)\widetilde{N}(\boldsymbol{x}_h)] = \mathrm{i}k_1^{-3}[R^{\mathrm{D}}(\boldsymbol{M}_t) + I(\boldsymbol{M}_t)], \tag{116}$$

which states that the two-dimensional Fourier transform $\mathcal{F}$ of $h\widetilde{N}$ is equal to the sum of incident plus reflected spectral amplitudes. (Note the k_1^{-3} factor goes away with scaled coordinates and height h). Similarly the sum of the two terms in (95) yields to first order in t

$$\mathcal{F}[\widetilde{N}(\boldsymbol{x}_h)] = k_1^{-2} m_1(\boldsymbol{M}_t)[R^{\mathrm{D}}(\boldsymbol{M}_t) - I(\boldsymbol{M}_t)], \tag{117}$$

i.e. the two-dimensional Fourier transform of the boundary unknown $\widetilde{N}$ is proportional to the difference of reflected and incident spectral amplitudes. If we take the inverse Fourier transform of both (116) and (117) and simply divide the results we get an approximate reconstruction of the surface h given by r which is

$$r(\boldsymbol{x}_t) = \mathrm{i}k_1^{-1} \frac{\mathcal{F}^{-1}[m_1(\boldsymbol{M}_t)(R^{\mathrm{D}}(\boldsymbol{M}_t) - I(\boldsymbol{M}_t))]}{\mathcal{F}^{-1}[R^{\mathrm{D}}(\boldsymbol{M}_t) + I(\boldsymbol{M}_t)]}. \tag{118}$$

Since we assumed our surface to be infinite the Fourier transforms are exact. For finite but nonperiodic surfaces, the surfaces must be tapered to zero outside some domain, and approximations must be introduced in addition to the use of perturbation theory. For periodic surfaces the results can be shown to reduce to Fourier series. Note that in this development, the boundary value $\widetilde{N}$ automatically dropped out in the division. The method has been implemented in one dimension with good results. Even limited data windows over the incident and scattered angles can produce useful results, so the method can be made practical.

The second method again for the Dirichlet problem assumes that the normal derivative boundary value $\widetilde{N}$ is approximated by the Kirchhoff approximation $\widetilde{N}_{\mathrm{KA}}$ which, for a single incident plane wave yields

$$\widetilde{N}_{\mathrm{KA}}(\boldsymbol{x}_h) = 2n_l M_l^{\mathrm{i}} \exp(\mathrm{i}k_1 \boldsymbol{M}_t^{\mathrm{i}} \cdot \boldsymbol{x}_t - \mathrm{i}k_1 M_3^{\mathrm{i}} h(\mathrm{x}_t)), \tag{119}$$

where the M_j^{i} are defined following (85). Using (94), (95), and (96) and integration by parts, the reflection spectral amplitude can be written as

$$R^D(\boldsymbol{M}_t) = k_1^2[m_1(\boldsymbol{M}_t)b(\boldsymbol{M}_t)]^{-1}[\boldsymbol{M}_t^i \cdot \boldsymbol{a}_t + M_3^i b]D(\boldsymbol{M}_t), \tag{120}$$

where

$$\boldsymbol{a}_t = \boldsymbol{M}_t - \boldsymbol{M}_t^i, \tag{121}$$

$$b(\boldsymbol{M}_t) = m_1(\boldsymbol{M}_t) + M_3^i, \tag{122}$$

and $\mathcal{D}$ is the "data" integral of the form

$$\mathcal{D}(\boldsymbol{M}_t) = \frac{1}{4\pi^2} \iint_{-\infty}^{\infty} \exp(-\mathrm{i}k_1 \boldsymbol{a}_t \cdot \boldsymbol{x}_t - \mathrm{i}k_1 b(\boldsymbol{M}_t) h(\boldsymbol{x}_t))\, \mathrm{d}\boldsymbol{x}_t. \tag{123}$$

This is not a Fourier transform since b is a function of $\boldsymbol{M}_t$, the "transform" variable. However, if we fix b we get a two-dimensional Fourier transform, which is restricted in $\boldsymbol{a}_t$. Its inverse transform yields an approximation to the surface $r(\boldsymbol{x}_t, b)$ which depends on b and which occurs in the phase

$$\exp(-\mathrm{i}k_1 b r(\boldsymbol{x}_t, b)) = \mathcal{F}^{-1}[\mathcal{D}(\boldsymbol{M}_t)], \tag{124}$$

and $r(\boldsymbol{x}_t, b)$ is given by the arctangent of the inverse transform. Depending on the choice of b, different surface reconstructions are possible. There is a constraint

$$|\boldsymbol{a}_t|^2 + b^2 \le 4, \tag{125}$$

which defines the Ewald sphere for these parameters. As b increases, the possible values of $|\boldsymbol{a}_t|$ decrease, which produces a low-pass filter on the data. Correspondingly, as b decreases, the set of transform values $|\boldsymbol{a}_t|$ increases, and it can be shown that this effectively includes more data for grazing incidence where the Kirchhoff approximation loses validity. The method has been successfully implemented in one dimension, and is able to reconstruct surface profiles with larger slopes than the perturbation method.[49]

Other methods to form an algorithm to reconstruct the surface are possible.[40,42] For example, there is an algorithm using a coupled system of equations to first solve for the boundary unknown $\widetilde{N}$ approximately, then uses a second equation which relates $\widetilde{N}$ to h. The equations are Volterra equations if only the forward propagating wave term is included. A second example uses a phase perturbation method of higher order. Other authors do not attempt to reconstruct the pointwise behavior of the surface, but rather settle for reconstructing specific surface parameters. If one includes time variability in the problem, the surface reflected wavefield can be recorded in one place, then time reversed and broadcast to a second place. The displacement in the signal produces a sensitivity to rms height and correlation function for a random surface, and this can be used to reconstruct these surface parameters.[37]

8.5. Solution Methods

All the methods we have discussed reduce to the solution of integral equations for the boundary unknowns. Briefly these reduce to the solution of equations of the form[2,11]

$$\lambda\, u(t) - \int_D K(t,s)u(s)\,ds = f(t), \tag{126}$$

where we consider only one-dimensional examples for simplicity, λ and $f(t)$ are assumed known, $K(t,s)$ is called the kernel and is known, u is the unknown, D is the domain, and $t \epsilon D$. For our cases the full vector corresponding to f was never zero since it represented the incident field or a function of the incident field, and $\lambda = 0$ (first-kind equation) or $\lambda = 1/2$ (second-kind equation).

If we pick a set of basis functions $\{\varphi_1, \varphi_2, \ldots, \varphi_n\}$ and assume that an approximate solution $u_n \approx u$ can be expanded in the basis as

$$u_n = b_j \varphi_j, \tag{127}$$

we can define the residual error as

$$r_n(t) = b_j(\lambda\varphi_j(t) - \int_D K(t,s)\varphi_j(s)\,ds) - f(t), \tag{128}$$

and the object is to make $r_n(t)$ as small as possible. For the method of collocation, n points or nodes $\{t_1, t_2, \ldots, t_n\}$ are chosen and we require that $r_n(t_i) = 0$. The method is simple and straightforward to implement, and only requires the dimension of the integrations to be equal to the dimensionality of the problem. For the Galerkin method, we require that

$$\langle r_n, \varphi_i \rangle = 0 \qquad i = 1, \ldots, n, \tag{129}$$

where $\langle \cdot, \cdot \rangle$ denotes an inner product. Again it is comparatively simple to implement and it generally yields better results than collocation although it requires the dimensions of the integrations to be twice the dimensionality of the problem. Various kinds of Galerkin and collocation schemes can be formulated depending on how the unknown is approximated over the domain, and rigorous error estimates are available[27].

Another popular approximation method is to write an iterative series of (126) with the first term approximating u under the integral as $\lambda^{-1} f(s)$. It is called the

Born approximation or field perturbation approximation in the physics literature. Higher order iterations are also possible, and in the mathematics literature it is referred to as a Neumann-type expansion. It yields an algorithm of the following type from (126)

$$\lambda u^{(n+1)}(t_k) = f(t_k) + \sum_{m=1}^{N} K(t_k, s_m) u^{(n)}(s_m), \tag{130}$$

where the superscripts refer to the iteration number and for $n = 0$

$$u^{(1)}(t_k) = \lambda^{-1} f(t_k). \tag{131}$$

If u under the integral is written as a complex phase term P as $u = \exp(\mathrm{i}P)$ and the phase term expressed in terms of a perturbation expansion, the resulting approximation u_{R}, called the Rytov approximation, can be written as

$$u_{\mathrm{R}}(t_k) = u^{(1)}(t_k) \exp\left(u^{(2)}(t_k)/u^{(1)}(t_k)\right), \tag{132}$$

which is effectively a partial summation of terms in (130).

More general iterative methods are also available. Equation (126) can be reduced to a matrix system of general form

$$\mathcal{Z} u = f, \tag{133}$$

where $\mathcal{Z}$ is called the impedance matrix. It can be composed of Green's functions and/or their normal derivatives. All these iterative methods split the impedance matrix in some way.[1,30] For example, choose a splitting matrix Q whose inverse exists. Multiply (130) by Q^{-1} and rewrite the result as

$$u = \mathcal{Z}_i u + Q^{-1} f, \tag{134}$$

where

$$\mathcal{Z}_i = I - Q^{-1} \mathcal{Z}. \tag{135}$$

The resulting algorithm is

$$u^{(n+1)} = \mathcal{Z}_i u^{(n)} + Q^{-1} f. \tag{136}$$

Some simple choices of Q are

$$Q = \frac{\omega}{2 - \omega} (\omega^{-1} D - L) D^{-1} (\omega^{-1} D - U), \tag{137}$$

where ω is the relaxation factor $0 < \omega < 2$ and

$$\mathcal{Z} = D - L - U, \tag{138}$$

where D is the diagonal part of $\mathcal{Z}$, and L and U are the remaining lower and upper triangular matrices. The resulting method is referred to as the symmetric successive over relaxation (SSOR) method. If Q is chosen as

$$Q = (I - L) D^{-1} (I - U), \tag{139}$$

so that

$$\mathcal{Z} = (I - L)(I - U) - LU, \tag{140}$$

then the iterative scheme is

$$u^{(n+1)} = (I - U)^{-1}(I - L)^{-1}LUu^{(n)} + (I - U)^{-1}(I - L)^{-1}f, \tag{141}$$

which is referred to as the forward–backward method (because U and L define source-field point combinations where the source is forward of the field point for U and backward for L) or the method of ordered multiple interractions (MOMI). Finally, the fast multipole method (FMM) uses a factorization of the impedance matrix of the form

$$\mathcal{Z} = \mathcal{Z}' + WTV, \tag{142}$$

where $\mathcal{Z}'$, W, T, and V are chosen to be sparse. Generally $\mathcal{Z}'$ is chosen as representing the nearest (matrix) neighbor interraction. There is also a factorization method which factors the impedance method into the sum of a sparse matrix, a block Toeplitz (flat surface) matrix and a remainder. The method is referred to as the sparse matrix flat surface iterative approach. The factorization, while maintaining its general form, is open to many variants.

Finally, it should be stressed that all these methods can be made to work with some degree of accuracy provided the problem itself is well conditioned. If not, then other more complicated methods must be found such as conditioning approximations, forms of least squares methods involving minimization techniques, or other regularization techniques.

8.6. Discussion

We briefly discuss some of the topics left out in the previous sections. We described very general integral equation methods in these sections but they were confined to scattering and transmission problems with a single deterministic infinite rough surface separating different but still homogeneous media. If the media are not homogeneous, then an additional volume integral term appears and the problems are those of combined surface-volume equations. The dimensionality of the problem thus increases. There are different combinations of this rough boundary plus volume problem. The pure rough surface problem can be mapped into a volume problem. One approach which uses partial differential equations rather than integral equations can be used to replace the rough surface with a rough equivalent layer which contains effective (constant) physical parameters such as density and index of refraction[35]. An example is an optical coating. This approach can also be useful when the problem is propagation in a rough-walled waveguide where the interior physical parameters vary. For these propagation problems what is often done is to replace the Helmholtz equation (an elliptic partial differential equation) with its forward propagating approximation called the parabolic approximation which is a parabolic equation and can be marched in the forward direction where the

Helmholtz equation cannot. Equivalent integral equations are Volterra equations whose iterates are known to converge. The disadvantage is that backscattered fields are ignored, and although this may be a small effect for each individual scattering, over long ranges and many multiple scatterings, this can become significant. An example occurs even when the angle of incidence is near grazing. The volume problem can also become important if the problem to model is an optical layer stack or a rough surface with a volume inhomogeneity buried below it.

We have presented results for infinite surfaces. Results for finite surfaces follow from these in different ways depending on what the word "finite" means. One possibility is to truncate the surface with a tapering of it to zero. The surface is still infinite but is mostly flat, and the flat surface integration can be extended to the full surface and approximated by usual flat surface results. If the surface is actually finite in extent then the diffraction problem simplifies for plane waves, for example, with the exterior diffraction integrals reducing to a plane wave since they are now over a full sphere rather than a hemisphere. The result is diffraction by a finite rough strip and the boundary values are related to the discontinuity of the field across the strip. For periodic surfaces [15] the reduction of the equations is straightforward. Again the plane wave integral terms reduce down to a single plane wave since there is integration only in a single periodic cell, side integrals cancel because of the Floquet periodicity, and the fields become infinite sum expansions in Bragg modes.[22,23] Rigorous methods for these problems can be found in [5, 6]. Within each period the surface could be a random variable, and the periodic surface reduction is the simplest way to treat finite random surfaces. The questions of how to treat surfaces which are "periodic" except for one or a finite number of the "periodic" cells removed, or for a surface "periodic" only over a semi-infinite line are much more difficult.

One of the concerns which increased the interest in rough surface scattering was the phenomenon known as backscatter enhancement.[8] This is an intensity peak in the retroreflection direction. There are two types of peaks which can occur, one narrow in angle which occurs when surface waves are present and the surface is shallow, the other broad in angle which occurs when the rms height is approximately a wavelength. The effect was first observed experimentally.

Finally, there are other approaches to the scattering problem. Transport equation methods have been applied to weakly random surfaces.[3] In the composite model the surface is decomposed into a long-scale part and a small-scale component.[13] The Kirchhoff approximation is used on the former and perturbation theory is used on the latter. Phase perturbation methods such as the Rytov approximation have been extended, as can perturbation methods involving the Green's function.[41] Boundary variation and analytic continuation can also be combined to yield scattering results.[7] There are also methods which directly use differential equations rather than formulate integral equations from them. Feynman diagram methods have also been used[50,14] as well as an expansion of the admittance operator using a Dirichlet-to-Neumann map.[33]

Many of the papers in our references below are themselves review papers which elaborate and extend our discussion above. They are [17, 24, 31, 26, 39, 44, 43], and [47]. These, along with some topical journal issues edited by Maradudin and Nieto-Vesperinas[32] and texts by Beckmann and Spizzichino,[4] Ogilvy,[36] and Voronovich[46] in conjunction with the papers in this book provide a thorough overview of the discipline.

References

1. R.J. Adams and G.S. Brown, "An iterative solution of one-dimensional rough surface scattering problems based on a factorization of the Helmholtz operator," *IEEE Trans. Antennas Propag.* **47**, 765–767 (1999).
2. K.E. Atkinson, *The Numerical Solution of Integral Equations of the Second Kind* (Cambridge University Press, Cambridge, 1997).
3. G. Bal, J.B. Keller, G. Papanicolaou, and L. Ryzhik, "Transport theory for acoustic waves with reflection and transmission at interfaces," *Wave Motion* **30**, 303–327 (1999).
4. P. Beckmann and A. Spizzichino: *The Scattering of Electromagnetic Waves from Rough Surfaces* (Pergamon, New York, 1963).
5. G. Bruckner and J. Elschner, "A two-step algorithm for the reconstruction of perfectly reflecting periodic profiles," *Inverse Problems* **19**, 315–329 (2003).
6. G. Bruckner and J. Elschner, "The numerical solution of an inverse periodic transmission problem," *Math. Methods Appl. Sci.* **28**, 757–778 (2005).
7. O. Bruno and F. Reitich, "Numerical solution of diffraction problems: a method of variation of boundaries," *J. Opt. Soc. Am.* **20**, 1168–1175 (1993).
8. V. Celli, A.A. Maradudin, A.M. Marvin and A.R. McGurn, "Some aspects of light scattering from a randomly rough metal surface," *J. Opt. Soc. Am. A* **2**, 2225–2239 (1985).
9. W.C. Chew, J.-M. Jin, C.-C. Lu, E. Michelssen, and J.M. Song, "Fast solution methods in electromagnetics," *IEEE Trans. Antennas Propag.* **45**, 533–543 (1997).
10. S.K. Cho, *Electromagnetic Scattering* (Springer, Berlin, 1990).
11. D. Colton and R. Kress, *Integral Equation Methods in Scattering Theory* (Wiley, New York, 1983).
12. H. Contopanagos, B. Dembart, M. Epton, J.J. Ottusch, V. Rokhlin, J.L. Visher, and S.M. Wandzura, "Well-conditioned boundary integral equations for three-dimensional electromagnetic scattering," *IEEE Trans. Antennas Propag.* **50**, 1824–1830 (2002).
13. R. Dashen and D. Wurmser, "A new theory for scattering from a surface," *J. Math. Phys.* **32**, 971–985 (1991).
14. J.A. DeSanto, "Green's function for electromagnetic scattering from a random rough surface," *J. Math. Phys.* **15**, 283–288 (1974).
15. J.A. DeSanto, "Scattering from a perfectly reflecting periodic surface: An exact theory," *Radio Sci.* **16**, 1315–1326 (1981).
16. J.A. DeSanto, "Exact spectral formalism for rough-surface scattering," *J. Opt. Soc. Am. A* **2**, 2202–2207 (1985).
17. J.A. DeSanto and G.S. Brown, "Analytical techniques for multiple scattering from rough surfaces," in *Progress in Optics*, vol. XXIII, E. Wolf ed., (North-Holland, Amsterdam, 1986), 1–62.
18. J.A. DeSanto, "A new formulation of electromagnetic scattering from rough dielectric interfaces," *J. Electr. Waves Appl.* **7**, 1293–1306 (1993).
19. J.A. DeSanto and P.A. Martin, "On angular-spectrum representations for scattering by infinite rough surfaces," *Wave Motion* **24**, 421–433 (1996).

20. J.A. DeSanto and P.A. Martin, "On the derivation of boundary integral equations for scattering by an infinite one-dimensional rough surface," J. *Acoust. Soc. Am.* **102**, 67–77 (1997).
21. J.A. DeSanto and P.A. Martin, "On the derivation of boundary integral equations for scattering by an infinite two-dimensional rough surface," J. *Math. Phys.* **39**, 894–912 (1998).
22. J.A. DeSanto, G. Erdmann, W. Hereman, B. Krause, M. Misra, and E. Swim, "Theoretical and computational aspects of scattering from periodic surfaces: two-dimensional perfectly reflecting surfaces using the spectral-coordinate method," *Waves Random Media* **11**, 455–487 (2001a).
23. J.A. DeSanto, G. Erdmann, W. Hereman, B. Krause, M. Misra, and E. Swim, "Theoretical and computational aspects of scattering from periodic surfaces: two-dimensional transmission surfaces using the spectral-coordinate method," *Waves Random Media* **11**, 489–526 (2001b).
24. J.A. DeSanto, "Scattering by Rough Surfaces," in *Scattering*, R. Pike and P. Sabatier, ed. (Academic, New York, 2002), 15–36.
25. J.A. DeSanto and A. Yuffa, "A new integral equation method for direct electromagnetic scattering in homogeneous media and its numerical confirmation," *Waves Random Complex Media* (submitted) (2005).
26. T.M. Elfouhaily and C-A. Guerin, "A critical survey of approximate scattering wave theories from random rough surfaces," *Waves Random Media* **14**, R1–R40 (2004).
27. G.C. Hsiao and R.E. Kleinman, "Mathematical foundations for error estimation in numerical solutions of integral equations in electromagnetics," *IEEE Trans. Antennas Propag.* **45**, 316–328 (1997).
28. A. Ishimaru, *Electromagnetic Wave Propagation, Radiation, and Scattering* (Prentice Hall, Englewood Cliffs, NJ, 1991).
29. D.S. Jones, *Methods in Electromagnetic Wave Propagation* (Clarendon Press, Oxford, 1994).
30. D.A. Kapp and G.S. Brown, "A new numerical method for rough-surface scattering calculations," *IEEE Trans. Antennas Propag.* **44**, 711–721 (1996).
31. A.A. Maradudin and E.R. Mendez, "Scattering by Surfaces and Phase Screens," in *Scattering*, R. Pike and P. Sabatier, ed. (Academic, New York, 2002), 864–894.
32. A.A. Maradudin and M. Nieto-Vesperinas (eds.), Topical issues on wave propagation and scattering from rough surfaces and related phenomena. *Waves Random Media* **7** (3) (1997) and **8** (3) (1998).
33. D.M. Milder, "An improved formulation of coherent forward scatter from random rough surfaces," *Waves Random Media* **8**, 67–78 (1998).
34. N. Morita, N. Kumagai, and J.R. Mautz, *Integral Equation Methods for Electromagnetics* (Artech House, Boston, 1990).
35. J. Nevard and J.B. Keller, "Homogenization of rough boundaries and interfaces," *SIAM J. Appl. Math.* **57**, 1660–1686 (1997).
36. J.A. Ogilvy, *Theory of Wave Scattering from Random Rough Surfaces* (Adam Hilger, Bristol, 1991).
37. J.H. Rose, M. Bilgen, P. Roux, and M. Fink, "Time-reversal mirrors and rough surfaces: Theory," *J. Acoust. Soc. Am.* **106**, 716–723 (1999).
38. M. Saillard and J.A. DeSanto, "A coordinate-spectral method for rough surface scattering," *Waves Random Media* **6**, 135–150 (1996).
39. M. Saillard and A. Sentenac, "Rigorous solutions for electromagnetic scattering from rough surfaces," *Waves Random Media* **11**, R103–R137 (2001).

40. A. Schatzberg and A.J. Devaney, "Rough surface inverse scattering within the Rytov approximation," *J. Opt. Soc. Am. A* **10**, 942–950 (1993).
41. W.T. Shaw and A.J. Dougan, "Green's function refinement as an approach to radar backscatter: General theory and application to LGA scattering from the ocean," *IEEE Trans. Antennas Propag.* **46**, 57–66 (1998).
42. C.J.R. Sheppard: "Imaging of random surfaces and inverse scattering in the Kirchhoff approximation," *Waves Random Media* **8**, 53–66 (1998).
43. A.V. Shchegrov, A.A. Maradudin, and E.R. Mendez, "Multiple scattering of light from randomly rough surfaces," in *Progress in Optics* vol. XLVI, E. Wolf, ed. (Elsevier, Amsterdam, 2004), 117–241.
44. A. Taurin, M. Fink, and A. Derode, "Multiple scattering of sound," *Waves Random Media* **10**, R31–R60 (2000).
45. P. Tran, V. Celli, and A.A. Maradudin, "Electromagnetic scattering from a two-dimensional randomly rough, perfectly conducting surface: iterative methods," *J. Opt. Soc. Am.* A **11**, 1686–1689 (1994).
46. A.G. Voronovich, *Wave Scattering from Rough Surfaces*, 2nd edn (Springer, Berlin Heidelberg, New York, 1999).
47. K.F. Warnick and W.C. Chew, "Numerical simulation methods for rough surface scattering," *Waves Random Media* **11**, R1–R30 (2001).
48. R.J. Wombell and J.A. DeSanto, "The reconstruction of shallow rough-surface profiles from scattered field data," *Inverse Problems* **7**, L1–L12 and 663 (Corrigenda) (1991a).
49. R.J. Wombell and J.A. DeSanto, "Reconstruction of rough-surface profiles with the Kirchhoff approximation," *J. Opt. Soc. Am.* A **8**, 1892–1897 (1991b).
50. G.G. Zipfel and J.A. DeSanto, "Scattering of a scalar wave from a random rough surface: A diagrammatic approach," *J. Math. Phys.* **13**, 1903–1911 (1972).

9
Experimental Studies of Scattering from Weakly Rough Metal Surfaces

K. A. O'DONNELL

División de Física Aplicada, Centro de Investigación Científica y de Educación Superior de Ensenada, Apartado Postal 2732, Ensenada, Baja California, 22800 México

9.1. Introduction

The surface of a weakly rough metal, with height fluctuations of a few nanometers, can produce remarkably strong and unusual optical effects under appropriate conditions. If the random roughness of a metal surface allows an incident light wave to launch surface plasmon polaritons, the diffuse scatter emitted by the surface will receive contributions when these surface waves are subsequently scattered from the surface. Under such conditions, it has been predicted that effects like backscattering enhancement may appear in the mean diffuse scatter emitted by the surface.[1,2] These theoretical works use sophisticated methods to account for multiple scattering processes involving plasmon–polariton excitation. In particular, the incident light wave may be roughness-coupled to surface waves, which may themselves be scattered many times within the surface, to finally be roughness-coupled out of the surface so as to contribute to the diffusely scattered light. Other theoretical works have used direct perturbation[3] or Monte Carlo[4] techniques to study such effects. This line of research has been extended to include theoretical studies of angular correlation functions[5] and of the generation of diffuse second harmonic light from weakly rough metals.[6,7]

There has been a shortage of related experimental works. Certainly, there has been considerable experimental effort directed toward scattering from the weak residual roughness of polished optical surfaces.[8] However, the intent of the work has often been to characterize the roughness or to better understand the various means of polishing surfaces. In any case, effects of plasmon–polariton excitation are not commonly seen in such work even for metallic optical surfaces, presumably because the surface roughness is not appropriate to produce significant surface wave excitation. There have also been many experiments done with strongly rough metal surfaces and, even though backscattering enhancement is sometimes observed, these effects are entirely unrelated to polariton excitation.[9] On the other hand, there have indeed been experimental observations of diffuse scatter arising from plasmon–polariton excitation on metallic gratings[10] or metal-coated coupling prisms[11] having incidental roughness. However, here the excitation is that of a specific surface wave mode, which occurs only for a particular illumination geometry.

This physical situation is very different from that of a randomly rough free space/metal interface where, over a wide range of incidence and scattering angles, the roughness simultaneously provides the coupling mechanisms for excitation, de-excitation, and multiple scattering of surface waves.

The purpose of this chapter is to describe experimental studies of the consequences of plasmon–polariton excitation on rough metal surfaces. It is possible that the lack of relevant experiments is due to difficulties in fabricating suitable surfaces; thus the work described here begins in Sect. 9.2 with a discussion of the essential surface wave coupling mechanisms and the lithographic surface fabrication methods employed to produce them. The fabrication method produces highly one-dimensional surface structures, with Gaussian height statistics and root-mean-square roughness of a few nanometers. The power spectrum of the roughness is of rectangular form and, to produce the effects of interest, covers a range of wavenumber that includes that of the surface plasmon polariton.

Section 9.3 presents results for the mean diffuse intensity scattered by these surfaces, which exhibit backscattering enhancement under a variety of conditions. This effect occurs only for p polarization, as expected for effects related to plasmon–polariton excitation on one-dimensional surfaces. It is seen that the rectangular spectrum allows one to either produce or suppress the backscattering peak and its associated distribution, according to the roughness couplings allowed by the bandwidth of the rectangular spectrum. Section 9.4 considers experiments in which the nonlinear response of the metal surface produces a diffuse scattering distribution of second-harmonic light. Here the rectangular spectrum is again useful in indicating the origin of the features seen in the distributions. With the power spectrum of the roughness centered on the wavenumber of the second-harmonic plasmon polariton, a backscattering effect is observed in the diffuse second-harmonic light. With the spectrum centered on the wavenumber of the fundamental plasmon–polariton, the effects are stronger and the observed features are attributed to a variety of nonlinear wave interactions. Finally, Sect. 9.5 returns to linear optical effects and considers the angular correlation functions of intensity scattered by the rough surface. Two distinct types of angular correlations are observed and the effects related to plasmon–polariton excitation are studied.

9.2. Experimental Methods

9.2.1. Essential Couplings

The lowest-order scattering processes producing polariton-related backscattering enhancement are shown in Fig. 9.1. The incident light wave of frequency ω is incident at angle θ_i upon a rough surface having dielectric constant $\varepsilon = \varepsilon_1 + i\varepsilon_2$. In path A of Fig. 9.1, it launches a plasmon polariton at point 1 traveling to, for example, the right along the surface. This polariton then reaches point 2, where it is scattered by the roughness to produce diffuse light escaping from the surface at an angle θ_s. The time-reversed process may also occur (path B), in which the

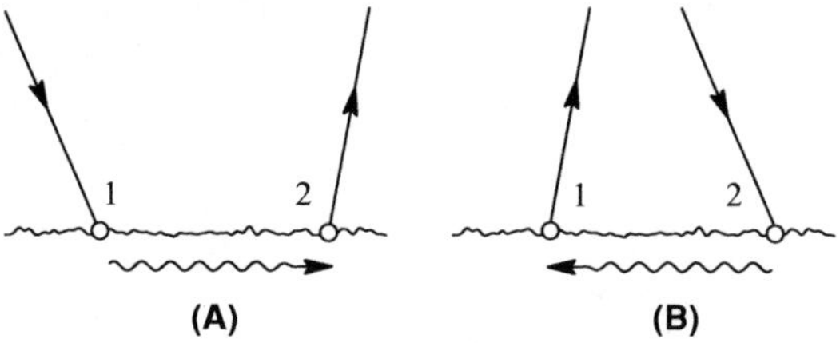

FIGURE 9.1. Multiple scattering paths on a rough metal surface.

two scattering events occur in the opposite order. Both paths share the same phase delay suffered by the surface waves; the phase difference $\Delta\phi$ between paths A and B instead arises in the free space and it is a straightforward geometrical exercise to show that $\Delta\phi = (\omega/c)(\sin\theta_i + \sin\theta_s)\Delta x$, where Δx is the distance along the mean surface from point 1 to point 2. It is obvious that $\Delta\phi$ vanishes at backscattering ($\theta_s = -\theta_i$) and the two contributions will interfere constructively and produce a backscattering peak. Said differently, paths A and B both produce broad scattering contributions in the upper half-space; these contributions are guaranteed to interfere constructively at backscattering, while in other directions they will, in general, interfere randomly. It is also notable that this same argument applies even if the surface waves are scattered an arbitrarily large number of times within the surface. Thus the initial surface wave launched from point 1 may be scattered into a long sequence of other surface waves before it reaches point 2, but for any such path 1→2 there will always be a time-reversed partner 2→1 for which the the phase difference will vanish at backscattering.

Although the previous arguments treat the couplings as occurring at points of the surface, it is useful to consider these couplings in terms of the wavenumbers k_r present in the power spectrum $G(k)$ of the random roughness. For example, the initial launching of surface waves traveling to the right and left may be expressed by the $\pm$1st-order grating equations for diffraction from roughness wavenumbers

$$+k_{\text{sp}} = k + k_{r1} \tag{1}$$

and

$$-k_{\text{sp}} = k - k_{r2}, \tag{2}$$

where $k = (\omega/c)\sin\theta_i$, k_{r1} and k_{r2} are positive roughness wavenumbers, and $\pm k_{\text{sp}} = \pm(\omega/c)\sqrt{\varepsilon_1/(\varepsilon_1 + 1)}$ are the wavenumbers of plasmon polaritons traveling to the right (+) or left (−). Similarly, with the plasmons $\pm k_{\text{sp}}$ now excited, they may themselves be scattered to a propagating scattered wave $q = (\omega/c)\sin\theta_s$ through two more coupling equations

$$q = +k_{\text{sp}} - k_{r3}, \tag{3}$$

and

$$q = -k_{\text{sp}} + k_{r4}, \tag{4}$$

where k_{r3} and k_{r4} are two more positive wavenumbers present in $G(k)$, and the $\mp$1st orders are chosen in Eqs. (3) and (4) to couple the evanescent waves $\pm k_{sp}$ to propagating waves. Equations (1)–(4) are indeed all couplings necessary to produce the backscattering effect.

It is obvious from inspection that, for small angles (i.e., small q and k), Eqs. (1)–(4) predict that all k_r's are near k_{sp}. More generally, for $|\theta_i|$ and $|\theta_s|$ less than some angle θ_{max}, it is readily shown that all couplings of Eqs. (1)–(4) are satisfied by a continuous range of roughness wavenumbers within $\pm(\omega/c)\sin\theta_{max}$ of k_{sp}.

9.2.2. Rectangular Spectra

The early theoretical approaches assumed that $G(k)$ was of Gaussian form, centered on zero wavenumber, and that the tail of $G(k)$ near k_{sp} was high enough to produce the polariton coupling just described.[1–4] However, even using sophisticated lithographic techniques, it has not been possible to fabricate such a spectrum and no comparable experimental results have appeared.

Fortunately, there is a different approach that has made experimental studies possible.[12] It has just been made clear that, for angles less than some θ_{max}, it is sufficient to have a spectrum $G(k)$ that is nonzero in a region of full width $2(\omega/c)\sin\theta_{max}$, centered on k_{sp}. Indeed, the simplest possibility is that $G(k)$ could be of constant height within this region and zero elsewhere.

It is possible to fabricate surfaces having this rectangular spectrum using lithography of photoresist. In particular, by exposing a photoresist-coated plate to many sinusoidal intensity patterns, with each pattern having a different wavenumber and phase, it has been possible to construct a satisfactory random surface profile. The mathematical justification of this procedure invokes a central limit theorem due to Rice[13] as follows. Photoresist responds by producing a surface relief $h(x)$ proportional to its total exposure and, assuming that a plate receives N sequential exposures, incoherently summed, we have

$$h(x) = \alpha \sum_{n=1}^{N} I_n[1 + M_n \cos(k_n x - \phi_n)]\Delta t_n\,, \tag{5}$$

where n denotes the nth exposure, I_n is the spatially averaged intensity, M_n is the modulation, k_n is the exposure wavenumber, ϕ_n is the phase, Δt_n is the exposure time, and α characterizes the material response. Assuming that ϕ_n is a random variable uniformly distributed on the range $(0, 2\pi)$, it is straightforward to show that

$$\langle h(x)\rangle = \alpha \sum_{n=1}^{N} I_n \Delta t_n\,, \tag{6}$$

so that

$$\Delta h(x) = \sum_{n=1}^{N} c_n \cos(k_n x - \phi_n)\,, \tag{7}$$

where $\langle\cdot\rangle$ denote an ensemble average, $\Delta h(x) \equiv h(x) - \langle h(x)\rangle$, and $c_n = \alpha I_n M_n \Delta t_n$. As discussed by Rice,[13] if $k_n = n\Delta k$, and if c_n is specified by

$$c_n = \sqrt{2\,G(k_n)\Delta k}\,, \tag{8}$$

it may be shown that $\Delta h(x)$ has the power spectrum $G(k)$ and Gaussian height statistics as $N \to \infty$ and $\Delta k \to 0$. Thus Rice's theorem is a method of constructing a random process $\Delta h(x)$ from its randomly-phased Fourier components, with appropriate amplitude weighting to produce the desired spectrum $G(k)$. The convergence to Gaussian statistics can be quite rapid and, for the case of interest here with c_n independent of n, $\Delta h(x)$ follows first-order statistics that appear Gaussian for N as small as 10.[14]

In the fabrication experiments, surfaces were each given 500 to 8000 sinusoidal exposures, which required a computer-controlled system. The photoresist plate was exposed to the sinusoidal interference pattern in the intersection of two beams of wavelength 442 nm produced by a HeCd laser, in a geometry similar to those used for holographic grating fabrication.[10] The maximum wavenumber $k_{\max}$ desired in the exposure series was obtained with the two exposing beams arriving at equal but opposite angles with respect to the plate normal. The minimum wavenumber required was produced by rotating the plate by an angle θ so that the two beams arrived asymmetrically with respect to the normal; the resulting wavenumber was $k_{\min} = k_{\max}\cos\theta$. Intermediate exposure wavenumbers were obtained by rotating the plate to intermediate angles under computer control. The random phase ϕ_n of Eq. (7) was produced by a glass wedge in one exposing beam which, under computer control, was positioned randomly for each exposure. A typical exposure sequence lasted 2 to 10 h, depending on the number of exposures. The surface standard deviation σ obtained varied between 3 and 30 nm, with σ decreasing with increasing N as the total exposure contrast was reduced.

There are a number of subtleties that are required to produce high quality surfaces with this technique. First, photoresist has high contrast and there is negligible response for low exposures followed by a critical region where there is a steeply sloped response with increasing exposure, which ends when all photoresist is removed from the substrate. Thus care was taken to place the mean exposure at the center of the steeply sloped response region. With uniform exposing times, the resulting spectrum $G(k)$ was found to be somewhat sloped; this was corrected by empirically varying Δt_n with wavenumber so as to flatten $G(k)$. Another concern is that surfaces produced with this method are periodic, as is clear from the Fourier series of Eq. (7), which would limit the spatial averaging for the reduction of speckle noise in the experiments. The periodicity was removed by randomly placing each exposure wavenumber in a uniformly probable range within Δk of its nominal value. Further, the surface one-dimensionality was slightly reduced by rotating the plate about its normal to a random angle (typically $<0.01°$) for each exposure, so as to produce a 1 mm correlation length in the direction in which the surface would have been of constant height. With these techniques, surfaces typically had a 35×25 mm area available for speckle averaging in the diffuse scattering experiments.

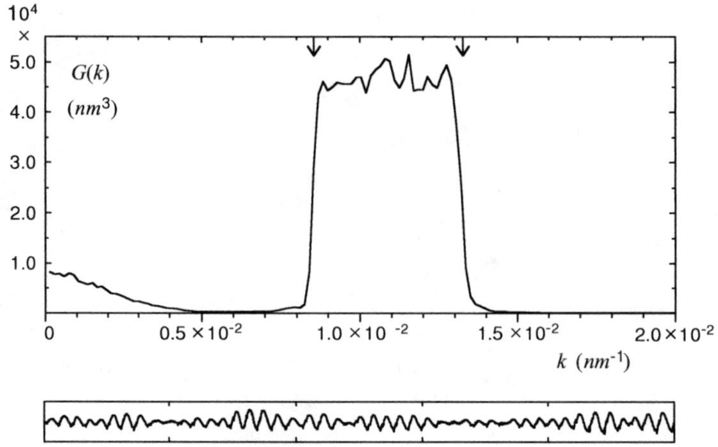

FIGURE 9.2. Upper plot: surface power spectrum $G(k)$ determined from profilometry of an experimental sample, normalized to have area σ^2. The arrows denote the desired $k_{\min}$ and $k_{\max}$. Lower plot: 25 μm segment of surface profilometer data; the vertical scale extends ± 70 nm.

At this point, the photoresist plate was coated with an optically thick layer of gold or silver using standard vacuum coating techniques. The surface was then measured with a Talystep mechanical profilometer and the surface height data were processed to determine $G(k)$. An example of such data is shown in Fig. 9.2, where a rectangular spectrum is clearly present. It is also seen that the surface profile resembles a sinusoid having a randomly modulated envelope, as expected for a narrowband random process.

9.2.3. Scattering Measurements

For a highly one-dimensional surface that is illuminated with a light wave perpendicular to the grooves of the surface, all diffuse scatter remains in the plane of incidence. The samples employed here were of sufficient quality that all diffuse scatter remained within 0.10° of the plane of incidence. The sample was mounted on a rotation stage to set the angle of incidence; a detector arm of length 70 cm was positioned by a concentric rotation stage to determine the angular dependence of the diffuse scatter. Both rotation stages were computer-controlled. A few mm diameter area of the rough surface was illuminated by the laser beam which, after reflection from the sample, focused at the position of the detector's field lens. Several low-power lasers were used as sources in the work of Sect. 9.3, while a pulsed laser was used to produce second harmonic generation in Sect. 9.4.

The detector's collection angle was typically set at 0.5° full width. This was found to resolve the narrow structures observed while being large enough to reduce speckle noise. To further reduce speckle noise, the sample was moved under computer control in a raster fashion while the detector signal was averaged to provide each plotted data point. Considerable effort has also been made to provide absolute normalization of all scattering data using procedures discussed elsewhere.[9] For an identically polarized incident wave of unit power, the diffuse intensity represents scattered power per unit planar angle in radians. All plotted results for diffuse scatter also contain a gap within 1° of the specular angle, where data could not be taken due to the glare of the specular reflection. Finally, scattering data have been taken through what appears in results as the backscattering direction by tilting the surface slightly so that the incident wave was no longer perpendicular to the surface groove direction. This small tilt (0.5°) was just sufficient for the scattered light to clear the final mirror of the illumination optics, and in cases checked no noticeable effect on the data has been found.

9.3. Experimental Results

9.3.1. Ideal Spectrum

Results are now presented for light scattering in the optimal case in which k_{sp} is centered in the rectangular region of $G(k)$.[12] The source employed was a HeNe laser of wavelength $\lambda = 612$ nm. The surface used was prepared in gold and from the dielectric constant it is estimated that $k_{sp} = 1.059\,(\omega/c) = 1.09 \times 10^{-2}\,\text{nm}^{-1}$. The surface was designed for a maximum coupling angle of $\theta_{max} = 13.5^\circ$, from which it follows that $k_{min} = 8.5 \times 10^{-3}\,\text{nm}^{-1}$ and $k_{max} = 1.33 \times 10^{-2}\,\text{nm}^{-1}$. Profilometry of the surface revealed a satisfactory rectangular spectrum and $\sigma = 10.9$ nm as the standard deviation of surface height. The skewness and kurtosis of the surface height are, respectively, -0.30 and 3.0, which are close to those of an ideal Gaussian variate (0 and 3).

The diffuse intensities $\langle I_p(\theta_s|\theta_i)\rangle$ and $\langle I_s(\theta_s|\theta_i)\rangle$ in p and s polarization are shown in Fig. 9.3 for three values of θ_i. First, in these results there is a considerable amount of diffuse scatter directed to large $|\theta_s|$. We attribute this to single scatter from the surface roughness consistent with the grating equation

$$q = k + n\,k_r, \tag{9}$$

where $q = (\omega/c)\sin\theta_s$, $k = (\omega/c)\sin\theta_i$, $n = \pm 1$, and k_r is a roughness wavenumber falling between k_{min} and k_{max}. With $n = 1$, Eq. (9) is consistent with diffuse scatter for $\theta_s \geq 56^\circ$ when $\theta_i = 0^\circ$, but all of these scattered modes become evanescent for $\theta_i \cong 10^\circ$ and remain so for larger θ_i. With $n = -1$, Eq. (9) predicts scattering for $\theta_s \leq 56^\circ$ when $\theta_i = 0^\circ$, $\theta_s \leq 41^\circ$ when $\theta_i = 10^\circ$, and for θ_s within the range $(-79^\circ, -31^\circ)$ when $\theta_i = 18^\circ$. All of these considerations are closely consistent with the behavior of $\langle I_p(\theta_s|\theta_i)\rangle$ and $\langle I_s(\theta_s|\theta_i)\rangle$ at large $|\theta_s|$ in Fig. 9.3. It is also worth noting that for $\theta_i = 0^\circ$ and 10°, the evanescent modes coupled to include the polariton resonances at $\pm k_{sp}$ as described by Eqs. (1) and (2).

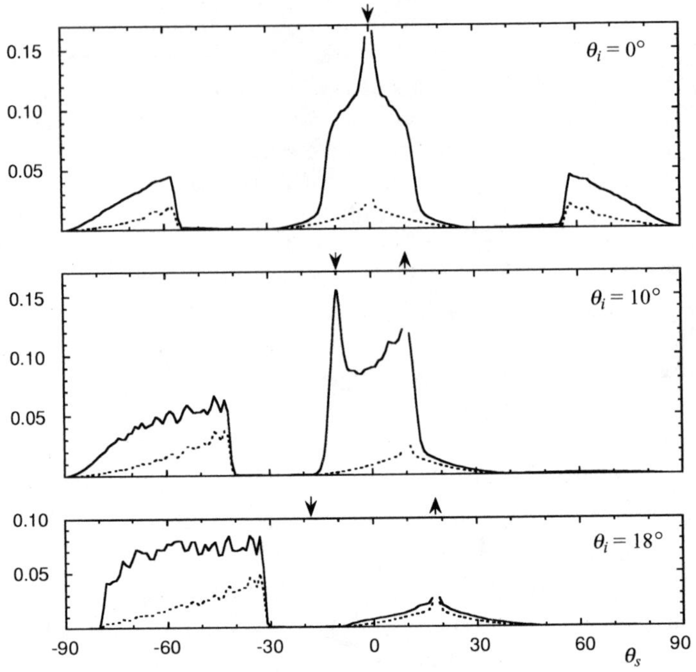

FIGURE 9.3. Scattering distributions $\langle I_p(\theta_s|\theta_i)\rangle$ (solid curves) and $\langle I_s(\theta_s|\theta_i)\rangle$ (dashed curves) for θ_i as noted. The arrows denotes the backscattering direction (downward arrows) and specular direction (upward arrows). The specular reflection is not shown.

Significantly different behavior is apparent in Fig. 9.3 at small $|\theta_s|$. For $\theta_i = 0°$ and $10°$, $\langle I_p(\theta_s|\theta_i)\rangle$ presents a compact shape with angular limits that remain fixed at $\theta_s = \pm 13°$, and a narrow peak of nearly twice the height of the remainder of this distribution appears at the backscattering position. However, at $\theta_i = 18°$ this scattering distribution has disappeared in $\langle I_p(\theta_s|\theta_i)\rangle$ and a much lower level of scatter is seen near the specular angle that is only slightly higher than $\langle I_s(\theta_s|\theta_i)\rangle$. In fact, $\langle I_s(\theta_s|\theta_i)\rangle$ always presents a low curve that does not change form significantly, and only proceeds into the direction of forward-scattering with increasing θ_i.

The behavior at small $|\theta_s|$ is consistent with backscattering enhancement in $\langle I_p(\theta_s|\theta_i)\rangle$ involving the surface polaritons. In particular, $G(k)$ has been constructed to allow the polaritons to outwardly couple to θ_s within angular limits at $\pm 13.5°$, for all θ_i allowing polariton excitation. The resulting distribution should remain within these fixed angular limits as θ_i is varied and, as seen in Fig. 9.3, may thus be easily distinguished from the motion of the single scatter consistent with Eq. (9). It appears likely that the low scattering levels seen in $\langle I_s(\theta_s|\theta_i)\rangle$ near the specular direction, as well as in $\langle I_p(\theta_s|\theta_i)\rangle$ for $\theta_i = 18°$ where the inward polariton coupling has broken down, correspond to single scattering (according to Eq. (9)) from the low levels in $G(k)$ at small wavenumber apparent in Fig. 9.2. The

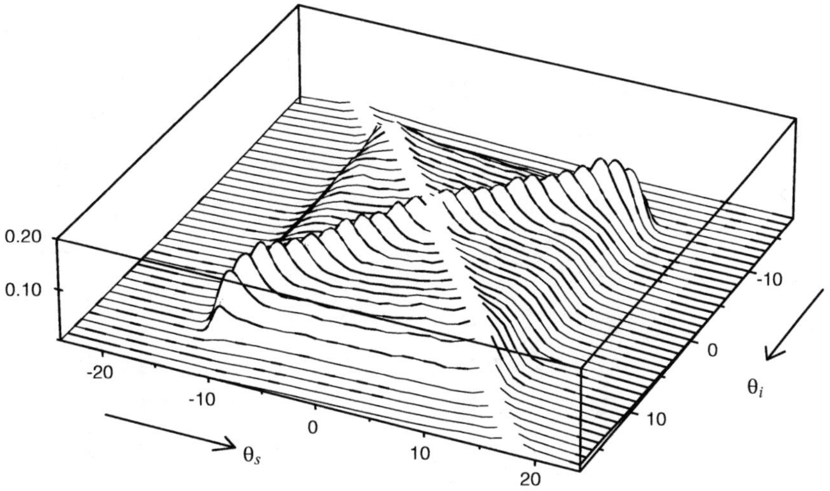

FIGURE 9.4. $\langle I_p(\theta_s|\theta_i)\rangle$ for $\lambda = 612$ nm and angles (θ_i, θ_s) as shown. Backscattering enhancement persists at $\theta_s = -\theta_i$. The break in each curve at $\theta_s = \theta_i$ surrounds the specular reflection that rises above the scale of the plot.

absolute level of $\langle I_p(\theta_s|\theta_i)\rangle$ in the region attributed to polariton excitation is also considerably stronger here than in theoretical studies.[1–4] It is likely that this strong scatter arises from the narrow width of $G(k)$, as the area of $G(k)$ within the width of the polariton resonance is far greater than would be found for a broad Gaussian spectrum with identical σ.

The claim that the observed effects arise from polariton excitation receives more support from more complete data sets. Figure 9.4 shows the diffuse intensity $\langle I_p(\theta_s|\theta_i)\rangle$ for all (θ_i, θ_s) within $(\pm 18°, \pm 24°)$. There is a remarkable square region in the (θ_i, θ_s) plane where $\langle I_p(\theta_s|\theta_i)\rangle$ abruptly rises from low levels. The boundaries of this region fall at θ_i or θ_s equal to $\pm 13°$ where the polariton coupling limits are expected; this indicates that *nearly all* of the diffuse scatter seen in Fig. 9.4 arises through the mechanisms involving the polariton described by Eqs. (1)–(4). A weak backscattering peak is visible in the scans at $\theta_i = \pm 13°$, and this peak quickly rises to become quite distinct for all intermediate θ_i. The power contained in the diffuse intensity $\langle I_p(\theta_s|\theta_i)\rangle$ for the range of θ_s shown in Fig. 9.4 is 0.0071 at $\theta_i = 18°$, although this rises to 0.026 at $\theta_i = 13°$, and then remains nearly fixed at 0.050 for $|\theta_i| \leq 11°$.

9.3.2. Detuned Case

The case is now considered when $G(k)$ still covers k_{sp}, but is no longer centered on it.[15] To achieve this, the values of k_{min} and k_{max} quoted earlier were kept, but the light source was now tuned to a frequency ω other than the optimal one, which

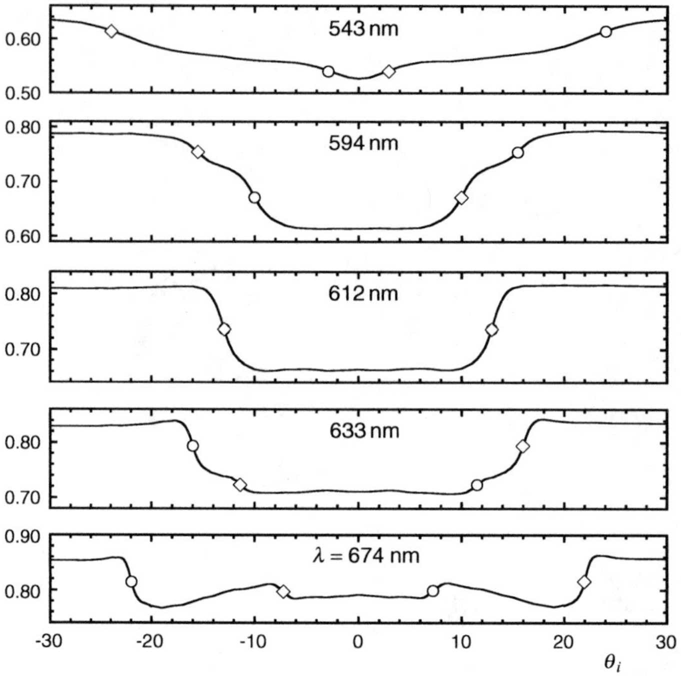

FIGURE 9.5. For unit incident power, the power of the specular reflection for the wavelengths λ indicated. The steeply sloped regions arising from the excitation of $+k_{sp}$ and $-k_{sp}$ are denoted, respectively, by circles and by diamonds. These regions overlap for $\lambda = 612$ nm.

moves k_{sp} away from the center of the rectangle of $G(k)$. The surface used here was fabricated in gold with $\sigma = 15.5$ nm and is indeed the sample of Fig. 9.2.

Figure 9.5 shows results for the surface's specular reflection in p polarization, which illustrate the inward couplings of Eqs. (1) and (2). For $\lambda = 674$ nm, a sharp decrease in the reflection appears at $\theta_i \cong -22.1°$. We associate this with the onset of the excitation of $+k_{sp} = 1.045\,(\omega/c)$ via k_{max} as in $+k_{sp} = (\omega/c)\sin\theta_i + k_{max}$. A second decrease occurs at $\theta_i \cong -7.3°$, which is associated with excitation of $-k_{sp} = -1.045\,(\omega/c)$ via k_{min} as in $-k_{sp} = (\omega/c)\sin\theta_i - k_{min}$. As θ_i increases further, there is a flat region where both $+k_{sp}$ and $-k_{sp}$ remain excited via wavenumbers in $G(k)$ between k_{min} and k_{max} until, symmetrically, at $\theta_i \cong 7.3°$ the excitation of $+k_{sp}$ ceases, and finally the excitation of $-k_{sp}$ breaks down at $\theta_i \cong 22.1°$. As λ decreases in Fig. 9.5, the excitation regions of $+k_{sp}$ and $-k_{sp}$ overlap more closely (633 nm), appear to coincide (612 nm, the optimum case of Figs. 9.3–9.4), and withdraw from one another (594 nm) until the flat region that indicates simultaneous $\pm k_{sp}$ excitation has disappeared (543 nm).

It is of interest to consider the diffuse scatter in one of the cases of Fig. 9.5 when the counterpropagating surface wave excitation is not necessarily simultaneous. Particularly interesting behavior is exhibited by $\langle I_p(\theta_s|\theta_i)\rangle$ with $\lambda = 674$ nm, for which results are shown in Fig. 9.6 for small angles. It is quite clear that two

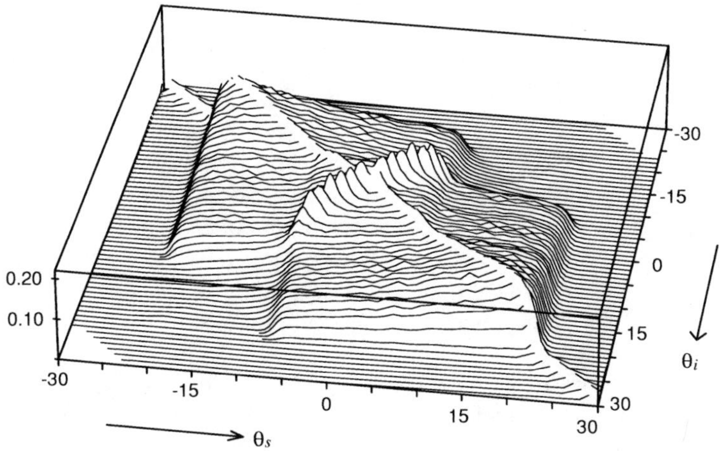

FIGURE 9.6. $\langle I_p(\theta_s|\theta_i)\rangle$ for $\lambda = 674$ nm and angles (θ_i, θ_s) as shown. The open region along the plot's diagonal for $\theta_s = \theta_i$ surrounds the specular reflection; backscattering enhancement is apparent as the peak in the central region for $\theta_s = -\theta_i$.

scattering distributions in the (θ_i, θ_s) plane are interacting with one another. One distribution apparently arises via $+k_{sp}$ because it is constrained to θ_i and θ_s within the range $(-22°, 7°)$, while a similar distribution arises via $-k_{sp}$ and is constrained to θ_i and θ_s within $(-7°, 22°)$. In the overlap of the two distributions a slightly raised region is seen ($|\theta_i| \leq 7°$ and $|\theta_s| \leq 7°$); it is only there that backscattering enhancement is clearly seen. The transition between regions is remarkably abrupt (occurring over roughly 3°) apparently because of the narrow polariton resonance and the sharp limits of $G(k)$.

The essential qualities of the backscattering effect are immediately obvious in this plot. First, when polariton-related scatter is insignificant (for $|\theta_i|$ or $|\theta_s| > 22°$), $\langle I_p(\theta_s|\theta_i)\rangle$ immediately falls to low levels. There are regions that receive contributions from surface waves traveling in only one direction (for example, with either $10° < \theta_i < 22°$ or $10° < \theta_s < 22°$, only $-k_{sp}$ contributes significantly) where scattering levels may be large but backscattering enhancement is not apparent. Only in the region of overlap, where contributions from $\pm k_{sp}$ are present, is backscattering enhancement clearly seen. It is of interest to consider that tuning λ has the effect of varying the amount of overlap between the two rectangular distributions present in Fig. 9.6; in the ideal case of Fig. 9.4 with $\lambda = 612$ nm, the two distributions have overlapped perfectly.

9.4. Second Harmonic Generation

It has been known for many years that a flat metal surface can produce low levels of second-harmonic light in a specularly reflected laser beam. Even though the bulk of an evaporated metal has inversion symmetry, there are contributions to the second-harmonic response that arise both from the surface and from the bulk of the medium.[16]

There are two pioneering theoretical studies that have considered related effects for a weakly rough metal in which a diffuse scattering distribution arises in the second harmonic.[6,7] Both works employed Gaussian roughness spectra with one-dimensional surface roughness. In a broad distribution of second-harmonic diffuse light, a pair of narrow peaks have been predicted that arise from plasmon–polariton excitation at the fundamental frequency.[6,7] Including multiple scattering effects, two additional peaks have been predicted that appear in the backscattering direction and in the direction normal to the surface.[7] The latter two peaks are of particular interest here because they are related to the effects of Sect. 9.3.

There have been few experimental works that have considered second-harmonic generation from a free space/metal interface. The reason for the lack of experiments is probably that detected signal levels have generally been too weak to provide meaningful results. There have been some works that have employed coupling prisms to enhance the effects;[17] however, as mentioned in Sect. 9.1 these systems are physically very different from a simple free space/metal interface. On the other hand, the surfaces having rectangular spectra employed in Sect. 9.3 produce unusually strong plasmon–polariton coupling, thus producing strong surface fields that may compensate for the weak surface nonlinearity. Still, it is to be anticipated that signal levels will be low.

The experimental geometry was as described in Sect. 9.2.3, but the laser source and the detection techniques were considerably different. A Nd:YAG laser/regenerative amplifier system produced pulses of wavelength 1064 nm, pulse length 100 ps, peak power 10^7 W, and repetition rate 1 kHz. The slightly convergent incident beam illuminated a 2 mm width of the sample and, as before, the sample was scanned to reduce speckle noise. The light scattered by the surface passed through an infrared-absorbing filter and a 532 nm interference filter, and was finally focused by a field lens onto a photon-counting photomultiplier. The incident laser beam was set to p polarization in the cases presented here, and it was verified that the second-harmonic scattering distributions were also p polarized. An electronic counter was gated to accept photoelectric counts within a 5 ns window coincident with each laser pulse and other photocounts were rejected. A number of procedures were performed to test the validity of the detected signal. For example, scanning the detector above the plane of incidence reduced the photocount rate to well under 1 s^{-1}. Similar effects were observed if the counter gate was set a few ns away from its synchronized position, or if the wavelength of the interference filter was changed slightly. Further, it was verified that the photocount rate depended quadratically on the fundamental laser power incident on the metal surface, as expected for second-harmonic generation.

The rectangular roughness spectrum may be centered on the polariton wavenumber at either the laser's frequency ω or the harmonic frequency 2ω. As will be seen in Sects. 9.4.1 and 9.4.2, these two cases provide quite different results. Most importantly, neither the backscattering peak nor the peak in the direction of the surface normal are present in the measurements presented here, and other unexpected phenomena appear.

9.4.1. Roughness Spectrum Centered on $k_{sp}^{2\omega}$

Results are now presented for the case with the harmonic plasmon wavenumber $k_{sp}^{2\omega}$ centered in the rectangular region of $G(k)$.[18] The surface was fabricated in silver, and we estimate from the dielectric constant that $k_{sp}^{2\omega} = 1.052\,(2\omega/c) = 1.24 \times 10^{-2}\,\text{nm}^{-1}$. The surface was designed for a maximum coupling angle of $\theta_{max} = 12.2°$, from which it follows that $k_{min} = 9.9 \times 10^{-3}\,\text{nm}^{-1}$ and $k_{max} = 1.49 \times 10^{-2}\,\text{nm}^{-1}$. Mechanical profilometry of the surface revealed a satisfactory rectangular spectrum with $\sigma = 11.1$ nm.

This surface should be capable of producing the couplings required for the backscattering peak predicted in [7]. The scattering mechanisms are as follows. First, the incident light wave of frequency ω interacts with the nonlinearity of the metal to produce a field at 2ω. The 2ω field is then scattered by the roughness to excite counterpropagating surface plasmon polaritons, which are then scattered by the roughness to emerge from the surface as propagating waves. As in the linear case of Sect. 9.3.1, the two distinct scattering processes involving the right- and left-traveling polaritons interfere constructively in the backscattering direction, producing the peak in the diffuse scatter. The nonlinearity is thus required only for the first surface interaction; the subsequent scattering processes are identical to those that would occur with the surface illuminated at 2ω.

The principal results for the 2ω scattering distribution $\langle I^{2\omega}(\theta_s|\theta_i)\rangle$ are shown in Fig. 9.7. It is seen that the distributions indeed resemble the results in the linear case of Fig. 9.3; this observation supports the statement that the situation should be similar to illuminating this surface at 2ω. In particular, the single scattering for large $|\theta_s|$ is similar to that of Fig. 9.3. Further, the polariton-related scatter is present for $|\theta_s| \leq \theta_{max}$ in Fig. 9.7 with $\theta_i = 3°$ and $10°$; this distribution again disappears as it should for $\theta_i = 17°$ as the polariton coupling ceases. However, there is also a striking difference between the linear and second-harmonic cases. In particular, instead of the backscattering peak of the linear case, the second-harmonic case produces remarkable minima there in Fig. 9.7. It is seen that the photon rates are quite low ($10\,\text{s}^{-1}$), but the signal appeared to be well isolated using the techniques described earlier. Considerable care has been taken in the normalization of the data and, while the procedure is lengthy, we estimate the absolute scale to be correct within a factor of 2. The area of each curve of Fig. 9.7 is approximately $5 \times 10^{-23}\,\text{cm}^2/\text{watt}$, which is far less than the specular second harmonic signal from a flat silver surface at oblique θ_i.[16]

Thus the striking result is the distinct backscattering minimum that is the opposite of the theoretical prediction. Our conditions differ from those of [7] which assume that $G(k)$ is of Gaussian form, and show backscattering results only for larger θ_i (35° and 45°). However, it is unclear how these differences could be

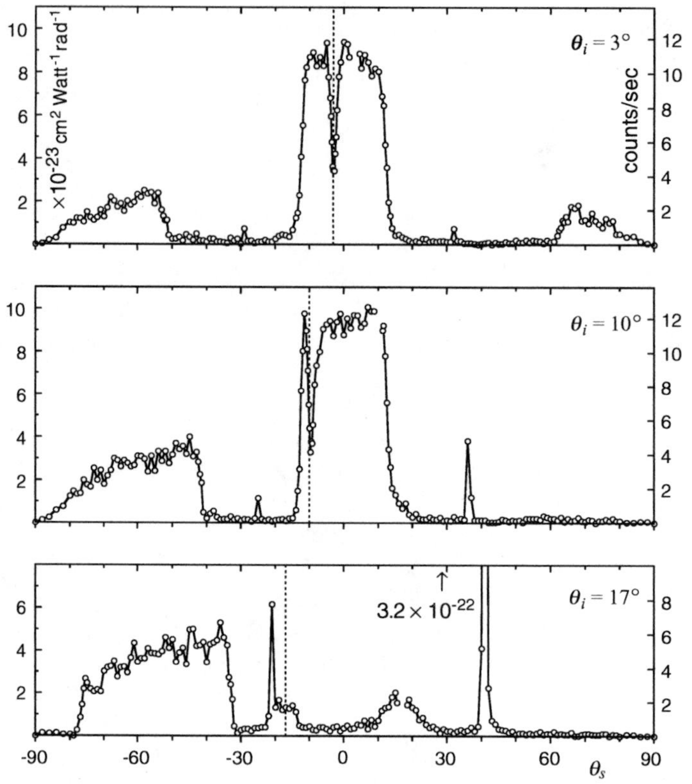

FIGURE 9.7. For the incident angles θ_i shown, the second-harmonic scattering distribution $\langle I^{2\omega}(\theta_s|\theta_i)\rangle$ in the case with $G(k)$ centered on the harmonic polariton wavenumber $k_{\rm sp}^{2\omega}$. Dashed vertical lines indicate the backscattering direction.

significant. Certainly, their term producing the backscattering effect (the last term of their Eq. (61)) is positive definite and could not produce a minimum. Monte Carlo calculations have been reported for a slightly wider rectangular spectrum;[19] a backscattering minimum appeared for $\theta_i = 0°$, which became a minimum of reduced depth for $\theta_i = 3°$ and finally a peak with $\theta_i = 15°$. There is thus some similarity to the experimental results, although there is little sign of variation of the depth of the minima in Fig. 9.7. Other theoretical results[17] even show backscattering peaks for non-normal incidence for one model of the surface nonlinearity, although calculations for another model exhibit persistent backscattering peaks for all θ_i shown and is thus consistent with Fig. 9.7.

These calculations demonstrate that much progress has been made toward understanding these unusual effects. Still, it may be said that the physical mechanism responsible for the minimum remains an open issue. The backscattering peak of linear optics occurs because the scattering contribution arising via the right-traveling polariton is random but identical to that of the left-traveling polariton at backscattering. To produce the backscattering minimum seen here at 2ω, one may modify this argument so as to introduce a phase shift between these two contributions, thus producing destructive interference. It appears most fruitful to look for a physical justification of a phase shift in the interaction required for the ω incident field to launch the 2ω polaritons, as this coupling appears to be the most significant difference between the linear and nonlinear scattering situations. It may also be relevant that a strongly rough metal surface is known to produce a minimum at backscattering in its second-harmonic distribution.[20] A plausible physical reason for this effect has been put forth,[20] although the scattering mechanisms are quite different in the case of strong roughness and do not involve surface plasmon polaritons.

9.4.2. Roughness Spectrum Centered on k_{sp}^{ω}

The second case considered here is that with the fundamental polariton wavenumber k_{sp}^{ω} centered in the rectangular region of $G(k)$.[21] This surface was also fabricated in silver, and we estimate from the dielectric constant that $k_{sp}^{\omega} = 1.01\,(\omega/c) = 5.96 \times 10^{-3}\,\text{nm}^{-1}$. The surface was designed for a maximum coupling angle of $\theta_{max} = 15°$, from which it follows that $k_{min} = 4.43 \times 10^{-3}\,\text{nm}^{-1}$ and $k_{max} = 7.49 \times 10^{-3}\,\text{nm}^{-1}$. Profilometry of the surface determined the surface roughness σ to be 28.3 nm.

This surface is capable of producing the couplings required by all other peaks of [6] and [7]. This may be demonstrated through consideration of the coupling principle implicit in [6] and [7]. In general, the surface nonlinearity acts upon a pair of ω-waves k_1^{ω} and k_2^{ω} to produce a scattered 2ω-wave $q^{2\omega}$ according to the rule

$$q^{2\omega} = k_1^{\omega} + k_2^{\omega}, \tag{10}$$

from which the emission angle θ_s of $q^{2\omega} = (2\omega/c)\sin\theta_s$ may be calculated. Any of the ω-waves present at the surface can be taken as k_1^{ω} or k_2^{ω}, although stronger waves should play more significant roles.

In particular, the two narrow peaks of [6] and [7] arise from the interaction of the incident wave $k_1^{\omega} = (\omega/c)\sin\theta_i$ with the fundamental plasmon polaritons $k_2^{\omega} = \pm k_{sp}^{\omega}$. Because the roughness strongly excites $\pm k_{sp}^{\omega}$, it is hoped that these peaks should appear in the experiments. The more interesting peak of [7] is in the direction normal to the mean surface. This was predicted to arise from the strongly excited fundamental plasmons $k_1^{\omega} = +k_{sp}^{\omega}$ and $k_2^{\omega} = -k_{sp}^{\omega}$. Obviously, Eq. (10) then predicts $q^{2\omega} = 0$, so this interaction would appear at the surface normal.

The scattering distribution $\langle I^{2\omega}(\theta_s|\theta_i)\rangle$ is shown in Fig. 9.8 for $\theta_i = 8°$, $13°$ and $17°$. The signals are much stronger than in Fig. 9.7 and, when there is excitation of k_{sp}^{ω} ($\theta_i = 8°$ and $13°$), there are two peaks that rise well above the scale. Indeed, these two peaks fall at the θ_s calculated for the incident wave/$\pm k_{sp}^{\omega}$ interaction described earlier, which indicates the origin of the peaks. However it is obvious that the normal peak at $\theta_s = 0°$ is completely absent. Instead, a broad distribution having a width of approximately 20° appears for small $|\theta_s|$, which is far wider than expected for the normal peak. In particular, the peak would have appeared

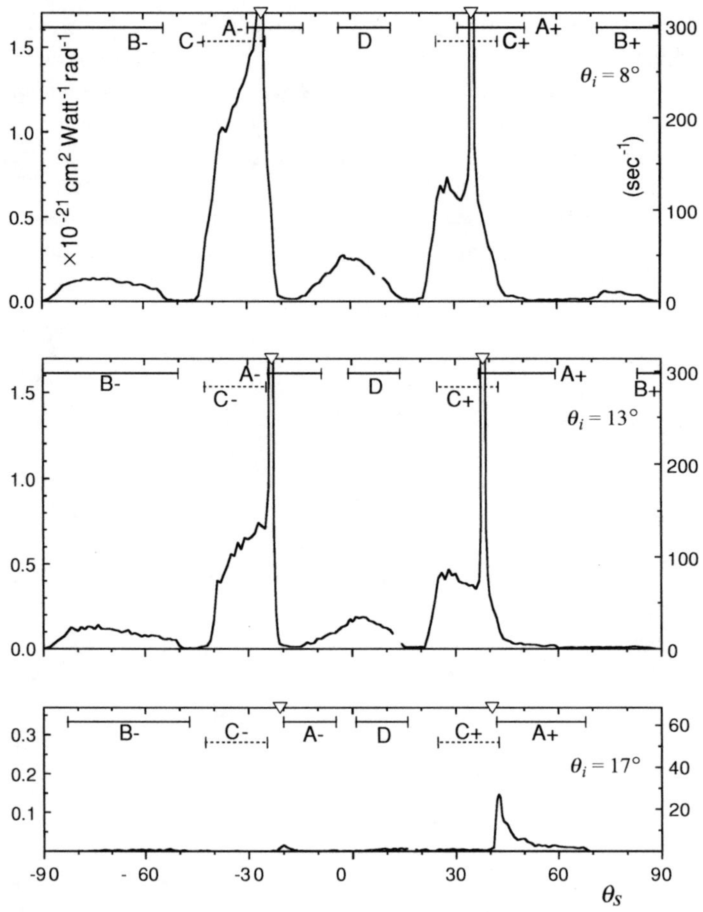

FIGURE 9.8. With k_{sp}^{ω} centered in the roughness spectrum $G(k)$, the distributions $\langle I^{2\omega}(\theta_s|\theta_i)\rangle$ for θ_i as shown. Inverted triangles denote the calculated position of the incident wave/$\pm k_{sp}^{\omega}$ interaction peaks, which rise to 2 to 4 times the figure scale.

with width similar to that of the two high peaks, as the widths are all governed by that of the resonance at k_{sp}^{ω}. Further data have been taken for other θ_i and σ;[21] the appearance is similar to Fig. 9.8 and a normal peak has never been observed. It may thus be summarized that all necessary conditions to produce the normal peak have been satisfied, but there is no evidence whatsoever of the peak.

Still, with the help of Eq. (10), it is possible to investigate the origin of the features seen in Fig. 9.8. For example, the interaction between the incident wave $k_1^{\omega} = (\omega/c)\sin\theta_i$ and the waves of the lowest-order linear scatter $k_2^{\omega} = (\omega/c)\sin\theta_i \pm k_r$ predicts that

$$q^{2\omega} = 2(\omega/c)\sin\theta_i \pm k_r\,, \tag{11}$$

where k_r can lie between $k_{\min}$ and $k_{\max}$. With $q^{2\omega} = (2\omega/c)\sin\theta_s$, it is then a simple matter to calculate the resulting range of of scattering angles θ_s to which coupling is possible, which are shown as A$\pm$ in Fig. 9.8. The coupling A+ seems to be apparent at low levels for $\theta_i = 13°$ and $17°$, although in other cases A$\pm$ overlaps with other scattering contributions.

Because all scattering contributions besides A+ disappear in $\langle I^{2\omega}(\theta_s|\theta_i)\rangle$ for $\theta_i = 17°$ when $\pm k_{\text{sp}}^{\omega}$ is not excited, it is likely that other contributions rely on excitation of $\pm k_{\text{sp}}^{\omega}$. Consider, for example, the nonlinear interaction between $k_1^{\omega} = \pm k_{\text{sp}}^{\omega}$ and, once again, the single scatter $k_2^{\omega} = (\omega/c)\sin\theta_i \pm k_r$. For k_r between $k_{\min}$ and $k_{\max}$, the coupling ranges in θ_s predicted by Eq. (10) are shown as B$\pm$ in Fig. 9.8. For $\theta_i = 8°$ the couplings B$\pm$ are clearly seen, at $\theta_i = 13°$ B$-$ is present while B+ is mostly evanescent, and at $\theta_i = 17°$ the scatter has disappeared, presumably because $\pm k_{\text{sp}}^{\omega}$ is not excited. These observations thus provide strong evidence that these wave interactions are occurring on the surface.

Some of the stronger signal levels in Fig. 9.8 may be attributed to the excitation of $\pm k_{\text{sp}}^{2\omega}$. In particular, $\pm k_{\text{sp}}^{2\omega}$ can couple to propagating waves by simple roughness scattering as in

$$q^{2\omega} = \pm k_{\text{sp}}^{2\omega} \mp k_r , \tag{12}$$

where the sign of the roughness coupling is reversed to make $q^{2\omega}$ a propagating wave. The coupling ranges produced for k_r between $k_{\min}$ and $k_{\max}$ are shown in Fig. 9.8 as C$\pm$ and coincide with some of the higher levels of $\langle I^{2\omega}(\theta_s|\theta_i)\rangle$, until the coupling also disappears for $\theta_i = 17°$. Thus there is excitation of $\pm k_{\text{sp}}^{2\omega}$ that is linked to fundamental plasmon excitation; this could arise in Eq. (10) with $k_1^{\omega} = \pm k_{\text{sp}}^{\omega}$ and $k_2^{\omega} = (\omega/c)\sin\theta_i \pm k_r$ so that $q^{2\omega} = k_{\text{sp}}^{2\omega}$, which is a special case of the B$\pm$ coupling described earlier.

The last interaction considered here, with coupling range denoted by D in Fig. 9.8, is with $k_1^{\omega} = (\omega/c)\sin\theta_i$ and with k_2^{ω} being part of the distribution associated with backscattering enhancement at frequency ω for $|\theta_s| \leq \theta_{\max}$. The range D spans small $|\theta_s|$ and, even though there are modest levels of scatter present, the distribution there is broader than the coupling limits and it is unclear if the observed scatter is related to the proposed process. However, for weaker σ, the experimental distribution does indeed produce scatter that is much more consistent with the interval D.[21]

Monte Carlo calculations have been recently reported for $\langle I^{2\omega}(\theta_s|\theta_i)\rangle$ with parameters as in the experimental results.[17] For two different models of the surface nonlinearity, the scattering levels differed considerably, but both models reproduced remarkably well the features of $\langle I^{2\omega}(\theta_s|\theta_i)\rangle$ seen in the experiments. In particular, the two high peaks are present, and the other components of $\langle I^{2\omega}(\theta_s|\theta_i)\rangle$ appear at similar positions and with similar relative heights. As in the experiments, a distinct peak at $\theta_s = 0°$ was not present throughout the results, although there was a minimum there only for $\theta_i = 0°$. Unfortunately, it has not been possible to take data under these conditions because of the glare of the specular reflection.

9.5. Angular Correlation Functions

In this section, we return to the case of linear scattering at a fixed ω, but with a different question: When will the scattered intensity for a given pair of angles $(\theta_{i1}, \theta_{s1})$ be correlated with a second intensity for different angles $(\theta_{i2}, \theta_{s2})$?

The discussion is restricted to the case in which the illuminated surface is large in the sense that many correlation cells of the surface are illuminated. The scattered amplitude then arises from many essentially independent contributions; it is permissible to invoke the central limit theorem and conclude that the scattered amplitude is jointly Gaussian. Applying the Gaussian moment theorem[22] it follows that

$$\langle \Delta I(\theta_{s1}|\theta_{i1})\, \Delta I(\theta_{s2}|\theta_{i2})\rangle = |\langle A(\theta_{s1}|\theta_{i1})\, A^*(\theta_{s2}|\theta_{i2})\rangle|^2 + |\langle A(\theta_{s1}|\theta_{i1})\, A(\theta_{s2}|\theta_{i2})\rangle|^2, \tag{13}$$

where $\Delta I \equiv I - \langle I\rangle$, and A is understood to be the diffusely scattered amplitude from which the specular reflection has been subtracted. Equation (13) is a generalization of the common circular Gaussian moment theorem[14] and has a second term that expresses the consequences of noncircular statistics. Thus, to investigate correlations of intensity, it is necessary to address the two fundamental amplitude correlations of Eq. (13).

A complete discussion would require a theoretical digression and thus the results are simply stated. It is found that $\langle A(\theta_{s1}|\theta_{i1})\, A^*(\theta_{s2}|\theta_{i2})\rangle$ is nonzero only when $(\omega/c)(\sin\theta_{s1} - \sin\theta_{i1}) = (\omega/c)(\sin\theta_{s2} - \sin\theta_{i2})$. Similarly, it is also found that $\langle A(\theta_{s1}|\theta_{i1})\, A(\theta_{s2}|\theta_{i2})\rangle$ is nonzero only when $(\omega/c)(\sin\theta_{s1} - \sin\theta_{i1}) = -(\omega/c)(\sin\theta_{s2} - \sin\theta_{i2})$. These statements have been justified rigorously in perturbation theory including all terms to fourth order in (σ/λ),[23] although this result ultimately relies on the statistical stationarity of the surface roughness and should then be valid for all orders.

The interpretation of Eq. (13) is thus unusual because the two terms contribute separately for different configurations. Defining Δ_{qk} as $(\omega/c)(\sin\theta_s - \sin\theta_i)$, only the first term contributes for two intensities that have equal Δ_{qk}, while only the second term contributes for two intensities that have opposite Δ_{qk}. For all other angular configurations, the intensity correlation vanishes. Physically, Δ_{qk} may be regarded as a measure of the deviation of geometry from the specular condition ($\Delta_{qk} = 0$). Thus the first term of Eq. (13) implies that a correlation may be present as long as this deviation is held constant, which may be regarded as the law of speckle motion as θ_i is varied. Further, the second term of Eq. (13) implies that there can be another correlation between intensities having opposite deviations from specular, which thus expresses the degree of symmetry of the speckle about specular. For brevity in the following discussions, the first and second terms of Eq. (13) are denoted by, respectively, $C^+_{\Delta I}$ and $C^-_{\Delta I}$.

To investigate these correlations, the experimental techniques of Sect. 9.2.3 were modified. The incident beam of wavelength 612 nm was focused to a waist at the surface having e^{-1} intensity diameter 67 μm, which served to make the speckle width large at the detector in the far field. To resolve the speckles, the detector aperture size was reduced to a slit much narrower than a speckle. Data were taken by placing the experiment in a configuration (θ_i, θ_s) and recording the intensity as the sample was translated to a series of spatial coordinates, thus producing 5.1×10^3 nearly independent realizations of scattered intensity. The geometry was then configured to new angles consistent with the correlation conditions discussed earlier, and the process was repeated. After data were taken for a total of 170 configurations, numerical computation of the correlations between the data sets revealed the correlation functions of interest. The surface employed was, as in Sect. 9.3.2, that of Fig. 9.2.

Data were taken for $\Delta_{qk}=0.04\,\omega/c$ which implies, for small angles, intensities approximately 2° away from the specular reflection. In Fig. 9.9, the results are plotted as a function of θ_{i2} for several values of θ_{i1}. For $\theta_{i1}=16.2°$, $C^+_{\Delta I}$ and $C^-_{\Delta I}$ both remain small because $\theta_{i1}>\theta_{\max}$. For smaller θ_{i1} producing polariton excitation, a nearly symmetric pair of peaks is seen in $C^+_{\Delta I}$ ($\theta_{i1}=6.3°$), which interact ($\theta_{i1}=0.6°$), and finally overlap ($\theta_{i1}=-1.1°$) to produce a single distinct peak with the geometry set at backscattering. On the other hand, $C^-_{\Delta I}$ is nonzero but remains at quite low levels throughout the measurements. In each curve of $C^+_{\Delta I}$, there is a point that represents the correlation between two identical intensities. This peak appears on the right and obviously implies perfect correlation. The peak on the left represents the correlation between two intensities related by reciprocity with $(\theta_{i2}, \theta_{s2})=(-\theta_{s1}, -\theta_{i1})$, for which perfect recorrelation is predicted. In the experiment the degree of correlation for the reciprocal point is 0.99 until it reaches unity when the reciprocal and autocorrelation points overlap at $\theta_{i1}=-1.1°$.

There are several notable aspects of these data. The first is how rapidly the intensity decorrelates; the peaks of $C^+_{\Delta I}$ in Fig. 9.9 are only approximately 3° wide. Further, it is seen that $C^-_{\Delta I}$ remains small throughout Fig. 9.9 and no peaks appear that are comparable to those of $C^+_{\Delta I}$. The degree of correlation implied by $C^-_{\Delta I}$ remains small, reaching at most 0.3, although this is not shown in Fig. 9.9. This situation is completely different for a weakly rough surface without significant plasmon–polariton excitation, where narrow peaks do not generally appear in $C^+_{\Delta I}$, the degree of correlation throughout both $C^+_{\Delta I}$ and $C^-_{\Delta I}$ is only slightly less than unity, and $C^+_{\Delta I}$ and $C^-_{\Delta I}$ are of similar height.[23] From an observational point of view, in the absence of polariton excitation the speckles can be seen moving into the forward direction with increasing θ_i, while maintaining their appearance. In this case the symmetry of the speckle about the specular reflection can be seen, and is even more obviously present as symmetric noise in data taken as a detector is scanned through specular. On the other hand, for a surface producing polariton excitation as in Fig. 9.9, the speckles boil or decorrelate extremely rapidly with increasing θ_i, to the extent that speckle motion is much less clearly seen. Further, the symmetry about specular is not apparent either visually or by scanning a detector through the specular region. Thus the physical behavior of the speckle pattern varies greatly, and these differences manifest themselves in $C^+_{\Delta I}$ and $C^-_{\Delta I}$.

Finally, it is notable that there remains more work to be done in the study of the correlation functions under a wide variety of conditions. It is stressed that $C^-_{\Delta I}$

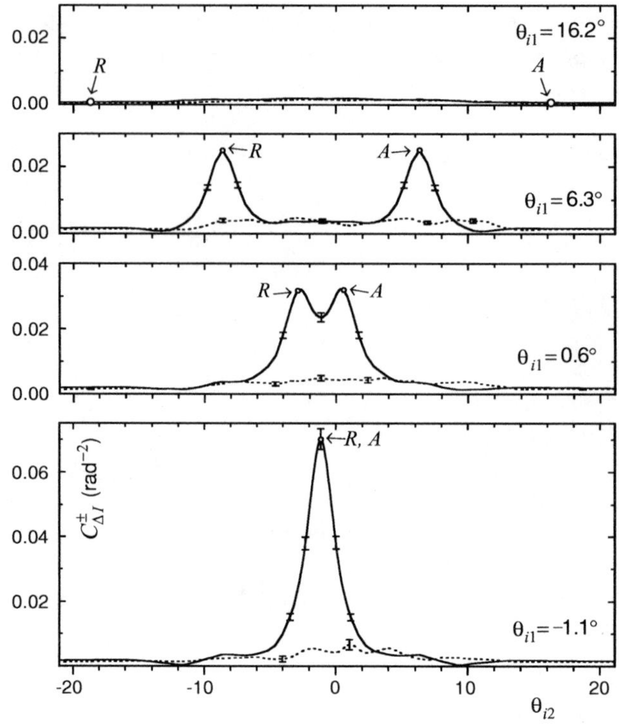

FIGURE 9.9. Intensity correlation functions $C^+_{\Delta I}$ (solid curves) and $C^-_{\Delta I}$ (dashed curves) for $\Delta_{qk} = 0.04\,\omega/c$ and θ_{i1} as indicated. The points A and R denote, respectively, autocorrelation and reciprocal configurations in $C^+_{\Delta I}$.

(or, in theoretical work $\langle A(\theta_{s1}|\theta_{i1})\, A(\theta_{s2}|\theta_{i2})\rangle$) remains poorly understood; it is unclear if multiple scattering can produce interesting effects in $C^-_{\Delta I}$, or even if it can be significant in cases other than that of extremely weak roughness. Also, it is notable that a variety of other intensity correlation functions have been predicted theoretically when the Gaussian moment theorem of Eq. (13) is violated due to a small surface illumination area.[5] Attempts to observe these in the experiments described here were unsuccessful, and using a smaller illuminated area was impractical due to alignment difficulties.

9.6. Conclusions

Experimental results have been presented for the diffuse scatter from randomly rough metal surfaces that produce significant plasmon–polariton excitation. To a large extent this work relies on the ability to fabricate and characterize such surfaces, and here a novel approach has been employed to produce metal surfaces with a rectangular roughness power spectrum that produces unusually strong effects.

The surfaces are fabricated using photolithography in which the total exposure is the sum of many sinusoidal interference patterns. The resulting roughness is consistent with a Gaussian random process and is highly one-dimensional, which simplifies the scattering experiments and allows comparison with theoretical works using this convenient surface model.

For these surfaces, the effects related to plasmon–polariton excitation are clear. In particular, with the rectangular spectrum centered on the plasmon–polariton wavenumber, the mean diffuse scatter contains a backscattering peak. If the plasmon–polariton wavenumber is shifted within the spectrum, the diffuse emission of the plasmon polaritons becomes more distinct, with the backscattering peak being present for angles of incidence when counterpropagating plasmon polaritons have been excited. Further, it has been shown that the form of the angular correlation functions of the diffuse intensity is strongly affected by plasmon–polariton excitation.

The intense surface fields produced also make possible the study of the emission of diffuse second harmonic light from the rough metal surfaces. For a rectangular spectrum centered on the wavenumber of the second harmonic plasmon–polariton, a backscattering effect appears as a distinct minimum in the second-harmonic diffuse scatter. With the rectangular spectrum centered on the fundamental plasmon polariton wavenumber, the second-harmonic diffuse scatter is considerably stronger and contains a number of nonlinear wave interactions that are discussed in detail. Such investigations serve to demonstrate the wide variety of nonlinear wave interactions possible on a metal surface and, ultimately, may lead to a better understanding of the nonlinear mechanisms themselves.

Acknowledgments. The author is grateful for the help of C. S. West, R. Torre, M. E. Knotts, and E. R. Méndez. He also acknowledges support from CICESE internal project 7329 and CONACYT grant 41947F.

References

1. A.R. McGurn, A.A. Maradudin, and V. Celli, "Localization effects in the scattering of light from a randomly rough grating," *Phys. Rev. B* **31**, 4866–4871 (1985).
2. V. Celli, A.A. Maradudin, A.M. Marvin, and A.R. McGurn, "Some aspects of light scattering from a randomly rough metal surface," *J. Opt. Soc. Am. A* **2**, 2225–2239 (1985).
3. A.A. Maradudin and E.R. Méndez, "Enhanced backscattering of light from weakly rough, random metal surfaces," *Appl. Opt.* **32**, 3335–3343 (1993).
4. T.R. Michel, "Resonant light scattering from weakly rough random surfaces and imperfect gratings," *J. Opt. Soc. Am. A* **11**, 1874–1885 (1994).
5. V. Malyshkin, A.R. McGurn, T.A. Leskova, A.A. Maradudin, and M. Nieto-Vesperinas, "Speckle correlations in the light scattered from weakly rough random metal surfaces," *Waves Random Media* **7**, 479–520 (1997).
6. R.T. Deck and R.K. Grygier, "Surface-plasmon enhanced harmonic generation at a rough metal surface," *Appl. Opt.* **23**, 3202–3213 (1984).

7. A.R. McGurn, T.A. Leskova, and V.M. Agranovich, "Weak-localization effects in the generation of second harmonics of light at a randomly rough vacuum-metal grating," *Phys. Rev. B* **44**, 11441–11456 (1991).
8. J.M. Bennett and L. Mattsson, *Introduction to Surface Roughness and Scattering* (Optical Society of America, Washington, D.C., 1989) and references therein.
9. M.E. Knotts, T.R. Michel, and K.A. O'Donnell, "Comparisons of theory and experiment in light scattering from a randomly rough surface," *J. Opt. Soc. Am. A* **10**, 928–941 (1993).
10. M.C. Hutley, *Diffraction Gratings* (Academic, London, 1982).
11. H. Raether, *Surface Plasmons on Smooth and Rough Surfaces and on Gratings* (Springer-Verlag, Berlin, 1988).
12. C.S. West and K.A. O'Donnell, "Observations of backscattering enhancement from polaritons on a rough metal surface," *J. Opt. Soc. Am. A* **12**, 390–397 (1995).
13. S.O. Rice, "Mathematical analysis of random noise," *Bell Syst. Tech. J.* **23**, 282–332 (1944).
14. J.W. Goodman, *Statistical Optics* (Wiley, New York, 1985).
15. C.S. West and K.A. O'Donnell, "Scattering by plasmon polaritons on a metal surface with a detuned roughness spectrum," *Opt. Lett.* **21**, 1–3 (1996).
16. K.A. O'Donnell and R. Torre, "Characterization of the second-harmonic response of a silver-air interface," *New J. Phys.* **7**, 154 (2005).
17. T.A. Leskova, A.A. Maradudin, and E.R. Méndez, "Multiple-scattering phenomena in the second-harmonic generation of light reflected from and transmitted through randomly rough metal surfaces," in *Optical Properties of Nanostructured Random Media*, V.M. Shalaev, ed. (Springer-Verlag, New York, 2002), 359–443.
18. K.A. O'Donnell, R. Torre, and C.S. West, "Observations of backscattering effects in second harmonic generation from a weakly rough metal surface," *Opt. Lett* **21**, 1738–1740 (1996).
19. M. Leyva-Lucero, E.R. Méndez, T.A. Leskova, A.A. Maradudin, and J.Q. Lu, "Multiple-scattering effects in the second-harmonic generation of light in reflection from a randomly rough surface," *Opt. Lett.* **21**, 1809–1811 (1996).
20. K.A. O'Donnell and R. Torre, "Second harmonic generation from a strongly rough metal surface," *Opt. Commun.* **138**, 341–344 (1997).
21. K.A. O'Donnell, R. Torre, and C.S. West, "Observations of second-harmonic generation from randomly rough metal surfaces," *Phys. Rev. B* **55**, 7985–7992 (1997).
22. H.M. Pedersen, "Object-roughness dependence of partially developed speckle patterns in coherent light," *Opt. Commun.* **16**, 63–67 (1976).
23. C.S. West and K.A. O'Donnell, "Angular correlation functions of light scattered from weakly rough metal surfaces," *Phys. Rev. B* **59**, 2393–2406 (1999).

10
Measuring Interfacial Roughness by Polarized Optical Scattering

THOMAS A. GERMER
Optical Technology Division, National Institute of Standards and Technology, Gaithersburg, MD 20899

10.1. Introduction

Optical scattering measurements are extremely sensitive for locating and characterizing defects, contamination, and roughness on smooth surfaces. Its sensitive, high-throughput, and nondestructive nature has made optical scattering the preferred method for inspecting many materials whose surfaces must be pristine before manufacturing devices on them, such as polished silicon for the semiconductor microelectronics industry, substrates for magnetic storage media, and glass used for information display systems.[1] Optical scattering often limits the performance of optics, such as those used for satellite telescopes or ring-laser gyroscopes. Understanding and being able to measure roughness can aid manufacturers in developing these materials. Lastly, the morphology of thin films is determined by the mechanisms of film growth and the interactions between the interfaces, and measurements of the relative roughness and correlation between the interfaces can yield significant information about the underlying physics in these systems.[2]

Since there may be a number of different sources of optical scatter in a material or thin film besides roughness, such as material inhomogeneity, subsurface defects, or particles, it may be important to distinguish among the different sources in order to properly interpret optical scattering measurements. Recent research has demonstrated that the polarization of light scattering can be instrumental in this application.[3–7] Theoretical and experimental results have shown that different scattering sources yield unique polarization signatures that can be used to distinguish scattering sources or validate the interpretation of intensity data.

This chapter will discuss the measurement and interpretation of roughness by angle-resolved optical scattering with an emphasis on utilizing information contained in the polarization. In Sect. 10.2, we will describe how to quantify scattered light in terms of its intensity and polarization properties. In Sect. 10.3, we will discuss measurement methods. In Sect. 10.4, we will describe roughness of a single interface, where we use the polarization information primarily to validate the interpretation. In Sect. 10.5, we will describe the scattering by the two layers of a dielectric film and show how the polarization can yield information about the relative amplitude and correlation between the roughness of the two interfaces. In the final section, Sect. 10.6, we will make some remarks about extending the technique to more interfaces.

10.2. Definitions

Angle-resolved measurements are concerned with quantifying light originating from a source direction, defined by a polar angle θ_i and azimuthal angle ϕ_i, and scattered into a reflected direction, defined by a polar angle θ_r and azimuthal angle ϕ_r. One quantity describing the directional dependence of scatter from a surface is the bidirectional reflectance distribution function (BRDF), defined historically[8] as the differential radiance $\mathrm{d}L_r$ (power per unit solid angle per unit projected surface area) scattered by a uniformly illuminated, homogeneous material per unit differential incident irradiance $\mathrm{d}E_i$ (power per unit surface area):

$$f_r(\theta_i, \phi_i, \theta_r, \phi_r) = \frac{\mathrm{d}L_r(\theta_i, \phi_i, \theta_r, \phi_r)}{\mathrm{d}E_i(\theta_i, \phi_i)}. \tag{1}$$

Eq. (1), while often quoted, is of little practical use, because most materials are not homogeneous and most illumination schemes are not uniform. That is, even if a material does not have appreciable variation across its surface, any real specimen has a finite extent. Likewise, while we can generate illumination that is approximately uniform within some region, that region of illumination must come to an end, if not simply at the specimen edge. The biggest problem with this definition is that diffuse materials often emit light outside of a finitely illuminated region. When we see Eq. (1), but ignore the words surrounding it, then we can easily find ourselves coping with an apparently infinite BRDF in certain regions of the sample.

An equivalent definition of the BRDF considers the average power $\langle\Phi_r\rangle$ scattered into a solid angle Ω for a given incident power Φ_i:

$$f_r(\theta_i, \phi_i, \theta_r, \phi_r) = \lim_{\Omega\to 0} \frac{\langle\Phi_r\rangle}{\Phi_i \Omega \cos\theta_r}. \tag{2}$$

That is, the BRDF is the average fraction of light scattered per *projected* solid angle for a finitely illuminated region. It is a distribution in the scattering direction and a function of incident direction. It is also a function of wavelength, polarization, and sample properties.

While the BRDF appears to have a term (i.e., $\cos\theta_r$) that might cause it to diverge for large scattering angles, most surface scattering sources behave in such a manner that the scattering per unit solid angle falls off in angle fast enough that the BRDF not only remains finite, but approaches zero for $\theta_r \to 90^\circ$. In real data, however, this may not be true. Rayleigh scatter by air within the field of view of the receiver, a primary background source in smooth surface scatter measurements, and other sources of stray light do not vanish at large scattering angles.[9] Furthermore, small uncertainties in the scattering angle can lead to apparent uncertainties in the BRDF, which do not translate to meaningful uncertainties for some applications. A convenient property of the BRDF is that it obeys Helmholtz reciprocity:

$$f_\mathrm{r}(\theta_\mathrm{i}, \phi_\mathrm{i}, \theta_\mathrm{r}, \phi_\mathrm{r}) = f_\mathrm{r}(\theta_\mathrm{r}, \phi_\mathrm{r}, \theta_\mathrm{i}, \phi_\mathrm{i}). \tag{3}$$

The definition of the BRDF given in Eq. (2) does not fully characterize the scatter properties of a material, though, because it fails to include any details of how scatter depends upon incident polarization or what polarization the scattered light is. The addition of information contained in the polarization makes measurements of the polarization attractive.

To characterize polarization states, the Mueller–Stokes formalism is convenient.[10] Any intensity-like quantity (e.g., power, radiance, and irradiance) can be quantified in terms of a Stokes vector, given by

$$\boldsymbol{\Phi} = \begin{pmatrix} \Phi_\mathrm{s} + \Phi_\mathrm{p} \\ \Phi_\mathrm{s} - \Phi_\mathrm{p} \\ \Phi_\mathrm{s+p} - \Phi_\mathrm{s-p} \\ \Phi_\mathrm{lcp} - \Phi_\mathrm{rcp} \end{pmatrix}, \tag{4}$$

where Φ_u is the power that we would measure if we used an analyzer that passes only u-polarized light: s indicating that the electric field is perpendicular to the plane defined by the direction of propagation and the surface normal, p indicating that the electric field is parallel to that plane, $\mathrm{s} \pm \mathrm{p}$ indicating that the electric field is $\pm 45^\circ$ with respect to those directions, and lcp and rcp indicating left- and right-circularly polarized light, respectively. We can describe how a material interacts with light, assuming that the material's effect upon the light is linear, using a 4×4 Mueller matrix, which relates an input Stokes vector to an output Stokes vector. A Mueller matrix BRDF, $\mathbf{F}_\mathrm{r}$, can then be defined as the Mueller matrix that relates the average scattered Stokes vector power $\langle \boldsymbol{\Phi}_\mathrm{r} \rangle$ to the incident Stokes vector power[11] $\boldsymbol{\Phi}_\mathrm{i}$:

$$\lim_{\Omega \to 0} \frac{\langle \boldsymbol{\Phi}_\mathrm{r} \rangle}{\Omega} = \mathbf{F}_\mathrm{r} \boldsymbol{\Phi}_\mathrm{i} \cos\theta_\mathrm{r}. \tag{5}$$

Note that we have rearranged Eq. (5), compared to Eq. (2), since one cannot divide a Stokes vector by another Stokes vector. Furthermore, we cannot measure $\mathbf{F}_\mathrm{r}$ with a single measurement.

In many applications, measurement of the full Mueller matrix BRDF is not necessary to yield the information we need. Rather, a specific incident polarization, which may depend upon incident and scattered directions, is chosen to maximize differentiation among scattering mechanisms, and we measure the Stokes vector of the scattered light. Sometimes, instead of reporting the Stokes vector elements, a different combination of them is reported. One set of parameters consists of the BRDF for the given incident polarization, f_r, the principal angle of the polarization ellipse, η, the degree of circular polarization, P_C, and the total degree of polarization P. These parameters are related to the measured Stokes vector intensity by

$$\begin{aligned} f_\mathrm{r} &= \langle \Phi_{\mathrm{r}0} \rangle / (\Omega\ \Phi_\mathrm{i} \cos\theta_\mathrm{r}) \\ \eta &= \tan^{-1}(\langle \Phi_{\mathrm{r}2} \rangle / \langle \Phi_{\mathrm{r}1} \rangle)/2 \\ P_\mathrm{C} &= \langle \Phi_{\mathrm{r}3} \rangle / \langle \Phi_{\mathrm{r}0} \rangle \\ P &= \sqrt{\langle \Phi_{\mathrm{r}1}^2 \rangle + \langle \Phi_{\mathrm{r}2}^2 \rangle + \langle \Phi_{\mathrm{r}3}^2 \rangle} / \langle \Phi_{\mathrm{r}0} \rangle, \end{aligned} \tag{6}$$

where $\Phi_{\mathrm{r}j}$ ($j = 0, 1, 2, 3$) is the j-th element of $\mathbf{\Phi}_\mathrm{r}$. One reason these parameters are particularly useful is that all of the information about the intensity is contained in one parameter (f_r), and all of the information about the randomness of the polarization is contained in another parameter (P). In many cases, the parameter η is all that is necessary to differentiate scattering mechanisms, since P_C is often predicted to be negligible.

While the Stokes vector analysis has been used to distinguish and quantify different scattering sources, at least for well chosen measurement conditions, the Mueller matrix analysis has not yet been found to yield additional information on isotropic surfaces that warrants the added difficulty that the measurements entail. However, it is not inconceivable that the Mueller matrix analysis will find itself useful for patterned or other anisotropic samples.

10.3. Measurement Methods

Figure 10.1 shows a schematic diagram of an instrument used for performing angle-resolved optical scatter measurements.[12] While there are a variety of guides that exist for developing BRDF instruments,[1,13,14] we will summarize a number of their most important features. We can divide these instruments into three parts: the source (elements (a)–(h) in Fig. 10.1), the sample holder and goniometer (element (i) in Fig. 10.1), and the receiver (elements (j)–(p) in Fig. 10.1). Details of the design of each of these parts depend upon our application. In this discussion, we will concentrate on those issues which are important for polarimetric measurements of nanoscale roughness.

The purpose of the source is to generate the beam of polarized incident light. The source consists of a laser (a), which is modulated by an optical chopper (b). The polarization state of the incident light is set by a fixed plane polarizer (c) and a rotatable linear retarder (d). Elements (a)–(d), plus the various mirrors that are needed to steer the beam around the table, generally impart some stray light on the beam. In order that the beam incident upon the sample have as good a beam profile as possible, the beam is spatially filtered with a lens and pinhole ((e) and (f)). Finally, a focusing element (g) focuses the beam onto the entrance aperture of the receiver (j). The focusing element can be a lens or a concave mirror. In order to have the least amount of stray light at small scattering angles, a high-quality concave mirror is usually preferred for this element. Finally, some baffling (h) or

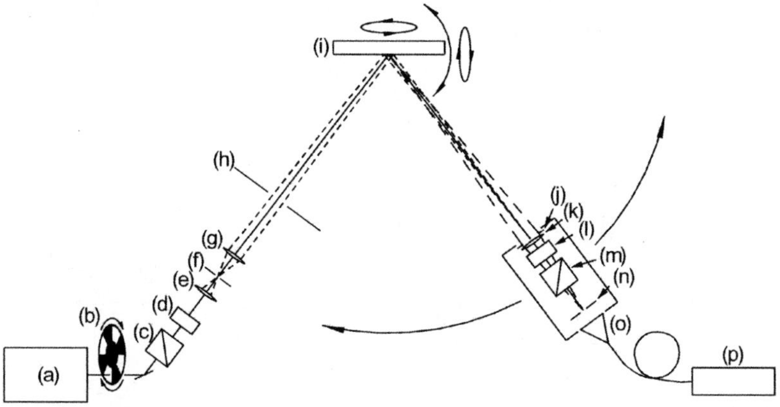

FIGURE 10.1. A schematic diagram of a polarization-resolved, angle-resolved scatterometer: (a) laser, (b) chopper, (c) source polarizer, (d) source rotating retarder, (e) spatial filter lens, (f) spatial filter, (g) primary lens or mirror, (h) baffling, (i) sample, (j) receiver aperture, (k) receiver lens, (l) receiver rotating retarder, (m) receiver polarizer, (n) field stop, (o) detector, and (p) lock-in amplifier.

enclosure is usually included. The baffling should not actually block any of the beam, since that will generally scatter strongly and increase the stray light in the system.

The sample holder and goniometer is designed to orient the sample (i) with respect to the source. A simple system may employ a single rotation axis, enabling scatter measurements in the plane of incidence. A more complex system with more axes of rotation enables measurements out of the plane of incidence and as functions of sample rotation. Lastly, linear translation of the sample may be needed to assess sample uniformity by obtaining scatter measurements from multiple spots on a sample.

The purpose of the receiver is to collect and analyze light over a known solid angle about a given direction. The receiver rotates about the illumination spot on the sample. The first element of the receiver is the receiver aperture (j). The area A of this aperture and its distance R from the sample determines the collected solid angle $\Omega = A/R^2$. A lens (k) in the receiver images the sample onto a field stop (n), so that the size of the field stop aperture determines the sample field of view. A smaller field of view reduces the amount of stray light accepted by the receiver. It should be set so that it is as small as possible, but always so that the field of view is larger than the illuminated area on the sample. The combination of a rotating retarder (l) and a fixed polarizer (m) selects a specific polarization state for analysis. A detector (o) is placed after the field stop, and a lock-in amplifier (p), synchronized with the chopper, is used for phase sensitive detection of the signal.

The dynamic range of angle-resolved light scattering instrumentation presents a challenge to accurate measurements of nanoscale roughness. Near the specular direction, the intensity can be very high, and the BRDF measurement is limited by the diffraction-limited spot size of the incident light on the receiver aperture. For a Gaussian beam focused onto the detector, the maximum measurable BRDF at normal incidence is[1]

$$f_r^{\max} = \frac{\pi D^2}{2\lambda^2}, \tag{7}$$

where D is the diameter of the illumination spot at the sample, and λ is the wavelength of the light. For $D = 5$ mm and $\lambda = 550$ nm, the maximum value of the BRDF is about 10^8 sr^{-1}. In the other extreme, the measured scatter signal is limited by the Rayleigh scatter by air surrounding the sample. That is, the field of view of the receiver will accept the scatter from the beam propagating through the air. In the absence of a sample, this quantity is given approximately by[9]

$$f_r^{\text{Rayleigh}} = \frac{4\pi^2(n-1)^2 l_{\text{FOV}}}{\lambda^4 N \sin\theta \cos\theta_r} \times \begin{cases} 1 & \text{for s-polarization} \\ \cos^2\theta & \text{for p-polarization,} \end{cases} \tag{8}$$

where θ is the viewing angle measured from the incident direction, l_{FOV} is the diameter of the field of view of the receiver, N is the number density of air, and n is the index of refraction of air. At 20 °C, standard atmospheric pressure, $\lambda = 550$ nm, viewing perpendicular to the beam propagation direction, s-polarized incident light, and for $l_{\text{FOV}} = 10$ mm, Eq. (8) yields a BRDF of approximately 1.5×10^{-8} sr^{-1}. With the reasonable assumption that we need to have signals above this level (although it is conceivable that one could subtract this scatter from data), the range of scatter levels extends a range of over 16 orders of magnitude.

The wide dynamic range can be obtained by a combination of multiple collection apertures and multiple detectors. The smallest aperture should be on the order of the beam diameter $2w_0$ at the detector,

$$2w_0 = \frac{2\lambda R}{\pi D}. \tag{9}$$

For $R = 500$ mm and $D = 5$ mm, the beam diameter is about 70 μm. As one varies the direction away from the specular direction, larger apertures need to be used, the largest typically spanning an angle from 0.5° to 2°. One cannot use just two apertures, however, for a number of reasons. First, the ratio of areas in these extremes is at least 5000, increasing the dynamic range requirements of the detector. More importantly, a well-designed instrument will have sufficiently low stray light at an angle where use of the small aperture encounters signals at the noise floor of the detector, opening up to the largest detector may actually accept the full specular beam to the detector. Therefore, any instrument should have a range of apertures, each varying by a factor of 4 to 7 from the next.

In addition to using multiple receiver apertures, multiple detectors can significantly expand the dynamic range of a scattering instrument. The detectors should have overlapping ranges of use, and the least sensitive detector should be capable of measuring the incident power. It is common to employ a silicon photodiode and a photomultiplier tube, for example, as the two detectors.

Figure 10.2 shows a representative measurement of an instrument signature, measured by scanning the receiver through the incident beam in the absence of a sample. Three regimes can be observed: the coherent incident beam at small

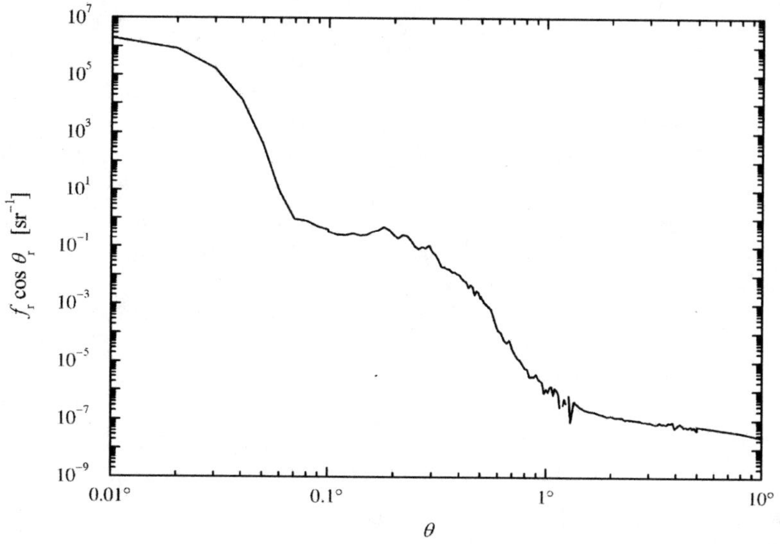

FIGURE 10.2. An instrument signature measured for an angle-resolved instrument at the author's institute.

angles ($\theta < 0.06°$), stray light at intermediate angles ($0.06° < \theta < 1°$), and Rayleigh scatter at large angles ($\theta > 1°$). The challenge of the instrument designer (and to some extent, the user, who must maintain the instrument) is to create an optical system that transitions to the Rayleigh scatter regime in as small an angle as possible. Low scatter optics, well-placed baffling, and reduction of the field of view of the detector to its minimum necessary size serve to reduce the stray light. The effect of an aperture change can be seen in Fig. 10.2 near $\theta = 1°$, where the noise level appears to rise before the next aperture can be used.

We must also consider the effect of laser speckle. For a spatially incoherent source of diameter D at the sample, the coherence length at the receiver aperture will be approximately given by the same expression as Eq. (9), within a small factor which depends upon the intensity profile over the source. If one considers a source of diameter D, an aperture of diameter D_{det}, and a source–aperture distance R, there will be approximately

$$N_{\text{speckle}} = \frac{\pi D_{\text{det}}^2/4}{\pi w_0^2} = \left(\frac{\pi D D_{\text{det}}}{2\lambda R}\right)^2 \tag{10}$$

speckles entering the receiver.[15] Assuming Poisson statistics, the relative standard deviation of the signals will be given by

$$\frac{\sigma_S}{\langle S \rangle} = \frac{1}{\sqrt{N_{\text{speckle}}}} = \frac{2\lambda R}{\pi D D_{\text{det}}} = \frac{\lambda}{D\sqrt{\pi \Omega}}. \tag{11}$$

For a $D = 5$ mm source, a $D_{\text{det}} = 5$ mm aperture, $R = 500$ mm, and $\lambda = 550$ nm, the estimated relative standard deviation would be 0.7%. While this is an acceptable value for many measurements, reduction of the illumination spot diameter or the detector aperture can easily place this value into a regime where laser speckle is by far the largest source of measurement uncertainty.

There are a number of ways to overcome speckle noise, of which some applications may be able to take advantage. The easiest in many applications is simply to make several measurements, either at different locations on the sample or at different sample rotations. Constant motion of the sample during the measurement can also effectively allow sampling over different surface realizations. Destroying the spatial coherence of the light source can be performed by a number of methods, including passing the beam through an ultrasonically vibrated multi-mode fiber. Such a beam, however, will necessarily have poorer focusing characteristics, and the angular resolution of the system will be degraded.

To perform measurements of the Mueller matrix BRDF, we must analyze the scattered polarization for a number of incident polarizations. A common method for obtaining the Mueller matrix is the $\omega-5\omega$ scheme, developed by Azzam,[16] whereby a quarter-wave retarder on the source ((d) in Fig. 10.1) is rotated at frequency ω, while another quarter-wave retarder on the receiver ((l) in Fig. 10.1) is rotated at frequency 5ω. The Mueller matrix elements are then linearly related to the Fourier components of the signal. An improvement on this scheme uses 0.37λ retarders instead of 0.25λ retarders, in order to improve the path on the Poincarré sphere taken by each of the rotating retarders.[17,18] Other methods employ the use of liquid crystal variable retarders[19] and photoelastic modulators.[20] Many measurements do not require full measurement of the Mueller matrix. For example, in most of the measurements presented in this chapter, only a single linearly polarized source is used, and the Stokes vector of the scattered light is measured. In that case, a rotating half-wave retarder can be used in the source to rotate the polarization into the desired angle.

10.4. Roughness of a Single Interface

10.4.1. Theory

We will begin our discussion of measurements with roughness of the interface between a condensed state (usually a solid, but could be a liquid) and a gas or vacuum. This case is the usual starting point for measurements of nanoscale roughness. If we assume that the surface height function $\Delta z(\boldsymbol{\rho})$ $[\boldsymbol{\rho} = (x, y)]$ is single valued, has zero mean, is much smaller than the wavelength of the light, and that the slopes $\partial \Delta z(\boldsymbol{\rho})/\partial x$ and $\partial \Delta z(\boldsymbol{\rho})/\partial y$ are much smaller than 1, the problem can be solved by first-order perturbation theory. The solution was first proposed by Rice[21] in 1951, while a more complete solution was developed by Barrick[22] in 1970. The Mueller matrix BRDF from a rough surface in the smooth surface approximation is given by

$$\mathbf{F}_\mathrm{r} = \frac{16\pi^2}{\lambda^4} \cos\theta_\mathrm{i} \cos\theta_\mathrm{r} \, \langle |Z(\boldsymbol{\kappa})|^2 \rangle \, \mathbf{Q}, \tag{12}$$

where the Mueller matrix $\mathbf{Q} = \mathbf{M}(\mathbf{q}, \mathbf{q}^\dagger)$ is nondepolarizing, and the function $\mathbf{M}$ is given in the appendix. The scattering matrix $\mathbf{q}$ has elements

$$\begin{aligned}
q_{\rm ss} &= (\epsilon - 1)k^2 \cos\phi_{\rm r}/[(k_{z\rm i} + k'_{z\rm i})(k_{z\rm r} + k'_{z\rm r})] \\
q_{\rm ps} &= -(\epsilon - 1)kk'_{z\rm r} \sin\phi_{\rm r}/[(k_{z\rm i} + k'_{z\rm i})(\epsilon k_{z\rm r} + k'_{z\rm r})] \\
q_{\rm sp} &= -(\epsilon - 1)kk'_{z\rm i} \sin\phi_{\rm r}/[(\epsilon k_{z\rm i} + k'_{z\rm i})(k_{z\rm r} + k'_{z\rm r})] \\
q_{\rm pp} &= (\epsilon - 1)(\epsilon k_{xy\rm i}k_{xy\rm r} - k'_{z\rm i}k'_{z\rm r} \cos\phi_{\rm r})/[(\epsilon k_{z\rm i} + k'_{z\rm i})(\epsilon k_{z\rm r} + k'_{z\rm r})],
\end{aligned} \tag{13}$$

where

$$\begin{aligned}
k'_{z\beta} &= k(\epsilon - \sin^2\theta_\beta)^{1/2} \\
k_{z\beta} &= k\cos\theta_\beta \\
k_{xy\beta} &= k\sin\theta_\beta \\
k &= 2\pi/\lambda
\end{aligned} \tag{14}$$

($\beta = \mathrm{i, r}$). $\langle |Z(\boldsymbol{\kappa})|^2 \rangle$ is the two-dimensional power spectral density (PSD) of the surface height function, where

$$Z(\boldsymbol{\kappa}) = \lim_{A\to\infty} \frac{1}{\sqrt{A}} \int_A \mathrm{d}^2\boldsymbol{\rho}\, \Delta z(\boldsymbol{\rho}) \exp(\mathrm{i}\,\boldsymbol{\kappa}\cdot\boldsymbol{\rho}). \tag{15}$$

The surface wavevector $\boldsymbol{\kappa}$ has elements κ_x and κ_y given by the diffraction equations

$$\begin{aligned}
\kappa_x &= k_{xy\rm r}\cos\phi_{\rm r} - k_{xy\rm i} \\
\kappa_y &= k_{xy\rm r}\sin\phi_{\rm r}.
\end{aligned} \tag{16}$$

In the above, we are ignoring, without loss of generality, the azimuthal angle of the source direction, $\phi_{\rm i}$. That is, we are defining our x-axis to be the intersection of the plane of incidence and the plane of the sample. The specular condition is $\theta_{\rm i} = \theta_{\rm r}$ and $\phi_{\rm r} = 0$. The surface wavevector (radians per unit length) is related to the spatial frequency (cycles per unit length) by a factor of 2π, so PSDs are often presented with respect to $|\boldsymbol{\kappa}|/(2\pi)$.

The limit in Eq. (15) does not exist for a randomly rough surface of infinite extent. It is precisely that issue that gives rise to the observed speckle pattern. In practice, however, a nonzero solid angle is collected. The limit of $|Z(\boldsymbol{\kappa})|^2$ integrated over a finite region of spatial frequencies does, in fact, exist. Notice that the Mueller matrix $\mathbf{Q}$ does not depend upon $Z(\boldsymbol{\kappa})$. So, while there will exist a speckle pattern in the intensity, that speckle pattern does not exist in the polarization.

The expressions in Eqs. (13) appear to be very similar to Fresnel reflection coefficients. In fact, $|q_{\rm ss}|^2$ is given by

$$|q_{\rm ss}|^2 = [R_{\rm s}(\theta_{\rm i})R_{\rm s}(\theta_{\rm r})]^{1/2}, \tag{17}$$

where $R_{\rm s}(\theta)$, is the specular reflectance of the substrate for s-polarization and incident angle θ. Thus, in the absence of specific values of the dielectric constant ϵ, it is relatively straightforward to make measurements that can be used to obtain $|q_{\rm ss}|^2$.

10.4.2. Limitations

An estimate of the maximum roughness amplitude that can be treated with the first-order approach can be found by considering diffraction from a sinusoidal grating.[23] Within the Kirchhoff approximation, the ratio of the second-order diffraction efficiency to the first-order diffraction efficiency is given by

$$[J_2(\delta)/J_1(\delta)]^2 \approx \delta^2, \tag{18}$$

where $\delta = ka(\cos\theta_\mathrm{i} + \cos\theta_\mathrm{r})$, and a is the amplitude of the sinusoid. If we ask the question, "When is the second-order diffraction intensity less than 5% of the first-order diffraction intensity?" we find a characteristic amplitude a. Since the rms roughness is given by $\sigma = a/\sqrt{2}$, we arrive at a practical estimate of the validity of the first-order theory:

$$\sigma < \frac{\lambda}{10(\cos\theta_\mathrm{i} + \cos\theta_\mathrm{r})}. \tag{19}$$

The expression shows that the perturbative approach can be valid for relatively large roughnesses, as long as the incident and scattering angles are large. This behavior can be easily observed by noticing that nearly all materials become specularly reflecting when viewed at grazing incidence. The measurement scheme where θ_i and θ_r are held equal and fixed at large angles, while ϕ_r is varied provides a means for measuring the roughness of surfaces approaching the wavelength of the light.[24]

10.4.3. The Inverse Problem

The proportionality between the BRDF and the surface PSD, given in Eq. (12), makes optical scattering an attractive method for measuring surface roughness in the smooth surface limit. One need only solve Eq. (12) for $\langle|Z(\boldsymbol{\kappa})|^2\rangle$, taking into account the incident polarization. For example, for s-polarized light, and performing measurements in the plane of incidence, we can use Eq. (17) to obtain the relatively simple expression

$$\langle|Z(\boldsymbol{\kappa})|^2\rangle = \frac{\lambda^4\langle\Phi_\mathrm{r}\rangle}{16\pi^2\Omega\Phi_\mathrm{i}\cos\theta_\mathrm{i}\cos\theta_\mathrm{r}^2[R_\mathrm{s}(\theta_\mathrm{i})R_\mathrm{s}(\theta_\mathrm{r})]^{1/2}}. \tag{20}$$

The range of surface wavevectors $|\boldsymbol{\kappa}|$ over which the measurement can be performed is limited at small $|\boldsymbol{\kappa}|$ by the range of angles over which the signal is sufficiently above the instrument signature and at large $|\boldsymbol{\kappa}|$ by the wavelength of the light. For example, if the minimum angle from the specular direction is 0.1°, and we are operating with $\lambda = 550$ nm and an incident angle of $\theta_\mathrm{i} = 60°$, the smallest spatial frequency is $|\boldsymbol{\kappa}|_{\min}/(2\pi) \approx 1.5\ \mathrm{mm}^{-1}$, and the largest spatial frequency is $|\boldsymbol{\kappa}|_{\max}/(2\pi) \approx 3\ \mu\mathrm{m}^{-1}$, a range spanning almost three orders of magnitude.

While roughness can contribute to elastic light scattering, many other sources can also contribute and interfere with the results. For example, Rayleigh scatter in the air, particles on the surface, subsurface defects, material inhomogeneities, and stray light can contribute to varying degrees. Therefore, if we want to apply

Eq. (12) to data to determine the roughness statistics, it is important that we test the basic hypothesis that roughness is indeed causing the scatter. There are a number of measurements we can make to check that the scatter is consistent with roughness.

The first of these consistency checks is angle scaling. That is, we can perform a scattering measurement using one incident angle, use Eq. (12) to estimate the PSD, then try the same measurement using a different incident angle. The problem with this method is that it can be quite deceptive. Scatter by material inhomogeneities, for example, has been shown to yield very similar results as roughness.[6,25,26] Just as the scattering by roughness is proportional to the PSD of the surface height function, the scattering by inhomogeneities is proportional to the PSD of the dielectric constant across the surface. Since the dependence upon direction in both cases depends upon those PSDs evaluated using the same diffraction equation, what we end up showing more than anything else is that the diffraction equation works.

A second consistency check that is often used is wavelength scaling.[1] Here, we perform the measurement at multiple wavelengths and, again, compare the estimated PSDs determined from Eq. (12). Subsurface defects and material inhomogeneities have scattering behaviors that are very similar, especially if the dielectric constant of the material does not change appreciably. For this reason, this method is often employed using very wide ranges of wavelengths, over which the dielectric constant of the material varies significantly. Over such a wide range of wavelengths, the measurement becomes substantially more difficult to perform, because we need to switch optical components and detectors. It also fails to tell us whether any of the scatter results from roughness, just that it does not result from roughness over the entire wavelength range. Despite these problems, it has been considered the mainstay for checking for roughness scatter, and was instrumental in helping recognize that many mirror materials, such as beryllium and aluminum, were inherently high scatterers, regardless of how smooth they were, because they tend to exhibit a high degree of scatter from material inhomogeneity.[27]

The third consistency check, polarization analysis, is much more powerful and relatively difficult to fool.[7] With this method, we are checking for consistency with the elements of the scattering matrix, Eq. (13). Scattering by other sources, such as particles, subsurface defects, and material inhomogeneity, yield scattering matrix elements that differ from those given in Eqs. (13). For example, the scattering matrix elements appropriate for subsurface defects or material inhomogeneity are given by[6,25,26] $q_{\mathrm{ss}}^{\mathrm{sub}} = q_{\mathrm{ss}}$, $q_{\mathrm{sp}}^{\mathrm{sub}} = q_{\mathrm{sp}}$, $q_{\mathrm{ps}}^{\mathrm{sub}} = q_{\mathrm{ps}}$, and

$$q_{\mathrm{pp}}^{\mathrm{sub}} = (\epsilon - 1)(k_{xy\mathrm{i}} k_{xy\mathrm{r}} - k'_{z\mathrm{i}} k'_{z\mathrm{r}} \cos\phi_{\mathrm{r}})/[(\epsilon k_{z\mathrm{i}} + k'_{z\mathrm{i}})(\epsilon k_{z\mathrm{r}} + k'_{z\mathrm{r}})]. \tag{21}$$

That is, only the pp-terms differ, and even then, only when θ_{i} and θ_{r} are both nonzero. Purely s-polarized incident light, for example, will yield no discrimination between these two scattering mechanisms. Thus, we must have some p-polarized light incident on the sample and use large incident and scattering polar angles. These restrictions argue for measurements out of the plane of incidence. For example, a useful measurement, which maps out the PSD over a wide range of surface wavevectors, is to set $\theta_{\mathrm{i}} = \theta_{\mathrm{r}}$ and vary ϕ_{r} from 0° to 180°. The incident polarization is linear at an angle η_{i} and continuously varied from 45° to 135° over this range, such that when $\phi_r = 90°$, the incident light is p-polarized ($\eta_i = 90°$). That is,

$$\eta_i = 45° + \phi_r/2. \tag{22}$$

The magnitude of the surface wavevector is then given by

$$|\kappa| = 2k \sin\theta_r \sin(\phi_r/2). \tag{23}$$

At each angle, the Stokes vector can be measured. In many instances, we only need to measure the linear components of the Stokes vector, because we expect little circular polarization from surface roughness, if the material is non- or weakly absorbing.

We can obtain reasonably good discrimination between scattering sources by performing the measurement in the plane of incidence by measuring the Stokes vector for 45° incident polarization. However, near the surface normal, that discrimination disappears, and we are left confirming the roughness hypothesis at the beginning and end of a scan, and hoping that at those angles near the surface normal the trend continues.

We can also perform Mueller matrix measurements to distinguish scattering sources. However, since s-polarized incident light has very little ability to discriminate sources, we would not expect that Mueller measurements would improve our confidence of the roughness hypothesis substantially from what we can obtain by optimizing the incident polarization to that which gives the largest discrimination.

It is interesting to note that, because the scatter by roughness in the smooth surface limit yields a deterministic polarization, the light scattered by such a rough surface does not depolarize the light.[28] While we observe speckle fluctuations in the intensity, especially if we use a small collection solid angle, little of those fluctuations are observed in the polarization state. Thus, polarization measurements often appear quite noise-free in comparison to their intensity counterparts.

10.4.4. Example

To demonstrate the methodology described in Sect. 10.4.3, we present data obtained from a thick metallic TiN layer grown on a silicon wafer. The thickness of the layer, 110 nm, is thick enough and the material absorbent enough that we can safely ignore any interfaces below the TiN. The light scattering measurement was carried out using $\lambda = 532$ nm light with $\theta_i = \theta_r = 60°$, varying the scattering azimuthal angle ϕ_r from near 0° to 170°. The polarization was varied as described in Eq. (22). The results of the measurements are shown in Fig. 10.3, in terms of the parameters given in Eqs. (6). The systematic uncertainties in the measurement are less than the size of the symbols, and the random uncertainties can be estimated by observing the variation of the data about a smooth curve.

We evaluated the polarization predicted by first-order vector perturbation theory. We chose $\epsilon = 1.6 + 4.6i$ so that the results matched in the specular direction, which is equivalent to using specular ellipsometry to determine its value. The measured polarization states agree very well with the predictions of the perturbation theory,

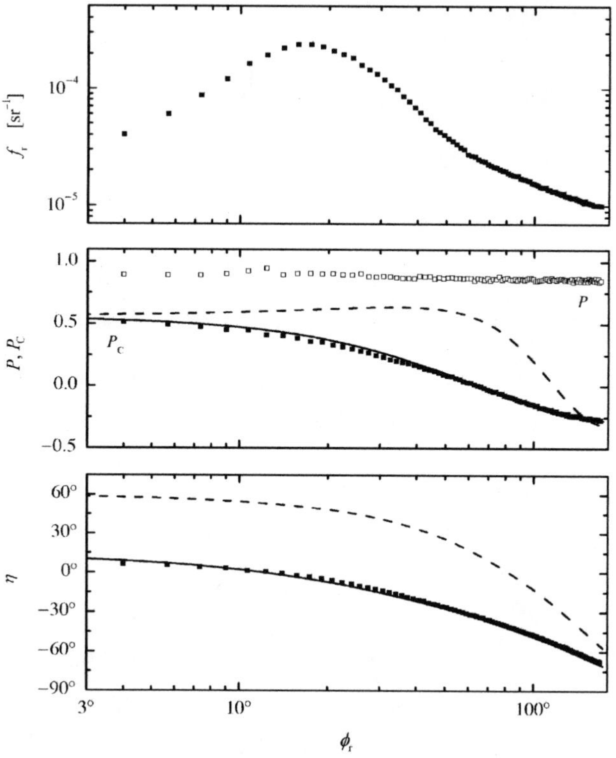

FIGURE 10.3. Results from out-of-plane ($\theta_i = \theta_r = 60°$) polarized light scattering measurements (symbols) for a TiN sample using 532 nm light: (top) the BRDF, f_r, (middle) the degrees of polarization and circular polarization, (open symbols) P and (closed symbols) P_C, respectively, and (bottom) the principal polarization angle η. The incident polarization was varied as described in Eq. (22) in the text. The curves represent the polarization states predicted for light scattered by (solid) a microrough surface and (dashed) material inhomogeneity.

which are shown as solid curves in Fig. 10.3. In particular, the measured P is close to 1, within about 15%, the deviations of which may be due to stray light in the experiment. The parameters P_C and η follow very closely to the curves.

Another likely scattering mechanism in metallic samples is scattering by material inhomogeneity. We show η and P_C predicted by Eq. (21) for this mechanism in Fig. 10.3, too. The data do not agree with such scattering. Scatter by particles, which depends upon particle size, yield different behaviors, as well.[6,29,30] It is clear from the comparison that the data agree very well with the microroughness theory and that the differentiation among the alternate scattering mechanisms is unambiguous. Thus, the polarization measurement establishes the validity of the microroughness interpretation, allowing us to convert the measured BRDF to the PSD of the surface height function.

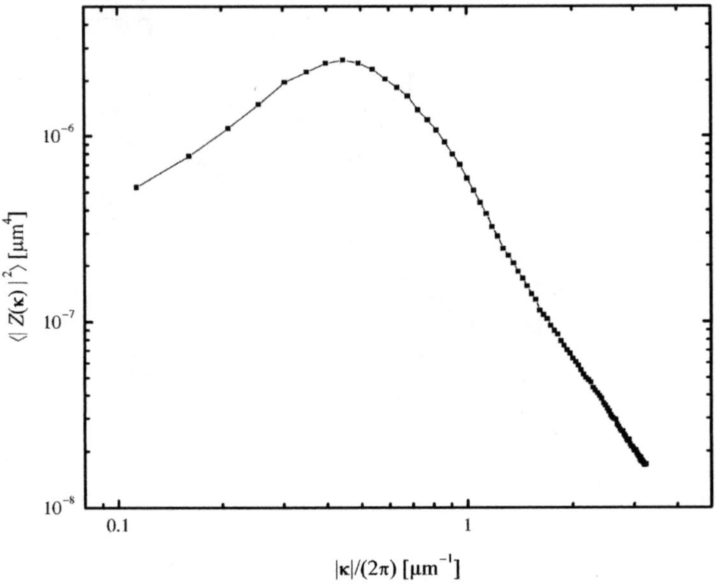

FIGURE 10.4. The PSD of the surface height function derived from the data shown in Fig. 10.3.

The result of converting the BRDF to PSD is shown in Fig. 10.4. The results show fractal behavior for large $\boldsymbol{\kappa}$, where $\langle|Z(\boldsymbol{\kappa})|^2\rangle \propto |\boldsymbol{\kappa}|^{-2.8}$, and a distinct peak in the power spectrum near $|\boldsymbol{\kappa}|/(2\pi) = 0.45\ \mu\text{m}^{-1}$. By integrating the two-dimensional PSD, we can obtain an estimate of the rms roughness. The total rms roughness over the bandwidth shown is about 2.6 nm.

The excellent agreement between the theory and experiment for microroughness implies that the polarization of light scattered by microroughness is not determined by the exact details of the surface height profile, but is a unique signature of the scattering mechanism. It therefore suggests that scatterometers can be designed to be blind to microroughness. For example, a device may be constructed with a number of detectors, each viewing a particular scattering direction, and each with a polarizer aligned to block the light from microroughness.[31] Such a device would collect light over a large solid angle, be microroughness-blind, and therefore be more sensitive to other sources of scatter, such as subsurface defects and particulate contamination.

10.5. Roughness of Two Interfaces

In Sect. 10.4, we described measurements that only used the polarization to validate the interpretation of the intensity. In this section, we will describe measurements in which the polarization is not used to validate the model, but is used to extract information about interface roughness. In this case, we are interested in the two interfaces of a dielectric film. The methodology that we describe parallels that used by specular ellipsometry[10,32] to determine film thickness. By performing an ellipsometric measurement of light diffusely scattered out of the specular direction, we move away from $\boldsymbol{\kappa} = 0$ and probe the *variations* in film thickness. That is, we measure the relative interface roughness and its degree of correlation.[33,34]

10.5.1. Theory

We now consider a film, having dielectric constant ϵ_f and mean thickness τ lying above a substrate of dielectric constant ϵ. The surface height functions of the buried and exposed interfaces are Δz_1 and Δz_2, respectively (leaving out the explicit dependence on $\boldsymbol{\rho}$). We apply first-order vector perturbation theory to this problem. The zero-order, unperturbed ($\Delta z_1 = 0$ and $\Delta z_2 = 0$) fields are found from the solution of the well-known problem of reflection from a dielectric film. The first-order calculation consists of expanding the electric and magnetic fields on both sides of each interface and the local surface normal to first order in the surface height functions Δz_j about their mean. The requirement that the tangential electric and magnetic fields be continuous across the boundary leads to relationships between zero-order and first-order fields. The theory self-consistently handles the multiple reflections that occur for both orders of the field. However, since it assumes that the film thickness is constant, it does not account for long-range nonconformal roughness, which has sufficient amplitude to substantially vary the local film thickness. In order for the theory to be valid, the modulations of the surface height functions, Δz_j, must be much less than the wavelength, λ, and the surface slope must be small.

Elson[35–39] described the solution to the first-order vector perturbation theory for scattering from interfacial microroughness in a dielectric stack. For the buried interface (1), the scattering matrix $\mathbf{q}^{(1)}$ to replace $\mathbf{q}$ in Eq. (12) has elements

$$q_{uv}^{(1)} = 4(\epsilon - \epsilon_\mathrm{f})k''_{z\mathrm{i}}k''_{z\mathrm{r}} \exp[\mathrm{i}(k''_{z\mathrm{i}} + k''_{z\mathrm{i}} - k_{z\mathrm{r}} - k_{z\mathrm{r}})\tau]s_{uv}^{(1)} \tag{24}$$

($u, v = \mathrm{s, p}$), where

$$\begin{aligned} s_{\mathrm{ss}}^{(1)} &= -k^2 \cos\phi_\mathrm{r}/(\Gamma_{\mathrm{si}}\Gamma_{\mathrm{sr}}) \\ s_{\mathrm{ps}}^{(1)} &= \epsilon_\mathrm{f} k k'_{z\mathrm{i}} \sin\phi_\mathrm{r}/(\Gamma_{\mathrm{pi}}\Gamma_{\mathrm{sr}}) \\ s_{\mathrm{sp}}^{(1)} &= \epsilon_\mathrm{f} k k'_{z\mathrm{r}} \sin\phi_\mathrm{r}/(\Gamma_{\mathrm{si}}\Gamma_{\mathrm{pr}}) \\ s_{\mathrm{pp}}^{(1)} &= -\epsilon_\mathrm{f}(\epsilon k_{xy\mathrm{i}}k_{xy\mathrm{r}} - \epsilon_\mathrm{f} k_{z\mathrm{i}}k_{z\mathrm{r}}\cos\phi_\mathrm{r})/(\Gamma_{\mathrm{pi}}\Gamma_{\mathrm{pr}}) \end{aligned} \tag{25}$$

$$\begin{aligned} \Gamma_{\mathrm{p}\beta} &= \epsilon_\mathrm{f} F_{\mathrm{p}\beta}^{(+)} k_{z\beta} - F_{\mathrm{p}\beta}^{(-)} k''_{z\beta} \\ \Gamma_{\mathrm{s}\beta} &= F_{\mathrm{s}\beta}^{(+)} k_{z\beta} - F_{\mathrm{s}\beta}^{(-)} k''_{z\beta} \end{aligned} \tag{26}$$

$$\begin{aligned} F_{\mathrm{p}\beta}^{(\pm)} &= \epsilon_\mathrm{f} K_\beta^{(\mp)} k'_\beta - \epsilon K_\beta^{(\pm)} k''_{z\beta} \\ F_{\mathrm{s}\beta}^{(\pm)} &= K_\beta^{(\mp)} k'_{z\beta} - K_\beta^{(\pm)} k''_{z\beta} \\ K_\beta^{(\pm)} &= \exp(2\mathrm{i}k''_{z\beta}\tau) \pm 1 \end{aligned} \tag{27}$$

$k''_{z\beta} = k(\epsilon_{\rm f} - \sin^2\theta_\beta)^{1/2}$ (β = i or r). The Fourier transform of the roughness of the m-th interface is given in Eq. (15), with Δz replaced with Δz_m. For the exposed interface (2), the scattering matrix $\mathbf{q}^{(2)}$ to replace $\mathbf{q}$ in Eq. (12) has elements

$$q_{uv}^{(2)} = (\epsilon_{\rm f} - 1)\exp[-{\rm i}(k_{z{\rm i}} + k_{z{\rm r}})\tau]s_{uv}^{(2)} \tag{28}$$

where

$$\begin{aligned} s_{\rm ss}^{(2)} &= -k^2 F_{\rm si}^{(+)} F_{\rm sr}^{(+)} \cos\phi_{\rm r}/(\Gamma_{\rm si}\Gamma_{\rm sr}) \\ s_{\rm ps}^{(2)} &= -kk''_{z{\rm i}} F_{\rm pi}^{(-)} F_{\rm sr}^{(+)} \sin\phi_{\rm r}/(\Gamma_{\rm pi}\Gamma_{\rm sr}) \\ s_{\rm sp}^{(2)} &= -kk''_{z{\rm r}} F_{\rm si}^{(+)} F_{\rm pr}^{(-)} \sin\phi_{\rm r}/(\Gamma_{\rm si}\Gamma_{\rm pr}) \\ s_{\rm pp}^{(2)} &= -(\epsilon_{\rm f} k_{xy{\rm i}} k_{xy{\rm r}} F_{\rm pi}^{(+)} F_{\rm pr}^{(+)} - k''_{z{\rm i}} k''_{z{\rm r}} F_{\rm pi}^{(-)} F_{\rm pr}^{(-)} \cos\phi_{\rm r})/(\Gamma_{\rm pi}\Gamma_{\rm pr}). \end{aligned} \tag{29}$$

Just as the matrix elements for scattering by single-interface roughness given in Eq. (13) are independent of the surface height function, those for scattering by the two interfaces of a dielectric film given in Eqs. (25) and (29) do not depend upon the respective surface height functions. Therefore, to first order, the scattering from a single rough interface will not depolarized light. Furthermore, the fields resulting from the scattering of each interface are independent of each other.

We can evaluate the special case of two interfaces that are totally conformal (correlated and equal roughness) by coherently adding the scattering matrices from each of them:

$$\mathbf{q}^{(\rm corr)} = \mathbf{q}^{(1)} + \mathbf{q}^{(2)}. \tag{30}$$

Similarly, if the two interfaces are equally rough, but have a random phase relationship between them (i.e., they are uncorrelated), then we can add the two sources incoherently:

$$\mathbf{Q}^{(\rm uncorr)} = \mathbf{M}(\mathbf{q}^{(1)}, \mathbf{q}^{(1)\dagger}) + \mathbf{M}(\mathbf{q}^{(2)}, \mathbf{q}^{(2)\dagger}). \tag{31}$$

In general, the surfaces may be neither correlated nor of equal roughness. In this case, we replace the factor $\mathbf{Q}\langle|Z|^2\rangle$ in Eq. (12) by

$$\langle|Z|^2\rangle\mathbf{Q} = \langle\mathbf{M}(Z_1\mathbf{q}^{(1)} + Z_2\mathbf{q}^{(2)}, Z_1^*\mathbf{q}^{(1)\dagger} + Z_2^*\mathbf{q}^{(2)\dagger})\rangle, \tag{32}$$

where we have dropped the explicit dependence of Z_m on $\boldsymbol{\kappa}$. Since the only random variables are Z_1 and Z_2, Eq. (32) can be simplified to

$$\begin{aligned} \langle|Z|^2\rangle\mathbf{Q} = {} & \langle|Z_1|^2\rangle\mathbf{M}(\mathbf{q}^{(1)}, \mathbf{q}^{(1)\dagger}) + \langle|Z_2|^2\rangle\mathbf{M}(\mathbf{q}^{(2)}, \mathbf{q}^{(2)\dagger}) \\ & + 2{\rm Re}\langle Z_1 Z_2^*\rangle{\rm Re}\mathbf{M}(\mathbf{q}^{(1)}, \mathbf{q}^{(2)\dagger}) \\ & - 2{\rm Im}\langle Z_1 Z_2^*\rangle{\rm Im}\mathbf{M}(\mathbf{q}^{(1)}, \mathbf{q}^{(2)\dagger}). \end{aligned} \tag{33}$$

10.5.2. The Inverse Problem

Equation (33) is an overdetermined equation in the PSD of the two interfaces, $\langle|Z_1|^2\rangle$ and $\langle|Z_2|^2\rangle$, and the cross-PSD, $\langle Z_1 Z_2^*\rangle$. That is, we can write Eq. (33) in the form $\langle |Z|^2 \rangle \mathbf{Q} = \mathbf{DZ}$, where $\mathbf{D}$ is a 16×4 matrix

$$\mathbf{D} = \begin{pmatrix} \mathbf{M}(\mathbf{q}^{(1)}, \mathbf{q}^{(1)\dagger}) \\ \mathbf{M}(\mathbf{q}^{(2)}, \mathbf{q}^{(2)\dagger}) \\ 2\mathrm{Re}\mathbf{M}(\mathbf{q}^{(1)}, \mathbf{q}^{(2)\dagger}) \\ -2\mathrm{Im}\mathbf{M}(\mathbf{q}^{(1)}, \mathbf{q}^{(2)\dagger}) \end{pmatrix}, \tag{34}$$

where each row consists of a flattened 4×4 matrix, and $\mathbf{Z}$ is a 4-element column vector

$$\mathbf{Z} = \begin{pmatrix} \langle |Z_1|^2 \rangle \\ \langle |Z_2|^2 \rangle \\ \mathrm{Re}\langle Z_1 Z_2^* \rangle \\ \mathrm{Im}\langle Z_1 Z_2^* \rangle \end{pmatrix}. \tag{35}$$

We can solve for $\mathbf{Z}$ in a least-squares sense by calculating the pseudoinverse, $\mathbf{D}^{-1} = (\mathbf{D}^{\mathrm{T}}\mathbf{D})^{-1}\mathbf{D}^{\mathrm{T}}$. Thus, we can determine the roughness statistics for the two interfaces of a thin film from the measured Mueller matrix BRDF $\mathbf{F}_{\mathrm{r}}$ from Eqs. (12),

$$\mathbf{Z} = \frac{\lambda^4}{16\pi^2 \cos\theta_{\mathrm{i}} \cos\theta_{\mathrm{r}}} (\mathbf{D}^{\mathrm{T}}\mathbf{D})^{-1}\mathbf{D}^{\mathrm{T}}\mathbf{F}_{\mathrm{r}}. \tag{36}$$

Let us consider the simpler case of a specific incident polarization state, specified by a unit intensity Stokes vector $\mathbf{S}$. The scattered Stokes vector will then be given by left-multiplying $\mathbf{S}$ by Eq. (33):

$$\begin{aligned} \langle |Z|^2 \rangle \mathbf{QS} = {} & \langle |Z_1|^2 \rangle \mathbf{M}(\mathbf{q}^{(1)}, \mathbf{q}^{(1)\dagger})\mathbf{S} \\ & + \langle |Z_2|^2 \rangle \mathbf{M}(\mathbf{q}^{(2)}, \mathbf{q}^{(2)\dagger})\mathbf{S} \\ & + 2\mathrm{Re}\langle Z_1 Z_2^* \rangle \mathrm{Re}\mathbf{M}(\mathbf{q}^{(1)}, \mathbf{q}^{(2)\dagger})\mathbf{S} \\ & - 2\mathrm{Im}\langle Z_1 Z_2^* \rangle \mathrm{Im}\mathbf{M}(\mathbf{q}^{(1)}, \mathbf{q}^{(2)\dagger})\mathbf{S}. \end{aligned} \tag{37}$$

Equation (37) is a fully determined equation in the roughness statistics. That is, Eq. (37) can be written as $\langle |Z|^2 \rangle \mathbf{QS} = \mathbf{D'Z}$, where $\mathbf{D}'$ is a 4×4 matrix

$$\mathbf{D}' = \begin{pmatrix} \mathbf{M}(\mathbf{q}^{(1)}, \mathbf{q}^{(1)\dagger})\mathbf{S} \\ \mathbf{M}(\mathbf{q}^{(2)}, \mathbf{q}^{(2)\dagger})\mathbf{S} \\ 2\mathrm{Re}\mathbf{M}(\mathbf{q}^{(1)}, \mathbf{q}^{(2)\dagger})\mathbf{S} \\ -2\mathrm{Im}\mathbf{M}(\mathbf{q}^{(1)}, \mathbf{q}^{(2)\dagger})\mathbf{S} \end{pmatrix}, \tag{38}$$

and each row is the transpose of a 4-element vector. Equation (38) can be inverted, provided the two surfaces scatter with different polarization states. Thus, the four degrees of freedom of a measured Stokes BRDF $\mathbf{f}_{\mathrm{r}}$ map onto the four degrees of freedom of the roughness statistics:

$$\mathbf{Z} = \frac{\lambda^4}{16\pi^2 \cos\theta_{\mathrm{i}} \cos\theta_{\mathrm{r}}} (\mathbf{D}')^{-1}\mathbf{f}_{\mathrm{r}}. \tag{39}$$

It is convenient for us to define a relative roughness

$$\chi = \sqrt{\langle |Z_1|^2 \rangle / \langle |Z_2|^2 \rangle} \tag{40}$$

and a complex correlation coefficient

$$c = \langle Z_1 Z_2^* \rangle / \sqrt{\langle |Z_1|^2 \rangle \langle |Z_2|^2 \rangle}. \tag{41}$$

Just as there is one constraint on a Stokes vector ($S_0^2 \geq S_1^2 + S_2^2 + S_3^2$), the parameter c must satisfy $|c| \leq 1$. For most realistic surfaces, c should have no imaginary component. It is interesting to note that the intensity and the polarization state of the scattered light separate in much the same way as for single interface roughness: the polarization state uniquely determines χ and c, while the intensity, once χ and c are known, determines the magnitude of the PSDs of the two interfaces. Another point to note is that when $|c| = 1$, we will observe no depolarization. In this case, there is no randomness in the ratio or relative phase of both sources, and so there is no randomness in their sum. Depolarization only occurs when there is incoherence between two sources.

10.5.3. Example

To demonstrate the application of the perturbation theory analysis for roughness of a dielectric film, we consider the behavior of $\lambda = 632.8$ nm light scattered by a 52 nm SiO_2 ($\epsilon_f = 2.13$) layer grown on a silicon ($\epsilon = 15.07 + 0.15i$) substrate. We let the incident angle be $\theta_i = 60°$ and scattering angle be $\theta_r = 60°$.

Before we present experimental results, we will make a number of observations about the theoretical predictions for four different limiting cases of interfacial roughness (roughness of each interface alone, correlated, and uncorrelated roughness). Figure 10.5 shows the scattered polarization state as a function of ϕ_r for s- and p-polarized incident light calculated for these cases. The results for s-polarized incident light (left column of Fig. 10.5) show only a small amount of differentiation between the roughness conditions, with none existing at $\phi_r = 0°$, $90°$, and $180°$. These results are similar to what we found for a single interface in Sect. 10.4. Symmetry dictates the polarization for $\phi_r = 0°, 90°$, and $180°$: for s-polarized light incident upon an isotropic sample in the static approximation, the scattered field must be antisymmetric about the incident plane and symmetric about the perpendicular plane. Therefore, in the plane of incidence ($\phi_r = 0°$ and $180°$), the scattered light must be s-polarized ($s_{sp} = s_{ps} = 0$), while for $\phi_r = 90°$, the scattered light must be p-polarized ($s_{ss} = 0$).

The results for p-polarized incident light (right column of Fig. 10.5) show a significantly greater differentiation between the different limiting cases, as long as we are sufficiently out of the plane of incidence (i.e., $\phi_r \neq 0°$ or $180°$). Again, symmetry requires that the scattered light be p-polarized in the plane of incidence. However, symmetry no longer exists about the perpendicular plane, so that for $\phi_r = 90°$, each case can yield a different polarization. Previous measurements have exploited this geometry to differentiate scattering from small particles, single rough surfaces, and subsurface defects.[7] We are often interested in extracting roughness statistics from data over as a wide range of surface wavevectors as possible. Since

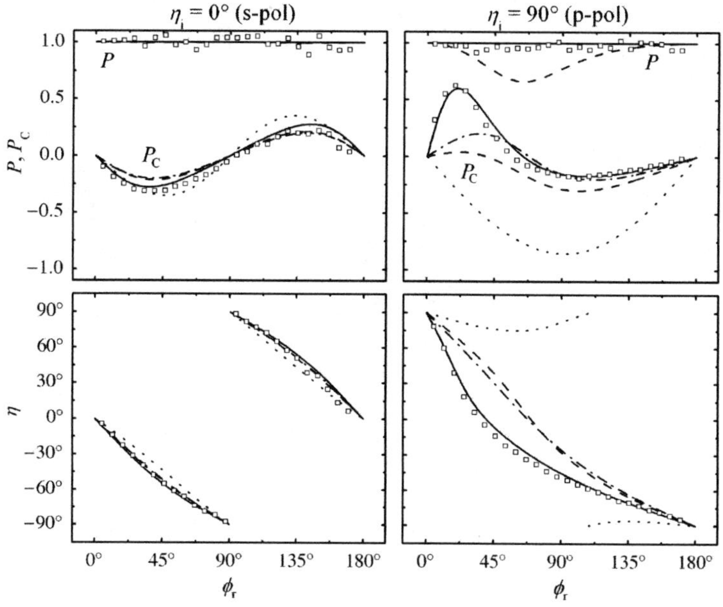

FIGURE 10.5. Polarization parameters P, P_C, and η for scattering out of the plane of incidence from (solid) correlated and equal roughness, (dashed) uncorrelated and equal roughness, (dotted) roughness of the exposed interface, (dash-dot) roughness of the buried interface, and (symbols) experimental results from a SiO_2 layer grown on microrough silicon. The incident light was (left column) s-polarized and (right column) p-polarized. Other parameters in the model are described in the text.

there is little differentiation between cases near $\phi_r = 0°$, the dynamic range of available spatial frequencies is limited.

Figure 10.6 presents two schemes that differentiate between interfacial roughness conditions for most scattering angles. One of these schemes uses circularly polarized incident light (left column of Fig. 10.6). Another scheme changes the incident polarization state as the viewing direction is varied. In the right column of Fig. 10.6, the incident light is linearly polarized, varied according to Eq. (22), as was done above for the single interface roughness measurements. We observe a reasonably good differentiation between the different roughness conditions at most scattering angles, using either of the two schemes, with somewhat better differentiation observed for the varying incident polarization scheme.

Because measurements out of the plane of incidence generally require more complicated instrumentation than those required for measurements in the plane of incidence, we include two schemes that work reasonably well in the plane of incidence. Figure 10.7 shows calculated polarization parameters for the different roughness conditions evaluated in the plane of incidence ($\theta_i = 60°, \phi_r = 0°$). Since the scattering matrices are diagonal for this geometry, we do not show results for s-polarized or p-polarized incident light. Incident light of either circular

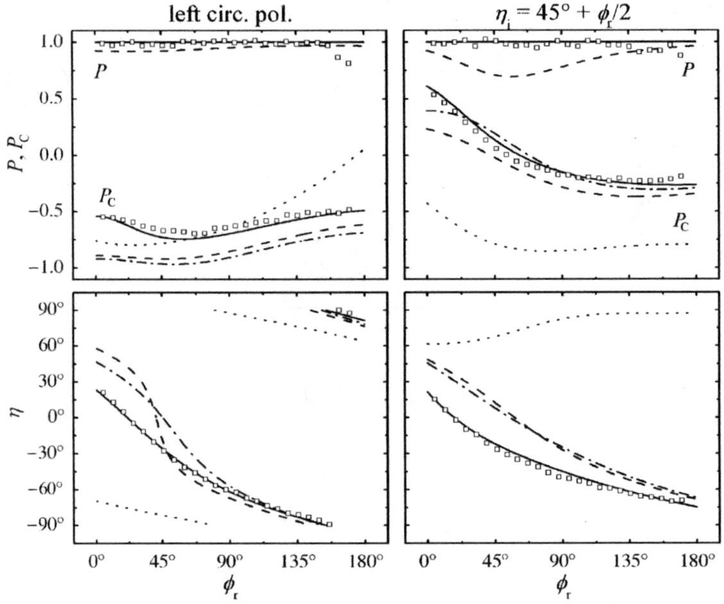

FIGURE 10.6. Polarization parameters P, P_C, and η for scattering out of the plane of incidence from (solid) correlated and equal roughness, (dashed) uncorrelated and equal roughness, (dotted) roughness of the exposed interface, (dash-dot) roughness of the buried interface, and (symbols) experimental results from a SiO_2 layer grown on microrough silicon. The incident light was (left column) left circularly polarized and (right column) linearly polarized at an angle $\eta_i = 45° + \phi_r/2$. Other parameters in the model are described in the text.

polarization or 45° linear polarization maps the four independent Mueller matrix elements onto the four Stokes vector elements. While we observe discrimination between the roughness cases in Fig. 10.7, it is relatively weak, with numerous curves crossing near $\theta_r = 0°$.

The results for polarized light scattering measurements from a 52 nm SiO_2 film thermally grown on a photolithographically produced microrough silicon surface are included in Figs. 10.5–10.7. The microrough surface consisted of a pseudorandom distribution of nominally 8 nm deep circular pits having diameters of nominally 1.31 μm and 1.76 μm.[40] Details of the experiment, its uncertainties, and the sample are given elsewhere.[12,33] This system should exhibit conformal roughness, at least for small surface wavevectors. The results shown in Figs. 10.5–10.7 indeed behave most like the equal roughness model for all incident polarizations, though a close inspection of the results reveals small discrepancies, which result from the buried interface being smoother than the exposed interface. The relative roughness of the two interfaces (χ) and the correlation coefficient c can be extracted using the technique outlined in Sect. 10.5.2. Figure 10.8 shows c and χ as functions

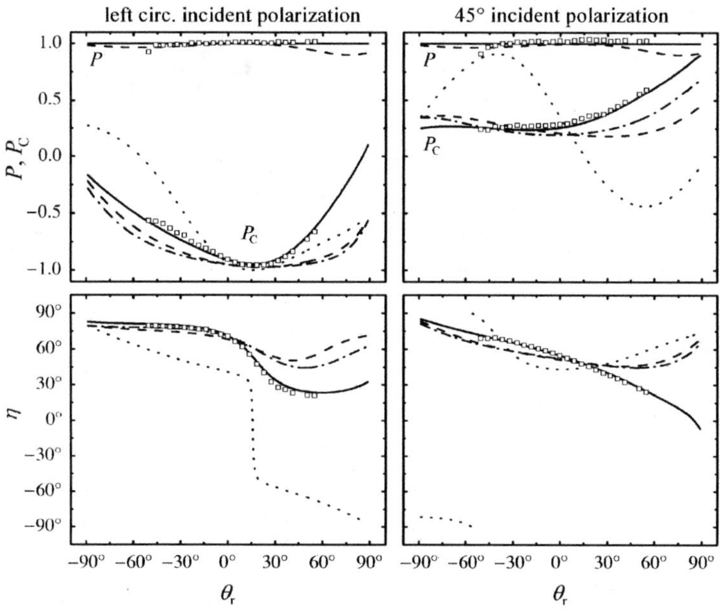

FIGURE 10.7. Polarization parameters P_C, P, and η for scattering in the plane of incidence from (solid) correlated and equal roughness, (dashed) uncorrelated and equal roughness, (dotted) roughness of the exposed interface, (dash-dot) roughness of the buried interface, and (symbols) experimental results from a SiO_2 layer grown on microrough silicon. The incident light was (left column) left circularly polarized and (right column) linearly polarized at an angle $\eta_i = 45°$. Other parameters in the model are described in the text.

of spatial frequency extracted from the data shown in Figs. 10.5–10.7. The indicated uncertainties represent single standard deviations of the extracted results obtained from the statistical uncertainties in the original data. The results obtained from all incident polarizations are consistent with each other, showing $\chi > 1$ and $c \approx 1$ for most spatial frequencies. Further validation of the method has been achieved by performing the measurements at multiple wavelengths and incident angles.[33] While measurements of the full Mueller matrix may allow different scattering mechanisms to be distinguished and quantified using the analysis given in Sect. 10.5.2, the results shown in Fig. 10.5 suggest that certain incident polarization states do not allow for much differentiation.

Figure 10.8 includes the results using data obtained in the plane of incidence. Large uncertainties and discrepancies result from the poor discrimination near 1 μm^{-1}. Comparison between the results of Figs. 10.5–10.7 suggest that maximum discrimination between different roughness conditions occurs in directions out of the plane of incidence. Other calculations show that such improvements also tend to occur for other scattering sources such as particles or subsurface defects.[6] While other researchers have performed light scattering ellipsometry measurements in the

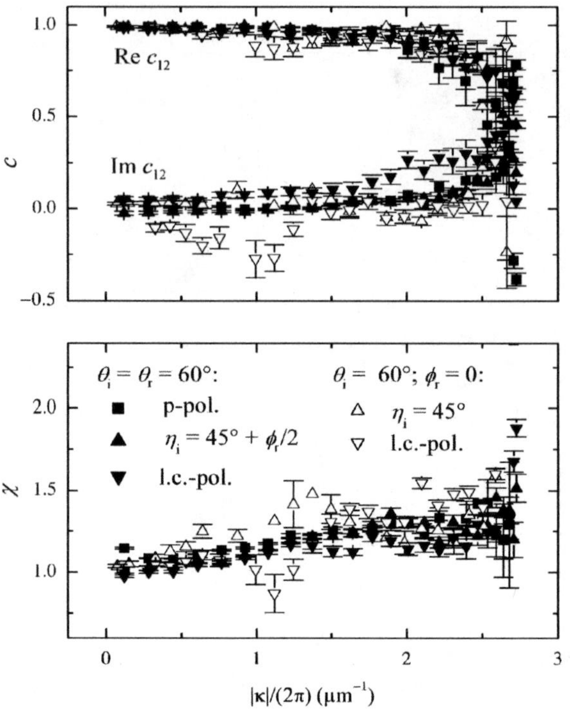

FIGURE 10.8. Roughness parameters extracted from polarized light scattering measurements from the 52 nm SiO_2 layer thermally grown on silicon. The results are obtained from measurements out of the plane of incidence (solid symbols) and in the plane of incidence (open symbols).

plane of incidence,[41] we chose to make full use of the polarization by performing such measurements in out-of-plane geometries.

It is noteworthy to point out that in Figs. 10.5–10.7, the theoretical predictions for buried interface roughness and uncorrelated roughness are the most poorly resolved. In both cases, the roughness of the bottom interface is present, and the top interface is incoherent with the bottom interface. When sources are incoherent, they add as intensities, rather than as fields, so that the smaller field has a correspondingly smaller effect. Hence, when the dielectric contrast between the substrate and the film is much larger than between film and the ambient environment, which is the case for our example, uncorrelated roughness of the top interface will be more difficult to observe in the presence of buried interface roughness.

In many realistic cases, any correlations between two interfaces are expected to be such that c is real and lies in the interval $0 \leq c \leq 1$. Any imaginary component to c implies a lateral offset in the roughness function. For this reason, it may be reasonable to use Eq. (39) to obtain a starting point for the roughness statistics, but to constrain Im $c = 0$ and perform a least-squares fit of the theory to the data.

We have also investigated a number of other systems, including a case of anti-correlated roughness (nominal $c = -1$) and a case of offset roughness (nominal $c = \exp(i\boldsymbol{\kappa} \cdot \mathbf{R})$, where $\mathbf{R}$ is a lateral offset in the two roughness functions).[42,43] These cases were much more complicated to analyze. While the amplitude of the roughness was small compared to the wavelength of the light, the lack of correlation caused unacceptably large variations in the thickness throughout the film. Thus, we find that the analysis presented here has much more rigid requirements in terms of the tolerable roughness amplitude over which the theory is valid.

10.6. Final Comments

It is worth considering, at the end, whether it is worth extending this methodology to three or more interfaces (that is, two or more films). After all, the space of valid Mueller matrices can be shown to be spanned by four scattering matrices. For example, we can decompose any valid Mueller matrix $\mathbf{M}$ into the sum

$$\mathbf{M} = \sum_{j=0}^{3} \sum_{k=0}^{3} a_{jk} \mathbf{M}(\boldsymbol{\sigma}_j, \boldsymbol{\sigma}_k^{\dagger}), \tag{42}$$

where $\boldsymbol{\sigma}_j$ are the Pauli matrices given in the appendix, and the 16 coefficients obey $a_{jk} = a_{kj}^*$. Therefore, one ought to be able to extract the roughness statistics for up to four interfaces from a Mueller matrix scattering measurement. However, the method would be very limited. In the specular direction, the scattering matrix for any interface will not have any off-diagonal elements, so only two of the four basis matrices are available. Since we cannot differentiate the different interfaces near the specular direction, the technique would therefore have a very narrow range of spatial frequencies over which to operate.

4 × 4 *Matrix Product of Two Scattering Matrices*

We define a 4 × 4 matrix product $\mathbf{M}(\mathbf{q}_1, \mathbf{q}_2)$ between two 2 × 2 scattering matrices, such that its elements are given by

$$\mathbf{M}(\mathbf{q}_1, \mathbf{q}_2)_{jk} = \tfrac{1}{2}\mathrm{Tr}(\mathbf{q}_1 \boldsymbol{\sigma}_k \mathbf{q}_2 \boldsymbol{\sigma}_j) \tag{43}$$

$(j, k = 0, 1, 2, 3)$ where the Pauli matrices are

$$\boldsymbol{\sigma}_0 = \begin{pmatrix} 1 & 0 \\ 0 & 1 \end{pmatrix}, \quad \boldsymbol{\sigma}_1 = \begin{pmatrix} 1 & 0 \\ 0 & -1 \end{pmatrix}, \quad \boldsymbol{\sigma}_2 = \begin{pmatrix} 0 & 1 \\ 1 & 0 \end{pmatrix}, \quad \boldsymbol{\sigma}_3 = \begin{pmatrix} 0 & -i \\ i & 0 \end{pmatrix}. \tag{44}$$

This operation is distributive with addition

$$\begin{aligned} \mathbf{M}(\mathbf{q}_1 + \mathbf{q}_2, \mathbf{q}_3) &= \mathbf{M}(\mathbf{q}_1, \mathbf{q}_3) + \mathbf{M}(\mathbf{q}_2, \mathbf{q}_3) \\ \mathbf{M}(\mathbf{q}_1, \mathbf{q}_2 + \mathbf{q}_3) &= \mathbf{M}(\mathbf{q}_1, \mathbf{q}_2) + \mathbf{M}(\mathbf{q}_1, \mathbf{q}_3) \end{aligned} \tag{45}$$

and associative with multiplication by a scalar

$$\mathbf{M}(k\mathbf{q}_1, \mathbf{q}_2) = k\mathbf{M}(\mathbf{q}_1, \mathbf{q}_2)$$
$$\mathbf{M}(\mathbf{q}_1, k\mathbf{q}_2) = k\mathbf{M}(\mathbf{q}_1, \mathbf{q}_2). \tag{46}$$

Although it is not commutative, the following relationship holds:

$$\mathbf{M}(\mathbf{q}_1, \mathbf{q}_2^\dagger) = [\mathbf{M}(\mathbf{q}_2, \mathbf{q}_1^\dagger)]^*. \tag{47}$$

If $\mathbf{q}_1 \neq \mathbf{q}_2^\dagger$, the matrix $\mathbf{M}(\mathbf{q}_1, \mathbf{q}_2^\dagger)$ is complex. The Mueller matrix $\mathbf{M}(\mathbf{q}, \mathbf{q}^\dagger)$, which is real, is the Mueller matrix equivalent of the scattering matrix $\mathbf{q}$.

References

1. J.C. Stover, *Optical Scattering: Measurement and Analysis*, 2nd edn (SPIE Optical Engineering Press, Bellingham, WA, 1995).
2. P. Müller-Buschbaum, J.S. Gutmann, C. Lorenz, T. Schmitt, and M. Stamm, "Decay of interface correlation in thin polymer films," *Macromolecules* **31**, 9265–9272 (1998).
3. W.S. Bickel, R.R. Zito, and V. Iafelice, "Polarized light scattering from metal surfaces," *J. Appl. Phys.* **61**, 5392–5398 (1987).
4. V.J. Iafelice and W.S. Bickel, "Polarized light-scattering matrix elements for select perfect and perturbed optical surfaces," *Appl. Opt.* **26**, 2410–2415 (1987).
5. G. Videen, J.-Y. Hsu, W.S. Bickel, and W.L. Wolfe, "Polarized light scattered from rough surfaces," *J. Opt. Soc. Am. A* **9**, 1111–1118 (1992).
6. T.A. Germer, "Angular dependence and polarization of out-of-plane optical scattering from particulate contamination, subsurface defects, and surface microroughness," *Appl. Opt.* **36**, 8798–8805 (1997).
7. T.A. Germer and C.C. Asmail, "Polarization of light scattered by microrough surfaces and subsurface defects," *J. Opt. Soc. Am. A* **16**, 1326–1332 (1999).
8. F.E. Nicodemus, J.C. Richmond, J.J. Hsia, I.W. Ginsberg, and T. Limperis, *Geometrical Considerations and Nomenclature for Reflectance*, NBS Monograph 160 (National Bureau of Standards, Gaithersburg, MD).
9. C. Asmail, J. Hsia, A. Parr, and J. Hoeft, "Rayleigh scattering limits for low-level bidirectional reflectance distribution function measurements," *Appl. Opt.* **33**, 6084–6091 (1994).
10. R.M.A. Azzam and N.M. Bashara, *Ellipsometry and Polarized Light* (Elsevier, Amsterdam, 1977).
11. D.S. Flynn and C. Alexander, "Polarized surface scattering expressed in terms of a bidirectional reflectance distribution function matrix," *Opt. Eng.* **34**, 1646–1650 (1995).
12. T.A. Germer and C.C. Asmail, "Goniometric optical scatter instrument for out-of-plane ellipsometry measurements," *Rev. Sci. Instrum.* **70**, 3688–3695 (1999).
13. SEMI ME1392, *Guide for Angle Resolved Optical Scatter Measurements on Specular or Diffuse Surfaces* (Semiconductor Equipment and Materials International, San José, CA, 2005).
14. ASTM E2387, *Standard Practice for Goniometric Optical Scatter Measurements* (ASTM International, West Conshohocken, PA, 2005).
15. J.W. Goodman, "Statistical properties of laser speckle patterns," in *Laser Speckle and Related Phenomena*, J.C. Dainty, ed. (Springer-Verlag, Berlin, 1984).

16. R.M.A. Azzam, "Photopolarimetric measurement of the Mueller matrix by Fourier analysis of a single detected signal," *Opt. Lett.* **2**, 148–150 (1978).
17. D.S. Sabatke, M.R. Descour, E.L. Dereniak, W.C. Sweatt, S.A. Kemme, and G.S. Phipps, "Optimization of retardance for a complete Stokes polarimeter," *Opt. Lett.* **25**, 802–804 (2000).
18. J.S. Tyo, "Design of optimal polarimeters: maximization of signal-to-noise ratio and minimization of systematic error," *Appl. Opt.* **41**, 619–630 (2002).
19. A. De Martino, Y.-K. Kim, E. Garcia-Caurel, B. Laude, B. Drévillon, "Optimized Mueller polarimeter with liquid crystals," *Opt. Lett.* **28**, 616–618 (2003).
20. G.E. Jellison and F.A. Modine, "Two-modulator generalized ellipsometry: theory," *Appl. Opt.* **36**, 8190–8198 (1997).
21. S.O. Rice, "Reflection of electromagnetic waves from slightly rough surfaces," *Commun. Pure Appl. Math.* **4**, 351–378 (1951).
22. D.E. Barrick, "Rough surfaces," in *Radar Cross Section Handbook*, G.T. Ruck, ed., (Plenum, New York, 1970).
23. T.V. Vorburger, E. Marx, and T.R. Lettieri, "Regimes of surface roughness measurable with light scattering," *Appl. Opt.* **32**, 3401–3408 (1993).
24. T. Karabacak, Y.-P. Zhao, M. Stowe, B. Quayle, G.-C. Wang, and T.-M. Lu, "Large angle in-plane light scattering from rough surfaces," *Appl. Opt.* **39**, 4658–4668 (2000).
25. J.M. Elson, "Theory of light scattering from a rough surface with an inhomogeneous dielectric permittivity," *Phys. Rev. B* **30**, 5460–5480 (1984).
26. J.M. Elson, "Characteristics of far-field scattering by means of surface roughness and variations in subsurface permittivity," *Waves Random Media* **7**, 303–317 (1997).
27. J.M. Elson, J.M. Bennett, and J.C. Stover, "Wavelength and angular dependence of light scattering from beryllium: comparison of theory and experiment," *Appl. Opt.* **32**, 3362–3376 (1993).
28. T.A. Germer, C.C. Asmail, and B.W. Scheer, "Polarization of out-of-plane scattering from microrough silicon," *Opt. Lett.* **22**, 1284–1286 (1997).
29. L. Sung, G.W. Mulholland, and T.A. Germer, "Polarized light-scattering measurements of dielectric spheres upon a silicon surface," *Opt. Lett.* **24**, 866–868 (1999).
30. J.H. Kim, S.H. Ehrman, G.W. Mulholland, and T.A. Germer, "Polarized light scattering by dielectric and metallic spheres on silicon wafers," *Appl. Opt.* **41**, 5405–5412 (2002).
31. T.A. Germer and C.C. Asmail, *Microroughness-blind optical scattering instrument* United States Patent #6,034,776 (2000).
32. H.G. Tompkins, *A User's Guide to Ellipsometry* (Academic Press, New York, 1993).
33. T.A. Germer, "Measurement of roughness of two interfaces of a dielectric film by scattering ellipsometry," *Phys. Rev. Lett.* **85**, 349–352 (2000).
34. T.A. Germer, "Polarized light scattering by microroughness and small defects in dielectric layers," *J. Opt. Soc. Am. A* **18**, 1279–1288 (2001).
35. J.M. Elson, "Light scattering from surfaces with a single dielectric overlayer," *J. Opt. Soc. Am.* **66**, 682–694 (1976).
36. J.M. Elson, "Infrared light scattering from surfaces covered with multiple dielectric overlayers," *Appl. Opt.* **16**, 2872–2881 (1977).
37. J.M. Elson, "Diffraction and diffuse scattering from dielectric multilayers," *J. Opt. Soc. Am.* **69**, 48–54 (1979).
38. J.M. Elson, J.P. Rahn, and J.M. Bennett, "Light scattering from multilayer optics: comparison of theory and experiment," *Appl. Opt.* **19**, 669–679 (1980).
39. J.M. Elson, "Multilayer-coated optics: guided-wave coupling and scattering by means of interface random roughness," *J. Opt. Soc. Am. A* **12**, 729–742 (1995).

40. B.W. Scheer, "Development of a physical haze and microroughness standard," in *Flatness, Roughness, and Discrete Defect Characterization for Computer Disks, Wafers, and Flat Panel Displays*, J.C. Stover, ed., *Proc. SPIE* **2862**, 78–95 (1996).
41. C. Deumié, H. Giovannini, and C. Amra, "Ellipsometry of light scattering from multilayer coatings," *Appl. Opt.* **35**, 5600–5608 (1996).
42. T.A. Germer, "Measurement of lithographic overlay by light scattering ellipsometry," in *Surface Scattering and Diffraction for Advanced Metrology II*, Z.-H. Gu and A.A. Maradudin, eds., *Proc. SPIE* **4780**, 72–79 (2002).
43. T.A. Germer and M.J. Fasolka, "Characterizing surface roughness of thin films by polarized light scattering," in *Advanced Characterization Techniques for Optics, Semiconductors, and Nanotechnologies*, A. Duparré and B. Singh, eds., *Proc. SPIE* **5188**, 264–275 (2003).

11
Scattering of Electromagnetic Waves from Nanostructured, Self-Affine Fractal Surfaces: Near-Field Enhancements

JOSÉ A. SÁNCHEZ-GIL,[†] JOSÉ V. GARCÍA-RAMOS,[†] VINCENZO GIANNINI,[†] AND EUGENIO R. MÉNDEZ[‡]

[†]*Instituto de Estructura de la Materia, Consejo Superior de Investigaciones Científicas, Madrid 28006 (Spain)* [‡]*Departamento de Óptica, División de Física Aplicada, Centro de Investigación Científica y de Educación Superior de Ensenada, Ensenada, Baja California 22800 (México)*

11.1. Introduction

Since the early days of fractality,[1] the scattering of electromagnetic (EM) waves from fractal surfaces has been a field of intense activity. Physical fractals appear ubiquitously in nature, possessing fractal properties within a broad, however finite, range of scales. Therefore the study of classical wave scattering from fractals is a problem of interest not only from a fundamental point of view, but also from the practical knowledge that probing technologies, such as surface optical characterization, remote sensing, radar, and sonar, can yield about a wide variety of systems. In fact, it is now well understood that many naturally occurring surfaces exhibit scale invariance, particularly in the form of self-affinity.[1–4]

There exists a large amount of theoretical works devoted to far-field wave scattering from fractal surfaces (cf. [5–13] and references therein). Most of them make use of approximations, such as the Kirchhoff approximation (KA) or perturbation methods, in order to obtain analytical expressions that are useful in certain regimes, thereby imposing a constraint on the length scales over which fractality might be present. Only very recent works are capable of dealing with arbitrarily rough metal or dielectric surfaces[7,8,11–13] on the basis of the Green's theorem integral equation formulation for rough surface scattering.[14–16]

In addition to far field scattering, the near EM field on self-affine fractals has attracted a great deal of attention.[13,17–27] Many theoretical works have exploited dipolar approximations (either retarded and nonretarded) to describe the optical response of fractal metal surface models.[17–23] Alternatively, full EM formulations have been recently employed, albeit typically one dimensional for the sake of numerical limitations.[20,24–28] To a large extent, the interest lies in the observation in the near field intensity distributions, through photon scanning tunneling microscopy (PSTM), of large concentrations of EM field intensity on bright spots.[22,29–34] The appearance of such localized optical modes is mediated by the roughness-induced excitation of surface-plasmon polaritons (SPPs) on nanostructured metal surfaces or nanoparticle aggregates. SPPs are surface electromagnetic waves bounded to a dielectric–metal interface and due to oscillations of the metal electron plasma.[35,36]

Recall that the occurrence of large surface EM fields is indeed crucial to the EM mechanism in surface-enhanced Raman scattering (SERS) and to other nonlinear surface optical processes.[18,20,21,23,24,26,34,37–40] Since the early days of SERS, it was established that SERS signals are enhanced orders of magnitude with respect to those of conventional Raman scattering,[37–39] thus being amply employed as a spectroscopic tool. The average enhancement factor, typically, $\sim 10^6$ stems from the combined action of two mechanisms of chemical (charge-transfer) and EM origin. The latter is widely accepted to be the most intense in several experimental configurations; it consists of surface-roughness-induced intensification of the EM field both at the pump frequency and at the Raman-shifted frequency,[38,39] mediated in most configurations by the above-mentioned excitation of SPPs. Furthermore, SERS single molecule probing has been recently reported,[41–44] claiming extremely large local enhancement factors. It is evident that large local EM fields must appear in the vicinity of the metal substrates such that a single, adsorbed molecule can be SERS detected, even if large effective resonant Raman cross sections are exploited.[44] SERS has been reported from a great variety of substrates: electrodes, colloids, silver islands, rough surfaces and films, etc. Interestingly, many SERS substrates posses scaling properties, physical fractality;[2–4] either self-similarity, as the widely employed colloidal aggregates,[2,45,46] or self-affinity, as in the case of deposited colloids, cold-deposited thin films, or evaporated or etched rough surfaces.[4,31,32,34,47]

Here we describe in detail the rigorous Green's theorem formulation to study the EM field scattered from nanostructured metal surfaces, restricted to one dimension for the sake of computational limitations. The scattering model is described in Sect. 11.2. We focus on fractal surfaces exhibiting self-affinity from tens of microns to the nanoscale. Actually, we analyze the influence on the near field of the size of the lower scale irregularities as is reduced from $\sim$50 nm to a few nanometers. Near-field calculations are presented in Sect. 11.3. The statistics of the surface-field enhancements is analyzed in Sect. 11.4, thereby discussing the impact of nanostructured self-affine fractals on SERS in Sect. 11.5. Finally, the main conclusions are summarized in Sect. 11.6.

11.2. Scattering Model

11.2.1. Scattering Geometry

The scattering geometry is depicted in Fig. 11.1. A monochromatic, linearly polarized incident beam of frequency ω impinges at an angle θ_0 on a randomly rough surface $z = \zeta(x)$. The rough interface separates a dielectric from a semi-infinite metal volume occupying the lower half-space $[z < \zeta(x)]$, characterized,

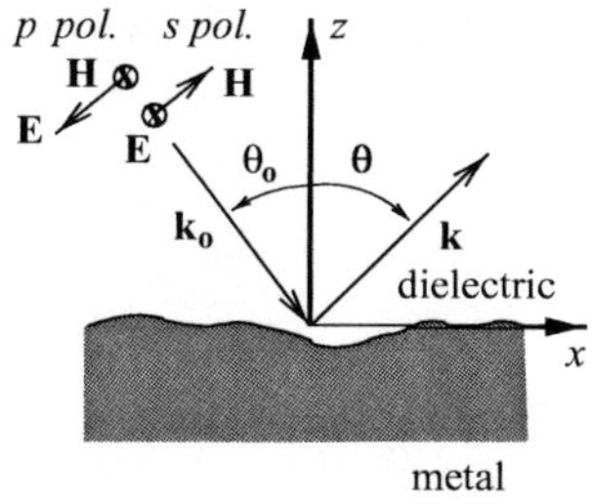

FIGURE 11.1. Illustration of the scattering geometry.

respectively, by isotropic, frequency-dependent homogeneous dielectric functions $\epsilon^{>}(\omega)$, $\epsilon^{<}(\omega)$.

With the aim of solving this scattering problem for arbitrarily large roughness parameters, we make use of the scattering integral equations based on the application of Green's second integral theorem.[15] It is well known that this fully vectorial formulation is considerably simplified when restricted to 1D surfaces and linear polarization,[14–16] thereby being reduced to four scalar integral equations for the only nonzero component of either the electric field amplitude

$$\mathbf{E}^{(s)} \equiv E^{(s)}(\mathbf{r}, \omega)\hat{\mathbf{y}},$$

with

$$\mathbf{H}^{(s)} \equiv H_x^{(s)}(\mathbf{r}, \omega)\hat{\mathbf{x}} + H_z^{(s)}(\mathbf{r}, \omega)\hat{\mathbf{z}},$$

for s polarization; or the magnetic field amplitude

$$\mathbf{H}^{(p)} \equiv H^{(p)}(\mathbf{r}, \omega)\hat{\mathbf{y}},$$

with

$$\mathbf{E}^{(p)} \equiv E_x^{(p)}(\mathbf{r}, \omega)\hat{\mathbf{x}} + E_z^{(p)}(\mathbf{r}, \omega)\hat{\mathbf{z}},$$

for p polarization.[14,16] From now on, since a time harmonic dependence $e^{-i\omega t}$ is assumed, the functional dependence on frequency will be omitted unless necessary for the sake of clarity.

11.2.2. Scattering Equations

The surface integral equations that fully describe the EM linear scattering problem, in the geometry of Fig. 11.1, are

$$\psi^{(i)}(\mathbf{r}) + \frac{1}{4\pi}\int_{-\infty}^{\infty} \gamma' \, dx' \left[\psi^{(>)}(\mathbf{r}') \frac{\partial G^{>}(\mathbf{r}, \mathbf{r}')}{\partial n'} - G^{>}(\mathbf{r}, \mathbf{r}') \frac{\partial \psi^{(>)}(\mathbf{r}')}{\partial n'} \right]$$

$$= \psi^{(>)}(\mathbf{r}), \; z > \zeta(x), \quad (1a)$$

$$= 0, \; z < \zeta(x); \quad (1b)$$

$$-\frac{1}{4\pi}\int_{-\infty}^{\infty}\gamma' \mathrm{d}x' \left[\psi^{(<)}(\mathbf{r}')\frac{\partial G^{<}(\mathbf{r},\mathbf{r}')}{\partial n'} - G^{<}(\mathbf{r},\mathbf{r}')\frac{\partial \psi^{(<)}(\mathbf{r}')}{\partial n'}\right]$$

$$= 0, z > \zeta(x), \qquad (1c)$$

$$= \psi^{(<)}(\mathbf{r}), z < \zeta(x). \qquad (1d)$$

The function $\psi(\mathbf{r})$ is defined as

$$\psi(\mathbf{r}) \equiv E^{(s)}(\mathbf{r}), \text{ for } s \text{ polarization}; \qquad (2a)$$

$$\equiv H^{(p)}(\mathbf{r}), \text{ for } p \text{ polarization}; \qquad (2b)$$

and the superscripts $>$ and $<$ denote the upper ($z > \zeta$) and lower ($z < \zeta$) semi-infinite half-spaces. The normal derivative is defined as $\partial/\partial n \equiv (\hat{\mathbf{n}} \cdot \nabla)$, with $\hat{\mathbf{n}} \equiv \gamma^{-1}(-\zeta'(x), 0, 1)$ and $\gamma = (1 + (\zeta'(x))^2)^{1/2}$. The 2D Green's function G is given by the zeroth-order Hankel function of the first kind $H_0^{(1)}$.

Our monochromatic incident field of frequency ω is a Gaussian beam of half-width W and incident angle θ_0 in the form

$$\psi^{(i)}(x,z) = \exp\{\imath k_\epsilon(x \sin\theta_0 - z\cos\theta_0) \times [1 + w(x,z)]\} \exp\left[-\frac{(x\cos\theta_0 + z\sin\theta_0)^2}{W^2}\right], \qquad (3a)$$

$$w(x,z) = \frac{1}{k_\epsilon^2 W^2}\left[\frac{2}{W^2}(x\cos\theta_0 + z\sin\theta_0)^2 - 1\right], \qquad (3b)$$

where $k_\epsilon = n_c^> \omega/c$ and $n_c^> = \sqrt{\epsilon^>}$.

In order to solve for the surface field and its normal derivative, two of the integral equations (note that they are not independent), typically Eqs. (1a) and (1c), are used as boundary conditions. Upon invoking the continuity conditions across the interface

$$E(x) = E^{(>,s)}(\mathbf{r})\mid_{z=\zeta^{(+)}(x)} = E^{(<,s)}(\mathbf{r})\mid_{z=\zeta^{(-)}(x)}, \qquad (4a)$$

$$\gamma^{-1}F(x) = \left[\frac{\partial E^{(>,s)}(\mathbf{r})}{\partial n}\right]_{z=\zeta^{(+)}(x)} = \left[\frac{\partial E^{(<,s)}(\mathbf{r})}{\partial n}\right]_{z=\zeta^{(-)}(x)}, \qquad (4b)$$

$$H(x) = H^{(>,p)}(\mathbf{r})\mid_{z=\zeta^{(+)}(x)} = H^{<,p)}(\mathbf{r})\mid_{z=\zeta^{(-)}(x)}, \qquad (4c)$$

$$\gamma^{-1}L(x) = \left[\frac{\partial H^{(>,p)}(\mathbf{r})}{\partial n}\right]_{z=\zeta^{(+)}(x)} = \frac{\epsilon^>}{\epsilon^<}\left[\frac{\partial H^{(<,p)}(\mathbf{r})}{\partial n}\right]_{z=\zeta^{(-)}(x)}, \qquad (4d)$$

with $\zeta^{(\pm)}(x) = \lim_{\varepsilon\to 0}(\zeta(x) \pm \varepsilon)$, two coupled integral equations are obtained for each polarization. The resulting system of integral equations can be numerically solved upon converting it into a system of linear equations through a quadrature scheme,[14,16] the unknowns being $E(x)$ and $F(x)$ for s polarization, and $H(x)$ and $L(x)$ for p polarization. Once these source functions are obtained, Eqs. (1a) and (1d) permit to calculate the scattered electric (magnetic) field amplitude for s (p) polarization at any point in the upper incident medium and inside the metal, respectively.

11.2.3. Near Field

In addition, from the latter integral equations, the corresponding expressions for the magnetic (electric) field amplitudes for s (p) polarization can be simply obtained by making use of Maxwell equations. These expressions can be useful in near electric or magnetic field calculations. This is indeed the situation in SERS where the near electric field locally excites the molecule vibrations that produce the Raman-shifted radiation that is detected. Thus we include below the electric field equations for p polarization.

In the incident medium, Eq. (1a) provides the only nonzero component of the magnetic field. Use of Maxwell's equation $\nabla \times \mathbf{H} = -\mathrm{i}\frac{\omega}{c}\epsilon \mathbf{E}$ leads to the following electric field components:

$$E_x^{(p,>)}(\mathbf{r}) = E_x^{(p,i)}(\mathbf{r}) - \mathrm{i}\frac{c}{4\pi\omega\epsilon^>}\int_{-\infty}^{\infty}\gamma'\mathrm{d}x'\left[H^{(p,>)}(\mathbf{r}')\frac{\partial^2 G^>(\mathbf{r},\mathbf{r}')}{\partial z\partial n'} - \frac{\partial G^>(\mathbf{r},\mathbf{r}')}{\partial z}\frac{\partial H^{(p,>)}(\mathbf{r}')}{\partial n'}\right] \tag{5a}$$

$$E_y^{(p,>)}(\mathbf{r}) = 0 \tag{5b}$$

$$E_z^{(p,>)}(\mathbf{r}) = E_z^{(p,i)}(\mathbf{r}) + \mathrm{i}\frac{c}{4\pi\omega\epsilon^>}\int_{-\infty}^{\infty}\gamma'\mathrm{d}x'\left[H^{(p,>)}(\mathbf{r}')\frac{\partial^2 G^>(\mathbf{r},\mathbf{r}')}{\partial x\partial n'} - \frac{\partial G^>(\mathbf{r},\mathbf{r}')}{\partial x}\frac{\partial H^{(p,>)}(\mathbf{r}')}{\partial n'}\right]. \tag{5c}$$

These equations can be rewritten in terms of the source functions $H(x)$ and $L(x)$ as follows:

$$\begin{aligned}E_x^{(p,>)}(\mathbf{r}) = E_x^{(p,i)}(\mathbf{r}) - \frac{\omega}{4c}\int_{-\infty}^{\infty}\gamma'\,\mathrm{d}x'\Bigg\{H(x')\\ \times\left[\frac{z-\zeta(x')}{|\mathbf{r}-\mathbf{r}'|^2}(\mathbf{n}\cdot(\mathbf{r}-\mathbf{r}'))H_2^{(1)}(k_\epsilon|\mathbf{r}-\mathbf{r}'|) - \frac{1}{\gamma' k_\epsilon|\mathbf{r}-\mathbf{r}'|}H_1^{(1)}(k_\epsilon|\mathbf{r}-\mathbf{r}'|)\right]\\ -L(x')\frac{z-\zeta(x')}{\gamma' k_\epsilon|\mathbf{r}-\mathbf{r}'|}H_1^{(1)}(k_\epsilon|\mathbf{r}-\mathbf{r}'|)\Bigg\}\end{aligned} \tag{6a}$$

$$E_y^{(p,>)}(\mathbf{r}) = 0 \tag{6b}$$

$$\begin{aligned}E_z^{(p,>)}(\mathbf{r}) = E_z^{(p,i)}(\mathbf{r}) - \frac{\omega}{4c}\int_{-\infty}^{\infty}\gamma'\mathrm{d}x'\Bigg\{H(x')\\ \times\left[\frac{-(x-x')}{|\mathbf{r}-\mathbf{r}'|^2}(\mathbf{n}\cdot(\mathbf{r}-\mathbf{r}'))H_2^{(1)}(k_\epsilon|\mathbf{r}-\mathbf{r}'|) - \frac{\zeta'(x')}{\gamma' k_\epsilon|\mathbf{r}-\mathbf{r}'|}H_1^{(1)}(k_\epsilon|\mathbf{r}-\mathbf{r}'|)\right]\\ +L(x')\frac{x-x'}{\gamma' k_\epsilon|\mathbf{r}-\mathbf{r}'|}H_1^{(1)}(k_\epsilon|\mathbf{r}-\mathbf{r}'|)\Bigg\},\end{aligned} \tag{6c}$$

where the explicit form of the Green's function has been taken into account, leading to the appearance of 1st and 2nd order Hankel functions of the first kind $H_1^{(1)}$, $H_2^{(1)}$. For the Gaussian incident field given by Eq. (3), the electric field components are

$$E_x^{(p,i)}(\mathbf{r}) = \frac{\mathrm{i}}{n_c^>} H^{(p,i)}(\mathbf{r}) \Big[\mathrm{i} \cos\theta_0 (1 + w(x,z)) - \left(\mathrm{i} \frac{4}{k_\epsilon^2 W^4} (x \sin\theta_0 - z \cos\theta_0) - \frac{2}{k_\epsilon W^2} \right) \sin\theta_0 (x \cos\theta_0 + z \sin\theta_0) \Big] \tag{7a}$$

$$E^{(p,i)}(\mathbf{r}) = 0 \tag{7b}$$

$$E_z^{(p,i)}(\mathbf{r}) = \frac{\mathrm{i}}{n_c^>} H^{(p,i)}(\mathbf{r}) \Big[\mathrm{i} \sin\theta_0 (1 + w(x,z)) + \left(\mathrm{i} \frac{4}{k_\epsilon^2 W^4} (x \sin\theta_0 - z \cos\theta_0) - \frac{2}{k_\epsilon W^2} \right) \cos\theta_0 (x \cos\theta_0 + z \sin\theta_0) \Big]. \tag{7c}$$

Equations (6) and (7) provide the electric field components in the incident medium of the resulting p-polarized EM field, incident plus scattered from the rough surface. The scattered electric field involves an additional surface integral in terms of the source functions, previously obtained from the above mentioned coupled integral equations. Analogous expressions, not shown here, for the corresponding electric field components inside the metal can be obtained from Eq. (1d). On the other hand, recall that a similar procedure can be straightforwardly developed to yield the magnetic field components in the case of s-polarized EM waves as surface integrals in terms of the surface electric field (y-component) and its normal derivative.

11.2.4. Surface Field

The calculation of the surface magnetic field intensity for s polarization can be done in a simple manner by exploiting the connection between its normal and tangential components on the upper (dielectric) medium

$$H_n^{(s)}(x) = -\frac{\mathrm{i}c}{\omega} \gamma^{-1} \frac{dE(x)}{dx}, \tag{8a}$$

$$H_t^{(s)}(x) = \frac{\mathrm{i}c}{\omega} \gamma^{-1} F(x). \tag{8b}$$

The corresponding expressions for the surface electric field components for p polarization are given by[24]

$$E_n^{(p)}(x) = \frac{\mathrm{i}c}{\omega \epsilon^>} \gamma^{-1} \frac{\mathrm{d}H(x)}{dx}, \tag{9a}$$

$$E_t^{(p)}(x) = -\frac{\mathrm{i}c}{\omega \epsilon^>} \gamma^{-1} L(x). \tag{9b}$$

As mentioned above, $E(x)$, $F(x)$, $H(x)$, $L(x)$ constitute the source functions of the integral equations in this scattering configuration.

11.2.5. Self-Affine Fractals

As a model describing many naturally occurring surface growth phenomena exhibiting self-affine fractality, we have chosen that given by the trace of a fractional Brownian motion through the Voss' algorithm.[3,48,49] In addition, this model yields self-affine fractal structures that resemble fairly well the properties of some SERS metal substrates.[31,32,34] The ensembles of realizations thus generated are characterized by their fractal dimension $D = 2 - \mathcal{H}$ ($\mathcal{H}$ being the Hurst exponent) and rms height δ. In order to avoid the inherent ambiguity in the definition of the rms height for self-affine fractals, δ refers in our calculations to the rms height defined over the entire fractal profile with length $L_f = 51.4$ μm. Recall that δ depends on the length Δx over which it is measured through[12] $\delta = l^{1-\mathcal{H}} \Delta x^{\mathcal{H}}$, l being the topothesy. From each generated fractal profile with N_f points and length L_f, sequences of N points (with constant N/N_f) are extracted to obtain similar profiles with identical properties except for the lower scale cutoff as determined by $\xi_L = L_f/N_f$. (Strictly speaking, ξ_L should be given by the minimum length scale above which the ensemble of such realizations exhibit self-affinity, resulting in a value typically larger than the mere discretization cutoff[3,20]).

From now on, we will focus on the fractal dimension $D = 1.9$, for the influence of decreasing ξ_L is more significant for larger fractal dimensions. The effect of varying D has been already studied on the far field[8] and the near field,[20] albeit for a relatively large lower cutoff ξ_L. The values of the lower scale cutoffs hitherto considered are $\xi_L = 51.4, 25.7, 12.85$, and 6.425 nm, resulting from sequences of $N = 102, 205, 410$, and 819 points extracted from profiles with $N_f = 1024, 2048, 4096$, and 8192 points. The length of all realizations is thus $L = L_f(N/N_f) = 5.14$ μm. Actually, the final number of sampling points per realization used in the numerical calculations is significantly higher for the sake of accuracy: $N_p = n_i N$ as obtained by introducing $n_i = 4 - 15$ cubic-splined interpolating points, the latter being chosen on the basis of numerical convergence tests.

Interestingly, this range of lower scale cutoff covers that of SERS substrates that can be obtained by depositing fractal colloidal aggregates of Ag particles with various diameters,[32,34,46] as well as the cutoff of evaporated rough surfaces,[31] approaching in the lower limit that of cold-deposited silver films. The fractal lower scale cutoff, beyond which fractality is preserved, is actually larger than ξ_L, though approximately subwavelength. The fractal upper scale cutoff is greater than the Gaussian beam width (a few microns). The results collected below restrict to vacuum and silver as (incident) dielectric and (scattering) metallic media, respectively. Similar considerations apply to other media, as revealed for water (dielectric) and other noble metals (Au and Cu) in [20]. Bulk dielectric function data for Ag are taken from [50]. It should be mentioned that in order to consider yet smaller lower scale cutoffs (with large δ), leading to surface irregularities narrower than ~10 nm,

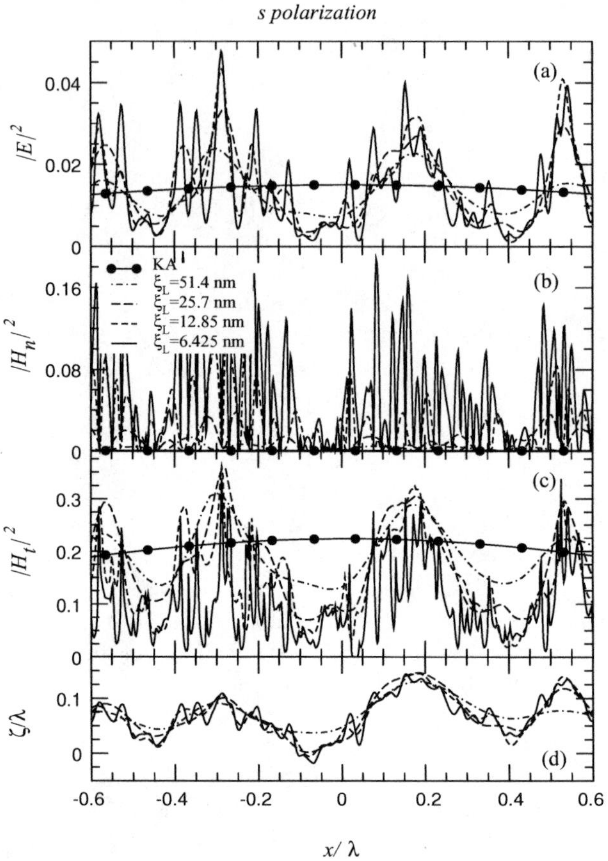

FIGURE 11.2. Surface electric (a) and magnetic (normal (b) and tangential (c)) field intensities on the center spot of the illuminated area ($L = 5.14\ \mu m$) for s-polarized scattering with $\theta_0 = -10^\circ$, $\lambda = 629.9$ nm, and $W = L/4\cos\theta_0$, from Ag fractal surfaces with $D = 1.9$, $\delta = 51.4$ nm, and $\xi_L = 51.4, 25.7, 12.85$, and 6.425 nm. The KA field intensity is also included (see text). (d) The corresponding surface profiles.

surface scattering effects yielding a larger imaginary part of the free electron (Drude) contribution to the dielectric function should be included.[51]

11.3. Near and Surface Field

We now turn to the investigation of the influence on the near EM field of the lower scale cutoff of the self-affine fractal surfaces.[13,25] Near field intensity distributions are relevant for the information they provide on the scattering process, and can be measured through near-field optical microscopy.

11.3.1. Surface Fields

First, we calculate the intensities of all the EM field components on the surface. The results for moderately rough fractals with $\delta = 51.4$ nm are shown in Figs. 11.2 and 11.3 for s and p polarization, respectively. The corresponding surface profiles are shown in the bottom, Figs. 11.2d and Figs. 11.3d. The nonzero components of the EM field being plotted are (cf. Sect. 11.2): the tangential, perpendicular to the incident plane, electric (respectively, magnetic) field and the normal and tangential (in the plane of incidence) magnetic (respectively, electric) field, in the case of s (respectively, p) polarization. For the sake of clarity, only the central part of the illuminated surface is shown.

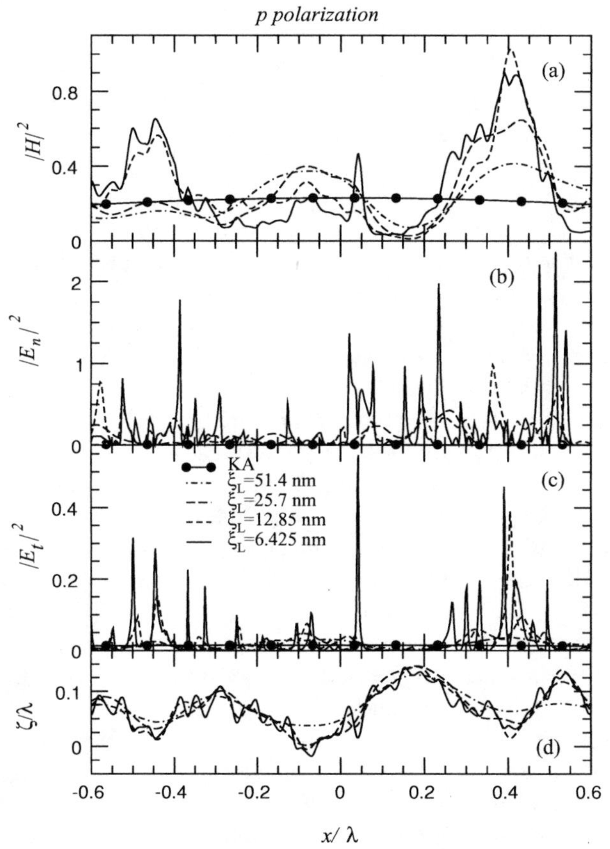

FIGURE 11.3. Surface magnetic (a) and electric (normal (b) and tangential (c)) field intensities on the center spot of the illuminated area ($L = 5.14$ μm) for p-polarized scattering with $\theta_0 = -10°$, $\lambda = 629.9$ nm, and $W = L/4 \cos\theta_0$, from Ag fractal surfaces with $D = 1.9$, $\delta = 51.4$ nm, and $\xi_L = 51.4$, 25.7, 12.85, and 6.425 nm. The KA field intensity is also included (see text). (d) The corresponding surface profiles.

Before analyzing the spatial distributions, let us recall what is expected within the KA, namely, the sum of the incident and the specularly (locally) reflected fields. Since the angular variation of the Fresnel coefficients for metals at this frequency is small, the local variations of the specular field can be neglected. Therefore, the KA field intensities are approximately given by those for a planar surface, namely

$$| E^{(s,\mathrm{KA})} |^2 \approx | 1 + \mathcal{R}_s |^2 | E^{(s,i)} |^2 \tag{10a}$$
$$| H_n^{(s,\mathrm{KA})} |^2 \approx \sin^2 \theta_0 \, | 1 + \mathcal{R}_s |^2 | H^{(s,i)} | \tag{10b}$$
$$| H_t^{(s,\mathrm{KA})} |^2 \approx \cos^2 \theta_0 \, | 1 - \mathcal{R}_s |^2 | H^{(s,i)} |, \tag{10c}$$

for s polarization; and

$$| H^{(p,\mathrm{KA})} |^2 \approx | 1 + \mathcal{R}_p |^2 | H^{(p,i)} |^2 \tag{11a}$$
$$| E_n^{(p,\mathrm{KA})} |^2 \approx \sin^2 \theta_0 \, | 1 + \mathcal{R}_p |^2 | E^{(p,i)} | \tag{11b}$$
$$| E_t^{(p,\mathrm{KA})} |^2 \approx \cos^2 \theta_0 \, | 1 - \mathcal{R}_p |^2 | E^{(p,i)} |, \tag{11c}$$

for p polarization, $\mathcal{R}_s$ and $\mathcal{R}_p$ being the corresponding Fresnel coefficients for θ_0 on a planar metal surface. The latter field intensities are plotted in Figs. 11.2 and 11.3. As expected, the KA does not hold even for the moderately rough surface profiles used therein, nor does perturbation theory (recall that the planar surface field intensities can be considered the zeroth-order approximation in the small-amplitude perturbation expansion of the field), but they provide the background about which the actual surface EM field intensities strongly vary.

In the case of s polarization, Fig. 11.2, significant variations appear in the surface EM field upon decreasing the lower scale cutoff due to the evanescent components. Fluctuations about the background distributions, Eqs. (10), become narrower and steeper the smaller is ξ_L, namely, the smaller are the surface features. This effect is more pronounced for the normal magnetic field, for which the expected KA background is indeed very small. Interestingly, it should be noted that the electric field intensity resembles with positive contrast the surface profile (the same is true for the total, tangential plus normal, magnetic field intensity), which is not the case in general.[52] Although not valid from a quantitative standpoint, small amplitude perturbation theory can provide a qualitative explanation. The first-order term of the surface electric field in powers of the surface height, which in turn gives the first-order correction to the planar surface background,[52,53] can be cast for this particular scattering geometry, polarization and near normal incidence in the form of a convolution integral involving the surface profile function. We have verified, though not shown here, that by simply increasing the angle of incidence up to $\theta_0 = 40^\circ$ the resemblance is slightly lost.

Upon illuminating with p-polarized light, see Fig. 11.3, the surface EM field intensity distributions become more complicated, with larger variations from one realization to another with diminishing ξ_L, and no resemblance whatsoever with the surface profiles. These variations are considerably larger than those for the far-field speckle patterns,[13] indicating the crucial role played by the evanescent components. Furthermore, the excitation of SPPs propagating along the surface and reradiating into vacuum mediates the scattering process. In fact, the spatial frequency of the large oscillations about the background in the surface EM field intensity is related to the SPP wavelength[20]; this is more easily observed in the surface magnetic field in Fig. 11.3a (also in the total electric field, not shown here) for the profiles with higher ξ_L. On the other hand, the surface normal electric field component for the smaller ξ_L (see Fig. 11.3b) reveals the appearance of narrow and bright spots where an enhancement of the field intensity of nearly two orders of magnitude occur, despite the relatively low value of δ. This clearly manifests the crucial role played by the lower scale cutoff in the excitation of localized optical modes.[25]

11.3.2. Near Field Map: Localized Surface-Plasmon Polaritons

It is thus seen that the effect of decreasing the nanoscale lower cutoff on self-affine fractals can lead to the appearance of p-polarized localized SPPs even for moderately rough rms heights. The effect can be quantitatively more relevant for very rough surfaces. Let us thus plot the EM field in the near vicinity of one such rough interface. This is done in Fig. 11.4, where the near-field intensity

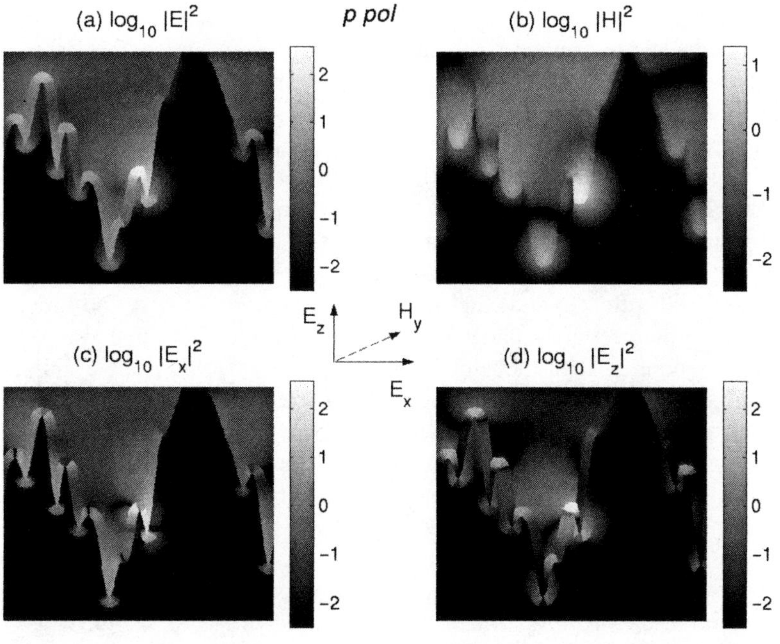

FIGURE 11.4. Near field intensity images in p polarization ((a) electric and (b) magnetic, the former split into (c) x, horizontal component, and (d) z, vertical component, all of them in a $\log_{10}$ scale) in an area of $0.5 \times 0.5\ \mu m^2$ close to a fractal surface with $D = 1.9$, $\delta = 257$ nm, and $\xi_L = 12.85$ nm, where a strong *localized optical mode* is observed.

maps in a logarithmic scale for *p* polarization are shown. All of the nonzero EM field components are included for the sake of completeness; recall that it has been reported that the magnetic field intensity can be also probed through PSTM for certain experimental configurations.[54] The actual surface profile, though not explicitly depicted, can be inferred from the fairly black, metallic regions due to the evanescent behavior of the EM fields inside metal with a small skin depth ($d \approx 25$ nm).

A very bright spot is found at the local maximum in the central part of the intensity map in Fig. 11.4a: The intensity field enhancement at such spot is $|E|^2 / |E^{(i)}|^2 \sim 10^4$. Note that there seems to be a bright, though weaker, magnetic field spot associated with the optical mode (see Fig. 11.4b), which is nonetheless slightly shifted to the right with respect to the electric field maximum. The electric field intensity decays rapidly upon moving into vacuum away from the bright spot, faster than expected for the evanescent decay of SPP propagating on a plane. This may help explain why localized optical modes experimentally observed through PSTM yield considerably smaller enhancement factors,[33] leaving aside the fact that no direct comparison with theoretical calculations for the actual experimental surface profile are available. With regard to the electric field orientation on the bright spot, Figs. 11.4c and 11.4d indicate that the normal electric field component is responsible for the bright spot; this component corresponds to $|E_z|^2$ in regions near surface maxima and minima (see Fig. 11.4d) and to $|E_x|^2$ near vertical surface walls (see Fig. 11.4c). It can be also observed that, as expected, the normal electric field is discontinuous across the interface, the tangential component being continuous. The continuity of the (tangential) magnetic field for *p* polarization is evident in Fig. 11.4a.

With regard to the physics underlying localized optical excitations, two mechanisms have been proposed: Anderson localization of SPP or SPP shape resonances. Theoretical results[27] rule out the former mechanism, whereas unequivocally connecting localized SPPs with SPP shape resonances occurring at either grooves or ridges. Upon examining the near-field patterns of localized SPP for many different ensembles (self-affine fractals and other Gaussian-correlated surfaces), striking qualitative similarities are found with those for isolated defects of analogous sizes, in strong correlation with their location at either grooves or ridges. This is evidenced in Fig. 11.5, where the *p*-polarized, near-electric-field map is shown for isolated, metal Gaussian defects with dimensions close to those of the typical grooves and ridges in the nanostructured surfaces being studied. In the case of ridges, Figs. 11.5b and 11.5d, see also Fig. 11.4 above, the electric field is fairly symmetric and normal to the surface, with a large peak at the very tip of the surface, resembling a *monopolar* configuration. This configuration is enforced by the boundary conditions, SPP polarization, and nonforbidden charge distribution at the very end of the metal tip at adjacent walls. In contrast, as seen in Figs. 11.5a and 11.5c, a SPP is trapped near the groove bottom with a *dipole*-like, opposite charge concentration on either groove wall.[27,28] This has been verified for a wide spectral range in the visible and near-IR, and should be applicable to any disordered nanostructured metal configuration supporting SPPs. It is not the case, however, of subwavelength nanoparticle aggregates, where localized optical excitations are

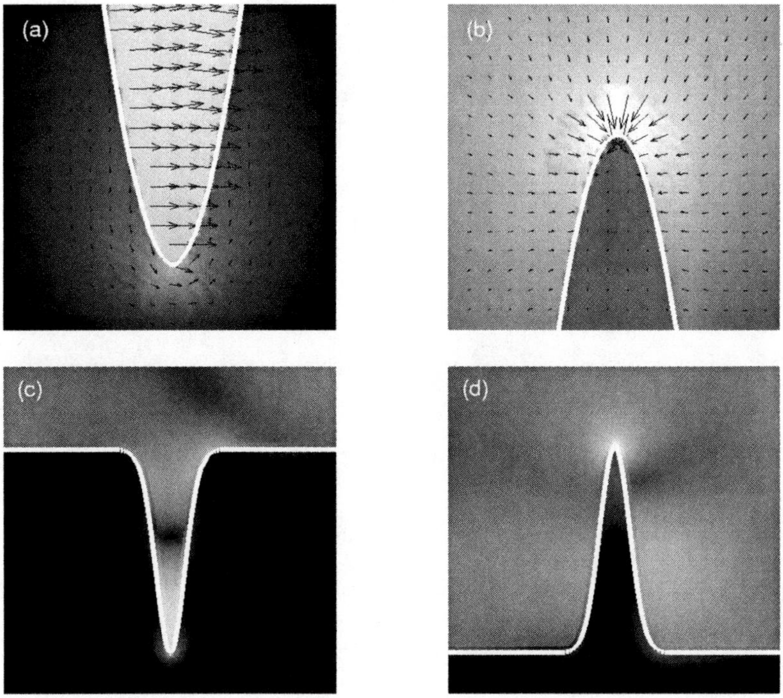

FIGURE 11.5. Near-field intensity images of the p-polarized electric field intensity (normalized to that of the incident field) for single Gaussian defects of $1/e$-width $A = 25.7$ nm and height $h = \pm\ 257$ nm ($\theta_0 = 40°$) in an area $0.5\times0.5\ \mu\text{m}^2$ close to either (c) a groove or (d) a ridge, zoomed in ($60 \times 60\ \text{nm}^2$) in (a) and (b), respectively. The surface profile is superimposed as a white curve. All gray scales span from $\log_{10}\sigma = -1$ (black) to 2 (white).

observed and interpreted in the quasistatic approach as Anderson-localized surface plasmons.[22]

11.4. Surface Field Enhancement: Statistics

We now analyze the enhancement factor of the surface electric field intensity:[25,26]

$$\sigma(\omega) = |\,\mathbf{E}(\omega)\,|^2 \;/\; |\,\mathbf{E}^{(\mathrm{i})}(\omega)\,|^2, \tag{12}$$

stemming from the excitation of localized SPPs as shown above, which play a decisive role in SERS and other surface nonlinear optical processes.[17,20,23,26,40]

The dependence on ξ_L of relevant statistical properties of the enhancement factor of the surface electric field intensity (mean $\langle\sigma\rangle$, and fluctuations $\Delta\sigma = (\langle\sigma^2\rangle/\langle\sigma\rangle^2 - 1)^{1/2}$) is analyzed from the calculations with normal incidence over ensembles of $N_{\text{rea}} = 100$ self-affine fractal realizations; typically, $N_{\text{data}} = N_{\text{rea}}N_p/2 \geq 10^5$ are used in the calculations. Moreover, several pump beam

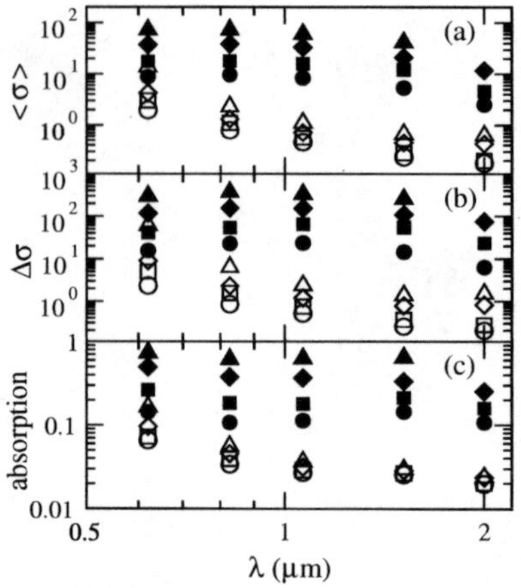

FIGURE 11.6. Spectral dependence of the enhancement factor of the surface electric field intensity (σ) obtained from the results for $N_{rea} = 100$ self-affine Ag fractal profiles with length $L = 5.14$ μm, $D = 1.9$, and $\delta = 51.4$ nm (hollow symbols) and $\delta = 257$ nm (filled symbols). (a) Mean, (b) fluctuations, and (c) normalized absorption. Circles: $\xi_L = 51.4$ nm; squares: $\xi_L = 25.7$ nm; diamonds: $\xi_L = 12.85$ nm; and triangles: $\xi_L = 6.425$ nm. (The result for $\delta = 257$ nm and $\xi_L = 6.425$ nm at $\lambda = 2$ μm is not shown for it does not satisfy numerical convergence with the available N_p.)

wavelengths in the visible and near IR are considered: $\lambda = 2\pi c/\omega$ =629.9, 826.6, 1064, 1512, and 2000 nm.

The results for the spectral dependence of $\langle\sigma\rangle$ and $\Delta\sigma$, along with the normalized absorption A, are presented in Fig. 11.6. First of all, note that throughout the visible and near IR regions studied, both the mean and fluctuations decrease with increasing ξ_L, as seen in Figs. 11.6a and 11.6b through the vertical variation for fixed λ; and as expected, σ is always larger for higher δ. The spectral response depends on the surface roughness parameters. For the rougher self-affine fractals, $\langle\sigma\rangle$ and $\Delta\sigma$ vary very smoothly over the frequency range shown, whereas maximum values are reached in the visible for the smoother self-affine fractals with a significant decrease into the near IR. Leaving aside the variations associated with large dispersion in the bulk dielectric function, as is the case of the strong damping due to the onset of interband transitions for Ag at $\lambda < 300$ nm, the spectral behavior of the enhancement factor can be roughly explained through the optimization of SPP excitation in the form of localized optical modes. The excitation mechanism is favored by surface features with small, nanoscale lateral dimensions and large heights. In terms of the increasing wavelength, light 'sees' surface features increasingly narrower, but smaller. This interplay may lead to the decay of σ in the IR, which occurs at higher λ for larger δ (only barely perceivable at $\lambda = 2$ μm for δ =257 nm). On the other hand, the total absorption spectra in Fig. 11.6c behave similarly to the spectra of the moments of the enhancement factor. It is worth noting that the small dissipative damping (less than 1%) occurring for planar Ag surfaces

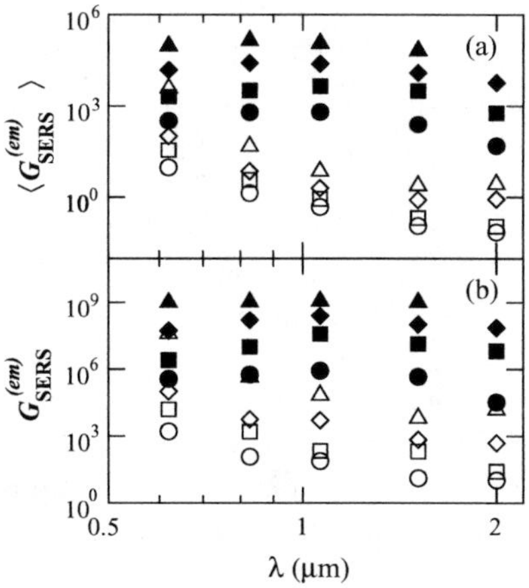

FIGURE 11.7. Spectral dependence of the SERS enhancement factor (EM mechanism) for the same self-affine Ag fractals as used in Fig. 11.6: (a) Average and (b) local maximum.

becomes huge (up to ~70%) for the rougher self-affine fractals with nanoscale cutoff due to the excitation and subsequent damping of SPP.

11.5. Surface-Enhanced Raman Scattering

The EM part of the SERS enhancement factor contains the contributions from σ at both the pump frequency ω and the Raman-shifted frequency ω_R, which we approximate by the square of the former,[38] namely

$$G_{\text{SERS}}^{(\text{em})} = \sigma(\omega)\sigma(\omega_R) \approx \sigma^2(\omega). \tag{13}$$

Roughly speaking, this approximation is justified provided that the Raman shift $\Delta\omega \equiv |\omega - \omega_R|$ is smaller than the typical line-width Γ_{SPP} of the localized SPP[24]; with regard to the results presented below,[20,26] this implies that even for the smaller line-widths found at the rougher surfaces, $\Gamma_{\text{SPP}} \gtrsim 0.2$ eV, the obtained SERS enhancement factors are valid for a reasonably wide range of Raman shifts $\Delta\omega \sim 1600$ nm^{-1}.

It is evident from Fig. 11.6 that mean enhancement factors can become large, but seemingly not exceeding $\langle\sigma\rangle \sim 10^2$. Fluctuations are in turn enhanced up to $\Delta\sigma \sim 10^3$, which is crucial in obtaining large enhancement factors in SERS and other nonlinear surface optical processes.[17,23,26] In Fig. 11.7a, the spectral dependence of the average SERS enhancement factor (EM mechanism), defined as $\langle G_{\text{SERS}}^{(\text{em})}\rangle = \langle\sigma^2\rangle$, is shown for the self-affine fractal Ag substrates considered here. Incidentally, similar results are obtained for other widely used metal substrates such as Au and Cu (except for wavelengths $\lambda < 600$ nm below the corresponding onset of interband transitions). Very large values are found, $\langle G_{\text{SERS}}^{(\text{em})}\rangle \lesssim 10^6$, on the order of those commonly inferred for SERS substrates.

In addition to large $\langle G_{\rm SERS}^{\rm (em)}\rangle$, it should be remarked that nanoscale cutoffs can lead to very large local values of $G_{\rm SERS}^{\rm (em)}$. This is shown in Fig. 11.7b, where the spectral variation of the maximum values of $G_{\rm SERS}^{\rm (em)}$ is plotted for the same self-affine Ag fractals as in Fig. 11.7a. The qualitative behavior of the local SERS enhancement factor with respect to pump frequency, surface RMS height and lower scale cutoff is analogous to that of the average SERS enhancement factor. Quantitatively, local enhancements can be considerably larger than average values, reaching $G_{\rm SERS}^{\rm (em)} \sim 10^9$. These huge local enhancement factors can be responsible for the bright spots at which SERS single-molecule detection has been claimed.[41–44] Nonetheless our electromagnetically induced calculated values of $G_{\rm SERS}^{\rm (em)}$ in Fig. 11.7b lie a few orders of magnitude below those allegedly needed in SERS single-molecule detection experiments;[41–43] in this regard, the chemical enhancement mechanism can account for the difference through resonance Raman.[44] Incidentally, these EM active sites can also enhance dramatically other nonlinear optical signals (fluorescence, second harmonic generation, etc.) of molecules adsorbed nearby, favoring single-molecule detection. Although narrow (nanoscale) surface features as high as possible are desirable, the roughness parameters for the actual SERS substrates are likely a compromise between δ and ξ_L; typically, lower ξ_L leads to proportionally smaller δ. This compromise yields different combinations of surface roughness parameters that could give rise to SERS activity, in agreement with the variety of SERS metal substrates widely employed.[32,46,47]

11.6. Concluding Remarks

To summarize, we have investigated in this chapter the influence of the nanoscale (in the region below a hundred nanometers) lower cutoff ξ_L on the scattering of light from one-dimensional, self-affine fractal Ag surfaces with large fractal dimension. Since no approximate methods are applicable for such roughness parameters, the rigorous EM scattering formulation has been described based on the Green's theorem integral equations, extended to account for all EM field components in the near field region.

Near-field distributions have been included for both *s* and *p* polarization (no depolarization takes place in this scattering geometry). It has been shown that nanoscale features have a strong impact on the near EM field distributions due to the relevant role played by the evanescent components. The *s*-polarized surface EM field intensity exhibits oscillations with higher frequency the smaller ξ_L is, but no significant electric field enhancements are found in this polarization. Drastic changes with decreasing ξ_L are found in the *p*-polarized, surface EM field stemming from the roughness-induced excitation of SPPs. Large, narrow peaks (called localized optical modes or localized SPPs) tend to appear with either increasing rms height or decreasing nanoscale, predominantly enhancing the normal electric field intensity (as expected from the SPP electric field orientation), and rapidly decaying into both vacuum and metal. Near field intensity maps around localized SPPs can help to interpret PSTM experimental results.

Localized SPPs have a strong impact on the statistics of the enhancement factor surface electric field intensity. Its average and fluctuations exhibit a remarkable increase upon diminishing the lower scale cutoff. In fact, the presence of strong nanoscale irregularities is crucial to the excitation of localized SPP and to the subsequent electric field enhancements; this has been investigated for very rough, Gaussian-correlated surfaces,[27,55] demonstrating that, strictly speaking, self-affinity is not critical. Large average SERS enhancement factors, which account fairly well for typically reported values of 4–6 orders of magnitude, are found over a wide spectral region in the visible and near IR. The inferred local SERS enhancement factors $\lesssim 10^9$ due to localized SPP are consistent with the values claimed to occur in SERS single molecule detection. In addition to SERS, the large field enhancements associated with such localized SPPs on self-affine fractal metal surfaces can shed light onto other surface nonlinear optical phenomena.

Acknowledgments. This work was supported by the Spanish Dirección General de Investigación (grants BFM2003-0427, FIS2004-0108, and FIS2006–07894) and Comunidad de Madrid (through grants GR/MAT/0425/2004 and MICROSERES, and through V. Giannini's predoctoral fellowship). The authors are grateful to I. Simonsen and S. Sánchez-Cortés for helpful discussions.

References

1. B.B. Mandelbrot, *Fractals* (Freeman, San Francisco, 1977); *The Fractal Geometry of Nature* (Freeman, San Francisco, 1982).
2. R. Jullien and R. Botet, *Aggregation and Fractal Aggregates* (World Scientific, Singapore, 1987).
3. J. Feder, *Fractals* (Plenum, New York, 1988).
4. A.-L. Barabási and H.E. Stanley, *Fractal Concepts in Surface Growth* (University Press, Cambridge, 1995).
5. M.V. Berry, "Difractals," *J. Appl. Phys.* **12**, 781–797 (1979).
6. B.J. West, "Sensing scaled scintillations," *J. Opt. Soc. Am. A* **7**, 1074–1100 (1990).
7. A. Mendoza-Suárez and E.R. Méndez, "Light scattering by a reentrant fractal surface," *Appl. Opt.* **36**, 3521–3531 (1997).
8. J.A. Sánchez-Gil and J.V. García-Ramos, "Far-field intensity of electromagnetic waves scattered from random, self-affine fractal metal surfaces," *Waves Random Media* **7**, 285–293 (1997).
9. I. Simonsen, A. Hansen, and O.M. Nes, "Determination of Hurst exponents by use of the wavelet transform," *Phys. Rev.* E **58**, 2779–2787 (1998).
10. Y.P. Zhao, G.C. Wang, and T.M. Lu, "Diffraction from non-Gaussian rough surfaces," *Phys. Rev. B* **55**, 13938–13952 (1997); Y.P. Zhao, C.F. Cheng, G.C. Wang, and T.M. Lu, "Power law behavior in diffraction from fractal surfaces," *Surf. Sci.* **409**, L703–L708 (1998).
11. G. Franceschetti, A. Iodice, and D. Riccio, "Scattering from dielectric random fractal surfaces via method of moments," *IEEE Trans. Geosci.* **38**, 1644–1655 (2000).
12. I. Simonsen, D. Vandembroucq, and S. Roux, "Wave scattering from self-affine surfaces," *Phys. Rev. E* **61**, 5914–5917 (2000); "Electromagnetic wave scattering from

conducting self-affine surfaces: an analytic and numerical study," *J. Opt. Soc. Am. A* **18**, 1101–1111 (2001).
13. J.A. Sánchez-Gil, J.V. García-Ramos, and E.R. Méndez, "Light scattering from self-affine fractal silver surfaces with nanoscale cutoff: far-field and near-field calculations,"*J. Opt. Soc. Am. A* **19**, 902–911 (2002).
14. A.A. Maradudin, T. Michel, A.R. McGurn, and E.R. Méndez, "Enhanced backscattering of light from a random grating," *Ann. Phys.* (*New York*) **203**, 255–307 (1990).
15. M. Nieto-Vesperinas, *Scattering and Diffraction in Physical Optics* (Wiley, New York, 1991).
16. J.A. Sánchez-Gil and M. Nieto-Vesperinas, "Light scattering from random rough dielectric surfaces," *J. Opt. Soc. Am. A* **8**, 1270–1286 (1991); "Resonance effects in multiple light scattering from statistically rough metallic surfaces," *Phys. Rev. B* **45**, 8623–8633 (1992).
17. V.M. Shalaev, "Electromagnetic properties of small-particle composites," *Phys. Rep.* **272**, 61–137 (1996).
18. E.Y. Poliakov, V.M. Shalaev, V.A. Markel, and R. Botet, "Enhanced Raman scattering from self-affine thin films," *Opt. Lett.* **21**, 1628–1630 (1996).
19. J.A. Sánchez-Gil and J.V. García-Ramos, "Strong surface field enhancements in the scattering of p-polarized light from fractal metal surfaces," *Opt. Commun.* **134**, 11–15 (1997).
20. J.A. Sánchez-Gil and J.V. García-Ramos, "Calculations of the direct electromagnetic enhancement in surface enhanced Raman scattering on random self-affine fractal metal surfaces," *J. Chem. Phys.* **108**, 317–325 (1998).
21. E.Y. Poliakov, V.A. Markel, V.M. Shalaev, and R. Botet, "Nonlinear optical phenomena on rough surfaces of metal thin films," *Phys. Rev. B* **57**, 14901–14913 (1998).
22. S. Grésillon, L. Aigouy, A.C. Boccara, J.C. Rivoal, X. Quelin, C. Desmaret, P. Gadenne, V.A. Shubin, A.K. Sarychev, and V.M. Shalaev, "Experimental observation of localized optical excitations in random metal-dielectric films," *Phys. Rev. Lett.* **82**, 4520–4523 (1999).
23. V.M. Shalaev, *Nonlinear Optics of Random Media* (Springer, Berlin, 2000).
24. J.A. Sánchez-Gil, J. V. García-Ramos, and E. R. Méndez, "Near-field electromagnetic wave scattering from random self-affine fractal metal surfaces: spectral dependence of local field enhancements and their statistics in connection with surface-enhanced Raman scattering," *Phys. Rev. B* **62**, 10515–10525 (2000).
25. J.A. Sánchez-Gil, J.V. García-Ramos, and E.R. Méndez, "Influence of nanoscale cutoff in random self-affine fractal silver surfaces on the excitation of localized optical modes," *Opt. Lett.* **26**, 1286–1288 (2001).
26. J.A. Sánchez-Gil and J.V. García-Ramos, "Local and average electromagnetic enhancement in surface-enhanced Raman scattering from self-affine fractal metal substrates with nanoscale irregularities," *Chem. Phys. Lett.* **367**, 361–366 (2003).
27. J.A. Sánchez-Gil, "Localized surface-plasmon polaritons in disordered nanostructured metal surfaces: Shape versus Anderson-localized resonances," *Phys. Rev. B* **68**, 113410 (2003).
28. F.J. García-Vidal and J.B. Pendry, "Collective theory for surface enhanced raman scattering," *Phys. Rev. Lett.* **77**, 1163–1166 (1996).
29. D.P. Tsai, J. Kovacs, Z. Wang, M. Moskovits, V.M. Shalaev, J. S. Suh, and R. Botet, "Photon scanning tunneling microscopy images of optical excitations of fractal metal colloid clusters," *Phys. Rev. Lett.* **72**, 4149–4152 (1994).

30. S. Bozhevolnyi, B. Vohnsen, I.I. Smolyaninov, and A.V. Zayats, "Direct observation of surface polariton localization caused by surface roughness," *Opt. Commun.* **117**, 417–423 (1995).
31. S. Bozhevolnyi, B. Vohnsen, A.V. Zayats, and I.I. Smolyaninov, "Fractal surface characterization: implications for plasmon polariton scattering," *Surf. Sci.* **356**, 268–274 (1996).
32. P. Zhang, T.L. Haslett, C. Douketis, and M. Moskovits, "Mode localization in self-affine fractal interfaces observed by near-field microscopy," *Phys. Rev. B* **57**, 15513–15518 (1998).
33. S. Bozhevolnyi, V.A. Markel, V. Coello, W. Kim, and V.M. Shalaev, "Direct observation of localized dipolar excitations on rough nanostructured surfaces," *Phys. Rev. B* **58**, 11441–11448 (1998).
34. V.A. Markel, V.M. Shalaev, P. Zhang, W. Huynh, L. Tay, T.L. Haslett, and M. Moskovits, "Near-field optical spectroscopy of individual surface-plasmon modes in colloid clusters," *Phys. Rev. B* **59**, 10903–10909 (1999).
35. H. Raether, *Surface Polaritons on Smooth and Rough Surfaces and on Gratings* (Springer-Verlag, Berlin, 1988).
36. W.L. Barnes, A. Dereux, and T.W. Ebbesen, "Surface plasmon sub-wavelength optics," *Nature* **424**, 824–826 (2003).
37. R.K. Chang and T.E. Furtak, *Surface Enhanced Raman Scattering* (Plenum, New York, 1982).
38. M. Moskovits, "Surface-enhanced spectroscopy," *Rev. Mod. Phys.* **57**, 783–826 (1985).
39. A. Wokaun, "Surface enhancement of optical-fields – mechanism and applications," *Mol. Phys.* **56**, 1–33 (1985).
40. K.A. O'Donnell, R. Torre, and C.S. West, "Observations of backscattering effects in second-harmonic generation from a weakly rough metal surface," *Opt. Lett.* **21**, 1738–1740 (1996); M. Leyva-Lucero, E.R. Méndez, T.A. Leskova, A.A. Maradudin, and J.Q. Lu, "Multiple-scattering effects in the second-harmonic generation of light in reflection from a randomly rough metal surface," *Opt. Lett.* **21**, 1809–1811 (1996).
41. S. Nie and S.R. Emory, "Probing single molecules and single nanoparticles by surface-enhanced Raman scattering," *Science* **275**, 1102–1106 (1997).
42. K. Kneipp, Y. Wang, H. Kneipp, L.T. Perlman, I. Itzkan, R.R. Dasari, and M.S. Feld, "Single molecule detection using surface-enhanced Raman scattering (SERS)," *Phys. Rev. Lett.* **78**, 1667–1670 (1997).
43. H. Xu, E.J. Bjerneld, M. Käll, and L. Börjesson, "Spectroscopy of single hemoglobin molecules by surface enhanced Raman scattering," *Phys. Rev. Lett.* **83**, 4357–4360 (1999).
44. C.J.L. Constantino, T. Lemma, P. Antunes, R. Aroca, "Single-molecule detection using surface-enhanced resonance raman scattering and langmuir-blodgett monolayers," *Anal. Chem.* **73**, 3674–3678 (2001).
45. M.I. Stockman, V.M. Shalaev, M. Moskovits, R. Botet, T.F. George, "Enhanced Raman scattering by fractal clusters: scale-invariant theory," *Phys. Rev. B* **46**, 2821–2830 (1992).
46. S. Sánchez-Cortés, J.V. García-Ramos, and G. Morcillo, "Morphological Study of Metal Colloids Employed as Substrate in the SERS Spectroscopy," *J. Colloid Interface Sci.* **167**, 428–436 (1994).
47. C. Douketis, Z. Wang, T.L. Haslett, and M. Moskovits, "Fractal character of cold-deposited silver films determined by low-temperature scanning tunneling microscopy," *Phys. Rev. B* **51**, 11022–11031 (1995).

48. R.F. Voss, in *The Science of Fractal Images*, H.-O. Peitgen and D. Saupe, eds. (Springer, Berlin, 1988).
49. R.F. Voss, "Random fractals: self-affinity in noise, mountains, and clouds," *Physica D* **38**, 362–371 (1989).
50. D.W. Lynch and W.R. Hunter, *Handbook of Optical Constants of Solids* E.D. Palik, ed. (Academic Press, New York, 1985), p. 356.
51. U. Kreibig and M. Vollmer, *Optical Properties of Metal Clusters* (Springer, Berlin, 1995).
52. R. Carminati and J.J. Greffet, "Influence of dielectric contrast and topography on the near field scattered by an inhomogeneous surface," *J. Opt. Soc. Am. A* **12**, 2716–2725 (1995).
53. J.J. Greffet, A. Sentenac, and R. Carminati, "Surface profile reconstruction using near-field data," *Opt. Commun.* **116**, 20–24 (1995).
54. E. Devaux, A. Dereux, E. Bourillot, J.-C. Weeber, Y. Lacroute, J.-P. Goudonnet, and C. Girard, "Local detection of the optical magnetic field in the near zone of dielectric samples," *Phys. Rev. B* **62**, 10504–10514 (2000).
55. J.A. Sánchez-Gil, J.V. García-Ramos, and E.R. Méndez, "Electromagnetic mechanism in surface-enhanced Raman scattering from Gaussian-correlated randomly rough metal substrates," *Opt. Express* **10**, 879–866 (2002).

12 Light Scattering by Particles on Substrates. Theory and Experiments

F. MORENO, J.M. SAIZ, AND F. GONZÁLEZ
Grupo de Óptica. Departamento de Física Aplicada. Universidad de Cantabria. 39005 Santander, Spain

12.1. General Introduction

During recent decades the study of the scattering of electromagnetic waves by rough surfaces has aroused the interest of many research groups and has been approached from different standpoints. Of the geometrical models developed either to reproduce experimental results or to understand the physics involved in the phenomena of scattering, particular mention should be made of the model consisting of ensembles of particles with simple geometry seeded onto flat surfaces. What makes these surfaces especially attractive is that they allow both controlled calculation for different sizes, shapes, densities, or optical properties, and the possibility of experimental testing of such systems. But these surfaces are also interesting for other reasons in certain specific areas, such as mirror degradation by particle contamination,[1] detection of surface defects in the semiconductor industry,[2] construction of biosensors,[3] optical particle sizing,[4] and near field optics.[5]

The problem of calculating the electromagnetic field scattered by these surfaces is not a simple one because of the presence of a substrate breaking the existing symmetry of the regular-shaped particle, and also because of the variability of the particles. In an attempt to solve the scattering problem associated with particles on flat surfaces, several theoretical approaches have been developed, each one corresponding to a particular set of conditions affecting the problem (small particles, nonhomogenous particles, specific optical properties, etc). The simplest system, in which the microstructure is much smaller than the wavelength, can be modeled by a dipole. This approach is valid when the field within the particle is nearly constant and holds when the scatterer is not too close to the interface. As the particle–substrate separation decreases, the dipole model begins to fail because higher order multipoles become significant through particle–substrate interaction. Some authors[6–8] have used exact image theory (EIT) to derive the scattering of electromagnetic waves from a small object above an interface separating two isotropic and homogeneous media. Most simple models assume that the dipole is illuminated by the superposition of the direct and reflected plane waves. The total scattered field from the dipole is the superposition of the direct and image scattered fields. Considering no multiple interactions between the dipole and the substrate,

Jakeman[9] showed that the analytical expressions for the field scattered from small particles (spheres, discs, and needles) distributed on a tilted interface gave first-order agreement with those of the Rayleigh–Rice theory. Videen[10] used similar expressions to study the polarization state of small spheres near substrates. Other authors have studied the multiple scattering effects between two small spherical metallic particles on a flat conducting surface[11] and the angular dependent polarization of out-of-plane optical scattering produced by dipole-like particles above and below a substrate.[12] For this kind of particle, several authors have performed an exact calculation in order to determine the effect of the substrate on scattering.[13–15] These studies use the exact solution for the emission of a dipole close to a plane interface and include the surface wave components of the scattered field, which in the case of metallic substrates are identified as the surface plasmon polaritons.

In a first attempt to model particles of finite size, many researchers have considered particles whose shape conforms to regular geometries such as a sphere, cylinder, and spheroid. For instance, Nahm and Wolfe[16] used a double interaction model to calculate the scattering by a sphere over a perfectly conducting mirror. In this model, the sphere is illuminated by the beam both directly and after specular reflection from the surface. This secondary reflected beam is affected by the reflection properties of the surface, and is partially obscured by the particle. In some approximations, further interaction between the sphere and the mirror can be assumed to be zero, so that the sphere scatters light as if it were isolated. Each of these two beams generates two contributions to the total scattered field: one direct and the other after being reflected by the mirror surface. Similar models include Fresnel reflection coefficients to handle nonperfectly conducting mirrors[17–19] and a geometrical shadowing factor to take into account the shadowing effect in both incoming and outgoing beams.[4] The latter correction has been applied to particle sizing under the name of the modified double interaction model (MDIM), and the assumptions and results will be discussed in detail in the third section.[20,21] Some authors[22,23] reduce the problem of light scattering by a sphere on a substrate to the problems of scattering by a sphere in a homogeneous medium and of the reflection of spherical waves by the substrate. They solve the first by using the Mie theory, whereas for the second one they use an extension of Weyl's method to calculate the reflection of dipole radiation by a flat surface. Videen expanded the interaction field about the image location to solve the scatter from a sphere in front of[24] and behind[25] a smooth arbitrary substrate, and later provided an exact theory for an arbitrary particle system in front of[26] and behind[27] a perfectly conducting substrate. Similar derivations for spheres resting on substrates have been provided by Johnson[28–30] and Fucile *et al.*[31] In this methodology, the boundary conditions at the particle and at the surface are satisfied simultaneously by projecting the fields in the half-space region not including the particle onto the half-space region including the particle. For nonperfectly conducting substrates, some numerical method or simplifying assumption must be used. Theoretical results for a cylinder above a substrate calculated using this method[32] were compared with experimental results.[33] Borghi *et al.*[34] presented a method for treating the two-dimensional scattering of a plane wave as an arbitrary configuration of perfectly conducting circular cylinders in front of a plane surface with general reflection properties. This is based on the plane-wave spectrum of cylindrical functions involved in the decomposition of the field scattered by a cylinder. The substrate can be dielectric, metallic, anisotropic or lossy media, or can be multilayered.

Much research on the problem of the scattering of electromagnetic waves by particulate surfaces has been formulated in terms of exact integral equations for the electromagnetic fields that are solved by standard numerical methods. It should be stressed that the formulation is exact in the sense that no additional physical assumptions or approximations are needed; i.e., all limitations arise from the numerical procedure and from the degree of similarity between the modeled and real (observed) systems. Integral equations are derived from integral theorems that combine differential Maxwell equations and appropriate boundary conditions. One of the methods most widely used is the extinction theorem (ET) of physical optics[35] which produces a surface integral equation relating both the incident field to the sources on the surface, and these to the scattered fields. Although the ET method was initially used to calculate the scattering by random rough surfaces, different authors have extended it to calculate the far-field[36] and near-field[37] scattering by small metallic particles on flat conducting substrates. The exact character of the formulation takes into account multiple interaction between particle and substrate. Saiz *et al.*[38] and Valle *et al.*[39] have studied the effect of particle size, particle mean distance and the effect of the optical constants on the light scattering. The method has also been used to study surface plasmon–polariton generation with small particles on real metallic substrates.[40] Madrazo and Nieto-Vesperinas[41–43] have established the ET method for multiply connected domains. This improvement allows numerical simulation of the scattering from systems composed of surfaces belonging to separated bodies of arbitrary shape and with different optical properties. A different integral method has been proposed by Greffet *et al.*,[44] who use adequate Green functions to obtain a volume integral extended to the particle volume, including the substrate. They calculated the far-field and near-field scattering produced by 2D particles deposited on a dielectric planar waveguide, paying special attention to particle interaction,[45] and the near field corresponding to a dielectric rod below a metallic surface under surface plasmon generation conditions.

Many other approaches to the problem of scattering from particles on surfaces can be found in the literature. Wriedt and Doicu[46] solved the scattering from an axisymmetric particle on or near a surface with a formalism based on the extended boundary condition method (EBCM) and the integral representation of spherical vector wavefunction over plane waves. Special attention was placed on the interaction between a particle and its image. Taubenblatt[47] used a modified version of the coupled dipole method (CDM) to calculate the far-field scattering intensity from a dielectric cylinder on a surface, when illuminated with a plane wave with field vectors along the cylinder axis. Using the same method, Schmehl *et al.*[48] analyzed the scattering by features acting as contaminants on surfaces. These studies reflect the considerable effort made to improve numerical techniques in order to accelerate the computation or to obtain good convergence rates.

Wojcik *et al.*[49] showed numerical solutions of Maxwell's equations for problems involving scattering from submicron objects on silicon wafers by considering time-domain finite elements. Kolbehdary *et al.*[50] analyzed scattering from a dielectric cylinder partially embedded in a perfectly conducting ground plane.

One simple and fairly successful approach is the ray-tracing solution, or, in other words, the application of the geometrical optics approximation. From a geometrical viewpoint, a plane wave incident on a metallic object is a beam of parallel rays of uniform density that is reflected by the sphere–substrate system. The scattered field is obtained as the coherent sum of the group of rays emerging from the surface with a common angle. This method has been shown to produce surprisingly good fits with experimental data when the observation is far from the specular direction (where the diffraction effects are too strong), even for particles whose size is of the order of the wavelength. For instance, Saiz and co-workers[51] have applied the method to the analysis of the backscattering from particles on a substrate and of the effect of different particle densities on scattering patterns.[52]

The experimental and theoretical work presented in this chapter attempts to provide an overview of the kind of results obtained in the analysis of light scattered by particles on surfaces. All these results come from different stages of the research performed by our group. Section 12.2 analyzes the near field scattered by a particle on a flat substrate for both 1D and 2D geometries. In the former, a cylindrical particle is selected and solutions are obtained by means of calculations based on the extinction theorem. In the latter, spherical particles are chosen with sizes much smaller than the incident wavelength, and calculations are made with a modified version of the numerical electromagnetic code (NEC). Section 12.3, which is devoted to the far-field case, describes several experimental techniques, all applied to metallic spherical particles or cylinders on conducting substrates, and presents different possibilities of obtaining information on the system from the experimental measurements, that is, the inverse problem.

12.2. Near Field of Particles on Substrates

12.2.1. Introduction

This section presents the near field scattered by a particle on a flat substrate and numerical analysis for both 1D and 2D geometries. For the 1D geometry, the particle will be an infinite cylinder with a section diameter smaller than the incident wavelength. The theoretical background introduced to solve the scattering problem associated with this geometry will prove useful for understanding the electromagnetic effects involved in the more general 2D problem. In this case, a spherical particle whose size is smaller than the incident wavelength will be assumed to lie on the substrate. Although the cylindrical and spherical particles are ideal shapes, the most important conclusions are valid for more complicated particle geometries. Numerical results obtained by applying the

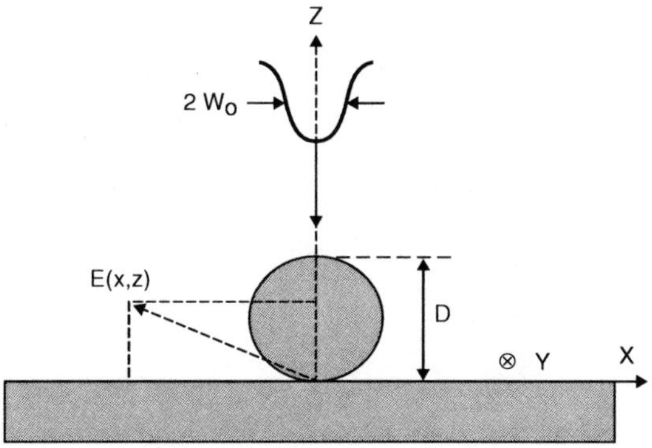

FIGURE 12.1. Scattering geometry.

extinction theorem,[35] finite element method[53] and a modification of the numerical electromagnetic code[14,15] will be shown.

According to the optical properties of the scattering system constituted by the particle and the substrate, four major combinations can be considered: metal–dielectric, metal–metal, dielectric–metal and dielectric–dielectric. All are of interest in the framework of the general electromagnetic problem of wave scattering by particles on substrates for both far- and near-field approximations. The latter presents interesting features when the substrate is metallic because of the generation of surface waves (surface plasmons),[54] which has led to what is termed subwavelength optics.[55] The case of metallic particles opens up the physics of localized plasmons to applications in SERS (surface enhanced Raman spectroscopy), biosensors, biomarkers, etc.[56] Consequently, most of the results presented in this section will correspond to the case of metallic substrate.

12.2.2. 1D Geometry

The scattering geometry is shown in Fig. 12.1. Both the cylinder and the substrate are assumed to be made of the same metallic material. The reference system is chosen so that the cylinder axis is parallel to the *Y*-direction and the flat substrate is parallel to the *XY*-plane. The incident wave vector $\vec{k}_0$ is perpendicular to the cylinder axis, and the components of the scattered electric field will be calculated at points on the plane *XZ* (scattering plane, which is also perpendicular to the cylinder axis). The calculations will be made for normal incidence, $\vec{k}_0 = (0, 0, -k_0)$.

For the 1D-geometry of Fig. 12.1, numerical calculation of Maxwell's integral equations derived from the application of the extinction theorem (see Sect. 7.11.1 of ref. 35) has been used to obtain the electric- and magnetic-field components of the scattered wave in the near field. The numerical treatment of the integral equations requires the surface to be discretized.[37] Furthermore, the surface length, L, has to be finite but long enough to avoid edge effects. For the same reason, a Gaussian

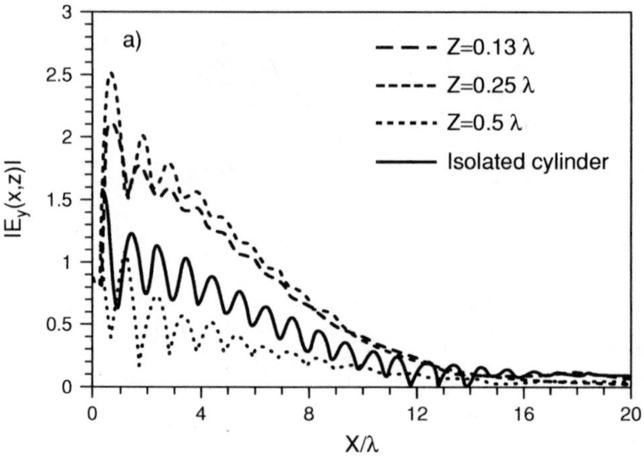

FIGURE 12.2. Evolution of $|E_y(x, z)|$ for S-incident polarization. Cylinder diameter $D = 0.5\lambda$. $\varepsilon = -11 + 1.5i$.

incident beam of suitable width (more than ten times the cylinder diameter) is chosen to illuminate the sample. The incident beam will be assumed to be linearly polarized perpendicular (S polarization) or parallel (P polarization) to the scattering plane, *XZ*. The cylinder diameter D will be restricted to values smaller than the incident wavelength. This is because (as seen later for P-incident polarization) the effects analyzed are more pronounced when $D < \lambda$ (nanosized particles in the visible range), since for these diameters the cylinder generates surface waves on the substrate more effectively.

In Fig. 12.2 the modulus of the total electric near-field component, $E_y(x, z)$, is plotted as a function of x for several values of z (height from the flat substrate) when the cylinder is illuminated by an S-polarized Gaussian beam. For the case of normal incidence, only one half of the scattering pattern is shown, since it is symmetric with respect to the position of the cylinder, which is assumed to be located at $x = 0$.

For comparison, we have also plotted in Fig. 12.2 the modulus of $E_y(x, z)$ (continuous line) when the cylinder is isolated and the direction of the incident beam is parallel to the Z-axis.[57] In this case, the total near electric field $E_y(x, z)$ has two contributions and can be expressed as

$$E_y(x, 0) = E_y^C(x, 0) + E_y^G(x, 0), \tag{1}$$

where E_y^C is the electric field of the wave scattered by the cylinder, and E_y^G is the electric field associated with the incident Gaussian beam. At normal incidence, the interference of these two terms produces a ripple whose spatial period is equal to the incident wavelength. For values of $x \gg \lambda$, where E_y^G is negligible, the only contribution comes from the wave scattered by the cylinder. This corresponds to the small background observed between 17λ and 20λ. When the cylinder is located on a flat substrate, Eq. (1) has to be rewritten as

$$E_y(x, z) = E_y^C(x, z) + E_y^G(x, z). \tag{2}$$

Now the term E_y^C of Eq. (2) is the result of the interference between the wave scattered directly from the cylinder and that scattered and reflected off the substrate. The term E_y^G comes from the interference between the incident Gaussian beam and its reflection on the flat substrate. As is evident, the pattern changes as z varies. The presence of the ripple is due to the interference between E_y^G and E_y^C in Eq. (2). A comparison of the near-field scattered patterns (broken lines) with that of the isolated cylinder (continuous line) shows that the interference ripple retains the same spatial period ($= \lambda$), but damps faster than in the isolated case. In fact, when $x \gg \lambda$ the background is hardly seen. This can be attributed to the almost destructive interference of the two contributions of E_y^C (the wave scattered from the cylinder and reflected off the substrate undergoes a phase shift of the order of π, and its amplitude is approximately equal to the incident wave). The presence of the flat substrate tends to weaken the range of the wave scattered by the cylinder.[11] As a result, interaction between two cylinders is more easily produced when they are isolated in space than when they are located on a flat substrate. In the latter case, the two cylinders have to be very close. This is indeed an important point when multiple scattering effects are analyzed for embossed surfaces.

In order to provide a complete picture of the near field for S-incident polarization and for the geometry of Fig. 12.1, Fig. 12.3(a) shows a 2D surface plot[53] of the modulus of the y-component of the scattered electric field. For comparison, Fig. 12.3(b) corresponds to the isolated cylinder case (no substrate underneath).

For P-incident polarization, Fig. 12.4 shows some near-field scattering patterns for the modulus of the x-component of the electric field for a cylinder of 0.5λ lying on a flat substrate together with the isolated cylinder case (continuous line). When the cylinder is isolated, it should be noted that the x-component of the scattered electric field is very weak. Only in close proximity to the cylinder can a very small interference ripple be observed. Its amplitude damps as x increases. For $x \gg \lambda$, the x-component of the scattered electric field has a negligible value.

When the cylinder is located on the flat substrate and $z \leq D$, the amplitude of the interference ripple increases considerably. This gives a constant background at points where the amplitude of the x-component of the incident field has a negligible value ($x > 16\lambda$). From these results, it seems reasonable to assume that the presence of the cylinder generates surface waves on the substrate (polaritons). Propagation of these waves along the x-direction is due to the coupling of non-radiative components of the scattered wave vector spectrum with the electronic plasma oscillations in the metal surface.[54] The background in Fig. 12.4 at points with $x > 16\lambda$ would correspond to the modulus of the electric field associated with these surface waves. Under this assumption, the x-component of the near total electric field can then be written as

$$E_x(x, z) = E_x^C(x, z) + E_x^G(x, z) + E_x^{\mathrm{sp}}(x, z), \tag{3}$$

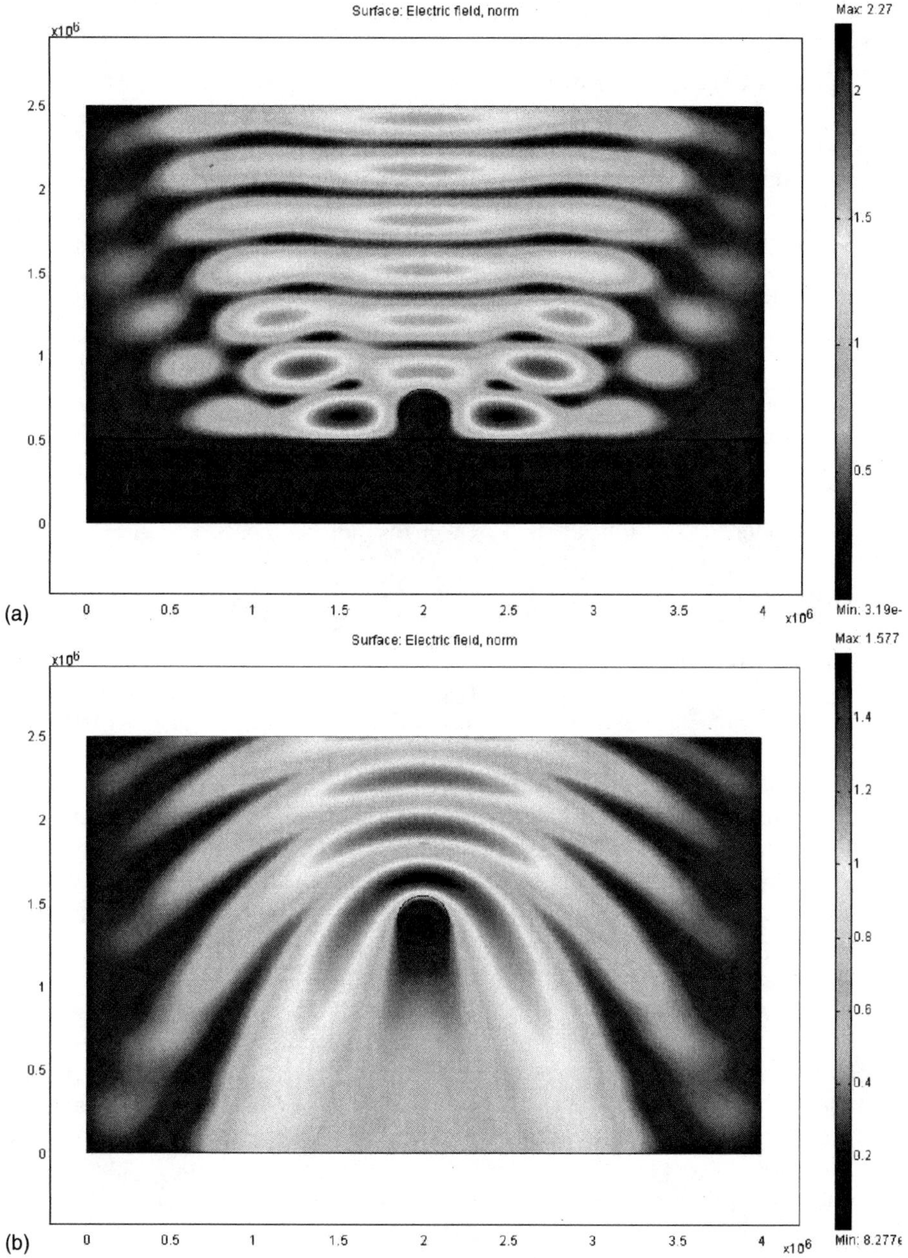

FIGURE 12.3. Two-dimensional surface plot of $|E_y(x, z)|$ for a cylinder on a flat substrate (a) and for an isolated cylinder (b) when the incident wave is S-polarized. Optical and geometrical parameters are the same as in Fig. 12.2. Scales are in microns.

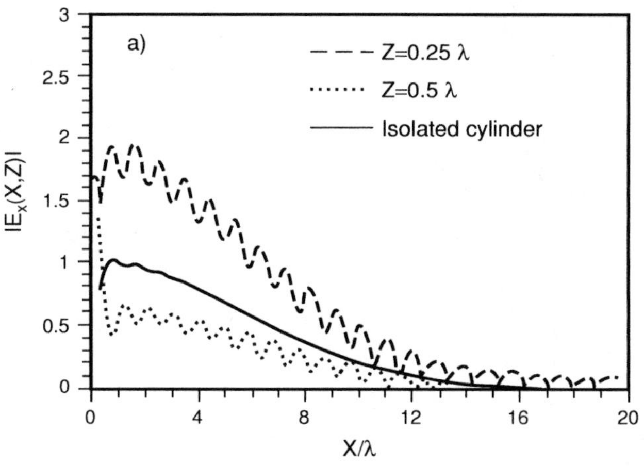

FIGURE 12.4. Evolution of $|E_x(x, z)|$ for P-incident polarization. Cylinder diameter $D = 0.5\lambda$. $\varepsilon = -11 + 1.5i$.

where the third term stands for the x-component of the surface wave electric field generated by the cylinder. For $z > 0$, the general expression for this term can be written as[54]

$$E_x^{sp} = E_{0x}^{sp} \exp(-k_z z) \exp[i(k_{sp} x)], \tag{4}$$

where E_{0x}^{sp} is the amplitude of the surface wave, k_{sp} is its propagation constant along the x-direction, and k_z is its damping constant along the z-direction.

From what has been stated for the isolated cylinder, the term $E_x^C(x, z)$ appears very small. In this case, a good approximation of Eq. (3) for $x > 5\lambda$ is

$$E_x(x, z) \cong E_x^G(x, z) + E_x^{sp}(x, z). \tag{5}$$

In Fig. 12.5 $|E_x(x, z)|$ as given by Eq. (5) is plotted together with the result obtained from the numerical calculation of the integral equations for $z = 0.25\lambda$. The agreement is very good. Only a few small reasonable discrepancies appear at points very close to, and far from, the cylinder. In the first case, they are due to the term $E_x^C(x, z)$ of Eq. (3) and neglected in Eq. (5); in the second case, they are due to the interference effect of the incident surface wave generated by the cylinder and that reflected at the edge of the surface.[58] This produces a small ripple, which slightly deforms the oscillations closer to the edge of the surface (this effect is not included in the previous equations).

As shown for the x-component, the z-component of the near total electric field can be written as

$$E_z(x, z) = E_z^C(x, z) + E_z^G(x, z) + E_z^{sp}(x, z). \tag{6}$$

In Fig. 12.6 $|E_z(x, z)|$ is represented for two values of $z < D$ together with the isolated cylinder case (continuous line). In the latter case, the ripple comes from

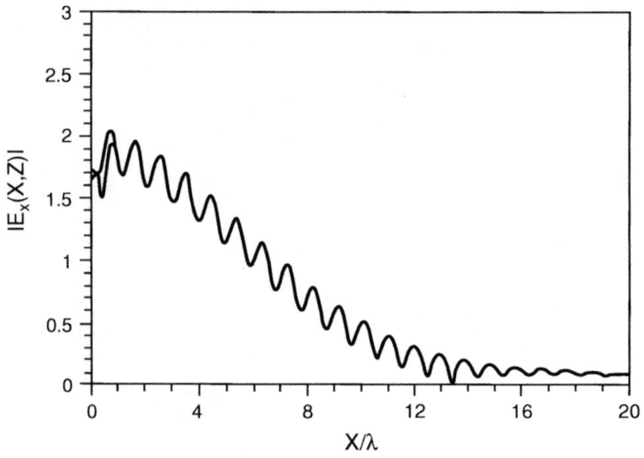

FIGURE 12.5. Comparison of $|E_x(x, z)|$ obtained through numerical calculation (thick line) with the analytical result of the modulus of Eq. (3) (thin line). Cylinder diameter $D = 0.5\lambda$. $\varepsilon = -11 + 1.5\text{i}$.

the interference of E_z^G and E_z^C ($E_z^{\text{sp}} = 0$). When the cylinder is on the substrate, the three terms of Eq. (6) contribute to the total scattered field. For the isolated cylinder, the z-component of the scattered field, unlike the x-component, has a nonnegligible value and reaches a constant value for $x \gg \lambda$. This value corresponds to the scattered field in the far-field approximation.

When the cylinder is located on the substrate and $x \gg \lambda$, the two contributions contained in $E_z^C(x, z)$ (direct and reflected off the substrate) tend to cancel each other out if the cylinder size is smaller than λ and E_z^C is evaluated at points with $z < \lambda$ (remember that $r_p(\theta) \to -1$ when $\theta \to \pi/2$). Let us assume that in this

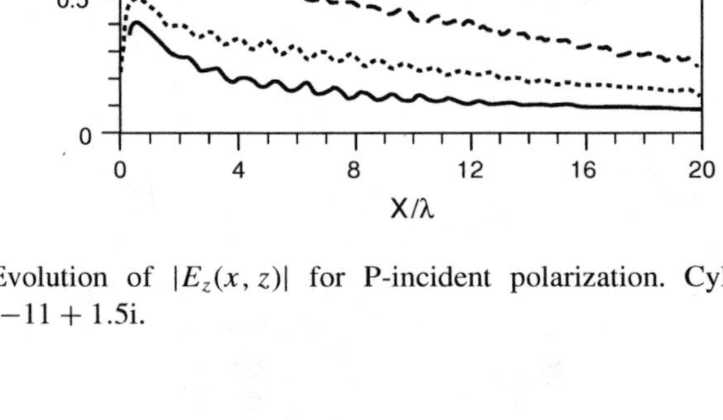

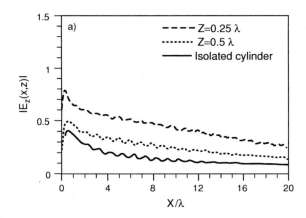

FIGURE 12.6. Evolution of $|E_z(x, z)|$ for P-incident polarization. Cylinder diameter $D = 0.5\lambda$. $\varepsilon = -11 + 1.5\text{i}$.

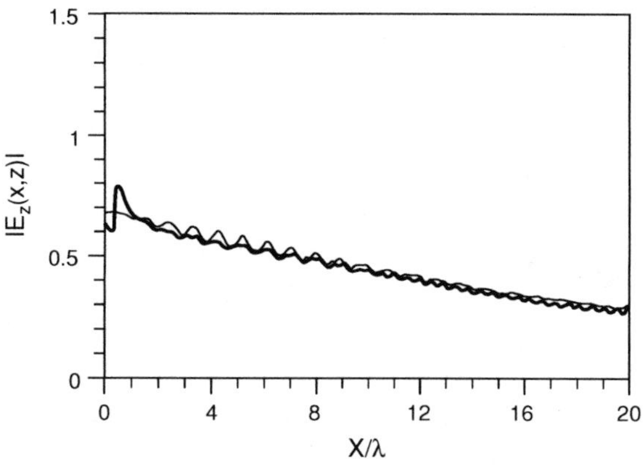

FIGURE 12.7. Comparison of $|E_z(x, z)|$ obtained through numerical calculation (thick line) with the analytical result of the modulus of Eq. (7) (thin line). Cylinder diameter $D = 0.5\lambda$. $\varepsilon = -11 + 1.5i$.

situation the following approximation holds:

$$E_z(x, z) \cong E_z^G(x, z) + E_z^{sp}(x, z). \tag{7}$$

In this equation $E_z^G(x, z)$ is given by

$$E_z^G(x, z) = \left[2x/i\left|\overrightarrow{k_i}\right|w_0^2\right] \exp\left[-x^2/w_0^2\right]\left\{\exp(-ik_{iz}z) + r_0 \exp(ik_{iz}z)\right\}, \tag{8}$$

where r_0 is the Fresnel reflection coefficient of the z-component of the incident Gaussian beam, and $E_z^{sp}(x, z)$ is given by an expression similar to Eq (4).

Figure 12.7 shows a comparison of $|E_z(x, z)|$ as calculated from Eq. (7) and that from the exact numerical methods for $z = 0.25\lambda$. As for the x-component, when the surface wave generated by the cylinder reaches the edge of the substrate, a reflected surface wave traveling in the opposite direction is generated. The interference between these two waves produces a periodical modulation in the x- and z-components of the electric field, whose spatial period is one half of the wavelength associated to the surface wave, $\lambda_{sp} = 2\pi/k_{sp}$.

Following the analysis of the local field, the evolution of the x- and z-components of the scattered electric field is studied when z is varied and the x-coordinate is kept fixed. As an example, a cylinder of diameter 0.2λ will be considered. In Fig. 12.8(a), a semilogarithmic plot of the evolution of $|E_x(x, z)|$ with z is shown in the range $[0.1\lambda$–$4\lambda]$ for four values of x: λ, 5λ, 10λ, and 15λ. For $x = \lambda$ and 5λ, it is difficult to see any sign of the presence of the surface wave. As z increases, only a periodic lobed structure appears. This is mainly due to the standing-wave behaviour of the resulting field due to the interference of the incident wave and that reflected off the substrate. For $x = 10\lambda$, and 15λ, $|E_x(x, z)|$ has a clear linear evolution in the range $[0.1\lambda - 0.5\lambda]$, which indicates the presence of the surface wave. The decay rate (the parameter k_z in Eq. (4)) gives a value of 2, which agrees with the theoretical result for k_z for a flat metallic cylinderless interface. For $z > 0.5\lambda$, the contribution

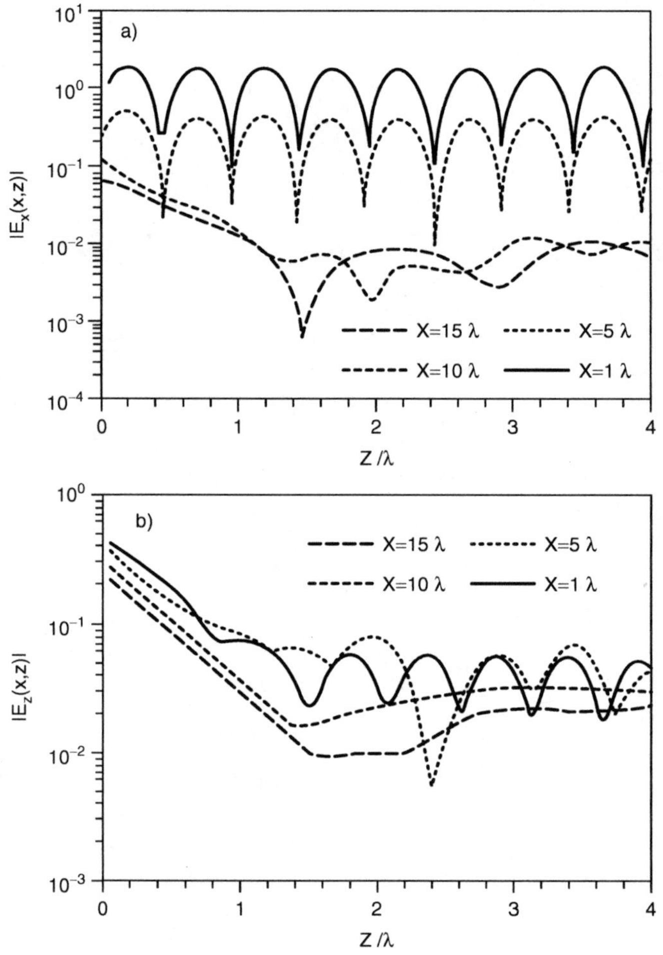

FIGURE 12.8. Evolution of (a) $|E_x(x, z)|$ and (b) $|E_z(x, z)|$ for P-incident polarization. Cylinder diameter $D = 0.2\lambda$. $\varepsilon = -11 + 1.5i$.

of the surface wave becomes negligible, and $E_x(x, z)$ is given approximately by the first term of Eq. (5), i.e., the scattering from the cylinder on the substrate. In Fig. 12.8(b) a semilogarithmic plot of $|E_z(x, z)|$ as a function of z is represented for the same values of x used in Fig. 12.8(a) (it should be remembered that the contribution of the incident field to the total z-component of the scattered wave field is very small). The linear behavior of $|E_z(x, z)|$ is maintained in the range $[0.1\lambda$–$0.5\lambda]$ for the four values of the x-coordinate. Therefore, the contribution of the z-component of the surface wave to the near field predominates. As is evident, this predominance is greater as x increases. For instance, for $x = 15\lambda$, the linear behavior of $|E_z(x, z)|$ is maintained until $z = \lambda$. For $z > \lambda$, the contribution from the cylinder of the z-component of the scattered electric field begins to dominate.

For P-incident waves, Fig. 12.9 shows a 2D surface plot of the near field generated by a metallic cylinder isolated (a) and on a flat metallic substrate (b).

12.2.3. 2D Geometry

All the results presented above for the 1D geometry serve as a basis for the analysis of the near field for the more realistic 2D case. In this section we present the near field produced by a small particle (as compared to the wavelength) located on a flat metallic substrate. As in the 1D case, if this system is illuminated by a monochromatic incident wave of wave vector $\vec{k}_0$, a component of the nonradiative part of the scattered wave vector spectrum ($k > k_0$) can couple with the electron–plasma oscillations of the metallic surface. For the 2D geometry, this coupling can excite a *cylindrical surface polariton* (CSP). Apart from the interest due to their applications,[56] this kind of surface wave is involved in the formation of periodical structures on the surfaces of some materials during the action of intense laser beams.[59] It should be noted that these cylindrical surface electromagnetic waves correspond to the surface waves appearing in Sommerfeld's solution to the classical problem of the field due to a dipole close to a surface.[60] Recently, Kosobukin[61] has developed a theory for the elastic scattering from insulating inhomogeneities in a medium accompanied by excitation of cylindrical surface phonon–polaritons, and his theory provides an asymptotic description of the fields.

For the analysis of the near field of a small particle on a flat metallic interface, a substrate made of gold ($\varepsilon = -11.95 + 1.33i$ at $\lambda = 633$ nm) with a small spherical isotropic gold particle (radius $= 0.05\lambda$) will be considered. A monochromatic ($\lambda = 633$ nm) linearly polarized plane wave, either parallel or perpendicular to the plane of incidence (*XZ* plane), is incident on the surface. In this case, the numerical calculation of the total electromagnetic field scattered by the surface has been carried out by means of a modification of the numerical electromagnetic code developed by Burke and Poggio.[62] The method is based on an integral representation of the scattered field related to the linear current distribution induced by the incident field, which in turn can be obtained by the application of the method of moments.[14]

From the above considerations, the small spherical particle is represented by three equal orthogonal thin wires of gold (length 0.1λ and diameter 0.01λ) centered at a point of height $h \geq$ the particle radius. The current elements excited on the conducting wires by the incident field can be regarded as the three components of the polarization induced in the particle.

Fig. 12.10 shows the squared modulus of the total electric field in a horizontal plane when a linear polarized plane wave is normally incident, with polarization along the Y-axis. The observation plane is parallel to the substrate and contains the center of the particle. The spatial distribution of the field on that plane will be of interest when other particles are present and interaction is to be considered. Besides the central peak, the interference of the incident field with that due to the particle is clearly seen. Two zones with different behavior can be distinguished in the interference pattern. In the first area, which corresponds to the angular

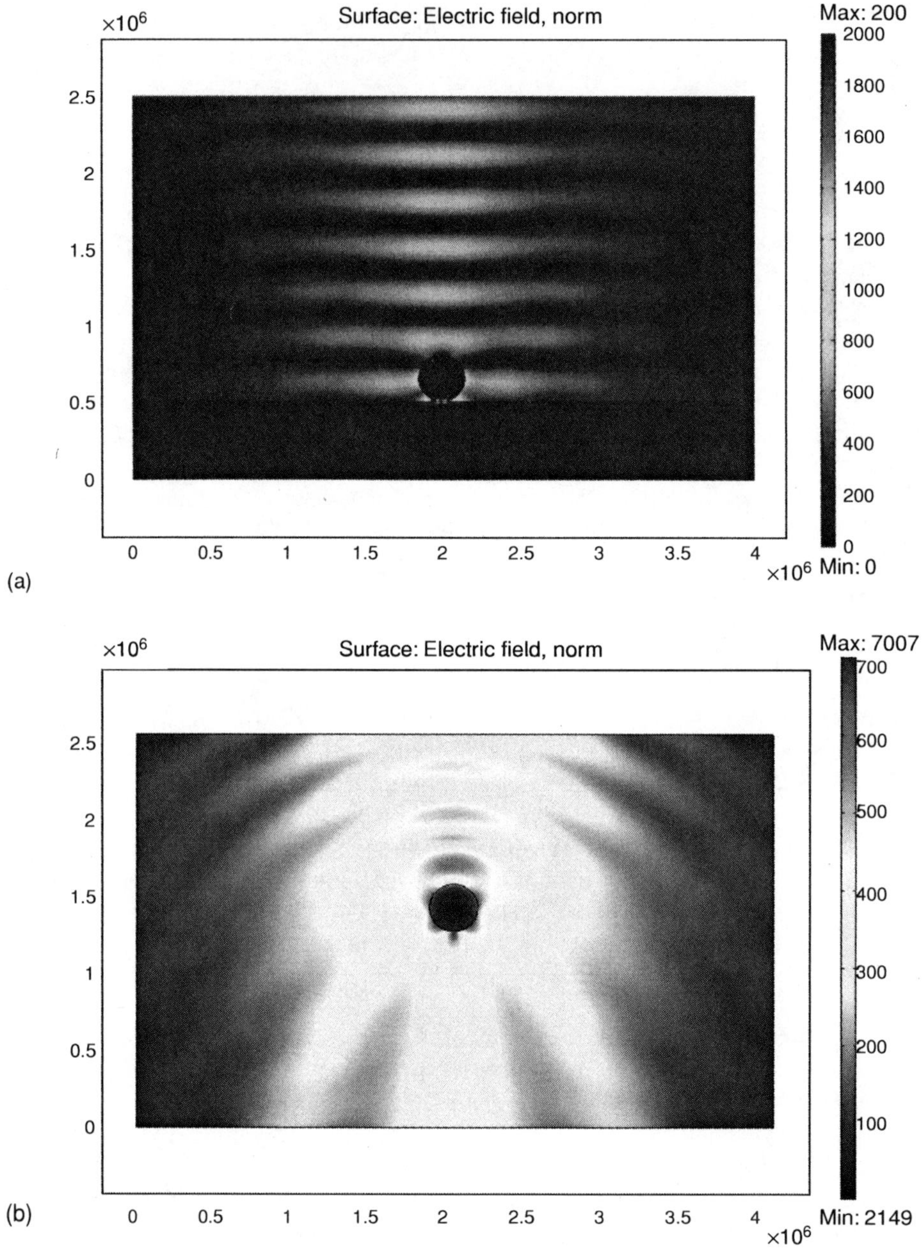

FIGURE 12.9. Two-dimensional surface plot of $\vec{E}_z(x, z)$ for a cylinder on a flat substrate (a) and for an isolated cylinder (b) when the incident wave is P-polarized. Optical and geometrical parameters are the same as in Fig. 12.2. Scales are in microns.

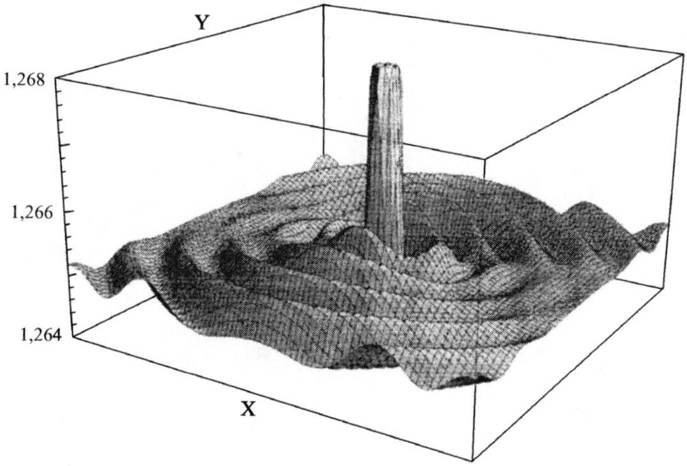

FIGURE 12.10. $|\vec{E}(x, z)|^2$ in a plane parallel to the substrate ($z \approx$ *particle radius*) for normal incidence and polarization parallel to the Y-axis. The plotted area is a $8\lambda \times 8\lambda$ square centered at the particle position.

interval $\pm 20^\circ$ around the X-axis in both directions, the interference contrast is low and decreases rapidly with distance. In the second, corresponding to the remaining plane, the contrast is higher and decreases slowly with distance. The maximum contrast is in the direction of the polarization induced in the particle by the incident wave (the Y-axis direction). If bulk waves are the dominant part of the field due to the particle, the visibility of the interference should be higher along the X-axis direction than along any other. Actually, the radiated field is attenuated close to the surface because of the destructive interference between the directly emitted field and that reflected off the substrate. Except for points very close to the particle, the angles of incidence of such reflection beams are quite large, almost causing a π phase shift.[11] On the other hand, along the Y-axis (maximum contrast) the only component of the radiated wave is the radial component. This decreases with distance as R^{-3}; it cannot, therefore, be responsible for the observed ripple. Thus, it can be assumed that the main contribution to the interference with the incident field comes from the surface waves.

In general, the electric field of CSP's has components normal to the surface, E_z, and radial components, E_r, parallel to the surface, described by Hankel functions with angular numbers $m = 0, 1$, depending on the polarization of the incident light and the angle of incidence.[61] The wave number k_{sp} is a function of the dielectric constants of the substrate (ε_2) and the medium above it (ε_1), with the same dependence as that found in Sect. 12.2.2 for the 1D case of plane surface plasmons.[54] When a linear polarized plane wave is normally incident, only CSP's with angular number $m = 1$ can be excited. This implies that the plasmon field has angular dependence $\sin(\varphi)$(φ is the cylindrical coordinate) and its interference with the incident field (along the Y-axis) takes the form $I' \approx \sin^2(\varphi)$ (I' being

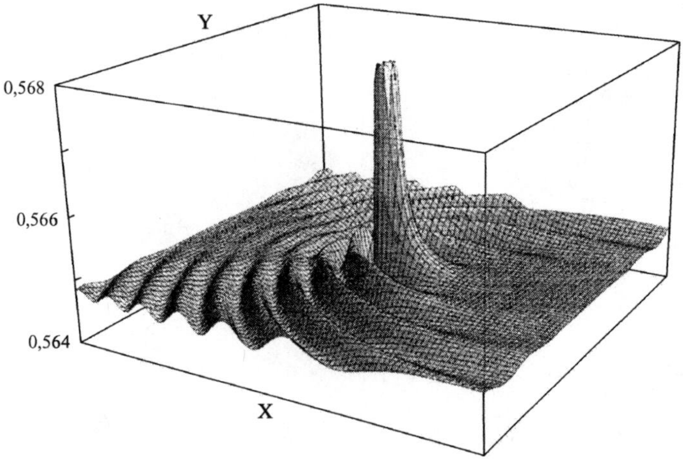

FIGURE 12.11. As in Fig. 12.10 but for an angle of incidence of 50°.

the interference term of the total intensity: $I' = 2Re(E_0^* \cdot E_s)$, where E_0 is the incident field and E_s is the field due to the particle). This can be seen in Fig. 12.10, where the maximum field due to the CSP is along the Y-axis direction ($\varphi = \pi/2$) parallel to the induced dipole moment, in contrast to what would happen with a dipole in free space. For $\varphi = 0$ or π, the plasmon field vanishes and the remaining interference is caused by the weak radiated component. Besides the amplitude, the two interference regions differ in their periods. For the radiated wave, the period is $\lambda = 2\pi/k_0$. In the plasmon case and for distances not too close to the particle, the separation of the maxima is $\lambda_{sp} = 2\pi/Re(k_{sp})$. With the same incident polarization (parallel to the Y-axis) but with an oblique incidence, the moment induced in the particle changes in strength but not in orientation. Therefore, the excited plasmon has the same spatial distribution and the change in the appearance of the interference pattern will only be due to the obliquity of the incident plane wave.

This can be observed in Fig. 12.11, which represents the squared modulus of the electric field for an angle of incidence of 50°.

A different situation occurs if the incident wave is P-polarized (electric field parallel to the XZ-plane). Fig. 12.12 shows the squared modulus of the total field for an angle of incidence of 50°. Here, the incident field excites a moment on the particle with two orthogonal components[14]

$$\begin{aligned} p_z &\approx \sin(\theta)[1 + r_p \exp(2ik_0 h \cos(\theta)], \\ p_x &\approx \cos(\theta)[1 - r_p \exp(2ik_0 h \cos(\theta)], \end{aligned} \tag{9}$$

r_p being the reflection coefficient of the substrate for P polarization. The radiated portion of the field is again negligible, for the same reason as before (except for points very close to the particle, where it produces the central peak). Figure 12.12

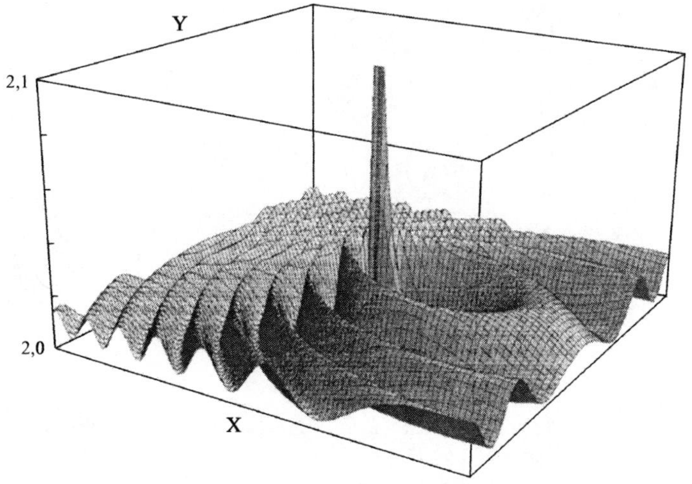

FIGURE 12.12. $|\vec{E}(x, z)|^2$ as in Fig. 12.10 but for an angle of incidence of 50° and incident polarization parallel (P-polarization) to the plane of incidence (XZ-plane).

is very similar to Fig. 12.4c of Ref. 59. The latter shows the ripple generated on a surface by the action of an intense laser radiation. A localized imperfection in the surface excited a cylindrical surface polariton (phonon–polariton in the case where the incident wavelength is resonant with the dielectric substrate) that interferes with the incident beam. The strong fields melt the surface with the shape of the interference ripple. The CSP excited by the vertical moment component has an angular number $m = 0$, as can be derived from symmetry considerations, while the CSP excited by the horizontal component has $m = 1$, as in the case of $\theta = 0°$. The first CSP implies circular symmetry; there are no directions along which the field associated with the plasmon vanishes. If the horizontal and vertical components of E_0, the incident electric field, and E_s due to the plasmon are explicitly taken into account, the interference term $I' = 2\mathrm{Re}(E_0^* \cdot E_s)$ can be separated into four terms[61]: I'_{xx}, I'_{zz}, I'_{xz}, I'_{zx} (the term I'_{xz}, for instance, would correspond to the E_x component excited by the p_z moment, interfering with the incident field). These terms hold the following angular dependences: $I'_{xx} \approx \cos^2(\varphi)$, I'_{xz}, $I'_{zx} \approx \cos(\varphi)$, $I'_{zz} \approx 1$. However, in Fig. 12.12, no direction with ripple extinction is observed, and so the term I'_{zz} clearly seems much greater than the others. Consequently, the polariton excited by the vertical induced moment is larger than the polariton excited by the horizontal moment. This is partly due to the dependence of the incident field on the angle of incidence that causes $p_z > p_x$ for $\theta = 50°$. Nonetheless, we have obtained the same behavior for angles of incidence where $p_z < p_x$. This means that the excitation of CSP is more effective for the vertical than for the horizontal dipole moment ($I'_{xz} < I'_{zz}$ because the ratio of the normal to the parallel component of the electric field for polaritons is[54] $E_z/E_x \approx \sqrt{\varepsilon_2}$ and, in our case, takes the value 3.5).

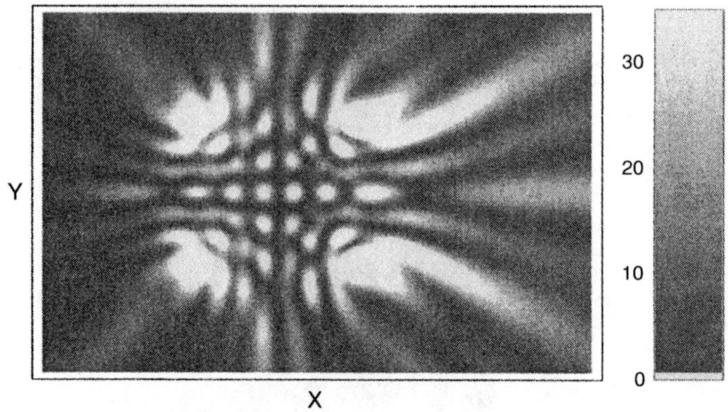

FIGURE 12.13. Scattered intensity distribution in a rectangle ($12\lambda \times 8\lambda$) for four scatterers on a metallic substrate separated by 3.6λ in the X-direction and 3.4λ in the Y-direction for an angle of incidence of 12° and P polarization (electric field parallel to the XZ-plane).

12.2.4. Concluding Remarks

Circular surface polaritons are surface waves whose wavelength is smaller than that corresponding to the excitation wave. They propagate along the metallic surface in the same way that conventional waves do in free space. Thus, they can be reflected and refracted and also undergo interference and diffraction.[56] Their physics has led to a new field known as *subwavelength optics* with many applications in several technological fields (photonic circuits, data storage, bio-photonics, etc.). As an example, Fig. 12.13 shows the near-field scattering pattern for a geometry of four scatterers on a metallic substrate located at the corners of a rectangle of dimensions 3.6λ in the X-direction and 3.4λ in the Y-direction. The rectangle is centered on the origin of the axes. The angle of incidence is 12° and the incident wave is S-polarized. Cavity effects due to CSP interference and diffraction are clearly visible.[14]

12.3. Far Field of Particles on Substrates

12.3.1. Introduction

The purpose of this section is to show how a system consisting of a particle located on a flat substrate scatters light under the far-field approximation. This will be done mainly from an experimental point of view although theoretical models will be introduced for comparison in order to extract the most important conclusions about the physics of this electromagnetic problem, whose near-field optics has been developed in the previous section.

The calculation of the far field produced by particulate surfaces requires the application of approximate or exact methods, each of which presents characteristics making them more or less suitable for certain given conditions. One important consequence of the accessibility of theoretical solutions is that the inverse problem can be approached experimentally. Any experimental scattering situation for which a solution can be calculated allows a systematic comparison. In this way, confrontation of theoretical and experimental results has led to a better understanding of the mechanism of the scattering process, and several ways of obtaining information about the scatters have been explored.

A wide variety of light scattering experiments involving particles on substrates have been described. This is mainly due to the following reasons:

- Experiments have been designed for very different purposes, including surface contamination research, particle sizing (both are parts of the so-called inverse problem), or modeling of real surfaces (planetary dust surfaces, sea surface among others)
- Experimental designs have very different requirements. This is true for near- and far-field geometries, each of which involves a different approach to design. With regard to far-field experiments we can find, for instance, measurements of the full angular variations (upper hemisphere), and variable incidence backscattering (or pure backscattering).
- The magnitudes experimentally measured vary from one experiment to another. Very often only scattered intensity is measured, sometimes with emphasis on a particular type of polarization, either in relation to the plane of incidence, defined by the surface normal to the incident light, or to the scattering plane, defined by the incident and observed directions, when light out of the incidence plane is studied. In addition, a specific polarization parameter may be analyzed, such as the degree of linear polarization or the depolarization ratios, but in some cases the whole Mueller matrix has been measured. Recently, one interesting approach chosen by some authors is the statistical analysis of time fluctuations of either the intensity or the polarization of scattered light.
- Experiments also vary with respect to the size, shape, optical properties, and surface density of the particles, protuberances or microstructures analyzed. This not only affects the characteristics of the experiment, but also the nature of the models used: single or multiple dipole models, models of the isolated Mie particle, rough surface models, etc.

This section presents a group of results, all of which correspond to substrates with metallic particles of size comparable to the wavelength. These results will be compared to others obtained with the theoretical models described in Sects. 12.1 and 12.2. Results are first presented for regular particles (spheres or cylinders with the axis normal to the scattering plane), followed by those for certain deviations from the simple regular case, namely irregularities (surface defects, buried particles), size variability (polydispersity), or high surface particle density (producing multiple scattering). Finally, a brief description is given of some results obtained in statistical studies.

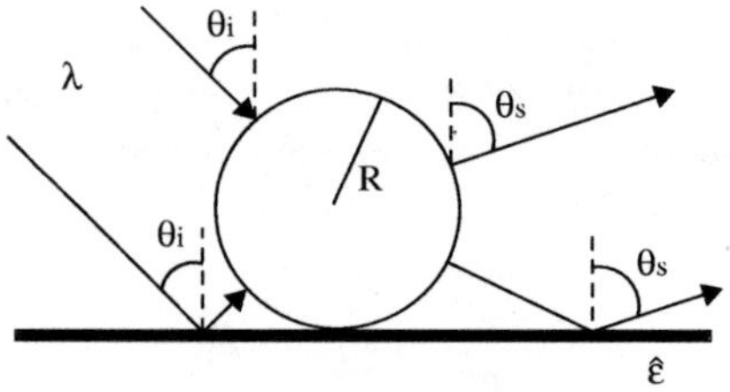

FIGURE 12.14. Geometry of the scatter system. λ: Incident wavelength; R: Radius of the particle; θ_i: Angle of incidence; θ_s: Scattering angle; ε: complex dielectric constant of the substrate and particle.

12.3.2. Regular Particles on Flat Substrates

When a particle whose size is of the order of the wavelength is located on a flat substrate, the angular dependence of the scattered light shows a characteristic lobed pattern with a maximum around the specular direction and a set of fairly visible maxima and minima whose angular positions depend strongly, but not exclusively, on the particle size.

A very useful model for this geometry was developed by Nahm and Wolfe[16] and a modified version[4] can be summarized as follows. A particle is illuminated by a direct beam of amplitude A_o and by its specular reflected beam, which is affected by the corresponding Fresnel reflection coefficient $\hat{r}(\theta_i)$ and by the phase shift $\delta(\theta_i)$ associated with the additional path length (see Fig. 12.14)

$$\delta(\theta_i) = (2\pi/\lambda)\, 2h \cos\theta_i, \tag{10}$$

where h is the distance from the center of the particle to the substrate, and $h = R$ for the contact case represented in Fig. 12.14. Each of the two incoming beams generates two contributions to the total scattered field. The first component in each case is that directly scattered from the protuberance, and the second is that scattered and reflected off the substrate, which is also affected by a Fresnel coefficient $\hat{r}(\theta_s)$ and a phase shift $\delta(\theta_s)$ similar to Eq.10.

The four components constituting the scattered field for an incident amplitude A_o can be written as

$$\begin{aligned}
[E_s]_1 &= A_o\hat{F}(\pi - \theta_i - \theta_s),\\
[E_s]_2 &= A_o\hat{F}(\theta_i - \theta_s)\,[1 - S(\theta_s)]^{1/2}\,\hat{r}(\theta_s)\exp\{i\delta(\theta_s)\},\\
[E_s]_3 &= A_o\hat{F}(\theta_i - \theta_s)\,[1 - S(\theta_i)]^{1/2}\,\hat{r}(\theta_i)\exp\{i\delta(\theta_i)\},\\
[E_s]_4 &= A_o\hat{F}(\pi - \theta_i - \theta_s)\,[1 - S(\theta_i)]^{1/2}\,[1 - S(\theta_s)]^{1/2}\,\hat{r}(\theta_i)\\
&\quad\times \exp\{\mathrm{i}\delta(\theta_i)\}\,\hat{r}(\theta_s)\exp\{\mathrm{i}\delta(\theta_s)\},
\end{aligned} \tag{11}$$

where $\hat{F}(\theta)$ is the complex far-field scattering amplitude for the isolated particle at an angle θ, and for the same radius, properties, and incident wavelength. These amplitudes can be evaluated by means of standard light scattering programs.[57] $S(\theta_i)$ and $S(\theta_s)$ are geometrical *shadowing factors* accounting for the fact that the

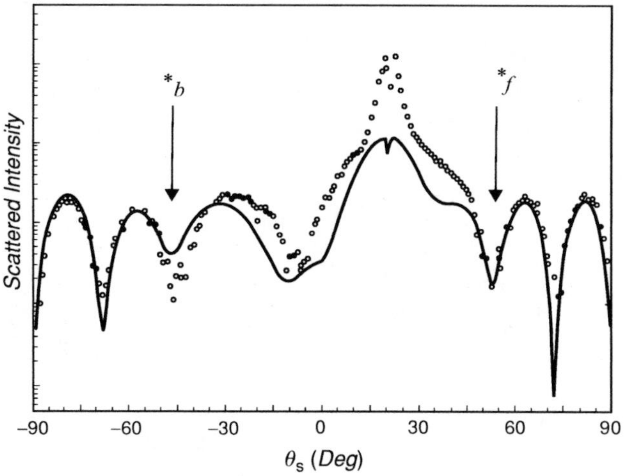

FIGURE 12.15. Scattering pattern corresponding to a cylinder on a substrate, both gold coated, of approximate diameter $D = 1.1\ \mu\text{m}$, for $\theta_i = 20°$. Circles: experimental results. Continuous line: results predicted by the model.

particle is obscuring its image both in the illuminating and scattering processes and is, therefore, contributing to the total energy conservation. These factors are easy to calculate for regular particles such as cylinders or spheres, and size the shadow cast by a particle onto its *image particle* for a given incidence.[4] Finally, the total scattered far field will be given by

$$E_T = \sum_{j=1}^{4} [E_s]_j. \tag{12}$$

A fast and accurate microsizing method[19] was proposed and developed for spherical and cylindrical metallic particles in 1996. This involves measuring the minima angular positions of the lobed S-polarized far-field scattering patterns at normal incidence. The particle mean diameter was fitted by comparing the theoretically predicted positions and the experimental minima. This method was later extended to the case of oblique incidence,[4] providing empirical expressions for sizing spheres and cylinders on substrates, based on the positions of their scattering minima. The method is reliable in the interval $\theta_i \in [0°–30°]$. Although the use of exact integral methods for the calculation of the scattering patterns could, in principle, increase this application interval, it should be noted that the flatness of the substrate surface becomes important as the angle of incidence increases. Therefore, any model used for sizing at high incidences should include information about surface flatness and defects.

Figure 12.15 shows the case of a cylinder 1.1 μm in diameter and for an angle of incidence $\theta_i = 20°$. The positions of the minima are strongly dependent on the size of the particle and also on the polarization chosen for the experiment.

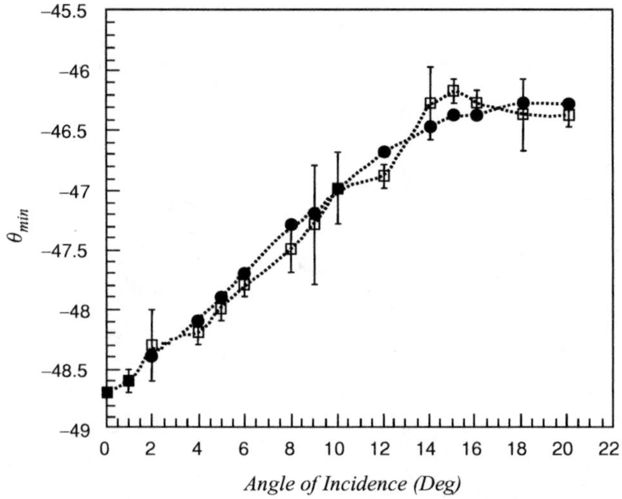

FIGURE 12.16. Position of the order $m = 1$ in the backward-scattering zone as a function of the angle of incidence for a cylinder of approximate diameter $D = 1.1\,\mu m$. Squares: experimental measurements (central values and error bars). Dots: positions predicted by the model.

This method was improved by simply tracking the evolution of a given minimum in relation to the angle of incidence. This dynamic procedure requires longer measuring times, but produces greater input of information, thereby increasing the accuracy of the method by an order of magnitude. In Fig. 12.16, the experimental tracking of the minimum labeled by (*b) in Fig. 12.15 is plotted together with the theoretical values obtained for the cylinders giving the best fit ($D = 1.077\,\mu m$, $\Delta D = 0.004\,\mu m$, for a cylinder of nominal diameter $1.1\,\mu m$). The minimum placed in the forward side and labeled (*f) yielded similar results ($D = 1.079\,\mu m$, $\Delta D = 0.005\,\mu m$).

An interesting point suggested by the nature of this model is the possibility of extending the method to other geometries. Once the scattered far field in the plane of incidence is known for an isolated particle of given size, shape, and orientation, and for a given wavelength, it can be implemented in Eqs. (11) used by the model in order to obtain its particular pattern.

12.3.3. Quasi-Regular Cases: Buried Particles and Surface Defects

The scattering produced by spherical particles or cylinders on substrates is very sensitive to the characteristics (roughness, inhomogeneity, defects) of the substrate itself. For example, when a silica fiber lying on a flat substrate is covered with gold, the substrate rises an amount that depends on the thickness of the coating. In some places the coating may produce a longitudinal folding of the gold layer. Here, we

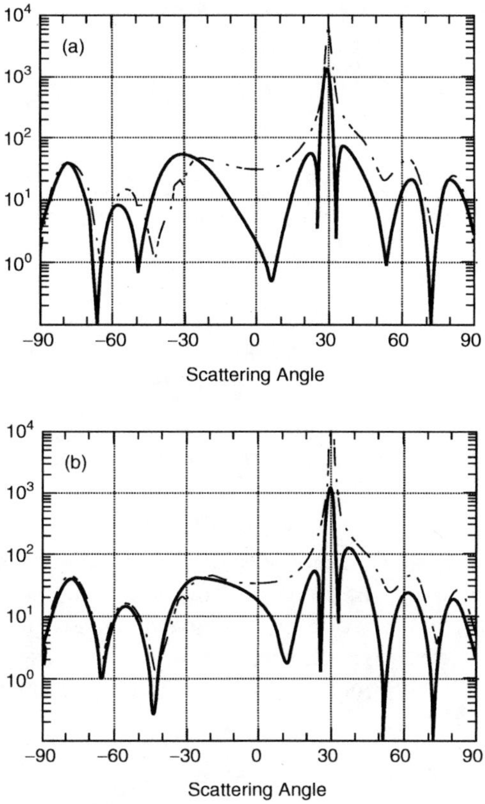

FIGURE 12.17. Scattering pattern corresponding to a cylinder of diameter $D = 1.1\,\mu m$ on a flat substrate for an incidence of $\theta_i = 30^\circ$ and s polarization. (a) Experiment (dashed) and calculated (continuous) curves. (b) The same, assuming a partially buried cylinder, by a fraction $\Delta = D/30$.

comment on two interesting examples: the buried particle and the particle with a neighboring protrusion, or *bumped* surface.

12.3.3.1. Buried Particles

Figures 12.17 show several scattering curves corresponding to a metallized fiber sized approximately $D = 1.1\,\mu m$ resting on a flat substrate. In both figures the dashed line corresponds to the experimental results obtained for S-polarized light with $\lambda = 0.633\,\mu m$ and $\theta_i = 30^\circ$. The continuous line represents the theoretical results obtained with numerical programs based on the extinction theorem[36] applied to the 1D case, i.e. a cylinder with light incident normal to the cylinder axis and scattering observed within the plane of incidence. While in Fig. 12.17.a the geometry introduced in the calculating program was exactly that of a cylinder lying on a substrate, in Fig. 12.17.b the calculation was performed for a surface in

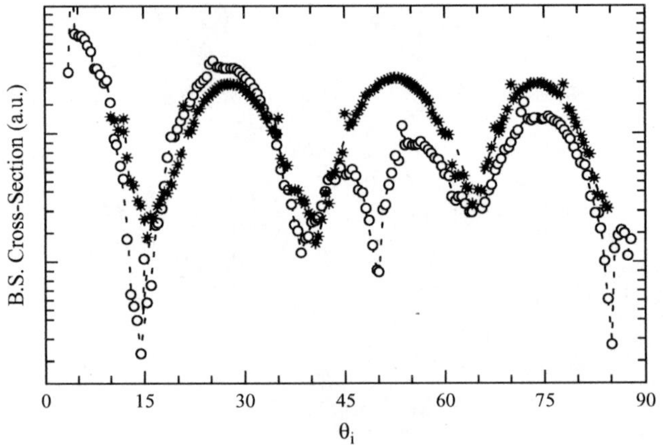

FIGURE 12.18. Experimental backscattering patterns obtained at different points of the fiber. See text for details.

which the cylinder is buried an amount equivalent to $D/30$, in agreement with the characteristics of the sputtering process applied to the sample.

The fit in the left-hand side of the plot (backward side) improves significantly with burying. With regard to this experiment, it is noteworthy that an observation in the far field provides information on a change in the geometry of a few tens of nanometers (20 nm in this case).

12.3.3.2. Particle with a Bumped Surface Nearby

This is an interesting example of a situation in which a solution to a particular inverse problem is provided by a combination of prior knowledge of the system and use of an appropriate model into which geometrical proposals may easily be introduced. In Fig. 12.18 two experimental backscattering plots are shown, corresponding to two different points of a fiber on a gold-coated substrate (in *pure* backscattering plots, $\theta_s = -\theta_i$, i.e. incidence and observation beams overlap). The line plotted in asterisks shows a perfectly regular and repetitive pattern, while the line plotted with circles is obtained at a given point where something produces a clear anomaly in the approximate interval [40°–60°]. A local distortion is observed in the form of an extra minimum, suggesting that, whatever is taking part in the scattering process affects a very particular angular range.

From Fig. 12.19 it is clear that each backscattering angle is associated with a point on the substrate surface where a particular feature (for instance, a neighboring bump (or dip) beside the cylinder) may strongly affect the light scattering in that angular interval.

Using the MDIM model, it is possible to obtain the scattering patterns for a surface containing a Gaussian-distributed height for each distance z to the base of the cylinder, thus producing a curved substrate, with different value h' for each

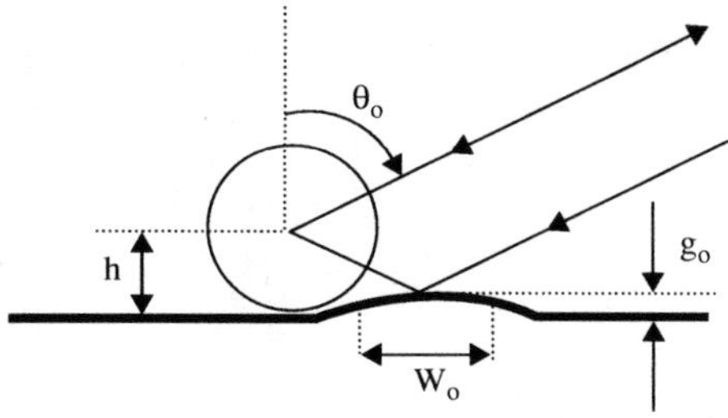

FIGURE 12.19. Scheme showing how MDIM implements the presence of a bump ($g_o < 0$ corresponds to a dip).

position z (or angle θ_i)

$$h'(z) \cong h - g_o \exp\left(-\frac{h^2(\tan\theta_i - \tan\theta_0)^2}{w_0^2}\right), \tag{13}$$

where θ_0 is the center of the affected angular interval, the parameter g_0 is the maximum height (or depth) of the bump ($g_0 > 0$) or dip ($g_0 < 0$), and w_o is its width. In Eq. 13 $\tan\theta_0 = z_0/h$, z_0 being the central point of the bump, or dip, measured from the cylinder[63].

The new variable h' produces a phase shift in some of the components involved in the MDIM model. Figure 12.20 shows the pattern produced for $w_o = 0.2\ \mu m$ and $g_0 = 0.4\ \mu m$.

The fit corresponds to a bump, which is in agreement with the fact that a gold-sputtered layer may easily rise above the lower substrate (this was confirmed

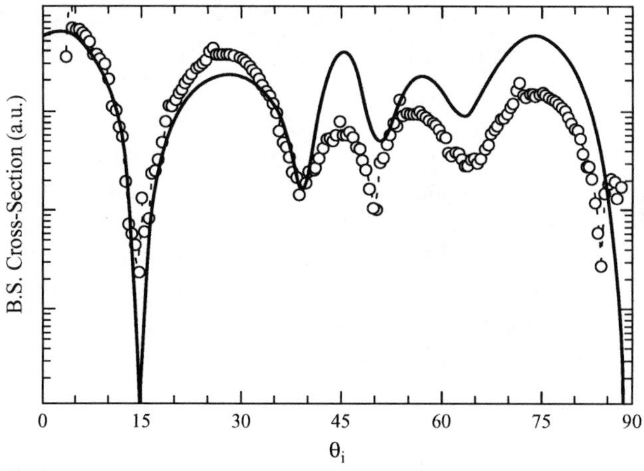

FIGURE 12.20. Agreement between the experimental backscattering pattern obtained at a point of the fiber (circles) and the results obtained with MDIM for a simulated bump of $w_o = 0.2\mu m$ and $g_0 = 0.4\mu m$.

by electron-microscope images). The presence of an extra minimum could be explained in this way, constituting one of several examples of inverse problems tackled with this technique.

12.3.4. Many Particles: Polydispersity, Shadowing and Multiple Scattering

12.3.4.1. Introduction

For surfaces sparsely seeded with identical particles, the scattering patterns correspond to those predicted for the isolated particle on the substrate. However, as the surface density increases, the role of multiple scattering becomes more important, and the pattern progressively loses the characteristic lobed structure. Eventually, when particles begin to stack in layers, the surface behaves very much like a rough surface, completely losing the single particle scattering signature.

The samples pictured in Figs. 12.21a and 12.21b represent this evolution, where the mean distance d is intuitively used as an inverse measure of the density of the sample, $d = (1/\rho)^{1/2}$, d being expressed in μm if ρ is expressed in particles/μm^2.

The loss of the single particle scattering signature can be observed in Fig. 12.22, for samples similar to those shown in Fig. 12.21, from $d = 1.5$ μm to $d = 4$ μm. In this case, the I_{ss} is measured , and plotted in log scale. The total scattering increases, but the characteristic lobed pattern is replaced by an increasingly uniform pattern.

Both evolution curves show that the importance of multiple scattering and its effects increases significantly as d decreases. In fact, it has been demonstrated[64,11] that d has to be shorter for particles on substrates than for particles in space in order to produce important multiple scattering effects. An explanation for this is the destructive interference between direct scattering and substrate reflected scattering from one particle to another.

Multiple scattering is not the only reason the lobed pattern is lost. In low surface density particle experiments, it is possible to observe that the visibility of the minima is lower than that calculated for the single scatterer on the flat substrate. One reason for this is the size polydispersity associated with an ensemble of similar, but not identical, particles. But other causes should be also mentioned: shape polydispersity, microirregularities in the substrate, presence of particle clusters, and also a persisting small amount of multiple scattering.

12.3.4.2. Polydispersity

A theoretical approach to the loss of visibility in the lobed pattern caused by polydispersity[65] leads to the conclusion that, in a first approximation, visibility depends on the order of the minimum and the degree of polydispersity, but not on particle size, so that approximate curves such as those shown in dashed lines with

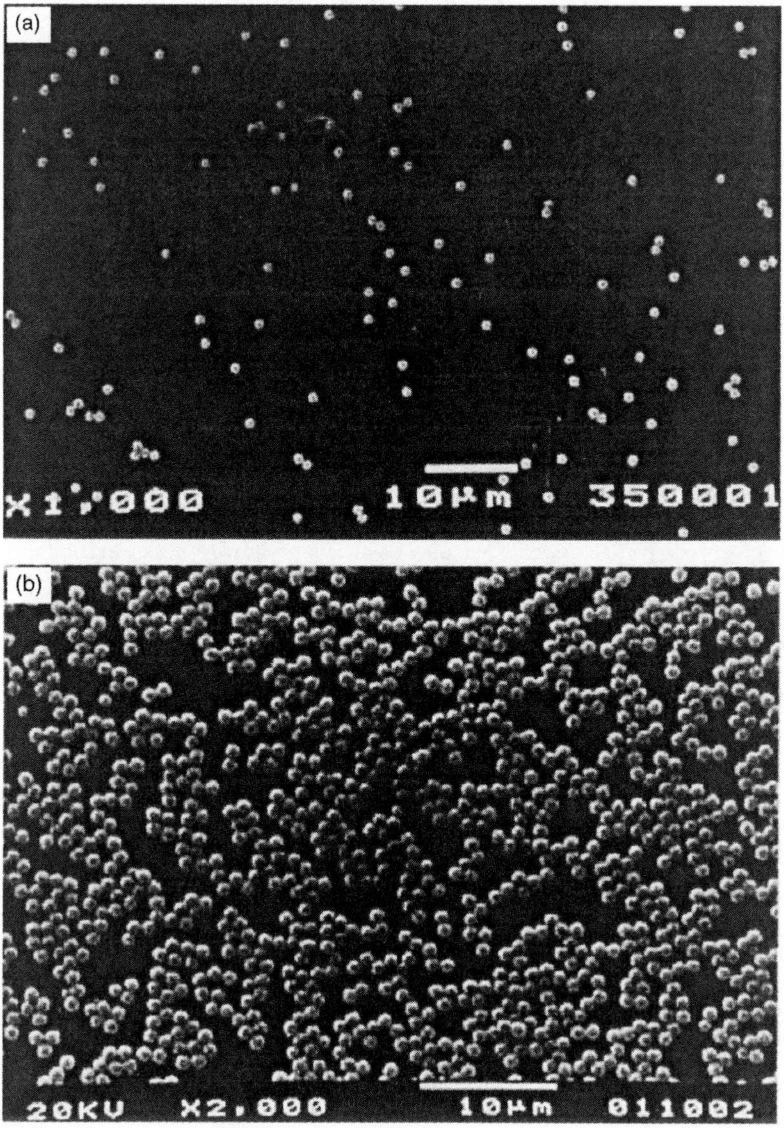

FIGURE 12.21. Scanning electron micrographs corresponding to (a) a sparse sample with $d = 6.5\mu m$ and (b) a dense sample with $d = 1.5\mu m$.

circles in Fig. 12.23 may be plotted not only for spheres but also for cylinders with incidence normal to the axis and observation within the plane of incidence.

In the case of a cylinder on the substrate, polydispersity must be understood as the averaging effect produced by size variations in the illuminated length.This polydispersity is the main cause of the loss of visibility for cylinders. Within this experimental framework, it is possible to understand the results obtained for

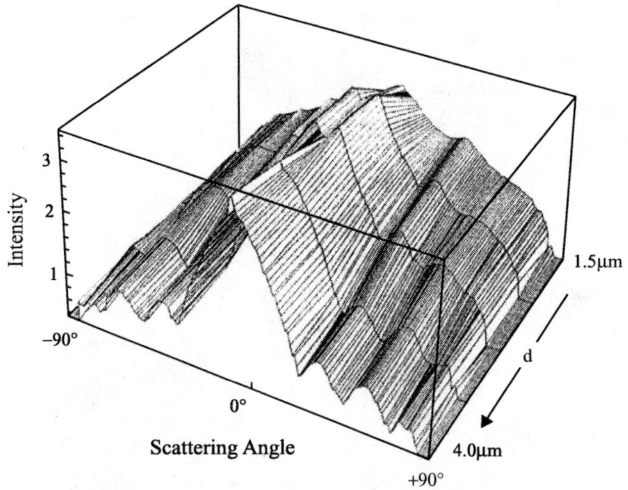

FIGURE 12.22. Evolution of a co-polarized scattering pattern as a function of the mean distance between particles. Dense samples produce a progressive loss of visibility in the lobed structure.

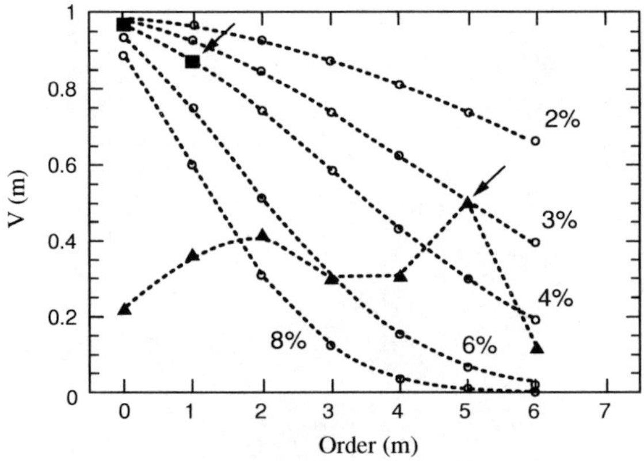

FIGURE 12.23. Evolution curves of the visibility $V(m)$. These curves are only measurable at the integer values (circles) and are obtained from an approximate calculation and assuming only one source for the loss of visibility: size polydispersity r, that is expressed in percent for each curve. Experimental values obtained for a cylinder sized $D = 1.1$ μm for $m = 0$ and $m = 1$ are plotted in squares. Experimental values for a sample of spheres sized $D = 3.2$ μm are plotted in triangles, showing evidence of other sources for the loss of visibility.

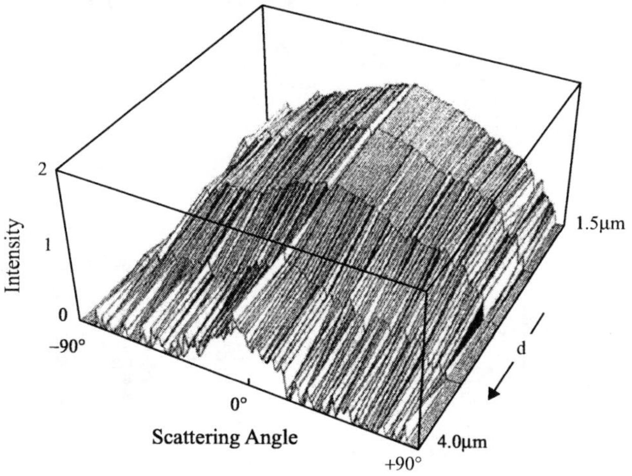

FIGURE 12.24. Evolution of the cross-polarized scattered intensity patterns for $\theta_i = 0°$ and s-incident wave as a function of the surface particle density, from the mean distance $d = 1.5\mu m$ to $d = 4\mu m$. Logarithmic scale.

a cylinder sized $D = 1.1$ μm (two minima) and for spheres sized $D = 3.2$ μm (seven minima). The visibilities found are plotted in Fig. 12.23 for the cylinder (squares), and for spheres (triangles). The number of orders available, depends, as mentioned above, on the particle size. The values found for the cylinder fit with one of the theoretical *only-polydispersity* curves, corresponding to $r = 0.04$ (4% polydispersity), which is in agreement with an estimate made by the manufacturers of the fiber for its size variations. For the spheres, because of the many causes of the visibility loss (each affecting different parts of the scattering patterns), we cannot expect the experimental values to follow a given curve. Instead, an upper-limit estimate of 3% (in the 5th order) can be established, provided that higher values would not allow to reach such values.

12.3.4.3. Multiple Scattering and Shadowing Effect

Fig. 12.24 is similar to Fig. 12.22, but here the measured cross-polarized intensity I_{sp} is plotted. This represents the increasing role of multiple scattering for this kind of sample, since the incident polarization (perpendicular to the plane of incidence) can be changed with respect to that plane (and measuring within that plane) only by multiple scattering.

One numerical method of evaluating the power of the surface to produce multiple scattering is to integrate the cross-polarized intensity over all scattering angles and to measure that with respect to the co-polarized intensity. This can be approximated by the following ratio in the plane of incidence (where specular light is not taken

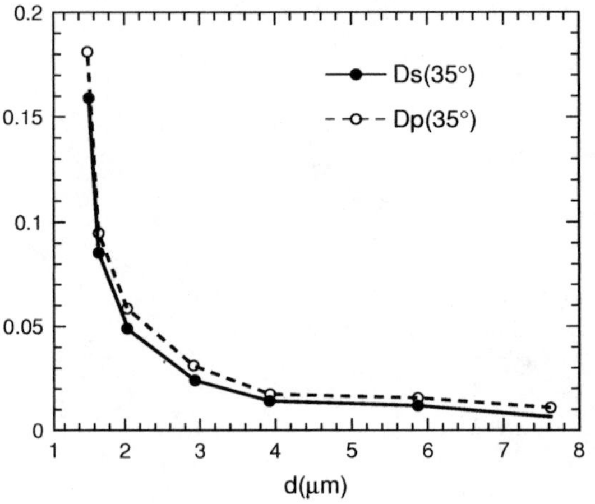

FIGURE 12.25. Evolution of the cross-polarization ratio for p-incident polarization as a function of the mean particle distance d.

into account):

$$D_s = \frac{\sum_{\theta_s=-90^\circ}^{\theta_s=+90^\circ} I_{\mathrm{sp}}(\theta_i, \theta_s)}{\sum_{\theta_s=-90^\circ}^{\theta_s=+90^\circ} I_{\mathrm{ss}}(\theta_i, \theta_s)}. \tag{14}$$

These *cross-polarization ratios* are obtained for different samples, and also for polarization parallel to the plane of incidence (D_p).

In Fig. 12.25 the ratios are plotted for a fixed angle of incidence ($\theta_i = 35^\circ$) as a function of the mean distance d, producing a characteristic shape of increasing interaction for decreasing mean distance. It should be noted that very short mean interparticle distances are required to obtain high values of the cross-polarization ratios.

An even more interesting dependence is discovered when the ratios are plotted as a function of the angle of incidence. In Fig. 12.26, D_s is plotted as a function of the angle of incidence and for samples with three different values of the surface particle density (mean distance values $d = 7.6$ μm, 2.9 μm and 1.5 μm for the dilute, intermediate and dense sample respectively). For dilute samples, D_s remain small, with almost no multiple scattering between particles and no depencedence on θ_i. For the intermediate sample, a significant increase is observed from $\theta_i = 60^\circ$; this is related to the way particles sized of the order of the wavelength scatter strongly in a wide forward lobe that can reach other particles for such incidences. From $\theta_i = 75^\circ$ onwards, cross-polarization effects tend to decay, suggesting that particles no longer increase the amount of light directed onto their neighbors through forward scattering, and receive less light instead. In other words, particles tend to be inside the shadow cast by other particles, producing what we call the

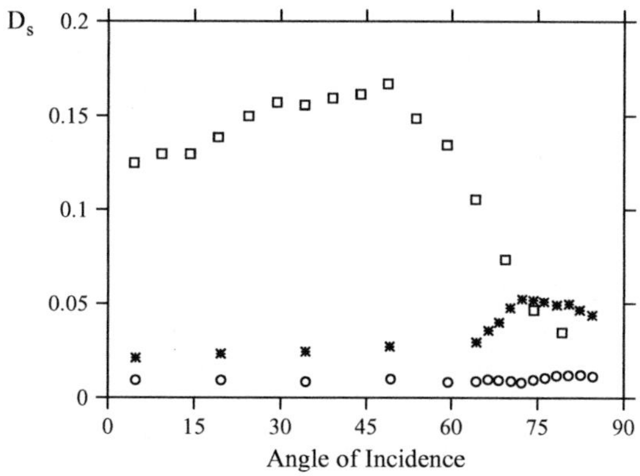

FIGURE 12.26. Cross-polarization ratios as a function of the angle of incidence for three different samples of spherical particles with particle mean distance $d = 7.6$ μm (circles), $d = 2.9$ μm (asterisks), and $d = 1.5$ μm (squares).

shadowing effect. In the dense sample, this effect is highly significant at relatively low incidences. The substrate plays an important role at low incidences, but is almost absent from the scattering process at high incidence angles, where the dense surface behaves more like a smooth rough surface.

12.3.5. Light Scattering Statistics

As mentioned above, an interesting approach to solving the inverse problem is statistical analysis of scattered light. The statistical techniques developed in this field are particularly applicable to the extraction of information about the samples in backscattering experiments. For particles on a flat or slightly rough surface, the particle surface density can be considered the main source of multiple scattering. For incident polarization, either parallel or perpendicular to the scattering plane, multiple scattering processes may produce changes in the polarization plane of the light detected within the plane of incidence. Consequently, a nonzero component in the direction perpendicular to that of the incident one (cross-polarized component) is observed when the average distance between particles, d, is short enough, i.e. particle surface density, $\rho = 1/d^2$, is high enough. Assuming a Gaussian regime (high number of scatterers), the probability of detecting zero when observing the cross-polarized component can be expressed as[66]

$$P(I_{\text{cross}} = 0) = \exp\left[-\frac{L\pi ab}{d^3}\right] = \exp(-\gamma ab), \tag{15}$$

where a and b are the semiaxes of the illuminated area (dependent on the width of the incident beam and the angle of incidence), and L is the distance at which multiple scattering between two particles can be neglected. The parameter γ includes only intrinsic dependences on the sample, whereas the presence of a and b shows the importance of the number of illuminated particles.

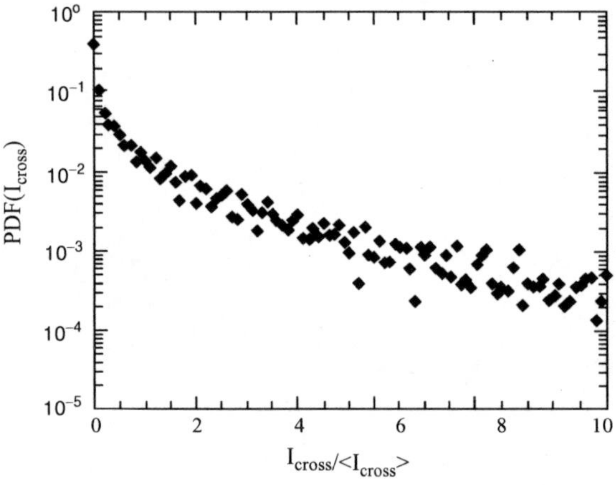

FIGURE 12.27. Probability density function for the cross-polarized intensity associated with light backscattered from a sample of spherical particles.

Experiments carried out to fit statistical expressions such as Eq.15, require a high number of measurements from the sample. This can be accomplished by rotating the sample and detecting the intensity fluctuations.[66] A probability density function (PDF) is found for either the co- and cross-polarized intensity.

In Fig. 12.27 a particular case is shown for backscattered cross-polarized detection.

The sample consisted of gold particles sized approximately 1 μm, seeded on a flat gold substrate.

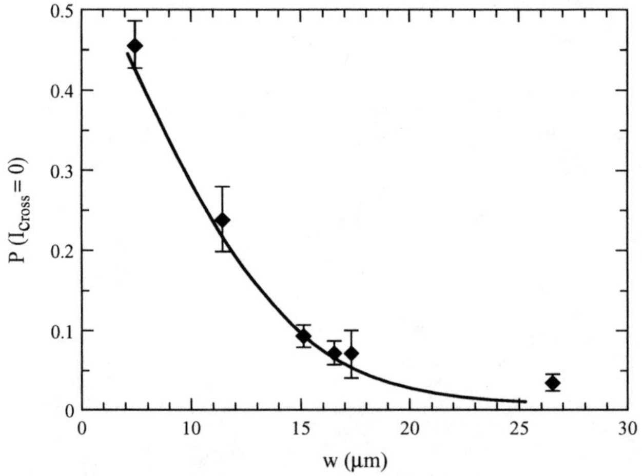

FIGURE 12.28. Dots show the values experimentally obtained for P($I_{cross} = 0$) for different values of the spot size w. The curve corresponds to the fit of such point to Eq. 15.

The probability value at $I_{cross} = 0$ (the highest value in the curve, i.e. the smallest relative error) can be assessed for different spot sizes, and γ can be obtained from Eq. 15. Figure 12.28 shows the evolution of $P(I_{cross} = 0)$ for six values of the spot size and the fit to Eq. 15 is plotted as a continuous line. From this fitting, $\gamma = 8.6\ 10^{-3}\ \text{mm}^{-2}$ is found, so that d can be obtained if L is known and vice versa. For $d = 8\ \mu\text{m}$, an estimate obtained from the electron microscope images, a value of $L \cong 1.25\ \mu\text{m}$ is obtained, slightly larger than the particle diameter, which is consistent for instance with results shown in Fig. 12.25 and with other experimental work.[66] Parameter γ is now seen as a measurement of the interacting capacity of the scatterers, given by their density and their ability to scatter light onto each other, and this finally contributes to the cross-polarized light in the backscattering direction.

12.3.6. Concluding Remarks

A system composed of a flat substrate seeded with particles can be analyzed in the far-field regime through different approaches. Some of these have been presented here. First, we presented the scattering patterns obtained when observing the angular dependences of the scattering intensity, and the evolution of some of their features, such as the minima positions. We then described the observation of backscattering as a function of the angle of incidence, which is highly sensitive to small changes in the system. Finally, we showed some results corresponding to a statistical analysis of the fluctuations associated with the scattered (or backscattered) signal. All these methods provide information about the scattering system, i.e. they are examples of how to tackle the more general inverse problem.

Acknowledgments. The authors thank the Dirección General de Enseñanza Superior for its financial support (projects BFM2001-1289 and FIS2004-06785) under which most of our research included in this chapter has been done.

References

1. P.R. Young, "Low-scatter mirror degradation by particle contamination," *Opt. Eng.* **15**, 516–520 (1976).
2. A.J. Pidduck, D.J. Robbins, I.M. Young, A.G. Cullis, and A.S.R. Martin, The formation of dislocations and their in-situ detection during silicon vapor phase epitaxi at reduced temperature, *Mater. Sci. Eng. B* **4**, 417–422 (1989).
3. P.N. Prasad, *Nanophotonics*, (John Wiley and Sons, 2004).
4. J.L de la Peña, F. González, J.M. Saiz, P.J. Valle, and F. Moreno, Sizing particles on substrates: a general method for oblique incidence, *J. Appl. Phys.* **85**, 432–438 (1999).
5. D.W. Pohl and D. Courjon (eds.), *Near Field Optics* (Kluwer Academic Publishers, London, 1993).
6. I.V. Lindell, A.H. Sihvola, K.O. Muinonen, and P.W. Barber, Scattering by a small object close to an interface. I. Exact-image theory formulation, *J. Opt. Soc. Am. A* **8**, 472–476 (1991).

7. K.O. Muinonen, A.H. Sihlova, I.V. Lindell, and K. A. Lumme, Scattering by a small object close to an interface. II. Study of backscattering, *J. Opt. Soc. Am. A* **8**, 477–482 (1991).
8. G. Videen, M.G. Turner, V.J. Iafelice, W.S. Bickel, and W.L. Wolfe, Scattering from a small sphere near a surface, *J. Opt. Soc. Am. A* **10**, 118–126 (1993).
9. E. Jakeman, Scattering by particles on an interface, *J. Phys. D: Appl. Phys.* **27**, 198–210 (1994).
10. G. Videen, W.L. Wolfe, and W.S. Bickel, The light-scattering Mueller matrix for a surface contaminated by a single particle in the Rayleigh limit, *Opt. Eng.* **31**, 341–349 (1992).
11. F. Moreno, J.M. Saiz, P.J. Valle and F. González, On the multiple scattering effects for small metallic particles on flat conducting substrates, *Waves Random Media* **5**, 73–88 (1995).
12. T.A. Germer, Angular dependence and polarization of out-of-plane optical scattering from particulate contamination, surface defects, and surface microroughness, *Appl. Opt.* **36**, 8798–8805 (1997).
13. E.M. Ortiz, P.J. Valle, J.M. Saiz, F. González, and F. Moreno, Multiscattering effects in the far-field region for two small particles on a flat conducting substrate, *Waves Random Media* **7**, 319–329 (1997).
14. E.M. Ortiz, P.J. Valle, J.M. Saiz, F. González, and F. Moreno, A detailed study of the scattered near field of nanoprotuberances on flat substrates, *J. Phys. D: Appl. Phys.* **31**, 3009–3019 (1998).
15. P.J. Valle, E.M. Ortiz and J.M. Saiz, Near field by subwavelength particles on metallic substrates with cylindrical surface plasmon excitation, *Opt. Commun.* **137**, 334–342 (1997).
16. K.B. Nahm and W.L. Wolfe, Light scattering models for spheres on a conducting plane: comparison with experiment, *Appl. Opt.* **26**, 2995–2999 (1987).
17. D.C. Weber and E.D. Hirleman, Light scattering signatures of individual spheres on optically smooth conducting surfaces, *Appl. Opt.* **27**, 4019–4026 (1998).
18. E.J. Bawolek and E.D Hirleman, *Particles on Surfaces 3*, K.L. Mittal, ed. (Plenum Press, New York, 1991).
19. F. Moreno, J.M. Saiz, P.J. Valle, and F. González, Metallic particle sizing on flat surfaces: application to conducting substrates, *Appl. Phys. Lett.* **68**, 3087–3089 (1996).
20. J.L. de la Peña, J.M. Saiz, and F. González, Profile of a fiber from backscattering measurements, *Opt. Lett.* **25**, 1699–1701 (2000).
21. J.L. de la Peña, J.M. Saiz, F. González, and F. Moreno, Detection and recognition of local defects in 1D structures, *Opt. Commun.* **196**, 33–39 (2001).
22. P.A. Bobbert and J. Vlieger, Light scattering by a sphere on a substrate, *Physica A* **137**, 209–242 (1986).
23. P.A. Bobbert, J. Vlieger, and R. Greef, Light reflection from a substrate sparsely seeded with spheres—comparison with an ellipsometric experiment, *Physica A* **137**, 243–257 (1986).
24. G. Videen, Light Scattering from a sphere on or near a surface, *J. Opt. Soc. Am. A* **8**, 483–489 (1991); errata, *J. Opt. Soc. Am. A* **9**, 844–845 (1992).
25. G. Videen, Light scattering from a sphere behind a surface, *J. Opt. Soc. Am. A* **10**, 110–117 (1993).
26. G. Videen, Light scattering from a particle on or near a perfectly conducting surface, *Opt. Commun.* **115**, 1–7 (1995).
27. G. Videen, Light scattering from an irregular particle behind a plane interface, *Opt. Commun.* **128**, 81–90 (1996).

28. B.R. Johnson, Light scattering from a spherical particle on a conducting plane: I. Normal incidence, *J. Opt. Soc. Am. A* **9**, 1341–1351 (1992); errata, *J. Opt. Soc. Am. A* **10**, 766 (1993).
29. B.R. Johnson, Calculation of light scattering from spherical particle on a surface by the multipole expansion method, *J. Opt. Soc. Am. A* **13**, 326–337 (1996).
30. B.R. Johnson, Light diffraction by a particle on an optically smooth surface, *Appl. Opt.* **36**, 240–246 (1997).
31. E. Fucile, P. Denti, E. Borghese, R. Saija, and O.I. Sindoni, Optical properties of a sphere in the vicinity of a plane surface, *J. Opt. Soc. Am. A* **14**, 1505–1514 (1997).
32. G. Videen and D. Ngo, Light scattering from a cylinder near a plane interface: theory and comparison with experimental data, *J. Opt. Soc. Am. A* **14**, 70–78 (1997).
33. G. Videen, W.S. Bickel, V.J. Iafelice, and D. Abromson, Experimental light-scattering Mueller matrix for a fiber on a reflecting optical surface as a function of incident angle, *J. Opt. Soc. Am. A* **9**, 312–315 (1992).
34. R. Borghi, F. Gori, M. Santarsiero, F. Frezza, and G. Schettini, Plane-wave scattering by a set of perfectly conducting circular cylinders in the presence of a plane surface, *J. Opt. Soc. Am. A* **13**, 2441–2452 (1996).
35. M. Nieto-Vesperinas, *Scattering and diffraction in Physical Optics* (Wiley-Interscience, New York, 1991).
36. P.J. Valle, F. González, and F. Moreno, Electromagnetic wave scattering from conducting cylindrical structures on flat substrates: study by means of the extinction theorem, *Appl. Opt.* **33**, 512–523 (1994).
37. P.J. Valle, F. Moreno, J.M. Saiz and F. González, Near-field scattering from subwavelength metallic protuberances on conducting flat substrates, *Phys. Rev. B* **51**, 13681–13690 (1995).
38. J.M. Saiz, P.J. Valle, F. González, E.M. Ortiz, and F. Moreno, Scattering by a metallic cylinder on a substrate: burying effects, *Opt. Lett.* **21**, 1330–1332 (1996).
39. P.J. Valle, F. Moreno, J.M. Saiz, and F. González, Electromagnetic interaction between two parallel circular cylinders on a planar interface, *IEEE Trans. Antennas Propag.* **44**, 321–325 (1996).
40. P.J. Valle, F. Moreno, and J.M. Saiz, Comparison of real- and perfect-conductor approaches for scattering by a cylinder on a flat substrate, *J. Opt. Soc. Am. A* **15**, 158–162 (1998).
41. A. Madrazo and M. Nieto-Vesperinas, Scattering of electromagnetic waves from a cylinder in front of a conducting plane, *J. Opt. Soc. Am. A* **12**, 1298–1309 (1995).
42. R. Carminati, A. Madrazo and M. Nieto-Vesperinas, Electromagnetic wave scattering from a cylinder in front of a conducting surface-relief grating, *Opt. Commun.* **111**, 26–33 (1994).
43. A. Madrazo and M. Nieto-Vesperinas, Scattering of light and other electromagnetic waves from a body buried beneath a highly rough random surface, *J. Opt. Soc. Am. A* **14**, 1859–1866 (1997).
44. J. -J. Greffet and F. -R Ladan, Comparison between theoretical and experimental scattering of an s-polarized electromagnetic wave by a two-dimensional obstacle on a surface, *J. Opt. Soc. Am. A* **8**, 1261–1269 (1991).
45. F. Pincemin, A.Sentenac, and J.-J.Greffet, Near field scattered by a dielectric rod below a metallic surface, *J. Opt. Soc. Am. A* **11**, 1117–1127 (1994).
46. T. Wriedt and A. Doicu, *Light Scattering by Microstructures*, F. Moreno and F. González, eds. (Lecture Notes in Physics Series, Springer-Verlag, Berlin, 2000), pp. 113–132.

47. M.A.Taubenblatt, Light scattering from cylindrical structures on surfaces, *Opt. Lett.* **15**, 255–257 (1990).
48. R. Schmehl, B.M. Nebeker, and E.D. Hirleman, Discrete-dipole approximation for scattering by features on surfaces by means of a two-dimensional fast Fourier transform technique, *J. Opt. Soc. Am. A* **14**, 3026–3036 (1997).
49. G.L. Wojcik, D.K. Vaughan, and L.K. Galbraith, Calculation of light scatter from structures of silicon surfaces, *Lasers in Microlithography*, SPIE **774**, 21–31 (1987).
50. M.A. Kolbehdari, H.A. Auda, and A.Z. Elsherbeni, Scattering from dielectric cylinder partially embedded in a perfectly conducting ground plane, *J. Electromagn. Waves Appl.* **3**, 531–554 (1989).
51. J.M. Saiz, F. González, F. Moreno, and P.J. Valle, Application of a ray-tracing model to the study of backscattering from surfaces with particles, *J. Phys. D: Appl. Phys.* **28**, 1040–1046 (1995).
52. F. González, J.M. Saiz, P.J. Valle and F. Moreno, Scattering from particulate metallic surfaces: effect of surface particle density, *Opt. Eng.* **34**, 1200–1207 (1995).
53. FEMLAB 3.0. *Electromagnetics Module User's Guide*. COMSOL AB (2004).
54. H. Raether, *Surface Plasmons on Smooth and Rough Surfaces and on Gratings* (Springer-Verlag, Berlin, 1988).
55. W.L. Barnes, A. Dereux, and T.W. Ebbesen, Surface plasmon subwavelength optics, *Nature* **424**, 824–830 (2003).
56. A.V. Zayats and I.I. Smolyaninov, Near-field photonics: surface plasmon polaritons and localized surface plasmons, *J. Opt. A: Pure Appl. Opt.* **5**, S16–S50 (2003).
57. P.W. Barber and S.C. Hill, *Light Scattering by Particles: Computational Methods* (World Scientific, Singapore, 1990).
58. P. Dawson, F. de Fornel, and J-P. Goudonnet, Imaging of surface plasmon propagation and edge interaction using a photon scanning tunnelling microscope, *Phys. Rev. Lett.* **72** (18), 2927–2930 (1994).
59. F. Keilmann and Y.H. Bai, Periodic surface structures frozen into CO_2 laser-melted quartz, *Appl. Phys. A* **29**, 9–18 (1982).
60. A. Sommerfeld, *Partial differential equations in physics* (Academic Press, New York, 1967) p. 236.
61. V.A. Kosobukin, Polarization and resonance effects in optical generation of cylindrical surface polaritons and surface structures, *Sov. Phys. – Solid State* **35**, 457–463 (1993).
62. G.J. Burke and A.J. Poggio, *Numerical Electromagnetics Code (NEC). Method of Moments*. Report UCID 18834 (Lawrence Livermore Lab, Livermore CA, 1981).
63. J.M. Saiz, J.L. de la Peña, F. González, and F. Moreno, Detection and recognition of local defects in 1D structures, *Opt. Commun.* **196**, 33–39 (2001).
64. F. González, J.M. Saiz, P.J. Valle, and F. Moreno, Multiple scattering in particulate surfaces: cross-polarization ratios and shadowing effects, *Opt. Commun.* **137**, 359–366 (1997).
65. J.L. de la Peña, J.M. Saiz, G.Videen, F. González, P.J. Valle, and F. Moreno, Scattering from particles on surfaes: visibility factor and polydispersity, *Opt.Lett.* **24**, 1451–1453 (1999).
66. E.M. Ortiz, F. González, J.M. Saiz, and F. Moreno, Experimental measurements of the statistics of the scattered intensity from particles on surfaces, *Opt. Express* **10**, 190–195 (2002).

13
Multiple Scattering of Waves by Random Distribution of Particles for Applications in Light Scattering by Metal Nanoparticles

KA KI TSE[†], LEUNG TSANG[‡], CHI HOU CHAN[†] AND KUNG-HAU DING[§]
[†] *Wireless Communications Research Centre, City University of Hong Kong, Hong Kong,*
[‡] *Department of Electrical Engineering, University of Washington, Seattle, WA 98195, USA*
[§] *Air Force Research Laboratory, Sensors Directorate, AFRL/SNHE, Hanscom AFB, MA 01731, USA*

13.1. Introduction

Electromagnetic scattering properties of particles are important issues for nanotechnology, terrestrial and planetary remote sensing, biomedical sensing and microscopy, wireless communication, astrophysics, and optical engineering. Scattering by a single particle is described by the Mie theory. For scattering by many particles, the classical theory assumes independent scattering in which the scattering intensity is set equal to the sum of scattering intensities from each particle. The independent scattering model is applied to calculate the phase matrix which is equal to the number density times the bistatic cross section of a single particle. The phase matrix is then used in radiative transfer theory to treat multiple incoherent scattering. The approach ignores the coherent wave interaction among the particles. The approach of independent scattering is particularly not valid for dense media when there is a high concentration of particles. For this case, particles are in close proximity of each other and the particles scatter collectively.

With the advent of computers, the electromagnetic problems are now often solved by numerical methods such as discrete dipole approximation (DDA)[1] and couple dipole (CD).[2–6] In DDA and CD, an object or a dense medium are divided by uniform grid line. Each small cubic box contains an electric dipole moment. Matrix equation is set up with the unknowns being the dipole moments of the cubes. This method can be used to solve the scattering problem of arbitrary geometry. However, it is computationally intensive because the entire geometry has to be discretized into fine cubic cells much smaller than a wavelength and the grid cells are arranged periodically. Multiple scattering can be formulated by using the Foldy Lax equation and using the T-matrix of Mie theory and the translation addition theorem. The Foldy Lax equations have been rigorously derived from Maxwell equations.[7] The Foldy Lax equations are solved numerically.

To distinguish such an approach from classical analytical theory, we call the approach numerical Maxwell model of 3-dimensional simulations (NMM3D). The advantage of Foldy Lax equations is that the vector spherical waves (partial waves) are used as basis functions rather than fine discretization of each particle. The unknowns are the partial wave coefficients for each particle. Thus the number of unknowns are many times less than that of fine discretization using cubic cells. Such an approach has been adopted by several types of wave scattering problem including light scattering and electron scattering.[8–12] In this chapter, we study multiple scattering based on numerical solutions of Foldy Lax equations putting particular emphasis on cases with high concentration of particles. We have also implemented the numerical method on parallel computing allowing us to solve cases of large number of particles. We used several thousand particles in our simulations.

Recently the optical properties of noble metal nanoparticles have attracted considerable interest. With the advances in nanotechnology, nanoparticles can be applied in a wide range of applications. The small size of nanoparticles allows them to penetrate into small objects, e.g. body cells. Their chemically inert properties, to avoid the problem of phototoxic and photo bleaching which existed in using organic fluorescent dye, make them applicable in biological and biomedical applications, such as optical spectroscopy,[15–16] biological sensing,[17–21] and plasmon resonance microscopy.[22–23] Their applications are also studied in optical engineering, e.g., integrated optics circuits[24] and optical data storage.[25] Experiments have been performed to investigate the plasmon resonant properties of nanoparticles.[26–31]

In this chapter, the approach is applied to nanotechnology in which we study plasmon resonance of scattering by metallic nanoparticles at optical frequency.

In Sect. 13.2, we cast Maxwell equation in the form of Foldy Lax equation. We describe steps for the numerical solution, and the computations of the extinction, the absorption and the phase matrix.

In Sect. 13.3, we study optical scattering by metallic nanoparticles. The plasmon resonance for various concentrations of gold nanoparticles is studied. It is desirable to have a high concentration of nanoparticles in applications. However, our results indicate the plasmon resonance can disappear at high concentrations.

In Sect. 13.4, we illustrate the results of the bistatic scattering properties. This relationship is described by the phase matrix. The phase matrix is obtained by solving Foldy Lax equations for many realizations and then averaging the scattered intensities over realizations.

In Sect. 13.5, we discuss the problem of optical scattering of nanoparticles above a rough surface or below a rough surface

13.2. Formulation for Foldy Lax Equations

Consider N spherical particles randomly located in free space. The particles are centered at $r_1, r_2, \ldots, r_N$ and are of radius a. With an incident electromagnetic plane wave with the incident direction $\hat{k}_i = \sin\theta_i \cos\phi_i \hat{x} + \sin\theta_i \sin\phi_i \hat{y} + \cos\theta \hat{z}$

and polarization $\overline{A}$, the incident wave is written as

$$\begin{aligned}\overline{E}^{\text{inc}}(\overline{r}) &= \overline{A}\,\mathrm{e}^{\mathrm{i}\overline{k}_i\cdot\overline{r}} \\ &= \mathrm{e}^{\mathrm{i}\overline{k}_i\cdot\overline{r}_l}\overline{A}\,\mathrm{e}^{\mathrm{i}\overline{k}_i\cdot(\overline{r}-\overline{r}_l)} \\ &= \mathrm{e}^{\mathrm{i}\overline{k}_i\cdot\overline{r}_l}\sum_{m,n}\left\{a_{mn}^{(M)}\overline{M}_{mn}^{(0)}(k\overline{r}\overline{r}_l) + a_{mn}^{(N)}\overline{N}_{mn}^{(0)}(k\overline{r}\overline{r}_l)\right\}. \end{aligned} \tag{1}$$

The exciting field coefficients of multiple scatterers can be expressed as the sum of incident field and the scattered field from other scatterers[13]:

$$\begin{aligned} w_{mn}^{(M)(l)} &= \mathrm{e}^{\mathrm{i}\overline{k}_i\cdot\overline{r}_l}a_{mn}^{(M)} + \sum_{\substack{j=1\\ j\neq 1}}^{N}\sum_{\mu\nu}\left[A_{mn\mu\nu}(k\overline{r_l r_j})T_{\nu}^{(M)(j)}w_{\mu\nu}^{(M)(j)}\right. \\ &\quad \left.+ B_{mn\mu\nu}(k\overline{r_l r_j})T_{\nu}^{(N)(j)}w_{\mu\nu}^{(N)(j)}\right] \end{aligned} \tag{2a}$$

$$\begin{aligned} w_{mn}^{(N)(l)} &= \mathrm{e}^{\mathrm{i}\overline{k}_i\cdot\overline{r}_l}a_{mn}^{(N)} + \sum_{\substack{j=1\\ j\neq 1}}^{N}\sum_{\mu\nu}\left[B_{mn\mu\nu}(k\overline{r_l r_j})T_{\nu}^{(M)(j)}w_{\mu\nu}^{(M)(j)}\right. \\ &\quad \left.+ A_{mn\mu\nu}(k\overline{r_l r_j})T_{\nu}^{(N)(j)}w_{\mu\nu}^{(N)(j)}\right], \end{aligned} \tag{2b}$$

where $A_{mn\mu\nu}$ and $B_{mn\mu\nu}$ are the translation addition coefficients; $w_{mn}^{(M)(l)}$ and $w_{mn}^{(N)(l)}$ denote the exciting field coefficients; $T_{\mu\nu}^{(M)(j)}$ and $T_{\mu\nu}^{(N)(j)}$ are the scattering T-matrix of particle (j). We can write Eq. (2) in matrix form. $\overline{\overline{A}}$ and $\overline{\overline{B}}$ are matrices of $A_{mn\mu\nu}$ and $B_{mn\mu\nu}$, respectively; $\overline{\overline{T}}^{(M)(j)}$ and $\overline{\overline{T}}^{(N)(j)}$ are matrices of $T_{\mu\nu}^{(M)(j)}$ and $T_{\mu\nu}^{(N)(j)}$, respectively; $\overline{a}^{(M)}$ and $\overline{a}^{(N)}$ are matrices of $a_{mn}^{(M)}$ and $a_{mn}^{(N)}$ in Eq. (1), respectively; and $\overline{w}^{(M)(l)}$ and $\overline{w}^{(N)(l)}$ are matrices of $w_{mn}^{(M)(l)}$ and $w_{mn}^{(N)(l)}$, respectively. $\overline{\overline{A}}$, $\overline{\overline{B}}$, $\overline{\overline{T}}^{(M)(j)}$ and $\overline{\overline{T}}^{(N)(j)}$ are square matrices with dimension $L_{\max}\times L_{\max}$, and $\overline{a}^{(M)}$, $\overline{a}^{(N)}$, $\overline{w}^{(M)(l)}$ and $\overline{w}^{(N)(l)}$ are column matrices with dimension $L_{\max}\times 1$. Then, we have

$$\overline{w}^{(M)(l)} = \mathrm{e}^{\mathrm{i}\overline{k}_i\cdot\overline{r}_l}\overline{a}^{(M)} + \sum_{\substack{j=1\\ j\neq 1}}^{N}\left[\overline{\overline{A}}(k\overline{r}_l)\overline{\overline{T}}^{(M)(j)}\overline{w}^{(M)(j)} + \overline{\overline{B}}(k\overline{r}_l)\overline{\overline{T}}^{(N)(j)}\overline{w}^{(N)(j)}\right] \tag{3a}$$

$$\overline{w}^{(N)(l)} = \mathrm{e}^{\mathrm{i}\overline{k}_i\cdot\overline{r}_l}\overline{a}^{(N)} + \sum_{\substack{j=1\\ j\neq 1}}^{N}\left[\overline{\overline{B}}(k\overline{r}_l)\overline{\overline{T}}^{(M)(j)}\overline{w}^{(M)(j)} + \overline{\overline{A}}(k\overline{r}_l)\overline{\overline{T}}^{(N)(j)}\overline{w}^{(N)(j)}\right]. \tag{3b}$$

We put equation (3a) and (3b) in a compact matrix form:

$$\overline{w}^{(l)} = \sum_{\substack{j=1\\ j\neq 1}}^{N}\overline{\overline{\sigma}}(k\overline{r_l r_j})\overline{\overline{T}}^{(j)}\overline{w}^{(j)} + \mathrm{e}^{\mathrm{i}\overline{k}_i\cdot\overline{r}_l}\overline{a}_{\text{in}} \tag{4a}$$

where

$$\overline{w}^{(l)} = \begin{bmatrix} \overline{w}^{(M)(l)} \\ \overline{w}^{(M)(l)} \end{bmatrix} \tag{4b}$$

$$\overline{a}_{\text{in}} = \begin{bmatrix} \overline{a}^{(M)} \\ \overline{a}^{(N)} \end{bmatrix} \tag{4c}$$

$$\overline{\overline{\sigma}}(k\overline{r}) = \begin{bmatrix} \overline{\overline{A}}(k\overline{r}) & \overline{\overline{B}}(k\overline{r}) \\ \overline{\overline{B}}(k\overline{r}) & \overline{\overline{A}}(k\overline{r}) \end{bmatrix} \tag{4d}$$

$$\overline{\overline{T}}^{(j)} = \begin{bmatrix} \overline{\overline{T}}^{(M)(j)} & 0 \\ 0 & \overline{\overline{T}}^{(N)(j)} \end{bmatrix}. \tag{4e}$$

In Eqs. (4a)–(4e), $\overline{w}^{(l)}$ and $\overline{a}_{\text{in}}$ are column matrices of dimension $2L_{\max} \times 1$ and $\overline{\overline{\sigma}}^{(i)}$ and $\overline{\overline{T}}^{(j)}$ are square matrices of dimension $2L_{\max} \times 2L_{\max}$.

We solve the matrix equation using iterative method. In iterative method, the matrix needs to be well conditioned. The matrix equations from Eq. (4) can be rewritten using scattered field coefficients and internal field coefficients. It has been shown that the condition number of the matrix equation with internal field coefficients has a better condition number.[14] Hence, the matrix equation with the internal field coefficients is used in this chapter. After solving the internal field coefficients, the scattered field coefficients can be found.

The relation between the exciting field coefficients and internal field coefficients is

$$\overline{c}^{(j)} = \begin{bmatrix} c^{(M)(j)} \\ c^{(N)(j)} \end{bmatrix} = \overline{\overline{B}}^{(j)} \overline{w}^{(j)}, \tag{5a}$$

where $\overline{c}^{(j)}$ is a column matrix of internal field coefficients, and $\overline{\overline{B}}^{(j)}$ is a diagonal matrix, i.e.

$$\overline{\overline{B}}^{(j)} = \begin{bmatrix} \overline{\overline{B}}^{(M)(j)} & 0 \\ 0 & \overline{\overline{B}}^{(N)(j)} \end{bmatrix} \tag{5b}$$

and

$$B_n^{(M)(j)} = \frac{\mathrm{i}}{ka} \frac{1}{j_n(k_j a)\left[kah_n^{(1)}(ka)\right]' - h_n^{(1)}(ka)[k_j a j_n(k_j a)]'} \tag{5c}$$

$$B_n^{(N)(j)} = \frac{\mathrm{i}k_j a}{(k_j a)^2\, j_n(k_j a)\left[kah_n^{(1)}(ka)\right]' - (ka)^2\, h_n^{(1)}(ka)[k_j a j_n(k_j a)]'}. \tag{5d}$$

Let

$$S_n^{(M)(j)} = -\mathrm{i}ka[j_n(ka)[k_j a j_n(k_j a)]' - j_n(k_j a)[ka j_n(ka)]']$$

$$S_n^{(M)(j)} = -\frac{\mathrm{i}}{k_j a}\left[k_j^2 a^2 j_n(k_j a)\,[ka j_n\,(ka)]' - k^2 a^2 j_n(ka)[k_j a j_n(k_j a)]'\right],$$

then,

$$c_{mn}^{(M)(l)} = e^{i\overline{k_i}\,\overline{r_q}} B_n^{(M)(l)} a_{mn}^{(M)} + B_n^{(M)(l)} \sum_{\substack{j=1\\ j\neq 1}}^{N} \sum_{\mu\nu} \left[A_{\mu\nu mn}(k\overline{r_l r_j}) S_\nu^{(M)(j)} c_{\mu\nu}^{(M)(j)} \right.$$
$$\left. + B_{\mu\nu mn}(k\overline{r_l r_j}) S_\nu^{(N)(l)} c_{\mu\nu}^{(N)(j)} \right] \tag{6a}$$

$$c_{mn}^{(N)(l)} = e^{i\overline{k_i}\,\overline{r_q}} B_n^{(N)(l)} a_{mn}^{(N)(l)} + B_n^{(N)(l)} \sum_{\substack{j=1\\ j\neq 1}}^{N} \sum_{\mu\nu} \left[B_{\mu\nu mn}(k\overline{r_l r_j}) S_\nu^{(M)(j)} c_{\mu\nu}^{(M)(j)} \right.$$
$$\left. + A_{\mu\nu mn}(k\overline{r_l r_j}) S_\nu^{(N)(j)} c_{\mu\nu}^{(N)(j)} \right]. \tag{6b}$$

After solving the internal field coefficients, the scattered field coefficients are calculated. In the observation direction $\hat{k}_s = \sin\theta_s \cos\phi_s \hat{x} + \sin\theta_s \sin\phi_s \hat{y} + \cos\theta_s \hat{z}$, the scattered field is

$$\overline{E}^S(\overline{r}) = \frac{e^{ikr}}{kr} \sum_{j=1}^{N} e^{-j\hat{k}\cdot\overline{r_j}} \sum_{m,n} \gamma_{mn} \left\{ a_{mn}^{S(M)(j)} \overline{C}_{mn}(\theta_s, \phi_s) i^{-n-1} \right.$$
$$\left. + a_{mn}^{S(N)(j)} \overline{B}_{mn}(\theta_s, \phi_s) i^{-n} \right\}. \tag{7}$$

13.3. Extinction and Absorption Efficiency of Metal Nanoparticles and Plasmon Resonance

In our simulations, the scattering of metal nanoparticles randomly located in a cubic volume are studied. We applied NMM3D to solve the Maxwell equation in the form of the Foldy Lax equation for the random medium. With the solved internal field coefficients, absorption is calculated by using the internal fields and extinction is calculated by applying optical theorem. The results for adhesive particles as well as various concentrations are studied.

13.3.1. Formulations

Consider N nanoparticles of radius 80 nm spread throughout evenly in a cubic box, which is shown in the Fig. 13.1. The optical constant of the gold particles is stated in [32].

A plane wave with the direction $\hat{k}_i = \sin\theta_i \cos\phi_i \hat{x} + \sin\theta_i \sin\phi_i \hat{y} + \cos\theta\hat{z}$ is incident onto the cubic box. The expression is given in (1). In the simulation, the incident angle is $\theta_i = 90°$ and $\phi_i = 270°$.

13.3.1.1. Extinction Cross Section

Extinction cross section represents the total power loss of the scatterers from the incident wave. The power loss is the sum of the scattered power and the absorbed power. The power loss is calculated from the imaginary part of the scattering amplitude in the forward direction by optical theorem. Therefore,

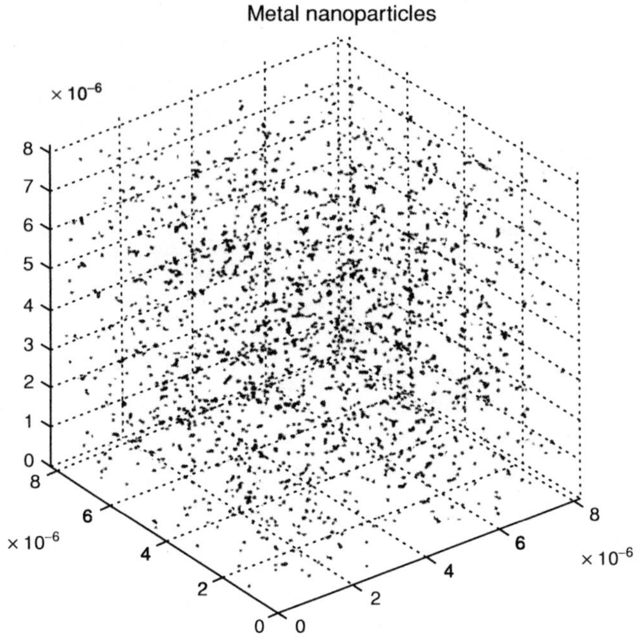

FIGURE 13.1. The simulated noble metal nanoparticles by Monte Carlo.

the total extinction cross section can be found by calculating the imaginary part of the scattering amplitude in the forward direction.

The scattered field can be expressed in terms of scattering dyadic:

$$\overline{E}_s(\overline{r}) = \frac{e^{ikr}}{r}\overline{\overline{F}}(\hat{k}_s, \hat{k}_i) \cdot \hat{e}_i E_0, \tag{8}$$

where $\hat{k}_i$ and $\hat{k}_s$ are the wave vector of the incident plane wave and the scattering field, respectively, and $\overline{\overline{F}}(\hat{k}_i, \hat{k}_i)$ is the scattering dyad in the forward direction. The relation between extinction cross section and scattering amplitude in the forward direction is

$$\sigma_e = \frac{4\pi}{k}\mathrm{Im}\big[\hat{e}_i \cdot \overline{\overline{F}}(\hat{k}_i, \hat{k}_i) \cdot \hat{e}_i\big]. \tag{9}$$

From Eqs. (7)–(8), we have

$$\overline{\overline{F}}\left(\hat{k}_s, \hat{k}_i\right) \cdot \hat{e}_i = \frac{1}{k}\sum_{j=1}^{N} e^{-ik\hat{k}_s \cdot \overline{r}_j} \sum_{mn} \gamma_{mn}\left[a_{mn}^{S(M)(j)}\overline{C}_{mn}(\theta_s, \phi_s)\, i^{-n-1} + a_{mn}^{S(N)(j)}\overline{B}_{mn}(\theta_s, \phi_s)\, i^{-n}\right]. \tag{10}$$

Substitute Eq. (10) into Eq. (9), the extinction cross section can be found:

$$\sigma_{eN} = \frac{4\pi}{k}\mathrm{Im}\left\{\frac{1}{k}\hat{e}_i \cdot \sum_{j=1}^{N} \mathrm{e}^{-ik\hat{k}_s\cdot\overline{r}_j} \sum_{mn} \gamma_{mn}\left[a_{mn}^{S(M)(j)}\overline{C}_{mn}(\theta_s,\phi_s)\,i^{-n-1} + a_{mn}^{S(N)(j)}\overline{B}_{mn}(\theta_s,\phi_s)\,i^{-n}\right]\right\}. \quad (11)$$

The extinction efficiency, $\tilde{\sigma}_{eN}$, is the ratio of the extinction cross section to the total cross section area of the particles, which is $\tilde{\sigma}_{eN} = \frac{\sigma_{eN}}{N_{\mathrm{part}}\pi a^2}$, where N_{part} is the number of particles and a is the radius of the particle.

13.3.1.2. Absorption Cross Section

After solving the internal field coefficient, the power absorbed by each particle is given by[14]

$$W_a^{(q)} = -\frac{1}{2\eta k^2}\sum_{mn}\left[\left|w_{mn}^{(M)(q)}\right|^2\left(ReT_n^{(M)} + \left|T_n^{(M)}\right|^2\right) + \left|w_{mn}^{(N)(q)}\right|^2\left(ReT_n^{(N)} + \left|T_n^{(N)}\right|^2\right)\right]. \quad (12)$$

The absorption cross section is

$$\sigma_{aN} = -\frac{1}{k^2}\sum_{j=1}^{N}\sum_{mn}\left\{\left|w_{mn}^{(M)(j)}\right|^2\left(ReT_n^{(M)(j)} + \left|T_n^{(M)(j)}\right|^2\right) + \left|w_{mn}^{(N)(j)}\right|^2\left(ReT_n^{(N)(j)} + \left|T_n^{(N)(j)}\right|^2\right)\right\}. \quad (13)$$

The absorption efficiency, $\tilde{\sigma}_{aN}$, is the ratio of the absorption cross section to the total cross section area of the particles, that is $\tilde{\sigma}_{aN} = \frac{\sigma_{aN}}{N_{\mathrm{part}}\pi a^2}$.

13.3.2. Results and Discussions

13.3.2.1. Convergence Test for Numerical Parameters

We first perform convergence test by changing $N_{\max}$, which is the truncation of the multipole expansion of n. A horizontally polarized plane wave where the unit vector of polarization $\overline{A} = \hat{x}$, propagating along the $\hat{y}$-axis, i.e. $\theta_i = 90°$ and $\phi_i = 270°$, is incident on 1000 gold nanoparticles with radius 80 nm in free space. The fractional volume of the nanoparticles is 1%. The refraction indices of gold are[32] shown in Table 13.1.

The error norm of extinction and absorption are 0.2% and 0.5%, respectively. Note that the error norm is defined as,

$$\mathrm{norm} = \sqrt{\frac{\sum_n \left(a_2^n - a_3^n\right)^2}{\sum_i \left(a_3^n\right)^2}},$$

TABLE 13.1. The refraction indices of gold.

Wavelength (nm)	$k = n + ip$	$\varepsilon = k^2$
400	1.658 + i1.956	−1.077 + i6.486
413.3	1.636 + i1.958	−1.157 + i6.407
427.5	1.616 + i1.94	−1.152 + i6.270
442.8	1.562 + i1.904	−1.185 + i5.948
459.2	1.426 + i1.846	−1.374 + i5.265
476.9	1.242 + i1.796	−1.683 + i4.461
495.9	0.916 + i1.84	−2.547 + i3.371
516.6	0.608 + i2.12	−4.125 + i2.578
539.1	0.402 + i2.54	−6.290 + i2.042
563.6	0.306 + i2.88	−8.201 + i1.763
652.6	0.166 + i3.15	−9.895 + i1.046
688.8	0.16 + i3.8	−14.414 + i1.216
729.3	0.164 + i4.35	−18.896 + i1.427
774.9	0.174 + i4.86	−23.589 + i1.691
826.6	0.188 + i5.39	−29.017 + i2.027

where a_i^n is the calculated value and the subscript i denotes the $N_{\max}$ for the truncation. Fig. 13.2 shows that using $N_{\max} = 2$ gives results of sufficient accuracy.

13.3.2.2. Extinction and Absorption of Two Particles with Various Orientations

Figure 13.3 shows the extinction and absorption cross section of two particles. The incident wave, $\overline{E}^{\text{inc}}(\overline{r}) = \overline{A}e^{i\overline{k}_i\cdot\overline{r}}$, where $\overline{A} = \hat{x}$ and $\bar{k}_i = -\hat{y}$, is incident horizontally toward the two particles. The two particles are touching each other. The touching particles are put in three different positions: (i) the two particles arranged parallel to the incident light polarization, i.e. they are put along the x-axis; (ii) the two particles arranged orthogonal to the incident light polarization, i.e. they are put along the z-axis; (iii) the two particles arranged one behind another one, i.e. they are along the y-axis.

All particles pairs are such that the point of touching is at the origin. The extinction and absorption cross section of single sphere is also shown for comparison. In (i), the extinction peak shows red shift. There is broadening of the extinction compared with those of single particle. In (ii), the extinction peak shows the blue shift. The resonance frequency shift for these two configurations is similar to that of silver nanocrystals in glass.[33] This phenomenon shows that the extinction is polarization dependent. In (iii), a sharpened extinction resonance is observed. This extinction peak is slightly shifted to the red. The absorption cross sections of (i) and (ii) are about the same. Absorption is polarization independent. In (iii), it is larger than (i) and (ii). Note that both extinction and absorption of (iii) are much larger than (i) and (ii). This implies that in this orientation more energy is dissipated.

13.3.2.3. Extinction and Absorption of Gold Nanoparticles with Various Fractional Volumes

Figure 13.4 shows the extinction and absorption of 2000 gold nanoparticles of radius 80 nm and different fractional volume. When the fractional volume is low,

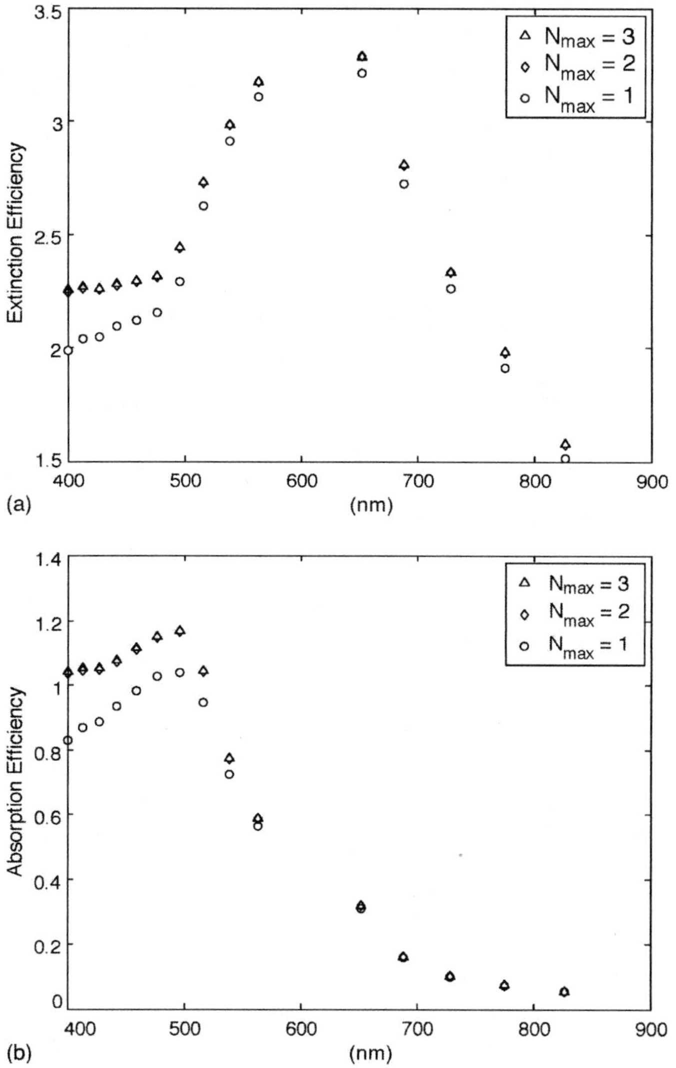

FIGURE 13.2. Convergence test against N_{max}, the maximum number of multipole expansions. The extinction efficiency (a) and absorption efficiency (b) are plotted against the wavelength. A thousand gold nanoparticles of radius 80 nm randomly located in the free space. Particles occupied 1% of volume.

i.e. less than 0.1%, the extinction shows a peak and the plasmon resonance behavior. When the fractional volume is 0.01%, the extinction is close to that of a single particle. When the fractional volume increases, the extinction gradually broadens. When the fractional volume further increases, there is a red shift. At 3% fractional volume, the plasmon resonance disappears. These physical features of dependence of extinction have been observed in experiments.[34] The corresponding extinction and wavelength of plasmon resonance are shown in Table 13.2. In Table 13.2, the

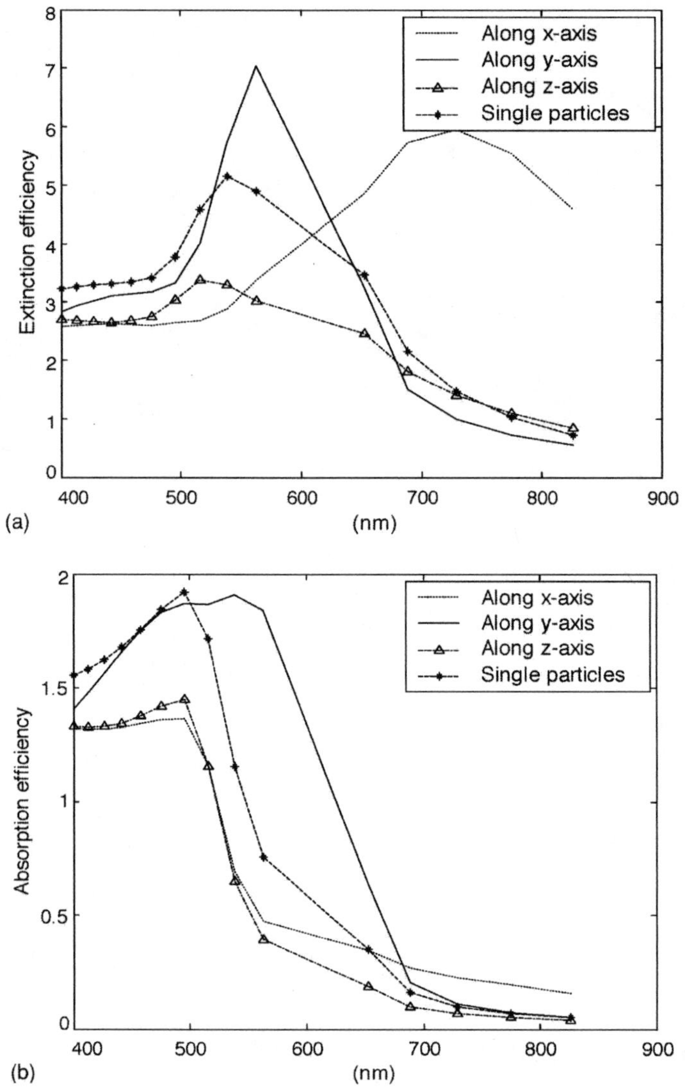

FIGURE 13.3. Comparison of extinction efficiency (a) and absorption efficiency (b) for two gold nanoparticles of radius 80 nm in free space with different locations, which are either put along the x-axis, y-axis or z-axis, and are plotted against the wavelength. The plane wave incident at an angle of $\theta_i = 90^\circ$ and $\phi_i = 270^\circ$ with horizontal polarization.

peak extinction and resonance frequency is constant for fractional volume $<0.1\%$. However, the peak extinction decreases more significantly when fractional volume exceeds 0.5%. At the same time, the extinction demonstrates the red shift. This phenomenon is more apparent for higher fractional volume. The red shift and

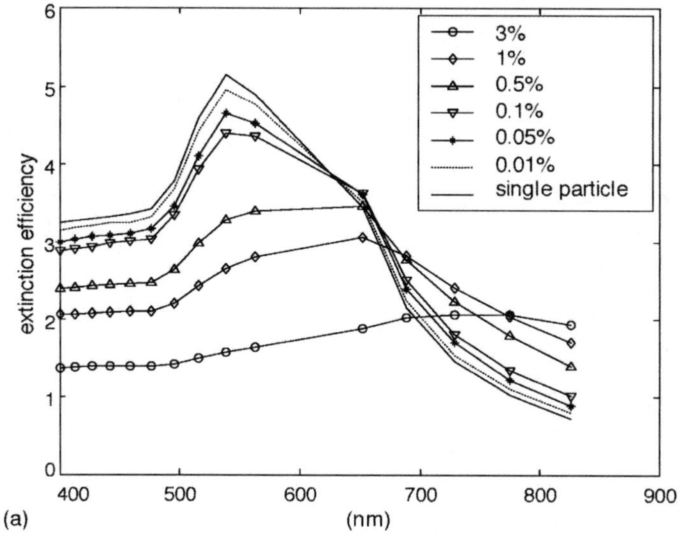

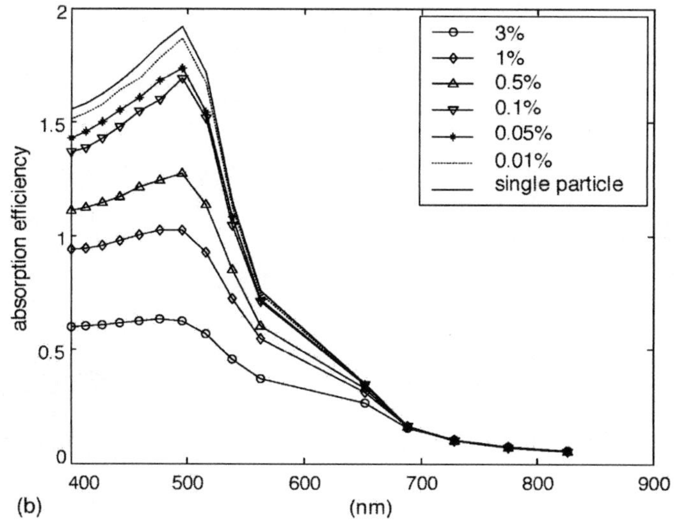

FIGURE 13.4. Extinction efficiency (a) and absorption efficiency (b) for 2000 gold nanoparticles of radius 80 nm with different fractional volume and extinction efficiency (c) and absorption efficiency (d) for 2000 gold nanoparticles of radius 80 nm with fractional volume 3% in free space with three cases (i) dipole mode only (ii) quadrupole only, and (iii) full spectrum plotted for comparison.

the decrease of extinction of concentration can have important consequences in applications.

Figure 13. 4(b) shows the absorption efficiency of the nanoparticles. The figure shows that the absorption efficiency decreases slightly for increasing fractional

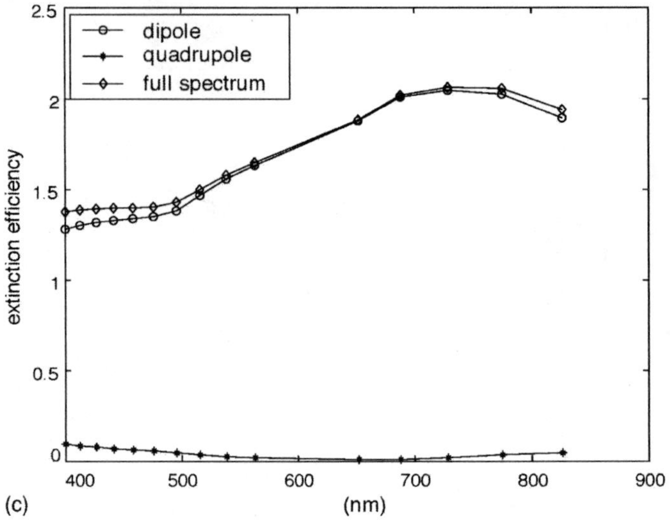

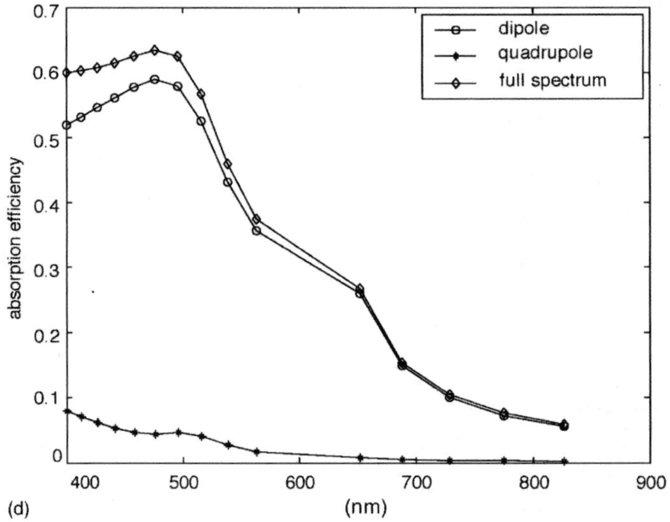

FIGURE 13.4. (*Continued*)

volume when fractional is less than 0.1%. However, it decreases considerably when the fractional volume exceeds 0.5%. When concentration increases, each particle will absorb less energy. It should be noted that the frequency of the peak absorption does not show any changes with the fractional volume.

Figure 13.4(c) shows extinction at fractional volume 3% using the dipole mode and the quadrupole mode. The extinction for the sum of dipole mode and quadrupole mode are also plotted. Result shows that the quadrupole mode only has a small effect and that the dipole mode is dominant in this region. Hence, the

TABLE 13.2. The corresponding extinction and wavelength of resonance (gold nanoparticles).

Fractional volume	Peak wavelength (nm)	Extinction
Single particle (for reference)	~539	5.147
0.01%	~539	4.965
0.05%	~539	4.659
0.1%	~539	4.400
0.5%	~653	3.468
1%	~653	3.058
3%	~729	2.063

red shift of the plasmon resonance and the broadening extinction is due to the dipole mode, the interaction of the electromagnetic modes. The absorption at fractional volume 3% using the dipole mode is shown in Fig. 13.4(d). It is similar to the extinction that the absorption of energy is mainly due to the dipole mode. The maximum absorption is obtained at $\lambda = 477$ nm, which is due to the dipole mode. It is noted that there is a small peak of absorption for quadrupole mode at $\lambda = 496$ nm.

13.3.2.4. Extinction and Absorption of Silver Nanoparticles with Different Fractional Volume

The extinction and absorption of 2000 silver nanoparticles with different fractional volume are shown Figs. 13.5(a) and 13.5(b). The radius of the particles is 40 nm. The refraction indices of silver are given in Table 13.3.[32]

The behavior of the plasmon resonance for silver nanoparticles is similar to that of the gold nanoparticles. The extinction is close to that of the single particle when the fractional volume is 0.01%. When the fractional volume is within 0.1%, the

TABLE 13.3. The refraction indices of silver.

Wavelength (nm)	$k = n + ip$	$\varepsilon = k^2$
332.4	0.321 + i0.902	−0.711 + i0.579
335.1	0.294 + i0.986	−0.886 + i0.580
339.7	0.259 + i1.12	−1.187+ i0.580
344.4	0.238 + i1.24	−1.481 + i0.590
354.2	0.209 + i1.44	−2.030 + i0.602
364.7	0.186 + i1.61	−2.558 + i0.599
375.7	0.2 + i1.67	−2.749 + i0.668
387.5	0.192 + i1.81	−3.239 + i0.695
400	0.173 + i1.95	−3.773 + i0.675
413.3	0.173 + i2.11	−4.422 + i0.730
427.5	0.16 + i2.26	−5.082 + i0.723
442.8	0.157 + i2.4	−5.735 + i0.754
459.2	0.144 + i2.72	−6.533 + i0.737
476.9	0.132 + i2.72	−7.381 + i0.718
495.9	0.13 + i2.88	−8.278 + i0.749

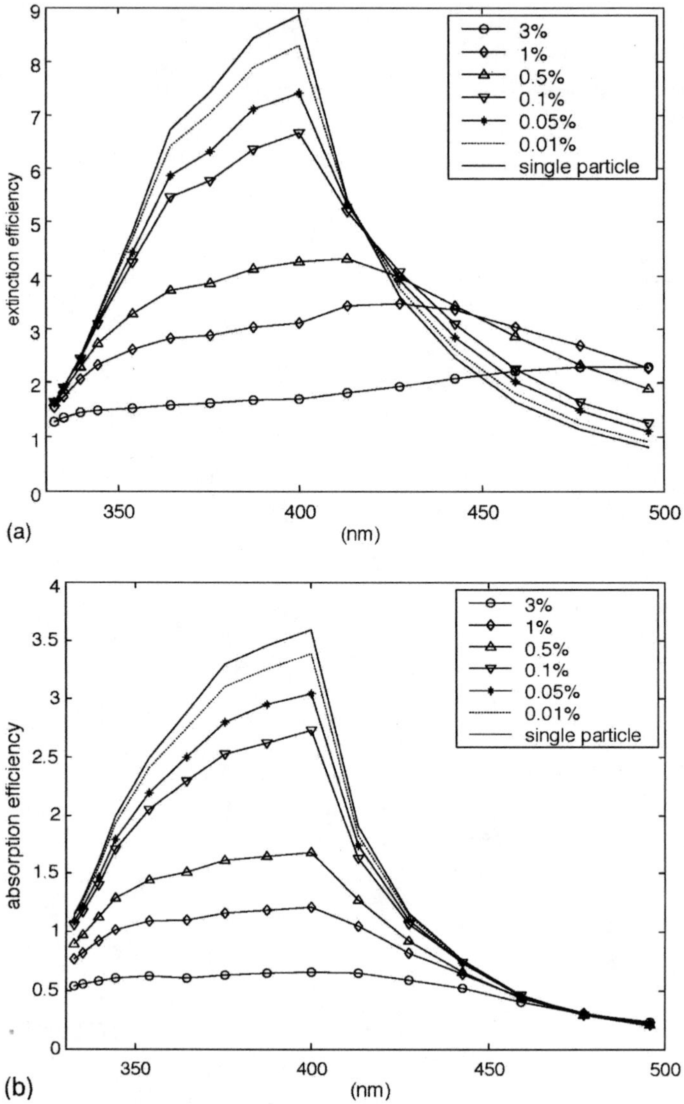

FIGURE 13.5. Extinction efficiency (a) and absorption efficiency (b) for 2000 silver nanoparticles of radius 40 nm with different fractional volume.

extinction peak decreases with increasing fractional volume. On the other hand, the extinction peak of gold nanoparticles only decreases slightly. When the fractional volume further increases, the extinction of the silver nanoparticles case gradually broadens and shows the red shift. The plasmon resonance disappeared when the fractional volume is 3%, which is the same for that of the gold nanoparticle. The extinction and the corresponding wavelength of plasmon resonance are shown in

TABLE 13.4. The corresponding extinction and wavelength of resonance (silver nanoparticles).

Fractional volume	Peak wavelength (nm)	Extinction
Single particle (for reference)	∼400	8.857
0.01%	∼400	8.304
0.05%	∼400	7.420
0.1%	∼400	6.664
0.5%	∼413	4.309
1%	∼428	3.474
3%	No resonance	

the Table 13.4. In Table 13.4, the resonance frequency is constant and the extinction peak decreases steadily when fractional volume does not exceed 0.1%. However, the peak extinction decreases more significantly and extinction demonstrates red shift when fractional volume further increases from 0.5%.

Figure 13.5(b) shows the absorption of the nanoparticles; it decreases steadily for increasing fractional volume when it does not exceed 0.1%. Nevertheless, it decreases considerably when the fractional volume exceeds 0.5%. It shows that each particle absorbs less energy for higher concentration. Similar to that of the gold nanoparticles, the frequency of the peak absorption does not change with changing fractional volume.

13.3.2.5. Energy Absorption of Each Particle in the Collection

We next show in Fig. 13.6 the absorption of each particle for the case of 200 particles randomly distributed over a cubic box of size 2.05 μm $\times$2.05 μm $\times$2.05 μm. A plane wave is incident onto the cubic box with incident angle at $\theta_i = 90°$ and $\phi_i = 270°$. The position of the particle is determined by their centre location (x, y, z) in the Cartesian coordinates system. We order the particles by its y-coordinates because the propagation direction of the incident wave is along the y-axis. It is possible to have two or more particles lying on the same y-plane, i.e. $y_1 = y_2 = \cdots$.

Figure 13.6(c) shows the average absorption efficiency of 200 particles of radius 80 nm for reference. The fractional volume of the sample is 5%. Two cases are studied: (i) the frequency with the maximum average absorption ($\lambda = 476.9$ nm, absorption efficiency = 0.938) and (ii) the frequency of the minimum average absorption ($\lambda = 826.6$ nm, absorption efficiency = 0.0642). The corresponding absorption efficiency for a single particle is 1.843 ($\lambda = 476.9$ nm) and 0.052 ($\lambda = 826.6$ nm).

Most of the particles in case (i) absorb more energy than those in (ii). It should be noted that the average absorption in (i) is about 15 times that of (ii). Consider case (i). The absorption of each particle depends on its y-position. The incident plane wave makes its entrance at the boundary $y \approx 2.05$ μm. Those particles that are near the boundary at $y \approx 2.05$ μm absorb the most energy. The particles in

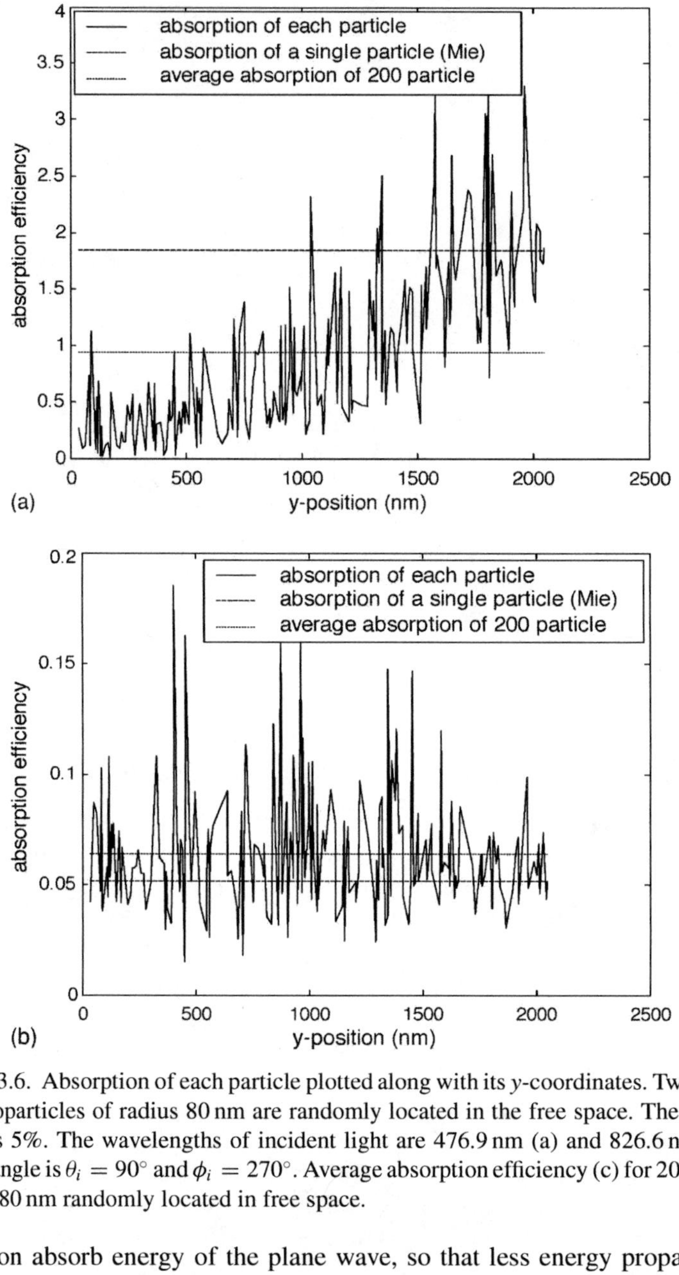

FIGURE 13.6. Absorption of each particle plotted along with its y-coordinates. Two hundred gold nanoparticles of radius 80 nm are randomly located in the free space. The fractional volume is 5%. The wavelengths of incident light are 476.9 nm (a) and 826.6 nm(b). The incident angle is $\theta_i = 90^\circ$ and $\phi_i = 270^\circ$. Average absorption efficiency (c) for 200 particles of radius 80 nm randomly located in free space.

this region absorb energy of the plane wave, so that less energy propagate into the interior region. Thus, the particles near the boundary at $y = 0$ absorb the least energy. About 55% of particles absorb less energy than the average absorption energy among all particles. Comparing the absorption efficiency to that of the single particle, only 10% of particles absorb more energy than the single particle case. This is because of the mutual wave interaction of the neighboring particles.

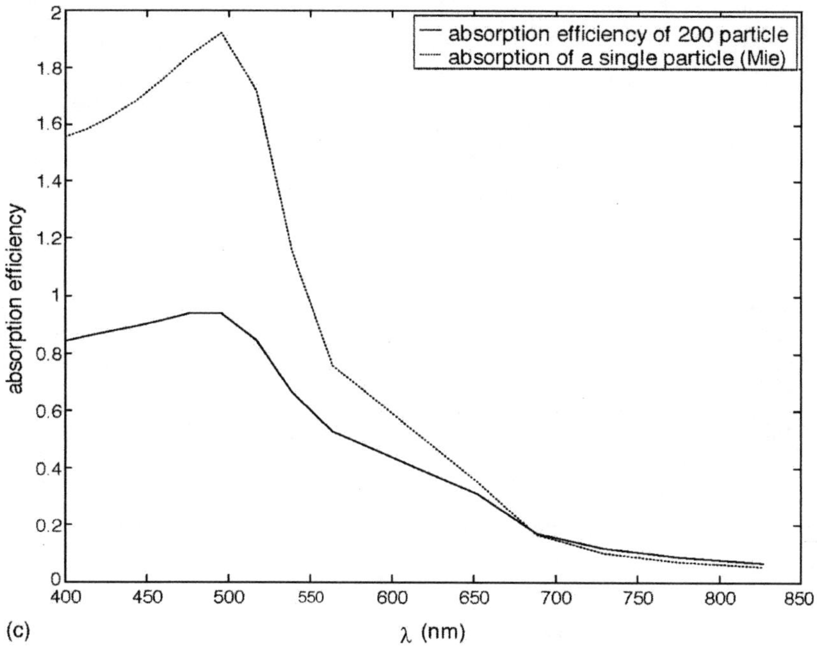

FIGURE 13.6. *(Continued)*

When the absorption efficiency is low, i.e. the absorption efficiency in (ii) is one fifteenth of that in (i), the phenomenon that appears in case (i) disappears in case (ii). For case (ii), the energy absorbed by each particle in all regions inside the cubic box is roughly the same. As each particle absorbs much less energy, most of the energy continues to propagate into the interior of the cubic box and all particles absorb roughly the same energy. Note that the absorption efficiency of 200 particles (0.064) is slightly higher than that of the single particle (0.052). This is caused by the excitation field in each particle that is enhanced by the scattered field of neighboring particles.

13.4. Phase Matrix of Light Scattering by Metal Nanoparticles

In this part, we will study the relationship between the incident wave intensity and the scattering field intensity as a function of the direction of the scattered field. As the particle size is larger than one-tenth of the wavelength, i.e. gold particle with diameter 160 nm at frequency with wavelength about 800 nm, it is not applicable to predict the Rayleigh scattering. Instead, we use NMM3D to formulate relationship of the incident field and the scattered field in the sense of 1–2 polarization frame system.

We consider a plane wave incident onto a linear, isotropic, homogeneous medium with randomly distributed nanoparticles. Since the random media is considered, averaging over realizations takes place. The plane wave is linear polarization and its polarized vector acts as $\hat{1}$ or $\hat{2}$ each time to calculate the response of the scattered field. We study the col-polarization of the phase matrix in the cases of 1 realization versus averaging over realizations, the denser medium versus less dense medium, as well as resonance versus non-resonance.

13.4.1. Formulation of Phase Matrix

Consider nanoparticles in a cubic box, which is shown in the Fig. 13.1. Following the steps of solving the multiple-scattering equations in part one, the internal field coefficients, $c_{mn}^{(M)(l)}$ and $c_{mn}^{(N)(l)}$. In Eq. (6), and hence, $a_{mn}^{S(M)(j)}$ and $a_{mn}^{S(N)(j)}$ are solved. Using the far-field expression for the $\overline{M}_{mn}$ and $\overline{N}_{mn}$, the scattered field $\overline{E}^S$ by N particles with the scattered direction $\hat{k}_s$, where $\hat{k}_s = (\sin\theta_s\cos\phi_s\hat{x} + \sin\theta_s\sin\phi_s y + \cos\theta_s\hat{z})$, is calculated using Eq. (7). The scattered field is averaged over realizations as

$$\left\langle \overline{E}_s(\theta_s, \phi_s) \right\rangle = \frac{1}{N_r}\sum_{\sigma=1}^{N_r} \overline{E}_s^{\sigma}(\theta_s, \phi_s), \tag{14}$$

where σ is the realization index, N_r denotes the number of realizations, and $\overline{E}_s^{\sigma}$ denotes the scattered field of the σth realization. The angular bracket stands for the ensemble average. After solving the scattered field, the phase matrix is best illustrated in the 1–2 polarization frame[13]

13.4.1.1. 1–2 Polarization Frame

Consider a volume containing many scatterers. The phase matrix is the matrix to describe the relation between each polarization of the scattered intensity due to the respective polarization of the incident wave. In the 1–2 system, they obey the relation:

$$\begin{bmatrix} |\overline{E}_{1s}|^2 \\ |\overline{E}_{2s}|^2 \end{bmatrix} = \begin{bmatrix} P_{11} & P_{12} \\ P_{21} & P_{22} \end{bmatrix} \begin{bmatrix} |\overline{E}_{1i}|^2 \\ |\overline{E}_{2i}|^2 \end{bmatrix}, \tag{15}$$

where the subscripts s and i denote the scattered field and the incident field. The subscripts 1 and 2 denote the polarization of the electric field. The phase matrix is the scattered cross-section normalized by the cross section area of all scatterers.

Let $\hat{k}_i$ be the propagation direction of incident field and $\hat{k}_s$ be the direction of the scattered field. $\hat{1}_i$ is the unit vector which is perpendicular to $\hat{k}_i$ and $\hat{k}_s$. Then $\hat{2}_i$ is perpendicular to both $\hat{k}_i$ and $\hat{1}_i$. Let $\hat{1}_i = \hat{1}_s$, then $\hat{2}_s$ is perpendicular to both $\hat{k}_s$ and $\hat{1}_s$. These relations can be written as

$$\hat{1}_i = \hat{1}_s = \frac{\hat{k}_s \times \hat{k}_i}{|\hat{k}_s \times \hat{k}_i|}; \quad \hat{2}_i = \hat{k}_i \times \hat{1}_i; \quad \hat{2}_s = \hat{k}_s \times \hat{1}_s.$$

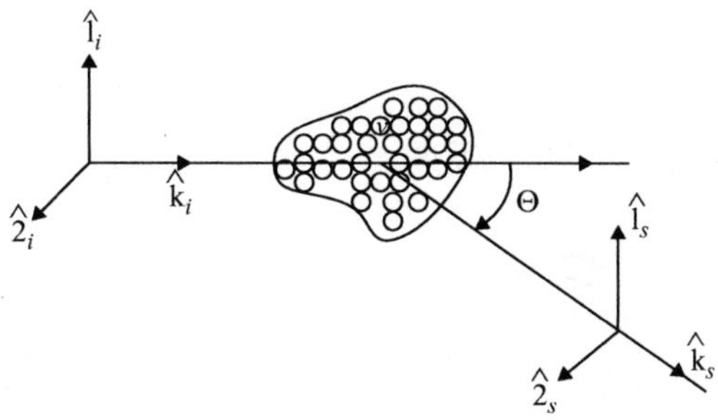

FIGURE 13.7. The orthogonal system for polarization of 1–2 system.

The $\hat{1}_i$, $\hat{2}_i$, $\hat{1}_s$, and $\hat{2}_s$ can be related to vertical and horizontal polarization of $\hat{v}_i$, $\hat{h}_i$, $\hat{v}_s$ and $\hat{h}_s$ by transformation, which will be shown in the latter part. The orthogonal system for polarization of 1–2 system is shown in Fig. 13.7.

13.4.1.2. Scattering Cross Section

To simulate the phase matrix, we let the incident electric field be propagating in the $\hat{z}$-direction with polarization of the electric field $\hat{e} = \hat{y}$, i.e. $\overline{E}_{\text{inc}}(\overline{r}) = \hat{y}e^{ikz}$, $\hat{k}_i = \hat{z}$, $\hat{v}_i = \hat{x}$, and $\hat{h}_i = \hat{y}$. Solving the Foldy Lax equation using internal field formulation by iterative method, the scattered field can be found using Eq. (7). For this case $\hat{h}_s = -\hat{1}_s$ and $\hat{v}_s = \hat{2}_s$, then $\tilde{E}_s = \tilde{E}_{vs}\hat{v} + \tilde{E}_{hs}\hat{h}$. Therefore, the scattering cross section is decomposed into vertical, $\hat{v}$, and horizontal, $\hat{h}$, polarization, i.e.

$$\sigma_{vb}(\hat{k}_s, \hat{k}_i) = r^2 \frac{\left\langle |\tilde{E}_{vs}(\theta_s, \phi_i)|^2 \right\rangle}{|\overline{E}^i|^2}; \quad \sigma_{hb}(\hat{k}_s, \hat{k}_i) = r^2 \frac{\left\langle |\tilde{E}_{hs}(\theta_s, \phi_i)|^2 \right\rangle}{|\overline{E}^i|^2}. \tag{16}$$

The phase matrix element can be computed by finding the scattering cross section with at $\hat{\phi}_s = 0°$ and $180°$ and $\hat{\phi}_s = 90°$ and $270°$.

- *Phase Matrix Elements Due to* $\hat{1}_i$

Consider the $\hat{x} - \hat{z}$ plane, i.e. $\phi_s = 0°$ and $180°$.

For $\phi_s = 0°$, $\hat{k}_s = \sin\theta_s\hat{x} + \cos\theta_s\hat{z}$, $\hat{v}_s = \cos\theta_s\hat{x} - \sin\theta_s\hat{z}$ and $\hat{h}_s = \hat{y}$. So that

$$\hat{1}_s = \hat{1}_i = \frac{(\hat{k}_s \times \hat{k}_i)}{|\hat{k}_s \times \hat{k}_i|} = -\hat{y} = -\hat{h}_s = -\hat{h}_i, \qquad \hat{2}_s = \hat{k}_s \times \hat{1}_s = \hat{v}_s$$

$$\text{and} \quad \hat{2}_i = \hat{k}_i \times \hat{1}_i = \hat{v}_i.$$

Similarly, the formulas for polarization vector for $\phi_s = 180°$ can be obtained. With these polarization vector relations, the phase matrix elements, P_{11} and P_{21}, become

$$P_{11}(\theta_s) = \frac{\sigma_{hb}}{N_{part}\pi a^2} \tag{17a}$$

$$P_{21}(\theta_s) = \frac{\sigma_{vb}}{N_{part}\pi a^2}, \tag{17b}$$

where σ_{hb} and σ_{vb} are found by Eq. (16) and $N_{part}\pi a^2$ is the cross section area of particles. P_{11} and P_{21} are the co-polarization and cross-polarization due to $\hat{1}_s$, respectively.

- *Phase Matrix Elements Due to $\hat{2}_i$*

Consider the $\hat{y}-\hat{z}$ plane, i.e. $\phi_s = 90^\circ$ and 270°

For $\phi_s = 90^\circ$, $\hat{k}_s = \sin\theta_s\hat{y} + \cos\theta_s\hat{z}$; $\hat{v}_s = \cos\theta_s\hat{y} - \sin\theta_s\hat{z}$ and $\hat{h}_s = -\hat{x}$. Then,

$$\hat{1}_s = \hat{1}_i = \frac{(\hat{k}_s \times \hat{k}_i)}{|\hat{k}_s \times \hat{k}_i|} = \hat{x} = -\hat{h}_s = \hat{v}_s; \qquad \hat{2}_i = \hat{k}_i \times \hat{1}_i = \hat{y} = \hat{h}_i;$$

$$\hat{2}_s = \hat{k}_s \times \hat{1}_s = \hat{v}_s.$$

The polarization of the incident wave, $\hat{e} = \hat{y}$, is the polarization of $\hat{2}_s$. With these polarization vector relations, the phase matrix elements, P_{12} and P_{22}, become

$$P_{12}(\theta_s) = \frac{\sigma_{hb}}{N_{part}\pi a^2} \tag{18a}$$

$$P_{22}(\theta_s) = \frac{\sigma_{vb}}{N_{part}\pi a^2}, \tag{18b}$$

where P_{22} and P_{12} are the co-polarization and cross-polarization due to $\hat{2}_s$, respectively.

13.4.2. *Results and Discussion*

13.4.2.1. Phase Matrices of Single Realization and Average Realizations

The phase matrices of a single realization and averaged over realizations are shown in Fig. 13.8. We used 50 realizations. There are 2000 gold nanoparticles of radius 80 nm randomly located in a cubic box. The wavelengths are 652 nm and 826 nm. Referring to Table 13.2, these are the resonant wavelengths and non-resonant wavelength, respectively. In the figure, the phase matrices of both single realization and averaged realizations give similar patterns. However, there are many ripples in the case of single realization. The scattering at different directions vary among different realizations. The results are smoothened by averaging over realizations. Note that the extinction and absorption efficiency do not vary as much over realizations. This is because the extinction and absorption efficiency involve summations over all direction and summations over all particles. These ripples do not occur in the forward direction, $\theta = 0^\circ$. It is because strong intensity propagates in this direction. The random phase effect due to the scattered waves from different positions of the particles becomes insignificant. In the figure, we see the scattering in different directions are uniform for the case of P_{11}. However, there is a minimum at about $\theta = 90^\circ$ in the case of P_{22}, The minimum is more obvious at wavelength 826 nm in spite of the strong intensity propagating in the forward direction. Note that polarized vector $\hat{1}$ of incident field and scattered field are the same and are perpendicular to both the propagation directions of incident field and scattered field. The polarized vector $\hat{2}$ of the scattered field is not the same as the $\hat{2}$ of incident field and is

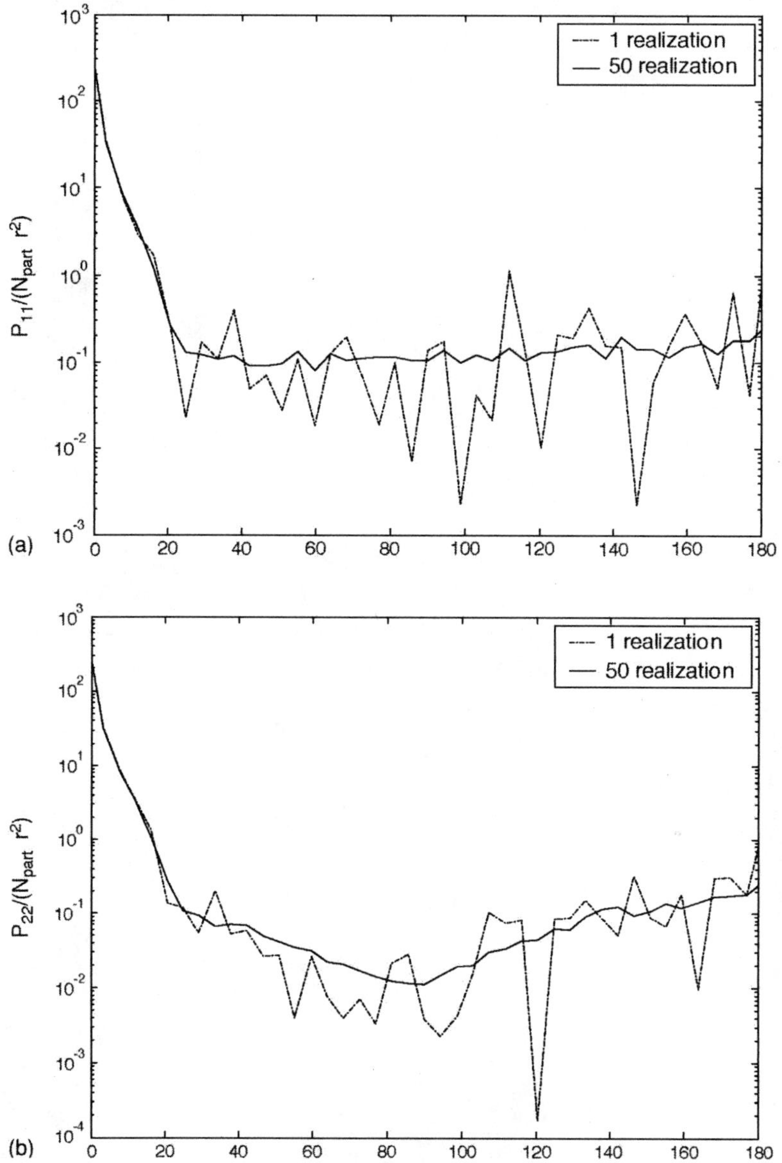

FIGURE 13.8. The phase matrices (a) P_{11} and (b) P_{22} at frequency with wavelength 652 nm and (c) P_{11} and (d) P_{22} at frequency with wavelength 826 nm are plotted against θ. The cubic box consists of 2000 gold nanoparticles of radius 80 nm. The particles occupied 1% in volume fraction.

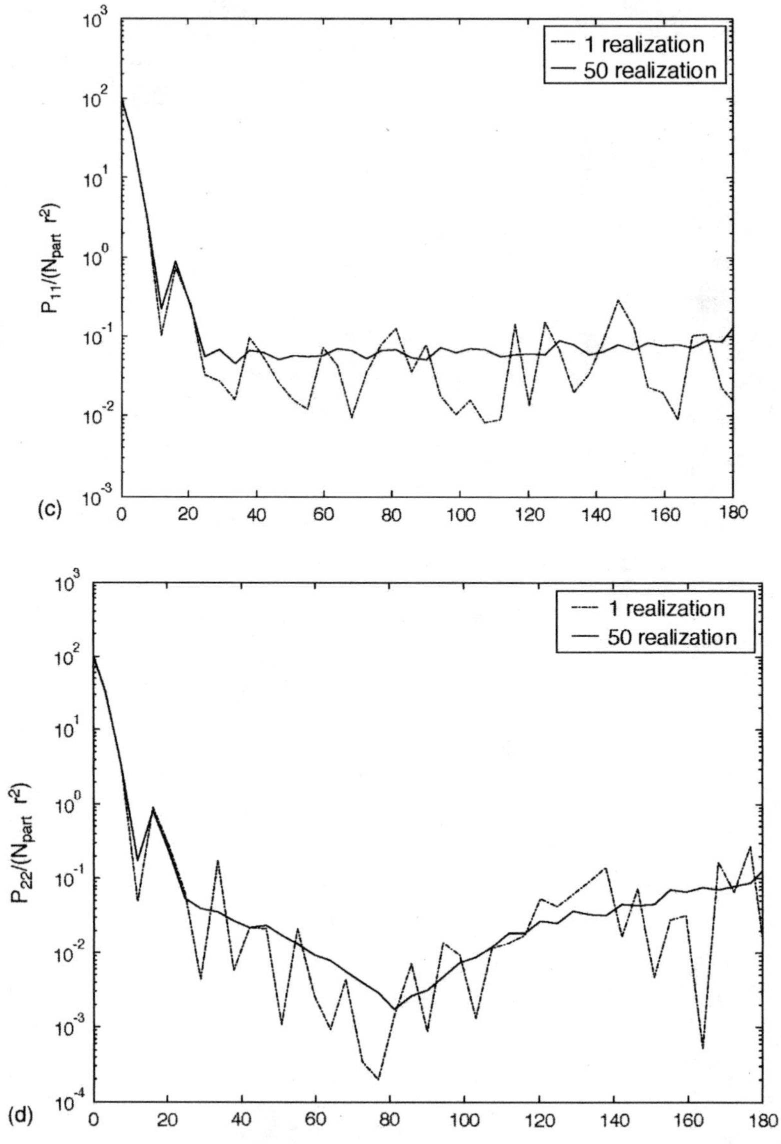

FIGURE 13.8. (*Continued*)

not perpendicular to the propagation direction of incident field. Large intensity propagates in the forward direction which is due to the coherent field.

13.4.2.2. Phase Matrices of 1% and 5%

The phase matrices of 2000 nanoparticles with radius 80 nm and fractional volume 1% and 5% are plotted in Fig. 13.9. Two frequencies are selected which is the

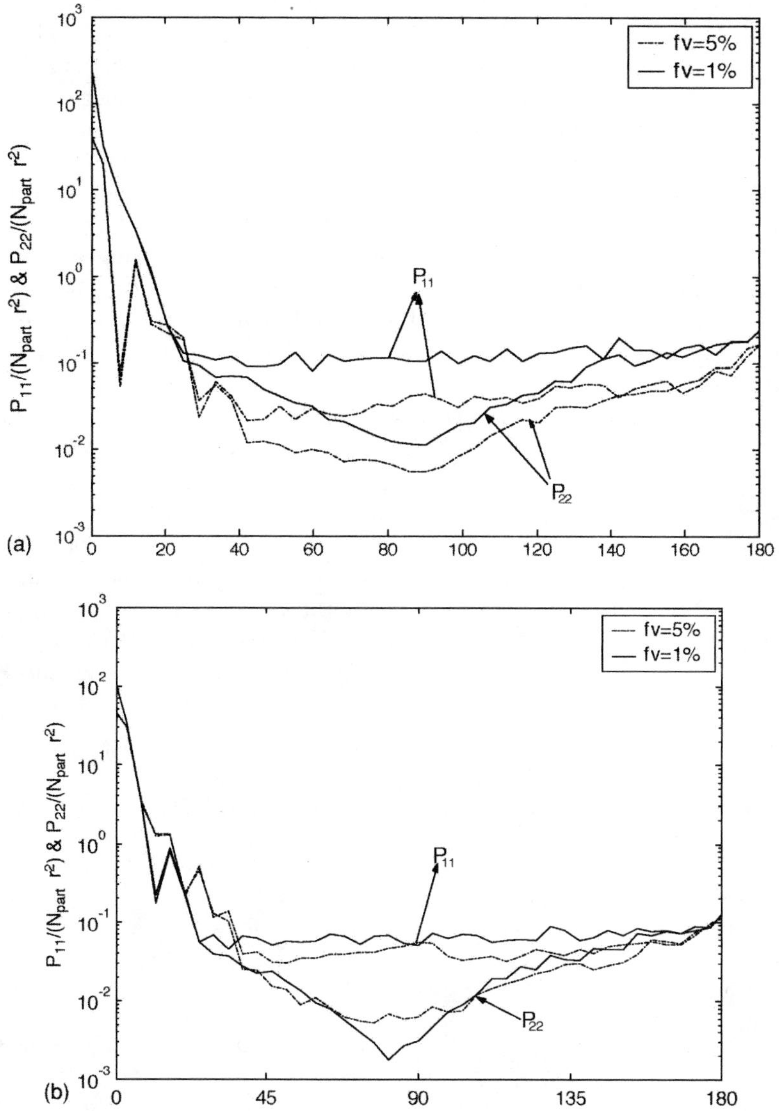

FIGURE 13.9. The phase matrices P_{11} and P_{22} at frequency with wavelength (a) 652 nm and (b) 826 nm are plotted against θ. The cubic box consists of 2000 gold nanoparticles of radius 80 nm. The particles occupied either 1% or 5% in volume fraction which are stated in the graphs.

resonant frequency (at $\lambda = 652$ nm) and nonresonant frequency (at $\lambda = 826$ nm) for the case of 1% fractional volume. In Fig. 13.9(a), the intensity level for the case of 1% volume fraction is much higher than that of 5% volume fraction at this frequency. Note that at this frequency, the nanoparticles with 1% volume fraction are in resonant state whereas that of with 5% volume fraction are not. It shows that the resonance of the system contributes to the scattering in all directions. The extinction efficiency for the 1% volume fraction is larger than that of 5% volume fraction. Note that the difference of the absorption efficiency between the two cases is not significant when compared with the extinction efficiency. The extinction efficiency for 1% volume fraction and 5% fraction at $\lambda = 652$ nm are 3 and 1.28, respectively. For the phase matrix, the strength of the scattering intensity for different angles at 1% volume fraction is about 2.3 times that of 5% volume fraction for P_{22} and is about 3 times for P_{11}. This means resonance of the system enhances more the scattering intensity for polarization $\hat{1}$. Note that the scattering intensity in the forward direction of 1% volume fraction is six times that of 5% volume fraction, which is about the same for both P_{11} and P_{22}.

When the frequency shifts to the lower frequency, i.e. $\lambda = 826$ nm, both systems of 1% volume fraction and 5% volume fraction are not in resonance state. The two results look alike, with similar intensity and similar pattern. The extinction efficiencies of 1% volume fraction and 5% volume fraction are 1.68 and 1.72, respectively, and are almost the same. The scattering intensity is about the same for P_{22} whereas the scattering intensity of 1% volume fraction is about 1.4 times that of 5% volume fraction for P_{11}. The scattering intensity in the forward direction of 1% volume fraction is about twice that of 5% volume fraction. The difference is larger than the other scattering direction. It means for lower density, more energy propagates through the nanoparticles in the forward direction than for high density.

13.4.2.3. Phase Matrix in Resonant Mode and Nonresonant Mode

The phase matrix for resonant mode (at $\lambda = 652$ nm) and nonresonant mode (at $\lambda = 826$ nm) for 1% volume fraction are shown in Fig. 13.10. The phase matrices for 5% volume fraction at both frequencies are also shown for comparison. Note that the scattering efficiency for 1% volume fraction at resonant mode and nonresonant mode are 2.7 and 1.6, respectively. In Figs. 13.10(a) and 13.10(b), the scattering intensities for P_{11} and P_{22} are different between resonant mode and nonresonant mode. The scattering intensities are higher at resonant mode than at nonresonant mode. The phase matrices of P_{11} and P_{22} for 5% volume fraction at these two frequencies are shown in Figs. 13.10(c) and 13.10(d). The scattering intensities are about the same. It means that resonance of the system contributes to the scattering intensities in all directions including the forward direction. Note that the ratio, $P_{11}(\lambda = 652\text{ nm}) / P_{11}(\lambda = 826\text{ nm})$, for 1% fractional volume is 2.4, and for 5% fractional volume is 0.9. The enhancement by the resonance in the forward direction is larger than the scattering efficiency. There is a small peak near 0°. This peak appears at $\lambda = 826$ nm for 1% volume fraction and at $\lambda = 652$ nm for 5%. The resonance occurs at $\lambda = 652$ nm for 1% volume fraction and this resonance mode disappears for 5% volume fraction. However, we can see the extinction efficiency for 5% volume fraction at $\lambda = 652$ nm and $\lambda = 826$ nm are 1.3 and 1.7,

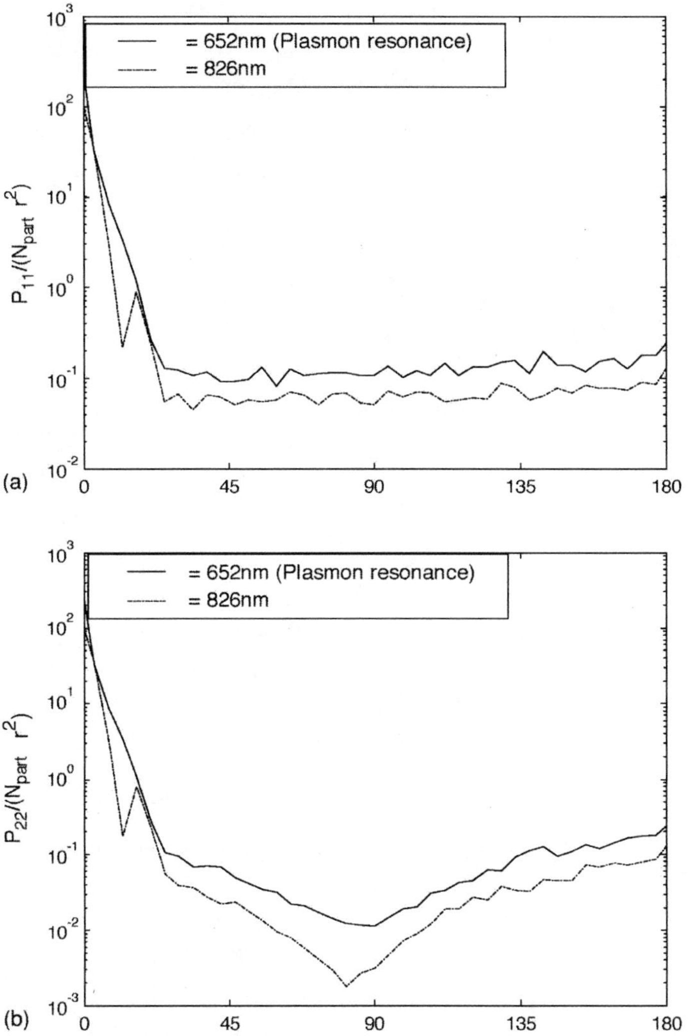

FIGURE 13.10. The phase matrices (a) P_{11} and (b) P_{22} for 1% volume fraction and (c) P_{11} and (d) P_{22} for 5% volume fraction are plotted against θ. The cubic box consists of 2000 gold nanoparticles of radius 80 nm.

respectively. Note that when the fraction volume increases, the resonance has a red shift and the resonant peak will be flattened. Thus, it is approaching resonant mode for 5% volume fraction at $\lambda = 826$ nm. The small peak near the forward direction occurs when the system is in nonresonant state, but it disappears when the frequency shift to the resonant frequency. The small peak is at 16° for 1% volume fraction and 12° for 5% volume fraction.

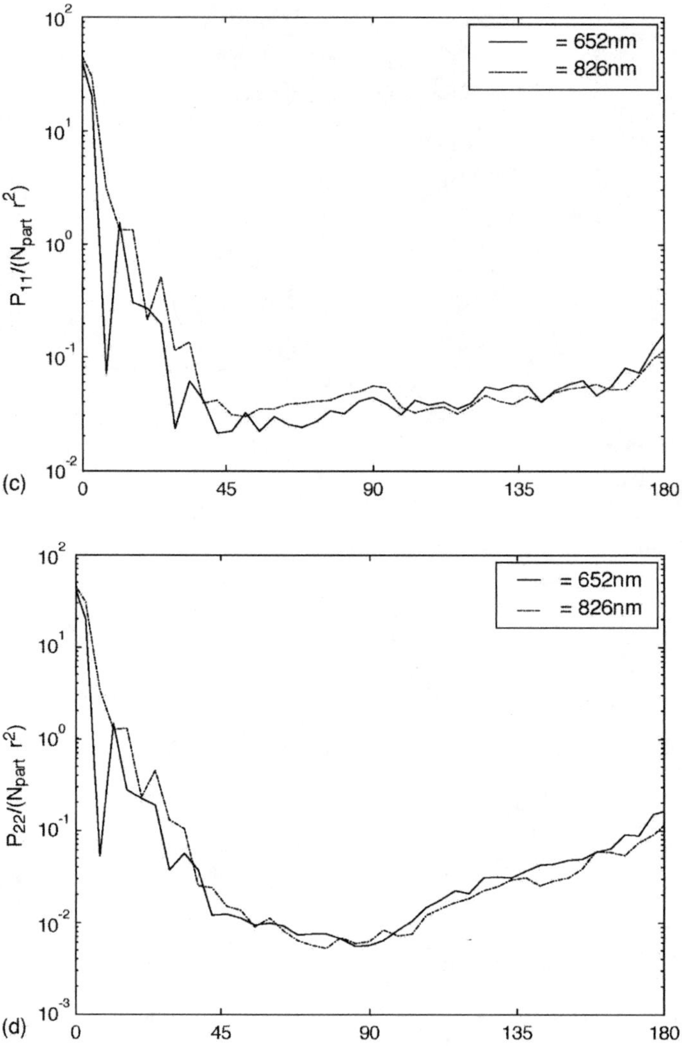

FIGURE 13.10. *(Continued.)*

13.5. Optical Scattering of Nanoparticles Below or Above a Random Rough Surface

Recent advances on nanotechnologies have made optical scattering of noble nanoparticles an important topic.[25,35–41] Optical scattering by nanoparticles has a wide range of applications, e.g., optical data storage,[25] confocal laser scanning microscopy,[35] near-field scanning optical microscopy,[36] and opto-plasmonic tweezers of micro/nano object.[37] Applications make use of optical enhancements on specific resonance wavelength that depends on the parameters of metal nanoparticles, such as particle size, particle shape, dielectric constant, and arrangement of the particles. The application of opto-plasmonic tweezers makes use of the radiation field from the nanoparticles that induce the light-induced dielectrophoresis force (L-DEP) on a dielectric object. The nanoparticles array is used instead of bulk or thin film because the direction of oscillating dipoles for such arrays can be controlled by the polarization of the incident light. The optical tweezers operate with the light frequency at the resonant mode to obtain the best performance. There are also studies of optical scattering by nanoparticles of nonspherical shapes.[38–39] In many optical scattering applications, the nanoparticles with different shapes, sizes and, arrangement are fabricated on a substrate. Jensen *et al.*[40] measured the extinction of silver nanoparticles arranged in a hexagonal array on a glass substrate. The extinction peak shows a red shift with a thicker substrate on top of the nanoparticles. These studies are useful for tuning the plasmonic resonance frequency of the nanoparticles. With larger particles of size up to one third of the wavelength, C. L. Haynes *et al.*[4] made detailed studies on nanoparticles that were arranged in either hexagonal or square arrays. The measured results show that the resonant wavelengths possess a linear dependence on the lattice spacing. An increase of interparticle distance gives an increase of resonant wavelength.

Often the nanoparticles are placed above a substrate or embedded in the substrate. The substrate usually has a random rough surface. Thus it is important to take into account the rough surface scattering effects in addition to the volume scattering by the nanoparticles. In addition, the interaction between rough surface scattering and volume scattering is also important.

To solve the scattering problem of particles on or buried in a substrate, we have used two approaches, the finite element method (FEM)[42] and the integral equation method[13]. In the FEM, periodic boundary condition is used in the horizontal direction. The problem is divided into two regions. Region I is the homogeneous region which is above the highest point of the rough surface. The scattered field in this region is expanded in upward traveling Floquet modes. Region II is the region below and consists of the rough surface as well as the medium below the rough surface. The medium below the rough surface consists of random-positioned particles. Region II is divided into a number of first-order triangular elements that can be adapted to the rough surface profile. Furthermore, since the field is expressed by the triangular elements, the permittivity ε_r can vary from element to element that takes into account the volume inhomogeneities. By matching the boundary conditions of the boundary between region I and region II, the unknown coefficients can be solved. We have shown the results of combined rough surface and volume scattering effects using this method.[41]

A second approach that we have used is the integral equation approach.[13] The region is divided into two regions. Region 0 is the region above the rough surface and region 1 is the region below the rough surface. The particles can be located either above the surface or below the surface. The fields of both region 0 and region 1 are expressed in the form of integral equations, which consist of the surface integral of all surfaces. The surfaces include the random rough surface and the surfaces of the particles. If the particles are above the rough surface, the expression of the fields of region 0 is the sum of the incident field and the surface integral of both the rough surface and the surfaces of the particles. If the particles are buried under the rough surface, the expression of field in region 1 is the summation of the surface integral of both the rough surface and that of the particles. By matching the boundary condition of all surfaces, coupled surface integral equations are obtained for rough surface scattering and particle scattering. The integral equations are put into coupled matrix equation by surface discretizations.

Both the FEM approach and the surface integral equation approach can be employed to solve the problem of optical scattering by nanoparticles above a substrate or embedded in a substrate with random rough surfaces.

Acknowledgments. The research in this book chapter was supported by Hong Kong RGC Central Allocation Grant 8730017.

References

1. T. Jensen, L. Kelly, A. Lazarides, and G.C. Schatz, "Electrodynamics of noble metal nanoparticles and nanoparticle clusters," *J. Cluster. Sci.* **10** (2), 295–317 (1999).
2. P.C. Chaumet, A. Rahmani, and G.W. Bryant, "Generalization of the coupled dipole method to periodic structures," *Phy. Rev. B* **67**, 165404 (2003.)
3. V.A. Markel, "Coupled-dipole approach to scattering of light from a one-dimensional periodic dipole structure," *J. Mod. Opt.* **40** (11), 2281–2291.
4. C.L. Haynes, A. MacFarland, L.L. Zhao, R.P.V. Duyne, G.C. Schatz, L. Gunnarsson, J. Prikulis, B. Kasemo, and M. Kall, "Nanoparticle optics: the importance of radiative dipole coupling in two-dimensional nanoparticle arrays," *J. Phys. Chem. B* **107**, 7337–7342 (2003).
5. S. Zou, N. Janel, and G.C. Schatz, "Silver nanoparticle array structures that produce remarkably narrow plasmon lineshapes," *J. Chem. Phys.* **120** (23), 10871–10875 (2004).
6. L.L. Zhao, K.L. Kelly, and G.C. Schatz, "The extinction spectra of silver nanoparticle arrays: influence of array structure on plasmon resonance wavelength and width," *J. Phys. Chem. B* **107**, 7343–7350 (2003).
7. L. Tsang, J.A. Kong, and R. Shin, *Theory of Microwave Remote Sensing*. (Wiley, New York, 1985).
8. S.Y. Tong, *Prog. Surf. Sci.* **7**, 1 (1975).
9. M.A. Van Hove and S.Y. Tong, *Surface Crystallography by LEED*. (Spinger-Verlag, Berlin, Heidelberg, New York, 1979).
10. M.A. Van Hove, W.H. Weinberg, and C.-M. Chan, *Low-energy electron diffraction: experiment, theory and structural determination*. (Springer-Verlag, Berlin, Heidelberg, New York, 1986).
11. J. Ng, C.T. Chan, P. Sheng, and Z.F. Lin, "Strong optical force induced by morphology-dependent resonances," *Opt. Lett.* **30**, 1956 (2005).
12. J. Ng, Z. Lin, C.T. Chan, and P. Sheng, "Photonic clusters formed by dielectric microspheres: numerical simulations," *Phys. Rev. B* **72**, 085130 (2005).
13. L. Tsang, J.A. Kong, K.H. Ding, and C.O. Ao, *Scattering of Electromagnetic Waves*, vol. 2: Numerical Simulations (Wiley, New York, 2001).

14. C.-T. Chen, L. Tsang, J. Guo, A.T.C. Chang, and K.-H. Ding, "Frequency dependence of scattering and extinction of dense media based on three-dimensional simulations of Maxwell's equations with applications to snow," *IEEE Trans. Geosci. Remote Sens.* **41**, 1844–1852 (2003).
15. C. Sonnichsen, S. Geier, N.E. Hecker, G. von Plessen, J. Feldmann, H. Ditlbacher, B. Lamprecht, J.R. Krenn, F.R. Aussenegg, V.Z-H Chan, J.P. Patz, and M. Moller, "Spectroscopy of single metallic nanoparticles using total internal reflection microscopy," *Appl. Phys. Lett.* **77**, 2949–2951 (2000).
16. H. Xu, E.J. Bjerneld, M. Kall, and L. Borjesson, "Spectroscopy of single hemoglobin molecules by surface enhanced Raman scattering," *Phys. Rev. Lett.* **83**, 4357–4360 (1999).
17. A.J. Haes and R.P. Van Duyne, "A nanoscale optical biosensor: Sensitivity and selectivity of an approach based on the localized surface plasmon resonance spectroscopy of triangular silver nanoparticles," *J. Am. Chem. Soc.* **124**, 10596–10604 (2002).
18. N. Nath and A. Chilkoti, "A colorimetric gold nanoparticle sensor to interrogate biomolecular interactions in real time on a surface," *Anal. Chem.* **74**, 504–509 (2002).
19. F. Frederix, J.-M. Friedt, K.-H. Choi, W. Laureyn, A. Campitelli, D. Mondelaers, G. Maes, and G. Borghs, "Biosensing based on light absorption of nanoscaled gold and silver particles," *Anal. Chem.* **75**, 6894–6900 (2003).
20. S. Schultz, D.R. Smith, J.J. Mock, and D.A. Schultz, "Single-target molecule detection ith nonbleaching multicolor optical immunolabels," *PNAS* **97**, 996–1001 (2000).
21. X. Hong and F.-J. Kao, "Microsurface plasmon resonance biosensing based on gold-nanoparticle film," *Appl. Opt. B* **43**, 2868–2873 (2004).
22. T. Zhang, H. Morgan, A.S.G. Curtis, and M. Riehle, "Measuring particle-substrate distance with surface plasmon resonance microscopy," *J. Opt. A: Pure Appl. Opt.* **3**, 333–337 (2001).
23. D. Yelin, D. Oron, S. Thiberge, E. Moses, and Y. Silberberg, "Multiphoton plasmon-resonance microscopy," *Opt. Express* **11**, 1385–1391 (2003).
24. M. Quinten, A. Leitner, J.R. Krenn, and F.T. Aussenegg, "Electromagnetic energy transport via linear chains of silver nanoparticles," *Opt. Lett.* **23**, 1331–1333 (1998).
25. H. Ditlbacher, J.R. Krenn, B. Lamprecht, A. Leitner, and F.R. Aussenegg, "Spectrally coded optical data storage by metal nanoparticles," *Opt. Lett.* **25**, 563–565 (2000).
26. J.P. Kottmann, O.J.F. Martin, D.R. Smith, and S. Schultz, "Dramatic localized electromagnetic enhancement in plasmon resonant nanowires," *Chem. Phys. Lett.* **341**, 1–6 (2001).
27. W. Rechberger, A. Hohenau, A. Leitner, J.R. Krenn, and B. Lamprecht, "Optical properties of two interacting gold nanoparticles," *Opt. Commun.* **220**, 137–141 (2003).
28. J.J. Mock, M. Barbic, D.R. Smith, D.A. Schultz, and S. Schultz, "Shape effects in plasmon resonance of individual colloidal silver nanoparticles," *J. Chem. Phys.* **116** (15), 6755–6759 (2002).
29. J.J. Mock, D.R. Smith, and S. Schultz, "Local refractive index dependence of plasmon resonance spectra from individual nanoparticles," *Nano Lett.* **3** (4), 485–491 (2003).
30. J.J. Mock, S.J. Oldenburg, D.R. Smith, D.A. Schultz, and S. Schultz, "Composite plasmon resonant nanowires," *Nano Lett.* **2** (5), 465–469 (2002).
31. M. Barbic, J.J. Mock, D.R. Smith, and S. Schultz, "Single crystal silver nanowires prepared by the metal amplification method," *J. Appl. Phys.* **91** (11), 9341–9345 (2002).
32. E.D. Palik, *Handbook of Optical Constants of Solids*, (Academic, Orlando, 1985).
33. J.J. Penninkhof, A. Polman, Luke A. Sweatlock, Stefan A. Maier, Harry A. Atwater, A.M. Vredenberg, and B.J. Kooi, "Mega-electron-volt ion beam induced anisotropic

plasmon resonance of silver nanocrystals in glass," *Appl. Phys. Lett.* **83** (20), 4137–4139 (2003).
34. X.-M. Duan, Z.-B. Sun, X.-M. Xia, X.-Z. Dong, and W.-Q. Chen, "Optical properties of metal and metal-semiconductor hybrid nanoparticles in polymer marix," *Progress in Electromagnetics Research Symposium 2005* (Hangzhou, China, 2005 Aug. 22–26).
35. M. Alschinger, M. Maniak, F. Stietz, T. Vartanyan, and F. Trager, "Application of metal nanoparticles in confocal laser scanning microscopy: improved resolution by optical field enhancement," *Appl. Phys. B* **76**, 771–774 (2003).
36. M. Gu and P.C. Ke, "Image enhancement in near-field scanning optical microscopy with laser-trapped metallic particles," *Opt. Lett.* **24** (2), 74–76 (1999).
37. X. Miao, H. Liao, and L.Y. Lin, "Opto-plasmon tweezers for rotation and manipulation of micro/nano objects," *International Optical MEMS Conference* (Oulu, Finland, 2005 Aug. 1–4).
38. W. Gotschy, K. Vonmetz, A. Leitner, and F.R. Aussenegg, "Thin films by regular patterns of metal nanoparticles: tailoring the optical properties by nanodesign," *Appl. Phys. B* **63**, 381–384 (1996).
39. W. Gotschy, K. Vonmetz, A. Leitner, and F.R. Aussenegg, "Optical dichroism of lithographically designed silver nanoparticle films," *Opt. Lett.* **21** (15), 1099–1101 (1996).
40. T.R. Jensen, M.D. Malinsky, C.L. Haynes, and R.P. Van Duyne, "Nanosphere lithography: tunable localized surface plasmon resonance spectra of silver nanoparticles," *J. Phys. Chem. B* **104**, 10549–10556 (2000).
41. K. Pak, L. Tsang, L. Li, and C.H. Chan, "Combined random rough surface and volume scattering based on Monte Carlo simulations of solutions of Maxwell's equations," *Radio Sci.* **28** (3), 331–338 (1993 May–Jun.).

14
Multiple-Scattering Effects in Angular Intensity Correlation Functions

TAMARA A. LESKOVA AND ALEXEI A. MARADUDIN
Department of Physics and Astronomy, and Institute for Surface and Interface Science, University of California, Irvine, CA 92697, U.S.A.

14.1. Introduction

The majority of the existing theoretical and experimental studies of multiple-scattering effects in the scattering of light from a randomly rough surface have been devoted to the reflectivity or transmissivity of such a surface or to its mean differential reflection or transmission coefficients, i.e. to the first and second moments of the scattered or transmitted field. Recently, however, attention has begun to be directed to the theoretical[1–20] and experimental[9,10,18,21–23] study of multiple-scattering effects in higher moments of the scattered field, in particular in angular intensity correlation functions. These correlation functions describe how the speckle pattern, formed by the interference of randomly scattered waves, changes when the angles of incidence and scattering are varied.

The interest in these correlation functions was stimulated by the expectation that, just as the inclusion of multiple-scattering processes in calculations of the angular dependence of the intensity of the light that has been scattered diffusely from, or transmitted diffusely through, a randomly rough surface, led to the prediction of enhanced backscattering[24] and enhanced transmission,[25] their inclusion in calculations of higher moments of the scattered or transmitted field would also lead to the prediction of new effects.

This interest was also stimulated by the results of earlier theoretical[26–30] and experimental[31–35] investigations of angular intensity correlation functions in the scattering of classical waves from volume disordered media. In the theoretical investigation[27] it was predicted that three types of correlations occur in such scattering, namely short-range correlations, long-range correlations, and infinite-range correlations. These were termed $C^{(1)}$, $C^{(2)}$, and $C^{(3)}$ correlation functions, respectively. The $C^{(1)}$ correlation function includes both the "memory effect"[28] and the "time-reversed memory effect,"[29,30] so named because of the wave vector conservation conditions they satisfy. The memory effect has been observed in volume scattering experiments,[32] as has the time-reversed memory effect.[34] The $C^{(2)}$ angular intensity correlation function has also been observed in volume scattering experiments,[33] as has the $C^{(2)}$ frequency intensity correlation function,[31] while the

$C^{(3)}$ frequency intensity correlation function has been observed in the transmission of microwaves through a long tube filled with polystyrene microspheres.[35]

In more recent work[36,37] a new angular intensity correlation function, the $C^{(0)}$ correlation function, was predicted, but it has yet to be studied experimentally.

Much of the theoretical work on multiple-scattering effects in angular intensity correlation functions in rough surface scattering has been carried out perturbatively for scattering from weakly rough random metal surfaces. Thus, Arsenieva and Feng[7] studied perturbatively the angular correlation function of the scattering amplitude in the cross-polarized scattering of light from a two-dimensional randomly rough metal surface. The squared modulus of this correlation function gives the analogue of the $C^{(1)}$ angular intensity function studied earlier in the context of scattering from volume disordered systems. By keeping only the contribution from ladder diagrams they obtained what is now called the memory effect peak in the envelope of the $C^{(1)}$ angular intensity correlation function, and related this peak to the enhanced backscattering peak in the angular dependence of the light scattered diffusely from the same surface.

The same angular amplitude correlation function was also studied perturbatively by Freilikher and Yurkevich.[8] By keeping only the contribution from maximally crossed diagrams they obtained what is now called the time-reversed memory effect peak in the envelope of the $C^{(1)}$ correlation function, and related it to the enhanced backscattering peak.

It should be noted, however, that the analogue of the $C^{(1)}$ correlation function in rough surface scattering had been predicted 18 years prior to the work of Arsenieva and Feng and of Freilikher and Yurkevich in single-scattering theories of the scattering of light from a randomly rough surface.[38,39] However, these single-scattering calculations could not capture the memory effect and time-reversed memory effect peaks predicted by the multiple-scattering calculations of this correlation function.

The analogs of the $C^{(2)}$ and $C^{(3)}$ angular intensity correlation functions in rough surface scattering were investigated theoretically by Malyshkin *et al.*[11,12] by small-amplitude perturbation theory for the scattering of p-polarized light from randomly rough metal surfaces whose roughness was sufficiently weak that these correlations were caused by the multiple scattering of the surface plasmon polaritons supported by vacuum–metal interface and excited through the roughness by the incident field.

The work of Malyshkin *et al.*[11,12] also revealed that in rough surface scattering the angular intensity correlation function possesses a contribution that they called the $C^{(10)}$ correlation that can be of the same order of magnitude as the $C^{(1)}$ correlation function. This correlation function had been overlooked in the earlier studies of Arsenieva and Feng,[7] and of Freilikher and Yurkevich[8] because the angular amplitude correlation function they investigated was equivalent to the use of the factorization approximation[26] in calculating the angular intensity correlation function. This approximation, which was not used in the work of Malyshkin *et al.*[11,12] is based on the assumption that the amplitude of the field scattered diffusely possesses a circular complex Gaussian joint probability density function.[40,41] It is good enough to describe the $C^{(1)}$ correlation function. However, the fact that it fails to predict the $C^{(10)}$ correlation function shows that the latter assumption is not universally valid. An indication that interesting correlation effects can occur when the factorization approximation is not used had been shown earlier by Nieto-Vesperinas and Sánchez-Gil[4–6] in predicting what they called enhanced long-range correlation functions.

Malyshkin *et al.*[11,12] also showed that there is an additional class of correlations between the $C^{(1)}$ and $C^{(2)}$ correlations in the order in the surface profile function in which they first appear in calculations of angular intensity correlation functions by means of small-amplitude perturbation theory. The lowest order contribution to the $C^{(1)}$ correlation function is of the fourth order in the surface profile function, while the lowest order contribution to the $C^{(2)}$ correlation function is of eighth order. The lowest order contribution to the additional class of correlations found by Malyshkin *et. al.* is of sixth order in the surface profile function, and these correlations were therefore named the $C^{(1.5)}$ correlations. They arise when the factorization approximation is not made, and are intimately connected with the roughness-induced excitation of forward and backward propagating surface plasmon polaritons at a dielectric–metal interface by the incident light, and introduce peaks into the angular intensity correlation function.

More recently, Leskova and Maradudin[19] studied theoretically the analogue in rough surface scattering of the $C^{(0)}$ correlation function that had earlier been studied in the context of the scattering of classical waves in volume disordered media.[36,37]

In this chapter we present an introduction to theoretical and experimental studies of multiple-scattering effects in angular intensity correlation functions in the context of the scattering of p- and s-polarized light from a one-dimensional, weakly rough, random surface. A study of multiple-scattering effects in the scattering of light from two-dimensional random surfaces has been presented by Malyshkin *et al.*[12] and the reader interested in this topic is referred to this paper. We will also restrict our discussion to surfaces whose profile functions obey Gaussian statistics. A brief discussion of angular intensity correlation functions of light scattered from non-Gaussian surfaces is presented in [42].

14.2. The Correlation Function $C(q, k|q', k')$ and Its Properties

For all of the scattering systems considered in this chapter, the illuminated surface is a one-dimensional random surface defined by $x_3 = \zeta(x_1)$. The region $x_3 > \zeta(x_1)$ is vacuum, while the region $x_3 < \zeta(x_1)$ is the scattering medium.

The surface profile function $\zeta(x_1)$ is assumed to be a single-valued function of x_1 that is differentiable, and that constitutes a stationary, zero-mean, Gaussian random process, defined by the properties

$$\langle \zeta(x_1) \rangle = 0 \tag{1}$$

$$\langle \zeta(x_1)\zeta(x_1') \rangle = \delta^2 W(|x_1 - x_1'|). \tag{2}$$

The angle brackets here and in all that follows denote an average over the ensemble of realizations of the surface profile function, and $\delta = \langle\zeta^2(x_1)\rangle^{\frac{1}{2}}$ is the rms height of the surface.

We will need the Fourier integral representation of $\zeta(x_1)$:

$$\zeta(x_1) = \int_{-\infty}^{\infty} \frac{\mathrm{d}Q}{2\pi}\hat{\zeta}(Q)\exp(\mathrm{i}Qx_1). \tag{3}$$

The Fourier coefficient $\hat{\zeta}(Q)$ is also a zero-mean Gaussian random process, and possesses the properties

$$\langle\hat{\zeta}(Q)\rangle = 0 \tag{4}$$

$$\langle\hat{\zeta}(Q)\hat{\zeta}(Q')\rangle = 2\pi\delta(Q+Q')\delta^2 g(|Q|), \tag{5}$$

where $g(|Q|)$ is the power spectrum of the surface roughness, and is the Fourier transform of the surface height autocorrelation function $W(|x_1|)$,

$$g(|Q|) = \int_{-\infty}^{\infty} \mathrm{d}x_1 W(|x_1|)\exp(-\mathrm{i}Qx_1). \tag{6}$$

In this chapter two forms of $W(|x_1|)$, and hence of $g(|Q|)$, will be considered. They are the Gaussian form

$$W(|x_1|) = \exp\left(-x_1^2/a^2\right) \tag{7a}$$

$$g(|Q|) = \sqrt{\pi}a\exp(-a^2Q^2/4), \tag{7b}$$

where the characteristic length a is the transverse correlation length of the surface roughness, and the West–O'Donnell[43] form

$$W(|x_1|) = \frac{\sin Q_{\max}x_1 - \sin Q_{\min}x_1}{(Q_{\max} - Q_{\min})x_1} \qquad Q_{\max} > Q_{\min} \tag{8a}$$

$$g(|Q|) = \frac{\pi}{Q_{\max} - Q_{\min}}[\theta(Q_{\max} - Q)\theta(Q - Q_{\min}) + \theta(Q_{\max} + Q)\theta(-Q - Q_{\min})], \tag{8b}$$

where $\theta(z)$ is the Heaviside unit step function. In the case of the latter power spectrum, if $Q_{\min}$ and $Q_{\max}$ are chosen to bracket the wave numbers of the surface or guided waves supported by the scattering system, light incident from the vacuum on a surface characterized by this power spectrum can preferentially excite forward and backward propagating waves of these types, which can then be efficiently converted back into electromagnetic volume waves in the vacuum. Multiple-scattering processes in which these waves play the role of intermediate states are strongly enhanced thereby.

The surface $x_3 = \zeta(x_1)$ is illuminated from the vacuum by a plane wave of frequency ω, whose plane of incidence is the x_1x_3 plane. The single nonzero component of the electromagnetic field in the region $x_3 > \zeta(x_1)_{\max}$ is given by

$$\Phi_\nu^>(x_1, x_3|\omega) = \exp[ikx_1 - i\alpha_0(k)x_3] + \int_{-\infty}^{\infty} \frac{dq}{2\pi} R_\nu(q|k) \exp[iqx_1 + i\alpha_0(q)x_3]. \quad (9)$$

Here $\Phi_\nu^>(x_1, x_3|\omega)$ is $H_2^>(x_1, x_3|\omega)$ in the case of p-polarization ($\nu = p$), and is $E_2^>(x_1, x_3|\omega)$ in the case of s-polarization ($\nu = s$), where $H_2^>(x_1, x_3|\omega)$ ($E_2^>(x_1, x_3|\omega)$) is the single nonzero component of the magnetic (electric) field in the vacuum region. The function $\alpha_0(q)$ entering Eq. (9) is defined by

$$\alpha_0(q) = ((\omega/c)^2 - q^2)^{\frac{1}{2}} \qquad |q| < \omega/c \quad (10a)$$

$$= i\left(q^2 - (\omega/c)^2\right)^{\frac{1}{2}} \qquad |q| > \omega/c. \quad (10b)$$

In writing Eq. (9) we have assumed a time dependence of the field of the form $\exp(-i\omega t)$, but have not indicated this explicitly. The angle of incidence θ_0, measured counterclockwise from the x_3-axis, and the scattering angle θ_s, measured clockwise from the x_3-axis, are related to the wavenumbers k and q by

$$k = (\omega/c)\sin\theta_0, \qquad q = (\omega/c)\sin\theta_s. \quad (11)$$

We are interested in the correlations that exist between the intensity fluctuations $\delta I(q|k) = I(q|k) - \langle I(q|k)\rangle$ and the intensity fluctuations $\delta I(q'|k') = I(q'|k') - \langle I(q'|k')\rangle$,

$$C(q, k|q', k') = \langle \delta I(q|k)\delta I(q'|k')\rangle \quad (12a)$$

$$= \langle I(q|k)I(q'|k')\rangle - \langle I(q|k)\rangle\langle I(q'|k')\rangle, \quad (12b)$$

where the intensity $I(q|k)$ entering these expressions is defined in terms of the scattering matrix $S(q|k)$ for the scattering of light of frequency ω from a one-dimensional random surface by[7]

$$I(q|k) = \frac{1}{L_1}\left(\frac{\omega}{c}\right)|S(q|k)|^2, \quad (13)$$

where L_1 is the length of the x_1-axis covered by the random surface. The scattering matrix $S(q|k)$ is related to the scattering amplitude $R(q|k)$ appearing in Eq. (9) by

$$S(q|k) = \frac{\alpha_0^{\frac{1}{2}}(q)}{\alpha_0^{\frac{1}{2}}(k)} R(q|k), \quad (14)$$

and possesses the property of *reciprocity*,

$$S(q|k) = S(-k|-q). \quad (15)$$

We note that in several studies of angular intensity correlation functions, e.g.[18] a normalized correlation function

$$\Xi(q, k|q', k') = \frac{\langle \delta I(q|k)\delta I(q'|k')\rangle}{\langle I(q|k)\rangle\langle I(q'|k')\rangle} \quad (16)$$

is considered. We have chosen to work with the nonnormalized correlation function (12) because the mean intensities $\langle I(q|k)\rangle$ and $\langle I(q'|k')\rangle$ appearing in the denominator of Eq. (16) themselves display a structure, e.g. peaks, as functions of q, k and q', k', and this structure unnecessarily complicates the structure of the correlation function $\langle \delta I(q|k)\delta I(q'|k')\rangle$ which is of primary interest here.

We can already obtain useful qualitative information about the correlation function $C(q,k|q',k')$ from Eqs. (12) and (13). Because the correlation of $\delta I(q|k)$ with itself should be stronger than the correlation of $\delta I(q|k)$ with $\delta I(q'|k')$ when $q' \neq q$ and $k' \neq k$, a peak in $C(q,k|q',k')$ is expected when $q' = q$ and $k' = k$. This peak has come to be called the *memory effect peak*. At the same time, since $S(q|k)$ is reciprocal, Eq. (15), a peak in $C(q,k|q',k')$ is also expected when $q' = -k$ and $k' = -q$. This peak is called the *reciprocal memory effect peak*. In earlier work this peak was called the time-reversed memory effect peak. This name had its origin in the fact that the time-reversed memory effect was first predicted to occur in random volume systems that were time-reversal invariant, e.g. aqueous suspensions of polystyrene microspheres.[29] In this case reciprocity and time-reversible invariance are equivalent. However, because this effect also appears, due to reciprocity, in systems that are not time-reversal invariant, such as the rough surface of a lossy metal, it seems more appropriate to call it the reciprocal memory effect, and we do so in this chapter.

In terms of the scattering matrix $S(q|k)$ the correlation function $C(q,k|q',k')$ takes the form

$$C(q,k|q',k') = \frac{1}{L_1^2}\left(\frac{\omega}{c}\right)^2 [\langle S(q|k)S^*(q|k)S(q'|k')S^*(q'|k')\rangle - \langle S(q|k)S^*(q|k)\rangle\langle S(q'|k')S^*(q'|k')\rangle]. \tag{17}$$

In the form given by Eq. (17) $C(q,k|q',k')$ contains purely specular contributions, i.e. terms proportional to $2\pi\delta(q-k)$ and/or $2\pi\delta(q'-k')$. Such terms are uninteresting. If we note that due to the stationarity of the surface profile function $\langle S(q|k)\rangle$ is diagonal in q and k, $\langle S(q|k)\rangle = 2\pi\delta(q-k)S(k)$, we can eliminate these terms by rewriting $C(q,k|q',k')$ in terms of $\delta S(q|k) = S(q|k) - \langle S(q|k)\rangle$. In addition, if we use the relations between averages of products of random functions and the corresponding cumulant averages,[44,45] we can finally write the contribution to $C(q,k|q',k')$ that is free from specular contributions as

$$\hat{C}(q,k|q',k') = \frac{1}{L^2}\left(\frac{\omega}{c}\right)^2 [|\langle \delta S(q|k)\delta S^*(q'|k')\rangle|^2 + |\langle \delta S(q|k)\delta S(q'|k')\rangle|^2 + \langle \delta S(q|k)\delta S^*(q|k)\delta S(q'|k')\delta S^*(q'|k')\rangle_c], \tag{18}$$

where $\langle\cdots\rangle_c$ denotes the cumulant average.

The result given by Eq. (18) is convenient for several reasons. Due to the stationarity of the surface profile function $\zeta(x_1)$, the amplitude correlation function $\langle \delta S(q|k)\delta S^*(q'|k')\rangle$ is proportional to $2\pi\delta(q-k-q'+k')$. It therefore gives rise to the contribution to $\hat{C}(q,k|q',k')$ denoted by $\hat{C}^{(1)}(q,k|q',k')$, which contains the memory and reciprocal memory effect peaks.[7,8] Since $2\pi\delta(0) = L_1$, when the argument of the delta function vanishes the $\hat{C}^{(1)}(q, k|q', k')$ correlation function is independent of the length of the surface, because it contains $[2\pi\delta(0)]^2$. The property of a speckle pattern that is reflected in the presence of the factor $2\pi\delta(q - k - q' + k')$ in $\hat{C}^{(1)}(q, k|q', k')$ is that if the angle of incidence is changed so that k goes into $k + \Delta k$, the entire speckle pattern shifts in such a way that any feature initially at q moves to $q' = q + \Delta k$. This is why the $\hat{C}^{(1)}$ correlation function was originally named the memory effect. In terms of the angles of incidence and scattering we have that if θ_0 is changed to $\theta_0' = \theta_0 + \Delta\theta_0$, any feature in the speckle pattern initially at θ_s is shifted to $\theta_s' = \theta_s + \Delta\theta_s$, where $\Delta\theta_s = (\cos\theta_0/\cos\theta_s)\Delta\theta_0$, to first order in $\Delta\theta_0$.

Due to the stationarity of the surface profile function, the amplitude correlation function $\langle\delta S(q|k)\delta S(q'|k')\rangle$ is proportional to $2\pi\delta(q - k + q' - k')$. It gives rise to the contribution to $\hat{C}(q, k|q', k')$ called $\hat{C}^{(10)}(q, k|q', k')$. When the argument of the delta function vanishes the $\hat{C}^{(10)}(q, k|q', k')$ correlation function is also independent of L_1. The property of a speckle pattern that is reflected in the presence of the factor $2\pi\delta(q - k + q' - k')$ in $\hat{C}^{(10)}(q, k|q', k')$ is that if the angle of incidence is changed in such a way that k goes into $k' = k + \Delta k$, a feature initially at $q = k - \Delta q$ will appear at $q' = k' + \Delta q$, i.e. at a point as much to one side of the new specular direction as the original point was on the other side of the original specular direction. For one and the same incident beam, $k' = k$, the $\hat{C}^{(10)}$ correlation function therefore reflects the symmetry of the speckle pattern with respect to the specular direction (in wave number space).

It should be emphasized that the preceding properties of the correlation functions $\langle\delta S(q|k)\delta S^*(q'|k')\rangle$ and $\langle\delta S(q|k)\delta S(q'|k')\rangle$, and their consequences for speckle patterns, are consequences of the assumed stationarity of the surface profile function, which implies a surface of infinite length, and are therefore independent of any approximations made in calculating these correlation functions. Thus, they will be present in the results obtained in the calculations based on a single-scattering approximation as well as in results obtained in calculations that take multiple-scattering processes into account.

In Figs. 14.1(a) and 14.1(b) we have plotted the differential reflection coefficient produced by the scattering of p-polarized light from a single realization of a one-dimensional randomly rough silver surface for angles of incidence $\theta_0 = 0^\circ$ and 10°, respectively. The curves were calculated using a rigorous computer simulation of the scattering problem.[46] The overall shift of the speckle pattern in Fig. 14.1(a) by about 10° is clearly visible in the speckle pattern plotted in Fig. 14.1(b). It is also seen in Figs. 14.1(a) and 14.1(b) that each speckle pattern is symmetric with respect to the specular direction.

The correlation function $\langle\delta S(q|k)\delta S^*(q'|k')\delta S(q'|k')\delta S^*(q'|k')\rangle_c$ appearing in Eq. (18) is proportional to $2\pi\delta(0) = L_1$, and gives rise to the long-range and infinite-range contributions to $\hat{C}(q, k|q', k')$ given by the sum $\hat{C}^{(1.5)}(q, k|q', k')$ $+\hat{C}^{(2)}(q, k|q', k') + \hat{C}^{(3)}(q, k|q', k')$. Its contribution to $\hat{C}(q, k|q', k')$ is of $O(L_1^{-1})$. Therefore, in the limit of a long surface or a large illumination area the long-range and infinite-range correlation functions are small compared to the

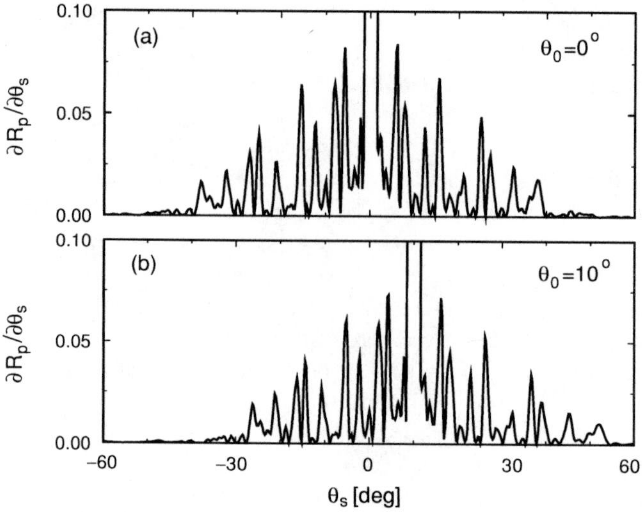

FIGURE 14.1. Computer generated speckle pattern produced with a single realization of a one-dimensional randomly rough silver surface defined by the Gaussian power spectrum (7b) with $\delta = 8$ nm, $a = 400$ nm, when the surface is illuminated by p-polarized light of wavelength $\lambda = 612.7$ nm ($\epsilon(\omega) = -17.2 + i0.498$). (a) $\theta_0 = 0°$; (b) $\theta_0 = 10°$.

short-range correlation function, and vanish in the limit of an infinitely long surface. Consequently, although they contain interesting multiple-scattering effects,[11,12] they are weak, and therefore will not be considered further here.

In conventional speckle theory[40,41] the surface profile function is assumed to be a stationary random process, and the random surface is assumed to be infinitely long. Under these conditions the scattering matrix $S(q|k)$ becomes the sum of a very large number of independent contributions from different points on the surface. According to the central limit theorem $\delta S(q|k)$ must therefore obey complex Gaussian statistics. In this case Eq. (18) becomes rigorously[47]

$$\begin{aligned}\hat{C}(q, k|q', k') &= \frac{1}{L_1^2}\left(\frac{\omega}{c}\right)^2 [|\langle\delta S(q|k)\delta S^*(q'|k')\rangle|^2 + |\langle\delta S(q|k)\delta S(q'|k')\rangle|^2] \\ &\equiv \hat{C}^{(1)}(q, k|q', k') + \hat{C}^{(10)}(q, k|q', k'),\end{aligned} \tag{19}$$

because all cumulant averages of products of more than two Gaussian random processes vanish. The third term on the right-hand side of Eq. (18) therefore gives the correction to the prediction of the central limit theorem due to the finite length of the surface.

If it is further assumed, as is done in conventional speckle theory, where the disorder is assumed to be strong, that $\delta S(q|k)$ obeys circular complex Gaussian statistics,[40,41] then $\langle \delta S(q|k)\delta S(q'|k')\rangle$ vanishes, and the expression for $\hat{C}(q, k|q', k')$ simplifies to

$$\hat{C}(q, k|q', k') = \frac{1}{L_1^2}\left(\frac{\omega}{c}\right)^2 |\langle \delta S(q|k)\delta S^*(q'|k')\rangle|^2 = \hat{C}^{(1)}(q, k|q', k'). \tag{20}$$

Therefore, theoretical calculations or experimental measurements of $\hat{C}(q, k|q', k')$ that reveal the existence of the $\hat{C}^{(10)}(q, k|q', k')$ correlation function, or its absence, indicate whether $\delta S(q|k)$ obeys complex Gaussian statistics or circular complex Gaussian statistics. The degree of surface roughness needed for $\delta S(q|k)$ to change from a complex Gaussian process to a circular complex Gaussian process, and hence for $\hat{C}^{(10)}(q, k|q', k')$ to vanish, will be discussed in Sect. 14.3.1.3.

The result given by Eq. (20) is often called the factorization approximation to $\hat{C}(q, k|q', k')$ [26]. We see that this approximation suffices if it is only the memory and reciprocal memory effects that are of interest. However, we also see from Eq. (18) what is omitted when this approximation is made, namely a correlation function $\hat{C}^{(10)}$ whose magnitude can be comparable to that of the correlation function $\hat{C}^{(1))}$ obtained with the use of the factorization approximation, and the correlations defined by the third term on the right-hand side of Eq. (18). In what follows the factorization approximation will not be used.

We now turn to a discussion of the several contributions to the correlation function $\hat{C}(q, k|q', k')$.

14.3. Determination of $\hat{C}(q, k|q', k')$

We have seen in the preceding section that the correlation function $\hat{C}(q, k|q', k')$, Eq. (18), is expressed in terms of the fluctuating part of the scattering matrix $\delta S(q|k) = S(q|k) - \langle S(q|k)\rangle$, where the scattering matrix is expressed in terms of the scattering amplitude $R(q|k)$ by $S(q|k) = (\alpha_0(q)/\alpha_0(k))^{\frac{1}{2}} R(q|k)$. We begin a description of the calculation of this correlation function by obtaining some useful general results that apply to any of the scattering structures considered in this chapter.

For all the scattering systems we will be concerned with, the scattering amplitude $R(q|k)$ is the solution of a reduced Rayleigh equation,[48]

$$\int_{-\infty}^{\infty} \frac{dq}{2\pi} M(p|q)R(q|k) = N(p|k), \tag{21}$$

where the forms of the functions $M(p|q)$ and $N(p|q)$ depend on the structure of the scattering system and on the polarization of the incident light. A solution of Eq. (21) is sought in the form[49]

$$R(q|k) = 2\pi\delta(q-k)R_0(k) - 2iG_0(q)T(q|k)G_0(k)\alpha_0(k), \tag{22}$$

where $R_0(k)$ is the Fresnel coefficient for the scattering of light from the scattering system in the absence of the surface roughness, while $G_0(k)$ is the electromagnetic Green's function for the scattering system in the absence of the surface roughness. The transition matrix $T(q|k)$ is postulated to satisfy the equation

$$T(q|k) = V(q|k) + \int_{-\infty}^{\infty} \frac{dp}{2\pi} V(q|p)G_0(p)T(p|k) \tag{23a}$$

$$= V(q|k) + \int_{-\infty}^{\infty} \frac{dp}{2\pi} T(q|p)G_0(p)V(p|k). \tag{23b}$$

On combining Eqs. (21), (22), and (23b) we find that the scattering potential $V(q|k)$ is the solution of the equation

$$\int_{-\infty}^{\infty} \frac{dq}{2\pi} \{M(p|q)[R_0(q) + 2iG_0(q)\alpha_0(q)] - N(p|q)\} \frac{V(q|k)}{2i\alpha_0(q)} = \frac{M(p|k)R_0(k) - N(p|k)}{2iG_0(k)\alpha_0(k)}. \tag{24}$$

It is convenient to define $G_0(q)$ by requiring that

$$R_0(q) + 2iG_0(q)\alpha_0(q) = -1, \tag{25}$$

so that

$$G_0(q) = i\frac{1 + R_0(q)}{2\alpha_0(q)}. \tag{26}$$

The equation satisfied by $V(q|k)$ then takes the form

$$\int_{-\infty}^{\infty} \frac{dq}{2\pi} [M(p|q) + N(p|q)] \frac{V(q|k)}{2i\alpha_0(q)} = \frac{N(p|k) - M(p|k)R_0(k)}{2iG_0(k)\alpha_0(k)}. \tag{27}$$

We now define the electromagnetic Green's function in the presence of the surface roughness as the solution of the equation

$$G(q|k) = 2\pi\delta(q-k)G_0(k) + G_0(q) \int_{-\infty}^{\infty} \frac{dp}{2\pi} V(q|p)G(p|k) \tag{28a}$$

$$= 2\pi\delta(q-k)G_0(k) + G_0(q)T(q|k)G_0(q), \tag{28b}$$

where $T(q|k)$ is the solution of Eq. (23). By combining Eqs. (22) and (28b), and making use of Eq. (25), we obtain the useful relation

$$R(q|k) = -2\pi\delta(q-k) - 2iG(q|k)\alpha_0(k). \tag{29}$$

The scattering matrix is therefore given by

$$S(q|k) = -2\pi\delta(q-k) - 2\mathrm{i}\alpha_0^{\frac{1}{2}}(q)G(q|k)\alpha_0^{\frac{1}{2}}(k), \tag{30}$$

so that

$$\delta S(q|k) = -2\mathrm{i}\alpha_0^{\frac{1}{2}}(q)[G(q|k) - \langle G(q|k)\rangle]\alpha_0^{\frac{1}{2}}(k). \tag{31}$$

The correlation function $\hat{\mathrm{C}}^{(1)}(q,k|q',k')$ is then obtained from Eq. (18) as

$$\hat{\mathrm{C}}^{(1)}(q,k|q',k') = \frac{16}{L_1^2}\left(\frac{\omega}{c}\right)^2 [\alpha_0(q)\alpha_0(k)\alpha_0(q')\alpha_0(k')] \times |[\langle G(q|k)G^*(q'|k')\rangle - \langle G(q|k)\rangle\langle G^*(q'|k')\rangle]|^2, \tag{32}$$

while $\hat{C}^{(10)}(q,k|q',k')$ is given by

$$\hat{\mathrm{C}}^{(10)}(q,k|q',k') = \frac{16}{L_1^2}\left(\frac{\omega}{c}\right)^2 [\alpha_0(q)\alpha_0(k)\alpha_0(q')\alpha_0(k')] \times |[\langle G(q|k)G(q'|k')\rangle - \langle G(q|k)\rangle\langle G(q'|k')\rangle]|^2. \tag{33}$$

Due to the stationarity of the surface profile function $\zeta(x_1)$, the ensemble-averaged Green's function $\langle G(q|k)\rangle$ is diagonal in q and k

$$\langle G(q|k)\rangle = 2\pi\delta(q-k)G(k). \tag{34}$$

where

$$G(k) = \frac{1}{G_0^{-1}(k) - M(k)}. \tag{35}$$

The function $M(k)$ in Eq. (35) is the averaged proper self-energy, which is obtained from the pair of equations[50]

$$\langle M(q|k)\rangle = 2\pi\delta(q-k)M(k) \tag{36a}$$

$$M(q|k) = V(q|k) + \int_{-\infty}^{\infty}\frac{\mathrm{d}p}{2\pi}\int_{-\infty}^{\infty}\frac{\mathrm{d}r}{2\pi}M(q|p)\langle G(p|r)\rangle[V(r|k) - \langle M(r|k)\rangle]. \tag{36b}$$

We now apply the preceding results to the determination of the $\hat{\mathrm{C}}^{(1)}(q,k|q',k')$ and $\hat{\mathrm{C}}^{(10)}(q,k|q',k')$ correlation functions for a system containing a single random surface between vacuum and a medium that supports a single surface electromagnetic wave, and for a film system with a random illuminated surface, that supports two or more guided electromagnetic waves.

14.3.1. Correlations in Single-Interface Systems

The physical system we consider in this section consists of vacuum in the region $x_3 > \zeta(x_1)$ and a metal characterized by an isotropic, complex, frequency-dependent, dielectric function $\epsilon(\omega) = \epsilon_1(\omega) + \mathrm{i}\epsilon_2(\omega)$ in the region $x_3 < \zeta(x_1)$. We are interested in the frequency range in which $\epsilon_1(\omega)$ is negative, while $\epsilon_2(\omega)$ is always nonnegative. This is the frequency range in which surface electromagnetic waves – surface plasmon polaritons – exist.[51]

This system is illuminated from the vacuum by a p-polarized plane wave of frequency ω, whose plane of incidence is the x_1x_3 plane, Eq. (9). The Fresnel reflection coefficient for the scattering of this wave from a planar vacuum–metal interface is

$$R_0(k) = \frac{\epsilon(\omega)\alpha_0(k) - \alpha(k)}{\epsilon(\omega)\alpha_0(k) + \alpha(k)}, \tag{37}$$

where

$$\alpha(k) = [\epsilon(\omega)(\omega/c)^2 - k^2], \qquad \mathrm{Re}\,\alpha(k) > 0, \quad \mathrm{Im}\alpha(k) > 0. \tag{38}$$

The Green's function $G_0(k)$, obtained from Eqs. (26) and (37), is

$$G_0(k) = \frac{\mathrm{i}\epsilon(\omega)}{\epsilon(\omega)\alpha_0(k) + \alpha(k)}. \tag{39}$$

Both $R_0(k)$ and $G_0(k)$ have simple poles at

$$\begin{aligned} k &= \pm\frac{\omega}{c}\left[\frac{\epsilon(\omega)}{\epsilon(\omega)+1}\right]^{\frac{1}{2}}, \\ &= \pm[k_{\mathrm{sp}}(\omega) + \mathrm{i}\Delta_\epsilon(\omega)] \end{aligned} \tag{40}$$

where, to the lowest order in $\epsilon_2(\omega)$,

$$k_{\mathrm{sp}}(\omega) = \frac{\omega}{c}\left(\frac{|\epsilon_1(\omega)|}{|\epsilon_1(\omega)| - 1}\right)^{\frac{1}{2}}, \qquad \Delta_\epsilon(\omega) = \frac{1}{2}\frac{\omega}{c}\frac{\epsilon_2(\omega)}{|\epsilon_1(\omega)|^{\frac{1}{2}}[|\epsilon_1(\omega)| - 1]^{3/2}}. \tag{41}$$

Here $k_{\mathrm{sp}}(\omega)$ is the wave number of the surface plasmon polariton of frequency ω supported by a planar vacuum–metal interface, while $\Delta_\epsilon(\omega)$ is the amplitude decay rate of the surface plasmon polariton due to ohmic losses as it propagates along the surface. The functions $M(p|q)$ and $N(p|q)$ entering Eq. (21) are

$$M(p|q) = \frac{[\alpha(p)\alpha_0(q) + pq]}{\alpha(p) - \alpha_0(q)} I(\alpha(p) - \alpha_0(q)|p - q) \tag{42a}$$

$$N(p|q) = \frac{[\alpha(p)\alpha_0(q) - pq]}{\alpha(p) + \alpha_0(q)} I(\alpha(p) + \alpha_0(q)|p - q), \tag{42b}$$

where

$$I(\gamma|Q) = \int_{-\infty}^{\infty} \mathrm{d}x_1 \exp(-\mathrm{i}Qx_1)\exp(-\mathrm{i}\gamma\zeta(x_1)). \tag{43}$$

The solution of Eq. (27) for the scattering potential $V(q|k)$ through terms linear in the surface profile function (the small-roughness approximation) is

$$V(q|k) = \frac{\epsilon(\omega) - 1}{\epsilon^2(\omega)}[\epsilon(\omega)qk - \alpha(q)\alpha(k)]\hat{\zeta}(q - k). \tag{44}$$

In what follows we will approximate $V(q|k)$ by an expression that is valid for small $|q|$ and $|k|$, namely

$$V(q|k) \cong \frac{1-\epsilon(\omega)}{\epsilon(\omega)} \frac{\omega^2}{c^2} \hat{\zeta}(q-k). \tag{45}$$

This is not an essential approximation, but it allows simpler expressions for $\hat{C}^{(1)}(q, k|q', k')$ and $\hat{C}^{(10)}(q, k|q', k')$ to be obtained than would be the case if it were not made, without sacrificing any of the physics of the results.

In the vicinity of its poles at $k = \pm[k_{sp}(\omega) + i\Delta_\epsilon(\omega)]$, the Green's function $G_0(k)$ can be written in the form

$$G_0(k) \cong \frac{C(\omega)}{k - k_{sp}(\omega) - i\Delta_\epsilon(\omega)} - \frac{C(\omega)}{k + k_{sp}(\omega) + i\Delta_\epsilon(\omega)}, \tag{46}$$

where the residue $C(\omega)$ at these poles is given by

$$C(\omega) = \frac{|\epsilon_1(\omega)|^{3/2}}{\epsilon_1^2(\omega) - 1}. \tag{47}$$

The results given by Eqs. (35) and (46) enable us to obtain a pole approximation for the averaged Green's function $G(k)$,

$$G(k) \cong \frac{C(\omega)}{p - k_{sp}(\omega) - i\Delta(\omega)} - \frac{C(\omega)}{p + k_{sp}(\omega) + i\Delta(\omega)}, \tag{48}$$

where we have defined the total amplitude damping rate of the surface plasmon polariton $\Delta(\omega)$ by

$$\Delta(\omega) = \Delta_\epsilon(\omega) + \Delta_{sp}(\omega), \tag{49}$$

with

$$\Delta_{sp}(\omega) = C(\omega) \operatorname{Im} M(k_{sp}(\omega)). \tag{50}$$

The function $\Delta_{sp}(\omega)$ is the amplitude decay rate of the surface plasmon polariton due to its roughness-induced conversion into volume electromagnetic waves in the vacuum above the metal surface and into other surface plasmon polaritons. In obtaining Eq. (48) we have neglected the small renormalization of the wavenumber $k_{sp}(\omega)$ arising from the real part of the proper self-energy $M(k_{sp}(\omega))$, but have taken into account that $\Delta_{sp}(\omega)$ can be comparable to, or larger than, $\Delta_\epsilon(\omega)$. The small renormalization of the residue $C(\omega)$ due to the surface roughness has also been neglected. In the small roughness approximation represented by Eq. (45) the decay rate $\Delta_{sp}(\omega)$ is given by (see Eqs. (36) and (50))

$$\begin{aligned}\Delta_{sp}(\omega) &= C(\omega) \operatorname{Im} \frac{1}{L_1} \int_{-\infty}^{\infty} \frac{dp}{2\pi} G(p) \langle V(k_{sp}(\omega)|p) V(p|k_{sp}(\omega)) \rangle \\ &\cong C(\omega) \delta^2 \left[\frac{|\epsilon_1(\omega)| + 1}{|\epsilon_1(\omega)|}\right]^2 \left(\frac{\omega}{c}\right)^4\end{aligned}$$

$$\times \operatorname{Im} \int_{-\infty}^{\infty} \frac{dp}{2\pi} g(|k_{sp}(\omega) - p|)C(\omega) \left[\frac{1}{p - k_{sp}(\omega) - i\Delta(\omega)} - \frac{1}{p + k_{sp}(\omega) + i\Delta(\omega)} \right]$$

$$\cong \frac{1}{2} C^2(\omega) \left[\frac{|\epsilon_1(\omega)| + 1}{|\epsilon_1(\omega)|} \right]^2 \left(\frac{\omega}{c} \right)^4 \delta^2 [g(0) + g(2k_{sp}(\omega)]. \tag{51}$$

These approximations will be useful in the interpretation of the results to be obtained in the remainder of this chapter.

Finally, we assume that the power spectrum of the surface roughness, $g(|q - k|)$, can be written in the separable form[52]

$$g(|q - k|) = \sum_{\ell=0}^{\infty} \phi_\ell(q)\phi_\ell(k), \tag{52}$$

where $\{\phi_\ell(q)\}$ are a suitably chosen set of functions. For example, if $g(|q - k|)$ has the Gaussian form

$$g(|q - k|) = \sqrt{\pi} a \exp[-a^2(q - k)^2/4], \tag{53}$$

$\phi_\ell(q)$ is given by[17]

$$\phi_\ell(q) = \left(\frac{\sqrt{\pi} a^{2\ell+1}}{2^\ell \ell!} \right)^{\frac{1}{2}} q^\ell \exp[-a^2 q^2/4]. \tag{54}$$

We now apply these results to the determination of $\hat{C}^{(1)}(q, k|q', k')$ and $\hat{C}^{(10)}(q, k|q', k')$.

14.3.1.1. The Correlation Function $\hat{C}^{(1)}(q, k|q', k')$

The difference $\langle G(q|k)G^*(q'|k')\rangle - \langle G(q|k)\rangle\langle G^*(q'|k')\rangle$ that enters the expression (32) for $\hat{C}^{(1)}(q, k|q', k')$ can be obtained from the solution of the Bethe–Salpeter equation,[50]

$$\langle G(q|k)G^*(q'|k')\rangle = \langle G(q|k)\rangle\langle G^*(q'|k')\rangle + \int_{-\infty}^{\infty} \frac{dr}{2\pi} \int_{-\infty}^{\infty} \frac{dr'}{2\pi} \int_{-\infty}^{\infty} \frac{ds}{2\pi} \int_{-\infty}^{\infty} \frac{ds'}{2\pi} \langle G(q|r)\rangle\langle G^* q'|r')\rangle\langle I^{(1)}(r, r'|s, s')\rangle \times \langle G(s|k)G^* s'|k')\rangle, \tag{55}$$

where $\langle I^{(1)}(r, r'|s, s')\rangle$ is the irreducible vertex function. The solution of Eq. (55) can be written formally as

$$\langle G(q|k)G^*(q'|k')\rangle = \langle G(q|k)\rangle\langle G^*(q'|k')\rangle + G(q)G^*(q')\langle\Gamma^{(1)}(q, q'|k, k')\rangle G(k)G^*(k'), \tag{56}$$

where $\langle \Gamma^{(1)}(q, q'|k, k')\rangle$ is the reducible vertex function. It is related to the irreducible vertex function through

$$\left\langle \Gamma^{(1)}(q, q'|k, k')\right\rangle = \langle I^{(1)}(q, q'|k, k')\rangle + \int_{-\infty}^{\infty} \frac{\mathrm{d}s}{2\pi} \int_{-\infty}^{\infty} \frac{\mathrm{d}s'}{2\pi} \left\langle I^{(1)}(q, q'|s, s')\right\rangle \times G(s)G^*(s')\left\langle \Gamma^{(1)}(s, s'|k, k')\right\rangle . \tag{57}$$

In writing Eqs. (56)–(57) we have used the fact that $\langle G(q|k)\rangle$ is diagonal in q and k, Eq. (34). On combining Eqs. (32) and (56) we find that

$$\hat{C}^{(1)}(q, k|q', k') = \frac{16}{L_1^2}\left(\frac{\omega}{c}\right)^2 [\alpha_0(q)\alpha_0(k)]|G(q)|^2|G(k)|^2 \times \left|\left\langle \Gamma^{(1)}(q, q'|k, k')\right\rangle\right|^2 |G(q')|^2|G(k')|^2[\alpha_0(q')\alpha_0(k')]. \tag{58}$$

In what follows we will focus on the determination of $\langle \Gamma^{(1)}(q, q'|k, k')\rangle$.

We will approximate $\langle I^{(1)}(q, q'|k, k')\rangle$ by the sum of the contributions from all maximally crossed diagrams in the small-roughness approximation. It is the contributions associated with these diagrams that describe the phase-coherent multiple-scattering processes that give rise to the effects we seek, namely the memory and reciprocal memory effect peaks. The definition of these diagrams and the rules for writing the contributions associated with them, in the case of scattering from one-dimensional randomly rough surfaces, are given explicitly in [53]. Due to the stationarity of the surface profile function $\zeta(x_1)$ each term in the expansion of $\langle I^{(1)}(q, q'|k, k')\rangle$ is proportional to $2\pi\,\delta(q - k - q' + k')$, so that $\langle I^{(1)}(q, q'|k, k')\rangle$ is given by

$$\begin{aligned}\left\langle I^{(1)}(q, q'|k, k')\right\rangle &= 2\pi\,\delta(q - k - q' + k') \\ &\times \Bigg\{ Wg(|q - k|) + \int_{-\infty}^{\infty} \frac{\mathrm{d}p_1}{2\pi} Wg(|q - p_1|)G(p_1)G^*(q' + k - p_1) \\ &\times Wg(|p_1 - k|) + \int_{-\infty}^{\infty} \frac{\mathrm{d}p_1}{2\pi} \int_{-\infty}^{\infty} \frac{\mathrm{d}p_2}{2\pi} Wg(|q - p_2|)G(p_2)G^*(q' + k - p_2) \\ &\times Wg(|p_2 - p_1|)G(p_1)G^*(q' + k - p_1)Wg(|p_1 - k|) + \cdots \Bigg\}\end{aligned} \tag{59}$$

where

$$W(\omega) = \delta^2 \left|\frac{1 - \epsilon(\omega)}{\epsilon(\omega)}\right|^2 \left(\frac{\omega}{c}\right)^4 . \tag{60}$$

To evaluate this sum we use Eq. (52). In numerical calculations the upper limit on the sum in Eq. (52) will be replaced by an integer N, which is increased until a convergent solution is obtained. With the use of this representation we obtain

$$\begin{aligned}\left\langle I^{(1)}(q, q'|k, k')\right\rangle &= 2\pi\delta(q - k - q' + k')\\ &\times\left\{Wg(|q-k|) + W^2\sum_{\ell=0}^{N}\sum_{\ell'=0}^{N}\phi_\ell(q)\right.\\ &\left.\times\{[\mathbf{I} - W\mathbf{K}(q'+k)]^{-1}\mathbf{K}(q'+k)\}_{\ell\ell'}\phi_{\ell'}(k)\right\},\end{aligned} \tag{61}$$

where the elements of the $(N+1)\times(N+1)$ matrix $\mathbf{K}(Q)$ are given by

$$K_{\ell\ell'}(Q) = \int_{-\infty}^{\infty}\frac{dp}{2\pi}\phi_\ell(p)G(p)G^*(Q-p)\phi_{\ell'}(p). \tag{62}$$

The result given by Eq. (61) now has to be substituted into Eq. (57), which is then solved by iteration. However in each of the integral terms in the iterative solution we keep only the contribution associated with the first term on the right-hand side of Eq. (61), and omit all contributions that contain the second term. The sum of the resulting integral terms is

$$\begin{aligned}&2\pi\delta(q-k-q'+k')\left\{\int_{-\infty}^{\infty}\frac{dp_1}{2\pi}Wg(|q-p_1|)G(p_1)G^*(q-q'-p_1)Wg(|p_1-k|)\right.\\ &+\int_{-\infty}^{\infty}\frac{dp_1}{2\pi}\int_{-\infty}^{\infty}\frac{dp_2}{2\pi}Wg(|q-p_2|)G(p_2)G^*(q'-q-p_2)\\ &\left.\times Wg(|p_2-p_1|)G(p_1)G^*(q-q'-p_1)Wg(|p_1-k|)+\cdots\right\}.\end{aligned} \tag{63}$$

This is just the sum of the contributions associated with all the ladder diagrams, starting with the two-rung ladder diagram.[53] With use of representation (52), this sum becomes

$$\begin{aligned}&2\pi\delta(q-k-q'+k')W^2\sum_{\ell=0}^{N}\sum_{\ell'=0}^{N}\\ &\times\phi_\ell(q)\{[\mathbf{I}-W\mathbf{K}(q-q')]^{-1}\mathbf{K}(q-q')\}_{\ell\ell'}\phi_{\ell'}(k).\end{aligned} \tag{64}$$

This contribution equals that of the second term on the right-hand side of Eq. (61) when $q' = -k'$. Therefore, we cannot neglect it in comparison with the latter contribution. On combining this result with the one given by Eq. (61), we finally obtain our approximation to $\langle\Gamma^{(1)}(q, q'|k, k')\rangle$:

$$\begin{aligned}\left\langle\Gamma^{(1)}(q, q'|k, k')\right\rangle &= 2\pi\delta(q-k-q'+k')\left\{Wg(|q-k|)\right.\\ &+W^2\sum_{\ell=0}^{N}\sum_{\ell'=0}^{N}\phi_\ell(q)\{[\mathbf{I}-W\mathbf{K}(q'+k)]^{-1}\mathbf{K}(q'+k)\}_{\ell\ell'}\phi_{\ell'}(k)\end{aligned}$$

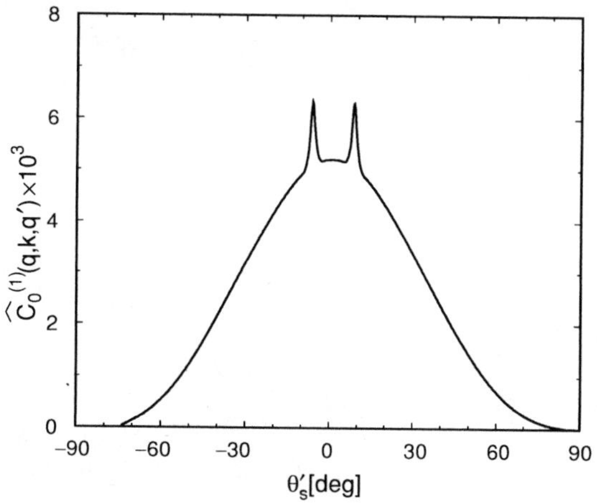

FIGURE 14.2. A plot of the envelope function $\hat{C}_0^{(1)}(q, k, q')$ as a function of θ_s' for $\theta_0 = 5.3°$ and $\theta_s = 8.6°$ when p-polarized light of wavelength $\lambda = 612.7$ nm is incident on a one-dimensional randomly rough gold surface ($\epsilon(\omega) = -9.00 + i1.29$). The power spectrum of the surface roughness has the Gaussian form (7b) with $\delta = 8$ nm and $a = 100$ nm.

$$+W^2 \sum_{\ell=0}^{N} \sum_{\ell'=0}^{N} \phi_\ell(q) \{[\mathbf{I} - W\mathbf{K}(q - q')]^{-1} \mathbf{K}(q - q')\}_{\ell\ell'} \phi_{\ell'}(k)\}. \tag{65}$$

The substitution of this result into Eq. (58) yields our approximation to $\hat{C}^{(1)}(q, k|q', k')$.

To illustrate the result obtained in this way it is convenient to represent $\hat{C}^{(1)}(q, k|q', k')$ in the form

$$\hat{C}^{(1)}(q, k|q', k') = 2\pi\delta(q - k - q' + k') \frac{1}{L_1} \hat{C}_0^{(1)}(q, k, q'), \tag{66}$$

where

$$\hat{C}_0^{(1)}(q, k, q') = \hat{C}^{(1)}(q, k|q', q' - q + k) \tag{67}$$

is the envelope of the correlation function $\hat{C}^{(1)}(q, k|q', k')$. It is independent of the length of the rough surface. It is a function of θ_s' for fixed values of θ_0 and θ_s, while θ_0' is determined from the constraint represented by the vanishing of the argument of the delta function in Eq. (66). In Fig. 14.2 we plot $\hat{C}_0^{(1)}(q, k, q')$ as a function of θ_s' for $\theta_0 = 6.3°$ and $\theta_s = 8.6°$ when p-polarized light of wavelength $\lambda = 612.7$ nm is incident on a randomly rough gold surface ($\epsilon(\omega) = -9.00 + i1.29$). The power spectrum of the surface roughness has the Gaussian form given by Eq. (7b), with $a = 100$ nm. The rms height of the surface is $\delta = 8$ nm. The envelope function $\hat{C}_0^{(1)}(q, k, q')$ displays peaks at $q' = q$ (the memory effect) and at $q' = -k$ (the reciprocal memory effect). These are multiple scattering effects arising from the third and second terms on the right-hand side of Eq. (65), respectively.

These results can be readily understood if we approximate the matrix $\mathbf{K}(Q)$, Eq. (62) by the element $K_{00}(Q)$, which can be quite a good approximation.[52] If, in addition, we use the pole approximation (48) for $G(p)$, which yields the result that

$$\begin{aligned} G(p)G^*(Q-p) &\cong \frac{C^2(\omega)}{2\mathrm{i}\Delta(\omega) - Q}\frac{2\mathrm{i}\Delta(\omega)}{(p-k_{\mathrm{sp}}(\omega))^2 + \Delta^2(\omega)} \\ &\quad + \frac{C^2(\omega)}{2\mathrm{i}\Delta(\omega) + Q}\frac{2\mathrm{i}\Delta(\omega)}{(p+k_{\mathrm{sp}}(\omega))^2 + \Delta^2(\omega)} \\ &\cong \frac{2\pi\mathrm{i}C^2}{2\mathrm{i}\Delta(\omega) - Q}\delta(p-k_{\mathrm{sp}}(\omega)) + \frac{2\pi\mathrm{i}C^2(\omega)}{2\mathrm{i}\Delta(\omega)+Q}\delta(p+k_{\mathrm{sp}}(\omega)), \end{aligned} \tag{68}$$

we find that

$$K_{00}(Q) = C^2(\omega)\phi_0^2(k_{\mathrm{sp}}(\omega))\frac{4\Delta(\omega)}{Q^2 + 4\Delta^2(\omega)}. \tag{69}$$

Since Q equals $q' + k$ in the second term on the right-hand side of Eq. (65), and it equals $q - q'$ in the third term, we see that $\langle \Gamma^{(1)}(q, q'|k, k')\rangle$, and hence $|\langle \Gamma^{(1)}(q, q'|k, k')\rangle|^2$, has peaks at $q' = -k$ and $q' = q$, as the result plotted in Fig. 14.2 shows.

The single-scattering contribution to $\hat{\mathrm{C}}_0^{(1)}(q, k, q')$, arising from the first term on the right-hand side of Eq. (65), is a structureless function of θ_s'.

The result obtained in this section indicates that an experimental determination of $\hat{\mathrm{C}}_0^{(1)}(q, k, q')$ in the vicinity of either the memory effect peak or the reciprocal memory effect peak yields essentially the same information as a measurement of the mean differential reflection coefficient in the vicinity of the enhanced backscattering peak, but without the experimental difficulties associated with placing a detector at the position of the source.[54] It is this feature of the correlation function $\hat{\mathrm{C}}^{(1)}(q, k|q', k')$ that emerged from the theoretical studies of it by Arsenieva and Feng[7] and by Freilikher and Yurkevich.[8]

14.3.1.2. The Correlation Function $\hat{C}^{(10)}(q, k|q'k')$

In exactly the same way as Eq. (55) was derived in [50], the Bethe–Salpeter equation for $\langle G(q|k)G(q'|k)\rangle$ can be derived, with the result that

$$\begin{aligned} \langle G(q|k)G(q'|k')\rangle &= \langle G(q|k)\rangle\langle G(q'|k')\rangle \\ &+ \int_{-\infty}^{\infty}\frac{\mathrm{d}r}{2\pi}\int_{-\infty}^{\infty}\frac{\mathrm{d}r'}{2\pi}\int_{-\infty}^{\infty}\frac{\mathrm{d}s}{2\pi}\int_{-\infty}^{\infty}\frac{\mathrm{d}s'}{2\pi}\langle G(q|r)\rangle\langle G(q'|r')\rangle\langle I^{(10)}(r, r'|s, s')\rangle \\ &\times\langle G(s|k)G(s'|k')\rangle. \end{aligned} \tag{70}$$

The solution of this equation can be written formally as

$$\begin{aligned} \langle G(q|k)G(q'|k')\rangle &= \langle G(q|k)\rangle\langle G(q'|k')\rangle \\ &\quad + G(q)G(q')\langle\Gamma^{(10)}(q, q'|k, k')\rangle G(k)G(k'), \end{aligned} \tag{71}$$

where the reducible vertex function $\langle \Gamma^{(10)}(q, q'|k, k')\rangle$ is related to the irreducible vertex function $\langle I^{(10)}(q, q'|k, k')\rangle$ by

$$\left\langle \Gamma^{(10)}(q, q'|k, k')\right\rangle = \left\langle I^{(10)}(q, q'|k, k')\right\rangle + \int_{-\infty}^{\infty} \frac{ds}{2\pi} \int_{-\infty}^{\infty} \frac{ds'}{2\pi} \left\langle I^{(10)}(q, q'|s, s')\right\rangle \times G(s)G(s')\left\langle \Gamma^{(10)}(s, s'|k, k')\right\rangle . \tag{72}$$

On combining Eqs. (33) and (71) we find that the correlation function $\hat{C}^{(10)}(q, k|q', k')$ is given by

$$\left\langle \hat{C}^{(10)}(q, k|q', k')\right\rangle = \frac{16}{L_1^2}\left(\frac{\omega}{c}\right)^2 [\alpha_0(q)\alpha_0(k)]|G(q)|^2|G(k)|^2 \times \left|\left\langle \Gamma^{(10)}(q, q'|k, k')\right\rangle\right|^2 |G(q')|^2|G(k')|^2[\alpha_0(q')\alpha_0(k')]. \tag{73}$$

In the present case the irreducible vertex function will also be approximated in the small roughness limit by the sum of the contributions from the maximally crossed diagrams. On evaluating these contributions in a standard manner,[53] we obtain for $\langle I^{(10)}(q, q'|k, k')\rangle$ the result

$$\begin{aligned}\left\langle I^{(10)}(q, q'|k, k')\right\rangle &= 2\pi\delta(q - k + q' - k')\Bigg\{ \tilde{W}g(|q - k|) \\ &+ \int_{-\infty}^{\infty} \frac{dp_1}{2\pi} \tilde{W}g(|q - p_1|)G(p_1)G(k - q' - p_1)\tilde{W}g(|p_1 - k|) \\ &+ \int_{-\infty}^{\infty} \frac{dp_1}{2\pi} \int_{-\infty}^{\infty} \frac{dp_2}{2\pi} \tilde{W}g(|q - p_2|)G(p_2)G(k - q' - p_2)\tilde{W}g(|p_2 - p_1|) \\ &\times G(p_1)G(k - q' - p_1)\tilde{W}g(|p_1 - k|) + \cdots \Bigg\},\end{aligned} \tag{74}$$

where, with the approximation of $V(q|k)$ given by Eq. (44),

$$\tilde{W}(\omega) = \delta^2 \left[\frac{1 - \epsilon(\omega)}{\epsilon(\omega)}\right]^2 \left(\frac{\omega}{c}\right)^4 . \tag{75}$$

We sum this series with the aid of the decomposition (52) and obtain

$$\begin{aligned}\left\langle I^{(10)}(q, q'|k, k')\right\rangle &= 2\pi\delta(q - k + q' - k')\Bigg\{ \tilde{W}g(|q - k|) \\ &+ \tilde{W}^2 \sum_{\ell=0}^{N}\sum_{\ell'=0}^{N} \phi_\ell(q)\{[\mathbf{I} - \tilde{W}\mathbf{L}(k - q')]^{-1}\mathbf{L}(k - q')\}_{\ell\ell'}\phi_{\ell'}(k)\Bigg\},\end{aligned} \tag{76}$$

where the elements of the $(N + 1) \times (N + 1)$ matrix $\mathbf{L}(Q)$ are given by

$$L_{\ell\ell'}(Q) = \int_{-\infty}^{\infty} \frac{dp}{2\pi} \phi_\ell(p)G(p)G(Q - p)\phi_{\ell'}(p). \tag{77}$$

As in the calculation of $\langle I^{(10)}(q, q'|k, k')\rangle$, when the result given by Eq. (76) is substituted into Eq. (72), and the resulting integral equation is solved by iteration, in each integral term in the resulting expansion only the contribution associated with the first term on the right-hand side of Eq. (76) is kept, and all contributions that contain the second are omitted. The sum of the resulting integral terms is given by

$$2\pi\delta(q-k+q'-k')\left\{\int_{-\infty}^{\infty}\frac{dp_1}{2\pi}\tilde{W}g(|q-p_1|)G(p_1)G(q+q'-p_1)\tilde{W}g(|p_1-k|)\right.$$

$$+\int_{-\infty}^{\infty}\frac{dp_1}{2\pi}\int_{-\infty}^{\infty}\frac{dp_2}{2\pi}\tilde{W}g(|q-p_2|)G(p_2)G(q+q'-p_2)\tilde{W}g(|p_2-p_1|)$$

$$\left.\times G(p_1)G(q+q'-p_1)\tilde{W}g(|p_1-k_1)+\cdots\right\}. \tag{78}$$

This is just the sum of the contributions associated with the ladder diagrams, starting with the two-rung ladder diagram. We sum the infinite series (78) with the aid of Eq. (52) to obtain the result that it is given by

$$2\pi\delta(q-k+q'-k')\tilde{W}^2\sum_{\ell=0}^{N}\sum_{\ell'=0}^{N}\phi_\ell(q)\{[\mathbf{I}-\tilde{W}\mathbf{L}(q+q')]^{-1}\mathbf{L}(q+q')\}_{\ell\ell'}\phi_{\ell'}(k). \tag{79}$$

Thus, our approximation to $\langle\Gamma^{(10)}(q, q'|k, k')\rangle$ is

$$\left\langle\Gamma^{(10)}(q,q'|k,k')\right\rangle = 2\pi\delta(q-k+q'-k')\left\{\tilde{W}g(|q-k|)\right.$$

$$+\tilde{W}^2\sum_{\ell=0}^{N}\sum_{\ell'=0}^{N}\phi_\ell(q)\{[\mathbf{I}-\tilde{W}\mathbf{L}(k-q')]^{-1}\mathbf{L}(k-q')\}_{\ell\ell'}\phi_{\ell'}(k)$$

$$\left.+\tilde{W}^2\sum_{\ell=0}^{N}\sum_{\ell'=0}^{N}\phi_\ell(q)\{[\mathbf{I}-\tilde{W}\mathbf{L}(q+q')]^{-1}\mathbf{L}(q+q')\}_{\ell\ell'}\phi_{\ell'}(k)\right\}. \tag{80}$$

The second and third terms on the right-hand side of this equation are equal when $q' = -k'$. Substitution of this expression for $\langle\Gamma^{(10)}(q, q'|k, k')\rangle$ into Eq. (73) yields our approximation to $\hat{C}^{(10)}(q, q'|k, k')$.

As in the case of $\hat{C}^{(1)}(q, q'|k, k')$, it is convenient to represent $\hat{C}^{(10)}(q, q'|k, k')$ in the form

$$\hat{C}^{(10)}(q,q'|k,k') = 2\pi\delta(q-k+q'-k')\frac{1}{L_1}\hat{C}_0^{(10)}(q,k,q'), \tag{81}$$

where $\hat{C}_0^{(10)}(q, k, q') = \hat{C}^{(10)}(q, k|q', q+q'-k)$ is the envelope of the correlation function $\hat{C}^{(10)}(q, q'|k, k')$. In Fig. 14.3 we plot $\hat{C}_0^{(10)}(q, k, q')$, as a function of θ_s'

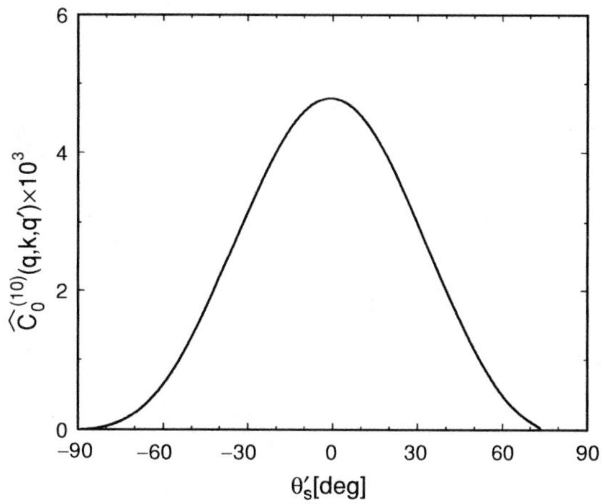

FIGURE 14.3. A plot of the envelope function $\hat{C}_0^{(10)}(q, k, q')$ as a function of θ_s' for $\theta_0 = 6.3°$ and $\theta_s = 8.6°$ when p-polarized light of wavelength $\lambda = 612.7$ nm is incident on a one-dimensional randomly rough gold surface ($\epsilon(\omega) = -9.00 + i1.29$). The power spectrum of the surface roughness has the Gaussian form (7b) with $\delta = 8$ nm and $a = 100$ nm.

for $\theta_0 = 6.3°$ and $\theta_s = 8.6°$, when p-polarized light of wavelength $\lambda = 612.7$ nm is incident on the same randomly rough gold surface that was assumed in obtaining the plot presented in Fig. 14.2. For the parameters of the scattering system assumed, $\hat{C}_0^{(10)}(q, k, q')$ is a structureless function of θ_s', and is of the same order of magnitude as $\hat{C}_0^{(1)}(q, k, q')$.

14.3.1.3. The Transition from Complex Gaussian to Circular Complex Gaussian Statistics

We have already noted, in Sect. 14.2, that when the surface profile function is assumed to be a stationary random process, and the random surface is assumed to be infinitely long, conventional speckle theory[40,41] assumes that the scattering matrix becomes the sum of a very large number of independent contributions from different points on the surface. The central limit theorem[55] then yields the result that $S(q|k)$ obeys complex Gaussian statistics. In this case the correlation function $\hat{C}(q, k|q', k')$ is rigorously given by $\hat{C}(q, k|q', k') = \hat{C}^{(1)}(q, k|q', k') + \hat{C}^{(10)}(q, k|q', k')$. As the strength of the surface roughness increases, it is assumed in conventional speckle theory that $S(q|k)$ obeys circular complex Gaussian statistics. In this case $\hat{C}(q, k|q', k')$ becomes rigorously $\hat{C}(q, k|q', k') = \hat{C}^{(1)}(q, k|q', k')$, i.e. $\hat{C}^{(10)}(q, k|q', k')$ vanishes. It is of interest to examine how rough the random surface has to be in order that $\hat{C}^{(10)}(q, k|q', k')$ is negligible. We can make an estimate of this degree of roughness in the following way.[20]

We consider the scattering of an s-polarized plane wave of frequency ω from a randomly rough, infinitely long, surface defined by $x_3 = \zeta(x_1)$. The region $x_3 > \zeta(x_1)$ is vacuum, while the region $x_3 < \zeta(x_1)$ is a perfect conductor.

The surface profile function $\zeta(x_1)$ is assumed to be a single valued function of x_1 that is differentiable, and that constitutes a stationary, zero-mean, Gaussian random process defined by Eqs. (1)–(2). We further assume that the surface height autocorrelation function $W(|x_1|)$ has the Gaussian form (7a).

The single nonzero component of the electric field in the vacuum region $x_3 > \zeta(x_1)_{\max}$ is the sum of an incoming incident field and a superposition of outgoing scattered waves

$$E_2(x_1, x_3|\omega) = \exp[ikx_1 - i\alpha_0(k)x_3] + \int_{-\infty}^{\infty} \frac{dq}{2\pi} R(q|k) \exp[iqx_1 + i\alpha_0(q)x_3]. \tag{82}$$

A reciprocal phase perturbation theory for the scattering matrix $S(q|k) = [\alpha_0(q)/\alpha_0(k)]^{\frac{1}{2}} R(q|k)$ was constructed in [56], with the result that

$$S(q|k) = \int_{-\infty}^{\infty} dx_1 \exp[-i(q-k)x_1] \exp[-2i\sqrt{\alpha_0(q)\alpha_0(k)}\zeta(x_1)], \tag{83}$$

when only the term linear in $\zeta(x_1)$ is kept in the exponent in the integrand. The average $\langle S(q|k)\rangle$ is given by

$$\langle S(q|k)\rangle = 2\pi\delta(q-k)\exp[-2\delta^2\alpha_0(q)\alpha_0(k)]. \tag{84}$$

The fluctuation $\delta S(q|k)$ can therefore be written as

$$\delta S(q|k) = \int_{-\infty}^{\infty} dx_1 \exp[-i(q-k)x_1] \times \{\exp[-2i\sqrt{\alpha_0(q)\alpha_0(k)}\zeta(x_1)] - \exp[-2\alpha_0(q)\alpha_0(k)\delta^2]\}. \tag{85}$$

By the use of this expression for $\delta S(q|k)$ we obtain for the correlators $\langle \delta S(q|k)\delta S^*(q'|k')\rangle$ and $\langle \delta S(q|k)\delta S(q'|k')\rangle$

$$\begin{aligned}\langle \delta S(q|k)\delta S^*(q'|k')\rangle &= 2\pi\delta(q-k-q'+k') \\ &\times \exp[-2(\alpha_0(q)\alpha_0(k) + \alpha_0(q')\alpha_0(k'))\delta^2] \\ &\times \int_{-\infty}^{\infty} du\{\exp[\delta^2\sqrt{\alpha_0(q)\alpha_0(q')\alpha_0(k)\alpha_0(k')}W(|u|)] - 1\} \\ &\times \exp[i(q'-k')u]\end{aligned} \tag{86}$$

$$\begin{aligned}\langle \delta S(q|k)\delta S(q'|k')\rangle &= 2\pi\delta(q-k+q'-k') \\ &\times \exp[-2(\alpha_0(q)\alpha_0(k) + \alpha_0(q')\alpha_0(k'))\delta^2] \\ &\times \int_{-\infty}^{\infty} du\{\exp[-\delta^2\sqrt{\alpha_0(q)\alpha_0(q')\alpha_0(k)\alpha_0(k')}W(|u|)] - 1\} \\ &\times \exp[-i(q'-k')u].\end{aligned} \tag{87}$$

It is seen from Eqs. (86) and (87) that in contrast to $\langle \delta S(q|k)\delta S^*(q'|k')\rangle$ the average $\langle \delta S(q|k)\delta S(q'|k')\rangle$ vanishes with increasing rms height δ for a fixed value of the

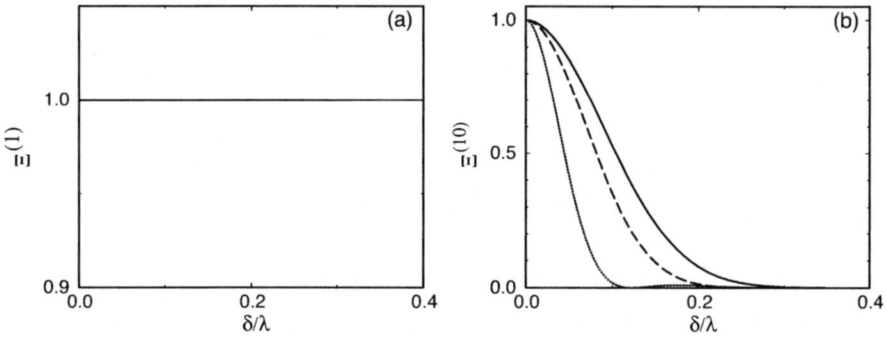

FIGURE 14.4. The normalized correlation functions $\Xi^{(1)}(a)$ and $\Xi^{(10)}(b)$, as functions of δ/λ for values of the transverse correlation length $a = 300$ nm, 500 nm, and 800 nm. The incident light is s-polarized and of wavelength $\lambda = 632.8$ nm. The scattering medium is a perfect conductor. Furthermore, $\theta_0 = 30°$, $\theta_s = 0°$, and $\theta'_s = 0°$. In (a) the results for the different correlation lengths could not be distinguished.[20]

transverse correlation length a, due to the negative exponential under the integral sign in the last line of Eq. (87).

In Fig. 14.4 we present plots of the normalized correlation function

$$\Xi^{(1)}(q, k|q'k') = |\langle \delta S(q|k)\delta S^*(q'|k')\rangle|^2/\langle \delta S(q|k)\delta S^*(q|k)\rangle\langle \delta S(q'|k')\delta S^*(q'|k')\rangle$$

and

$$\Xi^{(10)}(q, k|q', k') = |\langle \delta S(q|k)\delta S(q'|k')\rangle|^2/\langle \delta S(q|k)\delta S^*(q|k)\rangle\langle \delta S(q'|k')\delta S^*(q'|k')\rangle$$

as functions of δ/λ for several values of a. From the plots presented in Fig. 14.4 we see that $\Sigma^{(10)}(q, k|q'k')$ vanishes even for quite moderately weakly rough surfaces ($\delta/\lambda \cong 0.1$, $a = 800$ nm) for which $\Sigma^{(1)}(q, k|q'k')$ is close to unity.

In Fig. 14.5 we plot the envelopes of the correlation functions $\hat{C}^{(1)}(q, k|q', k')$ and $\hat{C}^{(10)}(q, k|q', k')$ as functions of δ/λ for several values of a. We see that the envelope of $\hat{C}^{(1)}(q, k|q', k')$ decreases with increasing δ/λ, as does the envelope of $\hat{C}^{(10)}(q, k|q', k')$, although the former function is still significantly larger than the latter as δ/λ increases.

In calculating the results plotted in Figs. 14.4 and 14.5, the value of q' was chosen to produce the same values of $\hat{\mathrm{C}}_0^{(1)}(q, k, q')$ and $\hat{\mathrm{C}}_0^{(10)}(q, k, q')$ in the limit of a weakly rough surface.

14.3.2. Correlations in Film Systems

Up to now in this chapter we have considered only angular intensity correlation functions arising in the scattering of (p-polarized) light from a one-dimensional randomly rough surface bounding a semi-infinite metal. Such a system supports only a single surface electromagnetic mode, and the peaks arising in $\hat{\mathrm{C}}_0^{(1)}(q, k, q')$ are associated with the excitation of this mode.

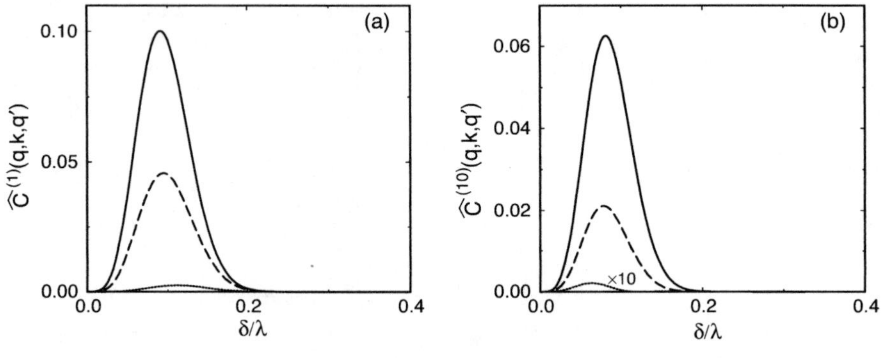

FIGURE 14.5. The envelopes $\hat{C}_0^{(1)}(a)$ and $\hat{C}_0^{(10)}(b)$ as functions of δ/λ for values of the transverse correlation length $a = 300$ nm, 500 nm, and 800 nm. The incident light is s polarized and of wavelength $\lambda = 632.8$ nm. The scattering medium is a perfect conductor. Furthermore, $\theta_0 = 30°$, $\theta_s = 0°$, and $\theta_s' = 0°$.[20]

However, structures exist that support two or more surface or guided modes. These include a metal film on a dielectric substrate, or a dielectric film on a metallic substrate. When one of the interfaces in such a structure is randomly rough, the existence of more than one mode of this type has been shown to give rise to satellite peaks bracketing the enhanced backscattering peak in the angular dependence of the intensity of light scattered diffusely from it.[57] It may be expected, therefore, that the angular intensity correlation function of the light scattered from one of these film structures will also display a richer structure than is present in this function in scattering from a system that supports only a single surface or guided wave. The possibility that this should be the case was first noted by Freilikher *et al.*[57] in the context of the $\hat{C}^{(1)}$ correlation function. In subsequent theoretical studies the angular intensity correlations of light scattered from a randomly rough dielectric film on a perfectly conducting substrate,[14,15] or reflected from or transmitted through a randomly rough, free-standing, thin metal film[16] were calculated. The resulting correlation functions indeed were found to have a richer structure than the correlation functions calculated for the light scattered from the randomly rough surface of a semi-infinite metal.

To illustrate the nature of the angular intensity correlation functions of light scattered from a film system that supports two or more guided modes, in this section we study the scattering of s-polarized light of frequency ω, whose plane of incidence is the x_1x_3 plane, from a system that consists of vacuum in the region $x_3 > \zeta(x_1)$, a dielectric film characterized by a complex dielectric constant ϵ_d in the region $-d < x_3 < \zeta(x_1)$, and a perfect conductor in the region $x_3 < -d$. The surface profile function $\zeta(x_1)$ is a single-valued function of x_1 that is differentiable, and constitutes a stationary, zero-mean, Gaussian random process defined by Eqs. (1)–(2). The random roughness of the surface is characterized by the Gaussian power spectrum (7b).

The matrices $M(p|q)$ and $N(p|q)$ entering Eq. (21) are found to be[58]

$$M(p|q) = \frac{e^{i\alpha_d(p)d}}{\alpha_0(q)+\alpha_d(p)} I(\alpha_0(q)+\alpha_d(p)|p-q) - \frac{e^{-i\alpha_d(p)d}}{\alpha_0(q)-\alpha_d(p)} I(\alpha_0(q)-\alpha_d(p)|p-q) \tag{88a}$$

$$N(p|q) = \frac{e^{i\alpha_d(p)d}}{\alpha_0(q)-\alpha_d(p)} I(\alpha_d(p)-\alpha_0(q)|p-q) - \frac{e^{-i\alpha_d(p)d}}{\alpha_0(q)+\alpha_d(p)} I(-\alpha_0(q)-\alpha_d(p)|p-q), \tag{88b}$$

where $\alpha_0(q)$ has been defined in Eq. (10), $\alpha_d(q) = [\epsilon_d(\omega^2/c^2) - q^2]^{\frac{1}{2}}$, with $\mathrm{Re}\alpha_d(q) > 0$, $\mathrm{Im}\alpha_d(q) > 0$, and

$$I(\gamma|Q) = \int_{-\infty}^{\infty} dx_1\, e^{-iQx_1} e^{i\gamma\zeta(x_1)}. \tag{89}$$

The reflection coefficient $R_0(k)$ is given by

$$R_0(k) = \frac{-i\alpha_d(k)\cos\alpha_d(k)d + \alpha_0(k)\sin\alpha_d(k)d}{i\alpha_d(k)\cos\alpha_d(k)d + \alpha_0(k)\sin\alpha_d(k)d}, \tag{90}$$

while the Green's function $G_0(k)$ is

$$G_0(k) = \frac{i\sin\alpha_d(k)d}{i\alpha_d(k)\cos\alpha_d(k)d + \alpha_0(k)\sin\alpha_d(k)d}. \tag{91}$$

Both $R_0(k)$ and $G_0(k)$ have simple poles at the wavenumbers $k = \pm k_1(\omega)$, $\pm k_2(\omega), \ldots, \pm k_N(\omega)$ of the guided waves supported by the structure in the absence of the surface roughness. For a given value of ϵ_d the number of these modes depends on the frequency ω and the thickness of the film. In the small roughness approximation the scattering potential $V(q|k)$ is given by

$$V(q|k) = (\epsilon_d - 1)\frac{\omega^2}{c^2}\hat{\zeta}(q-k). \tag{92}$$

The correlation function $\hat{C}^{(1)}(q,k|q',k')$ is still given by Eq. (58), where the reducible vertex function $\langle\Gamma^{(1)}(q,q'|k,k')\rangle$ is given formally by Eq. (65), with W now defined as

$$W = \delta^2|\epsilon_d - 1|^2\left(\frac{\omega^2}{c^2}\right)^4. \tag{93}$$

To make explicit the dependence of $\hat{C}^{(1)}(q,k|q',k')$ on its arguments we approximate the matrix $\mathbf{K}(Q)$ defined by Eq. (62) by the element $K_{00}(Q)$, which we will denote by $K(Q)$. The reducible vertex function $\langle\Gamma^{(1)}(q,q'|k,k')\rangle$ in this approximation becomes

$$\langle\Gamma^{(1)}(q,q'|k,k')\rangle = 2\pi\delta(q-k-q'+k')\Big\{Wg(|q-k|)$$
$$+W^2\phi_0(q)\left[\frac{K(q'+k)}{1-WK(q'+k)}+\frac{K(q-q')}{1-WK(q-q')}\right]\phi_0(k)\Big\}, \tag{94}$$

where

$$K(Q)=\int_{-\infty}^{\infty}\frac{\mathrm{d}p}{2\pi}\phi_0^2(p)G(p)G^*(Q-p). \tag{95}$$

The Green's function $G(p)$, Eq. (35), has simple poles at the wave numbers of the guided waves supported by the scattering structure, in the presence of the surface roughness, at the frequency of the incident light. We exploit this circumstance by making a pole approximation to $G(p)$ of the form[58]

$$G(p)=\sum_m \frac{C_m(\omega)}{p-p_m(\omega)-\mathrm{i}\Delta_m(\omega)}. \tag{96}$$

We will assume that the film supports two guided modes. The summation index n takes the values $-2,-1,1,2$ and $C_{-n}(\omega)=-C_n(\omega)$, $p_{-n}(\omega)=-p_n(\omega)$, and $\Delta_{-n}(\omega)=-\Delta_n(\omega)$. Here $p_1(\omega)$ and $p_2(\omega)$ are the wavenumbers of the two guided waves, while $\Delta_1(\omega)$ and $\Delta_2(\omega)$ are their decay rates, due to their roughness-induced conversion into volume electromagnetic waves and into each other. With the use of Eq. (96) we find that

$$G(p)G^*(Q-p)=G(p)G^*(p-Q)$$
$$=2\pi\mathrm{i}C_1^2\left\{\frac{\delta(p-q_1)}{2\mathrm{i}\Delta_1-Q}+\frac{\delta(p+q_1)}{2\mathrm{i}\Delta_1+Q}\right\}+2\pi\mathrm{i}C_2^2\left\{\frac{\delta(p-q_2)}{2\mathrm{i}\Delta_2-Q}+\frac{\delta(p+q_2)}{2\mathrm{i}\Delta_2+Q}\right\}$$
$$+2\pi\mathrm{i}C_1C_2\Big\{\frac{\delta(p-q_1)}{\mathrm{i}(\Delta_1+\Delta_2)+(q_1-q_2)-Q}+\frac{\delta(p+q_1)}{\mathrm{i}(\Delta_1+\Delta_2)+(q_1-q_2)+Q}$$
$$+\frac{\delta(p-q_2)}{\mathrm{i}(\Delta_1+\Delta_2)-(q_1-q_2)-Q}+\frac{\delta(p+q_2)}{\mathrm{i}(\Delta_1+\Delta_2)-(q_1-q_2)+Q}\Big\}, \tag{97}$$

where only the terms that are large for small Q have been kept, and the approximation

$$\frac{\Delta}{(p-q)^2+\Delta^2}\cong\pi\delta(p-q) \tag{98}$$

is used. It follows that the function $K(Q)$, Eq. (95), becomes

$$K(Q)=\sqrt{\pi}aW^2\Big\{C_1^2\,\mathrm{e}^{-\frac{a^2}{2}q_1^2}\frac{4\Delta_1}{Q^2+4\Delta_1^2}+C_2^2\,\mathrm{e}^{-\frac{a^2}{2}q_2^2}\frac{4\Delta_2}{Q^2+4\Delta_2^2}$$
$$+C_1C_2\left[\frac{\Delta_1+\Delta_2}{(q_1-q_2-Q)^2+(\Delta_1+\Delta_2)^2}+\frac{\Delta_1+\Delta_2}{(q_1-q_2+Q)^2+(\Delta_1+\Delta_2)^2}\right]$$
$$\times\left(\mathrm{e}^{-\frac{a^2}{2}q_1^2}+\mathrm{e}^{-\frac{a^2}{2}q_2^2}\right)\Big\}, \tag{99}$$

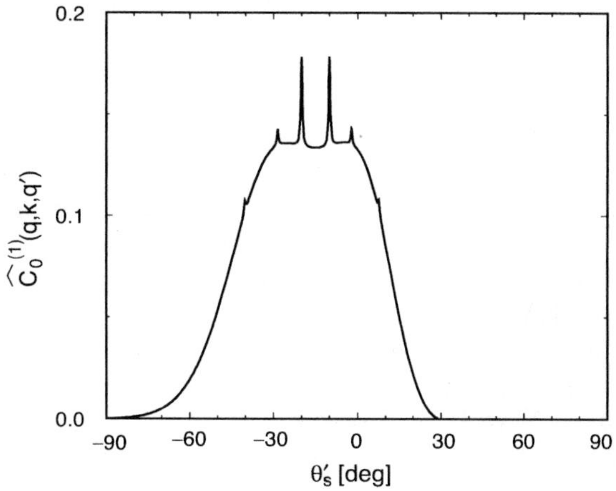

FIGURE 14.6. A plot of the envelope function $\hat{C}_0^{(1)}(q, k, q')$ as a function of θ_s' for $\theta_0 = 20^\circ$ and $\theta_s = -10^\circ$, when s-polarized light of wavelength $\lambda = 632.8$ nm is incident on a one-dimensional randomly rough surface of a dielectric film whose dielectric constant is $\epsilon_d = 2.6896 + i0.0075$. The film is deposited on the planar surface of a perfect conductor. Its mean thickness is $d = 500$ nm. The power spectrum of the surface roughness has the Gaussian form (7b) with $\delta = 15$ nm and $a = 100$ nm.

where we have used Eq. (54) and again have dropped all terms that are small when Q is small.

When the result given by Eq. (99) is substituted into Eq. (94), we find from the first two terms on the right-hand side of the former equation that the reducible vertex function $\langle \Gamma^{(1)}(q, q'|k, k')\rangle$ is large when $q' + k \cong 0$ and when $q - q' \cong 0$. This is the origin of the reciprocal memory effect peak, and the memory effect peak, respectively. However, we see from the third and fourth terms on the right-hand side of Eq. (99) that $\langle \Gamma^{(1)}(q, q'|k, k')\rangle$ is also large when $q' + k = \pm(q_1 - q_2)$ and when $q - q' = \pm(q_1 - q_2)$. Thus, we see that the correlation function $\hat{C}^{(1)}(q, k|q', k')$ acquires additional peaks whose angular positions depend on differences of the wave numbers of the guided or surface waves. That is, there are now additional memory effect peaks, and additional reciprocal memory effect peaks. Such peaks, of course, have no counterparts in the scattering of light from a semi-infinite metal.

In Fig. 14.6 we present a plot of the envelope function $\hat{C}_0^{(1)}(q, k, q')$ as a function of θ_s' for fixed values of θ_0 and θ_s, for s-polarized light scattered from the film system considered in this section. The wavelength of the incident light is $\lambda = 632.8$ nm, the dielectric constant of the film is $\epsilon_d = 2.6896 + i0.0075$, and the mean thickness of the film is $d = 500$ nm. The transverse correlation length of the surface roughness is $a = 100$ nm, while the rms height of the surface is $\delta = 15$ nm. The angle θ_0 is $\theta_0 = 20^\circ$, while $\theta_s = -10^\circ$. It is seen that in addition to the memory and reciprocal memory effect peaks $\hat{C}_0^{(1)}(q, k, q')$ displays the additional memory and reciprocal memory effect peaks predicted by Eq. (99).

14.4. Frequency Correlation Functions

In the discussion of the correlation function $C(q, k|q', k')$ in the preceding sections of this chapter, it was assumed that the incident and scattered field characterized by the wave numbers (k, q) and (k', q') possessed the same frequency ω. It is of some interest to relax this assumption, and to examine the situation where the fields characterized by (k, q) have the frequency ω, while the fields characterized by (k', q') have the frequency ω'. It might be thought that the intensities entering the correlation function will simply decorrelate when ω' is unequal to ω. However, we will see that in fact the angular intensity correlation function acquires a richer structure as a function of the scattering angle θ_s' for $\omega' \neq \omega$ than it possesses when $\omega' = \omega$. We examine this case in the present section.

Thus, the correlation function we consider is denoted by $C(q, k; \omega|q', k'; \omega')$, which is defined by

$$C(q, k; \omega|q', k'; \omega') = \langle I(q, k|\omega) I(q', k'|\omega')\rangle - \langle I(q, k|\omega)\rangle\langle I(q', k'|\omega')\rangle, \quad (100)$$

where the intensity $I(q, k|\omega)$ is

$$I(q, k|\omega) = \frac{1}{L_1}\frac{\omega}{c}|S(q, k|\omega)|^2, \quad (101)$$

and

$$k = (\omega/c)\sin\theta_0 \qquad q = (\omega/c)\sin\theta_s \quad (102a)$$

$$k' = (\omega'/c)\sin\theta_0' \qquad q' = (\omega'/c)\sin\theta_s', \quad (102b)$$

in terms of angles of incidence θ_0 and θ_0', and angles of scattering θ_s and θ_s'.

As before, we eliminate uninteresting specular terms by introducing the fluctuation $\delta S(q, k|\omega) = S(q, k|\omega) - \langle S(q, k|\omega)\rangle$ and working with the modified correlation function

$$\begin{aligned}\hat{C}(q, k; \omega|q', k'; \omega') = {} & \frac{1}{L_1^2}\frac{\omega}{c}\frac{\omega'}{c} \\ & \times\{\langle \delta S(q, k|\omega)\delta S^*(q, k|\omega)\delta S(q', k'|\omega')\delta S^*(q', k'|\omega')\rangle \\ & -\langle \delta S(q, k|\omega)\delta S^*(q, k|\omega)\rangle\langle \delta S(q', k'|\omega')\delta S^*(q', k'|\omega')\rangle\}. \quad (103)\end{aligned}$$

In what follows we will restrict our attention to the correlation function $\hat{C}^{(1)}(q, k; \omega|q', k'; \omega')$ since it is this contribution to $\hat{C}(q, k; \omega|q', k'; \omega')$ that displays the most dramatic consequences of including multiple-scattering processes in the calculation of the former correlation function. It is given by

$$\hat{C}^{(1)}(q, k; \omega|q', k'; \omega') = \frac{1}{L_1^2}\frac{\omega}{c}\frac{\omega'}{c}|\langle \delta S(q, k|\omega)\delta S^*(q', k'|\omega')\rangle|^2. \quad (104)$$

With the use of Eq. (31) this expression becomes

$$\hat{C}^{(1)}(q,k;\omega|q',k';\omega') = \frac{16}{L_1^2}\frac{\omega\omega'}{c^2}[\alpha_0(q,\omega)\alpha_0(k,\omega)\alpha_0(q',\omega')\alpha_0(k',\omega')] \times|[\langle G_\omega(q|k)G^*_{\omega'}(q'|k')\rangle - \langle G_\omega(q|k)\rangle\langle G^*_{\omega'}(q'|k')\rangle]|^2, \tag{105}$$

where $\alpha_0(q,\omega)$ is defined by Eqs. (10), and the Green's function $G_\omega(q|k)$ is again the solution of Eq. (28a), but now with its dependence on the frequency ω indicated explicitly.

The two-particle Green's function $\langle G_\omega(q|k)G^*_{\omega'}(q'|k')\rangle$ satisfies the Bethe–Salpeter equation

$$\begin{aligned}\langle G_\omega(q|k)G^*_{\omega'}(q'|k')\rangle &= \langle G_\omega(q|k)\rangle\langle G^*_{\omega'}(q'|k')\rangle \\ &+ \int_{-\infty}^{\infty}\frac{dr}{2\pi}\int_{-\infty}^{\infty}\frac{dr'}{2\pi}\int_{-\infty}^{\infty}\frac{ds}{2\pi}\int_{-\infty}^{\infty}\frac{ds'}{2\pi}\langle G_\omega(q|r)\rangle\langle G^*_{\omega'}(q'|r')\rangle\left\langle I^{(1)}_{\omega\omega'}(r,r'|s,s')\right\rangle \\ &\times\langle G_\omega(s|k)G^*_{\omega'}(s'|k')\rangle \\ &= 2\pi\delta(q-k)G_\omega(k)2\pi\delta(q'-k')G^*_{\omega'}(k') \\ &+ G_\omega(q)G^*_{\omega'}(q')\int_{-\infty}^{\infty}\frac{ds}{2\pi}\int_{-\infty}^{\infty}\frac{ds'}{2\pi}\left\langle I^{(1)}_{\omega\omega'}(q,q'|s,s')\right\rangle\left\langle G_\omega(s|k)G^*_{\omega'}(s'|k')\right\rangle.\end{aligned} \tag{106}$$

The solution of Eq. (106) can be written in terms of the reducible vertex function $\langle\Gamma^{(1)}_{\omega\omega'}(q,q'|k,k')\rangle$ as

$$\langle G_\omega(q|k)G^*_{\omega'}(q'|k')\rangle = \langle G_\omega(q|k)\rangle\langle G^*_{\omega'}(q'|k')\rangle + G_\omega(q)G^*_{\omega'}(q')\left\langle\Gamma^{(1)}_{\omega\omega'}(q,q'|k,k')\right\rangle G_\omega(k)G^*_{\omega'}(k'), \tag{107}$$

where $\langle\Gamma^{(1)}_{\omega\omega'}(q,q'|k,k')\rangle$ is the solution of

$$\left\langle\Gamma^{(1)}_{\omega\omega'}(q,q'|k,k')\right\rangle = \left\langle I^{(1)}_{\omega\omega'}(q,q'|k,k')\right\rangle + \int_{-\infty}^{\infty}\frac{ds}{2\pi}\int_{-\infty}^{\infty}\frac{ds'}{2\pi}\left\langle I^{(1)}_{\omega\omega'}(q,q'|s,s')\right\rangle G_\omega(s)G^*_{\omega'}(s')\left\langle\Gamma^{(1)}_{\omega\omega'}(s,s'|k,k')\right\rangle. \tag{108}$$

On combining Eqs. (105) and (107) we find that the correlation function $\hat{C}^{(1)}(q,k;\omega|q',k';\omega)$ is given by

$$\hat{C}^{(1)}(q,k;\omega|q',k';\omega') = \frac{16}{L_1^2}\frac{\omega\omega'}{c^2}[\alpha_0(q,\omega)\alpha_0(k,\omega)]|G_\omega(q)|^2|G_\omega(k)|^2 \times|\left\langle\Gamma^{(1)}_{\omega\omega'}(q,q'|k,k')\right\rangle|^2|G_{\omega'}(q')|^2|G_{\omega'}(k')|^2[\alpha_0(q',\omega')\alpha_0(k',\omega')]. \tag{109}$$

We again approximate the irreducible vertex function by the sum of the contributions from all maximally crossed diagrams. In evaluating this sum we approximate the scattering potential $V(q|k)$ by the expression given by Eq. (45). As a result, we obtain

$$\left\langle I^{(1)}_{\omega\omega'}(q,q'|k,k')\right\rangle = 2\pi\delta(q-k-q'+k')\Bigg\{W(\omega,\omega')g(|k-q|)$$

$$+W^2(\omega,\omega')\int_{-\infty}^{\infty}\frac{dp_1}{2\pi}g(|k-p_1|)G_\omega(p_1)G^*_{\omega'}(q'+k-p_1)g(|p_1-q|)$$

$$+W^3(\omega,\omega')\int_{-\infty}^{\infty}\frac{dp_1}{2\pi}\int_{-\infty}^{\infty}\frac{dp_2}{2\pi}g(|k-p_1|)G_\omega(p_1)G^*_{\omega'}(q'+k-p_1)g(|p_1-p_2|)$$

$$\times G_\omega(p_2)G^*_{\omega'}(q'+k-p_2)g(|p_2-q|)+\cdots\Bigg\}, \tag{110}$$

where

$$W(\omega,\omega') = \delta^2\left[\frac{1-\epsilon(\omega)}{\epsilon(\omega)}\right]\left[\frac{1-\epsilon(\omega')}{\epsilon(\omega')}\right]^*\frac{\omega^2\omega'^2}{c^4}. \tag{111}$$

To sum the series (110) we use representation (52) for the power spectrum of the surface roughness. The result can be written in the form

$$\left\langle I^{(1)}_{\omega\omega'}(q,q'|k,k')\right\rangle = 2\pi\delta(q-k-q'+k')\{W(\omega,\omega')g(|q-k|)$$

$$+W^2(\omega,\omega')\sum_{\ell=0}^{N}\sum_{\ell'=0}^{N}\phi_\ell(q)\{\mathbf{I}-W(\omega,\omega')\mathbf{K}(\omega,\omega'|q'+k)]^{-1}$$

$$\times\mathbf{K}(\omega,\omega'|q'+k)\}_{\ell\ell'}\phi_{\ell'}(k), \tag{112}$$

where

$$K_{\ell\ell'}(\omega,\omega'|Q) = \int_{-\infty}^{\infty}\frac{dp}{2\pi}\phi_\ell(p)G_\omega(p)G^*_{\omega'}(Q-p)\phi_{\ell'}(p). \tag{113}$$

When we substitute the result given by Eq. (112) into Eq. (108) and solve the resulting equation by iteration, where only the first term on the right-hand side of Eq. (112) is kept in the integral terms, we find that the reducible vertex function becomes

$$\left\langle \Gamma^{(1)}_{\omega\omega'}(q,q'|k,k')\right\rangle = \left\langle I^{(1)}_{\omega\omega'}(q,q'|k,k')\right\rangle + 2\pi\delta(q-k-q'+k')$$

$$\Bigg\{W^2(\omega,\omega')\int_{-\infty}^{\infty}\frac{ds_1}{2\pi}g(|q-s_1|)G_\omega(s_1)G^*_{\omega'}(q'-q+s_1)g(|s_1-k|)$$

$$+W^3(\omega,\omega')\int_{-\infty}^{\infty}\frac{ds_1}{2\pi}\int_{-\infty}^{\infty}\frac{ds_2}{2\pi}g(|q-s_1|)G_\omega(s_1)G^*_{\omega'}(q'-q+s_1)g(|s_1-s_2|)$$

$$\times G_\omega(s_2)G^*_{\omega'}(q'-q+s_2)g(|s_2-k|)+\cdots\Bigg\}. \tag{114}$$

The series in braces can be summed with the use of Eq. (52), with the result that

$$\begin{aligned}\left\langle\Gamma^{(1)}_{\omega\omega'}(q,q'|k,k')\right\rangle &= 2\pi\delta(q-k-q'+k')\{W(\omega,\omega')g(|q-k|)\\ &+W^2(\omega,\omega')\sum_{\ell=0}^{N}\sum_{\ell'=0}^{N}\phi_\ell(q)\{[\mathbf{I}-W(\omega,\omega')\mathbf{K}(\omega,\omega'|q'+k)]^{-1}\\ &\times\mathbf{K}(\omega,\omega'|q'+k)\}_{\ell\ell'}\phi_{\ell'}(k)\\ &+W^2(\omega,\omega')\sum_{\ell=0}^{N}\sum_{\ell'=0}^{N}\phi_\ell(q)\{[\mathbf{I}-W(\omega,\omega')\mathbf{K}(\omega,\omega'|q-q')]^{-1}\\ &\times\mathbf{K}(\omega,\omega'|q-q')\}_{\ell\ell'}\phi_{\ell'}(k). \end{aligned}\tag{115}$$

Substitution of this expression for $\langle\Gamma^{(1)}_{\omega\omega'}(q,q'|k,k')\rangle$ into Eq. (109) yields our approximation to $\hat{C}^{(1)}(q,k;\omega|q';k';\omega')$.

It is convenient to represent $\hat{C}^{(1)}(q,k;\omega|q',k';\omega)$ in the form

$$\hat{C}^{(1)}(q,k;\omega|q',k';\omega') = 2\pi\delta(q-k-q'+k')\frac{1}{L_1}\hat{C}^{(1)}_0(q,k;\omega|q';\omega'), \tag{116}$$

where

$$\hat{\mathrm{C}}^{(1)}_0(q,k;\omega|q';\omega') = \hat{\mathrm{C}}^{(1)}(q,k;\omega|q',q'-q+k;\omega'). \tag{117}$$

In Fig. 14.7 we plot $\hat{C}^{(1)}_0(q,k;\omega|q';\omega')$ for a silver surface whose roughness was characterized by the Gaussian power spectrum (7b) with the roughness parameters $\delta = 5$ nm and $a = 100$ nm. The incident light was p-polarized. The angles θ_0 and θ_s were fixed at $\theta_0 = 10^\circ$ and $\theta_s = -5^\circ$. The frequency ω corresponded to a vacuum wavelength of light $\lambda = 612.7$ nm. The envelope function is plotted as a function of θ_s' and $\delta\lambda = \lambda - \lambda' = (2\pi c/\omega) - (2\pi c/\omega')$. The values of the dielectric function of silver as a function of wavelength were obtained from the experimental results of Johnson and Christy.[59] When $\lambda' = \lambda$, $\hat{C}^{(1)}_0$ displays two peaks as a function of θ_s'. The one corresponding to $q' = q$, or $\theta_s' = \theta_s$, is the memory effect peak. The peak corresponding to $q' = -k$, or $\theta_s' = -\theta_0$, is the reciprocal memory effect peak. As $\delta\lambda$ increases from zero, each of these two peaks splits into two peaks, so that $\hat{C}^{(1)}_0$ now displays four peaks. As $\delta\lambda$ increases further, the two outer peaks shift farther apart, while the two inner peaks approach each other. At a critical value of $\delta\lambda$ the two inner peaks fuse, and $\hat{C}^{(1)}_0$ displays only three peaks. With a further increase of $\delta\lambda$, the central peak splits into two peaks, and $\hat{C}^{(1)}_0$ again displays four peaks.

This behavior can be understood in the following way. If we use the pole approximation (48) for the Green's function $G_\omega(k)$, we find that

$$\begin{aligned}G_\omega(p)G^*_{\omega'}(Q-p) &\simeq -\frac{C(\omega)C(\omega')}{Q+k_{\mathrm{sp}}(\omega')-k_{\mathrm{sp}}(\omega)-\mathrm{i}(\Delta(\omega)+\Delta(\omega'))}\\ &\times\left\{\frac{1}{p-k_{\mathrm{sp}}(\omega)-\mathrm{i}\Delta(\omega)}+\frac{1}{Q-p+k_{\mathrm{sp}}(\omega')-\mathrm{i}\Delta(\omega')}\right\}\end{aligned}$$

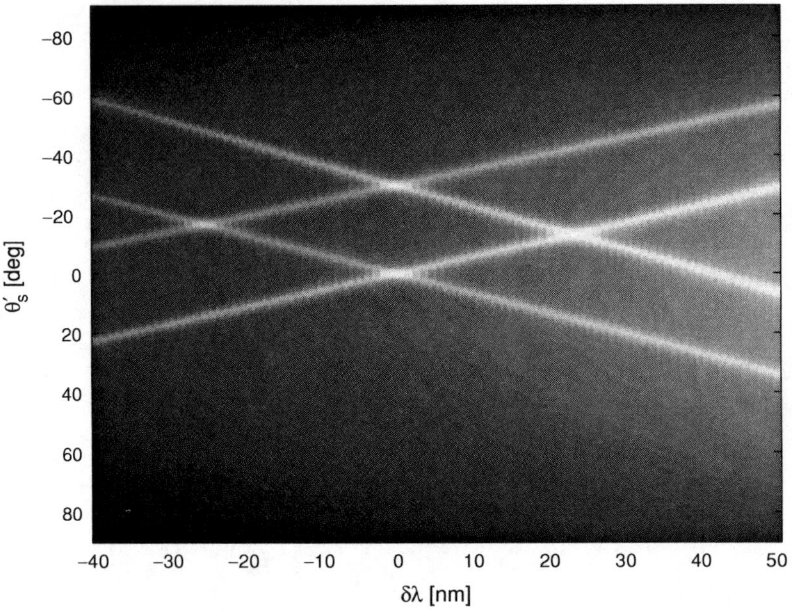

FIGURE 14.7. A plot of the envelope function $\hat{C}_0^{(1)}(q, k; \omega|q'; \omega')$ as a function of θ'_s and $\delta\lambda = \lambda - \lambda' = (2\pi c/\omega) - (2\pi c/\omega')$ for $\theta_0 = 10°$ and $\theta_s = -5°$, when p-polarized light is incident on a one-dimensional randomly rough silver surface characterized by the Gaussian power spectrum (7b) with $\delta = 5$ nm and $a = 100$ nm. The frequency ω corresponds to a vacuum wavelength $\lambda = 612.7$ nm.

$$-\frac{C(\omega)C(\omega')}{Q - k_{sp}(\omega') + k_{sp}(\omega) + \mathrm{i}(\Delta(\omega) + \Delta(\omega'))} \times \left\{ \frac{1}{p + k_{sp}(\omega) + \mathrm{i}\Delta(\omega)} + \frac{1}{Q - p - k_{sp}(\omega') + \mathrm{i}\Delta(\omega')} \right\}. \tag{118}$$

If we set $\omega' = \omega + \delta\omega$, then in the limit of small $\delta\omega$ and small Q Eq. (118) reduces to

$$\begin{aligned} G_\omega(p)G^*_{\omega'}(Q-p) &\simeq -\frac{C^2(\omega)}{Q + \delta\omega k'_{sp}(\omega) - \mathrm{i}2\Delta(\omega)} \frac{2\mathrm{i}\Delta(\omega)}{(p - k_{sp}(\omega))^2 + \Delta^2(\omega)} \\ &\quad - \frac{C^2(\omega)}{Q - \delta\omega k'_{sp}(\omega) + \mathrm{i}2\Delta(\omega)} \frac{2\mathrm{i}\Delta(\omega)}{(p + k_{sp}(\omega))^2 + \Delta^2(\omega)} \\ &\simeq -\frac{2\pi\mathrm{i}C^2(\omega)}{Q + \delta\omega k'_{sp}(\omega) - \mathrm{i}2\Delta(\omega)}\delta(p - k_{sp}(\omega)) \\ &\quad + \frac{2\pi\mathrm{i}C^2(\omega)}{Q - \delta\omega k'_{sp}(\omega) + \mathrm{i}2\Delta(\omega)}\delta(p + k_{sp}(\omega)), \end{aligned} \tag{119}$$

where the prime denotes differentiation with respect to ω.

If, in addition, we approximate the matrix $\mathbf{K}(\omega, \omega'|Q)$ by the element $K_{00}(\omega, \omega'|Q)$, we obtain from Eqs. (113) and (119) the result

$$K_{00}(\omega, \omega'|Q) = -\frac{\mathrm{i}C^2(\omega)\phi_0^2(k_{sp}(\omega))}{Q + \delta\omega k'_{sp}(\omega) - \mathrm{i}2\Delta(\omega)} + \frac{\mathrm{i}C^2(\omega)\phi_0^2(k_{sp}(\omega))}{Q - \delta\omega k'_{sp}(\omega) + \mathrm{i}2\Delta(\omega)}. \quad (120)$$

Now $Q = q' + k$ in the second term on the right-hand side of Eq. (115), which gives rise to the reciprocal memory effect peak, and $Q = q - q'$ in the third term on the right-hand side of this equation, which gives rise to the memory effect peak. From these results we find that $\hat{C}_0^{(1)}(q, k; \omega|q'; \omega')$ will now have reciprocal memory effect peaks at $q = -k \pm \delta\omega k'_{sp}(\omega)$, and memory effect peaks at $q' = q \pm \delta\omega k'_{sp}(\omega)$, for nonzero values of $\delta\omega = \omega' - \omega$.

14.5. Experimental Results

Several experimental studies of angular intensity correlation functions for light scattered from dielectric surfaces have been carried out.[21–23] In these studies, which are limited to the $\hat{C}^{(1)}$ correlation function, the existence of the memory effect and the reciprocal memory effect is demonstrated, but results for the envelope of $\hat{C}^{(1)}$ are not presented.

Experimental measurements of angular amplitude correlation functions of the type $\langle \delta S(q|k)\delta S^*(q'|k')\rangle$ have been carried out in the millimeter wave range for one-dimensional[10] and two-dimensional[9] randomly rough surfaces. The results display peaks that in $|\langle \delta S(q|k)\delta S^*(q'|k')\rangle|^2$ are the memory effect and reciprocal memory effect peaks.

Up to the present time there has been only a single experimental study of the correlation function $\hat{C}(q, k|q', k')$ that has revealed the existence of the memory effect and the reciprocal memory effect peaks in the $\hat{C}^{(1)}(q, k|q', k')$ correlation

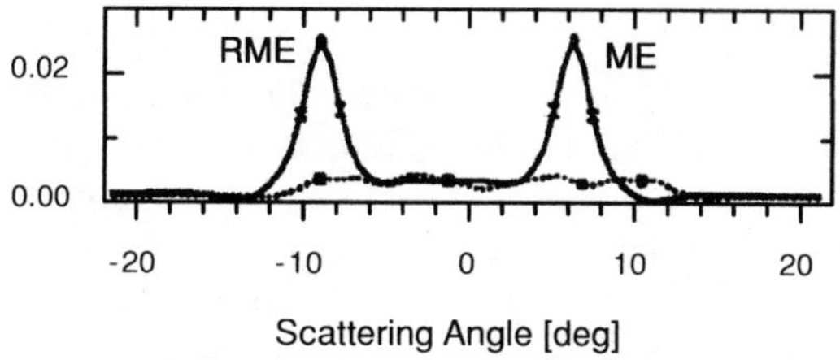

FIGURE 14.8. Experimental results for the envelope functions $\hat{C}_0^{(1)}(q, k, q')$ (solid curve) and $\hat{C}_0^{(10)}(q, k, q')$ (dotted curve) as functions of θ'_s for $\theta_0 = 6.3°$ and $\theta_s = 8.6°$ when p-polarized light of wavelength $\lambda = 612.7$ nm is incident on a one-dimensional randomly rough gold surface ($\epsilon(\omega) = -9.00 + \mathrm{i}1.29$). The surface roughness is characterized by the West–O'Donnell power spectrum (8b) with $\delta = 115.5$ nm, $k_{min} = 8.56 \times 10^{-3}$ nm^{-1} and $k_{max} = 13.3 \times 10^{-3}$ nm^{-1}. The peak in $\hat{C}_0^{(1)}$ labeled ME is the memory effect peak; the peak labeled RME is the reciprocal memory effect peak (after [18]).

function, and the existence of the $\hat{C}^{(10)}(q, k|q', k')$ correlation function. West and O'Donnell[18] have measured the envelope functions $\hat{C}_0^{(1)}(q, k, q')$ and $\hat{C}_0^{(10)}(q, k, q')$ in the scattering of p-polarized light from a weakly rough one-dimensional random gold surface.

To enhance the excitation of the surface plasmon polaritons at the vacuum–metal interface, West and O'Donnell fabricated their randomly rough gold surfaces on the basis of the power spectrum given by Eq. (8b) instead of the Gaussian power spectrum given by Eq. (7b). Their results for $\hat{C}_0^{(1)}(q, k, q')$ and $\hat{C}_0^{(10)}(q, k, q')$ are plotted in Fig. 14.8 as functions of θ_s' for fixed values of θ_0 and θ_s. The plot of $\hat{C}_0^{(1)}(q, k, q')$ displays two peaks. One of them is the memory effect peak, the other is the reciprocal memory effect peak. In contrast, the plot of $\hat{C}_0^{(10)}(q, k, q')$ is a structureless function of θ_s'. The existence of $\hat{C}_0^{(10)}(q, k, q')$ indicates that for the randomly rough surface used in the experiments of West and O'Donell the scattering matrix $S(q|k)$ obeys complex Gaussian statistics.

14.6. Conclusions

In this chapter we have presented an introduction to the existing theoretical and experimental studies of multiple-scattering effects in angular intensity correlation functions. The results of these investigations have provided information that complements that obtained in studies of features in the mean differential reflection and transmission coefficients, such as enhanced backscattering and enhanced transmission, and satellite peaks; they have predicted unsuspected symmetries in speckle patterns; and they have yielded information about the statistical properties of the scattering matrix.

Despite these advances in our understanding of angular intensity correlation functions, and of speckle patterns, there are interesting properties of these functions, associated with multiple-scattering processes, that remain to be explored. One is the existence of the $\hat{C}^{(1.5)}$, $\hat{C}^{(2)}$, $\hat{C}^{(3)}$, and $\hat{C}^{(0)}$ correlation functions. A combined theoretical and experimental effort will undoubtedly be required to elucidate the conditions under which any of these correlation functions can be observed, e.g., the length of the surface and the form of the incident field. Another is the nature of the correlation functions, and the statistical properties of the scattering matrix, when the surface profile function is no longer a Gaussian random process but instead obeys some form of non-Gaussian statistics. Some of these questions have been addressed on the basis of a random phase screen model,[42] but investigations that take multiple scattering into account have yet to be carried out. Yet another is the properties of higher-order correlation functions, such as $\langle \delta I(q|k)\delta I(q'|k')\delta I(q''|k'')\rangle$. While these three examples hardly exhaust the list of directions future studies of angular intensity correlation functions could take, they serve to indicate the possibilities afforded by this still developing sub-field of rough surface scattering.

Acknowledgments. This research was supported in part by Army Research Office Grant DAAD 19-02-1-0256.

References

1. A.R. McGurn and A.A. Maradudin, "Intensity correlation function for light elastically scattered from a randomly rough metallic grating," *Phys. Rev. B* **39**, 13160–13169 (1989).
2. T.R. Michel and K.A. O'Donnell, "Angular correlation functions of amplitudes scattered from a one-dimensional, perfectly conducting rough surface," *J. Opt. Soc. Am. A* **9**, 1374–1384 (1992).
3. M.E. Knotts, T.R. Michel, and K.A. O'Donnell, "Angular correlation functions of polarized intensities scattered from a one-dimensionally rough surface," *J. Opt. Soc. Am. A* **9**, 1822–1831 (1992).
4. M. Nieto-Vesperinas and J.A. Sánchez-Gil, "Enhanced long-range correlations of coherent waves reflected from disordered media," *Phys. Rev. B* **46** 3112–3115 (1992).
5. M. Nieto-Vesperinas and J.A. Sanchez-Gil, "Second-order statistics of non-Gaussian fluctuations of coherent waves reflected from disordered media," *Phys. Rev. B* **48**, 4132–4135 (1992).
6. M. Nieto-Vesperinas and J.A. Sánchez-Gil, "Intensity angular correlations of light multiply scattered from random rough surfaces," *J. Opt. Soc. Am. A* **10**, 150–157 (1993).
7. A.D. Arsenieva and S. Feng, "Correspondence between correlation functions and enhanced backscattering peak for scattering from smooth random surfaces," *Phys. Rev. B* **47**, 13047–13050 (1993).
8. V. Freilikher and I. Yurkevich, "Some aspects of electromagnetic wave scattering from a nonabsorbing rough surface," *Phys. Lett. A* **183**, 253–256 (1993).
9. Y. Kuga, C.T.C Le, A. Ishimaru, and L. Ailes-Sengers, "Analytical, experimental, and numerical studies of angular memory signatures of waves scattered from one-dimensional rough surfaces," *IEEE Trans. Geosci. Remote Sens.* **34**, 1300–1307 (1996).
10. C.T.C. Le, Y. Kuga, and A. Ishimaru, "Angular correlation function based on the second-order Kirchhoff approximation and comparison with experiments," *J. Opt. Soc. Am. A* **13**, 1057–1067 (1996).
11. V. Malyshkin, A.R. McGurn, T.A. Leskova, A.A. Maradudin, and M. Nieto-Vesperinas, "Speckle correlations in the light scattered from weakly rough one-dimensional random metal surfaces," *Opt. Lett.* **22**, 946–948 (1997).
12. V. Malyshkin, A.R. McGurn, T.A. Leskova, A.A. Maradudin, and M. Nieto-Vesperinas, "Speckle correlations in the light scattered from weakly rough random metal surfaces," *Waves Random Media* **7**, 479–520 (1997).
13. M. Nieto-Vesperinas, A.A. Maradudin, A.V. Schegrov, and A.R. McGurn, "Speckle patterns produced by weakly rough random surfaces: existence of a memory effect twin peak in the angular correlations and its consequences," *Opt. Commun.* **142**, 1–6 (1997).
14. J.A. Sánchez-Gil, " Degenerate optical memory effect in dielectric films with randomly rough surfaces," *Phys. Rev. B* **55**, 15928–15936 (1997).
15. A.R. McGurn and A.A. Maradudin, "Speckle correlations in the light scattered by a dielectric film with a rough surface: Guided wave effects," *Phys. Rev. B* **58**, 5022–5031 (1998).

16. A.R. McGurn and A.A. Maradudin, "Speckle correlations in the light reflected and transmitted by metal films with rough surfaces: surface wave effects," *Opt. Commun.* **155**, 79–90 (1998).
17. I. Simonsen, A.A. Maradudin, and T.A. Leskova, "The angular intensity correlation functions $C^{(1)}$ and $C^{(10)}$ for the scattering of s-polarized light from a one-dimensional randomly rough dielectric surface," *Proc.* SPIE **3784**, 218–231 (1999).
18. C.S. West and K.A. O'Donnell, "Angular correlation functions of light scattered from weakly rough metal surfaces," *Phys. Rev. B* **59**, 2393–2406 (1999).
19. T.A. Leskova and A.A. Maradudin, "The angular intensity correlation function C^0 in rough surface scattering," *Phil. Mag. B* **81**, 1289–1302 (2001).
20. I. Simonsen, T.A. Leskova, and A.A. Maradudin, "The angular intensity correlation function $C^{(1)}$ and $C^{(10)}$ for the scattering of light from randomly rough dielectric and metal surfaces," *Waves Random Media* **12**, 307–319 (2002).
21. J.Q. Lu and Zu-Han Gu, " Angular correlation function of speckle patterns scattered from a one-dimensional rough dielectric film on a glass substrate," *Appl. Opt.* **36**, 4562–4570 (1997).
22. Z.-H. Gu and J.Q. Lu, "Memory effect from a one-dimensional rough dielectric film on a glass substrate," *Proc. SPIE* **3141**, 269–280 (1997).
23. Z.-H. Gu and A.R. McGurn, "Dynamic behavior of speckles from rough surface scattering," *Proc. SPIE* **3784**, 285–295 (1999).
24. A.R. McGurn, A.A. Maradudin, and V. Celli, "Localization effects in the scattering of light from a randomly rough grating," *Phys. Rev. B* **31**, 4866–4871 (1985).
25. A.R. McGurn and A.A. Maradudin, "An analogue of enhanced backscattering in the transmission of light through a thin film with a randomly rough surface," *Opt. Commun.* **72**, 279–285 (1989).
26. B. Shapiro, "Large intensity fluctuations for wave propagation in random media, " *Phys. Rev. Lett.* **57**, 2168–2171 (1986).
27. S. Feng, C. Kane, P.A. Lee, and A.D. Stone, " Correlations and fluctuations of coherent wave transmission through disordered media," *Phys. Rev. Lett.* **61**, 834–837 (1988).
28. R. Berkovits, M. Kaveh, and S. Feng, "Memory effect of waves in disordered systems: a real-space approach," *Phys. Rev. B* **40**, 737–740 (1989).
29. R. Berkovits and M. Kaveh, "Time-reversed memory effect," *Phys. Rev. B* **41**, 2635–2638 (1990).
30. R. Berkovits and M. Kaveh, "The vector memory effect for waves," *Europhys. Lett.* **13**, 97–101 (1990).
31. N. Garcia and A.Z. Genack, "Crossover to strong intensity correlation for microwave radiation in random media," *Phys. Rev. Lett.* **63**, 1678–1681 (1989).
32. I. Freund and M. Rosenbluh, "Memory effects in propagation of optical waves through disordered media," *Phys. Rev. Lett.* **61**, 2328–2331 (1988).
33. M.P. van Albada, J.F. de Boer, and A. Lagendijk, "Observation of long-range intensity correlation in the transport of coherent light through a random medium," *Phys. Rev. Lett.* **64**, 2787–2790 (1990).
34. I. Freund and M. Rosenbluh, "Time reversal symmetry of multiply scattered speckle patterns," *Opt. Commun.* **82**, 362–369 (1991).
35. F. Scheffold and G. Maret, "Universal conductance fluctuations of light," *Phys. Rev. Lett.* **81**, 5800–5803 (1998).
36. B. Shapiro, "New type of intensity correlation in random media," *Phys. Rev. Lett.* **83**, 4733–4736 (1999).

37. S.E. Skipetrov and R. Maynard, "Nonuniversal correlations in multiple scattering," *Phys. Rev. B* **62**, 886–891 (2000).
38. H.M. Pedersen, "Second-order statistics of light diffracted from Gaussian, rough surfaces with applications to the roughness dependence of speckles," *Opt. Acta* **22**, 523–535 (1975).
39. D. Léger and J.C. Perrin, "Realtime measurement of surface roughness by correlation of speckle patterns," *J. Opt. Soc. Am.* **66**, 1210–1217 (1976). There is a typographical error in Eq. (14) of this reference: the correlation length T in the argument of the exponential should be squared.
40. J.W. Goodman, "Dependence of image speckle contrast on surface roughness," *Opt. Commun.* **14**, 324–327 (1975).
41. J.W. Goodman, *Statistical Optics* (Wiley, New York, 1985), Chapter 2.
42. A.A. Maradudin and E.R. Méndez, "Scattering by surfaces and phase screens," in *Scattering*, R. Pike and P. Sabatier eds. (Academic Press, New York, 2002), pp. 864–894.
43. C.S. West and K.A. O'Donnell, "Observations of backscattering enhancement from polaritons on a rough metal surface," *J. Opt. Soc. Am. A* **12**, 390–397 (1995).
44. R. Kubo, "Generalized cumulant expansion method," *J. Phys. Soc. Jpn.* **17**, 1100–1120 (1962).
45. A. Stuart and J. Keith Ord, *Kendall's Advanced Theory of Statistics*," vol. 1, 5th edn (Chares Griffin, London, 1987), p. 84ff.
46. A.A. Maradudin, T. Michel, A.R. McGurn, and E.R. Méndez, "Enhanced backscattering of light from a random grating," *Ann. Phys. (N.Y.)* **203**, 255–307 (1990).
47. B. Stoffregen, "Speckle statistics for general scattering objects: II. Mean, covariance and power spectrum of image speckle patterns," *Optik* **52**, 385–399 (1979).
48. F. Toigo, A. Marvin, V. Celli, and N.R. Hill, "Optical properties of rough surfaces: general theory and the small roughness limit," *Phys. Rev. B* **15**, 5618–5626 (1977).
49. G.C. Brown, V. Celli, M. Coopersmith, and M. Haller, "Unitary and reciprocal expansions in the theory of light scattering from a grating," *Surf. Sci.* **129**, 507–515 (1983).
50. G. Brown, V. Celli, M. Haller, A.A. Maradudin, and A. Marvin, "Resonant light scattering from a randomly rough surface," *Phys. Rev. B* **31**, 4993–5005 (1985), Appendix A.
51. E. Burstein, A. Hartstein, J. Schoenwald, A.A. Maradudin, D.L. Mills, and R.F. Wallis, "Surface polaritons-electromagnetic waves at interfaces," in *Polaritons*, E. Burstein and F. de Martini eds. (Pergamon, New York, 1974), pp. 89–110.
52. T.N. Antsygina, V.D. Freylikher, S.A. Gredeskul, L.A. Pastur, and V.A. Slusarev, "Backscattering enhancement by an impedance rough surface," *J. Electromagn. Waves Appl.* **5**, 873–884 (1991).
53. A.V. Zayats, I.I. Smolyaninov, and A.A. Maradudin, "Nanooptics of surface plasmon polaritons," *Phys. Rep.* **408**, 131–314 (2005).
54. A.V. Schegrov, A.A. Maradudin, and E.R. Mendez, "Multiple scattering of light from rough surfaces," in *Progress in Optics*, vol. 46, E. Wolf, ed. (Elsevier, Amsterdam, 2004), pp. 117–241.
55. Ref. [41], pp. 31–33.
56. J.A. Sánchez-Gil, A.A. Maradudin, and E.R. Méndez, "Limits of validity of three perturbation theories of the coherent scattering of light from one-dimensional randomly rough dielectric surfaces," *J. Opt. Soc. A* **12**, 1547–1558 (1995).

57. V. Freilikher, M. Pustilnik, and I. Yurkevich, "Wave scattering from a bounded medium with disorder," *Phys. Lett. A* **193**, 467–470 (1994).
58. J.A. Sánchez-Gil, A.A. Maradudin, Jun Q. Lu, V.D. Freilikher, M. Pustilnik, and I. Yurkevich, "Scattering of electromagnetic waves from a bounded medium with a random surface," *Phys. Rev. B* **50**, 15353–15368 (1994).
59. P.B. Johnson and R.W. Christy, "Optical constants of the noble metals," *Phys. Rev. B* **6**, 4370–4379 (1972).

15
Speckle Pattern in the Near Field

JEAN-JACQUES GREFFET AND RÉMI CARMINATI
Ecole Centrale Paris, CNRS Grande Voie des Vignes, 92295 Châtenay-Malabry Cedex, France

15.1. Introduction

Analysis of the speckle structure of a random field is a well-known topic.[1] With the development of near-field microscopy, the subject has been revisited in the last ten years. This paper addresses the structure of a random field in close proximity to an interface. First, we give a general overview of the differences between the structure of the field in the near field and in the far field. We emphasize the role of the evanescent waves in the near field. The second part of the paper reviews recent studies on field correlations in the near field for two cases: random thermal fields and light multiply scattered. The third part of the paper is devoted to the analysis of the speckle pattern above an interface in the single scattering regime. It is shown that in that case, the speckle pattern is nonuniversal and strongly related to the statistical properties of the surface. It is known that near-field images strongly depend on the specific properties of each tip. It cannot be assumed in general that the signal delivered by a near-field scanning microscope delivers a signal proportional to the square of the local electric field. In the last part of the paper, we derive from the reciprocity theorem a general form of the signal. In particular, we emphasize the role of polarization and the influence of the tip on the spectral response.

15.2. Role of Evanescent Waves in the Near Field

15.2.1. Angular Spectrum

In this section, we discuss the contribution of evanescent waves to the electric field in the half-space $z > 0$ above an interface. The upper half-space is filled with a medium with a dielectric constant ε_1 and a real refractive index $n_1 = \varepsilon_1^{1/2}$. To proceed, we introduce a plane-wave expansion of the electric field valid in the half-space $z > 0$:

$$\mathbf{E}(\mathbf{r}) = \int \frac{\mathrm{d}^2\mathbf{k}_\parallel}{(2\pi)^2} \mathbf{e}(\mathbf{k}_\parallel) \exp[\mathrm{i}(\mathbf{k}_\parallel \cdot \mathbf{r} + \gamma_1 z)], \tag{1}$$

where $\gamma_1 = (\varepsilon_1\omega^2/c^2 - \mathbf{k}_\parallel^2)^{1/2}$ with the determination $\text{Im}(\gamma_1) > 0$ and $\text{Re}(\gamma_1) > 0$. We have used the notation $\mathbf{k} = (k_x, k_y, k_z)$ and $\mathbf{k}_\parallel = (k_x, k_y, 0)$. The amplitude of the Fourier transform of the electric field in the plane $z = 0$ is denoted $\mathbf{e}(\mathbf{k}_\parallel)$. Since $\text{div}\mathbf{E} = 0$, we have $\mathbf{e}(\mathbf{k}_\parallel) \cdot \mathbf{k}_\parallel = 0$ so that only two components are needed to characterize $\mathbf{e}(\mathbf{k}_\parallel)$. We introduce two unit vectors that define the s- and p-polarizations:

$$\mathbf{e}(\mathbf{k}_\parallel) = e_s(\mathbf{k}_\parallel)\mathbf{a}_s(\mathbf{k}_\parallel) + e_p(\mathbf{k}_\parallel)\mathbf{a}_p(\mathbf{k}_\parallel), \tag{2}$$

where the unit vectors are given by

$$\mathbf{a}_s(\mathbf{k}_\parallel) = \mathbf{a}_z \times \frac{\mathbf{k}_\parallel}{|\mathbf{k}_\parallel|} \tag{3}$$

$$\mathbf{a}_p(\mathbf{k}_\parallel) = \mathbf{a}_s(\mathbf{k}_\parallel) \times \frac{\mathbf{k}_\parallel + \gamma_1\mathbf{a}_z}{k_1}, \tag{4}$$

where we have used $k_1 = n_1 k_0 = n_1\omega/c$ and $\mathbf{a}_z$ is the unit vector along the z-axis perpendicular to the interface. This plane wave expansion can be split into two contributions: the propagating waves with wavevectors with real z-components γ_1, corresponding to parallel wavevectors $|\mathbf{k}_\parallel| < k_1$, and evanescent waves with imaginary z-components, corresponding to large parallel wavevectors $|\mathbf{k}_\parallel| > k_1$. At distances larger than a wavelength from the interface, evanescent waves can be neglected. It follows that the spatial variations of the field have been filtered. Propagation is a low-pass filter with a cut-off given by $|\mathbf{k}_\parallel|_{\text{co}} = k_1$. We shall refer to the near-field region when dealing with distances to the interface much smaller than $\lambda_1 = 2\pi/k_1$ so that evanescent waves can be detected. The reader is referred to [2] for a comprehensive introduction to the basics of near-field optics.

15.2.2. Field and Intensity Correlations in the Near Field

Let us now study the correlation function of the field. We consider that the field is a random process, homogeneous in the plane (x, y) and spatially varying along the z-axis. This corresponds, e.g., to the case of a thermal field generated by an interface. This is also the case of a deterministic field scattered by a random medium limited by an interface $z = 0$, either a rough surface or a medium with a random dielectric constant such as a suspension of particles in water. The electromagnetic fields used in the experiments are usually microwave fields generated by an oscillator or optical fields generated by a laser. An interesting property of the fields is their spatial correlation function. We shall specifically examine the role of surface waves on the field correlation. Leaving aside polarization effects, we will consider a scalar field $\Psi(\mathbf{r})$.

From the Wiener–Khinchine theorem, the field correlation of a homogeneous field in the plane (x, y), known as the cross-spectral density in the context of coherence theory, can be written as the Fourier transform of the power spectral density of the field denoted by G_Ψ:

$$\langle\Psi(\mathbf{R} + \mathbf{r})\Psi^*(\mathbf{R})\rangle = \int G_\Psi(\mathbf{k}_\parallel)\exp[i\mathbf{k}_\parallel \cdot \mathbf{r}]\frac{d^2\mathbf{k}_\parallel}{(2\pi)^2}. \tag{5}$$

To define the power spectral density of the field, we first introduce a field Ψ_A equal to the field Ψ in the finite area A and null elsewhere. We can define its angular spectrum

$$\Psi_A(\mathbf{r}) = \int d^2\mathbf{k}_\parallel \Psi_A(\mathbf{k}_\parallel) \exp[i(\mathbf{k}_\parallel \cdot \mathbf{r} + \gamma_1 z)]. \tag{6}$$

The power spectral density is then given by

$$G_\Psi(\mathbf{k}_\parallel) = \lim_{A\to\infty} \frac{|\Psi_A(\mathbf{k}_\parallel)|^2}{A}. \tag{7}$$

To discuss the differences between the near field and the far field for the field correlation, let us first consider a constant power spectral density. In other words, the amplitudes $\Psi(\mathbf{k}_\parallel)$ in the plane $z = 0$ have a random phase and a constant modulus. It follows immediately from Eq. (5) that the cross-spectral density of the field is given by a delta of Dirac so that the field is delta-correlated. If we now consider the same field but at a distance from the surface larger than a wavelength, the evanescent waves can be neglected so that the correlation function is given by

$$\langle \Psi(\mathbf{R} + \mathbf{r})\Psi^*(\mathbf{R})\rangle = G_\Psi \int_0^{k_1} J_0(kr)\frac{k\mathrm{d}k}{2\pi} = G_\Psi \frac{k_1}{r} J_1(k_1 r), \tag{8}$$

where $k_1 = n_1\omega/c$. The typical length scale of this cross-spectral density is given by the wavelength. When dealing with blackbody radiation in three dimensions, one finds a cross-spectral density that varies like $\sin(k_1 r)/k_1 r$. For a Lambertian source, the cross-spectral density is also given by $\sin(k_1 r)/k_1 r$. The key difference between the near field and the far field is the possibility of observing a correlation length smaller than the wavelength in the near field. This is simply due to the presence of surface waves with large wavevectors. Whether these surface waves actually exist depends on the system. In what follows, we will consider different physical systems.

In many cases, the field can be viewed as a Gaussian statistical process. This is often observed for thermal light or for multiple scattering since the field is the result of the superposition of many different independent random fields. The intensity correlation is then simply related to the field correlation by virtue of the moment theorem for circular complex Gaussian variables.[3]

15.2.3. Speckle Patterns due to Random Thermal Fields

We discuss briefly in this section the transition between far field to near field of the cross-spectral density. We consider the cross-spectral density of the field emitted by a half-space ($z < 0$). It can be computed[4] accounting for the optical properties of the medium using stochastic electrodynamics.[5,6] For a material with local dielectric constant, the random current density is delta-correlated. It follows that the spatial frequency spectrum of the sources contains arbitrary high wavevectors. Thus the fields in the near field contain large wavevectors. A detailed asymptotic analysis is reported in [7]. It is found that in the near-field regime, the correlation length is controlled by the distance to the surface that plays the role of a cut-off frequency as already discussed.

Another important feature is the polarization dependence. The correlation functions of the parallel and perpendicular components of the field to the interface are different.[7] This is all the more important as electromagnetic modes along a surface such as surface plasmon polaritons or surface phonon polaritons are polarized. The polarization state cannot be described using two components perpendicular to a mean propagation direction when dealing with vectors in the near field. Three components are necessary. The scalar approximation is obviously rarely appropriate in the near field. A detailed study of polarization in near field has been reported recently.[8]

15.2.4. Multiple Scattering

Speckle patterns in the near field have been observed in the first images of surfaces taken with scanning near-field optical microscopes. Wavy fluctuations of the intensity are clearly seen, for instance, in images reported in [10, 11]. They are produced by the interference of waves scattered by the roughness of the surface. Their visibility strongly depends on the coherence of the illumination as first pointed out by de Fornel *et al.*[12,13]

A major application of the near-field scanning microscope has been the possibility of imaging surface plasmons. Leaving aside what is now called plasmonics, we will briefly review some particular experiments designed to study the structure of the intensity over a random rough surface. Two types of systems have been investigated: metallic surfaces with slightly rough surfaces and semicontinuous films that can be viewed as clusters of nanometric metallic particles. Localized states appearing as hot spots in the images have been observed by Tsai *et al.*,[14] Zhang *et al.*[15] and by Gresillon *et al.*[16] on semi continuous films. A theoretical treatment can be found in [17, 18]. Further studies have been reported recently by Chebanov *et al.*[19] Bozhevolnyi *et al.* have reported the observation of localized states on slightly rough surfaces.[20–22]

The enhancement of the field due to surface plasmon polaritons has also been studied in the context of surface enhanced Raman scattering. Numerical simulations have shown that fractal silver surfaces may produce very large enhancement of the field.[23,24] Very recently, Leskova *et al.*[25] have studied the spatial coherence of the field above a rough surface illuminated by a plane wave. In particular, they have found that the spatial coherence changes rapidly with the distance to the mean surface due to the role of evanescent waves.

More recently, near-field measurements of the speckle generated by volume random media have been reported. The statistics of the amplitude and phase of the field transmitted by a random medium has been studied.[26] Near-field measurements have been reported by Sebbah *et al.*[27] in the microwave range and by Emiliani *et al.*[28] in the optical range. In both results, the field correlation follows the usual far-field form first derived by Shapiro[29] proportional to $\sin(kr)\exp(-r/l)/kr$, where l is the mean free path. With scattering centers much smaller than the wavelength, one should expect smaller correlation lengths for measurements performed at a subwavelength distance from the medium. They were not observed in these experiments, presumably because of a lack of resolution of the tip or a distance between the tip and the surface larger than $\lambda/10$. The subwavelength correlation has been observed by Apostol and Dogariu.[30,31] Another remarkable result reported by Apostol and Dogariu is the non-Gaussian statistics observed in the near field under some illumination conditions.[32]

15.2.5. Experimental Difficulties

Near-field experiments are difficult to analyze quantitatively. The key importance of the tip used in near-field optical microscopy is experimentally observed by comparing different images of the same surface taken with different tips. Large differences are usually observed. It is thus difficult to analyze the signal without a model for the actual tip being used. There are three different experimental issues: (i) Is the tip a passive probe? (ii) Does the tip respond equally to all components of the electromagnetic field? (iii) Does the tip influence the spectral response?

A key issue to address is whether the tip is a passive probe or not. This amounts to assessing whether multiple scattering between the tip and the sample is negligible. When using a dielectric tip, this approximation can be justified.[9,33] By contrast, when detecting the field with a tip with a high dielectric constant and large volume, the approximation is usually not valid.[34] The passive probe assumption cannot be used when studying localized resonant modes with a high quality factor. Indeed, in these cases, the presence of the probe introduces radiative and nonradiative losses that may significantly alter the mode. It is therefore extremely difficult to measure quantitatively the enhancement of the fields in bright spots because the presence of the probe introduces losses that tend to decrease the resonance.

The second key issue when detecting electromagnetic fields in the near-field regime is the polarization sensitivity. Let us point out that the analysis of the detection process of near fields must take into account the polarization of the field and specific properties of the tip. The usual far-field assumption that the signal is proportional to the intensity is no longer valid, although it can be used in some cases.[9] Obviously, the polarization dependence of the tip is essential when detecting surface waves such as surface plasmon polaritons, which are polarized fields.

The third issue deals with the measurement of spectra in the near field. We will show that the spectral response of the tip has to be taken into account.

The last section of this paper will be devoted to the introduction of a model that allows us to include the tip characteristics in the modeling of the signal in the weak coupling regime.

15.3. Nonuniversal Speckle Pattern Produced by a Slightly Rough Surface

In this section, we will study the speckle pattern produced by a random rough surface. As discussed in the introduction, near-field images often show wavy structures even when imaging plane surfaces. We will show that these structures can be attributed to the intensity speckle produced by the residual roughness of the surface. This speckle pattern has two distinct properties as compared to the far-field developed speckle. It is nonuniversal and can be related to the statistical properties of the surface.[35] It is nonisotropic when the illumination is nonisotropic[35] . It has also been pointed out that the speckle pattern can be nonisotropic under normal incidence due to polarization effects.[36]

15.3.1. Statistical Description of a Random Rough Surface

In order to describe a random rough surface we assume that the height of the surface $z = S(\mathbf{r}_\parallel)$ at a given point $\mathbf{r} = (\mathbf{r}_\parallel, z)$ is a random variable with zero mean value, characterized by a Gaussian probability density function and a correlation function given by $\delta^2 C(|\mathbf{r}_\parallel - \mathbf{r}'_\parallel|)$, where δ is the roughness defined as the root mean square height:

$$\langle S(\mathbf{r}_\parallel)\rangle = 0 \tag{9}$$

$$\langle S(\mathbf{r}_\parallel)S(\mathbf{r}'_\parallel)\rangle = \delta^2 C(|\mathbf{r}_\parallel - \mathbf{r}'_\parallel|). \tag{10}$$

The brackets indicate the ensemble average. Let us emphasize that we assume an isotropic stationary random process as seen in the form of the correlation function which depends only on the distance $|\mathbf{r}_\parallel - \mathbf{r}'_\parallel|$. The correlation length a characterizes the correlation function $C(|\mathbf{r}_\parallel - \mathbf{r}'_\parallel|)$. Let us now introduce the Fourier transform of the restriction of the surface profile to a limited area A, $S_A(\mathbf{r}_\parallel)$ and the Fourier transform of the correlation function:

$$S_A(\mathbf{r}_\parallel) = \int \frac{\mathrm{d}^2\mathbf{k}_\parallel}{(2\pi)^2} S_A(\mathbf{k}_\parallel)\exp(\mathrm{i}\mathbf{k}_\parallel \cdot \mathbf{r}_\parallel), \tag{11}$$

$$C(|\mathbf{r}_\parallel|) = \int \frac{\mathrm{d}^2\mathbf{k}_\parallel}{(2\pi)^2} g(\mathbf{k}_\parallel)\exp(\mathrm{i}\mathbf{k}_\parallel \cdot \mathbf{r}_\parallel). \tag{12}$$

An important property is given by the Wiener–Khinchine theorem which relates the Fourier transform of the correlation function $g(\mathbf{k}_\parallel)$ to the power spectral density:

$$\lim_{A\to\infty} \frac{\langle |S_A(\mathbf{k}_\parallel)|^2\rangle}{A} = \delta^2 g(\mathbf{k}_\parallel). \tag{13}$$

15.3.2. *Amplitude of the Field Scattered by a Deterministic Slightly Rough Surface*

We consider a slightly rough surface separating two half-spaces filled with linear homogeneous media. The upper medium is characterized by the real dielectric constant ε_1 and the lower medium is characterized by the dielectric constant ε_2. The interface is illuminated in transmission from medium 2. The amplitude of the roughness is much smaller than the wavelength λ_1 so that the field in medium 1 can be computed by using a perturbative expansion in powers of δ/λ_1. Following [38], we seek a perturbative solution by writing the angular spectrum of the transmitted field in the form

$$\mathbf{e}_t(\mathbf{k}_\parallel) = \mathbf{e}_t^{(0)}(\mathbf{k}_\parallel) + \mathbf{e}_t^{(1)}(\mathbf{k}_\parallel) + \cdots. \tag{14}$$

Using the boundary conditions, one finds that the zero-order field is related to the incident field $\mathbf{e}_i = e_{i,s}\mathbf{a}_s(\mathbf{k}_\parallel^i) + e_{i,p}\mathbf{a}_p(\mathbf{k}_\parallel^i)$ by the Fresnel transmission factors:

$$\begin{bmatrix} e_{t,s}^{(0)} \\ e_{t,p}^{(0)} \end{bmatrix} = \begin{bmatrix} \frac{2\gamma_2^i}{\gamma_1^i+\gamma_2^i} & 0 \\ 0 & \frac{2n_1n_2\gamma_2^i}{\varepsilon_1\gamma_2^i+\varepsilon_2\gamma_1^i} \end{bmatrix} \begin{bmatrix} e_{i,s} \\ e_{i,p} \end{bmatrix}, \tag{15}$$

where $\gamma_j = [\varepsilon_j k_0^2 - \mathbf{k}_\parallel^2]^{1/2}$ with the determination $\mathrm{Im}(\gamma_j) > 0$ and $\mathrm{Re}(\gamma_j) > 0$ and where the superscript i indicates that the function is evaluated for the incident parallel wavevector $\mathbf{k}_\parallel^i$. The first-order amplitude of the scattered field is given by

$$\begin{bmatrix} e_{t,s}^{(1)} \\ e_{t,p}^{(1)} \end{bmatrix} = S(\mathbf{k}_\parallel^i - \mathbf{k}_\parallel)\mathbf{L}(\mathbf{k}_\parallel, \mathbf{k}_\parallel^i) \begin{bmatrix} e_{i,s} \\ e_{i,p} \end{bmatrix}, \tag{16}$$

where $\mathbf{L}$ is an operator defined by

$$\mathbf{L}(\mathbf{k}_\parallel, \mathbf{k}_\parallel^i) = i(\gamma_2 - \gamma_1) \begin{bmatrix} t_s^i\, \hat{\mathbf{k}}_\parallel.\hat{\mathbf{k}}_\parallel^{\mathrm{inc}} & t_p^i\, \frac{\gamma_1^{\mathrm{inc}}}{n_1k_0}\, \hat{\mathbf{k}}_\parallel^i.(\mathbf{a}_z \times \hat{\mathbf{k}}_\parallel) \\ \frac{t_s^i\, n_1k_0\,\gamma_2\, \hat{\mathbf{k}}_\parallel.(\mathbf{a}_z \times \hat{\mathbf{k}}_\parallel^i)}{\gamma_1\gamma_2 + |\mathbf{k}_\parallel|^2} & \frac{t_p^i\, \{\gamma_1^i\,\gamma_2\, \hat{\mathbf{k}}_\parallel^i.\hat{\mathbf{k}}_\parallel + |\mathbf{k}_\parallel|\, |\mathbf{k}_\parallel^i|\}}{\gamma_1\gamma_2 + |\mathbf{k}_\parallel|^2} \end{bmatrix}, \tag{17}$$

where t_p^i and t_s^i are the Fresnel transmission factors defined in Eq. (15) and the caret denotes a unit vector $\hat{\mathbf{x}} = \mathbf{x}/|\mathbf{x}|$. We can finally cast the scattered field to first order in the form

$$\mathbf{E}_t^{(1)}(\mathbf{r}) = \int \frac{\mathrm{d}^2\mathbf{k}_\parallel}{(2\pi)^2} S(\mathbf{k}_\parallel - \mathbf{k}_\parallel^i)\mathbf{L}(\mathbf{k}_\parallel, \mathbf{k}_\parallel^i)\mathbf{e}_i \exp(\mathrm{i}\gamma_1 z)\exp(\mathrm{i}\mathbf{k}_\parallel \cdot \mathbf{r}). \tag{18}$$

The first term of the equation shows that the spectrum of the scattered field is given by the spectrum of the surface translated by the parallel projection of the incident wavevector. This translation breaks the symmetry in the case of an illumination by a plane wave. The rotational symmetry can be restored by summing the contribution from all directions in the case of an isotropic illumination. The second term L accounts for polarization effects. The third term $\exp(\mathrm{i}\gamma_1 z)$ is a phase term for propagating terms but a filtering term for evanescent waves. This term is essential when studying the transition between the near field to the far field regime.

15.3.3. Speckle Pattern Generated by a Slightly Rough Surface in the Near Field

In what follows, we adopt a simplified point of view to describe the speckle pattern. We consider the speckle pattern generated by the transmitted field at the interface between two dielectric media. We study the intensity correlation function. Intensity is defined by the square modulus of the electric field. The key property of the speckle generated by a slightly rough surface is due to the fact that the field is a hologram of the surface as first introduced in [40]. This property stems from the interference structure of the intensity. Indeed, the intensity can be written as

$$I(\mathbf{r}_\parallel, z) = |\mathbf{E}_t^{(0)} + \mathbf{E}_t^{(1)}|^2. \tag{19}$$

To first order in δ/λ, we obtain

$$\begin{aligned} I(\mathbf{r}_\parallel, z) &= I^{(0)}(z) + I^{(1)}(\mathbf{r}_\parallel, z) \\ &= |\mathbf{E}_t^{(0)}(\mathbf{r}_\parallel, z)|^2 + 2Re[\mathbf{E}_t^{(0)*}(\mathbf{r}_\parallel, z)\cdot \mathbf{E}_t^{(1)}(\mathbf{r}_\parallel, z)]. \end{aligned} \tag{20}$$

Note that the zero-order term is the intensity when the roughness can be neglected. It is a constant term that does not contribute to the contrast in the image. It follows that the spatial modulation of the intensity is linearly related to the surface profile. We now turn to the study of the relationship between the structure of the speckle pattern and the statistical properties of the surface. Let us first study the intensity correlation function in the near field. From Eq. (20), we get

$$\langle I(\mathbf{r}_\parallel) I(\mathbf{r}_\parallel + \mathbf{R}_\parallel)\rangle = \langle I^{(0)}(\mathbf{r}_\parallel) I^{(0)}(\mathbf{r}_\parallel + \mathbf{R}_\parallel)\rangle + \langle I^{(1)}(\mathbf{r}_\parallel) I^{(1)}(\mathbf{r}_\parallel + \mathbf{R}_\parallel)\rangle, \tag{21}$$

where we have used the property $\langle I^{(1)}\rangle = 0$ that follows from $\langle S\rangle = 0$. We have not included the correlation between the second-order term and the zero-order term that involves only a constant contribution independent of $\mathbf{R}_\parallel$. Inserting Eqs. (18), (20) into Eq. (21), the intensity correlation function can be cast in the form

$$\langle I^{(1)}(\mathbf{r}_\parallel) I^{(1)}(\mathbf{r}_\parallel + \mathbf{R}_\parallel)\rangle = \mathrm{Re}\left\{\int \frac{d^2\mathbf{k}_\parallel}{(2\pi)^2} \delta^2 g(\mathbf{k}_\parallel) F(\mathbf{k}_\parallel) \exp(i\mathbf{k}_\parallel \cdot \mathbf{R}_\parallel)\right\}. \tag{22}$$

In other words, the power-spectral density of the near-field intensity $G_I(\mathbf{k}_\parallel)$ is related to the power spectral density of the roughness $\delta^2 g(\mathbf{k}_\parallel)$ by a linear filter

$$G_I(\mathbf{k}_\parallel) = \delta^2 g(\mathbf{k}_\parallel) F(\mathbf{k}_\parallel). \tag{23}$$

The filter F is linearly related to the operator $\mathbf{L}$ defined in Eq. (17). The explicit form of the filter F is given in [35]. We note that the filter depends on the illumination conditions. Thus it includes all the information on the coherence properties of the illuminating beam. These results are illustrated in Fig. 15.1 for a surface generated using the algorithm described in [39] and a field computed using the perturbative approach outlined above.

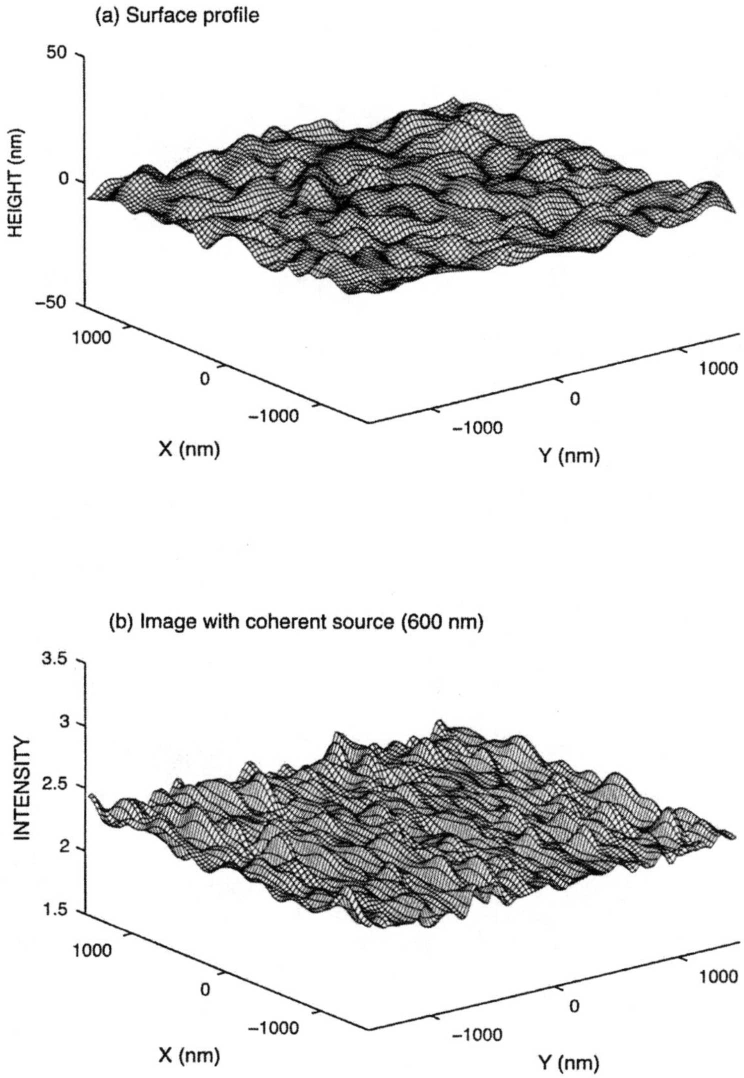

FIGURE 15.1. Near-field speckle pattern above a random rough surface with rms height $\delta = 0.005\lambda$, Gaussian correlation function $C(r) = \exp(-r^2/a^2)$ with $a = 0.5\lambda$. (a) Surface profile numerically generated. (b) Near-field intensity computed in a plane located at $\lambda/10$ above the mean surface. The surface is illuminated in total internal reflection with a wavelength $\lambda = 600$ nm and a field linearly polarized along the x-axis. It is clearly seen in the figure that the speckle pattern is not isotropic. Reprinted from [36].

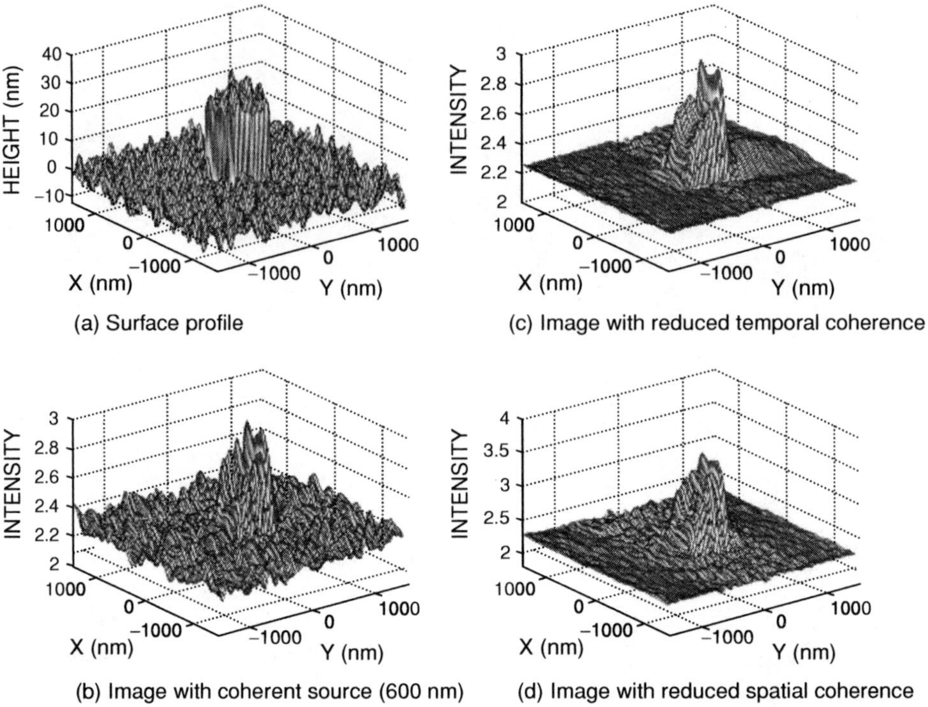

FIGURE 15.2. Near-field speckle pattern above a random rough surface with rms height $\delta = 1$ nm, correlation length $a = 120$ nm and a deterministic profile (a parallelepiped with height 15 nm, side 1200 nm, refractive index $n = 1.5$). The incident beam is in the yz plane. The field is linearly polarized along the x-axis. (a) Surface profile. (b) Near-field intensity computed in a plane located 30 nm above the mean surface with a monochromatic plane wave illumination. (c) Role of the temporal coherence: near-field intensity with a white light plane wave illumination with a flat spectrum in the interval [400, 800] nm. (d) Role of the spatial coherence. To reduce the spatial coherence of the illuminating beam we use a superposition of incoherent monochromatic plane waves with different incidence angles in the interval [45°, 65°]. Reprinted from [36].

Another interesting feature is the speckle contrast C_I. This quantity is defined as

$$C_I = \frac{\sqrt{\langle I^2 \rangle - \langle I \rangle^2}}{\langle I \rangle} = \frac{\sqrt{\langle (I^{(1)})^2 \rangle}}{I^{(0)}}. \tag{24}$$

It is seen that the speckle contrast is proportional to the rms-roughness δ.

In order to illustrate the influence of the coherence of the incident beam on the speckle structure in the near field, we have simulated the near-field intensity above a plane with a parallelepiped and a randomly rough component. Results are shown in Fig. 15.2.

15.4. Detection of Optical Near Fields

In this section we address the measurement of optical near fields in scanning near-field optical microscopy (SNOM). After pioneering works in the mid eighties,[41–43] SNOM techniques have developed rapidly and various schemes are now available (for an overview, see the following textbooks[44]). Current techniques allow surface structure imaging with subwavelength resolution, as well as measurements of confined electromagnetic fields, the latter making SNOM a particularly relevant instrument in nanooptics. Let us mention the detection and local excitation of surface plasmons on ordered[45] or disordered[22,46] surfaces, the local spectroscopy of isolated molecules,[47–49] the measurement of field modes in nanophotonics devices,[50] or the measurement of localized surface waves and speckle patterns on disordered surfaces.[16,51] The last two examples are of particular interest in the context of this book.

All SNOM techniques rely on the optical interaction between a sharp tip (local probe) and the near field generated by a sample with subwavelength structures (either by scattering or direct emission). For a given position of the tip, the far-field intensity radiated by the tip-sample system is collected, using, e.g., a conventional microscope objective or an optical fiber. Depending on the illumination geometry, several schemes have been proposed: far-field illumination of the sample and near-field collection of the signal by the local probe (etched fiber, apertureless tip, etc.), near-field illumination through a metal coated tip with a small aperture and far-field detection, near-field illumination and detection through the tip. In order to understand the contrast, the relationship between the measured signal and the local structure of the near field has to be known. It is the scope of this section to establish this relationship and to discuss the influence of several parameters (such as polarization or source spectrum) on the image formation.

Studying the contrast mechanism is a difficult problem because one has to handle an electromagnetic interaction in the near field, between objects whose length scales extend from the nanometer range (objects, tip apex) to macroscopic sizes (the entire probe). In the early nineties, the first theoretical models and numerical simulations were developed to understand the image formation and the underlying physics. Powerful numerical methods and analytical models have been developed, and have improved the understanding of the image formation and the influence of various parameters (polarization, illumination and detection conditions, coherence, etc.). For a review of the different methods and models, see [2, 52]. In this section, we shall show how a general and exact expression for the signal measured by a SNOM can be obtained. We will focus on the measurement of confined electromagnetic fields generated by scattering (nanostructures, gratings, rough surfaces, disordered films, etc.) or by direct emission from the sample itself (fluorescence, thermal emission, etc.). The fundamental tool is the reciprocity theorem of electromagnetism, whose formulations are given in the appendix. This theorem is widely used in the context of antenna theory.[53] It was used in the context of scanning-probe microscopy to model light emission by the scanning

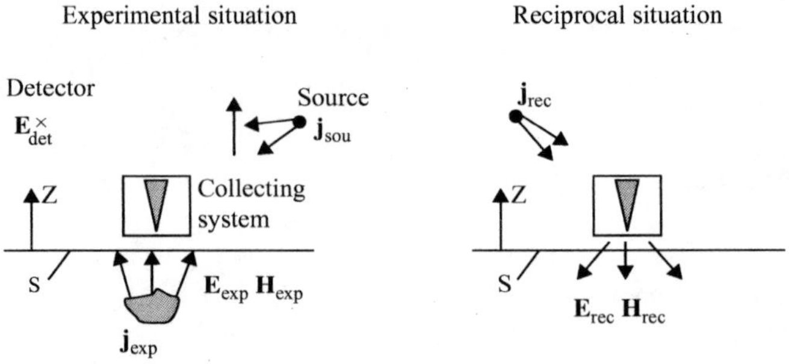

FIGURE 15.3. Left: Scheme of an experimental SNOM setup used for the near-field detection of an electromagnetic field. The source and detector positions are denoted by $\mathbf{r}_{sou}$ and $\mathbf{r}_{det}$, respectively. Right: reciprocal (fictitious) situation, in which the system (tip and illumination and detection optics) is illuminated by a point source located at the detector position.

tunneling microscope (STM),[54] and more recently to study the image formation in SNOM.[2,55,56]

The power of the method is to give an explicit expression for the measured signal in SNOM. The model leads naturally to the concept of a response function in near-field optical imaging. It allows a study of the influence of the relevant parameters. We will illustrate this point by explaining the polarization behavior and the spectral response of an apertureless SNOM using a sharp metallic tip.

15.4.1. General Expression for the Near-Field Optical Signal

We consider the near-field measurement of the electromagnetic field scattered (or directly emitted) by an arbitrary structure (e.g., a rough surface or a disordered layer on a substrate). The experimental situation is sketched in Fig. 15.3 (left). The radiating structure is described by a monochromatic current density $\mathbf{j}_{exp}(\omega)$. This current density represents either a current induced in the structure by an incident field, or a primary source (for example a fluorescent molecule, emitting at a frequency ω). The fields radiated by the current density $\mathbf{j}_{exp}$ will be denoted by $\mathbf{E}_{exp}$ (electric field) and $\mathbf{H}_{exp}$ (magnetic field). Note that these fields are the fields radiated *in the presence of the tip* and the illumination and detection optics (represented by the square box in Fig. 15.3).

In order to obtain an explicit expression for the field $\mathbf{E}_{det}$ at the detector position, we consider the fictitious situation sketched in Fig. 15.3 (right). In this situation, the tip and the illumination and detection optics are illuminated by a point source with current density $\mathbf{j}_{rec}(\omega)$, located at the position of the detector, *in the absence* of the structure generating the field to be measured. We shall refer to this situation as the *reciprocal situation*. The fields radiated at an arbitrary point in the reciprocal situation will be denoted by $\mathbf{E}_{rec}$ and $\mathbf{H}_{rec}$.

15.4.1.1. Reciprocity Theorem

Let us consider the situation in Fig. 15.3. The reciprocity theorem of electromagnetism[57–60] leads to (the general expression of this theorem is recalled in the appendix):

$$\mathbf{E}_{\text{det}} \cdot \mathbf{j}_{\text{rec}} = \mathbf{E}_{\text{rec}}(\mathbf{r}_{\text{sou}}) \cdot \mathbf{j}_{\text{sou}} + \int_S (\mathbf{E}_{\text{exp}} \times \mathbf{H}_{\text{rec}} - \mathbf{E}_{\text{rec}} \times \mathbf{H}_{\text{exp}}) \cdot \mathbf{a}_z \, \mathrm{d}^2\mathbf{R}, \tag{25}$$

where $\mathrm{d}^2\mathbf{r}_\parallel = \mathrm{d}x\,\mathrm{d}y$ and $\mathbf{a}_z$ is the unit vector along the z-axis. The integral is extended to a plane at a constant height z in the gap region (surface S represented by the horizontal line in Fig. 15.3). The left-hand side in Eq. (25) is a component of the electric field at the detector position. It is assumed that the detection is performed at a frequency ω, and uses an analyzer whose polarization direction is that of the vector $\mathbf{j}_{\text{rec}}$. The first term on the right-hand side represents a direct illumination of the detector by the external source (e.g. a laser source) which generates the incident field. The external source is represented by a point current density $\mathbf{j}_{\text{sou}}$ located at a point $\mathbf{r}_{\text{sou}}$.

15.4.1.2. Expression of the Detected Field

In order to simplify expression (25), we shall use the plane-wave representation (or angular spectrum) of the fields (see for example chapter 2 in [59]). In the region between the tip and the emitting structure (gap region), the electric field in the experimental situation can be written as

$$\mathbf{E}_{\text{exp}}(\mathbf{r}_\parallel, z) = \int \mathbf{e}^+_{\text{exp}}(\mathbf{k}_\parallel) \exp(\mathrm{i}\mathbf{k}_\parallel \cdot \mathbf{r}_\parallel + \mathrm{i}\gamma z) \frac{\mathrm{d}^2\mathbf{k}_\parallel}{4\pi^2} + \int \mathbf{e}^-_{\text{exp}}(\mathbf{k}_\parallel) \exp(\mathrm{i}\mathbf{k}_\parallel \cdot \mathbf{r}_\parallel - \mathrm{i}\gamma z) \frac{\mathrm{d}^2\mathbf{k}_\parallel}{4\pi^2}, \tag{26}$$

where $\mathbf{r} = (\mathbf{r}_\parallel, z)$ and $\gamma(\mathbf{k}_\parallel) = (k^2 - \mathbf{k}_\parallel^2)^{1/2}$, with $k = \omega/c$ and the determination $\mathrm{Re}(\gamma) > 0$ and $\mathrm{Im}(\gamma) > 0$. The integration on the wavevector $\mathbf{k}_\parallel$ is taken over the interval $0 < |\mathbf{k}_\parallel| < \infty$. The amplitudes of the plane waves propagating (or decaying) toward $z > 0$ are denoted by $\mathbf{e}^+_{\text{exp}}(\mathbf{k}_\parallel)$. The amplitudes of the plane waves propagating (or decaying) toward $z < 0$ are denoted by $\mathbf{e}^-_{\text{exp}}(\mathbf{k}_\parallel)$. All fields can be expanded in the same way. The expansions of the fields $\mathbf{E}_{\text{rec}}$ and $\mathbf{H}_{\text{rec}}$ in the reciprocal situation only have plane waves propagating (or decaying) along the direction $z < 0$, because the half-space $z < 0$ does not contain any (primary or induced) sources. Note that the presence of a flat substrate in the experimental situation could be accounted for. This slightly changes the definition of the reciprocal fields (the substrate has to be included in the reciprocal situation). For tutorial reasons, we have restricted the present work to the simplest situation. The complete situation is studied in [55].

Inserting the angular-spectrum representation of the experimental and reciprocal fields into Eq. (25) yields after tedious but straightforward algebra

$$\mathbf{E}_{\rm det} \cdot \mathbf{j}_{\rm rec} = \mathbf{E}_{\rm rec}(\mathbf{r}_{\rm sou}) \cdot \mathbf{j}_{\rm sou} - \frac{2}{\omega\mu_o} \int \gamma(\mathbf{k}_\parallel)\, \mathbf{e}_{\rm rec}(-\mathbf{k}_\parallel) \cdot \mathbf{e}^+_{\rm exp}(\mathbf{k}_\parallel)\, {\rm d}^2\mathbf{k}_\parallel \,, \tag{27}$$

where the summation extends from $0 < |\mathbf{k}_\parallel| < \infty$. An equivalent expression is obtained by transforming the fields back into direct space. It reads

$$\mathbf{E}_{\rm det} \cdot \mathbf{j}_{\rm rec} = \mathbf{E}_{\rm rec}(\mathbf{r}_{\rm sou}) \cdot \mathbf{j}_{\rm sou} - \frac{2{\rm i}}{\omega\mu_o} \int_S \left[\frac{\partial}{\partial z}\mathbf{E}_{\rm rec}(\mathbf{r}_\parallel, z)\right] \cdot \mathbf{E}^+_{\rm exp}(\mathbf{r}_\parallel, z)\, {\rm d}^2\mathbf{r}_\parallel \,, \tag{28}$$

where the summation is taken over a plane at a constant height z in the gap region (surface S represented by the horizontal line in Fig. 15.3). The field $\mathbf{E}^+_{\rm exp}$ corresponds to the first integral in Eq. (26). It describes the field illuminating the probe in the experimental situation, and contains only plane waves which propagate (or decay) toward $z > 0$.

Equations (27) and (28) are exact expressions for the electric field measured at the detector in the experimental situation. They both describe the coupling between this field and the field emitted by the structure under investigation ($\mathbf{E}_{\rm exp}$) by an integral relationship involving the field in the reciprocal situation. Depending on the situation under study, one expression or the other may be useful.

- Expression (28) (direct space) shows that *the detected field is given by an overlap integral between the experimental field and the z-derivative of the reciprocal field.* The latter can be identified with *the response function of the instrument*, describing the localization of the detection process as well as the influence of the experimental parameters (spectral or polarization response, for example).
- Expression (27) (Fourier space) describes how each spatial frequency $\mathbf{K}$ of the experimental field is detected. The coupling factor is proportional to $\mathbf{e}_{\rm rec}(-\mathbf{K})$, showing that the detection of a given spatial frequency $\mathbf{K}$ is efficient only if that frequency is present in the spectrum of the reciprocal field. In other words, a tip producing high-spatial frequencies in its scattered field when it is illuminated from the detector is able to detect the high spatial frequencies of a localized field, when it is used as a probe.

In optics, one usually measures the intensity of the fields, which is proportional to $|\mathbf{E}_{\rm det}|^2$. Therefore, for a polarized detection, the SNOM signal is given by the square modulus of Eq. (28). For an unpolarized detection one should add the intensities corresponding to two orthogonal directions of the fictitious reciprocal source $\mathbf{j}_{\rm rec}$. The form of the relationship between the detected field and the measured near field also allows us to study the measurement of near-field spatial correlations. This point was discussed in [61].

Finally, let us put forward that an expression similar to Eq. (28) was derived as a generalization of Bardeen's formalism, originally developed for electron tunneling between two weakly coupled electrodes. This result shows that SNOM and STM (scanning tunneling electron microscopy) can be handled with the same formalism. For more details on the derivation and its implications, see [62].

15.4.1.3. Calculation of the Response Function

The result in Eq. (28) shows that the near-field measurement can be described using a response function $(\partial/\partial z)\mathbf{E}_{\text{rec}}$. The key quantity is the reciprocal field. It characterizes the detection process, including the role of the tip. In general, this field is the solution of a difficult scattering problem, involving complex structures (probe, detection optics). Numerical simulations are in many cases the only way to solve this problem, at least in simplified geometries. Nevertheless, in some cases, analytical models can be used to describe the near field generated by the probe in the reciprocal situation. For example, in collection-mode techniques using an etched fiber with a metal coating and a small aperture (aperture SNOM), the reciprocal field is the field emitted by the aperture when it is illuminated by incident modes in the fiber. Such a field can be calculated numerically,[63] but is also approximately described by the Bethe–Bouwkamp model.[64,65] Another good example is the apertureless SNOM using sharp metallic tips.[66] In this case, the reciprocal field can be described by the field scattered by a perfectly conducting cone when it is illuminated by an incident plane wave. Such a model was introduced in SNOM by Cory *et al.*[67] Close to the tip of the cone, in the zone corresponding to $kr \ll 1$ where r is the distance from the tip (see Fig. 15.4), the electric field has an analytical expression. Such analytical (and approximate) expressions for near fields scattered by simple shapes can be found in refs.[68,69]

When the cone in Fig. 15.4 is illuminated by a point source located in the far field (the incident field being a plane wave in this case), the electric field close to the tip apex ($kr \ll 1$) is given by[68]

$$\mathbf{E} = k(kr)^{\nu-1} \sin\beta \left[\mathbf{u}_r + \frac{\mathbf{u}_\theta}{\nu} \frac{\partial}{\partial \theta} \right] a(\theta_o, \theta, \alpha)\,, \tag{29}$$

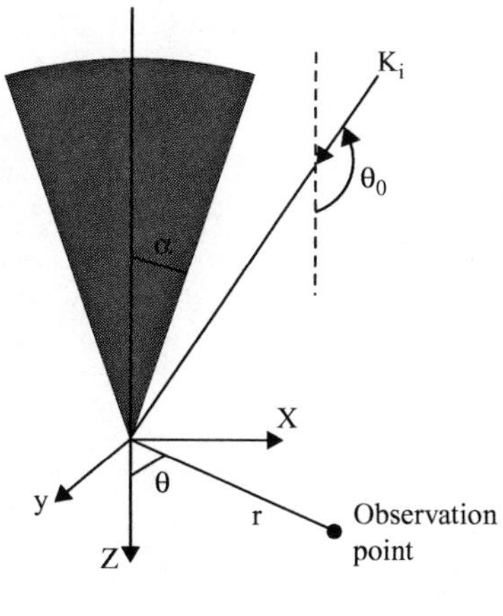

FIGURE 15.4. Tip shape used for modeling apertureless SNOM. Perfectly conducting cone, with semiangle α.

where a is a function of the incidence angle θ_o, of the observation angle θ and of the cone semiangle α. The other parameters are $k = \omega/c$, the polarization angle of the incident plane wave β ($\beta = 0$ for TE, or s-polarized illumination, $\beta = \pi/2$ for TM, or p-polarized illumination). $\mathbf{u}_r$ and $\mathbf{u}_\theta$ are the unit vectors in spherical coordinates. ν is a real and positive number, satisfying $\nu \leq 1$, and depending on the cone angle only. Note that the field close to the tip apex has a spatial structure which does not depend on the illumination conditions (direction of incidence and polarization). The parameters θ_o and β appear in an amplitude factor only. Moreover, let us point out that at larger distances from the tip apex (or when $\beta = 0$), the singular term in Eq. (29) does not describe the field by itself. Subsequent terms of higher order in kr have to be taken into account.[68]

Expression (29) describes the reciprocal field close to the tip. It can be used to calculate the response function $(\partial/\partial z)\mathbf{E}_{\text{rec}}$ appearing in Eq. (28). The spatial dependence of the z and x components of the response function is shown in Fig. 15.5.

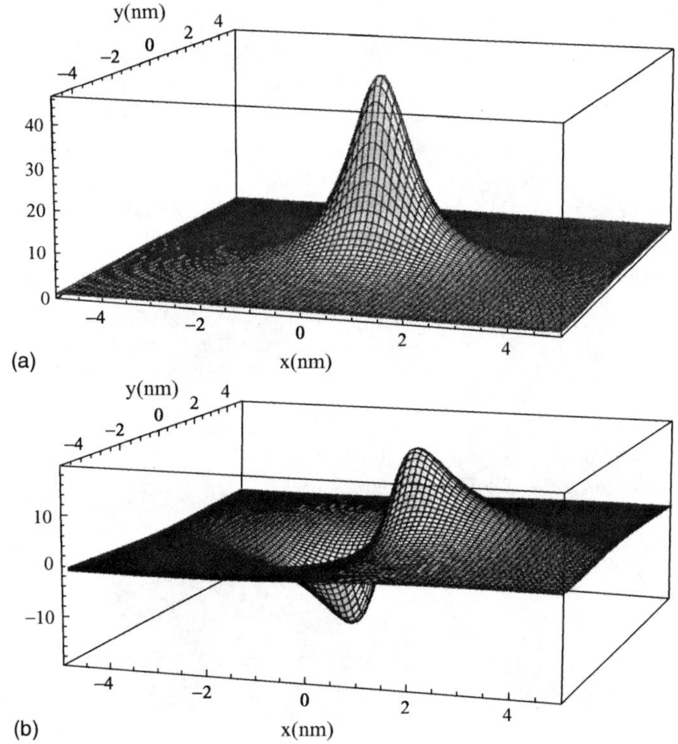

FIGURE 15.5. The z-component (left) and x-component (right) of the response function $(\partial/\partial z)\mathbf{E}_{\text{rec}}$ versus the lateral position $\rho = (x, y)$ in a plane perpendicular to the cone axis. Distance between the plane and the tip: $z = 1$ nm. The cone semi-angle is $\alpha = 3^o$.

First, we see that the z-component dominates in amplitude (see Fig. 15.5 (left)). As a consequence, the z-component of the experimental field will be detected with a higher signal level than the x-component (due to the dot product in Eq. (28)). Moreover, the full width at half maximum (FWHM) of the response function is a measure of the potential resolution of the system (although a precise discussion of the resolution should take into account the finite radius of curvature of the tip). We see in Fig. 15.5 that the FWHM of the response function is of the order of the distance between the observation plane and the tip. A consequence of this result is that multiple scattering is not necessary to attain a high resolution (i.e. at the nanometer scale).

15.4.2. Polarization Response

In order to illustrate the approach that has been developed, we will discuss the polarization behavior of an apertureless SNOM using a metallic tip, on the basis of experimental results.

Measurements of the signal versus the incident polarization, after reflection on a flat silicon substrate were reported by Aigouy *et al.*.[70] The detection is performed in the normal direction, i.e., in the direction of the tip axis. The polarization of the incident field is described by the angle β ($\beta = 0$ for s, or TE, polarization and $\beta = \pi/2$ for p, or TM, polarization). The result is presented in Fig. 15.6 (see [70] for details).

The result can be explained using Eq. (28). The field at the detector is described by the integral involving the enhanced field close to the tip apex (the first term on the right-hand side gives a negligible contribution in this experimental configuration). When the tip is at a few nanometers from the substrate, the experimental field (which illuminates the tip) is chiefly the enhanced tip field reflected by the substrate. Therefore, the field $\mathbf{E}^{+}_{\text{exp}}$ is proportional to the cone field in Eq. (29), which is itself proportional to $\sin\beta$. The measured signal, proportional to the square modulus of the field at the detector, is proportional to $\sin^2\beta$. This behavior is very close to the experimental result (see Fig. 15.6). Note that this is not the behavior that would be obtained by modeling the tip by a small sphere (dipole).

15.4.3. Spectral Response

Another interesting property of near-field detection is the spectral response. It was shown experimentally that, for apertureless SNOM using metallic tips, the spectral response in the visible is not flat, and depends on the tip geometry (cone angle for a conical tip).[71] Experimental results for two different tips are shown in Fig. 15.7. The measurement was performed in reflection, on a plane aluminum mirror (flat spectral response in the visible with a reflection factor $R = 0.9$), in a confocal configuration. The spectra are normalized by the spectra obtained with a far-field setup, i.e., without the tip (see [71] for details).

The spectral response can also be explained using Eq. (28). The field close to the tip apex is described by the cone model, Eq. (29). The frequency dependence

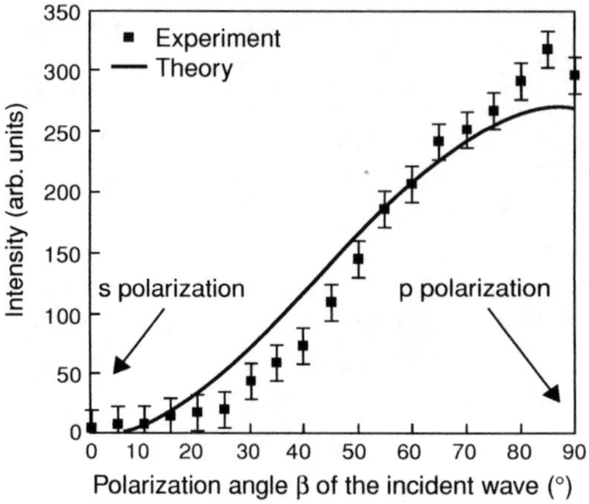

FIGURE 15.6. Response of an apertureless SNOM using a tungsten tip, versus the incident polarization (described by the angle β). The theoretical curve corresponds to a signal proportional to $\sin^2 \beta$. From [70].

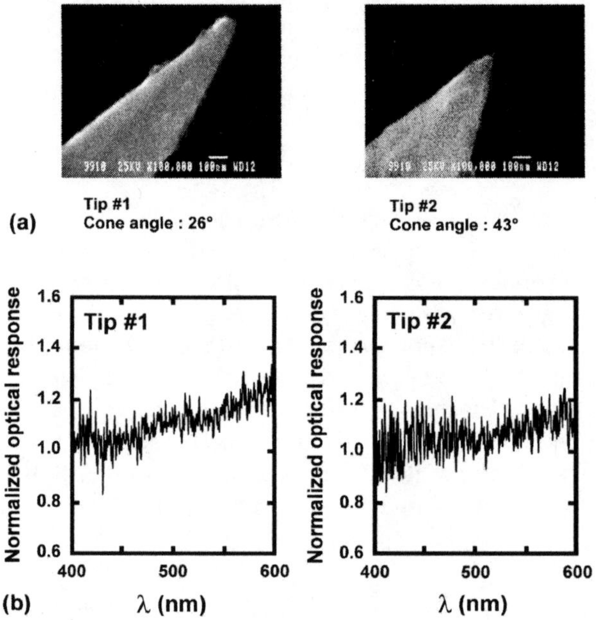

FIGURE 15.7. (*a*) SEM image of the two tungsten tips used to measure the spectral response of an apertureless microscope. (*b*) Spectral response in the visible regime, for both tips. From [71].

is ω^ν, where ν depends only on the cone semiangle.[68] This model is a good description of the reciprocal field $\mathbf{E}_{\text{rec}}$. The experimental field $\mathbf{E}_{\text{exp}}^{+}$ contains several contributions. One of them is the field reflected by the interface (without interaction with the tip) and the enhanced field at the tip apex reflected by the interface. This

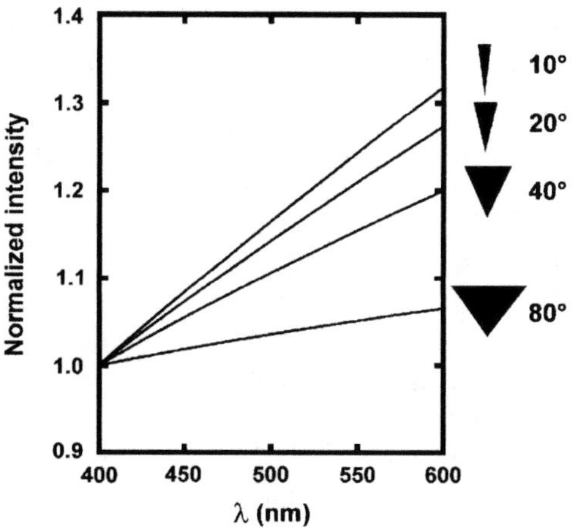

FIGURE 15.8. Theoretical spectral response for several semiangles of the conical tip. The signal is given by expression (28), with a reciprocal field described by the cone model (29). From [71].

last contribution is given by the product of expression (29) by a reflection factor which does not depend on the frequency, in this particular case. In a confocal configuration, as in [71] and in Fig. 15.7, the signal is given by the interference between these two contributions. Therefore, it is proportional to the integral term in Eq. (28), in which both $\mathbf{E}_{\text{rec}}$ and $\mathbf{E}^{+}_{\text{exp}}$ are described by the cone model Eq. (29). As a result, the signal is proportional to $\omega^{2\nu-1}$. The signal predicted by the model, for various semi-angles of the conical tip, is represented in Fig. 15.8, versus the wavelength ($\lambda = 2\pi c/\omega$). The agreement with the measurements is excellent.

Finally, let us point out that the spectral response cannot be studied using an electrostatic model for the field close to the tip apex, although such a model may be able to predict the spatial structure of the near field at small distance.[72]

Also note that the experimental results cannot be explained using a point-dipole model for the tip. Such a model would lead to a spectral dependence proportional to ω^4, in contradiction with the measurements. In view of these results, it seems that the small-sphere (dipole) model for the tip, which has been used to described apertureless setups,[73] does not reproduce (even qualitatively) the polarization and spectral response of such microscopes. The results in this section put forward the importance of a realistic tip for the study of near-field spectroscopy applications.

15.5. Conclusion

The advent of near-field microscopy has motivated new experimental studies of the speckle pattern in the near field. We have discussed the key issues involved in the analysis of the speckle patterns in the near field. A major difference is the fact that the length scale can be smaller than the wavelength. Unlike the far-field speckle pattern, the near-field speckle pattern is nonuniversal in a number of cases. It has been shown that there is a linear filter that relates the power spectral density of the surface profile and the near-field intensity. Information on the speckle pattern can thus be extracted from the intensity speckle pattern. Illumination conditions and polarization affect significantly the observed patterns.

The effects of the evanescent waves and the effects of the coherence of the illuminating field can be studied assuming that a near-field microscope yields a signal proportional to the near-field intensity. However, this assumption is not correct. Indeed, it does not take into account the properties of the tip used in the detection process. Within the weak coupling regime, the relation between the signal and the near field can be described by a linear response function characteristic of the tip and the illumination and detection geometry. It follows that different frequencies or different components of the field do not generate the same signal. In order to analyze quantitatively measured speckle patterns, one should use the appropriate response function.

Appendix: Reciprocity Theorem

Different expressions of the reciprocity theorem of electromagnetism can be found in the literature. We shall recall two of them here. For a derivation of these expressions, see [57–60].

Consider a scatterer, made of a linear material, and described by its constitutive dielectric and magnetic tensors $\overleftrightarrow{\epsilon}(\mathbf{r}, \omega)$ and $\overleftrightarrow{\mu}(\mathbf{r}, \omega)$. Both tensors are assumed to be *symmetric* (necessary condition for reciprocity).

In situation 1, the scatterer is illuminated by a monochromatic source occupying a volume V_1, and described by a current density $\mathbf{j}_1(\mathbf{r}, \omega)$. The resulting electric and magnetic fields at any point $\mathbf{r}$ are $\mathbf{E}_1(\mathbf{r}, \omega)$ and $\mathbf{H}_1(\mathbf{r}, \omega)$.

In situation 2 (another independent situation), the scatterer is illuminated by a monochromatic source occupying a volume V_2, and described by a current density $\mathbf{j}_2(\mathbf{r}, \omega)$. The resulting electric and magnetic fields at any point $\mathbf{r}$ are $\mathbf{E}_2(\mathbf{r}, \omega)$ and $\mathbf{H}_2(\mathbf{r}, \omega)$.

The Lorentz Reciprocity Theorem

For any closed surface Σ, with unit outward normal $\mathbf{n}$, closing a volume V containing the scatterer (and which may contain the sources or not), one has

$$\int_{\Sigma} (\mathbf{E}_1 \times \mathbf{H}_2 - \mathbf{E}_2 \times \mathbf{H}_1) \cdot \mathbf{n}\, \mathrm{d}\Sigma = \int_{V} (\mathbf{j}_1 \cdot \mathbf{E}_2 - \mathbf{j}_2 \cdot \mathbf{E}_1)\, \mathrm{d}V \,, \tag{A1}$$

in which the dependence on ω has been omitted in all fields. This expression was first introduced by Lorentz.[74] Equation (25) in the main text is a direct application of this formula to the geometry in Fig. 15.3.

Another Form of the Reciprocity Theorem

When the surface Σ is a sphere with radius $R \to \infty$, the surface integral vanishes (because the integrand vanishes in the far field). One obtains

$$\int_{V_1} \mathbf{j}_1(\mathbf{r}) \cdot \mathbf{E}_2(\mathbf{r})\, \mathrm{d}V = \int_{V_2} \mathbf{j}_2(\mathbf{r}) \cdot \mathbf{E}_1(\mathbf{r})\, \mathrm{d}V. \tag{A2}$$

For point-dipole sources placed at points $\mathbf{r}_1$ (situation 1) and $\mathbf{r}_2$ (situation 2), one has

$$\mathbf{j}_k(\mathbf{r}) = -\mathrm{i}\omega \mathbf{p}_k\, \delta(\mathbf{r} - \mathbf{r}_k)\,, \tag{A3}$$

where $k = 1, 2$ refers to the situation. This is another form of the reciprocity theorem, which is widely used, for example, in antenna theory.[53] Introducing these forms of the current densities into (A2) leads to

$$\mathbf{p}_1 \cdot \mathbf{E}_2(\mathbf{r}_1) = \mathbf{p}_2 \cdot \mathbf{E}_1(\mathbf{r}_2)\,. \tag{A4}$$

References

1. C. Dainty, *Laser speckle and related phenomena*, C. Dainty ed. (Springer Verlag, Heidelberg, 1984).
2. J.-J. Greffet and R. Carminati, "Image formation in near-field optics," *Prog. Surf. Sci.* **56**, 133–237 (1997).
3. J.W. Goodman, *Statistical Optics* (Wiley, New York, 1985).
4. R. Carminati and J.-J. Greffet, "Near-field effects in spatial coherence of thermal sources", *Phys. Rev. Lett.* **82**, 1660 (1999).
5. S.M. Rytov, Y.A. Kravtsov, and V.I. Tatarskii, *Principles of Statistical Radiophysics*, vol. 3 (Springer-Verlag, Berlin, 1989).
6. E.M. Lifshitz and L.P. Pitaevskii, *Statistical Physics*, vol. 9 (Pergamon Press, Oxford, 1980).
7. C. Henkel, K. Joulain, R. Carminati, and J.-J. Greffet, "Spatial coherence of thermal near fiels," *Opt. Commun.* **186**, 57 (2000).
8. T. Setälä , M. Kaivola, and A.T. Friberg, "Degree of polarization in near fields of thermal sources: Effects of surface waves", *Phys. Rev. Lett.* **88**, 123902 (2002).
9. R. Carminati and J. J. Greffet, "Two-dimensional numerical simulation of the photon scanning tunneling microscope. Concept of transfer function," *Opt. Commun.* **116**, 316–321 (1995).
10. N.F. van Hulst, F.B. Segerink, F. Achten, and B. Bölger, "Evanescent-field optical microscopy: effects of polarization, tip shape and radiative waves," *Ultramicroscopy* **42**, 416–421 (1992).
11. T.L. Ferrell, S.L. Sharp, and R.J. Warmack, "Progress in photon scanning tunneling microscopy (PSTM)," *Ultramicroscopy* **42**, 408–415 (1992).
12. F. de Fornel, P.M. Adam, L. Salomon, and J.P. Goudonnet, "Effect of coherence of the source on the images obtained with a photon scanning tunneling microscope", *Opt. Lett.* **19**, 1082–1084 (1994)
13. F. de Fornel, L. Salomon, J.C. Weeber, A. Rahmani, C. Pic, and A. Dazi, "Effects of the coherence in near-field microscopy," in *Optics at the nanometer scale*, M. Nieto-Vesperinas and N. García eds. (Kluwer Academic Publishers, Dordrecht, 1996).

14. D.P. Tsai, J. Kovacs, Z. Wang, M. Moskovits, V.M. Shalaev, J.S. Suh, and R. Botet, "Photon scanning tunneling microscopy images of optical excitations of fractal metal colloid clusters", *Phys. Rev. Lett.* **72**, 4149 (1994).
15. P. Zhang, T.L.Haslett, C. Douketis, and M. Moskovits, "Mode localization in self-affine fractal interfaces observed by near-field microscopy," *Phys. Rev. B* **57**, 15513 (1998).
16. S. Grésillon, L. Aigouy, A.C.Boccara, J.C. Rivoal, X. Quelin, C. Desmaret, P. Gadenne, V.A. Shubin, A.K. Sarychev, and V.M. Shalaev, "Experimental observation of localized optical excitations in random metal-dielectric films," *Phys. Rev. Lett.* **82**, 4520 (1999).
17. V.M. Shalaev, "Electromagnetic properties of small-particle composites," *Phys. Rep.* **272**, 61 (1996).
18. E.Y. Poliakov, V.A. Markel, V.M. Shalaev, and R. Botet, "Nonlinear optical phenomena on rough surfaces of metal thin films," *Phys. Rev. B* **57**, 14901 (1998).
19. K. Seal, A.K. Sarychev, H. Noh, D.A.Genov, A. Yamilov, V.M. Shalaev, Z.C. Ying, and H. Cao, "Near-field intensity correlations in semicontinuous metal-dielectric films," *Phys. Rev. Lett.* **94**, 226101 (2005).
20. S.I. Bozhevolnyi, I.I. Smolyaninov, and A.V. Zayats, "Near-field microscopy of surface-plasmon polaritons: Localization and internal interface imaging," *Phys. Rev. B* **51**, 17916 (1995).
21. S.I. Bozhevolnyi, B. Vohnsen, I.I. Smolyaninov, and A.V. Zayats, "Direct observation of surface polariton localization caused by surface roughness," *Opt. Commun.* **117**, 417 (1995).
22. S.I. Bozhevolnyi, V.S.Volkov, and K. Leosson, "Localization and waveguiding of surface plasmon polaritons in random nanostructures," *Phys. Rev. Lett.* **89**, 186801 (2002).
23. J.A. Sánchez-Gil, V. Garcia-Ramos, and E.R. Méndez, "Near-field electromagnetic wave scattering from random self-affine fractal metal surfaces: spectral dependence of local field enhancements and their statistics in connection with surface-enhanced Raman scattering," *Phys. Rev. B* **62** , 10515 (2000).
24. J.A. Sánchez-Gil, V. Garcia-Ramos, and E.R. Méndez, "Light scattering from self-affine fractal silver surfaces with nanoscale cutoff: far-field and near-field calculations," *J. Opt. Soc. Am. A* **19**, 902 (2002).
25. T.A. Leskova, A.A. Maradudin, and J. Muñoz-Lopez, "Coherence of light scattered from a randomly rough surface," *Phys. Rev. E* **71**, 036606 (2005).
26. A.A. Chabanov and A.Z. Genack, "Field distributions in the crossover from ballistic to diffusive wave propagation," *Phys. Rev. E* **56**, R1338 (1997).
27. P. Sebbah, B. Hu, A.Z. Genack, R. Pnini, and B. Shapiro, "Spatial-field correlation: The building block of mesoscopic fluctuations," *Phys. Rev. Lett.* **88**, 123901 (2002).
28. V. Emiliani, F. Intonti, D. Wiersma, M. Colocci, M. Cazayous, A. Lagendijk, and F. Aliev, "Near-field measurement of short-range correlation in optical waves transmitted through random media," *J. Micros.* **209**, 173 (2003).
29. B. Shapiro, "Large intensity fluctuations for wave propagation in random media," *Phys. Rev. Lett.* **57**, 2168 (1986).
30. A. Apostol and A. Dogariu, "Spatial correlations in the near field of random media," *Phys. Rev. Lett.* **91**, 093901 (2003).
31. A. Apostol and A. Dogariu, "First- and second-order statistics of optical near fields," *Opt. Lett.* **29**, 235 (2004).
32. A. Apostol and A. Dogariu, "Non-Gaussian statistics of optical near-fields," *Phys. Rev. E* **72**, 025602 (2005).

33. J.C. Weeber, F. de Fornel, and J.P. Goudonnet, "Numerical study of the tip-sample interaction in the photon scanning tunneling microscope," *Opt. Commun.* **126**, 285 (1996).
34. J.J. Greffet and R. Carminati,"Theory of imaging in near-field microscopy," in *Optics at the Nanometer Scale*, M. Nieto-Vesperinas and N. García, eds. (Kluwer Academic Press, Dordrecht, 1996).
35. J.J. Greffet and R. Carminati, "Relationship between the near-field speckle pattern and the statistical properties of a surface," *Ultramicroscopy* **61**, 43–50 (1995).
36. C. Liu and S.-H. Park, "Anisotropy of near-field speckle patterns," *Opt. Lett.* **30**, 1602–1604 (2005).
37. C. Cheng, C. Liu, X. Ren, M. Liu, S. Teng, and Z. Xu, "Near-field speckles produced by random self-affine surfaces and their contrast transitions," *Opt. Lett.* **28**, 1531–1533 (2003).
38. J.J. Greffet, "Scattering of electromagnetic waves by rough dielectric surfaces," *Phys. Rev. B* **37**, 6436(1988).
39. A.A. Maradudin, T. Michel, A.R. Mc Gurn, and E.R. Méndez, "Enhanced backscattering of light from a random grating," *Ann. Phys. (New York)* **203**, 255 (1990).
40. J.J. Greffet, A. Sentenac, and R. Carminati, "Surface profile reconstruction using near-field data," *Opt. Commun.* **116**, 20–24 (1995).
41. D.W. Pohl, W. Denk, and M. Lanz, "Optical stethoscopy: Image recording with resolution $\lambda/20$," *Appl. Phys. Lett.* **44**, 651 (1984).
42. E. Betzig *et al.*, "Near-field scanning optical microscopy (NSOM): development and biophysical applications," *Biophys. J.* **49**, 269 (1986).
43. R.C. Reddick, R.J. Warmack, and T.L. Ferrell, "New form of scanning optical microscopy," *Phys. Rev. B* **39**, 767 (1989); F. de Fornel, J.P. Goudonnet, L. Salomon, and E. Lesniewska, "An evanescent field optical microscope," *Proc. SPIE* **1**, 77 (1989); D. Courjon, K. Sarayeddine, and M. Spajer, "Scanning tunneling optical microscopy," *Opt. Commun.* **71**, 23 (1989).
44. D.W. Pohl and D. Courjon (eds.), *Near-Field Optics* (Kluwer, Dordrecht, 1993); O. Marti and R. Mller (eds.), *Photons and Local Probes*, (Kluwer, Dordrecht, 1995); N. García and M. Nieto-Vesperinas (eds.) *Optics at the Nanometer Scale*, (Kluwer, Dordrecht, 1996); M. Ohtsu (eds.) *Near-Field Nano/Atom Optics and Technology*, (Springer-Verlag, Tokyo, 1998).
45. O. Marti, O. Marti, H. Bielefeldt, B. Hecht, S. Herminghaus, P. Leiderer, and J. Mlynek, "Near-field optical measurement of the surface plasmon field," *Opt. Commun.* **96**, 225 (1993); B. Hecht, H. Bielefeldt, L. Novotny, Y. Inouye, and D. W. Pohl, "Local excitation, scattering, and interference of surface plasmons," *Phys. Rev. Lett.* **77**, 1889 (1996); J.R. Krenn *et al.*, "Squeezing the optical near-field zone by plasmon coupling of metallic nanoparticles," *Phys. Rev. Lett.* **82**, 2590 (1999).
46. S.I. Bozhevolnyi and F.A. Pudonin, "Two-dimensional micro-optics of surface plasmons," *Phys. Rev. Lett.* **78**, 2823 (1997); A.K. Sarychev and V.M. Shalaev, "Electromagnetic field fluctuations and optical nonlinearities in metal-dielectric composites," *Phys. Rep.* **335**, 275 (2000).
47. E. Betzig and J.K. Trautman, "Near-field optics: Microscopy, spectroscopy and surface modification beyond the difraction limit," *Science* **257**, 189 (1992).
48. E. Betzig and R. Chichester, "Single molecules observed by scanning near-field optical microscopy," *Science* **262**, 1422 (1993).
49. R.X. Bian, R.C. Dunn, X.S. Xie, and P.T. Leung, "Single molecule emission characteristics in near-field microscopy," *Phys. Rev. Lett.* **75**, 4772 (1995).

50. T. Guenther, V. Malyarchuk, J.W. Tomm, R. Müller, C. Lienau, and J. Luft, "Near-field photocurrent imaging of the optical mode profiles of semiconductor laser diodes," *Appl. Phys. Lett.* **78**, 1463 (2001).
51. S.I. Bozhevolnyi, "Localization phenomena in elastic surface-polariton scattering caused by surface roughness," *Phys. Rev. B* **54**, 8177 (1996).
52. C. Girard and A. Dereux, "Near-field optics theories," *Rep. Prog. Phys.* **59**, 657 (1996).
53. R.W.P. King, *Electromagnetic Engineering*, vol. 1 (McGraw Hill, New York, 1945), p. 311; D. S. Jones, *The Theory of Electromagnetism* (Pergamon, Oxford, 1964), pp. 59–65.
54. P. Johansson, R. Monreal, and P. Apell, "Theory for light emission from a scanning tunneling microscope," *Phys. Rev. B* **42**, 9210 (1990).
55. J.A. Porto, R. Carminati, and J.-J. Greffet, "Theory of electromagnetic field imaging and spectroscopy in scanning near-field optical microscopy," *J. Appl. Phys.* **88**, 4845 (2000).
56. J.N. Walford, J.A. Porto, R. Carminati, and J.-J. Greffet, "Theory of near-field magneto-optical imaging," *J. Opt. Soc. Am. A* **19**, 572 (2002).
57. L. Landau, E. Lifchitz, and L. Pitaevskii, *Electrodynamics of Continuous Media* (Pergamon Press, Oxford, 1984).
58. Ph. M. Morse and H. Feshbach, *Methods of Theoretical Physics* (McGraw-Hill, New York, 1953), Part I, sect. 7.5.
59. M. Nieto-Vesperinas, *Scattering and Diffraction in Physical Optics*, (Wiley, New York, 1991).
60. R. Carminati, M. Nieto-Vesperinas, and J.-J. Greffet, "Reciprocity of evanescent electromagnetic waves," *J. Opt. Soc. Am. A* **15**, 706–712 (1998).
61. K. Joulain, R. Carminati, J.-P. Mulet, and J.-J. Greffet, "Defintion and measurement of the local density of electromagnetic states close to an interface," *Phys. Rev. B* **68**, 245405 (2003).
62. R. Carminati and J.J. Sáenz, "Scattering theory of Bardeen's formalism for tunneling: New approach to near-field microscopy," *Phys. Rev. Lett.* **84**, 5156 (2000).
63. L. Novotny, D.W. Pohl, and P. Regli, "Light propagation through nanometer-sized structures: the two-dimensional-aperture scanning near-field optical microscope," *J. Opt. Soc. Am. A* **11**, 1768 (1994).
64. H.A. Bethe, "Theory of diffraction by small holes," *Phys. Rev.* **66**, 163 (1944); C.J. Bouwkamp, "Diffraction theory," *Rep. Prog. Phys.* **17**, 35 (1954).
65. D. Van Labeke, F. Baida, D. Barchiesi, and D. Courjon, "A theoretical model for the inverse scanning tunneling optical microscope (ISTOM)," *Opt. Commun.* **114**, 470 (1995).
66. F. Zenhausern, M.P. O'Boyle, and H. K. Wickramasinghe, "Apertureless near-field optical microscope," *Appl. Phys. Lett.* **65**, 1623 (1994); Y. Inouye and S. Kawata, "Near-field scanning optical microscope with a metallic probe tip," *Opt. Lett.* **19**, 159 (1994); P. Gleyzes, A.C. Boccara, and R. Bachelot, "Near field optical microscopy using a metallic vibrating tip," *Ultramicroscopy* **57**, 318 (1995).
67. H. Cory, A.C. Boccara, J.C. Rivoal, and A. Lahrech, "Electric field intensity variation in the vicinity of a perfectly conducting conical probe: application to near-field microscopy," *Microwave Opt. Technol. Lett.* **18**, 120 (1998).
68. J.J. Bowman, T.B.A. Senior, and P.L.E. Uslenghi (eds.), *Electromagnetic and Acoustic Scattering by Simple Shapes* (North-Holland, Amsterdam, 1969).
69. J. Van Bladel, *Singular Electromagnetic Fields and Sources* (Clarendon Press, Oxford, 1991).

70. L. Aigouy, A. Lahrech, S. Grésillon, H. Cory, A.C. Boccara, and J.C. Rivoal, "Polarization effects in apertureless scanning near-field optical microscopy: an experimental study," *Opt. Lett.* **24**, 187 (1999).
71. L. Aigouy, F.X. Andréani, A.C. Boccara, J.C. Rivoal, J.A. Porto, R. Carminati, J.-J. Greffet, and R. Mégy , "Near-field optical spectroscopy using an incoherent light source," *Appl. Phys. Lett.* **76**, 397 (2000).
72. W. Denk and D.W. Pohl, "Near-field optics: microscopy with nanometer-size fields," *J. Vac. Sci. Technol. B* **9**, 510 (1991).
73. F. Zenhausern, Y. Martin, and H.K. Wickramasinghe, "Scanning interferometric apertureless microscopy: Optical imaging at 10 angstroms resolution," *Science* **269**, 1083 (1995); C.J. Hill, P.M. Bridger, G.S. Picus, and T.C. McGill, "Scanning apertureless microscopy below the diffraction limit: comparisons between theory and experiment," *Appl. Phys. Lett.* **75**, 4022 (1999); R. Hillenbrand, T. Taubner, and F. Keilmann, "Phonon-enhanced light-matter interaction at the nanoscale," *Nature* **418**, 159 (2002).
74. H.A. Lorentz, *Collected Papers*, vol. III (Nijhoff, Den Haag, The Netherlands, 1936).

16
Inverse Problems in Optical Scattering

EUGENIO R. MÉNDEZ[†] AND DEMETRIO MACÍAS[‡]

[†]*División de Física Aplicada, Centro de Investigación Científica y de Educación Superior de Ensenada, Ensenada, B. C., México.* [‡]*Institut Charles Delaunay, Université de Technologie de Troyes, CNRS FRE 2848, Laboratoire de Nanotechnologie et d'Instrumentation Optique,12 rue Marie Curie, BP-2060 F-10010 Troyes CEDEX, France.*

16.1. Introduction

One of the main motivations for studying rough surface scattering problems consists in the desire to obtain information about the surface. The information obtained can be of a varied nature. One may be interested, for instance, on the surface profile function, on the optical properties of the surface, or, for random surfaces, on some statistical parameter of the height fluctuations. These are all inverse scattering problems. This chapter contains a review of some aspects of this broad field.

To begin, we consider the interaction of an electromagnetic wave propagating in air or vacuum with the irregular boundary of a homogeneous medium. The response of the system, in the form of scattered light in reflection and transmission, depends on the nature of the incident field (polarization, and spatial and temporal characteristics), on the shape of the boundary, and on the frequency dependent dielectric constant of the medium.

The incident field, as well as those reflected and transmitted, can be expressed as superpositions of plane waves. The linearity of the problem allows us to concentrate on the response of the system to an elementary excitation, in the form of a linearly polarized monochromatic plane wave. This response is given by the scattering amplitudes in reflection $R_{\alpha\beta}(\mathbf{q}_\parallel|\mathbf{k}_\parallel)$ and transmission $T_{\alpha\beta}(\mathbf{q}_\parallel|\mathbf{k}_\parallel)$, which represent the amplitudes of the plane wave components propagating with parallel wavevector $\mathbf{q}_\parallel$ in response to the excitation by a plane wave with parallel wavevector $\mathbf{k}_\parallel$. Here parallel refers to the plane of the surface in the absence of roughness. The subscripts α and β denote the linear polarization components of the incident and scattered light in which the electric field is parallel (p) or perpendicular (s) to the plane of incidence, defined by the incident wavevector and the normal to the surface.

The direct scattering problem consists in the determination of $R_{\alpha\beta}(\mathbf{q}_\parallel|\mathbf{k}_\parallel)$ and $T_{\alpha\beta}(\mathbf{q}_\parallel|\mathbf{k}_\parallel)$ for a given geometry and optical properties of the media involved. The classical approaches to solve the direct problem are described in [1]. Reviews on recent progress can be found in [2–5]. One important inverse problem, consists in the reconstruction of the surface profile function given these data. Optical detectors, however, are not sensitive to the phase of the field, and respond only to its time-averaged power. The inverse problem of recovering the surface profile function from intensity data is more challenging. Work on both of these problems will be reviewed in this chapter. We will only consider inverse problems with far-field data. Near-field scattering problems are treated elsewhere in this book.

Often, the surface is not known in detail, and can only be specified in statistical terms. The surface profile function is then considered to be a realization of a zero-mean stationary random process. It is then clear that the scattered fields are also random. Speckle theory deals with the statistics of these random fields.[6] Most of the work on speckle has been done with simple, approximate models for the interaction between the incident field and the surface. Rigorous work, on the other hand, has concentrated mainly on the calculation of the first few moments of the scattered field (e.g. $\langle R_{\alpha\beta}(\mathbf{q}_\parallel|\mathbf{k}_\parallel)\rangle$ or $\langle|R_{\alpha\beta}(\mathbf{q}_\parallel|\mathbf{k}_\parallel)|^2\rangle$). The inverse problem can consist, in this case, in the determination of a statistical property or parameter of the surface from an average property of the field. Reviews on these aspects of the inverse problem are given in [7, 8].

To solve inverse scattering problems, one needs to solve first the direct problem. This can be solved either approximately or with rigorous computer simulations. Our aim, in this review, is to provide a concise overview of the field. The literature is rather extensive and we will not attempt to provide a comprehensive account of this broad field. The review is organized as follows. In the first section, we present a simple method to treat the direct problem. The thin phase screen model will serve as the basis on which we illustrate the different inversion schemes. We then survey some established scattering techniques to estimate the standard deviation and the power spectral density of surface heights. The following two sections deal with the recovery of surface profile functions starting from far-field data. We consider first the case in which complex amplitude data are available, and then the case in which the phase information is lost. Finally, we present a brief discussion and our main conclusions.

16.2. The Scattering Amplitude

In this section, we consider the direct problem of the scattering of light by rough surfaces. To simplify the presentation, we center our attention on the case of surfaces with one-dimensional roughness. The discussion is based on a simplified model to describe the interaction of light with the rough surface. The inversion procedures surveyed here can all be illustrated with this model. Relevant more rigorous results, and results for the case of two-dimensional surfaces, will be mentioned or referenced.

Since beams, or other more general fields, can be expressed as superpositions of plane waves, we consider that the surface is illuminated by a monochromatic plane wave of frequency ω. A time dependence of the form $e^{-i\omega t}$ is assumed, but the explicit reference to it is suppressed. The medium of incidence is vacuum, and the surface is characterized by its profile and its complex, frequency-dependent dielectric functions $\epsilon(\omega)$ or, equivalently, by its complex refractive index $n_c = \sqrt{\epsilon(\omega)}$. The plane of incidence is the x_1x_3-plane, and the one-dimensional surface is invariant along x_2. Then, the scattering equations for p-polarized light, characterized by the magnetic field vector $\mathbf{H}(x_1, x_3) = (0, H_2(x_1, x_3), 0)$, and for s-polarized light, characterized by the electric field vector $\mathbf{E}(x_1, x_3) = (0, E_2(x_1, x_3), 0)$, are decoupled.

The electromagnetic field in the vacuum region $x_3 > \zeta(x_1)_{\max}$ is given by the sum of the incident plane wave and the scattered field:

$$\psi^>(x_1, x_3) = \psi^{\text{inc}}(x_1, x_3) + \int_{-\infty}^{\infty} \frac{dq}{2\pi} R(q|k)\, e^{iqx_1 + i\alpha_0(q)x_3}, \tag{1}$$

where $\psi^>(x_1, x_3) = H_2^>(x_1, x_3)$ for p-polarized light and $\psi^>(x_1, x_3) = E_2^>(x_1, x_3)$ for s-polarized light,

$$\psi^{\text{inc}}(x_1, x_3) = e^{ikx_1 - i\alpha_0(k)x_3}, \tag{2}$$

and $\alpha_0(q) = [(\omega/c)^2 - q^2]^{\frac{1}{2}}$, with $\operatorname{Re}\alpha_0(q) > 0$, $\operatorname{Im}\alpha_0(q) > 0$. The scattering amplitude $R(q|k)$ determines the amplitude of the wave scattered from the state with parallel wavenumber component $k = (\omega/c)\sin\theta_0$ into the state with parallel wavenumber component q. The states with $|q| < \omega/c$ represent propagating waves that are characterized by the scattering angle θ_s with $q = (\omega/c)\sin\theta_s$.

The differential reflection coefficient (DRC), which is defined as the fraction of the total flux incident onto the surface that is scattered into the angular interval $d\theta_s$ about the scattering direction defined by the angle θ_s, is given by

$$\left(\frac{\partial R}{\partial \theta_s}\right) = \frac{1}{2\pi L_1} \frac{\cos^2\theta_s}{\cos\theta_0} |R(q|\mathrm{k})|^2, \tag{3}$$

where L_1 denotes the length of the x_1-axis covered by the surface. The calculation of $R(q|k)$ is the central problem of rough surface scattering theory. We next outline a simple approximate method to solve the direct problem.

16.2.1. The Thin Phase Screen Model

A thin-phase screen may be visualized as a layer of negligible thickness that introduces upon reflection (transmission) phase variations in the reflected (transmitted) wave, without introducing any amplitude variations.[7,9] We now describe this approximate way of relating the height variations on the surface with the phase variations in the reflected (transmitted) wave.

Consider the reflection geometry depicted in Fig. 16.1(a). An incoming wave, traveling in a direction defined by the angle of incidence θ_0, is incident upon the scatterer, and is scattered in a direction defined by the angle θ_s. Suppose now that both, the interface and the scatterer, are displaced vertically by a distance ζ, as indicated in Fig. 16.1(b). We are interested in the phase difference introduced by this vertical displacement. The extra length traversed by the light when the scatterer is displaced is indicated with the short-dashed line segments in Fig. 16.1(b). Thus from the geometry shown in the figure we find that the change in the optical path

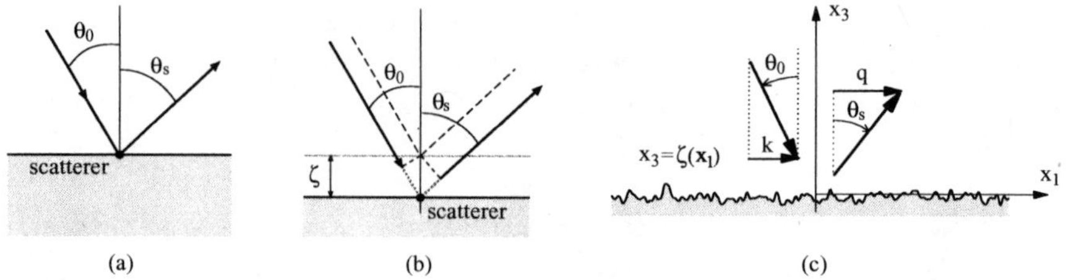

FIGURE 16.1. Illustration of the change in the optical path of a reflected ray upon a vertical displacement of a scatterer.

length due to this (negative) change in height is given by

$$\phi_b - \phi_a = \left(\frac{\omega}{c}\right)(\cos\theta_0 + \cos\theta_s)\zeta. \tag{4}$$

Note that for a positive displacement the sign of the phase change must be reversed. Thus, for a surface with height variations $\zeta(x_1)$ (see Fig. 16.1(c)) the random phase introduced upon reflection coincides with the result obtained with the Kirchhoff approximation.[1,10]

For the transmission geometry we consider the situation depicted in Fig. 16.2(a). A scatterer is placed on the interface separating two semiinfinite media with refractive indices n_d and n_0. The wave is incident from the dielectric medium with refractive index n_d in a direction given by the internal angle of incidence θ_d. After interacting with the scatterer, the wave is scattered into the medium (usually vacuum) with refractive index n_0, in a direction defined by the angle of scattering θ_s. Suppose now that both the interface and the scatterer are displaced vertically by a distance ζ. The segments that are responsible for the phase difference are denoted by the short-dashed lines in Fig. 16.2(b). From the figure we find that the optical path difference between these trajectories is given by

$$\phi_b - \phi_a = \left(\frac{\omega}{c}\right)(n_d\cos\theta_d - n_0\cos\theta_s)\zeta. \tag{5}$$

Often, in practical situations, the medium with index n_d is terminated by a lower flat interface with a medium of refractive index n_0. That is, the sample consists of a dielectric slab with a rough back surface [see Fig. 16.2(c)]. In such a situation, one must write the internal angle θ_d in terms of the angle of incidence θ_0 on the slab using Snell's law.

We can then conclude that, in the thin phase screen approximation, the scattering amplitude is given by the following expression:

$$R(q|k) = A_0 \int_{-\infty}^{\infty} dx_1\, e^{-iv_1x_1 - iv_3\zeta(x_1)}, \tag{6}$$

where $A_0 = \psi_0\kappa_0$, ψ_0 is the amplitude of the incident plane wave, κ_0 is a constant that accounts for the average reflectance or transmittance of the sample, and

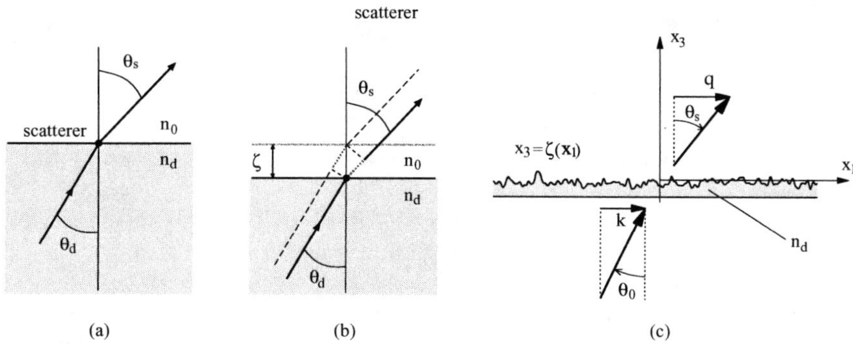

FIGURE 16.2. Illustration of the change in the optical path of a transmitted ray upon a vertical displacement of a scatterer.

$v_1 = (\omega/c)[\sin\theta_s - \sin\theta_0]$. For a reflection geometry

$$v_3 = (\omega/c)[\cos\theta_0 + \cos\theta_s], \tag{7}$$

while for the case of transmission through a dielectric slab with a rough surface

$$v_3 = \frac{\omega}{c}\left[n_d\sqrt{1-\left(\frac{n_0}{n_d}\right)^2\sin^2\theta_0} - n_0\cos\theta_s\right]. \tag{8}$$

This simple model provides a useful relationship between the surface height variations and the complex amplitude of the scattered field. The thin phase screen model represents a good approximation when the surface has only lateral features that are larger than the wavelength, small slopes, and does not produce significant amounts of multiple scattering.

The analysis of the inversion schemes presented in the following sections is based on the thin phase screen approximation. The model is adopted here not only for simplicity, but also because it permits the analysis of a variety of situations (such as transmission and reflection geometries) in a unified way. The extension of this model to two-dimensional surfaces is straightforward.

16.3. Estimation of Statistical Properties of Surfaces

In many practical problems, the profile of the surface is not known and one can only specify it in statistical terms. In dealing with the scattering of light by random surfaces, one is faced with a problem of an electromagnetic nature, compounded by the difficulties of modeling the statistical properties of the surface.

In this section we first present a brief review of the usual statistical models that characterize randomly rough surfaces. The surfaces are assumed planar in the absence of the roughness, and can be classified as two-dimensional and one-dimensional. In the former case, the departure of the surface from the plane (assumed to be the plane $x_3 = 0$) depends on both of the coordinates x_1 and x_2 in that plane. In the latter case it depends on only one of the coordinates in that plane, say x_1. For consistency with the rest of the review, we present here the case of one-dimensional surfaces. The two-dimensional case constitutes a simple extension of these results.

It is normally assumed that the surface profile can be represented by a continuos single-valued function of x_1. The surface profile function is then considered to be a realization of a stationary random process. Without loss of generality, it can also be assumed to be a zero-mean process. Before considering the estimation of the statistical properties of the surface, we present a phenomenological description of the random scattered field and of the kind of averages employed. We then proceed to evaluate the first few moments of the field and establish their connection with the statistical properties of the surface.

16.3.1. Statistical Characterization of Random Surfaces

The n-order joint probability density function (PDF) of surface heights is denoted by

$$P_Z(\zeta_1, \zeta_2, \ldots, \zeta_j, \ldots, \zeta_n),$$

where $\zeta_j = \zeta(x_1^{(j)})$ denote the surface heights at specified points in space. The characteristic function $M_Z(\omega_1, \ldots, \omega_n)$ is given by the n-order Fourier transform of $P_Z(\zeta_1, \ldots, \zeta_n)$.

A complete description of a random process would involve knowledge of the nth-order joint probability density function for all n (see, e.g., [11] p. 60). Normally, only a partial description of the process is possible. In some cases, this partial description is sufficient to solve a given scattering problem. For instance, with the usual experimental geometries and the theories based on the Kirchhof approximation or the thin phase screen model, knowledge of the second-order probability density function is enough to specify the scattered field in statistical terms.

A basic quantity for the description of the random process is the two-point height correlation function

$$\langle \zeta(x_1)\zeta(x_1') \rangle = \delta^2 W(|x_1 - x_1'|), \tag{9}$$

where the angled brackets represent an average over an ensemble of realizations of the surface, and

$$\delta^2 = \langle \zeta^2(x_1) \rangle. \tag{10}$$

The parameter δ represents the standard deviation or rms height of the surface. The fact that the autocorrelation function $W(|x_1 - x_1'|)$ depends on the coordinates x_1 and x_1' only through their difference is a reflection of the assumed stationarity of $\zeta(x_1)$.

It is also useful to introduce the Fourier integral representation of the surface profile function,

$$\zeta(x_1) = \int \frac{dk}{2\pi} e^{ikx_1} \hat{\zeta}(k), \tag{11}$$

where k is a wave vector. For stationary surfaces, the two-point correlation of the Fourier coefficient $\hat{\zeta}(k)$ id given by

$$\langle \hat{\zeta}(k)\hat{\zeta}(k') \rangle = 2\pi\delta(k + k')\delta^2 g(|k|), \tag{12}$$

where $\delta(k + k')$ represents a delta function. The function $g(|k|)$ appearing in Eq. (12) is called the power spectrum of the surface roughness, and is defined by

$$g(|k|) = \int dx_1 \, e^{-ikx_1} W(|x_1|). \tag{13}$$

The power spectrum $g(|k|)$ is a nonnegative function of $|k|$, and is normalized according to

$$\int \frac{dk}{2\pi} g(|k|) = 1. \tag{14}$$

A common assumption made in scattering theory is that the surface profile constitutes a realization of a Gaussian random process. For such processes, the joint probability density functions are known to all orders (see [11], p. 82) and, for zero-mean processes, are completely determined by the two-point height correlation function.

Another common assumption in scattering work is that the correlation function $W(|x_1|)$ is also Gaussian:

$$W(|x_1|) = e^{-x_1^2/a^2}. \tag{15}$$

The parameter a is known as the correlation length of the surface.

Other kinds of correlation functions and power spectra have been considered, including power law spectra and fractal surfaces (see e.g. [12–17]). On the other hand, although many surfaces of practical interest have non-Gaussian statistics, this problem has received less attention [18–21]. One of the chief difficulties in theoretical studies of non-Gaussian random surfaces is that there are not many random functions $\zeta(x_)$ for which the n-order joint PDF is known.

16.3.2. The Random Field and Its Averages

When coherent light interacts with a reflecting randomly rough surface the scattered field is also a (complex, in this case) random process. The reflected light contains, in general, a component that travels in the specular direction, and a diffuse component that can appear over the whole of the hemisphere. The field associated with the specular component has the same form and propagation behavior as the field reflected by a flat surface but, of course, since part of the incident power is taken

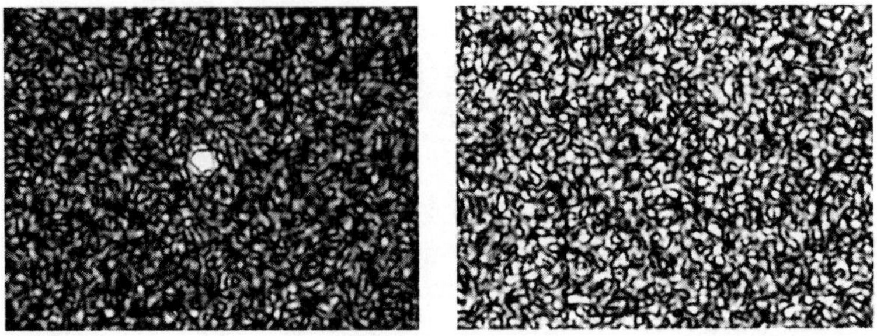

FIGURE 16.3. Photographs showing the speckle patterns produced by the interaction of light with a weak (a) and a strong (b) diffuser.

by the diffuse component, its power is diminished (see Fig. 16.3(a)). The diffuse component presents random intensity fluctuations; a phenomenon called speckle.[6] This random field is usually characterized in terms of averages and, at this point, it is worth discussing the nature of these averages.

One possibility is to consider averages over an ensemble of statistically equivalent surfaces. This kind of average is well-suited for theoretical work, but is impractical in the majority of experimental situations. Possible exceptions to this are surfaces that evolve with time and surfaces that are large enough to permit experiments with several statistically equivalent sections. In both cases, an ergodicity assumption must be invoked. In experimental work, one normally works with a single realization of a surface and, to estimate the mean intensity, the speckle fluctuations are smoothed by using a large area detector. Averages are then taken over an angular region that, ideally, is small compared with the angular extent over which the mean intensity varies, and yet, large compared with the speckle fluctuations. An attempt to justify this assumed equivalence has been given in [5].

In theoretical work, the mean intensity of the scattered light can be expressed as the sum of the intensity associated with the average field, the so-called coherent component, and another term that is called the diffuse component. Here, the averages are taken over an ensemble of statistically equivalent surfaces.

The coherent component is normally associated with the specular component and the incoherent component with the diffuse component. To our knowledge, however, there is no formal proof of the circumstances under which this is true; with nonstationary surfaces, one can find examples for wich this correspondence does not hold. However, since in most practical cases this correspondence appears to be valid, we assume it here for simplicity.

From Eq. (3), the mean differential reflection coefficient is given by

$$\left\langle \frac{\partial R}{\partial \theta_s} \right\rangle = \frac{1}{2\pi L_1} \frac{\omega}{c} \frac{\cos^2 \theta_s}{\cos \theta_0} \langle |R(q|k)|^2 \rangle. \tag{16}$$

Similarly, the coherent component is defined as

$$\left\langle \frac{\partial R}{\partial \theta_s} \right\rangle_{\text{coh}} = \frac{1}{2\pi L_1} \frac{\omega}{c} \frac{\cos^2 \theta_s}{\cos \theta_0} |\langle R(q|k) \rangle|^2$$
$$= \delta(\theta_s - \theta_0)\mathcal{R}(\theta_0)\,, \tag{17}$$

where we have assumed plane wave illumination and an infinite, stationary surface. Here, $\mathcal{R}(\theta_0)$ represents the reflectivity (transmissivity) of the random surface. For beams, $\mathcal{R}(\theta_0)$ represents the fraction of the incident power contained in the coherent component:

$$\mathcal{R}(\theta_0) = \int_{-\pi/2}^{\pi/2} \left\langle \frac{\partial R}{\partial \theta_s} \right\rangle_{\text{coh}} \mathrm{d}\theta_s\,. \tag{18}$$

The incoherent component is then defined as

$$\left\langle \frac{\partial R}{\partial \theta_s} \right\rangle_{\text{incoh}} = \frac{1}{2\pi L_1} \frac{\omega}{c} \frac{\cos^2 \theta_s}{\cos \theta_0} [\langle |R(q|k)|^2 \rangle - |\langle R(q|k) \rangle|^2]. \tag{19}$$

The visibility of the coherent component, i.e., its peak intensity in relation to the diffuse component in the neighborhood of the specular direction depends on (i) the size of the illuminated section of the surface and (ii) the standard deviation of heights, δ. The power contained in the coherent component, however, is independent of the size of the illumination and, for Gaussian processes at least, is determined solely by δ, providing a means for estimating this parameter.[22]

For the case of surfaces whose height variations are much smaller than the wavelength, the angular shape of the diffuse component is related to the power spectral density of the surface.[23,24] For rougher surfaces, the relation between the angular distribution of the diffuse component and the statistics of the surface is more complex. In the limit of surfaces with roughness parameter $\delta \gg \lambda$, the angular distribution represents a map of the distribution of slopes on the surface.

The speckle statistics convey very little information about the surface. Away from the specular direction, the intensity fluctuations obey a universal law[25] that is independent of the roughness parameters. The field correlation function (which defines, among other things, the average speckle size) is primarily determined by the shape of the illumination patch on the surface, and is also fairly independent of the roughness parameters. On the other hand, the motion and decorrelation of the speckle pattern as the angle of incidence is changed does depend on the roughness,[26,27] and can indeed be used to estimate δ.

16.3.3. The Coherent Component

From Eq. (6), the average scattering amplitude can be written as

$$\langle R(q|k) \rangle = A_0 \int_{-\infty}^{\infty} \mathrm{d}x_1\, \mathrm{e}^{-\mathrm{i}v_1 x_1} \langle \mathrm{e}^{-\mathrm{i}v_3 \zeta(x_1)} \rangle. \tag{20}$$

Since by assumption the surface is statistically stationary, the averaged quantity in the integrand is independent of x_1. Then, the average scattering amplitude can be written in the form

$$\langle R(q|k)\rangle = \left\langle e^{-iv_3\zeta(x_1)}\right\rangle R_F(q|k), \tag{21}$$

where $R_F(q|k)$ is the scattering amplitude corresponding to a flat surface of the same material. The term in angled brackets on the right-hand side of this expression can be calculated if the characteristic function associated with the height fluctuations is known. For the case of Gaussian phase fluctuations

$$\left\langle e^{-iv_3\zeta(x_1)}\right\rangle = e^{-v_3^2\delta^2/2}, \tag{22}$$

so that

$$|\langle R(q|k)\rangle|^2 = e^{-v_3^2\delta^2}|R_F(q|k)|^2. \tag{23}$$

Finally, normalizing the reflectivity of the rough sample by the reflectivity of a flat surface we obtain the result

$$\frac{\mathcal{R}(\theta_0)}{\mathcal{R}_F(\theta_0)} = e^{-v_3^2\delta^2}. \tag{24}$$

It should be mentioned that the same result is obtained for two-dimensional surfaces, and with treatments based on the Kirchhoff approximation.[22] It is also worth noting that the result represented by Eq. (24) is independent of the form of the height correlation function and, thus, independent of the correlation length. Departures from this result appear, gradually, as the lateral scale of the surface becomes smaller than the wavelength.[28]

Critical evaluations of several theoretical approaches of the coherent reflectance of randomly rough surfaces[29–31] indicate that, although the phase perturbation theory[28] provides a more accurate result for the case $a << \lambda$, the simple result given by expression (24) represents a good model for the reflectivity of a broad class of surfaces in the optical region of the spectrum. Equation (24) provides, then, an established practical method for the estimation[7,8,32,33] of δ. Although the result is based on the assumption of Gaussian height fluctuations, our experience indicates that the rms height of surfaces that are clearly non-Gaussian (e.g. surfaces with skewed histograms of heights) can still be estimated fairly stably by this method.

Nevertheless, we point out that it is possible to find or design surfaces whose reflectivity will show substantial departures from Eq. (24). Consider, for example, a surface with negative exponential statistics. Since $0 < \zeta(x_1, x_2) < \infty$, the points where $\zeta(x_1, x_2) = 0$ must be turning points of the process and have zero slope. The scattering from these specular points produces constructive interference in the specular direction (i.e., the phases are not random). Not surprisingly, for surfaces that belong to this statistical class, the coherent component cannot be extinguished by increasing the rms height of the surface[13].

For Gaussian (or "Gaussian-like") surfaces, as the rms height increases, the coherent component will, at some point, be indistinguishable from the diffuse background, precluding its measurement. In such circumstances, it is useful to turn a reflective strong scatterer into a weak scatterer by working at a large angle of incidence (i.e., reducing v_3). For the case of transmission, on the other hand, it is best to work in normal incidence and immerse the sample in a medium of nearly the same refractive index.

16.3.4. The Incoherent Component

We first consider the case $\delta \ll \lambda$. The exponential containing $\zeta(x_1)$ in Eq. (6) can be expanded in powers of this variable. Then, the quantity within square brackets appearing in the expression for the incoherent component (Eq. (19)) can be written in the form

$$\begin{aligned}\langle \Delta I(q|k)\rangle &= |A_0|^2 \left\langle \left| \int_{-\infty}^{\infty} \mathrm{d}x_1\, \mathrm{e}^{-\mathrm{i}v_1 x_1} \left[-\mathrm{i}v_3\zeta(x_1) - \frac{v_3^2}{2}\zeta^2(x_1) + \cdots \right] \right|^2 \right\rangle \\ &\approx 2\pi L_1 |A_0|^2 \delta^2 g(|q-k|), \end{aligned} \tag{25}$$

where only the terms up to second order in the surface profile function have been kept, and $g(|q|)$ represents the power spectral density of surface heights. Then, the incoherent component of the mean differential reflection coefficient can be written, approximately, as

$$\left\langle \frac{\partial R}{\partial \theta_s} \right\rangle_{\text{incoh}} = |A_0|^2 \, \frac{\omega}{c} \frac{\cos^2\theta_s}{\cos\theta_0} \, \delta^2 g(|q-k|). \tag{26}$$

This expression illustrates the fact that for weakly rough surfaces the angular scattering pattern provides a fairly direct map of the power spectral density of the heights of the surface. The same conclusion can be reached using more formal theories and two-dimensional surfaces.[23,24,32,33] With small amplitude perturbation, for instance, the incoherent component can be expressed as the product of the power spectral density (a "surface factor") and an "optical factor" that has a cosine dependence on the angle of incidence, a cosine-squared dependence on the angle of scattering, and also depends on the dielectric constant of the surface and the polarization combination of the incident and scattered waves.[33] Expressions for this optical factor for two-dimensional surfaces can be found, for example, in [34, 35]. Measurements of the angular distribution of the scattered light have become a standard method for characterizing optical surfaces, such as polished or diamond-turned mirrors [32, 33].

The simple relationship between the power spectral density and the scattering pattern is, essentially, a single scattering approximation. It should be noted, however, that even for surfaces with $\delta \ll \lambda$, multiple scattering effects can be significant.[36] An interesting approach to estimate the power spectral density in the presence of multiple scattering, based on a reverse Monte Carlo method, has been reported by Malyshkin *et al.*[37]

Consider now the case $\delta >> \lambda$. The mean intensity $\langle I(q|k)\rangle = \langle |R(q|k)|^2\rangle$ can be written as

$$\langle I(q|k)\rangle = |A_0|^2 \int_{-\infty}^{\infty} \mathrm{d}x_1 \int_{-\infty}^{\infty} \mathrm{d}x_1' \, \mathrm{e}^{-\mathrm{i}v_1(x_1-x_1')} \left\langle \mathrm{e}^{-\mathrm{i}v_3[\zeta(x_1)-\zeta(x_1')]} \right\rangle. \tag{27}$$

With the change of variables $x_1' = x_1 + u$, we can write

$$\langle I(q|k)\rangle = |A_0|^2 L_1 \int_{-\infty}^{\infty} \mathrm{d}u \, \mathrm{e}^{-\mathrm{i}v_1 u} \, g(u), \tag{28}$$

where

$$g(u) = \left\langle \mathrm{e}^{-\mathrm{i}v_3[\zeta(x_1)-\zeta(x_1+u)]} \right\rangle. \tag{29}$$

Due to the assumed stationarity of the surface, $g(u)$ is independent of x_1.

Since $(\omega/c)\delta \ll 1$, we can employ the approximation

$$g(u) = \left\langle \mathrm{e}^{-\mathrm{i}v_3 u\zeta'(x_1)} \right\rangle, \tag{30}$$

which is obtained by expressing the difference $\zeta(x_1) - \zeta(x_1 + u)$ in Eq. (29) in powers of u and retaining only the leading nonzero term. Since surface curvature effects are neglected, this is called the geometrical optics approximation.

This approximation leads to the result

$$\langle I(q|k)\rangle = |A_0|^2 \frac{2\pi L_1}{v_3} P_{\zeta'}\left(\frac{v_1}{v_3}\right), \tag{31}$$

where $P_{\zeta'}(x)$ represents the PDF of slopes on the surface. For the case of reflection from a surface, the mean DRC is then given by

$$\left\langle \frac{\partial R}{\partial \theta_s} \right\rangle = |A_0|^2 \frac{\cos^2\theta_s}{\cos\theta_0[\cos\theta_0 + \cos\theta_s]} P_{\zeta'}\left(\frac{\sin\theta_s - \sin\theta_0}{\cos\theta_s + \cos\theta_0}\right), \tag{32}$$

and for the important case of reflection from a Gaussian correlated, Gaussian surface, one has

$$\left\langle \frac{\partial R}{\partial \theta_s} \right\rangle = \frac{|A_0|^2}{\sqrt{2\pi}\,\sigma_{\zeta'}} \frac{\cos^2\theta_s}{\cos\theta_0[\cos\theta_0 + \cos\theta_s]} \mathrm{e}^{-\left(\frac{\sin\theta_s - \sin\theta_0}{\cos\theta_s + \cos\theta_0}\right)^2/\sigma_{\zeta'}^2}, \tag{33}$$

where $\sigma_{\zeta'} = \sqrt{2}\delta/a$ is the standard deviation of slopes on the surface. Apart from the angular prefactor, the scattering pattern constitutes a map of the probability density function of slopes of the surface. If, for instance, the rms height δ is determined by an independent method, the far-field angular distribution of the mean intensity can be used to determine the correlation length a.

For fully developed speckle patterns, the higher order intensity moments, such as $\langle I^2(q|k)\rangle$, are determined by $\langle I(q|k)\rangle$, and provide no further information on the scatterer.

16.3.5. Angular Correlations

The evolution of speckle patterns produced by randomly rough surfaces as the angle of incidence is changed does depend on the rms height. This information is contained in the angular intensity correlation of the speckle pattern. Speckle correlation techniques can be used to estimate δ for cases in which the coherent component is not detectable.[26,27]

Consider the amplitude correlation $C_R(q,k|q',k') = \langle R(q|k)R^*(q'|k')\rangle$. From Eq. (6), we can write

$$C_R(q,k|q',k') = |A_0|^2 \int_{-\infty}^{\infty} dx_1 \int_{-\infty}^{\infty} dx_1'\, e^{-i[v_1x_1 - v_1'x_1']} \left\langle e^{-i[v_3\zeta(x_1) - v_3'\zeta(x_1')]}\right\rangle, \quad (34)$$

where $v_1' = q' - k'$ and, for the reflection geometry assumed in this subsection, $v_3' = \alpha_0(q') + \alpha_0(k')$. With the change of variable $x_1' = x_1 + u$, we have

$$C_R(q,k|q',k') = |A_0|^2 2\pi\delta(v_1 - v_1') \int_{-\infty}^{\infty} du\, e^{iv_1'u} \left\langle e^{-i[v_3\zeta(x_1) - v_3'\zeta(x_1+u)]}\right\rangle. \quad (35)$$

The technique is particularly useful in cases in which $\langle R(q|k)\rangle$ is negligible (i.e. when $\delta > \lambda$). We then employ the geometrical optics approximation to write

$$\left\langle e^{-i[v_3\zeta(x_1) - v_3'\zeta(x_1+u)]}\right\rangle = \left\langle e^{iv_3'\zeta'(x_1)u]}\right\rangle \left\langle e^{-i[v_3 - v_3']\zeta(x_1)}\right\rangle, \quad (36)$$

where $\zeta'(x_1)$ denotes the derivative of $\zeta(x_1)$, and we have assumed that the heights and slopes on the surface are uncorrelated. This is true for the case of a Gaussian random process,[1] which is assumed here. With the use of Eqs. (28) and (30), the amplitude correlation can be written in the form

$$C_R(q,k|q',k') = \frac{2\pi}{L_1}\delta(v_1 - v_1')\, \langle I(q'|k')\rangle\, e^{-[v_3 - v_3']^2\delta^2/2}. \quad (37)$$

For a fully developed speckle pattern,[25] the intensity covariance defined as $C_{\Delta I}(q,k|q',k') = \langle I(q|k)I(q'|k')\rangle - \langle I(q|k)\rangle\langle I(q|k)\rangle$ can be expressed in terms of the amplitude correlation (see e.g. [11], p. 108):

$$C_{\Delta I}(q,k|q',k') = |C_R(q,k|q',k')|^2. \quad (38)$$

With the help of relations (4.1) in [38] and (24a) in [39], we can finally write

$$C_{\Delta I}(q,k|q',k') = \delta_{(q-k),(q'-k')}\langle I(q'|k')\rangle\, e^{-[v_3 - v_3']^2\delta^2}. \quad (39)$$

Expression (39) shows that, as the angle of incidence is changed, the speckle pattern translates angularly according to the law $(q - k) = (q' - k')$ (the so-called memory effect) and evolves or decorrelates through a factor that depends on δ. This property forms the basis of a method for the estimation of the rms height in the range[27] $1\ \mu\text{m} < \delta < 30\ \mu\text{m}$.

16.4. Estimation of the Surface Profile from Complex Amplitude Data

In the last section we introduced a simple scattering theory and illustrated the dependence of the first few statistical moments of the field on the statistical properties of the surface and, in particular, on its rms height. In the rest of the review, we deal with the recovery of the surface profile function starting from monochromatic far-field data. We begin, in this section, with the simpler problem of recovering the profile from amplitude data. It is assumed that angle-resolved complex amplitude data are available for several angles of incidence.

Inverse scattering procedures requiere the solution of the direct problem. Formal inversion schemes are based on approximate solutions of the direct scattering problem. In contrast to the large body of literature on the direct problem, only a few inversion schemes have been reported. An iterative algorithm based on the Kirchhoff approximation has been proposed by Wombell and DeSanto.[40] These authors have also studied an algorithm based on small amplitude perturbation theory.[41] Both of these algorithms consider data corresponding to only one angle of incidence, and seem to work well when the Kirchhoff approximation[40] or the small amplitude approximation[41] are valid.

Algorithms that make use of data corresponding to several angles of incidence have been proposed by Quartell and Sheppard.[42,43] These algorithms are inspired by the actions of a confocal microscope in a profiling mode.[44,45] They are based on the Kirchhoff approximation and seem to be more robust than the algorithms that use only a single angle of incidence. More recently, an algorithm based on similar notions has been studied by Macías, Méndez, and Ruiz-Cortés.[46] This "wavefront matching algorithm" is reviewed here.

16.4.1. *Inversion Algorithm*

As we have seen, for an incident plane wave (Eq. (2)), the response of the surface can be characterized by the scattering amplitude $R(q|k)$ (see Eq. (1)). Let us now consider a situation in which the surface is illuminated by a converging beam. We can write the incident field as a Debye integral[47]

$$\psi^{\text{inc}}(x_1 - \xi; x_3 - \eta) = \int_{-\infty}^{\infty} \frac{\mathrm{d}k}{2\pi} P_{\text{inc}}(k)\, \mathrm{e}^{\mathrm{i}k(x_1-\xi)}\, \mathrm{e}^{-\mathrm{i}\alpha_0(k)(x_3-\eta)}, \tag{40}$$

where the parameters (ξ, η) are the coordinates of the point of convergence of the beam, the pupil function $P_{\text{inc}}(k)$ is assumed to be given by

$$P_{\text{inc}}(k) = \text{rect}\left(\frac{k}{2k_{\text{max}}}\right), \tag{41}$$

rect (x) represents the rectangle function, $k_{\text{max}} = (\omega/c)\sin\theta_{0m}$, and θ_{0m} is the maximum value attained by the angles of incidence.

Due to the linearity of the problem, it is clear that the scattered field $\psi^{sc}(x_1, x_3; \xi, \eta)$ can be written as a superposition of plane wave solutions:

$$\psi^{sc}(x_1, x_3; \xi, \eta) = \int_{-\infty}^{\infty} \frac{dk}{2\pi} P_{inc}(k)\, e^{-ik\xi + i\alpha_0(k)\eta} \int_{-\infty}^{\infty} \frac{dq}{2\pi} R(q|k)\, e^{iqx_1 + i\alpha_0(q)x_3}. \tag{42}$$

In other words, if $R(q|k)$ is known for all q and k, the response of the system to an arbitrary incident field can be calculated.

Consider now a fictitious reference beam that arises from a diffraction-limited spot centered on the point of convergence of the incident field (ξ, η). This reference field is defined by the expression

$$\psi^{ref}(x_1 - \xi; x_3 - \eta) = \int_{-\infty}^{\infty} \frac{dp}{2\pi} P_{ref}(p)\, e^{ip(x_1-\xi)}\, e^{i\alpha_0(p)(x_3-\eta)}, \tag{43}$$

where

$$P_{ref}(p) = \mathrm{rect}\left(\frac{p}{2p_{max}}\right), \tag{44}$$

$p_{max} = (\omega/c)\sin\theta_{sm}$, and θ_{sm} is the maximum value considered for the angles of the reference field. This angle will be identified with the maximum angle of detection (or scattering).

The wavefront matching algorithm considers the interference between the reference and scattered fields, integrated over a plane above the surface. The interference term is given by twice the real part of the function

$$U(\xi, \eta) = \int_{-\infty}^{\infty} \psi^{sc}(x_1, x_3; \xi, \eta)\, [\psi^{ref}(x_1 - \xi; x_3 - \eta)]^* \, dx_1. \tag{45}$$

It is postulated that $\mathrm{Re}\{U(\xi, \eta)\}$ is an extremum when $(\xi, \eta) = (x_1, \zeta(x_1))$. The idea behind this is illustrated in Fig. 16.4, where we depict two different situations; in Fig. 16.4(a), the focused beam is reflected by a flat, horizontal mirror located on the focusing plane. In Fig. 16.4(b), on the other hand, the mirror is located in an out-of-focus plane. The reference beam emanates from the point $(x_1, x_3) = (0, 0)$. It is clear that in the first situation the wavefronts of the reflected and reference beams have the same curvature (they are perfectly matched), while in the second one the curvatures are different. At least from an intuitive point of view, when the two fields are in phase, the interference is constructive everywhere and the integral should be maximized. On the other hand, when the wavefront curvatures are different, some cancellation occurs. In this approximation, for a more general surface, the function $\mathrm{Re}\{U(\xi, \eta)\}$ should also be a maximum when the coordinates of the focusing point (ξ, η) coincide with a surface point.

Substitution of Eqs. (42) and (43) into Eq. (45) yields:[46]

$$U(\xi, \eta) = \int_{-\infty}^{\infty} \frac{dk}{2\pi} \int_{-\infty}^{\infty} \frac{dq}{2\pi} P_{inc}(k) P^*_{ref}(q) R(q|k) \Phi(q, k|\xi, \eta), \tag{46}$$

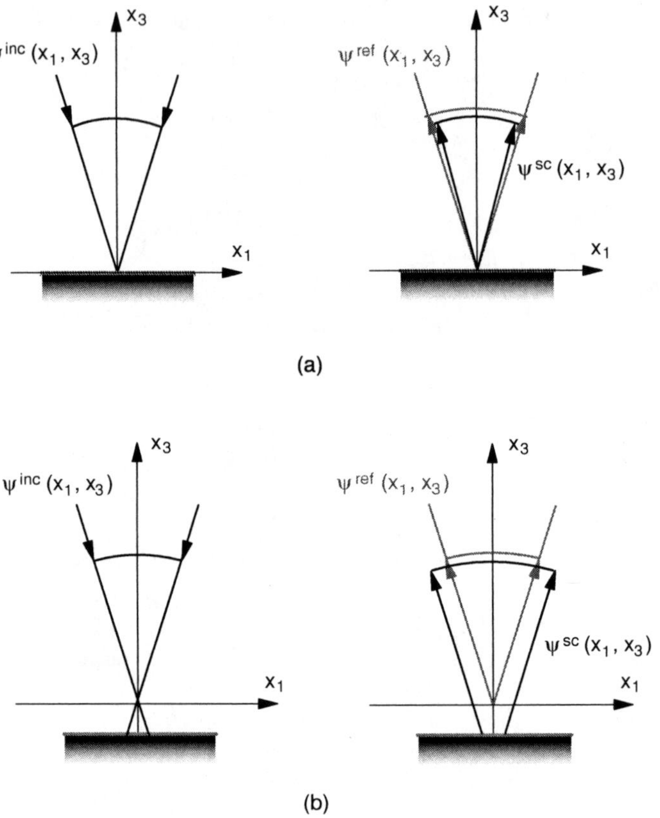

FIGURE 16.4. Illustration of the reference and scattered field in the reflection of a focused beam; (a) mirror in focus and (b) mirror out of focus.

where

$$\Phi(q, k|\xi, \eta) = \mathrm{e}^{-\mathrm{i}(k-q)\xi + \mathrm{i}[\alpha_0(k)+\alpha_0(q)]\eta} . \tag{47}$$

The function $U(\xi, \eta)$ can be visualized as some kind of potential that is obtained through a double integral transform of the scattering amplitude $R(q|k)$.

For the reconstruction of the surface profile function, the wavefront matching algorithm searches of the extrema of the function

$$F^{(\mathrm{wm})}(\xi, \eta) = 2\Re e\{U(\xi, \eta)\}, \tag{48}$$

with $U(\xi, \eta)$ given by Eq. (46). The basic idea is to use the scattering data to calculate $F^{(\mathrm{wm})}(\xi, \eta)$ and display it as a function of the two variables (ξ, η).

An alternative way of visualizing the effect of exploring the parameter space (ξ, η) is in terms of the incident and reference beams. Since the point (ξ, η) represents the focal point of the incident field, and the central position of the diffraction limited spot that gives rise to the reference field, the situation can be understood in

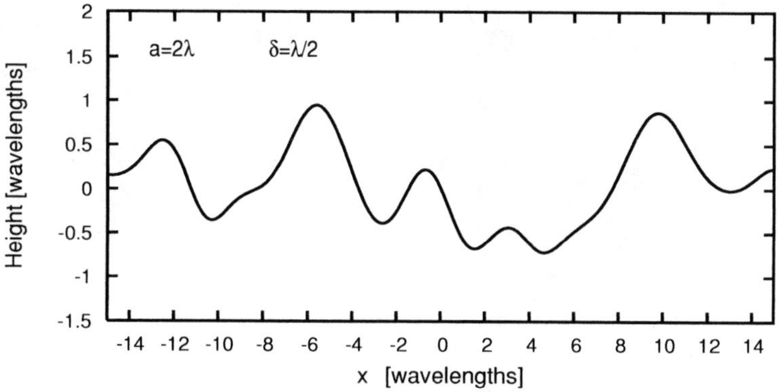

FIGURE 16.5. Realization of a Gaussian random process with the correlation length $a = 2\lambda$ and standard deviation of heights $\delta = \lambda/2$.

terms of the confocal scanning optical microscope.[42,44,45] The function $F^{(wm)}(\xi, \eta)$ is then related to the signal obtained with an interferometric confocal scanning optical microscope.[43,48] It is also worth mentioning that in our formalism, the standard confocal signal $F^{(conf)}(\xi, \eta)$ is given by the expression

$$F^{(conf)}(\xi, \eta) = |U(\xi, \eta)|^2. \tag{49}$$

16.4.2. Numerical Example

We now consider the specific example of the reconstruction of the profile of a perfectly conducting surface. The surface, of length $L_1 = 30\lambda$, is assumed to be a section of a realization of a Gaussian random process with a Gaussian correlation function. It was sampled at intervals of $\lambda/10$ so that, in total, it is represented by an array of $N = 300$ points. The statistical parameters assumed are, a correlation length $a = 2\lambda$ and a standard deviation of heights $\delta = 0.5\lambda$. The random profile considered is shown in Fig. 16.5.

The scattering data were calculated by rigorous numerical methods based on the solution of an integral equation[49] for a chosen set of angles of incidence and scattering. In this particular example we assume that complex amplitude scattering data corresponding to 61 angles of incidence and 61 angles of scattering, equally spaced in q and k, between -80° and 80° are available. So, the scattering amplitude $R(q|k)$ is sampled to produce a 61×61 matrix. The sampling was chosen in accord with the criterion given in [46]. The scattering amplitude calculated with this surface for the case of p-polarization is illustrated in Fig. 16.6 as a function of both k and q.

In Fig. 16.7(a), we show the function $F^{(wm)}(\xi, \eta)$, defined by expressions (46)–(48), calculated with the scattering data $R(q|k)$ shown in Fig. 16.6. A series of fringes resembling the surface profile can be observed. In order to facilitate the

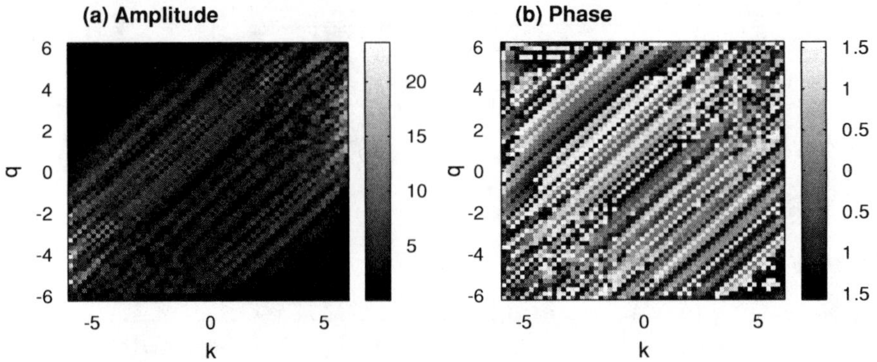

FIGURE 16.6. The scattering amplitude $R(q|k)$ as a function of q and k corresponding to the surface profile shown in Fig. 16.5 for the case of p-polarization.

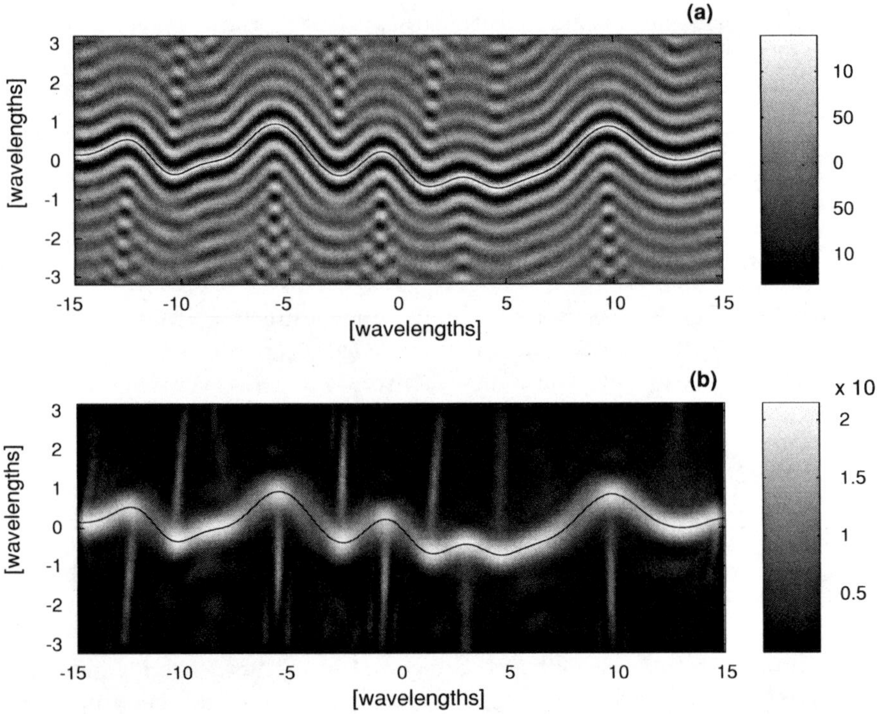

FIGURE 16.7. Gray scale maps of the functions $F^{(\mathrm{wm})}(\xi, \eta)$ (a) and $F^{(\mathrm{conf})}(\xi, \eta)$ (b). In both cases, the surface was illuminated with p-polarized light. The solid lines in reversed contrast show the original profile.

visualization of the results, the actual profile used in this calculations is shown using a solid black line. One can see that the brightest fringe coincides with the surface profile function.

For comparison, the confocal signal $F^{(\mathrm{conf})}$, defined in Eq. (49), is shown in Fig. 16.7(b). As expected, the profile falls fairly well within the bright regions of the figure. Unwanted effects appear in the form of the white vertical streaks that can be observed in the neighborhood of the turning points of the profile.[50] From any one of these two maps, the surface profile function can be estimated fairly reliably.

Similar results are obtained for the case of s polarization, the main difference being that the contrast of the map of Fig.16.7(a) is reversed. The algorithm can be applied not only to the recovery of perfectly conducting surface profiles, but also to dielectric and metallic ones and even multilayer systems.[51] So far, only one-dimensional surfaces have been studied but, in principle, the extension to surfaces with two-dimensional roughness should be possible.

Despite the fact that the algorithm is based on simple notions about the interaction of the light and the surface, it works well in a variety of situations. It produces excellent reconstructions in cases in which single scattering is dominant. The reconstructions deteriorate gradually as multiple scattering effects become more important, and are sensitive to phase noise.[46]

16.5. Estimation of the Surface Profile from Intensity Data

Since optical detectors are not phase sensitive, usually one only has access to intensity data. This means that the phase information present in the scattered field (arguably the most important) has been lost, and one cannot use the kind of strategy reviewed in the previous section. Among other things, the solution of the direct problem is no longer unique and the inversion strategies must be radically different.

In this section, we review some recent investigations on the possibility of recovering the surface profile function from far-field intensity data. The problem is approached as a nonlinear least-squares bounds-constrained optimization problem. Two kinds of representation of the objective variables have been considered; a spectral representation,[52] and a representation based on spline curves.[53] For simplicity, and to illustrate the approach, we review only the spectral representation.

It is assumed that we have access to far-field angle-resolved scattered intensity data corresponding to several angles of incidence. The goal is to retrieve the unknown surface profile function from these data. Some constraints on the kind of surface that we seek are introduced in order to reduce the search space.

The inverse scattering problem can be reformulated in terms the fitness (objective) functional

$$f\left[\zeta^{(c)}(x_1)\right] = \min\left\{\sum_{i=1}^{N_{\mathrm{ang}}} \left\| I^{(m)}(q|k_i) - I^{(c)}\left(q|k_i;\zeta^{(c)}(x_1)\right)\right\|_2^2\right\}, \tag{50}$$

where the symbol $\|\cdot\|_2$ represents the Euclidean norm, and N_{ang} is the number of angles of incidence considered. Also, $I^{(m)}(q|k_i)$ is an angle-resolved far-field scattered intensity pattern of the surface of interest (measured or calculated) and $I^{(c)}(q|k_i;\zeta^{(c)}(x_1))$ is a calculated intensity pattern obtained by solving the direct problem with a trial surface profile $\zeta^{(c)}(x_1)$. The functional $f[\zeta^{(c)}(x_1)]$ can be interpreted as an assessment of the closeness between the angular distributions of intensity $I^{(m)}(q|k)$ and $I^{(c)}(q|k;\zeta^{(c)}(x_1))$. The goal then would be to find a surface for which the condition $I^{(c)}(q|k) = I^{(m)}(q|k)$ is satisfied. When this happens, and if the solution to the problem is unique, the original profile has been retrieved.

Note that in our definition of the fitness functional we require that the proposed surface reproduces the "measured" scattering data for several angles of incidence. The satisfaction of these constraints should reduce the number of possible solutions and, hopefully, produce a unique one. The inverse scattering problem can be viewed, now, as the problem of minimizing $f[\zeta^c(x_1)]$.

To deal with the scattering problem numerically, the surface must be sampled. From the preceding discussion it seems natural to choose, as the parameters of interest, the surface heights evaluated at the sampling points. Changing these numbers independently, however, would lead to surfaces with abrupt height changes, which does not correspond to the physical situation of interest. One way to avoid this problem is to restrict the search space to randomly rough surfaces that belong to a certain class. We are, thus, faced with a problem of constrained optimization.

Consider the case in which the target surface as a realization of a stationary, zero-mean one-dimensional Gaussian random process. With this assumption, the random process is completely characterized by its two point correlation function, which we also assume to be Gaussian. Surfaces belonging to this class can be generated numerically with the spectral method described in [49]. Correlated random numbers that represent the surface heights at the sampling points can be obtained through the expression

$$\zeta_n = \frac{\delta}{\sqrt{L_1}} \sum_{j=-N/2}^{N/2-1} \frac{[M_j + \mathrm{i}N_j]}{\sqrt{2}} \left[\sqrt{\pi}a \exp\left\{-\left(\frac{aq_j}{2}\right)^2\right\}\right]^{1/2} \exp\left\{\mathrm{i}q_j\chi_n\right\}. \tag{51}$$

Here, N represents the total number of points on the surface, L_1 represents its length, $\chi_n = -L_1/2 + (n - 0.5)\Delta x$ are the sampling points spaced by Δx along x_1, $q_j = -\pi/\Delta x + 2\pi(j - 0.5)/L_1$ are the sampling points in the Fourier space, and $\zeta_n = \zeta(\chi_n)$. The random sets $\{M_j\}$ and $\{N_j\}$ contain statistically independent random Gaussian variables with zero mean and unit standard deviation. In order to produce a set of real random numbers $\{\zeta_n\}$, it is required that the complex array $\{M_j + \mathrm{i}N_j\}$ be Hermitian. The first and second derivatives of the surface profile function, which are required for the direct rigorous scattering calculations, can be obtained by differentiation of Eq. (51).

16.5.1. Evolutionary Inversion Procedure

At least in principle, any of the optimization techniques reported in the literature could be employed to minimize Eq. (50). However, the form of this equation and the constraint imposed by the representation scheme of Eq. (51) suggest the use of an algorithm belonging to the class of "direct search methods."[54] The main characteristic of this kind of technique is that, throughout the optimization process, one only needs to know the values of the fitness function and not its derivatives.

Evolutionary algorithms are a relatively recent set of direct search methods that have been successful in the solution of ill-posed inverse problems in different scientific disciplines. They are based on the Darwinian principle of variation and selection. Examples of these heuristic population-based techniques are the genetic algorithms,[55] the evolution strategies,[56] and the genetic[57] and evolutionary programming.[58] Given the characteristics of the inverse problem studied here, the evolutionary strategies seem to be the best-suited evolutionary algorithms for this task.

Evolutionary strategies follow the canonical structure shown in the flow diagram of Fig. 16.8. The starting point of the optimization process is the generation of a random set $P_\mu|_{g=0}$ of μ possible solutions to the problem which, in the present context, are the set of Gaussian randomly rough one-dimensional surfaces $\{\zeta_n\}$ generated through Eq. (51). A secondary population P_λ of λ elements is generated through the application of the "genetic" operations of recombination and mutation over the elements of the initial population P_μ. This represents the start of the main evolutionary loop.

It should be mentioned that we have previously used λ to denote the wavelength of the light, which is the usual notation in optical work. Due to the different context in which the two quantities are employed, use of the same symbol to denote both should not lead to much confusion.

Recombination exploits the search space through the exchange of information between different elements of the population. Mutation, on the other hand, explores the search space through the introduction of random variations in the newly recombined elements.

Once the secondary population has been generated, one needs to evaluate the quality of its elements. For this, the direct problem must be solved for each of the surfaces in the secondary population. A fitness value is associated to each surface ζ_n^j on the basis of Eq. (50). Only those elements of the secondary population leading to promising regions of the search space will be retained, through some selection scheme, as part of the population for the next iteration of the evolutionary loop. The procedure is repeated until a defined termination criterion has been reached. The respective sizes of the initial and the secondary populations remain constant throughout the search process.

In the scheme reviewed here, mutations are introduced by changing some of the elements of the Hermitian array employed in the generation of a given surface (see Fig. 16.9). Provided that the new numbers, M_j and N_j, are zero-mean Gaussian-distributed random numbers with unit standard deviation, and the Hermiticity of

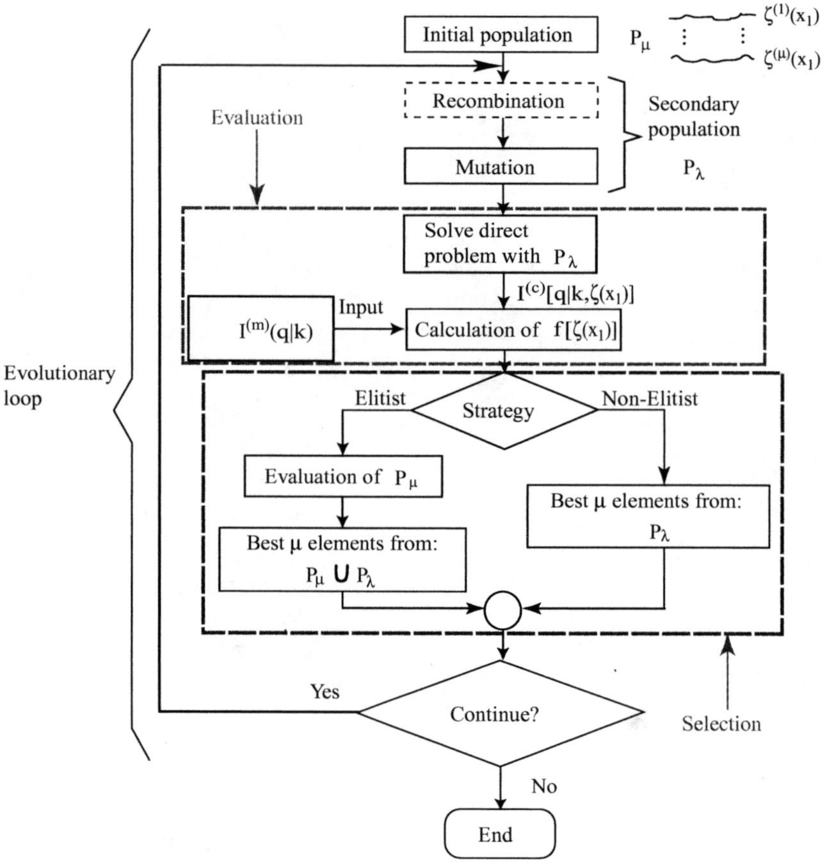

FIGURE 16.8. Flow diagram of the evolutionary loop.

the array is conserved, the new surface will belong to the statistical class specified for the search space.

The selection operator generates, through a deterministic process, the set of surfaces that will serve as the population for the next iteration of the algorithm. There are two selection procedures employed in evolution strategies. The first one is known as the "elitist" or $(\mu + \lambda)$ strategy, whereas the second one is called the "nonelitist" or (μ, λ) strategy. In the (μ, λ) scheme, the elements to be selected belong, exclusively, to the secondary population. An important consequence of this is the possibility that the best elements of the new population are less fit than the best element of the previous population. This possible deterioration of the fitness values helps the algorithm avoid regions of attraction that could lead to premature convergence to a local minimum. Of course, if the deterioration persists, the algorithm diverges.

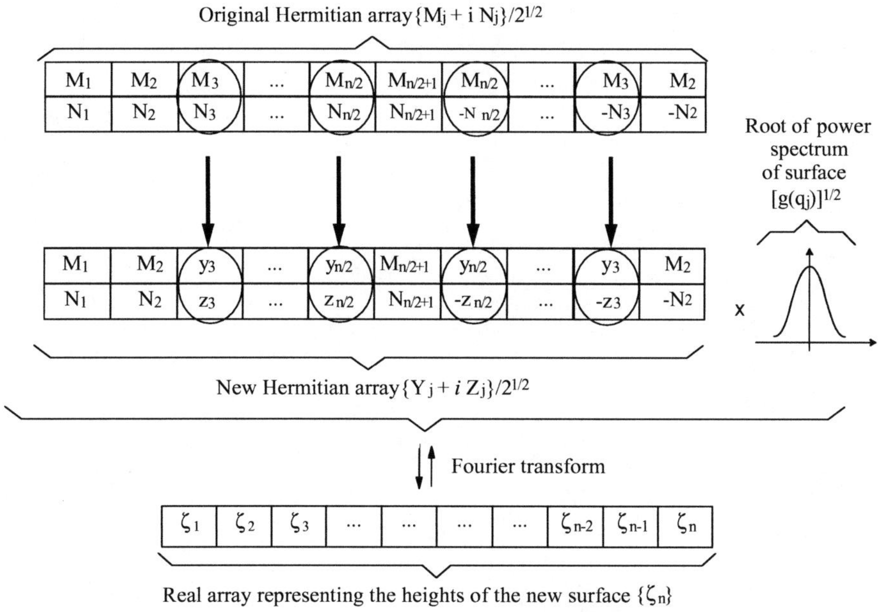

FIGURE 16.9. Schematic representation of the mutation operation.

16.5.2. Results of a Numerical Experiment

The scattering data that serve as the input to the algorithm were obtained through a rigorous numerical solution of the direct scattering problem.[49] Since the time of computation required to find the optimum increases with the number of sampling points on the surface, and the direct problem needs to be solved many times, in order to keep the problem to a manageable size we chose a surface with $N = 128$ sampling points. The surface profile used to generate these scattering data is shown in Fig. 16.10. It constitutes a realization of a zero-mean stationary Gaussian-correlated Gaussian random process with a $1/e$-value of the correlation function $a = 2\lambda$ and standard deviation of heights $\delta = 0.5\lambda$. The surface was sampled at intervals $\Delta x = \lambda/10$. Far-field intensity data corresponding to four different angles of incidence ($\theta_0 = -60^\circ$, -30°, 0°, and 40°) were available. In Fig. 16.11, we show the scattering pattern produced by the surface shown in Fig. 16.10 for the case of normal incidence.

To start the evolutionary loop, a set of surfaces belonging to the same statistical class as the original surface (zero-mean stationary Gaussian-correlated Gaussian random process with $a = 2\lambda$ and $\delta = 0.5\lambda$) were generated. This set constitutes the initial population. We chose the typical values[56] $\mu = 10$ and $\lambda = 100$. For brevity, we only present results obtained with the elitist strategy. Results obtained with the nonelitist one are similar. The maximum number of iterations was $g = 300$, which also provided the termination criterion.

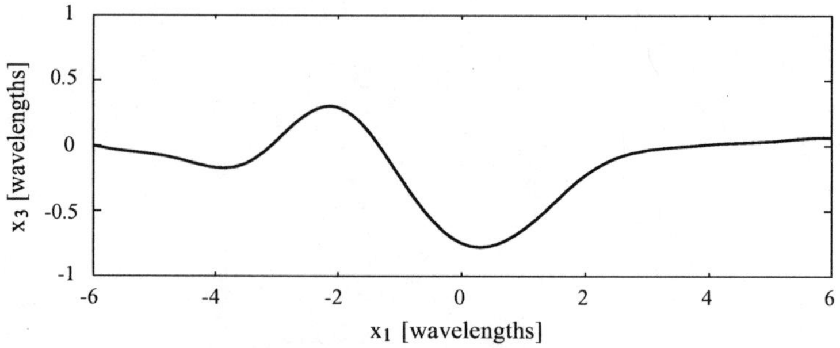

FIGURE 16.10. Profile used in the generation of the scattering data.

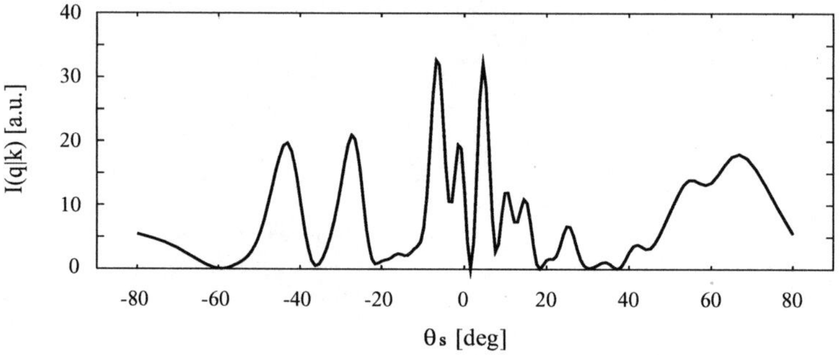

FIGURE 16.11. Scattered intensity produced by the surface depicted in Fig. 16.10 for the case of normal incidence.

The target profile was searched starting from 30 different, and randomly chosen, initial states. Not in all of these attempts to recover the profile the algorithms converged to the target surface. However, we found that a low value of $f(\zeta_n)$ corresponded, in most cases, to a profile that was close to the original one. So, the final value of $f(\zeta_n)$ was used as the criterion to decide whether the function profile had been reconstructed or not.

In Fig. 16.12, we present results obtained without recombination using the elitist strategy. The original profile is shown with the dotted curve, while the profile retrieved with the elitist strategy is shown with the solid curve. We can see that, in this case, the target profile has been retrieved quite well. It should be mentioned that, often, the technique recovers profiles that are displaced horizontally or vertically with respect of the original surface. These displacements of the recovered profile are understandable, as the far-field intensity is insensitive to such shifts. On the other hand, such displacements are unimportant for practical profilometric applications.

An interesting result that demonstrates the lack of uniqueness of the solution when intensity data are used is shown in Fig. 16.13. The dotted curve represents the original profile, while the reconstruction obtained with the elitist strategy is

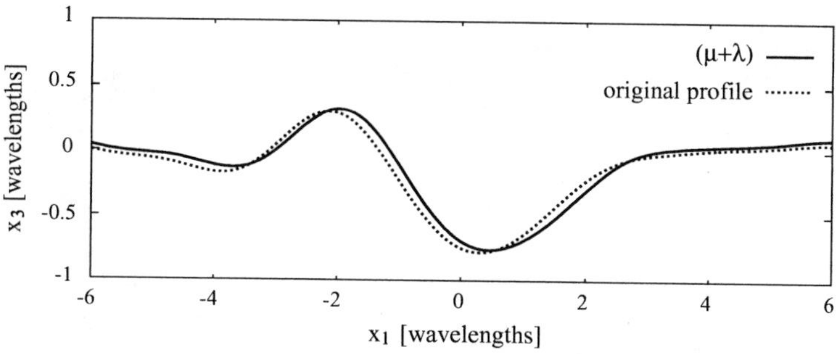

FIGURE 16.12. Reconstruction of the surface profile using the $(\mu + \lambda)$ (solid line) strategy. The original profile is plotted with a dotted line.

represented with the solid line. For this numerical experiment, the initial population was different from the one used in the example of Fig. 16.12. One can see that, in this case, the recovered profile does not resemble the original one. As explained below, these results illustrate a curious symmetry property of the scattering problem for situation in which single scattering is dominant.

The profile recovered in Fig. 16.13 resembles the sought profile if we reflect it with respect to the x_1 and x_3 axes; that is, if we replace $\zeta^{(c)}(x_1)$ by $-\zeta^{(c)}(-x_1)$. This is illustrated in Fig. 16.14. To better understand this property, let us consider the direct scattering problem in the thin phase screen approximation.

The far-field scattering amplitude $R(q|k)$ is given by Eq. (6). Consider now the profile $z(x_1)$, defined as $z(x_1) = -\zeta(-x_1)$. The scattering amplitude obtained with this profile can be written as

$$[R(q|k)]_z = A_0 \int_{-\infty}^{\infty} dx_1 \, e^{-iv_1 x_1 + iv_3 \zeta(-x_1)}. \tag{52}$$

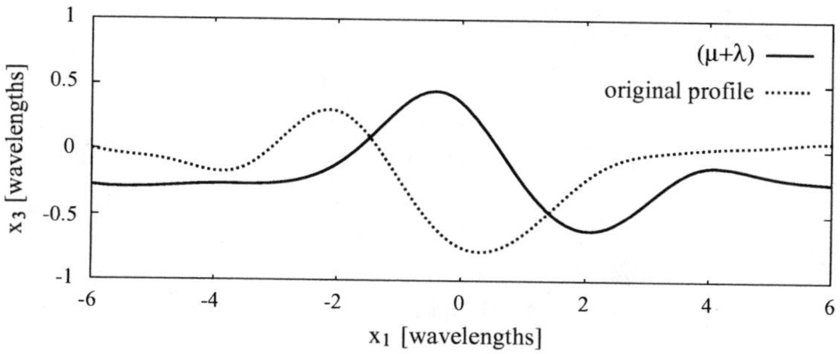

FIGURE 16.13. Reconstruction of the surface profile using the $(\mu + \lambda)$ (solid line) strategy, starting from a different initial population than in Fig. 16.12. The original profile is plotted with a dotted line.

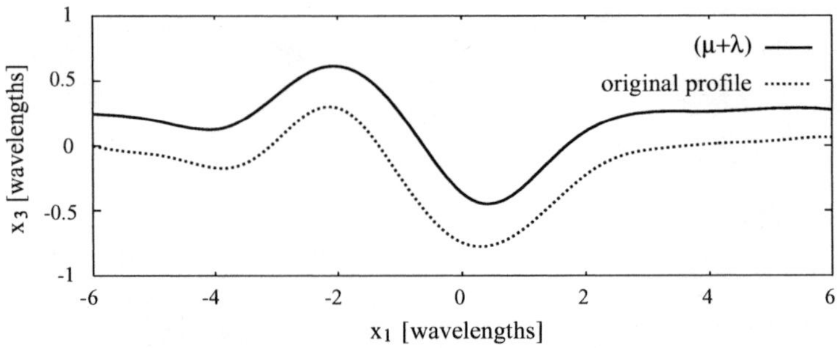

FIGURE 16.14. Reconstructed profile shown in Fig. 16.13, reflected on both axes. The original profile is plotted with a dotted line.

With the change of variable $u = -x_1$, it can be readily shown that the intensity patterns corresponding to $\zeta(x_1)$ and $z(x_1)$ are equal:

$$I(q|k) = [I(q|k)]_z. \tag{53}$$

Thus, the validity of the thin phase screen approximation leads to multiple solutions of the inverse scattering problem. It should be mentioned that, in general, the rigorous solution of the direct problem does not have this kind of symmetry. It is thus tempting to think that multiple scattering effects reduce the number of possible solutions of the inverse problem.

16.6. Discussion and Conclusions

In this review, we have considered the retrieval of surface profile information from monochromatic far-field angle-resolved data. After a brief discussion of the direct problem, we reviewed some methods to estimate statistical parameters of the surface, such as the rms height and the height correlation length.

A simple and practical method to estimate δ is based on the determination of the strength of the coherent component. More elaborate techniques need to be used for cases in which the coherent component cannot be detected. Information on the lateral scale of the surface is contained in the angular distribution of the mean intensity. For weakly scattering surfaces this distribution is, essentially, a map of the power spectral density of surface heights, whereas in the limit of very rough single scattering surfaces it is related to the PDF of slopes on the surface. A more complex relation occurs in more general cases.

We have also considered the recovery of the surface profile function itself. Two problems of different nature were discussed. The first, and simpler one, assumes the availability of complex amplitude data, while the second one assumes that only intensity data are available.

The approach taken to solve the first problem is based on single scattering assumptions about the interaction of the light with the surface. The resulting procedure is simple, and should also work with two-dimensional surfaces. The processing of the complex amplitude data involves, essentially, an integral transform of the scattering amplitude. The method, however, breaks down when multiple scattering is important, and is also sensitive to phase noise in the input data.

In the second case, when only intensity data are given, the phase information is lost and the solution of the inverse problem is not unique. A radically different strategy must be employed. In our approach, the problem is visualized as a problem of constrained optimization and solved using evolutionary strategies. These algorithms are heuristic, involve random search strategies, and there is no guarantee that they will converge to the correct solution. Nevertheless, even though numerical experiments have only been conducted with quite short one-dimensional surfaces, the results are encouraging. These methods are able to retrieve surface profile functions in a variety of situations, including cases in which multiple scattering is important.

An example that illustrates the potential of these heuristic procedures to solve inverse scattering problems in the presence of multiple scattering is shown in Fig. 16.15. The surface contains a triangular groove whose half-width and depth are equal to λ. In the region of the groove, the surface slopes are ± 1, which leads to significant amounts of multiple scattering. Let us first look at the results obtained with the wavefront matching algorithm for this surface. A gray scale map of $F^{(\mathrm{wm})}(\xi, \eta)$ for the case of normally incident s-polarized illumination is shown in Fig. 16.15(a). A dark fringe is only clearly defined in the region where the surface is horizontal. As expected, in the region where multiple scattering dominates, it is difficult to infer the position of the interface.

An attempt to reconstruct the surface profile function from these data by searching for the minimum at each position x_1 is shown in Fig. 16.15(b) (small circles); the failure of the algorithm in the region where multiple scattering occurs is evident. In the same figure, we show the results obtained with a nonelitist hybrid evolutionary strategy in which the surface was represented in terms of spline curves[53] (crosses). The algorithm is hybrid because it couples an evolutionary strategy with the downhill simplex method of Nelder and Mead,[59] which is a local search algorithm. One can see that the reconstruction obtained with the hybrid algorithm is reasonably close to the target profile.

Although the numerical studies carried out so far are limited in many ways and the evolutionary strategies lack mathematical formality, we have found that these algorithms are able to find the correct solution in many cases. They have thus the potential of becoming a practical tool for solving inverse problems when phase information is unavailable.

Acknowledgments. The work of ERM has been supported by the Consejo Nacional de Ciencia y Tecnoligía, through grant 47712-F. DM gratefully acknowledges financial support of the Conseil Régional de Champagne-Ardenne.

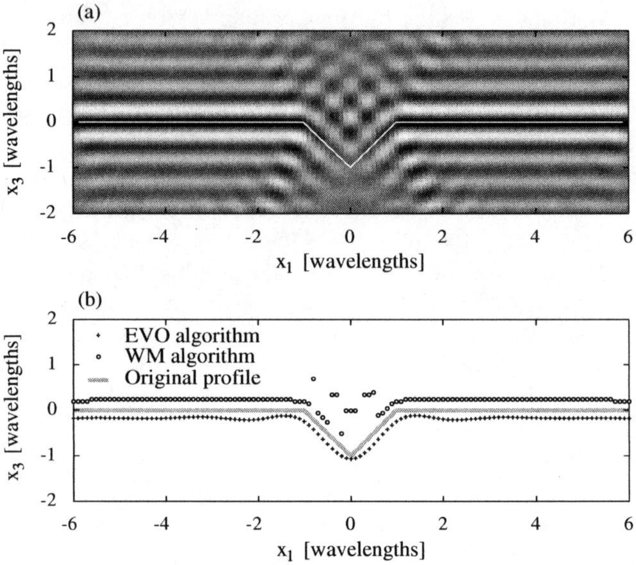

FIGURE 16.15. Comparison of reconstructed profiles using the wavefront matching algorithm and a hybrid algorithm that uses an evolutionary strategy coupled with the downhill simplex method. (a) Gray scale map of $F^{(\mathrm{wm})}(\xi, \eta)$ for the case of normal incidence and s-polarized waves. The original profile is shown with a white line. (b) Estimated profiles. To facilitate the visualization, the curves corresponding to the hybrid and wavefront matching algorithms have been slightly shifted downwards and upwards, respectively. The original profile is shown with a thick gray line.

References

1. J.A. Ogilvy, *Theory of Wave Scattering from Random Rough Surfaces* (Institute of Physics Publishing, Bristol, UK, 1991).
2. M. Saillard and A. Sentenac, "Rigorous solutions for electromagnetic scattering from rough surfaces," *Waves Random Media* **11**, R103–R137 (2001).
3. A.A. Maradudin and E.R. Méndez, "Scattering by surfaces and phase screens," in *Scattering: Scattering and Inverse Scattering in Pure and Applied Science*. Roy Pike and Pierre Sabatier, ed., (Academic Press, San Diego, 2002), pp. 864–894.
4. T.M. Elfouhaily and C.-A. Guérin, "A critical survey of approximate scattering wave tehories from random rough surfaces," *Waves Random Media* **14**, R1–R40 (2004).
5. A.V. Shchegrov, A.A. Maradudin, and E.R. Méndez, "Multiple Scattering of Light from Randomly Rough Surfaces," in *Progress in Optics*, E. Wolf, ed. (Elsevier, Amsterdam, 2004), pp. 117–241.
6. J.C. Dainty (ed.), *Laser Speckle and Related Phenomena*, 2nd edn (Springer-Verlag, Berlin, 1984).
7. W.T. Welford, "Optical estimation of statistics of surface roughness from light scattering measurements," *Opt. Quantum Electron.* **9**, 269–387 (1977).
8. W.T. Welford, "Laser speckle and surface roughness," *Contemp. Phys.* **21**, 401–412 (1980).

9. J.A. Ratcliffe, "Some aspects of diffraction theory and their application to the ionosphere," *Rep. Prog. Phys.* **19**, 188–267 (1956).
10. P. Beckmann and A. Spizzichino, *The Scattering of Electromagnetic Waves from Rough Surfaces* (Pergamon Press, London, 1963).
11. J.W. Goodman, *Statistical Optics* (Wiley New York, 1985).
12. M.V. Berry, "Diffractals," *J. Phys. A: Math. Gen.* **12**, 781–97 (1979).
13. E. Jakeman, "Scattering by a corrugated random surface with fractal slope," *J. Phys. A: Math. Gen.* **15**, L55–L59 (1982);
14. E. Jakeman, "Fraunhofer scattering by a sub-fractal diffuser," *Opt. Acta* **30**, 1207–1212 (1983).
15. A. Mendoza-Suárez and E.R. Méndez, "Light scattering by reentrant fractal surfaces," *Appl. Opt.* **36**, 3521–3531 (1997).
16. C.A. Guérin, M. Holschneider, and M. Saillard, "Electromagnetic scattering from multiscale rough surfaces," *Waves Random Media* **7**, 331–349 (1997).
17. J.A. Sánchez-Gil, J.V. García-Ramos, and E.R. Méndez, "Influence of nanoscale cutoff in random self-aff ine fractal silver surfaces on the excitation of localized optical modes," *Opt. Lett.* **26**, 1286–1288 (2001).
18. P. Beckmann, "Scattering by non-Gaussian surfaces," *IEEE Trans. Antennas Propag.* **AP-21**, 169–175 (1973).
19. M.J. Kim, E.R. Méndez, and K.A. O'Donnell, "Scattering from gamma-distributed surfaces," *J. Mod. Opt.* 34, 1107–1119 (1987).
20. V.I. Tatarskii, "Characteristic functional for one class of non-Gaussian random functions," *Waves Random Media* **5**, 243–252 (1995).
21. V.V. Tatarskii and V.I. Tatarskii, "Non-Gaussian statistical model of the ocean surface for wave-scattering theories," *Waves Random Media* **6**, 419–435 (1996).
22. H.E. Bennett and J.O. Porteus, "Relation between surface roughness and specular reflectance at normal incidence," *J. Opt. Soc. Am.* **51**, 123–129 (1961).
23. J.M. Elson and J.M. Bennett, "Relation between the angular dependence of scattering and the statistical properties of optical surfaces," *J. Opt. Soc. Am.* **69**, 31–47 (1979).
24. J.C. Stover, S.A. Serati, and C.H. Guillespie, "Calculation of surface statistics from light scatter," *Opt.Eng.* **23**, 406–412 (1984).
25. J.W. Goodman, "Statistical properties of speckle patterns," in *Laser Speckle and Related Phenomena*, 2nd edn (Springer-Verlag, Berlin, 1984). Pp. 9–75.
26. D. Léger, E. Mathieu, and J.C. Perrin, "Optical surface roughness determination using speckle correlation techniques," *Appl. Opt.* **14**, 872–877 (1975).
27. D. Léger and J.C. Perrin, "Real-time measurement of surface roughness by correlation of speckle patterns," *J. Opt. Soc. Am.* **66**, 1210–1217 (1976).
28. J. Shen and A.A. Maradudin, "Multiple scattering of waves from random rough surfaces," *Phys. Rev. B* **22**, 4234–4240 (1980).
29. J.A. Sánchez-Gil, A.A. Maradudin, and E.R. Méndez, "Limits of validity of three perturbation theories of the coherent scattering of light from a one-dimensional randomly rough dielectric surface," *J. Opt. Soc. Am. A* **12**, 1547–1558 (1995).
30. E.I. Chaikina, A.G. Navarrete, E.R. Méndez, A. Martínez, and A.A. Maradudin, "Coherent scattering by one-dimensional randomly rough metallic surfaces," *Appl. Opt.* **37**, 1110–1121 (1998).
31. A.G. Navarrete, E.I. Chaikina, E.R. Méndez, and T.A. Leskova, "Experimental study of the reflectance of two-dimensional metal surfaces with a random roughness distribution," *J. Opt. Technol.* **69**, 71–76 (2002).

32. J.C. Stover, *Optical Scattering: Measurement and Analysis*, 2nd edn (SPIE Optical Engineering Press, Bellingham, Washington, D.C., 1995).
33. J.M. Bennett and L. Mattsson, *Introduction to Surface Roughness and Scattering*, 2nd edn (Optical Society of America, Washington, D.C., 1999).
34. J.J. Greffet, "Scattering of electromagnetic waves by rough dielectric surfaces," *Phys. Rev. B* **37**, 6436–6441 (1988).
35. M.U. González, J.A. Sánchez-Gil, Y. González, L. González, and E.R. Méndez, "Polarized laser light scattering applied to surface morphology characterization of epitaxial III-V semiconductor layers," *J. Vac. Sci. Technol. B* **18**, 1980–1990 (2000).
36. C.S. West and K.A. O'Donnell, "Observations of backscattering enhancement from polaritons on a rough metal surface," *J. Opt. Soc. Am. A* **12**, 390–397 (1995).
37. V. Malyshkin, S. Simeonov, A.R. McGurn, and A.A. Maradudin, "Determination of surface profile statistics from electromagnetic scattering data," *Opt. Lett.* **22**, 58–60 (1997).
38. A.A. Maradudin and E.R. Méndez, "Enhanced backscattering of light from a weakly rough, random metal surface," *Appl. Opt.* **32**, 3335–3343 (1993).
39. C.S. West and K.A. O'Donnell, "Angular correlation functions of light scattered from weakly rough metal surfaces," *Phys. Rev. B* **59**, 2393–2406 (1999).
40. R.J. Wombell and J.A. DeSanto, "Reconstruction of rough-surface profiles with the Kirchhoff approximation," *J. Opt. Soc. Am. A* **8**, 1892–1897 (1991).
41. R.J. Wombell and J.A. DeSanto, "The reconstruction of shallow rough-surfaces profiles from scattered field data," *Inverse Problems* **7**, L7–L12 (1991).
42. J.C. Quartel and C.J.R. Sheppard, "Surface reconstruction using an algorithm based on confocal imaging," *J. Mod. Opt.* **43**, 469–486 (1996).
43. J.C. Quartel and C.J.R. Sheppard, "A surface reconstruction algorithm based on confocal interferometric profiling," *J. Mod. Opt.* **43**, 591–605 (1996).
44. C.J.R. Sheppard and A. Choudhury, "Image formation in the scanning microscope," *Opt. Acta*, **24**, 1051–1073 (1977).
45. T. Wilson and C.J.R. Sheppard, *Theory and Practice of Scanning Optical Microscopy* (Academic Press, New York, 1984).
46. D. Macías, E.R. Méndez, and V. Ruiz-Cortés, "Inverse scattering with a wave-front-matching algorithm," *J. Opt. Soc. Am. A* 19, 2064–2073 (2002).
47. E. Born and E. Wolf, *Principles of Optics* (Cambridge University Press, Cambridge, 1999), p. 485.
48. D.K. Hamilton and C.J.R. Sheppard, "A confocal interference microscope," *Opt. Acta* **29**, 1573–1577 (1982).
49. A.A. Maradudin, T. Michel, A.R. McGurn, and E.R. Méndez, "Enhanced backscattering of light from a random grating," *Ann. Phys. (N.Y.)* **203**, 255–307 (1990).
50. J.F. Aguilar, M. Lera, and C.J.R. Sheppard, "Imaging of spheres and surface profiling by confocal microscopy," *Appl. Opt.* **39**, 4621–4628 (2000).
51. Macías, D., *Estudios numéricos de esparcimiento inverso de ondas electromagnéticas*, PhD Thesis (Centro de Investigación Científica y de Educación Superior de Ensenada, México, 2003).
52. D. Macías, G. Olague, and E.R. Méndez, "Surface profile reconstruction from scattered intensity data using evolutionary strategies," S. Cagnoni *et al.*, eds. (EvoWorkshops: Springer LNCS **2279**, Berlin, 2002), pp. 233–244.
53. D. Macías, G. Olague and E.R. Méndez, "*Hybrid evolution strategy-downhill simplex algorithm for inverse light scattering problems*," S. Cagnoni *et al.*, eds. (EvoWorkshops: Springer LNCS **2611**, Berlin, 2003), pp. 339–409.

54. T.G. Kold, R.M. Lewis, and V. Torczon, "Optimization by direct search: new perspectives on some classical and modern methods," *SIAM Rev.* **45**, 385–482 (2003).
55. J.H. Holland, *Adaption in Natural and Artificial Systems* (MIT Press/Bradford Books, Cambridge, Massachusetts, 1992).
56. H.P. Schwefel, *Evolution and Optimum Seeking* (Wiley, New York, 1995).
57. J.R. Koza, "Genetic programming," in *Encyclopedia of Computer Science and Technology*, J.G. Williams and A. Kent, eds. (Marcel-Dekker, US, 1998), pp. 29–43.
58. L.J. Fogel, "Autonomous automata," *Ind. Res.* **4**, 14–19 (1962).
59. J. Nelder and R. Mead, "A simplex method for function optimization," *Comput. J.* **7**, 308–313 (1965).

17
The Design of Randomly Rough Surfaces That Scatter Waves in a Specified Manner

ALEXEI A. MARADUDIN
Department of Physics and Astronomy, and Institute for Surface and Interface Science, University of California, Irvine, CA 92697, U.S.A.

17.1. Introduction

It is our aim in this chapter to provide an introduction to a form of the inverse problem in rough surface scattering, and methods for its solution, that differs somewhat from the usual form of this problem. In the usual formulation of this problem scattering data, such as the angular and polarization dependence of the intensity of the scattered field, are provided by experimentalists, and the surface profile function, or some statistical properties of it, such as the power spectrum of the surface roughness, or just the rms height of the surface, is extracted from these data. The type of inverse problem we consider here is how to design a one- or two-dimensional randomly rough surface that scatters in a specified manner a wave or a beam incident on it. We consider two different cases: (i) the scattered field is required to have a prescribed angular dependence of its mean intensity; and (ii) it is required to have a specified wavelength dependence of its mean intensity at a fixed scattering angle. Applications of each of these types of surfaces will be presented.

We have chosen to work with randomly rough surfaces in designing optical diffusers with specified scattering properties, rather than with deterministic surfaces, because, as we will see, the use of such surfaces leads to precise algorithms for designing them, something that we have found more difficult to find in dealing with deterministic surfaces. Moreover, it is more convenient, and far more elegant, to specify a whole class of surfaces that have the desired properties.

An experimentalist fabricates only a single realization of such a random surface from an ensemble of an infinite number of possible realizations of that surface. Does this mean that a deterministic surface has in fact been produced? Although this is a bit of a philosophical question, we would argue that a deterministic surface has not been fabricated. The fact that we know its profile does not mean that it is not random. For example, few people would argue with the statement that a piece of ground glass has a random profile. Is it made deterministic by a measurement of it? Surely not. Consequently, we regard surfaces fabricated by the use of the algorithms developed in our work as random. Their profiles are realizations of a random process. We just happen to know the details of a particular realization.

A more subtle question arises from the fact that in a theoretical study of scattering from a randomly rough surface the property of the scattered field of interest, for example its intensity, is averaged over the ensemble of realizations of the surface profile function. In contrast, in an experimental study of such scattering the measurement is made on a single realization of the surface profile function drawn from that ensemble. It is generally believed that the experimental result will agree with the theoretical result if the surface is large enough, and the incident field illuminates a large enough area of that surface. In this case, it is argued, the mean intensity of the scattered field can be regarded as a spatial average of the actual scattered intensity over an area that contains many realizations of the surface roughness. The assumption that this spatial average is equivalent to the ensemble average used in the design of our random surfaces is difficult to prove rigorously, because some of the surfaces generated are not stationary. The validity of this assumption is a very important question, as it arises in virtually all experimental work on rough surface scattering. A partial answer to it can be found in [1], but it is a question that should be pursued in greater depth in the future.

17.2. A Surface That Produces a Scattered Field with a Specified Angular Dependence of Its Mean Intensity

The earliest examples of efforts to design randomly rough surfaces appear to be the efforts to devise a nonabsorbing optical diffuser that scatters light uniformly within a specified range of scattering angles, and produces no scattering outside this range. We will call such an optical element a *band-limited uniform diffuser*. Such an element could have applications, for example, in projection systems, where one wishes the illumination of a projection display to be concentrated in a specific area, where the viewers are, rather than scattered over a wide area. Band-limited uniform diffusers can also be useful in microscope illumination systems, in the fabrication of displays and projection screens, and in Fourier transform holography. A brief survey of early efforts to design and fabricate one- and two-dimensional surfaces that act as band-limited uniform diffusers can be found in [2].

In this section we present a new approach[3,4] to the design of a two-dimensional randomly rough surface that, when illuminated at normal incidence by a scalar plane wave, produces a scattered field with an essentially arbitrary prescribed angular distribution of intensity within an essentially arbitrary domain of scattering angles.

The physical system we consider consists of vacuum in the region $x_3 > \zeta(\mathbf{x}_\parallel)$, where $\mathbf{x}_\parallel = (x_1, x_2, 0)$ is a position vector in the plane $x_3 = 0$, and the scattering medium in the region $x_3 < \zeta(\mathbf{x}_\parallel)$. The surface profile function $\zeta(\mathbf{x}_\parallel)$ is assumed to be a continuous, single-valued function of $\mathbf{x}_\parallel$ that is differentiable with respect to x_1 and x_2, and constitutes a random process, but not necessarily a stationary one. The surface $x_3 = \zeta(\mathbf{x}_\parallel)$ is illuminated from the vacuum by a scalar plane wave of frequency ω, and it is assumed that the Dirichlet boundary condition is satisfied on this surface.

The differential reflection coefficient $(\partial R/\partial \Omega_s)$ is defined in such a way that $(\partial R/\partial \Omega_s)\, \mathrm{d}\Omega_s$ is the fraction of the total time-averaged incident flux that is scattered into the element of solid angle $\mathrm{d}\Omega_s$ about a given scattering direction. Since the surface defined by the profile function $\zeta(\mathbf{x}_\parallel)$ is randomly rough, it is the mean differential reflection coefficient that is of interest to us. In the geometrical optics limit of the Kirchhoff approximation, which we adopt because of its simplicity, it is given by[2]

$$\left\langle \frac{\partial R}{\partial \Omega_s} \right\rangle = \frac{1}{S} \left(\frac{\omega}{2\pi c} \right)^2 \int \mathrm{d}^2 u_\parallel \exp(-\mathrm{i}\mathbf{q}_\parallel \cdot \mathbf{u}_\parallel) \times \int \mathrm{d}^2 x_\parallel \langle \exp[-\mathrm{i}a\mathbf{u}_\parallel \cdot \nabla \zeta(\mathbf{x}_\parallel)] \rangle \tag{1}$$

in the case of normal incidence. In this expression S is the area of the plane $x_3 = 0$ covered by the random surface; $\mathbf{q}_\parallel = (\omega/c) \sin\theta_s(\cos\phi_s, \sin\phi_s, 0)$; θ_s and ϕ_s are the polar and azimuthal scattering angles, respectively; and $a = (\omega/c)(1 + \cos\theta_s)$. The angle brackets here and in the rest of this chapter denote an average over the ensemble of realizations of the surface profile function $\zeta(\mathbf{x}_\parallel)$. Our goal is to find the function $\zeta(\mathbf{x}_\parallel)$ that produces a specified form of $\langle \partial R/\partial \Omega_s \rangle$.

We begin by covering the x_1x_2 plane by equilateral triangles of edge b (Fig. 17.1). The vertices of these triangles are given by the vectors $\mathbf{x}_\parallel(m, n) = m\mathbf{a}_1 + n\mathbf{a}_2$, where $m, n = 0, \pm 1, \pm 2, \ldots$, and the basis vectors are $\mathbf{a}_1 = (b, 0)$ and $\mathbf{a}_2 = (b/2, \sqrt{3}b/2)$. Each triangle is labeled by the coordinates of its center of gravity. These are given by the mean values of the coordinates of its three vertices. Thus, the triangle defined by the vertices (m, n), $(m + 1, n)$, and $(m, n + 1)$ is

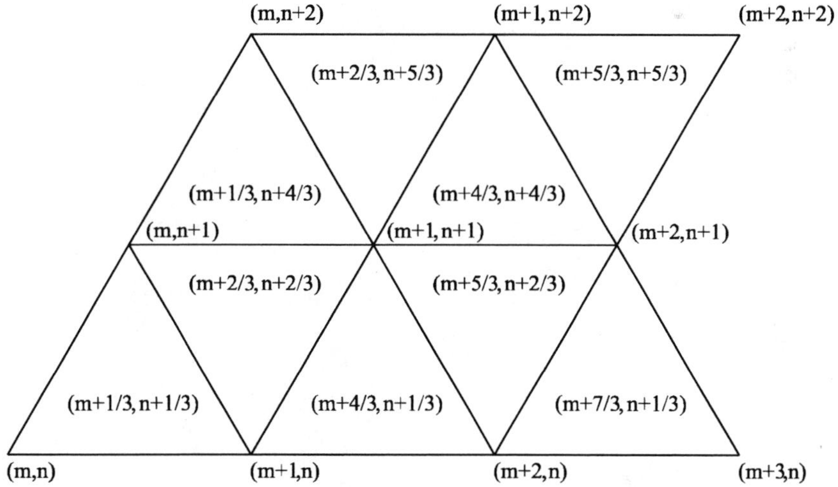

FIGURE 17.1. A segment of the x_1x_2 plane showing the equilateral triangles above which the triangular facets are placed that generate the two-dimensional randomly rough surfaces defined by Eqs. (2) and (3).

the $(m + 1/3, n + 1/3)$ triangle. Similarly, the triangle defined by the vertices $(m + 1, n)$, $(m + 1, n + 1)$, and $(m, n + 1)$ is the $(m + 2/3, n + 2/3)$ triangle. As m and n run through the values $0, \pm 1, \pm 2, \ldots$, the $(m + 1/3, n + 1/3)$ triangles generated constitute the subset of triangles with horizontal bases and points at the top, while the $(m + 2/3, n + 2/3)$ triangles generated constitute the subset of triangles with points at the bottom and horizontal tops. Together these two sets of triangles cover the $x_1 x_2$ plane.

For $\mathbf{x}_\parallel$ within the triangle $(m + 1/3, n + 1/3)$ the surface profile function is assumed to be

$$\zeta(\mathbf{x}_\parallel) = b^{(0)}_{m+\frac{1}{3},n+\frac{1}{3}} + a^{(1)}_{m+\frac{1}{3},n+\frac{1}{3}} x_1 + a^{(2)}_{m+\frac{1}{3},n+\frac{1}{3}} x_2, \tag{2}$$

while for $\mathbf{x}_\parallel$ within the triangle $(m + 2/3, n + 2/3)$ the surface profile function is

$$\zeta(\mathbf{x}_\parallel) = b^{(0)}_{m+\frac{2}{3},n+\frac{2}{3}} + a^{(1)}_{m+\frac{2}{3},n+\frac{2}{3}} x_1 + a^{(2)}_{m+\frac{2}{3},n+\frac{2}{3}} x_2. \tag{3}$$

The coefficients $a^{(1,2)}_{m+\frac{1}{3},n+\frac{1}{3}}$ and $a^{(1,2)}_{m+\frac{2}{3},n+\frac{2}{3}}$ are assumed to be independent identically distributed random deviates. Therefore, the joint probability density function (pdf) of the two coefficients associated with a given triangle,

$$\begin{aligned} &\left\langle \delta\left(s_1 - a^{(1)}_{m+\frac{1}{3},n+\frac{1}{3}}\right) \delta\left(s_2 - a^{(2)}_{m+\frac{1}{3},n+\frac{1}{3}}\right)\right\rangle \\ &\quad = \left\langle \delta\left(s_1 - a^{(1)}_{m+\frac{2}{3},n+\frac{2}{3}}\right) \delta\left(s_2 - a^{(2)}_{m+\frac{2}{3},n+\frac{2}{3}}\right)\right\rangle \\ &\quad = f(s_1, s_2), \end{aligned} \tag{4}$$

is independent of the coordinates labeling the triangle.

It is now straightforward to show that Eq. (1) becomes

$$\left\langle \frac{\partial R}{\partial \Omega_s} \right\rangle (q_1, q_2) = \left(\frac{\omega}{ac}\right)^2 f\left(-\frac{q_1}{a}, -\frac{q_2}{a}\right), \tag{5}$$

where $\langle \partial R/\partial \Omega_s \rangle (q_1, q_2)$ is the mean differential reflection coefficient expressed in terms of the components of the wave vector $\mathbf{q}_\parallel$. We invert this equation to obtain

$$f\left(\frac{q_1}{a}, \frac{q_2}{a}\right) = (1 + \cos\theta_s)^2 \left\langle \frac{\partial R}{\partial \Omega_s} \right\rangle (-q_1, -q_2). \tag{6}$$

With the changes of variables

$$\frac{q_1}{a} = s_1, \qquad \frac{q_2}{a} = s_2, \tag{7}$$

we find that

$$\sin\theta_s = \frac{2 s_\parallel}{1 + s_\parallel^2}, \qquad \cos\theta_s = \frac{1 - s_\parallel^2}{1 + s_\parallel^2}, \tag{8}$$

where $s_\parallel = (s_1^2 + s_2^2)^{\frac{1}{2}}$. Equation (6) then becomes

$$f(s_1, s_2) = \frac{4}{(1 + s_\parallel^2)^2} \left\langle \frac{\partial R}{\partial \Omega_s} \right\rangle \left(-\frac{2(\omega/c) s_1}{1 + s_\parallel^2}, -\frac{2(\omega/c) s_2}{1 + s_\parallel^2} \right). \tag{9}$$

For generation of the surface profile function we also need the (marginal) pdf of $a^{(1)}_{m+\frac{1}{3},n+\frac{1}{3}}$ and of $a^{(1)}_{m+\frac{2}{3},n+\frac{2}{3}}$,

$$f(s_1) = \int_{-\infty}^{\infty} f(s_1, s_2)\, \mathrm{d}s_2, \tag{10}$$

and the conditional pdf of $a^{(2)}_{m+\frac{1}{3},n+\frac{1}{3}}$ ($a^{(2)}_{m+\frac{2}{3},n+\frac{2}{3}}$) given $a^{(1)}_{m+\frac{1}{3},n+\frac{1}{3}}$ ($a^{(1)}_{m+\frac{2}{3},n+\frac{2}{3}}$),

$$f(s_2 \mid s_1) = \frac{f(s_1, s_2)}{f(s_1)}. \tag{11}$$

It is seen from Eqs. (2) and (3) that the coefficients $a^{(1,2)}_{m+\frac{1}{3},n+\frac{1}{3}}$ ($a^{(1,2)}_{m+\frac{2}{3},n+\frac{2}{3}}$) are the partial derivatives of $\zeta(\mathbf{x}_\|)$ along the horizontal edge of the triangle $(m+1/3, n+1/3)$ $((m+2/3, n+2/3))$ and along the normal to this edge inside the triangle. For generation of the surface profile function we also need the joint pdf of the derivatives of $\zeta(\mathbf{x}_\|)$ along the other edges of the triangles and along the normals to these edges within the triangles, and the corresponding marginal and conditional pdfs. Thus, if we denote the derivatives of $\zeta(\mathbf{x}_\|)$ along the left edge of the triangle $(m+1/3, n+1/3)$ and along the normal to this edge inside the triangle by $\alpha^{(1)}_{m+\frac{1}{3},n+\frac{1}{3}}$ and $\alpha^{(2)}_{m+\frac{1}{3},n+\frac{1}{3}}$, respectively, where $\alpha^{(1)}_{m+\frac{1}{3},n+\frac{1}{3}} = \frac{1}{2}a^{(1)}_{m+\frac{1}{3},n+\frac{1}{3}} + \frac{\sqrt{3}}{2}a^{(2)}_{m+\frac{1}{3},n+\frac{1}{3}}$, $\alpha^{(2)}_{m+\frac{1}{3},n+\frac{1}{3}} = \frac{\sqrt{3}}{2}a^{(1)}_{m+\frac{1}{3},n+\frac{1}{3}} - \frac{1}{2}a^{(2)}_{m+\frac{1}{3},n+\frac{1}{3}}$, the joint pdf of $\alpha^{(1)}_{m+\frac{1}{3},n+\frac{1}{3}}$ and $\alpha^{(2)}_{m+\frac{1}{3},n+\frac{1}{3}}$ is given by

$$\begin{aligned}\tilde{f}(s_1, s_2) &= \left\langle \delta\left(s_1 - \alpha^{(1)}_{m+\frac{1}{3},n+\frac{1}{3}}\right)\delta\left(s_2 - \alpha^{(2)}_{m+\frac{1}{3},n+\frac{1}{3}}\right)\right\rangle \\ &= f\left(\frac{1}{2}s_1 + \frac{\sqrt{3}}{2}s_2, \frac{\sqrt{3}}{2}s_1 - \frac{1}{2}s_2\right).\end{aligned} \tag{12}$$

The marginal pdf $\tilde{f}(s_1) = \langle \delta(s_1 - \alpha^{(1)}_{m+\frac{1}{3},n+\frac{1}{3}})\rangle$ is then given by

$$\tilde{f}(s_1) = \int_{-\infty}^{\infty} f\left(\frac{1}{2}s_1 + \frac{\sqrt{3}}{2}s_2, \frac{\sqrt{3}}{2}s_1 - \frac{1}{2}s_2\right) \mathrm{d}s_2, \tag{13}$$

and the conditional pdf of $\alpha^{(2)}_{m+\frac{1}{3},n+\frac{1}{3}}$ given $\alpha^{(1)}_{m+\frac{1}{3},n+\frac{1}{3}}$ is

$$\tilde{f}(s_2 \mid s_1) = \frac{\tilde{f}(s_1, s_2)}{\tilde{f}(s_1)}. \tag{14}$$

In a similar fashion, if we denote the derivatives of $\zeta(\mathbf{x}_\|)$ along the left edge of the triangle $(m+2/3, n+2/3)$ and along the normal to this edge within the triangle by $\beta^{(1)}_{m+\frac{2}{3},n+\frac{2}{3}}$ and $\beta^{(2)}_{m+\frac{2}{3},n+\frac{2}{3}}$, respectively, where $\beta^{(1)}_{m+\frac{2}{3},n+\frac{2}{3}} = -\frac{1}{2}a^{(1)}_{m+\frac{2}{3},n+\frac{2}{3}} +$

$\frac{\sqrt{3}}{2}a^{(2)}_{m+\frac{2}{3},n+\frac{2}{3}}$, $\beta^{(2)}_{m+\frac{2}{3},n+\frac{2}{3}} = \frac{\sqrt{3}}{2}a^{(1)}_{m+\frac{2}{3},n+\frac{2}{3}} + \frac{1}{2}a^{(2)}_{m+\frac{2}{3},n+\frac{2}{3}}$, the joint pdf of $\beta^{(1)}_{m+\frac{2}{3},n+\frac{2}{3}}$ and $\beta^{(2)}_{m+\frac{2}{3},n+\frac{2}{3}}$ is given by

$$\begin{aligned}\tilde{\tilde{f}}(s_1, s_2) &= \left\langle \delta\left(s_1 - \beta^{(1)}_{m+\frac{2}{3},n+\frac{2}{3}}\right)\delta\left(s_2 - \beta^{(2)}_{m+\frac{2}{3},n+\frac{2}{3}}\right)\right\rangle \\ &= f\left(-\frac{1}{2}s_1 + \frac{\sqrt{3}}{2}s_2, \frac{\sqrt{3}}{2}s_1 + \frac{1}{2}s_2\right). \end{aligned} \tag{15}$$

The marginal pdf $\tilde{\tilde{f}}(s_1) = \langle\delta(s_1 - \beta^{(1)}_{m+\frac{2}{3},n+\frac{2}{3}})\rangle$ is then given by

$$\tilde{\tilde{f}}(s_1) = \int_{-\infty}^{\infty} f\left(-\frac{1}{2}s_1 + \frac{\sqrt{3}}{2}s_2, \frac{\sqrt{3}}{2}s_1 + \frac{1}{2}s_2\right) ds_2, \tag{16}$$

and the conditional pdf of $\beta^{(2)}_{m+\frac{2}{3},n+\frac{2}{3}}$ given $\beta^{(1)}_{m+\frac{2}{3},n+\frac{2}{3}}$ is

$$\tilde{\tilde{f}}(s_2 \mid s_1) = \frac{\tilde{\tilde{f}}(s_1, s_2)}{\tilde{\tilde{f}}(s_1)}. \tag{17}$$

Finally, if we denote the heights of the surface at the three vertices of triangle $(m + 1/3, n + 1/3)$ by $h_{m,n}$, $h_{m+1,n}$, $h_{m,n+1}$, we find that

$$b^{(0)}_{m+\frac{1}{3},n+\frac{1}{3}} = (m + n + 1)h_{m,n} - mh_{m+1,n} - nh_{m,n+1}. \tag{18}$$

Similarly, if we denote the heights of the surface at the three vertices of triangle $(m + 2/3, n + 2/3)$ by $h_{m+1,n}$, $h_{m+1,n+1}$, $h_{m,n+1}$, we find that

$$b^{(0)}_{m+\frac{2}{3},n+\frac{2}{3}} = (n + 1)h_{m+1,n} - (m + n + 1)h_{m+1,n+1} + (m + 1)h_{m,n+1}. \tag{19}$$

The construction of a realization of the surface profile function proceeds in a sequential fashion. We begin by assuming, with no loss of generality, that the height of the surface at the vertex (m, n) is zero, $h_{m,n} = 0$. The marginal pdf $f(s_1)$ is used with the rejection method[5] to obtain the coefficient $a^{(1)}_{m+\frac{1}{3},n+\frac{1}{3}}$, which yields the height $h_{m+1,n}$. The conditional pdf $f(s_2|s_1)$ is used with this value of $a^{(1)}_{m+\frac{1}{3},n+\frac{1}{3}}$ as s_1 and the rejection method to obtain the coefficient $a^{(2)}_{m+\frac{1}{3},n+\frac{1}{3}}$, and hence the height $h_{m,n+1}$. The coefficient $b^{(0)}_{m+\frac{1}{3},n+\frac{1}{3}}$ is determined from $h_{m,n}$, $h_{m+1,n}$, and $h_{m,n+1}$ according to Eq. (18). The surface above the triangle $(m + 1/3, n + 1/3)$ is now specified.

The slope of the edge joining the vertices $(m + 1, n)$ and $(m, n + 1)$, namely $(h_{m,n+1} - h_{m+1,n})/b$, is $\beta^{(1)}_{m+\frac{2}{3},n+\frac{2}{3}}$. If this value of $\beta^{(1)}_{m+\frac{2}{3},n+\frac{2}{3}}$ is used as s_1 in the marginal pdf $\tilde{\tilde{f}}(s_1)$ and in the joint pdf $\tilde{\tilde{f}}(s_1, s_2)$, the use of the conditional pdf

$\tilde{\tilde{f}}(s_2|s_1)$ with the rejection method gives the value of the slope $\beta^{(2)}_{m+\frac{2}{3},n+\frac{2}{3}}$ in the direction normal to the edge joining $(m+1,n)$ and $(m,n+1)$, and hence the height $h_{m+1,n+1}$. The values of $a^{(1,2)}_{m+\frac{2}{3},n+\frac{2}{3}}$ are then obtained from the values of $\beta^{(1,2)}_{m+\frac{2}{3},n+\frac{2}{3}}$. From the heights $h_{m+1,n}$, $h_{m+1,n+1}$, and $h_{m,n+1}$ the coefficient $b^{(0)}_{m+\frac{2}{3},n+\frac{2}{3}}$ is determined from Eq. (19). The surface above the triangle $(m+2/3, n+2/3)$ is now specified.

Continuing the construction of the surface, the slope of the edge joining the vertices $(m+1,n)$ and $(m+1,n+1)$, namely $(h_{m+1,n+1}-h_{m+1,n})/b$, is $\alpha^{(1)}_{m+\frac{4}{3},n+\frac{1}{3}}$. If this value of $\alpha^{(1)}_{m+\frac{4}{3},n+\frac{1}{3}}$ is used as the value of s_1 in the marginal pdf $\tilde{f}(s_1)$ and in the joint pdf $\tilde{f}(s_1,s_2)$, the use of the conditional pdf $\tilde{f}(s_2|s_1)$ together with the rejection method gives the value of the slope $\alpha^{(2)}_{m+\frac{4}{3},n+\frac{1}{3}}$ in the direction normal to the edge joining $(m+1,n)$ and $(m+1,n+1)$, and hence the height $h_{m+2,n}$. The values of $a^{(1,2)}_{m+\frac{4}{3},n+\frac{1}{3}}$ are then obtained from these values of $\alpha^{(1,2)}_{m+\frac{4}{3},n+\frac{1}{3}}$. From the heights $h_{m+1,n}$, $h_{m+2,n}$, and $h_{m+1,n+1}$, the coefficient $b^{(0)}_{m+\frac{4}{3},n+\frac{1}{3}}$ is obtained from Eq. (18). The surface above the triangle $(m+4/3, n+1,3)$ is now specified.

By continuing in this fashion the surface profile function $\zeta(\mathbf{x}_\parallel)$ above the first row of equilateral triangles is constructed.

The surface profile function above the second, third, ..., row of equilateral triangles is determined in the same fashion. In this way a single realization of a two-dimensional rough surface is constructed. By construction it is a single-valued function of $\zeta(\mathbf{x}_\parallel)$. It is also a continuous function of $\mathbf{x}_\parallel$ across each edge joining two nearest neighbor vertices, because that edge is shared by two neighboring triangles, and the surface profile function is therefore a continuous function of $\zeta(\mathbf{x}_\parallel)$ in its entirety, whose statistical properties are defined by the joint pdf $f(s_1, s_2)$.

To determine how well the angular distribution of the intensity of the field scattered from the random surface generated by the method just described agrees with the mean differential reflection coefficient $\langle \partial R/\partial\Omega_s \rangle$ used as the input to this method, a large number N_p of realizations of the random surface is generated, and for each realization the scattering problem for a scalar plane wave incident normally on it is solved. The differential reflection coefficient $\partial R/\partial\Omega_s$ is calculated from each solution, and an arithmetic average of the N_p results for $\partial R/\partial\Omega_s$ yields the mean differential reflection coefficient $\langle \partial R/\partial\Omega_s \rangle$.

Rigorous computer simulation calculations of scattering from two-dimensional random surfaces are computationally intensive and time consuming.[6–8] We therefore solve the scattering problem for each realization of our random surface in the Kirchhoff approximation, but without passing to the geometrical optics limit of it. At normal incidence the mean differential reflection coefficient in this approximation is given by[4]

$$\left\langle \frac{\partial R}{\partial \Omega_s} \right\rangle = \frac{1}{S}\left(\frac{\omega}{2\pi c}\right)^2 \langle |r_1(\mathbf{q}_\parallel) + r_2(\mathbf{q}_\parallel)|^2 \rangle, \tag{20}$$

where

$$r_1(\mathbf{q}_\parallel) = \mathrm{i}\frac{\sqrt{3}b}{2}\sum_{m=-N}^{N-1}\sum_{n=-N}^{N-1}\frac{\exp\left(-\mathrm{i}ab^{(0)}_{m+\frac{1}{3},n+\frac{1}{3}}\right)}{q_1 + aa^{(1)}_{m+\frac{1}{3},n+\frac{1}{3}}}$$

$$\times \exp\left[-\mathrm{i}\left(q_1 + aa^{(1)}_{m+\frac{1}{3},n+\frac{1}{3}}\right)\left(m + \frac{n}{2} + \frac{1}{2}\right)b\right.$$

$$\left. - \mathrm{i}\frac{\sqrt{3}}{2}\left(q_2 + aa^{(2)}_{m+\frac{1}{3},n+\frac{1}{3}}\right)\left(n + \frac{1}{2}\right)b\right]$$

$$\times \left\{\exp\left[-\mathrm{i}\left(q_1 + aa^{(1)}_{m+\frac{1}{3},n+\frac{1}{3}}\right)\frac{b}{4}\right]\mathrm{sinc}\left[\frac{\sqrt{3}b}{4}\left(q_2 + aa^{(2)}_{m+\frac{1}{3},n+\frac{1}{3}}\right)\right.\right.$$

$$\left. - \frac{b}{4}\left(q_1 + aa^{(1)}_{m+\frac{1}{3},n+\frac{1}{3}}\right)\right]$$

$$- \exp\left[\mathrm{i}\left(q_1 + aa^{(1)}_{m+\frac{1}{3},n+\frac{1}{3}}\right)\frac{b}{4}\right]\mathrm{sinc}\left[\frac{\sqrt{3}b}{4}\left(q_2 + aa^{(2)}_{m+\frac{1}{3},n+\frac{1}{3}}\right)\right.$$

$$\left.\left. + \frac{b}{4}\left(q_1 + aa^{(1)}_{m+\frac{1}{3},n+\frac{1}{3}}\right)\right]\right\}, \tag{21a}$$

$$r_2(\mathbf{q}_\parallel) = \mathrm{i}\frac{\sqrt{3}b}{2}\sum_{m=-N}^{N-1}\sum_{n=-N}^{N-1}\frac{\exp\left(-\mathrm{i}ab^{(0)}_{m+\frac{2}{3},n+\frac{2}{3}}\right)}{q_1 + aa^{(1)}_{m+\frac{2}{3},n+\frac{2}{3}}}$$

$$\times \exp\left[-\mathrm{i}\left(q_1 + aa^{(1)}_{m+\frac{2}{3},n+\frac{2}{3}}\right)\left(m + \frac{n}{2} + 1\right)b\right.$$

$$\left. - \mathrm{i}\frac{\sqrt{3}}{2}\left(q_2 + aa^{(2)}_{m+\frac{2}{3},n+\frac{2}{3}}\right)\left(n + \frac{1}{2}\right)b\right]$$

$$\times \left\{\exp\left[-\mathrm{i}\left(q_1 + aa^{(1)}_{m+\frac{2}{3},n+\frac{2}{3}}\right)\frac{b}{4}\right]\mathrm{sinc}\left[\frac{\sqrt{3}b}{4}\left(q_2 + aa^{(2)}_{m+\frac{2}{3},n+\frac{2}{3}}\right)\right.\right.$$

$$\left. + \frac{b}{4}\left(q_1 + aa^{(1)}_{m+\frac{2}{3},n+\frac{2}{3}}\right)\right]$$

$$- \exp\left[\mathrm{i}\left(q_1 + aa^{(1)}_{m+\frac{2}{3},n+\frac{2}{3}}\right)\frac{b}{4}\right]\mathrm{sinc}\left[\frac{\sqrt{3}b}{4}\left(q_2 + aa^{(2)}_{m+\frac{2}{3},n+\frac{2}{3}}\right)\right.$$

$$\left.\left. - \frac{b}{4}\left(q_1 + aa^{(1)}_{m+\frac{2}{3},n+\frac{2}{3}}\right)\right]\right\}, \tag{21b}$$

with $a = (\omega/c)(1 + \cos\theta_s)$, $S = 8N^2(\sqrt{3}b^2/4)$, and $\mathrm{sinc}x = \sin x/x$.

To illustrate the method for generating a two-dimensional randomly rough Dirichlet surface that scatters in a specified manner a plane wave incident normally on it, we apply it to the design of a surface that acts as a band-limited uniform diffuser within a rectangular region of scattering angles. The mean differential reflection coefficient we seek in this case is

$$\left\langle \frac{\partial R}{\partial \Omega_s} \right\rangle (q_1, q_2) = \frac{\theta(q_{1m} - |q_1|)}{2(cq_{1m}/\omega)} \frac{\theta(q_{2m} - |q_2|)}{2(cq_{2m}/\omega)}, \tag{22}$$

so that from Eq. (9) we find that

$$f(s_1, s_2) = \frac{4}{(1 + s_\parallel^2)^2} \frac{\theta(s_{1m}(1 + s_\parallel^2) - |s_1|)}{4s_{1m}} \frac{\theta(s_{2m}(1 + s_\parallel^2) - |s_2|)}{4s_{2m}}, \tag{23}$$

where $s_{1m} = (cq_{1m}/2\omega)$ and $s_{2m} = (cq_{2m}/2\omega)$. This result simplifies greatly when the range of variation of s_1 and s_2 is small enough that $s_\parallel^2$ can be neglected with respect to unity with little error, e.g. when $|s_{1m}| \leq 0.15$ and $|s_{2m}| < 0.15$. When this is the case Eq. (23) becomes

$$f(s_1, s_2) = \frac{\theta(s_{1m} - |s_1|)}{2s_{1m}} \frac{\theta(s_{2m} - |s_2|)}{2s_{2m}}. \tag{24}$$

The marginal pdf $f(s_1)$ is readily found to be

$$f(s_1) = \frac{\theta(s_{1m} - |s_1|)}{2s_{1m}}, \tag{25}$$

while the conditional pdf $f(s_2 \mid s_1)$ is

$$f(s_2 \mid s_1) = \frac{\theta(s_{2m} - |s_2|)}{2s_{2m}}. \tag{26}$$

In Fig. 17.2 we present a segment of a single realization of the surface profile function $\zeta(\mathbf{x}_\parallel)$ determined by the approach described in this section for the case of the band-limited uniform diffuser within a rectangular region of scattering angles, for which $f(s_1, s_2)$ is given by Eq. (24).

In Fig. 17.3 we present a plot of the corresponding $\langle \partial R/\partial \Omega_s \rangle$ calculated on the basis of the Kirchhoff approximation represented by Eqs. (20)–(21).

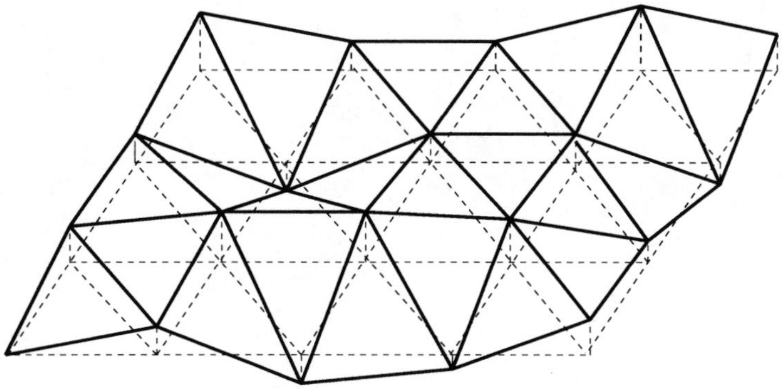

FIGURE 17.2. A segment of a single realization of a numerically generated surface profile function for the band-limited uniform diffuser within a rectangular domain of scattering angles, for which $f(s_1, s_2)$ is given by Eq. (24).

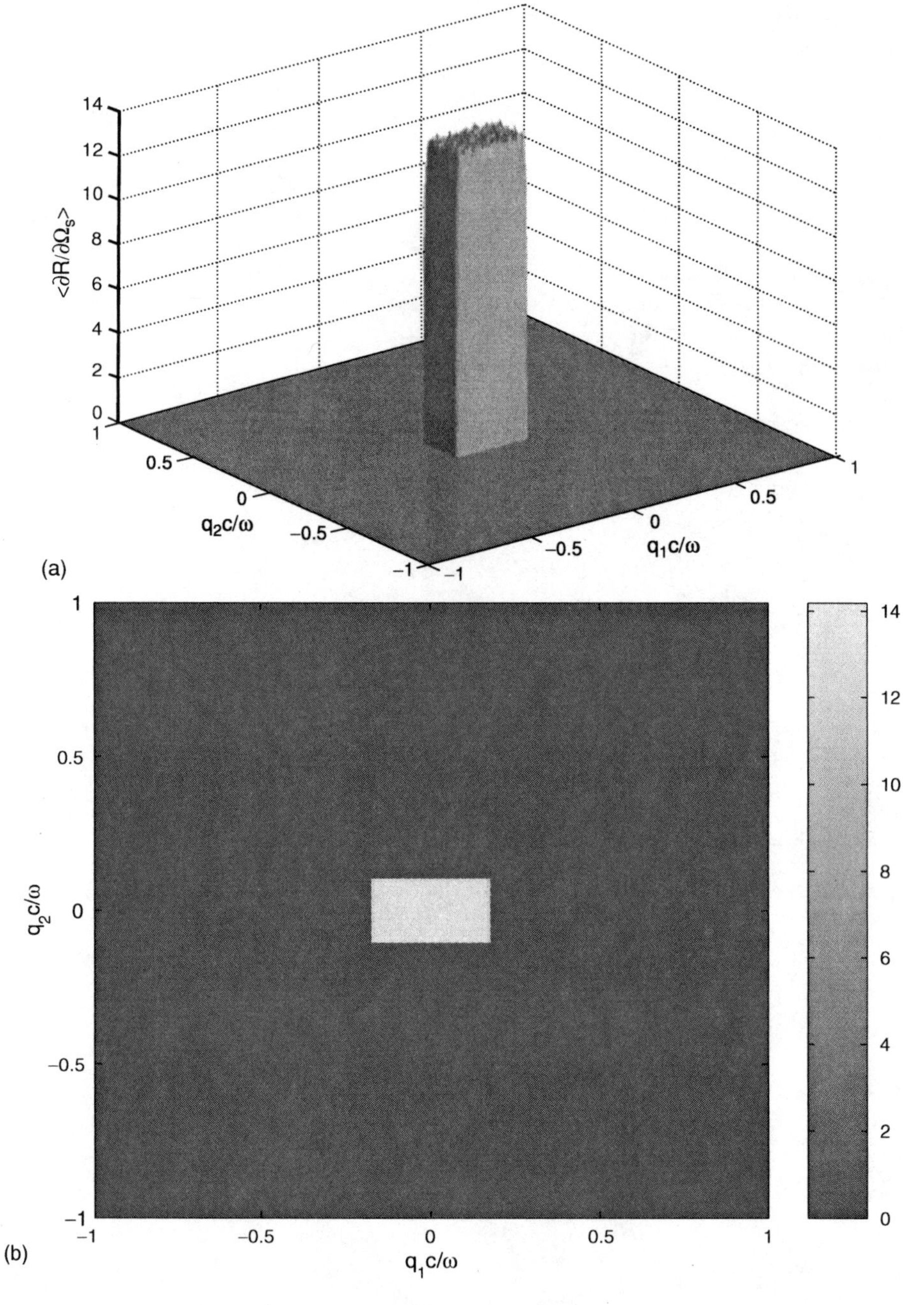

FIGURE 17.3. The mean differential reflection coefficient $\langle \partial R/\partial \Omega_s \rangle$ for scattering from a two-dimensional randomly rough surface that acts as a band-limited uniform diffuser within a rectangular domain of scattering angles. The cross sections corresponding to $q_2 = 0$ and $q_1 = 0$ are also presented.

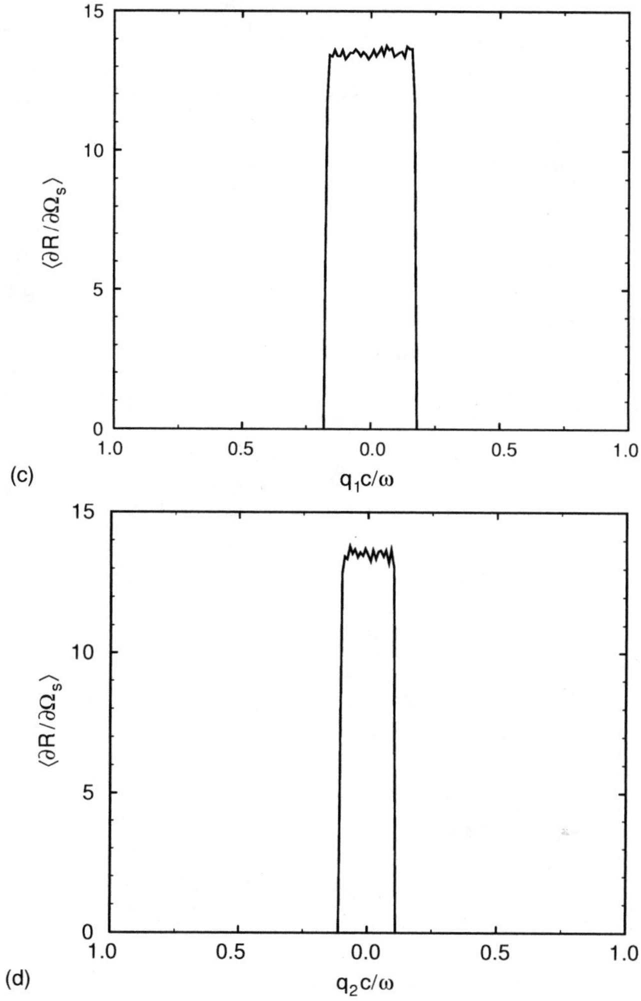

FIGURE 17.3. *(Continued)*

The wavelength of the incident field was $\lambda = 632.8$ nm. The roughness parameters assumed were $b = 20\ \mu$m, $q_{1m} = (\omega/c)\sin 10^\circ \cong 0.1736\ (\omega/c)$, $q_{2m} = (\omega/c)\sin 6^\circ \cong 0.1045\ (\omega/c)$. The results of $N_p = 3000$ realizations of the surface profile function were averaged to obtain the average indicated in Eq. (20). The results presented show that the angular dependence of the intensity of the scattered field is in very good agreement with that defined in Eq. (22). The intensity of the scattered field vanishes for scattering angles outside the domain defined by Eq. (22), and the cutoff is very sharp. For scattering angles within this rectangular domain the intensity of the scattered field is very nearly constant.

A surface designed to produce the mean differential reflection coefficient given by Eq. (22) with $q_{1m} = q_{2m}$ was fabricated on photoresist, and the angular dependence of the intensity of light transmitted through it was measured. The surface was fabricated by superposing two orthogonal one-dimensional surfaces defined by pdfs of the form given by Eq. (25), and generated by the approach described in [9]. A surface designed to act as a band-limited uniform diffuser within a square domain of scattering angles in scattering also acts as a band-limited uniform diffuser within a square region of angles of transmission in transmission. However, due to effects of refraction within the photoresist film, the value of $q_{1m} = q_{2m}$ defining the square domain of angles of transmission within which the mean differential transmission coefficient is constant is different from the value it has in scattering. In Fig. 17.4 we present a gray-level plot of the angular dependence of the intensity of the light transmitted through the surface fabricated in the manner described. It was obtained by the use of collimated white light. The use of white light for illumination is a way of averaging over the speckles that would be present in the intensity distribution of the transmitted light if monochromatic light were used for illumination. It is an experimental equivalent of the ensemble averaging used in obtaining Eq. (9). The resulting intensity distribution is seen to be strongly band-limited within a square region of the (q_1, q_2) plane, although the corners are a bit rounded, and is constant within the region.

Two-dimensional randomly rough surfaces that act as band-limited uniform diffusers within circular,[3] triangular,[3] and elliptical[3,4] regions of scattering angles, as well as a surface that acts as a Lambertian diffuser,[4] have also been designed by the approach presented in this section, demonstrating the versatility of this approach.

17.3. A Surface That Synthesizes the Infrared Spectrum of a Known Compound

In correlation spectroscopy[10–12] the degree of correlation between the transmission or reflection of an unknown sample and that of a reference cell containing a known compound is determined over a fixed spectral range, as a means of identifying the unknown sample. In the case that the known compound in the reference cell is toxic and/or corrosive, or is short-lived, it is useful to have an optical element that synthesizes the infrared spectrum of the compound for use in a correlation spectrometer.

The production of a synthetic spectrum requires the design of a one-dimensional rough surface with the property that at a fixed scattering angle the spectrum of the light scattered from it accurately reproduces a desired spectrum.

The existing theoretical approaches to the solution of the design problem[13–15] have been based on surfaces in the form of grating-like structures of N lines, each of width Δ, whose depths relative to a base level are adjusted by an iterative procedure to produce a wavelength dependence of the intensity of the scattered

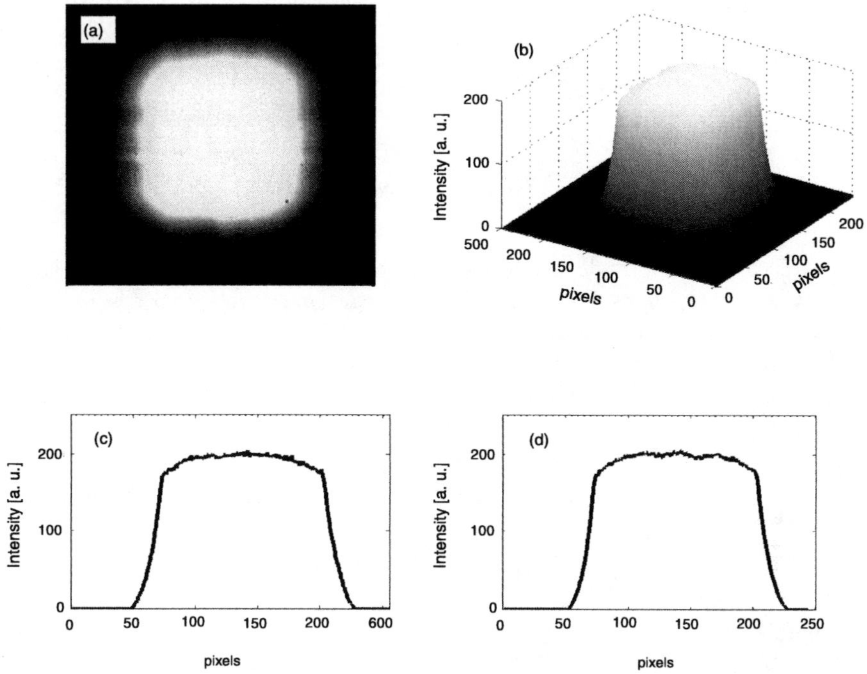

FIGURE 17.4. A CCD camera image (a) and its three-dimensional (b) depiction of an experimental result for the mean differential transmission coefficient of white light transmitted through a two-dimensional randomly rough photoresist surface that has been designed to act as a band-limited uniform diffuser within a square-domain of scattering angles; (c) and (d) are vertical cross sections of the mean differential transmission coefficient in two perpendicular directions [courtesy of E. E. García-Guerrero and E. R. Méndez].

field at the prescribed scattering angle that matches the infrared spectrum of a given compound throughout the spectral range of interest.

In contrast to the deterministic approach to the design of a one-dimensional rough surface that synthesizes a specified experimental infrared spectrum adopted in,[13–15] in this section we present an alternative, probabilistic, approach to the solution of this problem.

The physical system we consider initially consists of vacuum in the region $x_3 > \zeta(x_1)$ and a perfect conductor in the region $x_3 < \zeta(x_1)$. The surface profile function $\zeta(x_1)$ is assumed to be a continuous single-valued function of x_1 that constitutes a random process, but not necessarily a stationary one. The surface $x_3 = \zeta(x_1)$ is illuminated at normal incidence by an s-polarized plane wave of frequency ω, whose plane of incidence is the x_1x_3 plane. The intensity of the scattered field is measured as a function of ω in the far field at a fixed scattering angle.

The starting point for our analysis is the expression for the single nonzero component of the scattered electric field in the vacuum obtained in the Fraunhofer approximation,[16] which we write in the form

$$E_2^>(x_1, x_3|\omega)_{sc} = -E_0\left(\frac{\omega}{2\pi cr}\right)^{\frac{1}{2}} \exp\{[\mathrm{i}(\omega/c)r - \pi/4)]$$
$$\times \int_{-\infty}^{\infty} \mathrm{d}x_1' \exp[-\mathrm{i}(\omega/c)x_1' \sin\theta_s - \mathrm{i}(\omega/c)(1+\cos\theta_s)\zeta(x_1')], \tag{27}$$

where E_0 is the amplitude of the incident field, $r = (x_1^2 + x_3^2)^{\frac{1}{2}}$, $x_1/r = \sin\theta_s$, $x_3/r = \cos\theta_s$, and θ_s is the fixed scattering angle, measured clockwise from the x_3-axis. The squared modulus of the scattered field averaged over the ensemble of realizations of the surface profile function is

$$\left\langle |E_2^>(x_1, x_3|\omega)_{sc}|^2\right\rangle = |E_0|^2\left(\frac{\omega}{2\pi cr}\right)$$
$$\times \int_{-\infty}^{\infty} \mathrm{d}x_1' \int_{-\infty}^{\infty} \mathrm{d}x_1'' \exp[-\mathrm{i}(\omega/c)\sin\theta_s(x_1' - x_1'')]$$
$$\times \langle \exp[-\mathrm{i}(\omega/c)(1+\cos\theta_s)(\zeta(x_1') - \zeta(x_1''))]\rangle. \tag{28}$$

Our aim is to find a surface profile function $\zeta(x_1)$ for which the right-hand side of this equation reproduces the frequency dependence of the expression on the left-hand side, which we assume is given.

To this end we assume that $\zeta(x_1)$ has the form

$$\zeta(x_1) = \alpha x_1 + b_1 d_n, \qquad nb < x_1 < (n+1)b, \qquad n = -N, -N+1, \ldots, N-1, \tag{29}$$

where N is a large integer, the $\{d_n\}$ are independent identically distributed random deviates, b_1 and b are characteristic lengths, and α is a characteristic slope of the surface that will be determined later. Because the $\{d_n\}$ are independent and identically distributed random deviates, the probability density function (pdf) of d_n, $f(\gamma) = \langle\delta(\gamma - d_n)\rangle$, is therefore independent of n. The form of $\zeta(x_1)$ given by Eq. (29) violates our initial assumption that it is a continuous function of x_1. However, we will see that the discontinuities of the surface profile function at $x_1 = nb$, where $n = 0, \pm1, \pm2\ldots$, do not affect our determination of $\zeta(x_1)$.

With the form of $\zeta(x_1)$ given by Eq. (29) it is straightforward to show that

$$\left\langle |E_2^>(x_1, x_3|\omega)_{sc}|^2\right\rangle = |E_0|^2\left(\frac{\omega}{2\pi cr}\right)\{2Nb^2 \operatorname{sinc}^2((\omega b/2c)(\sin\theta_s$$
$$+\alpha(1+\cos\theta_s))[1 - |F((\omega/c)(1+\cos\theta_s)b_1)|^2]$$
$$+(2Nb)^2 \operatorname{sinc}^2(N(\omega b/c)(\sin\theta_s$$
$$+\alpha(1+\cos\theta_s)))|F((\omega/c)(1+\cos\theta_s)b_1)|^2, \tag{30}$$

where

$$F(\nu) = \int_{-\infty}^{\infty} \mathrm{d}\gamma f(\gamma) \exp(-\mathrm{i}\nu\gamma). \tag{31}$$

We now choose the slope α such that the argument of each sinc function in Eq. (30) vanishes:

$$\alpha = -\frac{\sin\theta_s}{1+\cos\theta_s} = -\tan\left(\frac{\theta_s}{2}\right). \tag{32}$$

This result means that the direction of observation is the direction of specular reflection from each segment of the surface.

Since $\mathrm{sinc}(0) = 1$, we have obtained the result that

$$\begin{aligned}\left\langle |E_2^>(x_1, x_3|\omega)_{\mathrm{sc}}|^2 \right\rangle &= |E_0|^2 \left(\frac{\omega}{2\pi c r}\right) 2N(2N-1)b^2 \\ &\times \left[|F((\omega/c)(1+\cos\theta_s)b_1)|^2 + \frac{1}{2N-1} \right]. \end{aligned}\tag{33}$$

We define the experimental intensity $I(\omega)$ by

$$\left\langle |E_2^>(x_1, x_3|\omega)_{\mathrm{sc}}|^2 \right\rangle = 2N(2N-1)b^2 |E_0|^2 \left(\frac{\omega}{2\pi c r}\right) I(\omega). \tag{34}$$

It follows from Eqs. (33) and (34) that

$$I(\omega) = \left[|F((\omega/c)(1+\cos\theta_s)b_1)|^2 + \frac{1}{2N-1} \right], \tag{35}$$

so that

$$|F((\omega/c)(1+\cos\theta_s)b_1)| = \left[I(\omega) - \frac{1}{2N-1} \right]^{\frac{1}{2}}. \tag{36}$$

The problem of determining the surface profile function $\zeta(x_1)$ thus reduces to obtaining the pdf of d_n, $f(\gamma)$, from a knowledge of the modulus of its Fourier transform $F(\nu)$. Once the function $f(\gamma)$ has been determined, a long sequence of $\{d_n\}$ is obtained from it by, e.g., the rejection method,[5] and a realization of the surface profile function is constructed on the basis of Eq. (29).

If $F(\nu)$ were known, the problem of obtaining $f(\gamma)$ would reduce to evaluating a Fourier integral,

$$f(\gamma) = \int_{-\infty}^{\infty} \frac{\mathrm{d}\nu}{2\pi} F(\nu) \exp(\mathrm{i}\gamma\nu). \tag{37}$$

However, we know only $|F(\nu)|$, which is given by Eq. (36). We are therefore forced to use an iterative approach to obtain $f(\gamma)$ from $|F(\nu)|$. In fact, we use a modified Gerchberg–Saxton algorithm[17,18] for this purpose.

The function $|F(\nu)|$ is known only for positive values of ν, in a range that we denote by $0 < \nu < \nu_{\max}$. However, it is convenient for what follows to assume, with no loss of generality, that $|F(\nu)|$ exists in the interval $-\nu_{\max} < \nu < \nu_{\max}$, and is an even function of ν. Moreover, instead of working with Fourier integral transforms, we will use discrete Fourier transforms. Thus, we introduce the definitions

$$T_0 = 2\nu_{\max} \tag{38}$$

$$\nu_k = \frac{T_0}{2M}k \qquad k = -M, -M+1, \ldots, M \tag{39}$$

$$\gamma_n = \frac{2\pi}{T_0}n \qquad n = -M, -M+1, \ldots, M, \tag{40}$$

and $F(\nu_k) \equiv F_k$, $f(\gamma_n) \equiv f_n$. It follows from Eqs. (38) and (40) that the domain of existence of $f(\gamma)$ is $-\pi M/\nu_{\max} < \gamma < \pi M/\nu_{\max}$. The discrete analogues of Eqs. (31) and (37) are

$$F_k = \frac{2\pi}{T_0} \sum_{n=-M}^{M} f_n \exp(-i\pi kn/M) \qquad k = -M, -M+1, \ldots, M \tag{41}$$

$$f_n = \frac{T_0}{4\pi M} \sum_{k=-M}^{M} c_k F_k \exp(i\pi nk/M) \qquad n = -M, -M+1, \ldots, M, \tag{42}$$

where

$$c_k = \begin{cases} \frac{1}{2} & k = -M, M \\ 1 & k \neq -M, M \end{cases}. \tag{43}$$

Because it is a probability density function, $f(\gamma)$ must be real, nonnegative, and normalized to unity. The first of these conditions requires that the phase $\chi(\nu)$ of the function $F(\nu) = |F(\nu)| \exp[i\chi(\nu)]$ must be an odd function of ν, $\chi(-\nu) = -\chi(\nu)$. The last of these conditions requires that $\chi(0) = 0$.

The iterative determination of $f(\gamma)$ proceeds as follows. We generate, for positive values of k, a sequence of real random numbers $\{\chi_0(\nu_k)\}$, drawn from a uniform probability density function in the interval $(-\pi, \pi)$, and define $\chi_0(\nu_{-k}) = \chi_0(-\nu_k) = -\chi_0(\nu_k)$, with $\chi_0(0) = 0$. We then construct the complex functions $F_k^{(0)} = |F(\nu_k)| \exp[i\chi_0(\nu_k)]$, and use them to evaluate the sum

$$f_n^{(0)} = \frac{T_0}{4\pi M} \sum_{k=-M}^{M} c_k F_k^{(0)} \exp(i\pi nk/M), \tag{44}$$

which is real. To implement the constraint that $f(\gamma)$ be nonnegative each negative coefficient $f_n^{(0)}$ is set equal to zero, while each positive $f_n^{(0)}$ is left unchanged. The resulting set of values of $f_n^{(0)}$ is denoted by $\{\tilde{f}_n^{(0)}\}$ after it has been scaled to satisfy the normalization condition

$$\frac{2\pi}{T_0} \sum_{n=-M}^{M} \tilde{f}_n^{(0)} = 1. \tag{45}$$

The $\{\tilde{f}_n^{(0)}\}$ are then used to generate a set of values $\{F_k^{(1)}\}$,

$$F_k^{(1)} = \frac{2\pi}{T_0} \sum_{n=-M}^{M} \tilde{f}_n^{(0)} \exp(-i\pi kn/M) = \left|F_k^{(1)}\right| \exp[i\chi_1(\nu_k)]. \tag{46}$$

From the set $\{F_k^{(1)}\}$ we generate a new set $\{\tilde{F}_k^{(1)}\}$ according to $\tilde{F}_k^{(1)} = |F(\nu_k)| \times \exp[i\chi_1(\nu_k)]$. From this set we obtain the set $\{f_n^{(1)}\}$ as

$$f_n^{(1)} = \frac{T_0}{4\pi M} \sum_{k=-M}^{M} c_k \tilde{F}_k^{(1)} \exp(i\pi nk/M). \tag{47}$$

Each negative coefficient $f_n^{(1)}$ is equated to zero, while each positive $f_n^{(1)}$ is left unchanged. After the resulting set of coefficients is scaled so that it is normalized to unity, it is denoted by $\{\tilde{f}_n^{(1)}\}$. The $\{\tilde{f}_n^{(1)}\}$ are used to generate a set

$$F_k^{(2)} = \frac{2\pi}{T_0} \sum_{n=-M}^{M} \tilde{f}_n^{(1)} \exp(-i\pi kn/M) = \left|F_k^{(2)}\right| \exp[i\chi_2(\nu_k)], \tag{48}$$

from which we construct a new set $\{\tilde{F}_k^{(2)}\}$ according to $F_k^{(2)} = |F(\nu_k)| \exp[i\chi_2(\nu_k)]$. The iteration scheme then proceeds as before. Eight to ten iterations usually suffice to obtain a good result for $f(\gamma)$.

To test how well a surface designed in this way scatters light with an intensity whose frequency dependence reproduces the input spectrum $I(\omega)$, we calculate $I(\omega)$ on the basis of the Kirchhoff approximation,[2]

$$I(\omega) = \frac{1}{2N(2N-1)} \left(\frac{\omega r}{2\pi c}\right) \langle |R(r, \theta_s)|^2 \rangle, \tag{49}$$

where

$$\begin{aligned} R(r, \theta_s) = \sum_{n=-N}^{N-1} \int_{-\frac{\pi}{2}}^{\frac{\pi}{2}} d\theta \exp\Bigg\{ - i(\omega/c) \Bigg[r \cos(\theta - \theta_s) - (\sin\theta + \alpha(1 + \cos\theta)) \\ \times \left(n + \frac{1}{2}\right) b - d_n b_1 (1 + \cos\theta) \Bigg] \Bigg\} \\ \times \operatorname{sinc}((\omega b/2c)(\sin\theta + \alpha(1 + \cos\theta))). \end{aligned} \tag{50}$$

The integral in Eq. (50) has to be evaluated numerically.

To illustrate the approach outlined here for the design of a one-dimensional random surface that synthesizes an experimental spectrum, we consider the synthesis

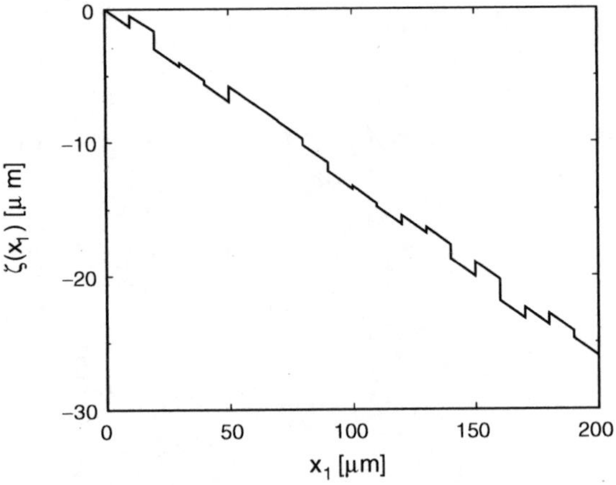

FIGURE 17.5. A segment of a single realization of a numerically generated one-dimensional random surface that has been designed to synthesize the infrared absorption spectrum of HF.

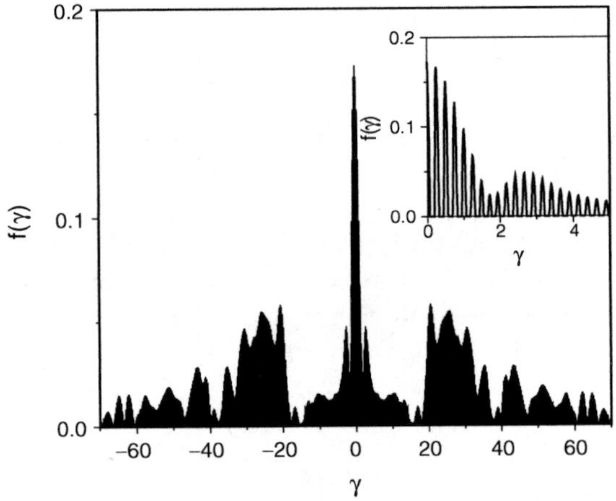

FIGURE 17.6. The probability density function $f(\gamma)$ used to generate the surface profile function presented in Fig. 17.5.

of the infrared spectrum of HF in the region 3600–4300 cm^{-1}. We designed a surface that synthesizes this spectrum at a scattering angle $\theta_s = 15°$. It consists of $2N = 5000$ segments, each of length $b = 10\,\mu\text{m}$, for a total length of 5 cm. The characteristic depth b_1 is chosen to be $b_1 = 0.1\,\mu\text{m}$. A segment of one realization of the resulting surface profile function $\zeta(x_1)$ is presented in Fig. 17.5. The probability density function of d_n, $f(\gamma)$, used to generate this surface is presented in Fig. 17.6.

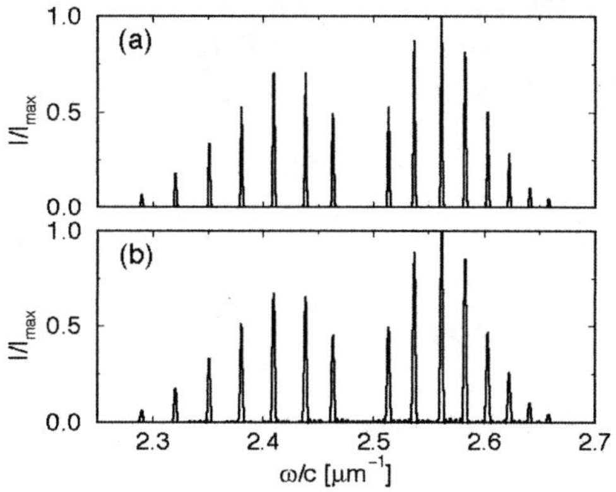

FIGURE 17.7. (a) The infrared absorption spectrum of HF from 3600 to 4300 cm^{-1}. (b) A theoretical synthetic spectrum of HF calculated by the approach presented in this chapter.

The infrared spectrum of HF in the region 3600–4300 cm^{-1} consists of fifteen sharp rotational lines superimposed on a broad vibrational background.[19] It is shown in Fig. 17.7(a). The intensity $I(\omega)$ calculated on the basis of Eqs. (49)–(50) for scattering from a *single* realization of the surface profile function, so that no ensemble averaging was carried out, is plotted in Fig. 17.7(b). It displays peaks at the correct wavelengths, whose relative amplitudes are close to those of the peaks in the experimental spectrum. The peaks in the theoretical spectrum are somewhat wider than those in the experimental spectrum.

It may seem surprising that scattering from a single realization of a random surface that was constructed by requiring that the *mean* intensity of the scattered field, i.e. the intensity of the scattered field averaged over the ensemble of realizations of the surface profile function, has a specified dependence on the wavelength of the incident field, yields a theoretical spectrum in such close agreement with the input experimental spectrum. The reason that it is not necessary to generate an ensemble of realizations of the surface profile function, and evaluate the arithmetic average of the intensity of the field scattered from each realization, is that the intensity is calculated as a function of the wavelength at a fixed scattering angle that coincides with the angle of specular reflection from each segment of the surface. In this case no speckles arise that have to be averaged over, as is the case if the intensity is calculated as a function of the scattering angle at a fixed wavelength. Averaging over the ensemble of realizations of the surface profile function only smoothes out the background noise that is clearly seen in Fig. 17.7(b). However, the random surface has to be long enough in order that scattering from a single realization of it produces a spectrum in good agreement with the experimental spectrum. The more complicated the spectrum that we seek to reproduce is, the more complicated is the resulting pdf $f(\gamma)$ and, as a result, the longer the surface should be to represent well the statistics of the surface that are required. In the calculations carried out here, we found that a value of $2N = 5000$ sufficed for this purpose.

It should also be emphasized that any realization of the surface profile function obtained by the rejection method from the pdf $f(\gamma)$ plotted in Fig. 17.6 produces a spectrum with the aid of Eqs. (49) and (50) that coincides with the one presented in Fig. 17.7(b).

Realizations of the random surfaces considered in this section can be fabricated by the method employed for this purpose in [15] and by the method proposed in [9].

Thus, the results of this section indicate that treating one-dimensional rough surfaces as randomly rough surfaces is an effective approach to the design of optical elements that synthesize infrared spectra of known compounds.

17.4. Conclusions

The two types of randomly rough surfaces considered in this chapter, and the applications for which they were designed, while hardly exhaustive, indicate the breadth of the technological opportunities presented by the ability to design surfaces with specified scattering properties. The probabilistic approach to the design problem described here, in which the inverse problem is formulated as a search for the joint probability density function of the slopes of the triangular facets from which a two-dimensional random surface is constructed, or for the probability density function for the slopes or depths of the segments from which a one-dimensional random surface is constructed, not only yields algorithms by means of which surfaces with specified scattering properties can be designed, but enables the statistical properties of these surfaces to be studied as well.

An outstanding unsolved problem is that of fabricating, e.g. on photoresist, randomly rough surfaces of the kind designed in Sect. 17.2. Methods exist for the fabrication of surfaces of the kind designed in Sect. 17.3, and have been cited in that section, but they are not applicable to the surfaces generated in Sect. 17.2. A different approach has to be found.

The focus in this chapter has been exclusively on the design of randomly rough surfaces with specified scattering properties. However, the methods used to solve this design problem can also be applied to the design of randomly rough surfaces that transmit light in a specified manner. Although some steps in this direction have been taken,[20,21] they have been limited to one-dimensional randomly rough surfaces. The design of two-dimensional randomly rough surfaces with prescribed transmission properties is an unexplored problem. It is an important one because experimental studies of designer surfaces, and applications of such surfaces, are more easily carried out in transmission, where the source and detector are on opposite sides of the surface, than in reflection where they are on the same side of the surface.

The design of surfaces with specified scattering or transmission properties is thus still a work in progress. It is hoped that the brief review of accomplishments to date, and opportunities for future studies presented here, will stimulate further development of this new type of inverse scattering problem.

Acknowledgments. The work presented in this chapter has been carried out in collaboration with Drs. E. E. García-Guerrero, T. A. Leskova, M. Leyva-Lucero, E. R. Méndez, and J. Muñoz-Lopez. I am grateful to all of them for their help. This research was supported in part by Army Research Office grant DAAD 19-02-1-0256.

References

1. A.V. Shchegrov, A.A. Maradudin, and E.R. Méndez, "Multiple scattering of light from randomly rough surfaces," in *Progress in Optics*, vol. 46, E. Wolf, ed. (Elsevier, Amsterdam, 2004), pp. 117–241. See, in particular, Section 4.3.2.
2. T.A. Leskova, A.A. Maradudin, E.R. Méndez, and J. Muñoz-Lopez, "The design and photofabrication of random achromatic optical diffusers for uniform illumination," in *Wave Propagation in Periodic and Random Media*, P. Kuchment, ed. (American Mathematical Society, Providence, RI, 2003), pp. 117–140.
3. E.R. Méndez, T.A. Leskova, A.A. Maradudin, and J. Muñoz-Lopez, "Design of two-dimensional random surfaces with specified scattering properties," *Opt. Lett.* **29**, 2917–2919 (2004).
4. E.R. Méndez, T.A. Leskova, A.A. Maradudin, M. Leyva-Lucero, and J. Muñoz-Lopez, "The design of two-dimensional random surfaces with specified scattering properties," *J. Opt. A* **7**, S141–S151 (2005).
5. W.H. Press, S.A. Teukolsky, W.T. Vetterling, and B.P. Flannery, *Numerical Recipes in Fortran*, 2nd edn (Cambridge University Press, New York, 1992), pp. 281–282.
6. P. Tran and A.A. Maradudin, "Scattering of a scalar beam from a two-dimensional randomly rough hard wall: Enhanced backscattering," *Phys. Rev. B* **45**, 3936–3939 (1992).
7. C. Macaskill, and B.J. Kachoyan, "Iterative approach for the numerical simulation of scattering from one- and two-dimensional rough surfaces," *Appl. Opt.* **32**, 2839–2847 (1993).
8. P. Tran and A.A. Maradudin, "Scattering of a scalar beam from a two-dimensional randomly rough hard wall: Dirichlet and Neumann boundary conditions," *Appl. Opt.* **32**, 2848–2851 (1993).
9. E.R. Méndez, E.E. García-Guerrero, T.A. Leskova, A.A. Maradudin, J. Muñoz-Lopez, and I. Simonsen, "The design of one-dimensional random surfaces with specified scattering properties," *Appl. Phys. Lett.* **81**, 798–800 (2002).
10. R. Goody, "Cross-correlating spectrometer," *J. Opt. Soc. Am.* **58**, 900–908 (1968).
11. H.O. Edwards and J.P. Daikin, "Gas sensors using correlation spectroscopy compatible with fibre-optic operation," Sensors Actuators **11**, 9–19 (1993).
12. F.W. Taylor, J.T. Houghton, G.D. Peskett, C.D. Rogers, and E.J. Williamson, "Radiometer for remote sounding of the upper atmosphere," *Appl. Opt.* **11**, 135–141 (1972).
13. M.B. Sinclair, M.A. Butler, A.J. Ricco, and S.D. Senturia, "Synthetic spectra: a tool for correlation spectroscopy," *Appl. Opt.* **36**, 3342–3348 (1997).
14. M.B. Sinclair, M.A. Butler, S.H. Kravitz, W.J. Zabrzycki, and A.J. Ricco, "Synthetic infrared spectra," *Opt. Lett.* **22**, 1036–1038 (1997).
15. G. Zhou, F.E.H. Tay, and F.S. Chau, "Design of the diffractive optical elements for synthetic spectra," *Opt. Express* **11**, 1392–1399 (2003).
16. J.W. Goodman, *Introduction to Fourier Optics* (McGraw-Hill, New York, 1968), pp. 57–74.

17. R.W. Gerchberg and W.O. Saxton, "A practical algorithm for the determination of phase from image and diffraction plane pictures," *Optik* **35**, 237–246 (1972).
18. J.R. Fienup, "Phase retrieval algorithms: a comparison," *Appl. Opt.* **21**, 2758–2769 (1982).
19. http://vpl.ipac.caltech.edu/spectra/.
20. A.A. Maradudin, T.A. Leskova, E.R. Méndez, and I. Simonsen, "Band–limited uniform diffuser in transmission," *Proc. SPIE* **4100**, 113–123 (2000).
21. A.A. Maradudin, E.R. Méndez, T.A. Leskova, and I. Simonsen, "A band-limited uniform diffuser in transmission. II. A multilayer system," *Proc. SPIE* **4447**, 130–139 (2001).

Index

Printed in the USA